Advances in Renewable Energy Systems

Advances in Renewable Energy Systems

Editors

Ferdinando Salata
Virgilio Ciancio

MDPI

Basel • Beijing • Wuhan • Barcelona • Belgrade • Novi Sad • Cluj • Manchester

Editors

Ferdinando Salata
University of Rome
"Sapienza"
Rome
Italy

Virgilio Ciancio
University of Rome
"Sapienza"
Rome
Italy

Editorial Office
MDPI AG
Grosspeteranlage 5
4052 Basel, Switzerland

This is a reprint of articles from the Special Issue published online in the open access journal *Processes* (ISSN 2227-9717) (available at: https://www.mdpi.com/journal/processes/special_issues/renewable_systems).

For citation purposes, cite each article independently as indicated on the article page online and as indicated below:

Lastname, A.A.; Lastname, B.B. Article Title. *Journal Name* **Year**, *Volume Number*, Page Range.

ISBN 978-3-7258-2693-3 (Hbk)
ISBN 978-3-7258-2694-0 (PDF)
doi.org/10.3390/books978-3-7258-2694-0

Contents

About the Editors

Ferdinando Salata

Ferdinando Salata graduated in Mechanical Engineering with specialisation in Energy at the University of Rome "Sapienza" in 2003. He completed his PhD in "Technical Physics" in 2008 at the same university. In recent years, he has carried out in-depth research into the optimisation of energy demand in buildings; cogeneration systems; urban microclimate; thermal comfort in the outdoor environment; energy optimisation and reliability of air conditioning and lighting systems; natural ventilation in buildings; desalination using absorption machines; and thermal conductivity problems in soils. In recent years, he has carried out in-depth research into the optimisation of energy demand in buildings; cogeneration systems; urban microclimate; thermal comfort in the outdoor environment; energy optimisation and reliability of air conditioning and lighting systems; natural ventilation in buildings; desalination using absorption machines; and thermal conductivity problems in soils.

Virgilio Ciancio

Virgilio Ciancio graduated in Architecture with a specialization in Architectural Restoration from the University of Roma Tre in 2013. He completed his PhD in "Technical Physics" in 2019 at the University of Rome "Sapienza". In recent years, he has conducted in-depth research on dynamic building energy consumption simulations. He has focused on the analysis of urban microclimate as well as acoustic and lighting comfort in outdoor environments. He has been part of research groups dedicated to airport noise reduction and street lighting studies.

processes MDPI

Editorial

Special Issue: Advances in Renewable Energy Systems

Ferdinando Salata * and Virgilio Ciancio *

Department of Astronautics, Electrical and Energetics Engineering, University of Rome "Sapienza", 00184 Rome, Italy
* Correspondence: ferdinando.salata@uniroma1.it (F.S.); virgilio.ciancio@uniroma1.it (V.C.)

The greatest challenge that the industrialized world faces today is converting its energy production systems into those that utilize renewable sources, which are more environmentally friendly compared to traditional systems [1]. In recent years, this challenge has become twofold: on the one hand, it is necessary to develop new technologies capable of freeing themselves from traditional energy sources, which are responsible for the emission of greenhouse gases; on the other hand, it is urgent to accelerate this energy transition to mitigate global warming and its consequences on the natural environment as quickly as possible [2]. Ecosystems around the world have been stressed by obsolete energy production schemes in an attempt (which now clearly appears unsustainable) to pursue the paradigm of constant growth, characterized by the assumption of available resources and an environment capable of absorbing and accepting changes introduced by humans. The international scientific community has the task of guiding industrial systems towards energy generation processes that respect the planet through the use of sources, with medium- to long-term sustainable production processes, capable of producing economic returns that justify investments for the energy transition [3]. Energy production must undergo a period of profound change that leads to the adoption of energy systems that do not cause further damage to the climate [4].

To achieve these results, it is necessary to completely review the energy production structure, with a particular focus on the energy production systems implemented in the first industrialization countries, but, above all, in the emerging countries experiencing stronger economic and demographic growth. These systems must be able to adequately exploit every available energy source, with zero emissions and with respect for the environment [5]. The electrification of energy end uses will change the entire energy production chain, as well as the energy supply network [6]. Therefore, the sustainability of the entire energy supply chain needs new management and design paradigms on which to reinvent itself and evolve in the near future [7]. Energy storage systems constitute one of the most relevant aspects on which scientific research will have to develop new technical solutions and technologically advanced materials to solve the known problems of the unpredictability of renewable sources [8].

The Special Issue "Advances in Renewable Energy Systems" is a collection of some interesting scientific works on these aspects, which have enriched the current state of the art in the scientific landscape. Experts from many research fields have shared their ideas and experiences in a multidisciplinary way on these topics. The following 18 arguments have been collected:

- Heat Pump (MAHP) to Produce Electrical Energy and Revaluated Heat [9].
- A Different Approach to Develop a District Heating Grid Based on the Optimization of Building Clusters [10].
- Impact of Electric Vehicles on Energy Efficiency with Energy Boosters in Coordination for Sustainable Energy in Smart Cities [11].
- Game Analysis of the Evolution of Energy Structure Transition Considering Low-Carbon Sentiment of the Decision-Makers in the Context of Carbon Neutrality [12].

Citation: Salata, F.; Ciancio, V. Special Issue: Advances in Renewable Energy Systems. *Processes* **2024**, *12*, 1583. https://doi.org/10.3390/pr12081583

Received: 18 July 2024
Accepted: 26 July 2024
Published: 29 July 2024

- Coordination of Configurations of Technologically Integrated "European Green Deal" Projects [13].
- Analysis of Enhanced Heat Transfer Characteristics of Coaxial Borehole Heat Exchanger [14].
- How Does Public Sentiment Affect the Socially Responsible Behavior of Construction Enterprises? [15].
- Dynamic Performance Assessment of PMSG and DFIG-Based WECS with the Support of Manta Ray Foraging Optimizer Considering MPPT, Pitch Control, and FRT Capability Issues [16].
- Thermal, Lighting and IAQ Control System for Energy Saving and Comfort Management [17].
- CFD Modeling of an H-Type Darrieus VAWT under High Winds: The Vorticity Index and the Imminent Vortex Separation Condition [18].
- Life Cycle Assessment and Cumulative Energy Demand Analyses of a Photovoltaic/Thermal System with MWCNT/Water and GNP/Water Nanofluids [19].
- Modeling and Multi-Stage Planning of Cement-IIES Considering Carbon-Green Certificate Trading [20].
- Evaluating the Efficacy of Intelligent Methods for Maximum Power Point Tracking in Wind Energy Harvesting Systems [21].
- Existing Stature and Possible Outlook of Renewable Power in Comprehensive Electricity Market [22].
- Desiccant Technologies for Improving Air Quality: An Overview of the Brazilian Scenario and Comparison of Available Design Software for Manufactured Desiccant Wheels [23].
- Fast-Frequency-Response Control Method for Electrode Boilers Supporting New Energy Accommodation [24].
- Simulation Experiment Design and Control Strategy Analysis in Teaching of Hydrogen–Electric Coupling System [25].
- Research on Multi-Objective Energy Management of Renewable Energy Power Plant with Electrolytic Hydrogen Production [26].

The ultimate goal of this Special Issue is to encourage readers of these scientific contributions to initiate a debate on future scenarios related to climate change and its effects on energy systems, which must shift towards sustainability and the preferential use of renewable energy.

I must congratulate the Authors and Reviewers for their exceptional work and thank the Editors, Assistants, and the entire MDPI staff for the quality of their work.

A sustainable future is closer than we can imagine.

Funding: This research received no external funding.

Conflicts of Interest: The authors declare no conflicts of interest.

References

1. Guo, S.; Liu, Q.; Sun, J.; Jin, H. A Review on the Utilization of Hybrid Renewable Energy. *Renew. Sustain. Energy Rev.* **2018**, *91*, 1121–1147. [CrossRef]
2. Chicco, G.; Mancarella, P. Distributed Multi-Generation: A Comprehensive View. *Renew. Sustain. Energy Rev.* **2009**, *13*, 535–551. [CrossRef]
3. Bocken, N.M.P.; Short, S.W.; Rana, P.; Evans, S. A Literature and Practice Review to Develop Sustainable Business Model Archetypes. *J. Clean. Prod.* **2014**, *65*, 42–56. [CrossRef]
4. Jacobson, M.Z. Review of Solutions to Global Warming, Air Pollution, and Energy Security. *Energy Environ. Sci.* **2009**, *2*, 148–173. [CrossRef]
5. Blanco, H.; Faaij, A. A Review at the Role of Storage in Energy Systems with a Focus on Power to Gas and Long-Term Storage. *Renew. Sustain. Energy Rev.* **2018**, *81*, 1049–1086. [CrossRef]
6. Shahsavari, A.; Akbari, M. Potential of Solar Energy in Developing Countries for Reducing Energy-Related Emissions. *Renew. Sustain. Energy Rev.* **2018**, *90*, 275–291. [CrossRef]
7. Dincer, I.; Acar, C. A Review on Clean Energy Solutions for Better Sustainability. *Int. J. Energy Res.* **2015**, *39*, 585–606. [CrossRef]

8. Tan, K.M.; Babu, T.S.; Ramachandaramurthy, V.K.; Kasinathan, P.; Solanki, S.G.; Raveendran, S.K. Empowering Smart Grid: A Comprehensive Review of Energy Storage Technology and Application with Renewable Energy Integration. *J. Energy Storage* **2021**, *39*, 102591. [CrossRef]
9. Hernández-Magallanes, J.A.; Domínguez-Inzunza, L.A.; Lugo-Loredo, S.; Sanal, K.C.; Cerdán-Pasarán, A.; Tututi-Avila, S.; Morales, L.I. Energy and Exergy Analysis of a Modified Absorption Heat Pump (MAHP) to Produce Electrical Energy and Revaluated Heat. *Processes* **2022**, *10*, 1567. [CrossRef]
10. Pompei, L.; Mannhardt, J.; Nardecchia, F.; Pastore, L.M.; de Santoli, L. A Different Approach to Develop a District Heating Grid Based on the Optimization of Building Clusters. *Processes* **2022**, *10*, 1575. [CrossRef]
11. Kumar, P.; Nikolovski, S.; Ali, I.; Thomas, M.S.; Ahuja, H. Impact of Electric Vehicles on Energy Efficiency with Energy Boosters in Coordination for Sustainable Energy in Smart Cities. *Processes* **2022**, *10*, 1593. [CrossRef]
12. Wang, X.; Guo, Z.; Zhang, Z.; Li, B.; Su, C.; Sun, L.; Wang, S. Game Analysis of the Evolution of Energy Structure Transition Considering Low-Carbon Sentiment of the Decision-Makers in the Context of Carbon Neutrality. *Processes* **2022**, *10*, 1650. [CrossRef]
13. Tryhuba, A.; Mudryk, K.; Tryhuba, I.; Hutsol, T.; Glowacki, S.; Faichuk, O.; Kovalenko, N.; Shevtsova, A.; Ratajski, A.; Janaszek-Mankowska, M.; et al. Coordination of Configurations of Technologically Integrated "European Green Deal" Projects. *Processes* **2022**, *10*, 1768. [CrossRef]
14. Sun, L.; Fu, B.; Wei, M.; Zhang, S. Analysis of Enhanced Heat Transfer Characteristics of Coaxial Borehole Heat Exchanger. *Processes* **2022**, *10*, 2057. [CrossRef]
15. Dang, X.; Wang, S.; Deng, X.; Zhang, Z.; Zhang, N.; Mao, H. How Does Public Sentiment Affect the Socially Responsible Behavior of Construction Enterprises? *Processes* **2022**, *10*, 2403. [CrossRef]
16. Mahmoud, M.M.; Atia, B.S.; Abdelaziz, A.Y.; Aldin, N.A.N. Dynamic Performance Assessment of PMSG and DFIG-Based WECS with the Support of Manta Ray Foraging Optimizer Considering MPPT, Pitch Control, and FRT Capability Issues. *Processes* **2022**, *10*, 2723. [CrossRef]
17. Zanoli, S.M.; Pepe, C. Thermal, Lighting and IAQ Control System for Energy Saving and Comfort Management. *Processes* **2023**, *11*, 222. [CrossRef]
18. Acosta-López, J.G.; Blasetti, A.P.; Lopez-Zamora, S.; de Lasa, H. CFD Modeling of an H-Type Darrieus VAWT under High Winds: The Vorticity Index and the Imminent Vortex Separation Condition. *Processes* **2023**, *11*, 644. [CrossRef]
19. Dolgun, G.K.; Koşan, M.; Kayfeci, M.; Georgiev, A.G.; Keçebaş, A. Life Cycle Assessment and Cumulative Energy Demand Analyses of a Photovoltaic/Thermal System with MWCNT/Water and GNP/Water Nanofluids. *Processes* **2023**, *11*, 832. [CrossRef]
20. Guo, Z.; Zhou, S. Modeling and Multi-Stage Planning of Cement-IIES Considering Carbon-Green Certificate Trading. *Processes* **2023**, *11*, 1219. [CrossRef]
21. Umar, D.A.; Alkawsi, G.; Jailani, N.L.M.; Alomari, M.A.; Baashar, Y.; Alkahtani, A.A.; Capretz, L.F.; Tiong, S.K. Evaluating the Efficacy of Intelligent Methods for Maximum Power Point Tracking in Wind Energy Harvesting Systems. *Processes* **2023**, *11*, 1420. [CrossRef]
22. Das, S.S.; Kumar, J.; Dawn, S.; Salata, F. Existing Stature and Possible Outlook of Renewable Power in Comprehensive Electricity Market. *Processes* **2023**, *11*, 1849. [CrossRef]
23. Castillo Santiago, Y.; Nunes, B.G.; Fontana, G.S.; Busanello, D.; Santos, A.F.; Santos, S.M.D.; de Mello, E.N.; Sphaier, L.A. Desiccant Technologies for Improving Air Quality: An Overview of the Brazilian Scenario and Comparison of Available Design Software for Manufactured Desiccant Wheels. *Processes* **2023**, *11*, 2031. [CrossRef]
24. Shi, T.; Chen, Z.; Guo, S.; Li, D. Fast-Frequency-Response Control Method for Electrode Boilers Supporting New Energy Accommodation. *Processes* **2023**, *11*, 3098. [CrossRef]
25. Shi, T.; Sheng, J.; Chen, Z.; Zhou, H. Simulation Experiment Design and Control Strategy Analysis in Teaching of Hydrogen-Electric Coupling System. *Processes* **2024**, *12*, 138. [CrossRef]
26. Shi, T.; Gu, L.; Xu, Z.; Sheng, J. Research on Multi-Objective Energy Management of Renewable Energy Power Plant with Electrolytic Hydrogen Production. *Processes* **2024**, *12*, 541. [CrossRef]

processes

Article

CFD Modeling of an H-Type Darrieus VAWT under High Winds: The Vorticity Index and the Imminent Vortex Separation Condition

Jansen Gabriel Acosta-López [1], Alberto Pedro Blasetti [2], Sandra Lopez-Zamora [1] and Hugo de Lasa [1,*]

[1] Chemical Reactor Engineering Centre, Department of Chemical and Biochemical Engineering, The University of Western Ontario, London, ON N6A 3K7, Canada
[2] Departamento de Ingeniería Química, Universidad de Nacional de la Patagonia San Juan Bosco, Comodoro Rivadavia U9005, Chubut, Argentina
* Correspondence: hdelasa@uwo.ca; Tel.: +1-5196612144

Abstract: This study introduces a Vorticity Index (VI) and an Imminent Vortex Separation Condition (IVSC), which are considered valuable indicators to quantify the vorticity impact on vertical axis wind turbines (VAWTs) operation. The VI and IVSC are specifically applied to a H-Darrieus vertical axis wind turbine (VAWT). Findings show that these two parameters display a direct relationship with the aerodynamic forces that govern the performance of this type of VAWT. This analysis is accomplished via 2D-CFD simulations of a H-Darrieus with a symmetrical NACA 0018, powered by high winds (8 and 20 m/s), by using a Shear Stress Transport SST k-ω model. The 2D model used is validated for Class II winds (8 m/s), for tip speed ratios (λ) ranging from 0.4 to 0.9. Power coefficients (C_p) predictions are close to those obtained with both 3D simulations and with experimental data, reported in the technical literature. It is found with the numerical simulations developed, that despite the significant increase of the average rotor overall torque values, when the wind speed is augmented from 8 m/s to 20 m/s, the energy extracted by the rotor seems to be moderately lessened by the amplified turbulence and vorticity.

Keywords: H-Darrieus; high winds; vorticity index; vortex separation

Citation: Acosta-López, J.G.; Blasetti, A.P.; Lopez-Zamora, S.; de Lasa, H. CFD Modeling of an H-Type Darrieus VAWT under High Winds: The Vorticity Index and the Imminent Vortex Separation Condition. *Processes* **2023**, *11*, 644. https://doi.org/10.3390/pr11020644

Academic Editors: Ferdinando Salata and Virgilio Ciancio

Received: 26 January 2023
Revised: 10 February 2023
Accepted: 11 February 2023
Published: 20 February 2023

1. Introduction

The continuous decrease of natural resources available for energy production and the unwanted CO_2 emissions produced from fossil fuels combustion, have generated great interest in the development of alternative and renewable energies [1]. According to the latest report from the global company BP, considering an accelerated scenario, by the year 2050 alternative and renewable energies might even become the largest source of energy in the world, comprising approximately 60% of the global market [2]. These types of energies consist of solar, biomass, geothermal, hydroelectric and wind [3]. Being wind energy a favored option for planners and national governments [3–5]. Compared to conventional energy sources, wind power is positioned as one of the most important sources of renewable energy and the high potential of its use in electric power generation [5–7].

Furthermore, wind energy is considered to provide unique opportunities for the implementation of small scale and decentralized hydrogen production technologies [8,9]. Wind energy contributes to the first step in an integrated process, where it can be converted into electrical power. Electrical energy can be used later for the production of hydrogen via photocatalysis or electrolysis [10].

Given these incentives, in the last decades, the extraction of energy from the wind through wind turbines has seen an accelerated development. The growing investment in researching these technologies has resulted in the development of increasingly powerful turbines designs at more affordable prices. These designs include both horizontal axis wind turbines (HAWTs) as well as vertical axis wind turbines (VAWTs) [11].

In spite of most implemented wind turbines have been the HAWT designs, given their greater efficiency while compared to VAWTs [12,13], the HAWTs still show certain disadvantages such as (a) high maintenance costs, (b) location requirements [4,14] and (c) wind direction dependence [15]. In this respect, VAWTs can be considered a better option than HAWTs under certain operating conditions because of their: (i) variable wind direction potential, (ii) adaptability for domestic installations, (iii) compactness, and (iv) capability to operate under severe turbulent flow conditions [14,16], normally present under high winds. The main differences between VAWT and HAWT configurations are summarized in Table 1.

Table 1. Comparison of Parameters between VAWT and HAWT.

	Vertical Axis Wind Turbine (VAWT)	Horizontal Axis Wind Turbine (HAWT)
Efficiency	Lower	Higher
Space Efficiency [a,b]	Higher	Lower
Wind Direction [a,b]	Independent	Dependent
Yaw mechanism [b]	No	Yes
Self-Starting [b]	No	Yes
Height from ground [b]	Small	Large
Tower Sway [b]	Small	Large
Generator location [b]	Ground Level	Not Ground Level Required
Installation Cost [a,b]	Lower	Higher
Maintenance Cost [a,b]	Lower	Higher
Shadow Flickering [b]	Less	More
Noise [a,b]	Low	High
Bird/bat safety [a,b]	High	Low

[a] [17], [b] [18].

Given their reported advantages, VAWTs have become increasingly relevant in the field of wind energy research, as can be seen in the increasing number of bibliographic publications available for each of these configurations (Figure 1). In general, the research areas of primary interest regarding VAWTs are concerned with improving turbine rotor efficiency, through aerodynamic analysis and wind tunnel experimentation [17]. In this regard, different designs have been evaluated to reduce the problems that affect their performance. However, aerodynamics analysis and experiments in wind tunnels can be costly and require significant research time [19].

Thus, computational models that allow establishing the behavior of VAWTs, have become one of the preferred tools of both industry and researchers in this field [20,21]. These computational methods can be used to establish vorticity parameters, which can significantly help to quantify the influence of vorticity on VAWT performance. To accomplish this the main objectives of this work are the following:

- To implement a Vorticity Index (VI), defined as the ratio between the leading-edge vorticity (LEV) and the trailing-edge vorticity (TEV), to quantify vorticity in the VAWTs.
- To develop this study using a 2D model validated with experimental data reported in the literature [22].
- To establish a relationship between the Vorticity Index (VI) and the Imminent Vortex Separation Condition (IVSC) with the VAWT extracted energy, for a VAWT functioning at 8 m/s and 20 m/s.

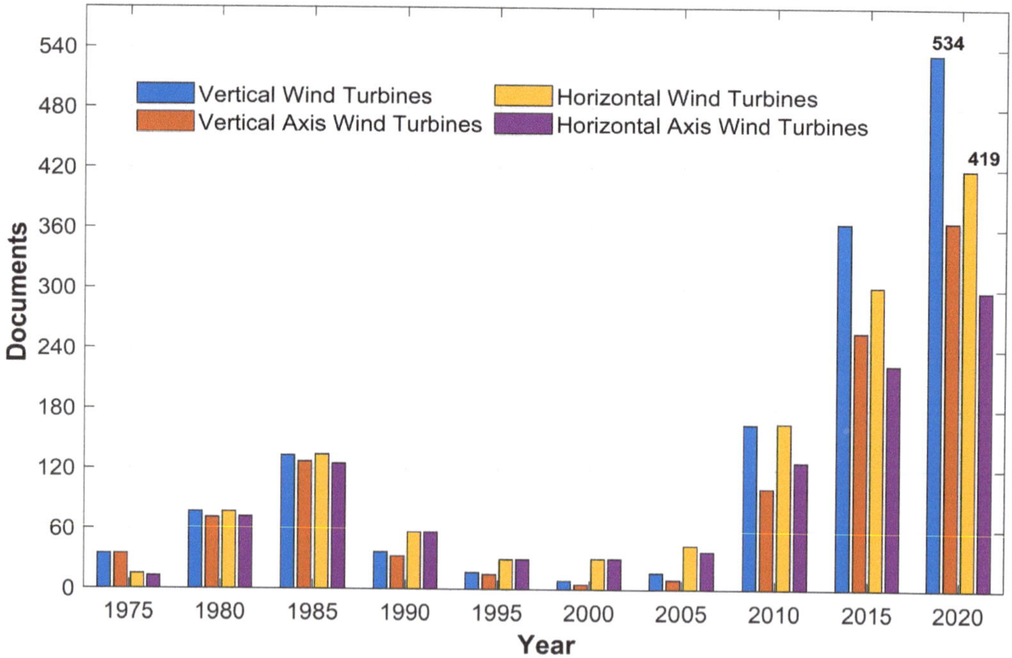

Figure 1. Studies Concerning Wind Turbines in the 1970–2020 Period (2021 Scopus's Database).

2. Numerical Simulation

2.1. Physical Model

One of the most studied VAWT designs is the Darrieus type H (H-Darrieus) wind turbine. This design mainly consists of 3 aerodynamic profile blades attached to a vertical rotating shaft, which are driven by the force of the wind that hits them [23]. In recent years, different CFD simulations of H-Darrieus wind turbines have been carried out implementing 2D models [24–27] and 3D models [6,28,29]. In the case of most of these simulations, the shaft is usually not included in the geometric model, as shown in Figure 2.

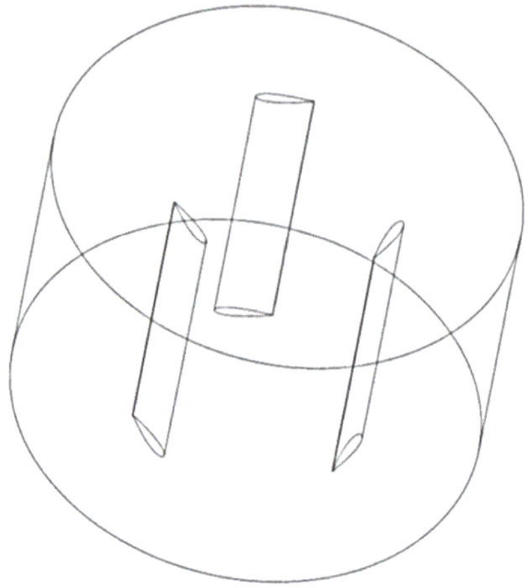

Figure 2. Schematic View of the H-Darrieus Wind Turbine Geometry Used for CFD Calculations.

As reported in Table 2, the design parameters and dimensions of the geometric model implemented in this work corresponded to the dimensions of a H-Darrieus wind turbine that was experimentally tested by Elkhoury et al. [22] using a NACA 0018 blade. This specific blade configuration was selected given the abundant experimental and simulation data for VAWTs with NACA 0018.

Table 2. Design Parameters of the H-Darrieus Wind Turbine [22].

Parameter	Symbol	Value
Rotor Diameter [m]	D	0.8
Blade Airfoil	-	NACA 0018
Blade Shape	-	Straight
Chord Length [m]	c	0.2
Rotor Height [m]	H	0.8 m (1 m adopted for 2D simulation)
Blades Number	N	3
Solidity	σ	0.75

Note: Solidity, $\sigma = \frac{N\,c}{D}$.

2.2. Computational Domain

For H-Darrieus wind turbine CFD simulations, two domains of interest can normally be considered: (i) a fixed domain which represents the control volume and (ii) a rotating domain inside of which, the rotor is accounted for. In this work, the fixed domain corresponds to a rectangle of 20 m × 8 m, and the rotating domain involves a 1.6 m or a $2D$ diameter circle (Figure 3). It can be noticed that the center of the rotor was located at $10D$ from the inlet boundary and at $15D$ from the outlet boundary. This was determined in order to ensure the correct development of the wake effect [6], which may lead to an underestimation of the fluid velocity and an overestimation of the turbulence [30].

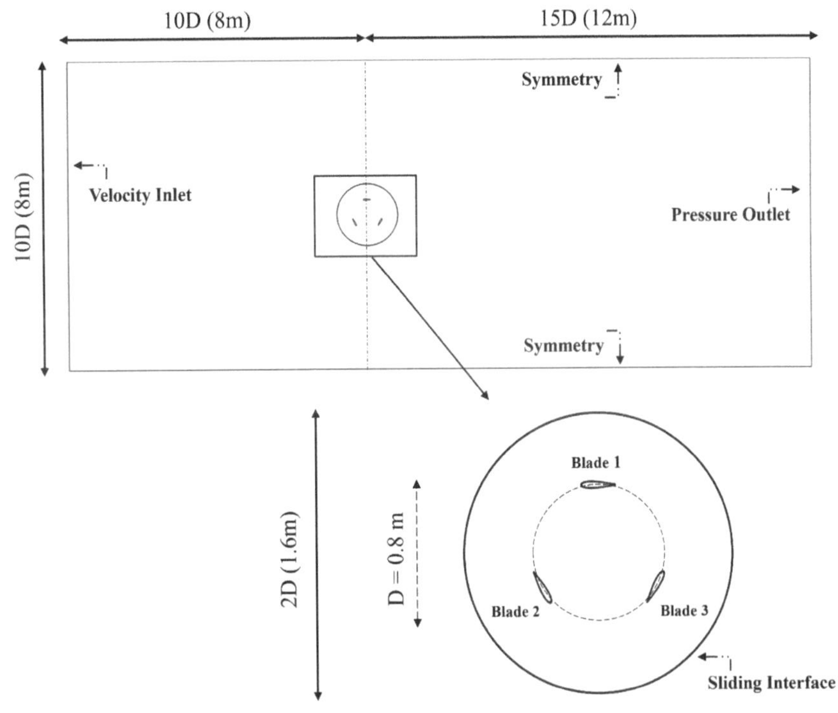

Figure 3. Description of the Computational Domain and Boundary Conditions Used for Calculations.

The boundary conditions: inlet, outlet, lateral sides, blades, and contact region, for rotating and fixed domains were set by using the velocity inlet, the pressure outlet, the symmetry plane, non-slip walls, and sliding interface, respectively.

2.3. Setting Up

In this work, the commercial software ANSYS 19.1 Fluent was selected as the CFD simulation package, to solve the flow equations. The continuity Equation (1) and the momentum Equation (2) used to solve URANS simulations in the ANSYS Fluent software are shown below [31,32]:

$$\frac{\partial \overline{u}_i}{\partial x_i} = 0 \quad (1)$$

$$\frac{\partial \overline{u}_i}{\partial t} + \frac{\partial}{\partial x_j}\left(\overline{u}_i \overline{u}_j \right) = -\frac{1}{\rho}\frac{\partial \overline{p}}{\partial x_i} + \frac{\partial}{\partial x_j}\left(v\,\frac{\partial \overline{u}_i}{\partial x_j} - \overline{u}'_i \overline{u}'_j \right) \quad (2)$$

where, $\overline{u}_i$ and $\overline{u}_j$ are the mean values of velocity in the axial system of coordinates, u'_i and u'_j its fluctuating components, $\overline{p}$ is the mean pressure, v is the kinematic viscosity, ρ is the density of the fluid and t is the time. It is important to point out that the term $\overline{u}'_i\overline{u}'_j$, commonly represents the Reynolds stress tensor, which depends on the turbulence model selected [32].

For URANS calculations, the pressure-based solver was utilized while the pressure-velocity coupling was handled using coupled algorithm. The second-order upwind scheme was used for the spatial discretization of the pressure, the turbulence model, and the momentum equations. The main settings implemented for all the simulations are listed in Table 3.

Table 3. Details of Numerical Set-Up.

Parameter	Symbol	Value
Viscous Model	SST k-ω	k-ω Shear Stress Transport
Air Density	ρ	1.225 kg/m^3
Air Viscosity	μ	1.79×10^{-5} Pa s
Air Velocity	U_∞	8 m/s, 20 m/s
Turbulent Intensity		1%
Tip Speed Ratio	λ	0.5–1.5
Solver Type		Pressure-Based
Calculation algorithm		Coupled
Spatial Discretization		2nd
Time discretization		According to λ, calculated to achieve 2° of rotation per time step
Residuals		1×10^{-4}

Note: Tip Speed Ratio, $\lambda = \frac{R\,\omega}{U_\infty}$.

Turbulence Model

During RANS and URANS simulations, different numerical models can be employed to determine the eddy viscosity turbulence [21,33]. These turbulence models used for CFD simulation include: (a) a one-equation Spalart-Allmaras model, (b) two-equation models such as k-ε (RNG), Realizable k-ε (RKE) and SST k-ω models, (c) three equations model such SST k-ω with intermittency (SSTI). Comparisons of their results with experimental data have been reported by [17,34,35]. In agreement with the previous research [31,36], the Shear Stress Transport (SST) based models have been demonstrated as being the most accurate. Among them, the SST k-ω model was shown to be effective for VAWT CFD simulations. This is mainly because the SST k-ω model performs well in the treatment of the boundary layer over the blades and works well in the air free-flowing areas [16,37,38]. Therefore, the SST k-ω model was chosen for this study.

2.4. Meshing

The computational mesh used in the present study was developed using the ANSYS 19.1 meshing tool. To determine the best mesh size in terms of computational cost and accuracy, the influence of the computational grid on the main calculations was evaluated. In this study, three different unstructured computational meshes were tested for a set wind speed (U_∞) of 8 m/s at a tip speed ratio (λ) of 1. Following this analysis, the medium mesh was selected as the optimal one. This was determined after obtaining an average power coefficient (C_p) value difference lower than 1.25% when testing the finest mesh. The number of elements within the computational domains, for each mesh, and their respective average C_p values, are reported in Table 4.

Table 4. Mesh Independence Test for λ = 1.

		Mesh Set		
		Coarse	Medium	Fine
Face Sizing	Fixed Domain Element Size [m]	0.120	0.080	0.060
	Rotor Domain Element Size [m]	0.012	0.0073	0.0053
Edge Sizing	Interface Element Size [m]	0.011	0.0069	0.0049
	Number of Divisions around Blades Surface	480	600	600
Elements Number	Fixed Domain	29,008	75,226	143,289
	Rotor Domain	62,416	125,564	216,843
	Total	91,424	200,790	360,132
Power Coefficient	C_p	0.196	0.200	0.202
	%Difference of C_p with respect to the Fine Mesh	−3.06%	−1.23%	/

Similar approaches for grid dimensioning were used for the three meshes considered in the present study, as follows: (a) First, the triangle method was applied for both domains, (b) Then, the rotating domain was refined by using a face sizing, (c) Following this step, the interface between the fixed domain and the rotating domain was refined by using an edge sizing, (d) Finally, the blades' surfaces were grid sized by increasing the number of elements around them and by creating an adequate inflation number for each of them. This was done to guarantee a $y+$ value < 1. This approach was considered to capture the viscous sublayer, as suggested by [20,30]. It is important to mention that the selected medium mesh size (Figure 4), complies with two conditions: (a) the 0.36 lowest orthogonal quality for the whole mesh is larger than 0.1 and (b) the 0.79 maximum skewness for the whole mesh is smaller than the 0.95 maximum advised [38].

2.5. Angular Marching Step

The timestep size is an important parameter when developing unsteady simulations, since it has a great influence on the final values obtained [37]. In this regard, the simulation for U_∞ = 8 m/s at λ = 1 was considered as the base case, with three different time steps being evaluated during ten consecutive H-Darrieus wind turbine rotor rotations. This approach was judged to be reliable in order determine the adequate time step to generate the most consistent C_p results, at the lowest computational cost. As a result, the three timesteps evaluated for the base case were 0.003491 s, 0.001745 s and 0.000873 s, which corresponded to angular marching steps ($\Delta\alpha$) of 4°, 2° and 1°, respectively.

As shown in Figure 5, for each $\Delta\alpha$, the C_p value stabilization required a different number of simulated rotations, being the case when $\Delta\alpha$ = 4 the one that required the smallest number of simulated rotations. On the other hand, although when $\Delta\alpha$ = 2°, a higher number of simulated rotations was required to provide stabilized C_p values than when using $\Delta\alpha$ = 4°, its final C_p value showed a 3% difference only versus the C_p value calculated when $\Delta\alpha$ = 1°, as reported in Table 5. Thus, $\Delta\alpha$ = 2° was considered a good compromise as an angular marching step, reducing computational cost without considerably compromising the accuracy of the results.

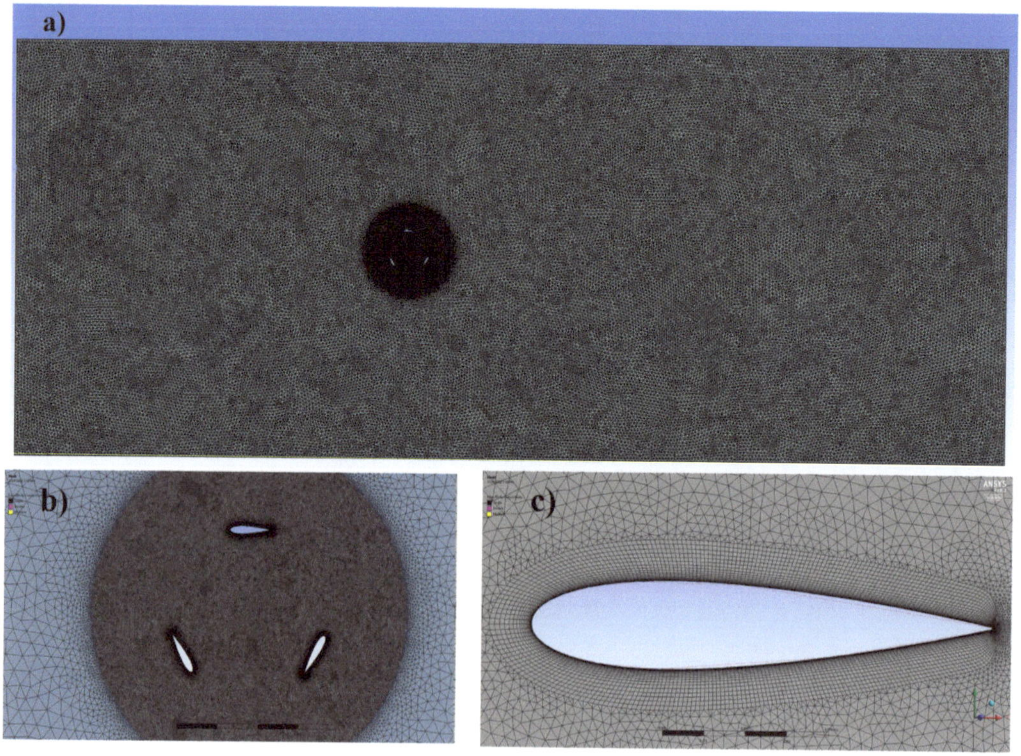

Figure 4. Medium Mesh Detailed Grids: (**a**) Grids in the Entire Computational Domain, (**b**) Grids in the Rotating Domain, (**c**) Grids around Blade 1.

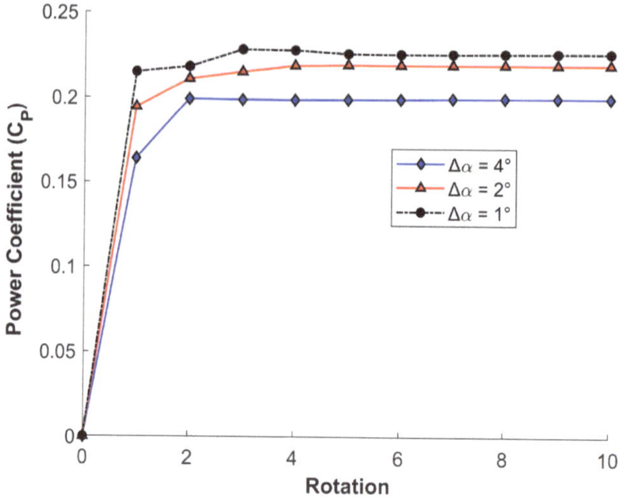

Figure 5. Power Coefficient Changes with the Number of Rotations for U_∞ = 8 m/s when adopting Different $\Delta\alpha$ at λ = 1.

Table 5. Timestep Independence Test for λ = 1.

Time Step	Angular Marching Step ($\Delta\alpha$)	C_m	C_p	%Difference of C_p with Respect to Smallest Time Step
0.003491 s	4°	0.200	0.200	−12%
0.001745 s	2°	0.220	0.220	−3%
0.000873 s	1°	0.227	0.227	<0.1%

3. Results and Discussion

In this work, all calculations were performed using the MATLAB R2021a Software, based on the values generated with the ANSYS 19.1 Fluent solver. The power coefficient (3) was calculated using the torque results obtained via CFD simulations. A minimum of eight rotor rotations were considered for all the simulations. This to guarantee the consistency of numerical calculations between rotations, as recommended in the technical literature [20]. The average C_p values were then estimated and presented in this section, for U_∞ = 8 m/s and U_∞ = 20 m/s, at various tip speed ratios (λ) of interest.

$$C_p = \frac{P_{turbine}}{P_{wind}} = \frac{T\,\omega}{\frac{1}{2}\rho\, U_\infty^3 (2R\,H)} \quad (3)$$

where T represents the mechanical torque, ω denotes the rotational speed of the rotor, U_∞ stands for the free wind speed, ρ is the air density and, R and H are radius and height of the rotor, respectively.

Considering the convergence of the results between rotations, the behavior of the torque, the power coefficient, and the drag and lift forces were reported for the last simulated H-Darrieus wind turbine rotation for both wind speeds, and this for two different λ values of 0.5 and 0.9. This was done in order to analyze the influence of the wind speed on the general aerodynamics of the proposed H-Darrieus wind turbine. Finally, graphical contours were produced at various stages of the H-Darrieus wind turbine operation, to illustrate the effects of vortex shedding on the performance of the wind turbine and to relate them to the vorticity index proposed in this study.

3.1. Model Validation

To validate the 2D CFD simulations of the H-Darrieus wind turbine carried out in this study, the first section of this article compares the calculated average C_p values with the values reported by Ma et al. [37], at various tip speed ratios (λ). In their work, the authors evaluated the same rotor configuration for a wind speed (U_∞) of 8 m/s with runs developed in a wind tunnel and by using 3D CFD simulations. Although 2D simulations can generate certain discrepancies with respect to the experimental results, such simulations can provide a reliable representation of the flow in the mid-plane of the turbine, at a considerably lower computational cost than 3D simulations and even 2.5D simulations [17,27]. The differences obtained in each type of CFD simulation with respect to the data obtained experimentally, are reported in Table 6.

As shown in Figure 6, simulations predictions of average C_p values for a 2D model were greater than those obtained in 3D simulation. While 2D simulations may provide a somewhat limited description of the complex dynamics of the flow, around the wind turbine [19,29,39], for tip speed ratios ranging from 0.4 to 0.9, the average relative error is about +/−9.27%, with a maximum relative error value of 27.5%. Therefore, this approach is acceptable for the validation of the vorticity index (VI) within the expected level of confidence of the experimental data, when tip speed ratios are below 0.9 values. For λ values larger than 0.9, observed 2D simulations results could not adequately represent the experimental data, and therefore were not considered in the present analysis.

Table 6. Power Coefficient Comparison of 2D SST k-ω Simulation with Experimental Data and 3D SST k-ω Simulation Results.

λ	Power Coefficient (C_p)				
	Experimental Results [37]	2D SST k-ω (This Work)	Relative Error 2D vs. Exp	3D SST k-ω [37]	Relative Error 3D vs. Exp
0.4	0.028	0.030	5.9%	0.030	5.5%
0.5	0.040	0.036	−10.6%	0.048	19.6%
0.7	0.073	0.053	−27.5%	0.079	8.6%
0.8	0.101	0.108	7.2%	0.105	4.7%
0.9	0.128	0.157	22.9%	0.132	3.1%
1.0	0.167	0.220	31.5%	0.159	−4.7%
1.2	0.190	0.326	71.4%	0.184	−3.3%
1.3	0.195	0.355	81.9%	0.191	−2.0%
1.4	0.186	0.375	101.5%	0.191	2.5%
1.5	0.169	0.385	128.0%	0.180	6.0%

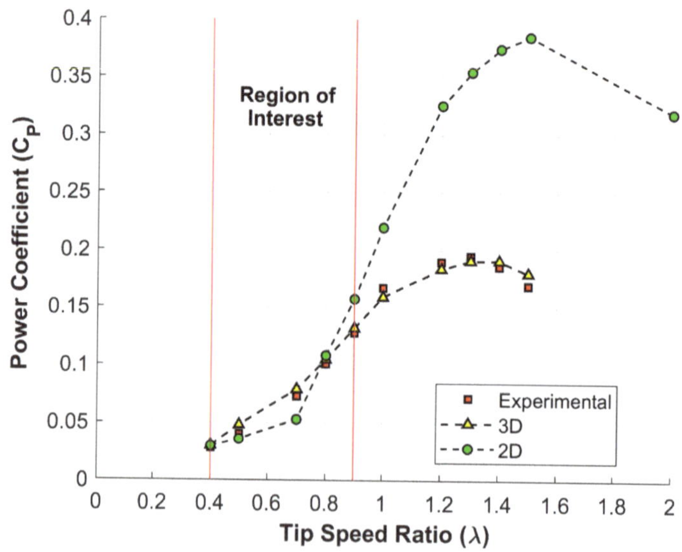

Figure 6. Power Coefficient Changes with Tip Speed Ratio for U_∞ = 8 m/s. (o) 2D Model Used in the Present Study, (Δ) 3D Data from [37], (□) Experimental Data from [37].

3.2. Influence of High Winds

3.2.1. Wind Speed Effect on Overall Torque—Convergence Criteria

The stability criteria for the average rotor overall torque values, following two consecutive revolutions, was achieved for both low (8 m/s) and high (20 m/s) wind speeds. As shown in Figure 7, a difference smaller than 1% in the average toque value between two subsequent rotations was achieved for all the tip speed ratios evaluated, as suggested by [40]. In this respect, a significant increase of about seven to eight times, in the average rotor overall torque values was noticed when the wind speed increased from 8 m/s to 20 m/s. In the case of U_∞ = 8 m/s simulations (Figure 7a), this value varied between 0.93 N m (for λ = 0.4) and 2.2 N m (for λ = 0.9), while in the case of the U_∞ = 20 m/s simulations (Figure 7b), the average rotor overall torque varied between 7.2 N m (for λ = 0.4) and 16.7 N m (for λ = 0.9).

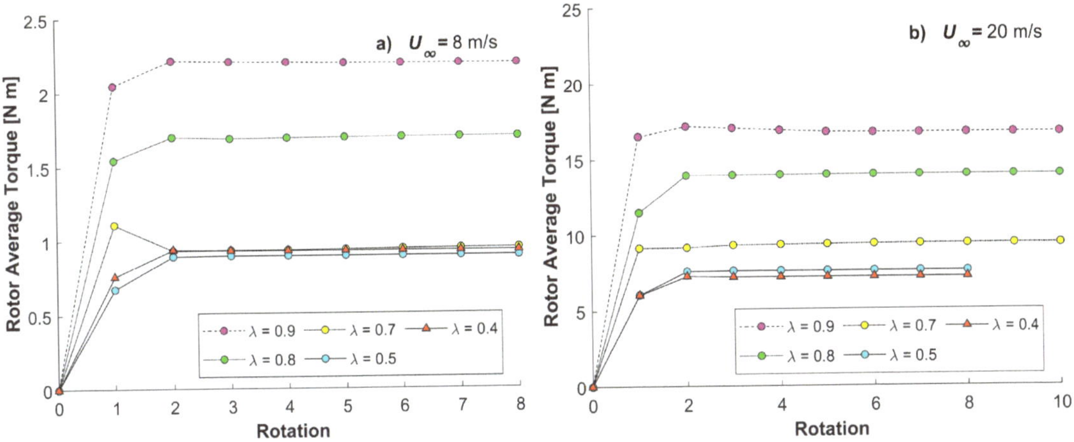

Figure 7. Average Rotor Overall Torque at Different λ for (**a**) U_∞ = 8 m/s and (**b**) U_∞ = 20 m/s, Using the 2D Model of the Present Study. Note: Rotor Average Torque = $\frac{1}{2\pi}\int_0^{2\pi} T\, d\theta$.

Regarding VAWTs, the upcoming sections of the present article, reports the influences of drag and lift forces on the performance of the H-Darrieus wind turbine's rotor. This is accomplished by considering that the overall torque of the turbine largely depends on the positions of the blades and the dominant forces on the rotor during its rotations, with this either being the lift force or the drag force [32,39].

3.2.2. Wind Speed Effect on Drag and Lift Forces

One of the main advantages of VAWTs is that they do not require a specific wind orientation mechanism to operate (Table 1), which is of great importance in regions of high rotating winds. It is, however, also well known that the aerodynamic forces during VAWTs operation are cyclic and contribute to fatigue and low power coefficients (C_p) while compared to the ones obtained in HAWTs.

In the case of an H-Darrieus wind turbine, the aerodynamic forces generated on the sectional profile of the blades as consequence of airflow through the turbine rotor, are significantly influenced by the azimuthal angle (θ) [27] and the angle of attack α of the wind. These forces, generally referred to as the lift forces (FL) and drag forces (FD) correspond to the force perpendicular to the flow and the force in the direction of the wind relative flow velocity (W), respectively [41]. Thus, based on FL and FD, the accountable forces of the torque such as normal forces and the tangential forces, can be established as described in Figure 8.

Figures 9 and 10 report the drag and the lift forces on the 3 blades of the H-Darrieus wind turbine for the same wind speed (U_∞ = 8 m/s) at two different tip speed ratios: λ = 0.5 and λ = 0.9. Furthermore, these figures show not only the displacements of the maximum and minimum values of these aerodynamic forces but also the variations in their cyclical behavior, both in terms of magnitude, and in terms of distribution over a rotational period. Then, one can notice that in the case for the drag forces, the maximum and minimum values on the main blade at λ = 0.5 (Figure 9a) are reached at θ = 60° and θ = 176°, respectively. At λ = 0.9 (Figure 9b), there is a peak displacement, and these values are reached at θ = 84° and θ = 162°, respectively. On the other hand, in the case for lift forces, the displacement effect is small, and the maximum and minimum values on the main blade are reached at θ = 118° and θ = 50°, respectively when λ = 0.5 (Figure 10a), and at θ = 114° and θ = 54°, respectively when λ = 0.9 (Figure 10b). Thus, it can be observed that at higher λ, the drag forces on the blades increase significantly (Figure 9). Furthermore, in the case of the lift forces (Figure 10), an effect of λ on the second negative peak of the distribution in the θ = 60°–180° azimuthal range is evidenced. One can see that this second minimum (−20 N) is attained at an earlier azimuthal angle, as compared to the angle reached when $\lambda = 0.5$ (−10 N).

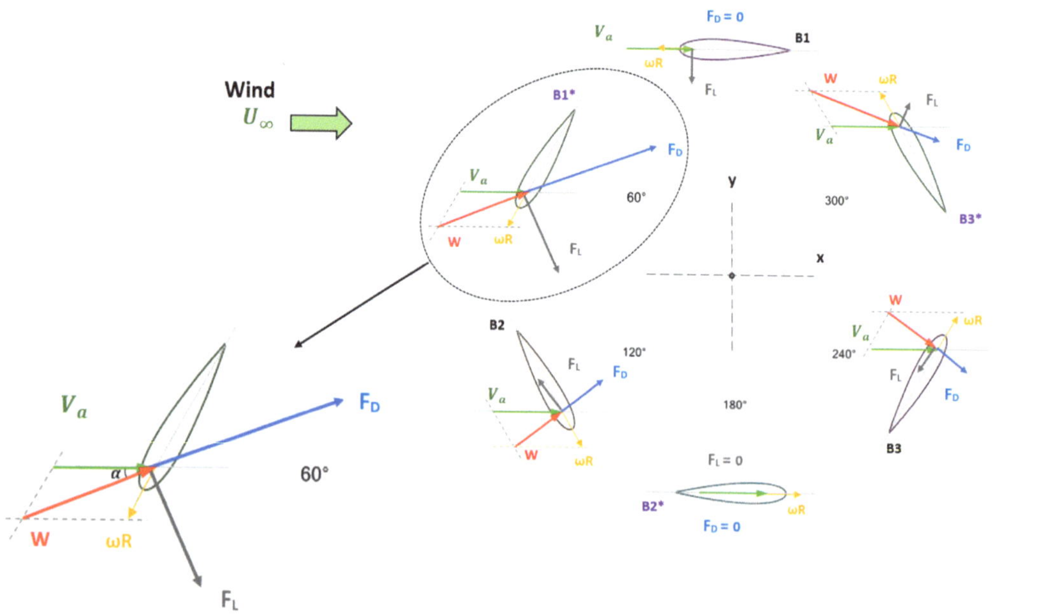

Figure 8. Wind and Blades Velocity Components with Forces acting on H-Darrieus' blades at Various Azimuthal Positions, with FL, and FD representing the lift and drag forces vectors and W and V_a representing the relative flow velocity and the induced velocity of wind.

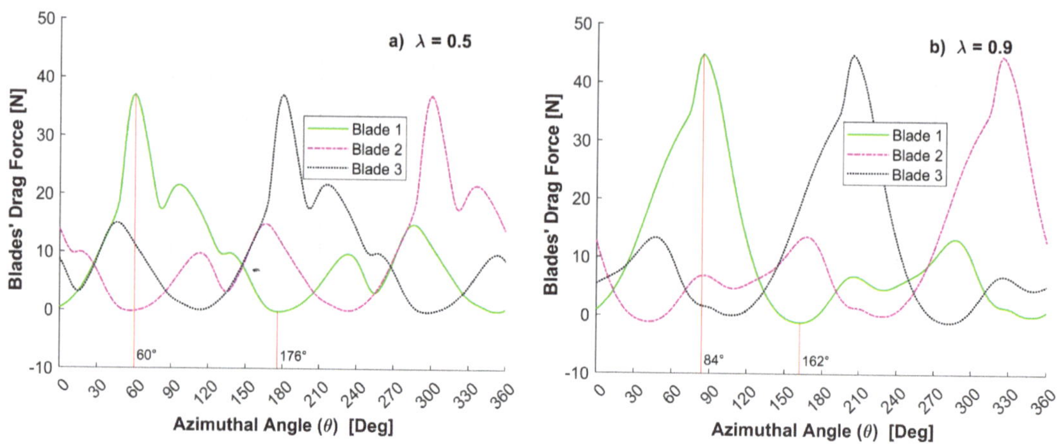

Figure 9. Drag Force for Blades 1, 2 and 3 at Various Azimuthal Angles, for U_∞ = 8 m/s at (**a**) $\lambda = 0.5$ and (**b**) $\lambda = 0.9$, Using the 2D Model of the Present Study.

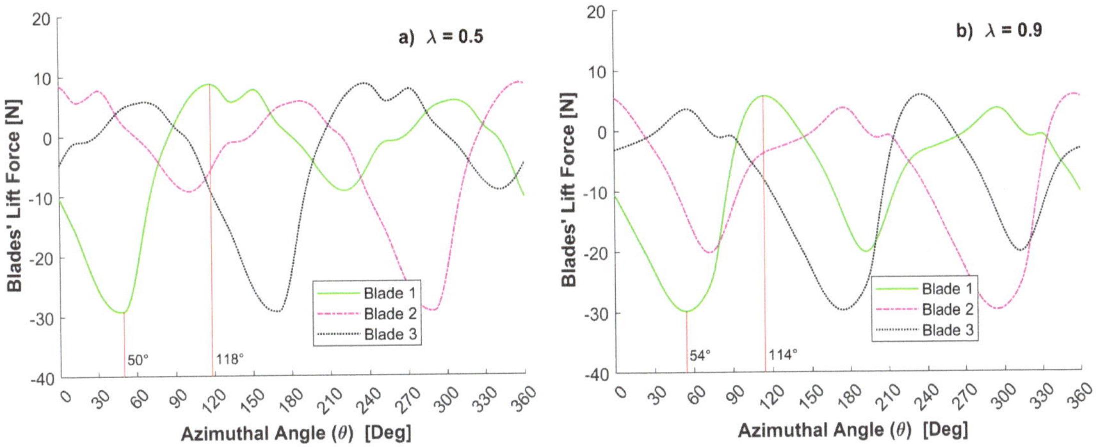

Figure 10. Lift Force for Blades 1, 2 and 3 at Various Azimuthal Angles, for U_∞ = 8 m/s at (**a**) λ = 0.5 and (**b**) λ = 0.9, Using the 2D Model of the Present Study.

On the other hand, when the behavior of the drag force (Figure 11) and the lift force (Figure 12) on the main blade for both wind speeds are compared, they show practically identical distributions, with the maximum and the minimum values of these two forces on the main blade being reached approximately at the same azimuthal angles (θ).

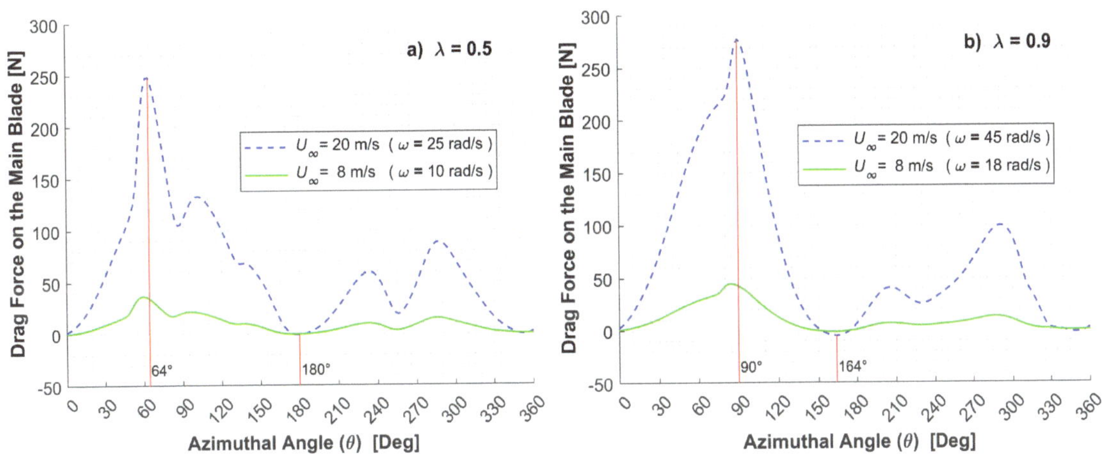

Figure 11. Drag Force for Blade 1at Various Azimuthal Angles, for U_∞ = 8 m/s and U_∞ = 20 m/s at: (**a**) λ = 0.5 and (**b**) λ = 0.9, Using the 2D Model of the Present Study.

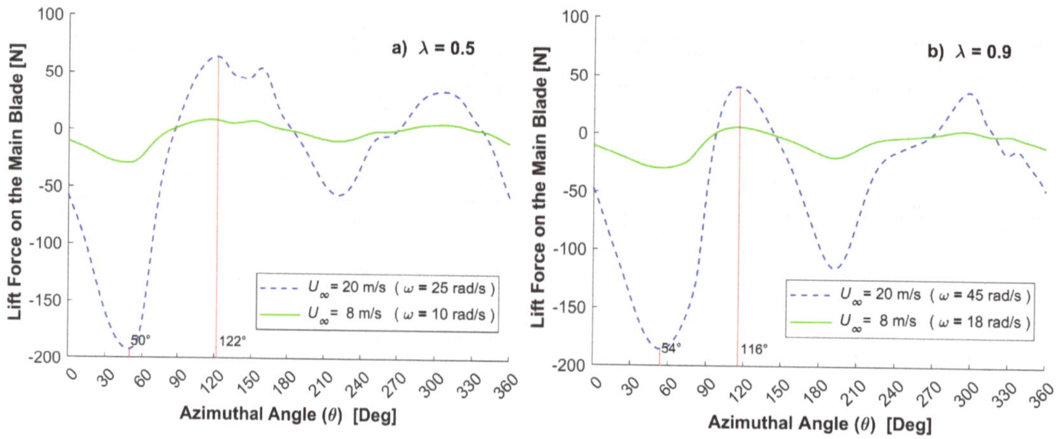

Figure 12. Lift Forces for Blade 1at Various Azimuthal Angles, for U_∞ = 8 m/s and U_∞ = 20 m/s at: (**a**) λ = 0.5 and (**b**) λ = 0.9, Using the 2D Model of the Present Study. Note: The direction of the lift forces is established using as a reference the vertical "y" axis.

In the case of a fixed pitch blade, as the angle of attack increases, "static stall" conditions are reached, with flow separation and draft forces being greater than lift forces. However, due to the rotation of a VAWT, there is a pitching oscillation of the blades that results in a "dynamic stall" condition. This condition is characterized by a vortex generation that starts at the leading edge of the blade and is shed later, with lift and drag forces fluctuating [25] as the angle of attack rapidly changes. Thus, as the wind turbine rotate, different azimuthal angles are reached and the angle of attack (α) on each blade influences drag and lift forces. The influence of λ and the azimuthal position (θ) on the angle of attack (α) can be described as proposed by Laneville & Vittecoq [42]:

$$tan\ (\alpha) = \frac{Sin\ (\theta)}{(Cos\ (\theta) + \lambda)} \tag{4}$$

Then, when the main blade (Blade 1) starts to turn counterclockwise from the $\theta = 0°$ angular position (Figure 8), a "dynamic stalling" effect seems to start to develop with lift forces decreasing and drag forces increasing progressively. One can observe that for Blade 1, lift and drag force minimum and maximum values take place between $\theta = 50°$ and $\theta = 64°$ with a lift force magnitude change from −30 N for U_∞ = 8 m/s to −190 N for U_∞ = 20 m/s (Figure 12a). There are no significant changes in this azimuthal angle, when λ increases from 0.5 to 0.9 (Figure 12b). In addition, and at the same conditions, drag forces on Blade 1 show a peak increase from 38 N for U_∞ = 8 m/s to 250 N for U_∞ = 20 m/s (Figure 11a) 280 N for U_∞ = 20 m/s is even reached when λ increases from 0.5 to 0.9 (Figure 11b).

A summary of both the maximum and the minimum drag and lift forces for Blade 1 and their corresponding angles of attack, for both U_∞ = 8 m/s and U_∞ = 20 m/s at the two λ of interest, is reported in Table 7. One can observe, on this basis, that the maximum position of the drag forces for U_∞ = 8 m/s shifts from $\theta = 60°$ to $\theta = 84°$ as the tip speed ratio increases from 0.5 to 0.9. This change can also be seen when U_∞ = 20 m/s is used, with the maximum position of the drag force moving from $\theta = 64°$ to $\theta = 90°$. In the case of the lift forces, the displacement effect is negligible. Maximum lift force peaks for Blade 1, for both wind speeds, are observed at λ = 0.5 and λ = 0.9, in the θ = 118°–120° and θ = 114°–116° azimuthal range, respectively.

Table 7. Summary of Maximum and Minimum Drag and Lift Forces on the Blade 1 for U_∞ = 8 m/s and U_∞ = 20 m/s at both λ = 0.5 and λ = 0.9.

	λ	θ [Degrees]	α [Degrees]	
Figure 9 Drag Forces U_∞ = 8 m/s	0.5	60	40.9	Max. Drag
		176	−8.0	Min. Drag
	0.9	84	44.7	Max. Drag
		162	−80.6	Min. Drag
Figure 10 Lift Forces U_∞ = 8 m/s	0.5	50	33.8	Min. Lift
		118	88.0	Max. Lift
	0.9	54	28.5	Min. Lift
		114	61.6	Max. Lift
Figure 11 Drag Forces U_∞ = 20 m/s	0.5	64	43.8	Max. Drag
		180	0.0	Min. Drag
	0.9	90	48	Max. Drag
		164	−77.5	Min. Drag
Figure 12 Lift Forces U_∞ = 20 m/s	0.5	50	33.8	Min. Lift
		120	90	Max. Lift
	0.9	54	28.5	Min. Lift
		116	62.8	Max. Lift

3.2.3. Wind Speed Effect on Individual Torque

Regarding the torque values obtained on each of the H-Darrieus wind turbine blades, for U_∞ = 8 m/s one can observe that they present a cyclic pattern at both λ = 0.5 and λ = 0.9 (Figure 13), with the torque being related to the position of the individual blades, at different azimuthal angles. One can note that the peak torque values for each blade show a close to 26° azimuthal angle difference, when λ increases from 0.5 to 0.9. This can be assigned to the substantial change in the angle of incidence of each blade, according to the rotor position.

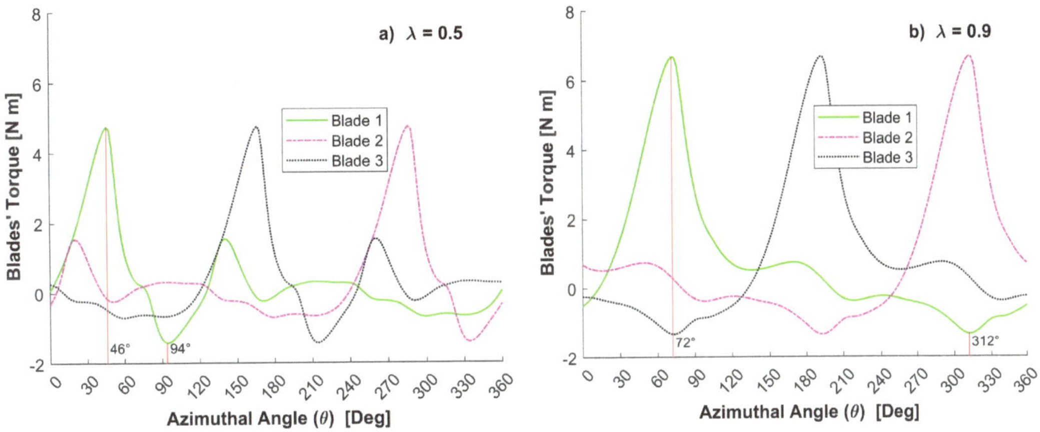

Figure 13. Torque on Blades 1, 2 and 3 at Various Azimuthal Angles, for U_∞ = 8 m/s at: (**a**) λ = 0.5 and (**b**) λ = 0.9, Using the 2D Model of the Present Study.

On this basis, it is possible to envision how an increment in the rotor speed increases the torque on the blades and shifts their peaks. Furthermore, when comparing the behavior of the torque on Blade 1, for a complete rotation of the rotor, and using both U_∞ = 8 m/s and U_∞ = 20 m/s (Figure 14), one can notice that the maximum and minimum torque values on the main blade are reached at close θ values for both wind speeds. It can also be noticed that torque values generated at U_∞ = 20 m/s are ten times larger than those produced at U_∞ = 8 m/s. However, given that for both U_∞ the peak and low torque distribution functions show a similar trend, one can conclude that both a substantial torque increase and oscillations can be expected at higher wind velocities. On the other hand, for a $U_\infty = 20$ m/s at $\lambda = 0.5$ (Figure 14a), maximum and minimum torque values on Blade 1 are reached at $\theta = 50°$ ($\alpha = 33.8°$) and $\theta = 100°$, respectively. While for $U_\infty = 20$ m/s at $\lambda = 0.9$ (Figure 14b) the maximum torque is reached at $\theta = 78°$ ($\alpha = 41.4°$) and the minimum torque shifts to $\theta = 316°$.

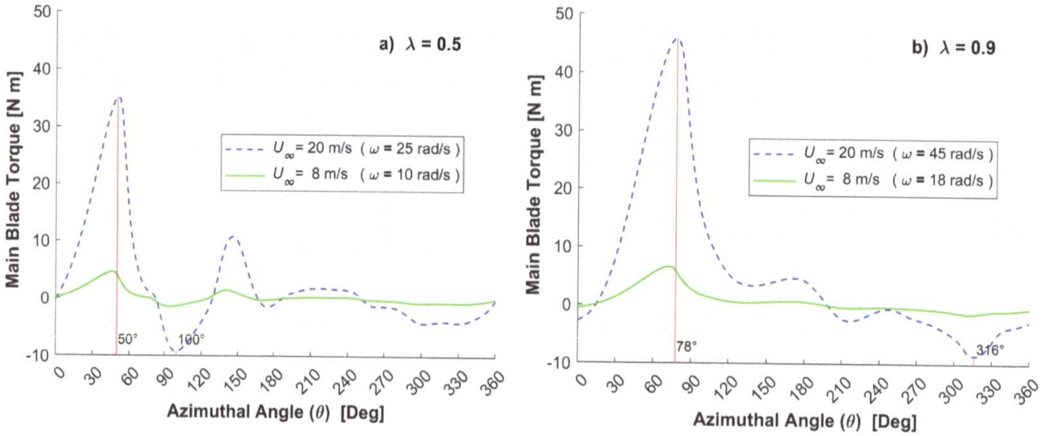

Figure 14. Torque on Blade 1 at Various Azimuthal Angles, for $U_\infty = 8$ m/s and $U_\infty = 20$ m/s at: (**a**) $\lambda = 0.5$ and (**b**) $\lambda = 0.9$, Using the 2D Model of the Present Study.

By relating the position of the peak drag force on Blade 1 (Figure 9) to the maximum torque on it (Figure 13), one can conclude that the drag forces have a greater influence on torque peak than lift forces, having Blade 1 under dynamic stall conditions in the $\theta = 0°$ to $\theta = 60°$ azimuthal position range. This is also valid for $U_\infty = 20$ m/s, with maximum torque values being obtained when $\theta = 50°$ and $\theta = 78°$ (Figure 14). Likewise, as shown in Figure 8, when Blade 1 is in the $\theta = 60°$ to $\theta = 120°$ azimuthal position range, it will experience lift force vectors impacted by an asynchronous system of forces that affect both torque and power. This effect has been anticipated for a H-Darrieus wind turbine, where mechanical power is the result of the overall torque, mainly attributed to aerodynamic lift forces [43].

Additionally, from the torque patterns observed at different azimuthal positions, for both wind speeds at $\lambda = 0.5$ and $\lambda = 0.9$ (Figure 14), it can be observed that a positive torque value on the main blade is reached before $\theta = 90°$. However, beyond this point, the torque is relatively small or negative. Similar torque patterns were recorded at both λ, with torque magnitudes differences speeds between both wind speeds 30% larger for $\lambda = 0.9$ than for $\lambda = 0.5$. Regarding the results obtained in this work, torque pulsation reductions as a consequence of tip-speed-ratio increases can be explained by an increased propensity of the flow to stay attached to the blade surface. This has been analyzed by Ahmedov & Ebrahimi [25] through simulations of a 4 blade H-Darrieus wind turbine under turbulent winds. This appears to be consistent with torque results presented in Figure 14 for both, low (8 m/s) and high (20 m/s) wind speeds.

3.2.4. Wind Speed Effect on the Power Coefficient

As previously mentioned, at high wind speeds, the torque on the three blades increases considerably, with the average C_p values remaining at similar levels (Figure 6). Likewise, the C_p changes during a complete rotor's rotation are almost the same for both wind speeds, either at $\lambda = 0.5$ or at $\lambda = 0.9$, with minimal differences in their magnitudes, as shown in Figure 15.

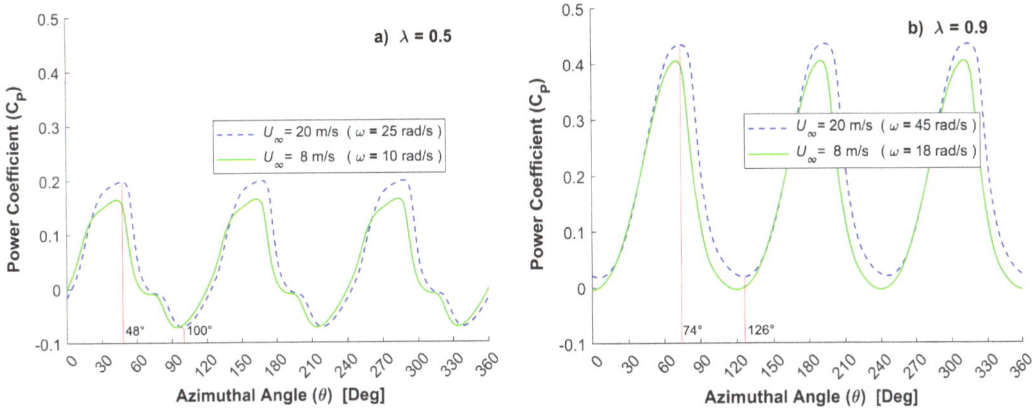

Figure 15. Power Coefficient Changes with Azimuthal Angles for U_∞ = 8 m/s and U_∞ = 20 m/s at: (**a**) λ = 0.5 and (**b**) λ = 0.9, Using the 2D Model of the Present Study.

According to (3), it is apparent that the maximum value of C_p is obtained at a maximum overall torque (T). This occurs at azimuthal angles (θ) that are close to those of the torque peak values for each one of the blades. In the case of the main blade, when U_∞ = 8 m/s, the azimuthal angles for the maximum C_p were θ = 48° and θ = 74° at λ = 0.5 as at λ = 0.9. Torque peak values were obtained at θ = 46° and θ = 72°, respectively. Similar behavior was observed for U_∞ = 20 m/s.

3.3. Vorticity Index Results

Dynamic stall is a process that results in an increase of lift forces, due to rapid changes of the angle of attack of the airfoil. This is normally characterized by a delayed flow separation over an aerodynamic airfoil, which is beyond the steady-state stall angle [44]. As shown in Figure 12, it produces an increased lift that is also followed by a lift decay related to vortex separation conditions, during rotor rotation. According to Ahmedov & Ebrahimi [25], the occurrence and intensity of the dynamic stall effect can be related to the reduced frequency (k^*) parameter. In a VAWT, the k^* represents the ratio between the time required for the wind to pass over a blade and the time needed for the angle of attack to change from positive to negative, with this representing he degree of unsteadiness of the airfoil [45]. In order to further address this issue, and as shown in Figure 16, for λ = 0.5, and in Appendix A for λ = 0.9, the vorticity magnitude [46] and its changes over the blade surface, have to be considered. It can be observed that vorticity magnitude reaches maximum values, either at the blade leading edge (LE) or the blade trailing edge (TE), at different azimuthal angles.

When analyzing these vorticity magnitude values at the leading edge (LE) and the trailing edge (TE) of Blade 1, for U_∞ = 8 m/s and U_∞ = 20 m/s, it was found that the azimuthal angle of the maximum lift force and torque can be predicted. This is done by relating the vorticity levels at both the LE and the TE of the blade. Therefore, a "Vorticity Index (VI)" was defined as the ratio between the leading edge vorticity (LEV) and the trailing edge vorticity (TEV), for different azimuthal positions:

$$Vorticity\ Index\ (VI) = \frac{Leading\ Edge\ Vorticity\ (LEV)}{Trailin\ Edge\ Vorticity\ (TEV)} \tag{5}$$

It was found that maximum torque values shown in Figures 13 and 14 are related to this definition. Vorticity Index results for U_∞ = 8 m/s and U_∞ = 20 m/s are available in the Appendices B and C, respectively.

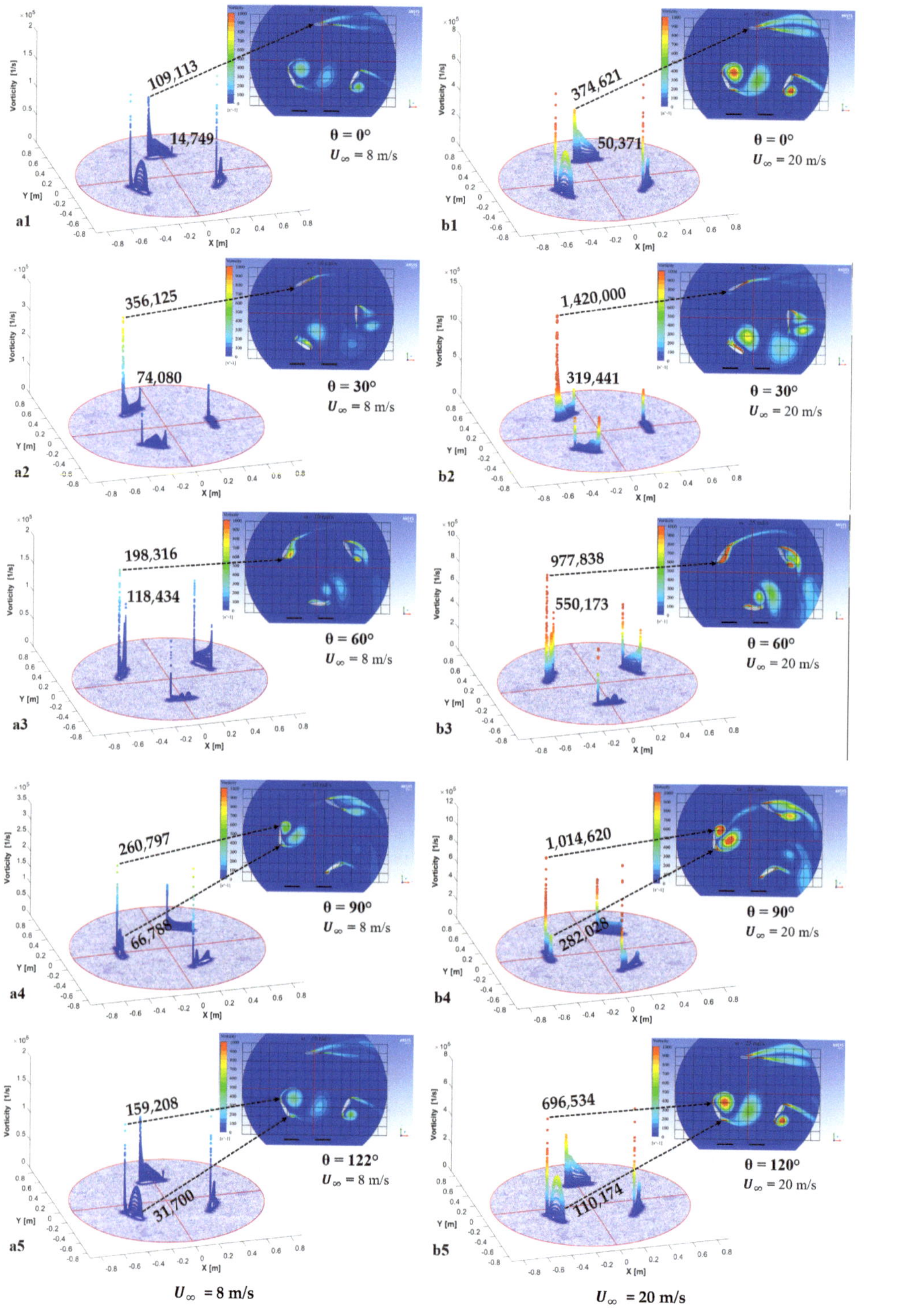

Figure 16. Vorticity Values at the LE and TE of Blade 1 at λ =0.5 for (**a1–a5**) U_∞ = 8 m/s and (**b1–b5**) U_∞ = 20 m/s.

As shown in Figure 17, by plotting vorticity values at the LE and the VI results with the azimuthal position of Blade 1 at λ = 0.5 and λ = 0.9 for U_∞ = 8 m/s, it is observed that LEV increases until reaching a maximum value at θ = 40° and at θ = 70° for λ = 0.5 and λ = 0.9, respectively. Concerning the vorticity index plot, a VI reduction is observed for λ = 0.9, occurring from the θ = 0° to θ = 40° azimuthal position range. Afterwards, from θ = 40° to θ = 70°, VI values remain close to constant. For azimuthal angles larger than θ = 70°, VI starts decreasing rapidly until minimum values are reached. Similar results were observed for λ = 0.5, with a smaller range of constant VI values from the θ = 25° to θ = 40° azimuthal angles. It was also found that maximum torque values shown in Figure 13, are located at about the same azimuthal angles where the LEV for Blade 1 reaches a maximum value and the VI starts to decrease. These critical azimuthal angles are associated with an average VI $\approx$ 4.2 either at λ = 0.5 or λ = 0.9, for both U_∞ = 8 m/s (Figure 17) and U_∞ = 20 m/s (Figure 18).

Then, considering Figure 19a,b, it can also be confirmed that a highest and consistent 4.2 VI value is reached at maximum torques. This is the case for U_∞ = 8 m/s, with a maximum torque being located at θ = 40° when λ = 0.5 and at θ = 70° when λ = 0.9. Similarly, the maximum torque position for U_∞ = 20 m/s is located at θ = 50° when λ = 0.5 and at θ = 80° when λ = 0.9. Thus, reported results show that maximum torque conditions take place when VI close to 4, with the boundary layer starting to separate from the blade LE surface without a vortex formation.

As previously shown in Figure 19, the vorticity index (VI) is a valuable indicator to identify the azimuthal angle where a maximum torque is reached. Furthermore, analyzing the variations of the vorticity as shown in Figures 17 and 18, it was found that there is a relationship between the VI and maximum drag and lift forces as shown in Figures 20 and 21. This condition can be used to identify a so-called "Imminent Vortex Separation Condition (IVSC)".

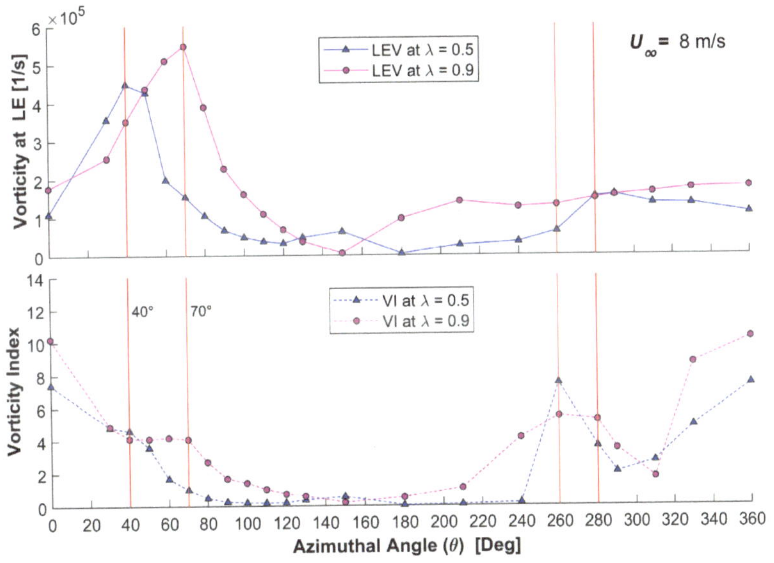

Figure 17. LE Vorticity and VI Values for Blade 1 at Various Azimuthal Angles for U_∞ = 8 m/s at λ =0.5 and λ =0.9.

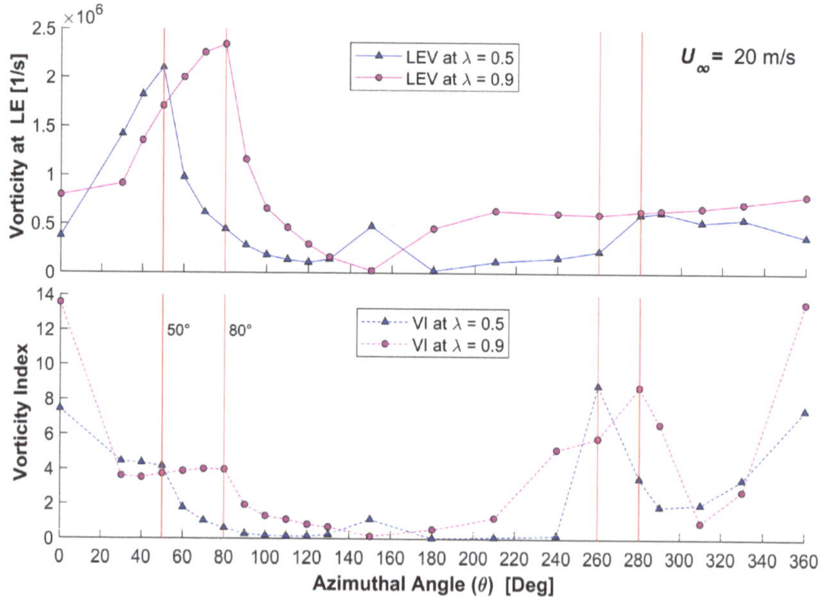

Figure 18. LE Vorticity and VI Values for Blade 1 at Various Azimuthal Angles for U_∞ = 20 m/s at λ =0.5 and λ =0.9.

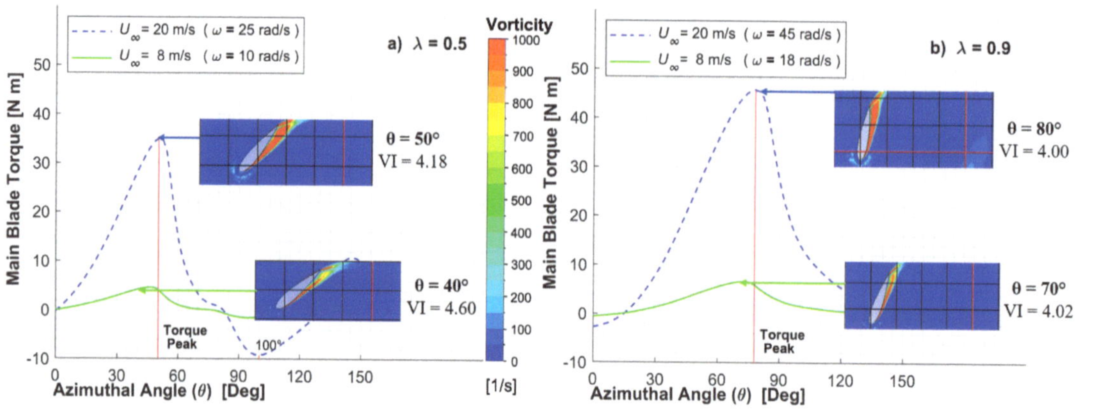

Figure 19. Torque and VI for Blade 1 with Azimuthal Position for U_∞ = 8 m/s and U_∞ = 20 m/s at: (**a**) λ = 0.5 and (**b**) λ = 0.9, Using the 2D Model of the Present Study.

It is noticed that VIs are related to the azimuthal angle position of the maximum drag force, and this occurs when the vorticity magnitude at the leading edge reaches a maximum values and starts decreasing once the dynamic stall vortex (DSV) detaches from the blade. This condition designated as the Imminent Vortex Separation Condition (IVSC), is determined by the magnitude of the VIs at the LE and TE changing from values greater than 1 to values close to 1. This condition is consistently followed by another IVSC, with VIs changing now, from the TE to the LE with VIs reaching 0.18 levels.

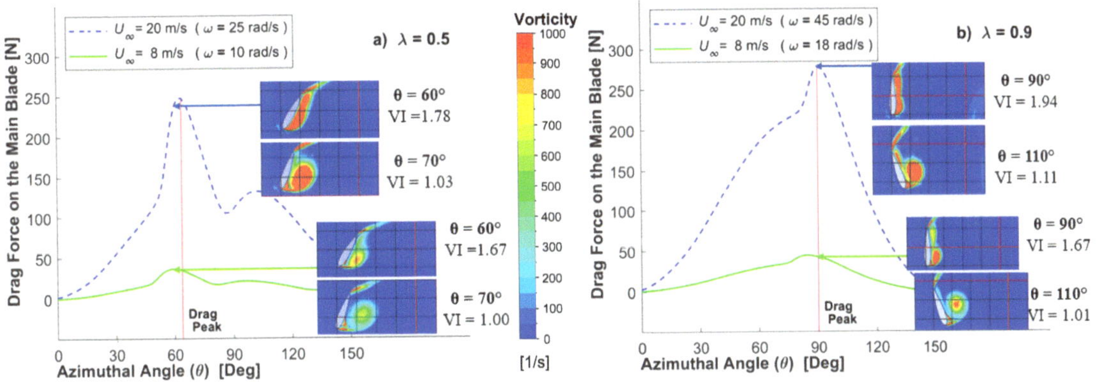

Figure 20. Drag Force and VI for Blade 1 with Azimuthal Position for U_∞ = 8 m/s and U_∞ = 20 m/s at: (**a**) λ = 0.5 and (**b**) λ = 0.9, Using the 2D Model of the Present Study.

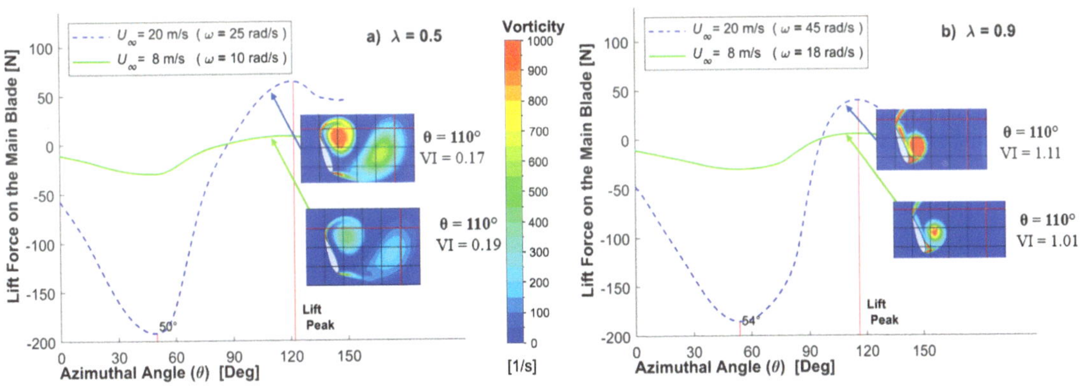

Figure 21. Lift Force and VI for Blade 1 and VI with Azimuthal Position for U_∞ = 8 m/s and U_∞ = 20 m/s at: (**a**) λ = 0.5 and (**b**) λ = 0.9, Using the 2D Model of the Present Study.

As shown in Figure 20, it can be noticed that the azimuthal position of the maximum drag force is independent of the wind speed and is only a function of the λ values. Nevertheless, when the wind speed increases from 8 to 20 m/s, it leads to a 30° difference in the azimuthal angle of the maximum drag force and this for both λ values. Furthermore, when λ equals 0.5 (Figure 20a), a maximum drag force is obtained at θ = 60° for both wind speed velocities with VI being 1.73 on average. Likewise, when λ equals to 0.9 (Figure 20b), the maximum drag force is obtained at θ = 90° for both wind speed velocities with VI being 1.8 on average. Under these conditions, it was identified that a VI change from a value greater than 1 to a close to 1, and this is a typical example of the IVSC condition. This is justifiable given VI is defined as the ratio between the vorticity at the LE and the TE. For instance, a VI of 1 is obtained for Blade 1 for both U_∞ = 8 m/s and U_∞ = 20 m/s at θ = 70° when λ = 0.5 and at θ = 110° when λ = 0.9.

Regarding lift forces, it is claimed that the dynamic stall process and the formation of a Dynamic Stall Vortex (DVS) with changes in the angle of attack, indicate that maximum lift forces are exerted when the flow separation is completed [44]. This is followed by a sudden decrease of lift forces. In agreement with this and as reported in Figure 21a, it can be observed that at λ = 0.5, maximum lift forces, appear at vortex separation from the TE of Blade 1, for both U_∞ = 8 m/s and U_∞ = 20 m/s, having VI values of close to 0.17, with this providing another example of IVSC.

Furthermore, Figure 21a, reports those findings, with a vortex separation moving from the TE to LE for $\lambda = 0.5$. However, when the tip speed ratio increases to 0.9 (Figure 21b), the maximum lift forces are reached at $\theta = 110^{\circ}$ for both wind speeds, displaying VI close to 1 related to a vortex separation from the LE of the blade. The reported data confirms then, that maximum lift forces for Blade 1 are related to a VI of 1 and this when an IVSC condition occurs.

Additionally, as shown in Figure 22a,b,e,f, it can be observed that an IVSC from the LE of Blade 1 is reached when the VI value is equal to 1.0, with this being the case for both wind speeds and at different tip speed ratios. Therefore, an IVSC from the LE to the TE of Blade 1 can be expected when VI = 1. This means that following this event, there is vortex separation on the suction side of the blade. Furthermore, and as shown in Figure 22c,d, when $\lambda = 0.5$, there is a correspondence of an IVSC from the trailing edge of the blade at $\theta = 110^{\circ}$. This is coincident with the maximum lift values (Figure 21a), when the VI is in the 0.15–0.19 range. These results indicate that any time the vorticity index is close to 1, the vortex separation condition from the LE to the TE of the blade will take place and when the VI is close to 0.18 value, the IVSC from the TE towards the LE of the blade will take place. Therefore, according to the VI values, the points of maximum lift and maximum torque can be related to the azimuthal position of each blade, and to the influence of the dynamic stall effect. Under this condition, the vortex separation occurring from the LE to the TE of the blade, and vice versa, is responsible for VAWT performance. Beyond the azimuthal position of maximum torque, there is a marked reduction in the vorticity, which is characterized by a drop in the VI from a constant ratio of about 4, to very low values of about 0.15.

Thus, during the rotation of Blade 1, having a fixed angle of attack and varying azimuthal angles, there are two conditions at which the VI approaches 1. One is when the drag force reaches a maximum value, and a second is when the lift force displays a maximum level, with this being true at any wind speed considered. Both conditions are characterized by an IVSC starting at the LE and moving towards the TE of the Blade 1. The IVSCs are determined by the angle of attack of the blade, a parameter that is a function of the azimuthal blade position. It can be concluded that a maximum drag force occurs during the first quarter and a maximum lift force at the second quarter of the blade circumferential path. In addition, when the VI approaches a value of around 0.20, a reverse imminent vortex separation condition from the TE to the LE is expected. Here, there is a very modest influence of the maximum lift on the overall wind turbine torque. One should notice that in the present study, this occurs at an azimuthal angle of 110 degrees.

Furthermore, by combining a visual interpretation of the air flow and vorticity patterns through the CFD simulation analysis, one can conclude that the VI represents a significant parameter that describes the vorticity and dynamic stall influence on lift forces and torque. As a result, and to improve the performance of a VAWT, the turbulence and vorticity have to be reduced through the development of innovative configurations and design parameters, including blade geometry, pitching cycles or/and solidity ratios. Thus, it is expected that in this respect, the vorticity index can provide a quantity that can be used to characterize more objectively different VAWT configurations in board ranges of wind speeds and dimensionless tip speed ratios and this in order to address different strategies for improving them with more favorable better C_p coefficients.

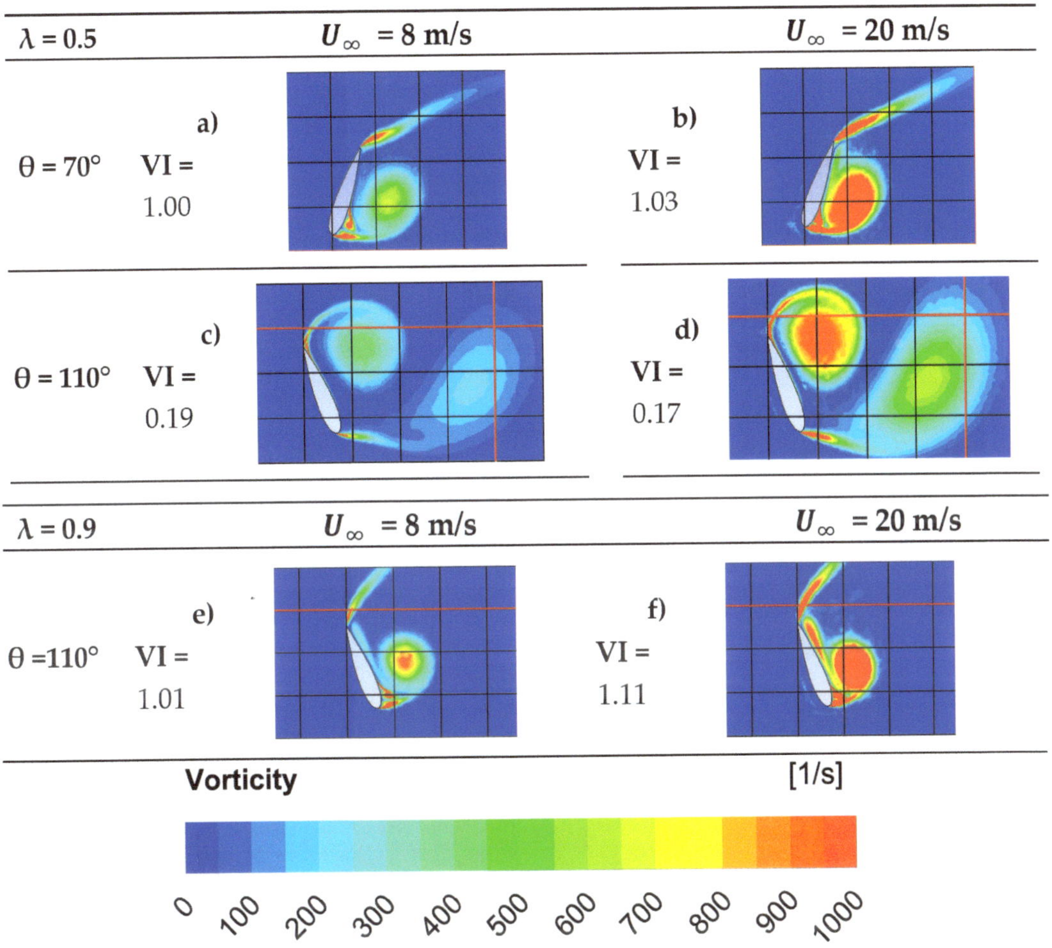

Figure 22. Vorticity Contours for Imminent Vortex Separation Condition (IVSC) for U_∞ = 8 m/s and for U_∞ = 20 m/s, at (**a–d**) λ = 0.5 and (**e,f**) λ = 0.9.

4. Conclusions

In this work, Unsteady Reynolds Averaged Navier-Stokes (URANS) simulations are successfully developed using the ANSYS Fluent software to evaluate the performance of a vertical H-Darrieus wind turbine configuration at wind conditions of 8 and 20 m/s. The wind turbine was first simulated by means of a 2D CFD model. This 2D model was then validated by comparing it with a 3D CFD model and experimental data, for a wind speed (U_∞) of 8 m/s. This evaluation showed that the 2D model was more precise in determining tip speed ratios (λ) of up to 1. Additionally, the 2D model was able to predict the vorticity distribution on each blade, at two different tip speed ratios: λ = 0.5 and λ = 0.9. Furthermore, URANS simulations of the H-Darrieus wind turbine, showed the following:

- The average overall torque values for a complete rotor rotation period are found to provide good numerical convergence criteria [20] for high (20 m/s) wind speeds.
- The maximum drag forces for 8 and 20 m/s wind speeds, are obtained for an azimuthal angle (θ) range of 65° to 85°. This corresponds to an angle of attack (α) close to 45°.

- The maximum torque on the main blade, for 8 and 20 m/s wind speeds is delivered at the following azimuthal positions: between $\theta = 45^\circ$ and $\theta = 50^\circ$ ($\alpha = 31.1$–33.8°) for $\lambda = 0.5$, and between $\theta = 72^\circ$ and $\theta = 78^\circ$ ($\alpha = 38.2$–41.4°) for $\lambda = 0.9$.
- Much higher torques are delivered at 20m/s, versus the ones produced at 8m/s. Nonetheless, high wind speeds showed just a moderate influence on the final average power coefficient value of a 3-bladed H-Darrieus VAWT. The overall gains are lessened by the increased turbulence and vorticity, which reduce the energy extraction by the rotor during the wind turbine operation, within the same range of tip speed ratios. This matter may require further studies.
- The torque is significantly reduced for blade azimuthal angles $\theta > 90^\circ$. This even yielded negative torque values, which are attributed to flow separation and strong vorticity interactions with the blades.
- The flow vorticity has a noticeable relation with the turbine performance. The leading edge vorticity (LEV) increases until the blade rotation reaches the azimuthal angle of maximum torque. The proposed vorticity index (VI) displays a constant value around 4 before reaching this maximum torque.
- The VI can be used to quantitatively assess vortex separation conditions. After the maximum torque azimuthal position, VI starts decreasing rapidly until minimum values reached because of vortex generation at the LE of the blade and vorticity accumulation at its TE. Furthermore, when the VI attains a value of 1, the Imminent Vortex Separation Condition (IVSC) takes place with the vortex formed by dynamic stall condition almost detaching from the LE of the blade. This occurs at comparable azimuthal angles for equal tip speed ratios, with this being independent of the wind speed.

Author Contributions: Conceptualization, J.G.A.-L., A.P.B. and H.d.L.; methodology, J.G.A.-L., A.P.B., S.L.-Z. and H.d.L.; software, J.G.A.-L.; validation, J.G.A.-L., S.L.-Z., A.P.B. and H.d.L.; investigation, J.G.A.-L., A.P.B. and S.L.-Z.; resources, H.d.L.; writing—original draft preparation, J.G.A.-L. and A.P.B.; writing—review and editing, H.d.L.; supervision, A.P.B. and H.d.L.; funding acquisition, H.d.L. All authors have read and agreed to the published version of the manuscript.

Funding: This research was funded by Natural Sciences and Engineering Research Council, Canada: HdL Discovery Grant; ELAP Scholarship Program: Jansen Gabriel Acosta-Lopez Scholarship; Compute Canada, Universidad Nacional de La Patagonia, Comodoro Rivadavia, Argentina, Universidad Nacional de Colombia, Manizales-Campus, Colombia.

Data Availability Statement: The data that support the findings of this study are available from the corresponding author, H.d.L., upon reasonable request.

Acknowledgments: The authors acknowledge the computational resources provided by the Shared Hierarchical Academic Research Computing Network (SHARCNET) of Compute Canada and the financial support via the NSERC (Natural Sciences and Engineering Research Council) Discovery Grant awarded to Hugo de Lasa. We also wish to express our deepest recognition to Beatriz Helena Aristizábal Zuluaga, who unfortunately passed away during the development of this study. As a member of the Grupo de Trabajo Académico en Ingeniería Hidráulica y Ambiental (GTA IHA) from Universidad Nacional de Colombia sede Manizales, Aristizábal Zuluaga was a resourceful researcher who dedicated part of her life to dealing with environmental pollution issues, especially air pollution. It is largely thanks to Beatriz and her initiative to promote academic exchange between universities that this work began in the first place. We would also like to thank Florencia de Lasa who helped with the editing of this article.

Conflicts of Interest: The authors declare no conflict of interest.

Notation

c	Chord Length [m]
C_p	Power Coefficient
D	Rotor diameter [m]
F_D	Drag force [N]
F_L	Lift force [N]
H	Blade span [m]
k^*	Reduced frequency
N	Number of blades
P	Power [W]
$\overline{p}$	Mean pressure [Pa]
q	Dynamic pressure [Pa]
R	Rotor Radius [m]
t	Time [s]
T	Torque [N m]
$\overline{u}$	Mean fluid velocity [m s^{-1}]
u'	Fluctuating fluid velocity [m s^{-1}]
U_∞	Wind speed [m s^{-1}]
V_a	Wind induced velocity [m s^{-1}]
W	Wind relative flow velocity [m s^{-1}]
$y+$	Non-dimensional first cell wall distance
Greek Symbols	
α	Angle of attack [deg]
$\Delta\alpha$	Angular marching step [Δdeg]
θ	Azimuthal angle [deg]
λ	Tip speed ratio = $\frac{R\,\omega}{U_\infty}$ [-]
μ	Fluid viscosity [Pa s]
v	Kinematic viscosity [m^2/s]
ρ	Fluid density [kg m^{-3}]
σ	Solidity = $\frac{N\,c}{D}$ [-]
ω	Angular velocity [rad s^{-1}]
Abbreviations	
CFD	Computational Fluid Dynamics
DVS	Dynamic Stall Vortex
HAWT	Horizontal Axis Wind Turbine
IVSC	Imminent Vortex Separation Condition
LE	Leading Edge
LEV	Leading Edge Vorticity
RANS	Reynolds Averaged Navier-Stokes method
SRS	Scale Resolving Simulation
SST	Shear Stress Transport
TE	Trailing Edge
TEV	Trailing Edge Vorticity
TSR	Tip Speed Ratio
URANS	Unsteady Reynolds Averaged Navier–Stokes
VAWT	Vertical Axis Wind Turbine
VI	Vorticity Index

Appendix A

This appendix reports the vorticity values at the leading edge (LE) and the trailing edge (TE) of Blade 1, for U_∞ = 8 m/s and U_∞ = 20 m/s, at λ = 0.9 and for different azimuthal positions.

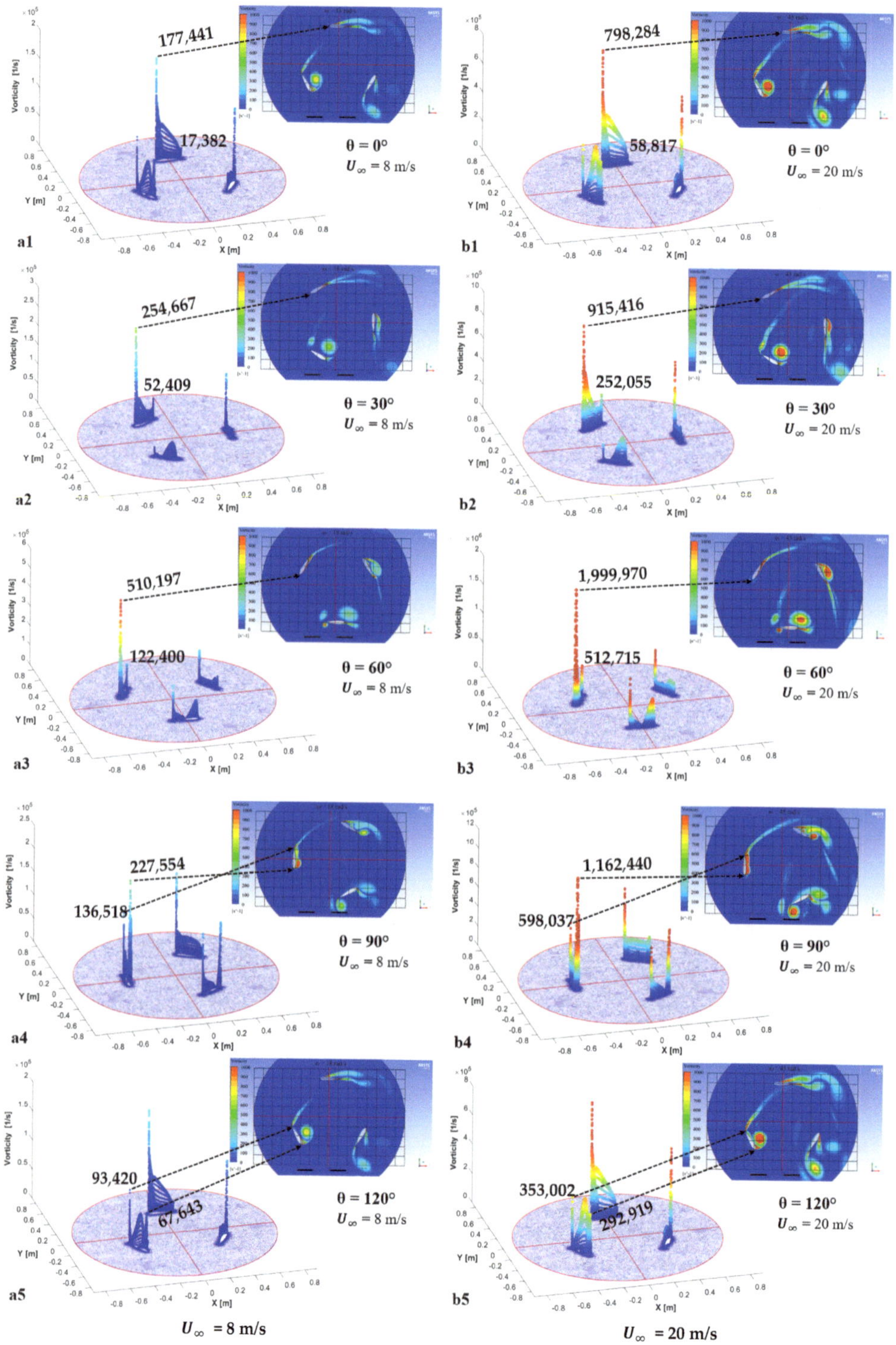

Figure A1. Vorticity Values at LE and TE for Blade 1 at λ = 0.9 for (**a1–a5**) U_∞ = 8 m/s and (**b1–b5**) U_∞ = 20 m/s.

Appendix B

This appendix documents all the vorticity values obtained from the 2D CFD model simulations, for $U_\infty = 8$ m/s, at both $\lambda = 0.5$ and $\lambda = 0.9$.

Table A1. Vorticity at the Leading Edge (LE) and Trailing Edge (TE) and VI for Blade 1 at Various Azimuthal Angles for $U_\infty = 8$ m/s.

θ (Deg)	λ = 0.5 LEV (1/s)	λ = 0.5 TEV (1/s)	λ = 0.5 VI	λ = 0.9 LEV (1/s)	λ = 0.9 TEV (1/s)	λ = 0.9 VI
0	109,113	14,749	7.40	177,441	17,382	10.21
30	356,125	74,080	4.81	254,667	52,409	4.86
40	447,702	97,372	4.60	351,808	85,683	4.11
50	425,774	119,716	3.56	435,962	106,290	4.10
60	198,316	118,434	1.67	510,197	122,400	4.17
70	153,440	152,854	1.00	547,857	134,543	4.07
80	104,366	214,936	0.49	388,710	144,981	2.68
90	66,788	260,797	0.26	227,554	136,518	1.67
100	48,225	229,827	0.21	160,139	114,489	1.40
110	37,120	197,066	0.19	108,520	107,670	1.01
120	31,700	159,208	0.20	67,643	93,420	0.72
130	47,272	131,531	0.36	36,521	62,172	0.59
150	61,833	112,285	0.55	5273	26,617	0.20
180	3525	109,692	0.03	95,543	186,638	0.51
210	25,936	305,729	0.08	139,748	132,173	1.06
240	34,814	199,575	0.17	125,114	30,306	4.13
260	62,373	8341	7.48	130,162	23,938	5.44
280	151,299	42,150	3.59	147,775	28,452	5.19
290	156,562	76,667	2.04	155,332	44,992	3.45
310	134,699	50,313	2.68	163,063	94,642	1.72
330	133,190	27,380	4.86	173,659	19,959	8.70
360	109,113	14,749	7.40	177,441	17,382	10.21

Appendix C

This appendix documents all the vorticity values obtained from the 2D CFD model simulations, for $U_\infty = 20$ m/s, at both $\lambda = 0.5$ and $\lambda = 0.9$.

Table A2. Vorticity at the Leading Edge (LE) and Trailing Edge (TE) and VI for Blade 1 at Various Azimuthal Angles for $U_\infty = 20$ m/s.

θ (Deg)	λ = 0.5 LEV (1/s)	λ = 0.5 TEV (1/s)	λ = 0.5 VI	λ 0.9 LEV (1/s)	λ 0.9 TEV (1/s)	λ 0.9 VI
0	374,621	50371	7.44	798,284	58,817	13.57
30	1,420,000	319,441	4.45	915,416	252,055	3.63
40	1,822,510	415,671	4.38	1,354,560	382,740	3.54
50	2,102,810	503,018	4.18	1,708,160	455,701	3.75
60	977,838	550,173	1.78	1,999,970	512,715	3.90
70	618,315	602,600	1.03	2,261,180	561,845	4.02
80	448,167	740,703	0.61	2,345,430	586,044	4.00
90	282,028	1,014,620	0.28	1,162,440	598,037	1.94
100	183,997	972,682	0.19	661,746	504,683	1.31
110	137,458	830,830	0.17	463,944	416,734	1.11
120	110,174	696,534	0.16	292,919	353,002	0.83
130	142,477	564,754	0.25	163,461	240,015	0.68
150	479,418	422,522	1.13	25,712	147,822	0.17
180	25,802	505,638	0.05	456,111	851,410	0.54
210	119,269	1,124,200	0.11	642,122	532,874	1.21
240	158,636	766,737	0.21	614,202	119,143	5.16
260	226,673	25,705	8.82	603,590	103,893	5.81
280	603,043	171,088	3.52	637,702	72,947	8.74
290	628,803	336,567	1.87	646,205	97,522	6.63
310	523,672	263,367	1.99	669,941	705,871	0.95
330	555,391	160,771	3.45	715,332	258,413	2.77
360	374,621	50,371	7.44	798,284	58,817	13.57

References

1. Zayats, I. The historical aspect of windmills architectural forms transformation. *Procedia Eng.* **2015**, *117*, 685–695. [CrossRef]
2. BP plc. bp Energy Outlook 2022 Edition. 2022. Available online: https://www.bp.com/content/dam/bp/business-sites/en/global/corporate/pdfs/energy-economics/energy-outlook/bp-energy-outlook-2022.pdf (accessed on 30 June 2022).
3. Zhao, D.; Han, N.; Goh, E.; Cater, J.; Reinecke, A. *Wind Turbines and Aerodynamics Energy Harvesters*, 1st ed.; Academic Press: Cambridge, MA, USA, 2019.
4. Dai, K.; Bergot, A.; Liang, C.; Xiang, W.-N.; Huang, Z. Environmental issues associated with wind energy—A review. *Renew Energy* **2015**, *75*, 911–921. [CrossRef]
5. Maalouly, M.; Souaiby, M.; ElCheikh, A.; Issa, J.; Elkhoury, M. Transient analysis of H-type Vertical Axis Wind Turbines using CFD. *Energy Rep.* **2022**, *8*, 4570–4588. [CrossRef]
6. He, J.; Jin, X.; Xie, S.; Cao, L.; Wang, Y.; Lin, Y.; Wang, N. CFD modeling of varying complexity for aerodynamic analysis of H-vertical axis wind turbines. *Renew Energy* **2020**, *145*, 2658–2670. [CrossRef]
7. Hossain, M.; Ali, M.H. Future research directions for the wind turbine generator system. *Renew. Sustain. Energy Rev.* **2015**, *49*, 481–489. [CrossRef]
8. Al-Sharafi, A.; Sahin, A.Z.; Ayar, T.; Yilbas, B.S. Techno-economic analysis and optimization of solar and wind energy systems for power generation and hydrogen production in Saudi Arabia. *Renew. Sustain. Energy Rev.* **2017**, *69*, 33–49. [CrossRef]
9. Aiche-Hamane, L.; Belhamel, M.; Benyoucef, B.; Hamane, M. Feasibility study of hydrogen production from wind power in the region of Ghardaia. *Int. J. Hydrogen Energy* **2009**, *34*, 4947–4952. [CrossRef]
10. Benghanem, M.; Mellit, A.; Almohamadi, H.; Haddad, S.; Chettibi, N.; Alanazi, A.M.; Dasalla, D.; Alzahrani, A. Hydrogen Production Methods Based on Solar and Wind Energy: A Review. *Energies* **2023**, *16*, 757. [CrossRef]
11. Etemadeasl, V.; Esmaelnajad, R.; Dizaji, F.F.; Farzaneh, B. A novel configuration for improving the aerodynamic performance of Savonius rotors. *Proc. Inst. Mech. Eng. Part A J. Power Energy* **2019**, *233*, 751–761. [CrossRef]
12. Borg, M.; Shires, A.; Collu, M. Offshore floating vertical axis wind turbines, dynamics modelling state of the art. part I: Aerodynamics. *Renew. Sustain. Energy Rev.* **2014**, *39*, 1214–1225. [CrossRef]
13. Schubel, P.J.; Crossley, R.J. Wind turbine blade design. *Energies* **2012**, *5*, 3425–3449. [CrossRef]
14. Souaissa, K.; Ghiss, M.; Chrigui, M.; Bentaher, H.; Maalej, A. A comprehensive analysis of aerodynamic flow around H-Darrieus rotor with camber-bladed profile. *Wind Eng.* **2019**, *43*, 459–475. [CrossRef]
15. Ferreira, C.J.S.; Bijl, H.; Van Bussel, G.; Van Kuik, G. Simulating Dynamic Stall in a 2D VAWT: Modeling strategy, verification and validation with Particle Image Velocimetry data. *J. Phys. Conf. Ser.* **2007**, *75*, 012023. [CrossRef]
16. Sridhar, S.; Zuber, M.; Shenoy B., S.; Kumar, A.; Ng, E.Y.K.; Radhakrishnan, J. Aerodynamic comparison of slotted and non-slotted diffuser casings for Diffuser Augmented Wind Turbines (DAWT). *Renew. Sustain. Energy Rev.* **2022**, *161*, 112316. [CrossRef]
17. Rezaeiha, A.; Montazeri, H.; Blocken, B. On the accuracy of turbulence models for CFD simulations of vertical axis wind turbines. *Energy* **2019**, *180*, 838–857. [CrossRef]
18. Kanyako, F.; Janajreh, I. Vertical Axis Wind Turbine performance prediction for low wind speed environment. In Proceedings of the 2014 IEEE Innovations in Technology Conference, Warwick, RI, USA, 16 May 2014. [CrossRef]
19. Lanzafame, R.; Mauro, S.; Messina, M. 2D CFD modeling of H-Darrieus Wind Turbines using a transition turbulence model. *Energy Procedia* **2014**, *45*, 131–140. [CrossRef]
20. Balduzzi, F.; Bianchini, A.; Maleci, R.; Ferrara, G.; Ferrari, L. Critical issues in the CFD simulation of Darrieus wind turbines. *Renew Energy* **2016**, *85*, 419–435. [CrossRef]
21. Michna, J.; Rogowski, K. Numerical Study of the Effect of the Reynolds Number and the Turbulence Intensity on the Performance of the NACA 0018 Airfoil at the Low Reynolds Number Regime. *Processes* **2022**, *10*, 1004. [CrossRef]
22. Elkhoury, M.; Kiwata, T.; Aoun, E. Experimental and numerical investigation of a three-dimensional vertical-axis wind turbine with variable-pitch. *J. Wind Eng. Ind. Aerodyn.* **2015**, *139*, 111–123. [CrossRef]
23. EMöllerström, E.; Gipe, P.; Beurskens, J.; Ottermo, F. A historical review of vertical axis wind turbines rated 100 kW and above. *Renew. Sustain. Energy Rev.* **2019**, *105*, 1–13. [CrossRef]
24. Mohamed, O.S.; Ibrahim, A.A.; Etman, A.K.; Abdelfatah, A.A.; Elbaz, A.M. Numerical investigation of Darrieus wind turbine with slotted airfoil blades. *Energy Convers. Manag. X* **2020**, *5*, 100026. [CrossRef]
25. Ahmedov, A.; Ebrahimi, K.M. Numerical Modelling of an H-type Darrieus Wind Turbine Performance under Turbulent Wind. *Am. J. Energy Res.* **2017**, *5*, 63–78. [CrossRef]
26. Trivellato, F.; Castelli, M.R. On the Courant-Friedrichs-Lewy criterion of rotating grids in 2D vertical-axis wind turbine analysis. *Renew Energy* **2014**, *62*, 53–62. [CrossRef]
27. Nguyen, M.T.; Balduzzi, F.; Bianchini, A.; Ferrara, G.; Goude, A. Evaluation of the unsteady aerodynamic forces acting on a vertical-axis turbine by means of numerical simulations and open site experiments. *J. Wind. Eng. Ind. Aerodyn.* **2020**, *198*, 104093. [CrossRef]
28. Alaimo, A.; Esposito, A.; Messineo, A.; Orlando, C.; Tumino, D. 3D CFD analysis of a vertical axis wind turbine. *Energies* **2015**, *8*, 3013–3033. [CrossRef]
29. Orlandi, A.; Collu, M.; Zanforlin, S.; Shires, A. 3D URANS analysis of a vertical axis wind turbine in skewed flows. *J. Wind Eng. Ind. Aerodyn.* **2015**, *147*, 77–84. [CrossRef]

30. Lam, H.; Peng, H. Study of wake characteristics of a vertical axis wind turbine by two- and three-dimensional computational fluid dynamics simulations. *Renew Energy* **2016**, *90*, 386–398. [CrossRef]
31. Hosseini, A.; Goudarzi, N. Design and CFD study of a hybrid vertical-axis wind turbine by employing a combined Bach-type and H-Darrieus rotor systems. *Energy Convers. Manag.* **2019**, *189*, 49–59. [CrossRef]
32. Lee, J.-H.; Lee, Y.-T.; Lim, H.-C. Effect of twist angle on the performance of Savonius wind turbine. *Renew Energy* **2016**, *89*, 231–244. [CrossRef]
33. Alfonsi, G. Reynolds-averaged Navier-Stokes equations for turbulence modeling. *Appl. Mech. Rev.* **2009**, *62*, 040802. [CrossRef]
34. Geng, F.; Kalkman, I.; Suiker, A.; Blocken, B. Sensitivity analysis of airfoil aerodynamics during pitching motion at a Reynolds number of 1.35×105. *J. Wind Eng. Ind. Aerodyn.* **2018**, *183*, 315–332. [CrossRef]
35. Ghasemian, M.; Ashrafi, Z.N.; Sedaghat, A. A review on computational fluid dynamic simulation techniques for Darrieus vertical axis wind turbines. *Energy Convers. Manag.* **2017**, *149*, 87–100. [CrossRef]
36. Barnes, A.; Marshall-Cross, D.; Hughes, B.R. Validation and comparison of turbulence models for predicting wakes of vertical axis wind turbines. *J. Ocean Eng. Mar. Energy* **2021**, *7*, 339–362. [CrossRef]
37. Ma, N.; Lei, H.; Han, Z.; Zhou, D.; Bao, Y.; Zhang, K.; Zhou, L.; Chen, C. Airfoil optimization to improve power performance of a high-solidity vertical axis wind turbine at a moderate tip speed ratio. *Energy* **2018**, *150*, 236–252. [CrossRef]
38. ANSYS Inc. *ANSYS Fluent Theory Guide 15*; ANSYS, Inc.: Canonsburg, PA, USA, 2013.
39. Celik, Y.; Ma, L.; Ingham, D.; Pourkashanian, M. Aerodynamic investigation of the start-up process of H-type vertical axis wind turbines using CFD. *J. Wind. Eng. Ind. Aerodyn.* **2020**, *204*, 104252. [CrossRef]
40. Song, C.; Wu, G.; Zhu, W.; Zhang, X.; Zhao, J. Numerical investigation on the effects of airfoil leading edge radius on the aerodynamic performance of H-rotor Darrieus vertical axis wind turbine. *Energies* **2019**, *12*, 3794. [CrossRef]
41. AlQurashi, F.; Mohamed, M.H. Aerodynamic forces affecting the H-rotor darrieus wind turbine. *Model. Simul. Eng.* **2020**, *2020*, 1368369. [CrossRef]
42. Laneville, A.; Vittecoq, P. Dynamic Stall: The Case of the Vertical Axis Wind Turbine. *J. Sol. Energy Eng.* **1986**, *108*, 140–145. Available online: http://solarenergyengineering.asmedigitalcollection.asme.org/ (accessed on 9 July 2022). [CrossRef]
43. Dessoky, A.; Bangga, G.; Lutz, T.; Krämer, E. Aerodynamic and aeroacoustic performance assessment of H-rotor darrieus VAWT equipped with wind-lens technology. *Energy* **2019**, *175*, 76–97. [CrossRef]
44. Choudhry, A.; Arjomandi, M.; Kelso, R. Methods to control dynamic stall for wind turbine applications. *Renew Energy* **2016**, *86*, 26–37. [CrossRef]
45. Li, S.; Zhang, L.; Yang, K.; Xu, J.; Li, X. Aerodynamic performance of wind turbine airfoil DU 91-W2-250 under dynamic stall. *Appl. Sci.* **2018**, *8*, 1111. [CrossRef]
46. Menter, F.R. Best Practice: Scale-Resolving Simulations in ANSYS CFD. 2015. Available online: www.ansys.com (accessed on 29 October 2022).

processes

Review

Desiccant Technologies for Improving Air Quality: An Overview of the Brazilian Scenario and Comparison of Available Design Software for Manufactured Desiccant Wheels

York Castillo Santiago [1,*], Bruno Gomes Nunes [2], Geovani Souza Fontana [2], Daiane Busanello [2], Alexandre Fernandes Santos [2], Samuel Moreira Duarte Santos [1,3], Estefania Neiva de Mello [3] and Leandro A. Sphaier [1]

1 Laboratory of Thermal Sciences (LATERMO), Mechanical Engineering Department (TEM/PGMEC), Fluminense Federal University, Rua Passo da Pátria 156, Niterói 24210-240, RJ, Brazil; samuel.m.santos@eletrobras.com (S.M.D.S.); lasphaier@id.uff.br (L.A.S.)

2 Escola Técnica Profissional, Grupo ETP, Rua Eng. Rebouças 2213, Curitiba 80230-040, PR, Brazil; projetos.etp@gmail.com

3 Centrais Elétricas Brasileiras S.A (ELETROBRAS), Rua da Quitanda 196, Rio de Janeiro 20091-005, RJ, Brazil

* Correspondence: yorkcastillo@id.uff.br

Abstract: A review of desiccant dehumidification technologies for improving air quality is presented, mainly focusing on alternatives for air conditioning systems for minimizing Sick Building Syndrome. The principles and types of desiccant wheels, as well as the existing selection software for these types of equipment, were reviewed and comparatively evaluated. The study focused on the Brazilian context; thus, information about this country's air conditioning systems and laws were evaluated. Possible applications of desiccant wheels, such as their integration into cooling cycles and the sensible heat wheel, were also analyzed. Finally, several examples of commercial desiccant wheel selection software that are useful in many situations were evaluated. Nevertheless, it was evidenced that the available software could not perform an operation analysis for only a specific period. Therefore, creating computational tools to select desiccant wheels is essential when considering the data from the different Brazilian regions for a year.

Keywords: dehumidification; desiccant wheels; air quality; air conditioning systems

Citation: Castillo Santiago, Y.; Nunes, B.G.; Fontana, G.S.; Busanello, D.; Santos, A.F.; Santos, S.M.D.; Mello, E.N.d.; Sphaier, L.A. Desiccant Technologies for Improving Air Quality: An Overview of the Brazilian Scenario and Comparison of Available Design Software for Manufactured Desiccant Wheels. *Processes* **2023**, *11*, 2031. https://doi.org/10.3390/pr11072031

Academic Editors: Ferdinando Salata and Virgilio Ciancio

Received: 17 June 2023
Revised: 27 June 2023
Accepted: 4 July 2023
Published: 7 July 2023

1. Introduction

In air conditioning design, it is essential to consider the addition of external air to closed environments such as offices, schools, hospitals, and malls, among others. The American Society for Heating, Refrigeration and Air-Conditioning Engineering (ASHRAE) established that if a building has more than 20% of people with symptoms of some disease or discomfort when exposed to this environment, it is considered a building that fits a Sick Building Syndrome (SBS) situation [1].

The lack of indoor air renewal causes the accumulation of chemical pollutants (carbon monoxide and dioxide, sulfur dioxide and formaldehyde, and ammonia, among others) and biological contaminants (fungi, algae, protozoa, bacteria, and mites), which makes the air hazardous for human health and is a risk factor for people with respiratory diseases [2]. Environments with high levels of these pollutants are conventionally called SBS; the World Health Organization (WHO) recognized this problem in 1982 [3].

In the 1970s, an energy crisis spread globally due to conflicts between oil-exporting countries in the Middle East. As a result, oil suffered a reduction in production and a price increase, reaching 400% in 1973. In the same period, engineers looking for alternatives to reduce electric energy consumption [4] reduced the external air supply to buildings [5,6]. Consequently, the first reports of SBS [7] were seen.

Fungal exposures are receiving increasing attention as an occupational and public health problem due to the high prevalence of fungal contamination in buildings [8]. Moisture and moisture-related problems are significant sources of fungal contamination in homes, buildings, and schools [9]. Indoor fungi can be inhaled, and the presence of such particles has been linked to many illnesses and symptoms (infections, allergic reactions, or toxic responses) among occupants of moisture-damaged buildings [10]. On the other hand, moisture damage in buildings could alter various components of building materials, which can be another source of deterioration in indoor air quality (IAQ) [11]. Fungal growth is associated with moisture content, defined as the ratio between a material's free water and dry weight [12].

Some studies indicated a relation between humidity, indoor-measured visible biological contaminant growth, and its effects on human health. Sterling et al. [13] analyzed the humidity impact on contaminants (such as viruses, bacteria, and fungi) and the possible impacts on human health. At room temperature, the authors observed that the required humidity range for minimizing risks to human health by biological contaminants was between 40 and 60%, as shown in Figure 1.

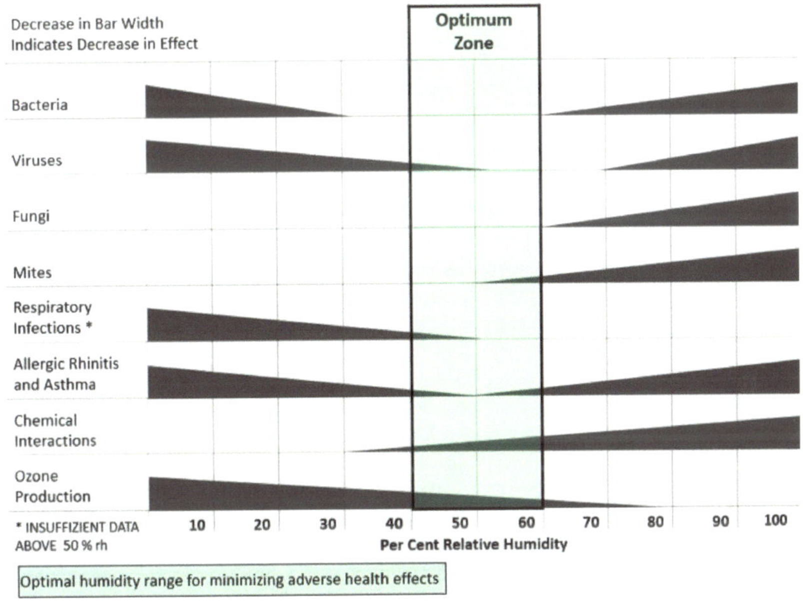

Figure 1. Optimal humidity range for minimizing risks to human health. Source: [13].

In 1984, the WHO estimated that up to 30% of buildings could be affected by SBS. In the early 1980s, SBS was correlated with environmental problems. Table 1 shows some of the main symptoms associated with SBS. However, the concept is evolving, including in terms of psychological aspects, economic implications, energy savings, and climate change [14].

Brasche [15] observed that women suffer more from SBS when compared to men, with 44.3% of women suffering from SBS while only 26.2% of men faced this problem, as shown in Table 2. However, this percentage difference between genders was due to position within the company; in the same study, when men and women were in environments with the same IAQ, the rates of occurrence of SBS were similar.

Table 1. Symptoms associated with SBS. Source: [14].

Symptoms	Occurrence (%)
Lethargy	57
Stuffy nose	47
Dry throat	46
Headache	43
Itchy eyes	28
Dry eyes	27
Runny nose	23
Flu	23
Breathing difficulty	9
Chest pain	9

Table 2. Gender-related differences in SBS. Source: [15].

	Women		Men	
	People	%	People	%
With symptoms	393	44.3	151	26.2
Without symptoms	495	55.7	426	73.8
Total	888	100	577	100

In the mid-1990s, when the issue of indoor air quality began to gain importance in Brazil, air conditioning systems became a main subject for maintaining cleanliness and conservation. In 1998, with the death of the then-Minister of Communications Sérgio Motta, attributed mainly to the bacterium Legionella, the then-Minister of Health José Serra announced the formation of a study group to publish legislation that obliged those responsible for air conditioning systems to maintain clean systems, therefore preventing diseases in the Brazilian population [16].

The Ministry of Health published in 1998 the ORDINANCE N°3523 [17], which defines values and definitions to maintain IAQ as described in Art. 4, the definition of Sick Building Syndrome. Based on Art. 5, it is possible to mention the need to meet a minimum level of air renewal (27 $m^3/h/person$) and a minimum filtration of class G1. On the other hand, the Brazilian Ministry of Health published in 2003 the RESOLUTION No. 09 (RE-09) [18], which presents additional information from ORDINANCE No. 3523/98. One of the definitions given in RE-09 is the concept of air conditioning, which corresponds to the air-treatment process intended to maintain the IAQ requirements of the conditioned space by controlling variables such as temperature, humidity, speed, particulate matter, biological particles, and carbon dioxide content.

RE No. 09 also presents parameters related to the maximum recommended value for microbiological contamination, which is 750 cfu/m^3 (colony-forming unit/cubic meter), and chemical contamination, which is 1000 ppm for the concentration of carbon dioxide (CO_2) and 80 $\mu g/m^3$ of aero dispersoids. In addition to these parameters, there are recommended operating ranges for temperature, humidity, air speed, air renewal rate, and the degree of air purity that must comply with the Brazilian Regulatory Standard NBR 16401-3 for indoor air quality [19].

An essential point for maintaining IAQ is to keep the equipment in full operation; for this, it was necessary to create a Maintenance, Operation, and Control Plan (MOCP). Considering that in maintaining IAQ requirements in situations with a considerable latent heat load, as commonly found in HVAC applications, a significant amount of moisture removal is required, desiccant technologies have a significant advantage over the traditional vapor compression-based (also termed mechanical dehumidification) alternatives. This is because on the one hand, vapor compression systems need to cool the process airstream to temperatures below the dewpoint to obtain the dehumidification effect, which occurs due to regular condensation. Naturally, this involves an additional high-grade energy expense.

On the other hand, when desiccant systems perform dehumidification, the subcooling becomes unnecessary, thus saving a considerable portion of high-grade energy.

In Brazil, the Civil House published Law No. 13,589 of 2018 [20], which obliges public and collective buildings to keep the MOCP active, thus guaranteeing the ranges established for a good IAQ. Increasing air renewal in buildings without proper treatment does not solve all the problems associated with IAQ. According to Che et al. [21], this methodology can increase indoor humidity, which can also substantially raise the potential for microbiological growth, as this increase in ventilation can augment the concentration of pollutants in the environment. However, the practice of increasing the rates of air renewal has established itself among designers [22]. Outdoor air recommendations are stipulated for each type of design for designers seeking to increase the speed of outdoor air to the maximum allowed according to NBR 16401-3 and obtain systems that provide temperature, humidity, and specific filtration control for the condition of the environment in question. It could consider using total heat exchangers as enthalpy wheels, which can provide energy savings to the building.

According to Pargeter [23], the energy consumption required by conventional air treatment practices for air conditioning in commercial buildings represents an average of 10% of all energy spent in the United States. The author also noted that a large part of the energy consumption for air conditioning in these commercial buildings is due to air renewal, which is different from the internal comfort conditions. In addition to electricity consumption, the higher air renewal rate leads to greater investments and maintenance costs for such Heating, Ventilation, and Air Conditioning (HVAC) systems.

One alternative to conventional refrigeration systems is the so-called desiccant cooling cycle [24,25], which employs active desiccant rotors for dehumidifying the outdoor air and running the produced dry air through evaporative coolers to provide the refrigeration effect. The main advantage of such technology is that no special refrigerant fluids are required and it is mainly driven by thermal energy, which means that solar or thermal waste energy can power these cycles. On the other hand, the COP of this cycle is quite low, which could not be a significant issue when considering that a renewable source or waste energy is being used as a source of power. These alternatives can also be used alongside other cooling systems, as previously demonstrated [26,27].

This research assessed alternatives for air HVAC systems to minimize Sick Building Syndrome and improve air quality while considering international programs/standards. For this purpose, an alternative technology known as desiccant wheels was studied by analyzing their principles and types when the existing selection software for these types of equipment was performed. In addition, energy-efficiency programs worldwide and in the Brazilian context were analyzed while aiming at implementing strategies in which desiccant wheels are appropriate. Finally, some examples of commercial software for desiccant wheels were compared with an aim to identify the different tools available in the air conditioning market.

2. Energy Efficiency Programs Worldwide

The International Energy Agency [28] estimated that about 2/3 of homes worldwide will have an air conditioner installed by the year 2050, representing sales of 10 air conditioners per second within the next 30 years. There is an expectation that air cooling systems could account for more than 20% of the world's electrical energy demand growth (provided that no alternative technologies for air conditioning become available). According to the Brazilian Association of Refrigeration, Air Conditioning, Ventilation, and Heating (ABRAVA) [29], the Brazilian air conditioning and refrigeration sector in the year 2021 presented a percentage growth of 9.8% by the year 2020, while the amount of split-type air conditioners that were produced in 2021 exceeded 3.5 million units. This significant increase in sales in the HVAC sector resulted in an increase in energy consumption. Therefore, programs aimed at energy efficiency were created to obtain better results regarding air conditioning systems, increases in renewal air, and electricity consumption.

2.1. World

The first climate zoning initiatives for building energy-efficiency programs were mainly undertaken in countries (Sweden and Norway) dominated by heating with extreme weather conditions [30]. Nowadays, several countries, such as Australia [31], China [32], Germany [33], India [34], Japan [35], Saudi Arabia [36], the United Kingdom [37], and the United States [38], are subject to climate zoning for energy-efficiency analysis in buildings. In Brazil, the first definition of climatic zones was presented by Roriz et al. [39]. These climate zones are used for various purposes and support thermal regulations with prescription-based and performance-based requirements, standardized data for building energy calculation, energy standards, voluntary labeling programs, and design guidelines [40].

The number of zones needed to characterize a country is essential for the success of building energy-efficiency programs [41]. An excess of zones leads to complicated energy-efficiency programs for buildings, making their use and adoption difficult. A low number leads to extensive zones with significant climatic variations, making them unsuitable for any building energy-efficiency program [42].

The United States of America (USA) has a national building code applicable to its total territory but also has state-level building regulations that address smaller fractions of its territory [43]. According to the International Energy Conservation Code [44] and Standard 90.1-2019 of ASHRAE [38], the USA is divided into 17 zones. The low-resolution national climate zoning of the USA is contrasted with the climate zoning of the state of California, where significant energy-efficiency efforts have recently been implemented. California has a higher resolution climate zoning containing 16 zones, with climate zones defined by energy use [38].

On the other hand, Tunisia has two climate classifications (one for thermal regulation purposes and another for passive construction design guidelines) based on performance metrics such as energy consumption by HVAC and thermal comfort in buildings without HVAC. However, climate zoning developed for thermal-regulation purposes has minor sensitivity to some climatic variables, such as wind speed and direction [45]. Climate zoning is usually adopted to guide requirements for civil construction, which has a high economic impact. However, in some countries, such as Brazil, climate zoning took more time to become part of building requirements due to the slow transition from voluntary to mandatory requirements [46]. Another factor that complicates the implementation of climate zoning is an arbitrary number of zones, especially in the case of neighboring areas with similar climates but located on opposite sides of the boundaries between adjacent zones. For example, on the border between Argentina, Brazil, and Uruguay, recommended values for coverage based on climate zoning range from 0.5 to 2 W/m^2K depending on which side of the border the building is located (despite negligible climatic variations in this region) [40].

The European Union has developed two strategies to combat CO_2 emissions and promote building energy efficiency. Initially, energy performance certification was created in the 1990s as an essential method for reducing energy use and CO_2 emissions. Subsequently, countries adopted an Energy Policy Strategy called Horizon 20-20 ("H2020") to reduce greenhouse gas emissions. European Union countries have been implementing various methods, laws, and projects to achieve the targets prescribed by H2020; however, the main focus has been on building efficient new buildings and renovating older ones [47]. In China, mandatory minimum standards for home appliances have been adopted, and an increase in regulation and enforcement of these standards is noticeable. This country promotes a voluntary energy-efficiency labeling program for devices and uses EU-adopted labeling conventions. Corporate income tax incentives encourage energy-efficient technologies and measures [48].

2.2. Brazil

Due to the importance of air conditioning and the associated energy consumption, programs were created to optimize the electricity consumption of buildings, especially

HVAC systems. Thus, programs have emerged aimed at studying technologies to improve the energy efficiency of an air treatment system. According to Xu et al. [49], in 2001, Brazil suffered the largest energy crisis, leading the population and institutions to mobilize and seek alternatives to save electricity. Therefore, different energy-efficiency programs were implemented in Brazil.

2.2.1. National Electric Energy Conservation Program

With the increasing concern about energy consumption, a discussion in Brazil on the energy issue began in 1984 conducted by The National Institute of Metrology, Standardization, and Industrial Quality (INMETRO). This institution contributed to creating the Brazilian Labeling Program (PBE in Portuguese), which works through informative labels to inform the consumer about the energy efficiency of the item in question [50]. The PBE gained strength by adding two partners: the National Energy Conservation Program (PROCEL in Portuguese) and the National Program for the Rationalization of the Use of Petroleum and Natural Gas Derivatives (CONPET in Portuguese) [51].

In 1993, the PROCEL Seal was created to inform consumers about more efficient equipment (mainly air conditioners, freezers, and refrigerators). Thus, it serves to identify the level of energy efficiency of appliances based on tests in laboratories certified by INMETRO and classified according to their level of energy consumption. Those that are classified as A are the ones that have better energy efficiency; that is, they consume less energy [52].

Considering buildings consume approximately 50% of Brazil's energy, in 2003 the federal government created the PROCEL EDIFICA seal. This strategy seeks to evaluate commercial, residential, and public service building parameters to prepare buildings with the lowest possible energy expenditure by considering the facade, lighting, and air conditioning system [53]. In buildings that are structured from their conception to have the PBE EDIFICA seal, they present an energy reduction of up to 50% compared to buildings that have not adopted the PBE EDIFICA methodology. In buildings that have undergone significant retrofits, the percentage of energy reduction rises to 30% [54].

Figures 2 and 3 present the results of annual investments and historical results obtained by PROCEL from 2008 to 2020 showing that there was an exponential increase in energy savings while investments after 2011 decreased.

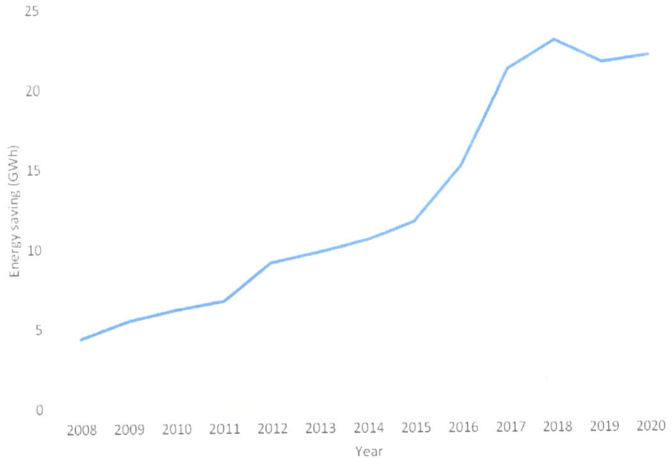

Figure 2. Energy savings generated by PROCEL from 2008 to 2020. Source: adapted from [55].

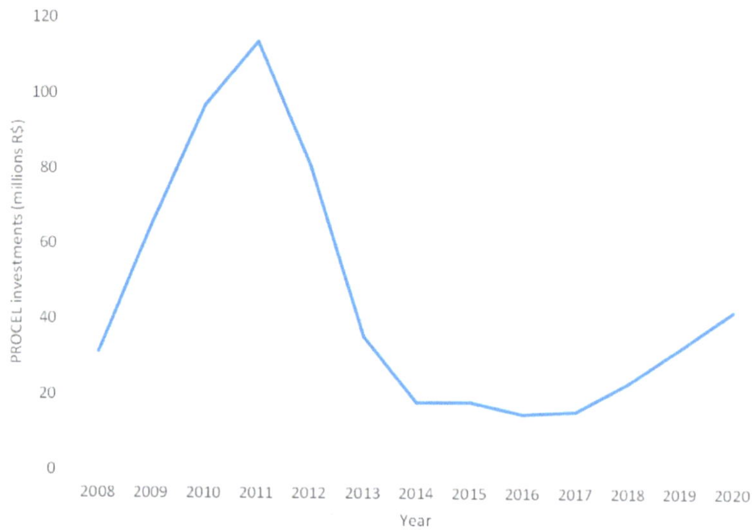

Figure 3. The investments made in PROCEL. Source: adapted from [55].

2.2.2. Other Programs

In addition to PROCEL, other programs and mechanisms were created with the support of the Ministry of Mines and Energy, as well as laws and decrees in order to promote energy efficiency and energy conservation at the national level. Among all these, the main programs adopted are briefly presented:

- National Program for the Rationalization of the Use of Petroleum and Natural Gas Derivatives (CONPET): CONPET is a Ministry of Mines and Energy program that Petrobras subsidizes through technical, administrative, and financial resources. It is a program instituted by decree in 1991 that is responsible for designing, operationalizing strategies, promoting institutional articulation, and disseminating the program's actions. Initially, it aimed to encourage the efficient use of non-renewable sources in the economy;
- Law No. 9478 of 6 August 1997: establishes the principles and objectives of the National Energy Policy, which aims to protect the environment and promote energy conservation. In addition, this law also created the National Energy Policy Council (CNPE). CNPE seeks to generate the rational use of the country's energy resources. It will support regulatory agencies such as the National Electric Energy Agency (ANEEL) and the National Agency of Petroleum, Natural Gas, and Biofuels (ANP);
- Law No. 9991 of 24 July 2000: establishes percentages of net operating revenue (NOR) of electric energy distributors to encourage the development of energy-efficiency projects aimed at use at the final point of consumption. In addition, it seeks the transformation of the electric energy market through equipment and new technologies;
- Law No. 10,295 of 17 October 2001 (regulated by Decree No. 4059 of 19 December 2001): Law No. 10,295, also known as the Energy Efficiency Law, establishes the procedures to determine the electrical consumption of machines and energy-consuming appliances (whether they are manufactured or only marketed in the country) and target programs for each type of equipment. The main objective is to promote structural transformations in the market of energy-consuming equipment;
- Decree No. 4059 also defines the procedures and responsibilities to establish the indicators and levels of energy efficiency. For this purpose, it instituted the Management Committee of Indicators and Energy Efficiency Levels (CGIEE) composed of the main institutions in the energy sector, such as the Ministry of Mines and Energy; the Ministry

of Development, Industry, and Foreign Trade; the Ministry of Science and Technology and Innovation; ANEEL; and ANP, among others;

- Law No. 13,280 of 3 May 2016: reserves 20% of the resources of electric energy companies destined for energy efficiency for application in PROCEL. This law also created the Energy Efficiency Management Committee, which transferred to ANEEL the power to define the collection schedule, fines, penalties, and payment method for the resources invested in PROCEL;
- Sectoral Funds: these were created to provide financial resources to research, development, and innovation projects in Brazil and to contribute to the national advancement of technology, science, and innovation in their areas of activity, such as energy, water, mineral, oil, and natural gas. Sectoral funds use resources from the National Fund for Scientific and Technological Development (FNDCT), which was created in 1969;
- PROESCO: this is intended to finance energy-efficiency projects and was approved in 2006 by the National Bank for Economic and Social Development. The program also supports implementing projects that present evidence that will contribute to energy savings and focuses mainly on lighting, compressed air, air conditioning and ventilation, refrigeration and cooling, and other sectors.

3. Desiccant Dehumidifiers

According to Kavanaugh [56], several improvements have already been implemented over the years in terms of the energy efficiency of air conditioning equipment. However, other technologies can reduce energy consumption, including desiccant wheels, enthalpy wheels, and cross-flow heat exchangers used to treat external air in buildings.

3.1. Background

In 1951, Carl Munters filed a patent for a desiccant-based drying system. He realized the potential for attracting water molecules and materials such as silica gel. From this concept, the process and development of drying technology using desiccant wheels began [57].

Desiccant wheels (Figure 4) are based on a wheel with desiccant crystals impregnated and grown on a fiberglass substrate. The lightweight wheel has a high surface-area-to-airflow ratio [58]. The desiccant wheel is a passive desiccant wheel or enthalpy wheel when there is no regeneration air heater. At the same time, it is called an active desiccant wheel when it is provided with an air heater, and the regeneration and process air sides are slatted apart. The wheel is installed with thermal insulation and air-proof material so that there is no mass and energy exchange with the surroundings [59].

During operation, the wheel rotates continuously and the desiccant cycles through adsorption, regeneration, and cooling every 4–5 min. Drying air is constantly regenerated in a closed circuit. The hot regeneration air passes through the desiccant medium, releasing the released moisture into the atmosphere [60]. No ambient air is introduced into the process because desiccant cooling is conducted using dry air. The desired dew point is achieved by changing the spin speed and other dryer variables without excessively drying the thermally sensitive materials [61]. Desiccant wheels are used in various heating and residential environments to prevent the growth of mold and mildew. As conventional air conditioning systems are limited, desiccants remove moisture (latent energy) in hot and humid climates [62].

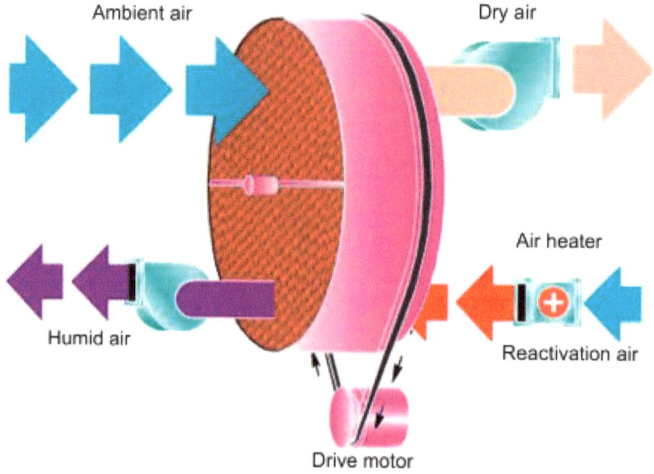

Figure 4. Desiccant wheel [59].

3.2. *Types of Desiccant Dehumidification Systems*

Generally speaking, there are five types of desiccant dehumidification systems; however, the most widespread technology is the vertical desiccant rotor. Each of these technologies is described below.

3.2.1. Spray Drying Tower

The spray drying tower (Figure 5) has two tanks corresponding to the condenser and the regenerator. In this system, the humid air enters the segment of the condenser and passes through a saline fog that will capture the humidity of the air, while the dry air is inflated for the process. At the bottom of the condenser tank, the saline solution is pumped to the second tank (regenerator), through which the second flow of high-temperature external air passes, causing the saline solution to lose moisture to the regeneration air. Thus, the hygroscopic material returns to the condenser tank [63].

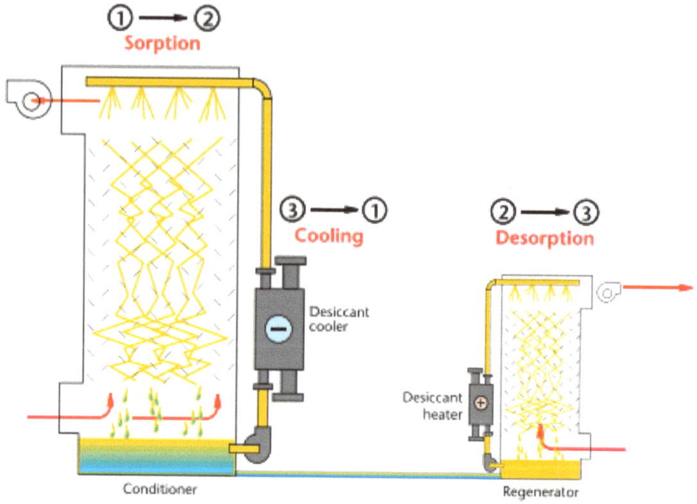

Figure 5. Spray drying tower scheme. Source: [64].

3.2.2. Dual-Tower Desiccant Dryers

The dual-tower desiccant dryers have two vessels (one for process and another for regeneration) as shown in Figure 6; each vessel has solid silica gel inside. Moist air passes through the first pressurized cylinder, trapping moisture in the silica gel until it saturates the hygroscopic material. After the saturation of the first cylinder, the direction of the process air to the second cylinder is reversed, the first being regenerated with external and heated air [65].

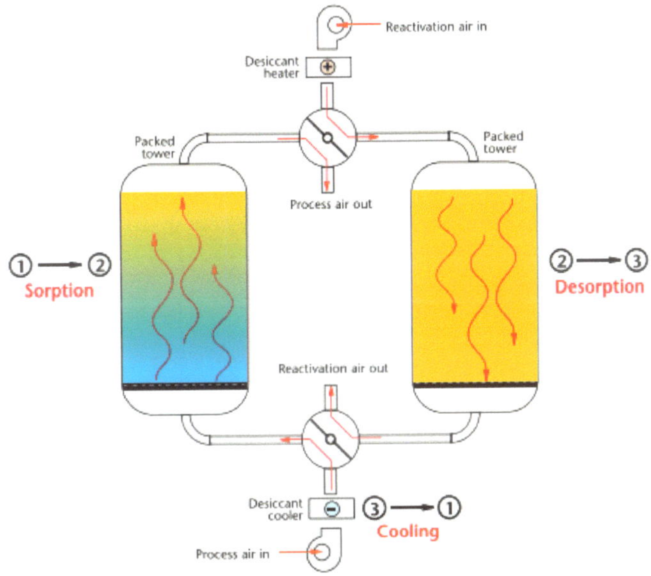

Figure 6. Dual-tower desiccant dryer scheme. Source: [64].

3.2.3. Tray Dryer

The spray dryer has two air flows (process and regeneration) as presented in Figure 7. The hygroscopic material responsible for capturing the humidity from the air is arranged in trays; the humid air passes through the process air sector, and the external air of the regeneration passes through 1/4 of the trays. The trays rotate so the silica gel passes from the process to the regeneration sector [66].

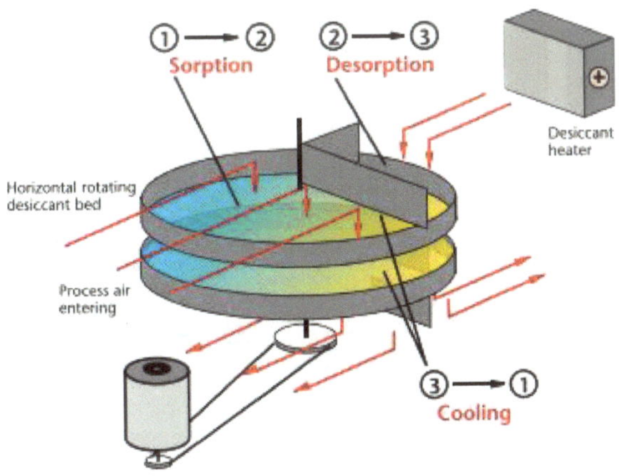

Figure 7. Tray dryer scheme. Source: [64].

3.2.4. Multi-Belt Dryer

The multi-belt dryer has two streams of trays arranged vertically (Figure 8) in which the granular hygroscopic material is installed, and the humid process air passes through the trays. After the silica gel saturation, the tray rotates to the regeneration sector, where the moisture impregnated on the silica gel is eliminated [67].

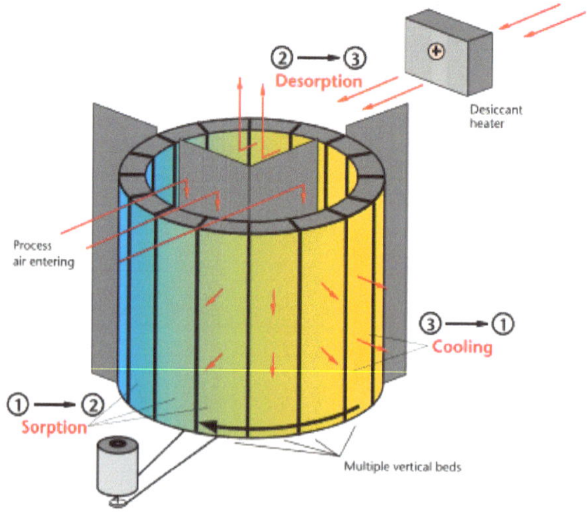

Figure 8. Multi-belt dryer scheme. Source: [64].

3.2.5. Desiccant Dehumidifier

Figure 9 shows the working principle of the desiccant dehumidifier, which has two air flows; the first occupies $\frac{3}{4}$ of the wheel and is responsible for adsorbing all excess moisture from the air. After the air passes through the rotor, it is inflated with low moisture content [68].

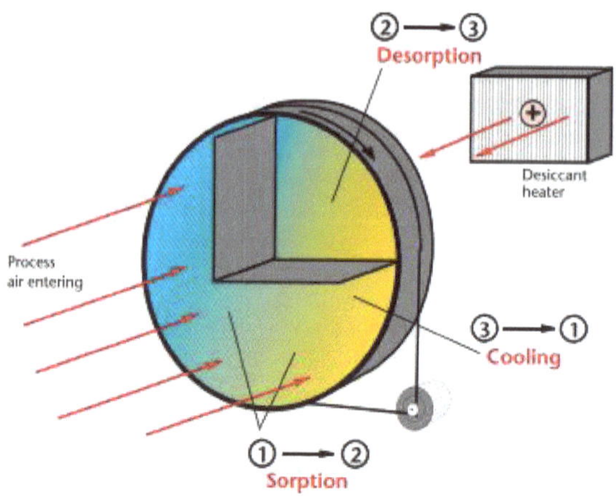

Figure 9. Desiccant dehumidifier scheme. Source: [64].

The second airflow, called regeneration air, occupies $\frac{1}{4}$ of the rotor and is responsible for extracting moisture in a vapor state from the desiccant rotor using a flow of heated air. The rotor structure is constructed in the shape of a beehive in order to have the largest possible area of hygroscopic material in contact with the air [69]. The hygroscopic material may be silica gel or a mixture of silica gel and zeolites [70].

3.3. Desiccant Dehumidification by Heat-Recovery Wheel

Heat-recovery wheels have different operating characteristics from desiccant wheels; their main objective is to promote heat transfer and humidity between the two existing air flows in the equipment [71]. These types of equipment also can minimize cross-contamination due to exhaust and supply airflows and provide high-efficiency systems with a low-pressure drop in the airflows [72]. Thus, some technologies such as cross-flow heat exchangers and enthalpy wheels were developed [73].

3.3.1. Enthalpy Wheels

The rotary heat exchanger (Figure 10), commonly known as an enthalpy wheel, is an "air-air"-type heat exchanger that works on the principle of sensible heat (temperature) and latent heat (humidity) exchange between the flow of renewal air and the flow of exhaust air [74]. Enthalpy wheels can be classified as sensible heat wheels and enthalpy wheels. The denomination for these two types is due to their application, where the sensible heat wheels act with the function of only exchanging the sensible heat, while the enthalpy wheel works completely between the energies of the air (the sensible heat and the latent heat). These two processes occur between the two air masses that travel through the equipment [75].

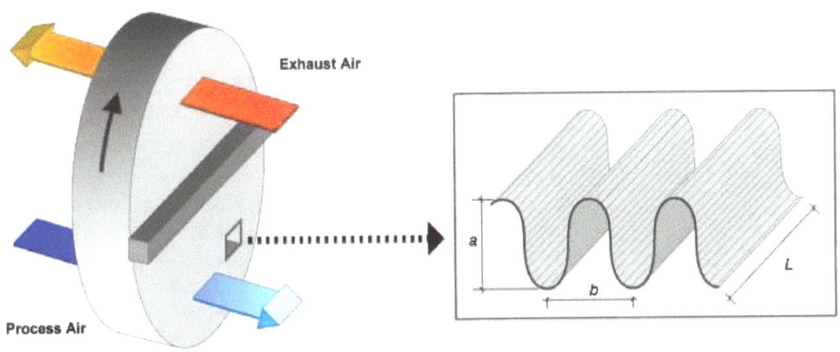

Figure 10. Enthalpy wheel [74].

The enthalpy wheel is an essential piece of equipment for building air-treatment systems that has some advantages and disadvantages. As an advantage, it can be mentioned that it is a technology that reaches efficiencies of around 50% to 80% [76]. Its operation and high operating flows provide energy savings since the cooling system that works after the enthalpy wheel can have a lower cooling load than the conventional one (without using the enthalpy wheel) [77]. For any design to meet the Green Building certifications, some minimum assumptions must be met, including the rate of air renewal of the central systems. Thus, it is possible to reduce the size of the cooling or heating coils of the air-treatment units [78]. Its main disadvantage is the initial investment in the system; however, this value has a short payback period [79].

3.3.2. Cross-Flow Heat Exchangers

The cross-flow heat exchanger is an "air-to-air" static heat exchanger. Like the enthalpy wheel, the cross-flow heat exchanger can be classified into two types [80]. The sensible cross-flow heat exchanger performs the function of only sensible heat exchange between the two air masses, and the conventional cross-flow heat exchanger (Figure 11) exchanges sensible heat and latent heat during the thermal exchange process of the equipment [81].

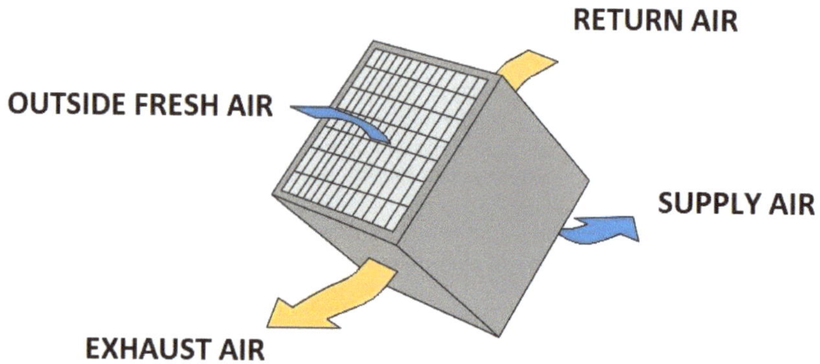

Figure 11. Cross-flow heat exchangers.

Cross-flow heat exchangers are generally used for gas-to-liquid heat-transfer applications in which the gas is on the outside and the liquid is on the tube side of the heat exchanger. Most of these exchangers have fins that increase the surface area and improve heat transfer. In most sensible heat-transfer applications, the external fluid is treated as unmixed flow, while the tube-side fluid is treated as mixed flow [82]. Nasif [83] found that cross-flow heat exchangers significantly reduce energy consumption. They reduces the latent load in hot and humid environments and provide 100% external air to the environment in question.

4. Software for Selecting Desiccant Wheels

Selecting a suitable desiccant requires that the involved dehumidification process be simulated. This can be quite time-consuming depending on the level of detail necessary for a specific simulation and a considerable amount of operating and construction data related to the wheel, as shown in previous works [84–87]. As a result, commercial software became available to facilitate the selection process of dehumidifiers while considering human breathing. Thus, Dong et al. [88] developed a plate dehumidifier model to analyze the shrinkage shape and the variable film thickness of the falling film. The results indicated that the moisture removal rates were augmented from 2.0 g/kg to 2.56 g/kg, the average film thickness decreased from 0.952 mm to 0.889 mm as the contact angles diminished from 85° to 5°, while the wetting area increased from 0.145 m^2 to 0.176 m^2.

Sun et al. [89] compared pre- and post-operative nasal airway models based on a 3-year-old nasal cavity model with apparent adenoid hypertrophy. The results showed that nasopharynx obstruction can induce significantly biased flow distribution in the central nasal passage even if the obstruction site is downstream of the nasal airway. Some review papers offered several works; for example, Tian and Ahmadi [90] studied the current state of computational modeling of micron and nano-elongated particle dynamics and their application to fiber transport and deposition in human respiratory airways; the most versatile and practical approach was the Eulerian–Lagrangian. Ma et al. [91] analyzed the parameters that affect the thermal performance of latent heat thermal energy storage systems (LHTES), including the inlet temperature of the heat transfer fluid, heat transfer rate, properties of the phase-change materials (PCMs), phase-change temperature, geometric parameters, and water tube layout, among others. Finally, the authors discussed the climatic conditions for air-LHTES and found that the higher ambient temperature led to incomplete solidification of the PCMs during the limited nighttime in air-LHTES.

This section presented software for selecting desiccant wheels provided by manufacturers, and some characteristics were analyzed. It is worth noting that there are other programs available for desiccant wheel simulations (TRNSYS, ESP, and Modelica, among others); however, this paper focused on desiccant manufacturer software; therefore, the analysis of other software was left for future studies.

4.1. Novel Aire

Novel Aire Technologies, based in the United States, is a company that has been operating in the dehumidification market for approximately 20 years with energy conservation products and desiccant wheels that has more than 500 customers in 25 countries around the world. The characteristics of Novel Aire software are presented in Table 3.

Table 3. Novel Aire software.

Airflow	340 m^3/h Up to 59,500 m^3/h
Diameter of rotors Main input data	250 mm up to 3050 mm Process airflow

4.2. Munters

Munters was founded by Carl Munters in 1955 and is based in Sweden. It functions in the dehumidification and air conditioning market and serves thousands of customers with about 3500 employees, 17 factories worldwide, and an installed base of 320,000 air-treatment systems. The characteristics of Munters software are presented in Table 4.

Table 4. Munters software.

Airflow	120 m^3/h Up to 170,000 m^3/h
Diameter of rotors Main input data	N/A Process airflow Regeneration airflow

4.3. Rotor Source

Rotor Source was founded in 1999 and is headquartered in the United States. It currently operates the dehumidification market and focuses on selling desiccant wheels to manufacturers of dehumidification equipment. The characteristics of Rotor Source software are presented in Table 5.

Table 5. Rotor Source software.

Airflow	200 m^3/h Up to 150,000 m^3/h
Diameter of rotors Main input data	220 mm up to 3300 mm Process airflow Regeneration airflow

4.4. Puresci

Puresci is a China-based company with over 10 years of experience in the desiccant rotor field. It has projects with more than 1000 clients around the world. The characteristics of the rotor source Puresci are presented in Table 6.

Table 6. Puresci software.

Airflow	N/A
Diameter of rotors Main input data	350 mm up to 3050 mm Process airflow Regeneration airflow

4.5. Comparison between Software

Several companies that develop heat-recovery technologies using desiccant wheels are recognized by international organizations such as Eurovent and AHR Exhibition and by

Brazilian agencies such as ABRAVA and the South Brazilian Association of Refrigeration, Air Conditioning, Heating and Ventilation (ASBRAV) were studied.

The research on commercial software for the technologies studied (Table 7) does not present a careful analysis of the operating results over a year; in most cases, the software only shows the efficiency and output data. In addition, it is necessary that for all software, the user must have external knowledge of thermal engineering to be able to analyze a project.

Table 7. Software comparison.

	Novel Aire	Munters	Rotor Source	Puresci
	Input data			
Process inlet dry bulb temperature	X	X	X	X
Absolute humidity entering the process	X	X	X	X
Regeneration inlet dry bulb temperature	X	X	X	X
Absolute inlet humidity in regeneration	X	X	X	X
Dry bulb temperature after heating in regeneration	X	X	X	X
	Output data			
Output dry bulb temperature in the process	X	X	X	X
Output absolute humidity in the process	X	X	X	X
Dry bulb temperature after regeneration output	X	X	X	X
Output absolute humidity at regeneration output	X	X	X	X
Moisture removal charge	X	X		

Therefore, the software should incorporate an updated database, such as the ASHRAE weather data viewer, into its source code to cover as many locations as possible with updated data. It is also necessary for the software to be able to perform a dynamic analysis of the total hours of the year and not just a one-off analysis.

5. Conclusions

This work presented options for reducing electrical energy consumption in air conditioning and refrigeration systems using desiccant dehumidification technologies. Initially, the importance of air renewal in buildings was analyzed, and it was observed that the circulation lack of renewed air could contribute to the environment being a favorable place for spreading respiratory diseases, which is a risk factor for a worker in any organization. Thus, a building with a high level of pollutants falls under the Sick Building Syndrome.

Subsequently, it was observed that the consumption of conventional air conditioning systems in countries such as the United States and Brazil may require up to 10% of the electricity demand of these countries. Therefore, programs aimed at energy efficiency were created or strengthened to obtain better results when referring to air conditioning systems, increases in renewal air, and electric energy consumption. For example, European Union countries adopted an energy policy known as Horizon 20-20 to reduce greenhouse gas emissions and implemented various laws and projects to achieve the Horizon 20-20 objectives.

The Brazilian Labeling Program has been implemented in Brazil since 1984 to indicate an item's energy efficiency through informative labels. This program was strengthened by adding two policies: the National Energy Conservation Program and the National Program Rationalization of the Use of Petroleum Derivatives and Natural Gas. Other guidelines, such as Law No. 9478 of 1997, Law No. 9991 of 2000, and Law No. 10,295 of 2001, were implemented to promote efficiency and energy conservation in Brazil. All these programs were created to meet the current and future markets that demand high energy consumption. However, it is necessary to explore new technologies such as desiccant dehumidification, which have shown promising results in terms of energy efficiency.

It is worth noting that a desiccant dehumidification system has a coating applied for airflow dehumidification. As the wheel turns, the desiccant alternately passes through the incoming air (where humidity is adsorbed) and through a regeneration zone (where the desiccant is dried and humidity is expelled). The wheel continues to rotate, and the adsorption process is repeated. Several variables can affect desiccant performance; these include the inlet dry bulb temperature, absolute inlet humidity, and speed of the desiccant face, among others. Hence, selecting desiccant wheels is fundamental depending on the project to be developed. Thus, several examples of desiccant wheel selection software were investigated to list and present the main characteristics of each one. It was observed that the multiple programs studied did not offer a detailed analysis of the annual operation and usually only showed the output data. Therefore, the selection software could be improved by incorporating a database update and considering ASHRAE weather data viewer to cover more locations, especially in Brazil.

For future studies, it is suggested to develop software focused on measuring energy and economic benefits when using the technology of desiccant/enthalpic wheels in the recovery of energy from the renewal air and that contemplates improvements concerning the software researched in this work. In addition, other examples of commercial software for desiccant wheel modeling can be assessed.

Author Contributions: Conceptualization, Y.C.S. and D.B.; methodology, Y.C.S.; software, B.G.N. and G.S.F.; validation, Y.C.S., A.F.S. and S.M.D.S.; formal analysis, L.A.S.; investigation, Y.C.S. and D.B.; resources, A.F.S.; data curation, S.M.D.S.; writing—original draft preparation, Y.C.S.; writing—review and editing, S.M.D.S. and L.A.S.; visualization, E.N.d.M.; supervision, E.N.d.M.; project administration, D.B.; funding acquisition, A.F.S. All authors have read and agreed to the published version of the manuscript.

Funding: For financial support, the authors acknowledge ELETROBRAS; FAPERJ (Grant: 210.805/2021, 210.061/2020, 211.731/2021, and 200.480/2023); and the National Agency of Petroleum, Natural Gas and Biofuels (PRH-ANP No 51.1).

Data Availability Statement: The data presented in this study are available upon request from the corresponding author. The data are not publicly available due to project privacy.

Conflicts of Interest: The authors declare no conflict of interest.

References

1. Gül, H. Sick Building Syndrome from the Perspective of Occupational and Public Health. In *Sick Building Syndrome: In Public Buildings and Workplaces*; Abdul-Wahab, S.A., Ed.; Springer: Berlin/Heidelberg, Germany, 2011; pp. 89–104. ISBN 978-3-642-17919-8.
2. Wang, X. Analysis on the Influence of Indoor Air Pollution on Human Health and Prevention Measures. *IOP Conf. Ser. Earth Environ. Sci.* **2019**, *300*, 032059. [CrossRef]
3. World Health Organization. *WHO Global Air Quality Guidelines*; World Health Organization: Geneva, Switzerland, 2021.
4. Suárez Useche, M.A.; Castillo Santiago, Y.; Restrepo, J.B.; Albis Arrieta, A.R.; Agámez Salgado, K.P. Evaluation of the Zinc Sulfate Catalytic Effect in Empty Fruit Bunches Pyrolysis. *Processes* **2022**, *10*, 1748. [CrossRef]
5. Valencia Ochoa, G.; Castillo Santiago, Y.; Duarte Forero, J.; Restrepo, J.B.; Albis Arrieta, A.R. A Comprehensive Comparative Analysis of Energetic and Exergetic Performance of Different Solar-Based Organic Rankine Cycles. *Energies* **2023**, *16*, 2724. [CrossRef]
6. Castillo Santiago, Y.; Henao, N.C.; Venturini, O.J.; Sphaier, L.A.; Duarte, S.V.; de Rezende, T.T.; Ochoa, G.V. Techno-Economic Assessment of Producer Gas from Heavy Oil and Biomass Co-Gasification Aiming Electricity Generation in Rankine Cycle. *Processes* **2022**, *10*, 2358. [CrossRef]
7. Graf, R. *Oil and Sovereignty Petro-Knowledge and Energy Policy in the United States and Western Europe in the 1970s*, 1st ed.; Graf, R., Ed.; Berghahn Books: New York, NY, USA, 2018; ISBN 9781785338069.
8. Mensah-Attipoe, J.; Toyinbo, O. Fungal Growth and Aerosolization from Various Conditions and Materials. In *Fungal Infection*; de Loreto, É.S., Tondolo, J.S.M., Eds.; IntechOpen: Rijeka, Croatia, 2019; pp. 1–13. ISBN 978-1-83880-469-5.
9. Norbäck, D.; Zock, J.-P.; Plana, E.; Heinrich, J.; Svanes, C.; Sunyer, J.; Künzli, N.; Villani, S.; Olivieri, M.; Soon, A.; et al. Mould and Dampness in Dwelling Places, and Onset of Asthma: The Population-Based Cohort ECRHS. *Occup. Environ. Med.* **2013**, *70*, 325–331. [CrossRef]
10. Wolkoff, P. Indoor Air Humidity, Air Quality, and Health—An Overview. *Int. J. Hyg. Environ. Health* **2018**, *221*, 376–390. [CrossRef]
11. Makul, N.; Fediuk, R.; Szelag, M. Advanced Interactions of Cement-Based Materials with Microorganisms: A Review and Future Perspective. *J. Build. Eng.* **2022**, *45*, 103458. [CrossRef]

12. Guerra, F.L.; Lopes, W.; Cazarolli, J.C.; Lobato, M.; Masuero, A.B.; Dal Molin, D.C.C.; Bento, F.M.; Schrank, A.; Vainstein, M.H. Biodeterioration of Mortar Coating in Historical Buildings: Microclimatic Characterization, Material, and Fungal Community. *Build. Environ.* **2019**, *155*, 195–209. [CrossRef]
13. Sterling, E.M.; Arundel, A.; Sterling, T.D. Criteria for Human Exposure to Humidity in Occupied Buildings. *ASHRAE Trans.* **1985**, *91*, 611–622.
14. Israeli, E.; Pardo, A. The Sick Building Syndrome as a Part of the Autoimmune (Auto-Inflammatory) Syndrome Induced by Adjuvants. *Mod. Rheumatol.* **2011**, *21*, 235–239. [CrossRef] [PubMed]
15. Brasche, S. Why Do Women Suffer from Sick Building Syndrome More Often than Men?—Subjective Higher Sensitivity versus Objective Causes. *Indoor Air* **2001**, *11*, 217–222. [CrossRef]
16. AERIS Brasil: Legionella e o Ar Condicionado. Available online: https://www.aerisrs.com.br/legionella-no-brasil-e-o-ar-condicionado (accessed on 29 September 2022).
17. BRASIL. Ministério da Saúde, Gabinete do Ministro. PORTARIA N° 3.523, DE 28 DE AGOSTO DE 1998. Available online: https://bvsms.saude.gov.br/bvs/saudelegis/gm/1998/prt3523_28_08_1998.html (accessed on 30 September 2022).
18. BRASIL. RESOLUÇÃO-RE Nº 9, de 16 de JANEIRO de 2003. Revisa e atualiza a RE/ANVISAnº 176, de 24 de Outubro de 2000, sobre Padrões Referenciais de Qualidade do Ar Interior em Ambientes Climatizados Artificialmente de Uso Público e Coletivo. Available online: https://bvsms.saude.gov.br/bvs/saudelegis/anvisa/2003/rdc0009_16_01_2003.html (accessed on 30 September 2022).
19. ABNT. NBR 16401-3-2008—Sistemas Centrais e Unitários, parte 3: Qualidade do ar interior. *Assoc. Bras. Normas Técnicas* **2008**, *24*. Available online: http://www.caramuru.com.br/pdf/NBR_16401-3_2008%20(1).pdf (accessed on 30 September 2022).
20. BRASIL. Presidência da República, Secretaria-Geral, Subchefia para Assuntos Jurídicos, LEI Nº 13.589, DE 4 DE JANEIRO DE 2018. Available online: https://www.planalto.gov.br/ccivil_03/_ato2015-2018/2018/lei/l13589.htm (accessed on 30 September 2022).
21. Che, W.W.; Tso, C.Y.; Sun, L.; Ip, D.Y.K.; Lee, H.; Chao, C.Y.H.; Lau, A.K.H. Energy Consumption, Indoor Thermal Comfort and Air Quality in a Commercial Office with Retrofitted Heat, Ventilation and Air Conditioning (HVAC) System. *Energy Build.* **2019**, *201*, 202–215. [CrossRef]
22. ABRAVA. Revista ABRAVA Climatização e Refrigeração—edição Dezembro de 2021. Available online: https://abrava.com.br/revista-abrava-refrigeracao-climatizacao-edicao-dezembro-2021-disponivel-para-download/ (accessed on 30 September 2022).
23. Pargeter, S. Reducing Building HVAC Costs with Site-Recovery Energy. *Facil. Manag.* **2012**, *28*, 24–28.
24. Sphaier, L.A.; Nóbrega, C.E.L. Parametric Analysis of Components Effectiveness on Desiccant Cooling System Performance. *Energy* **2012**, *38*, 157–166. [CrossRef]
25. Sphaier, L.A.; Nóbrega, C.E.L. Desiccant Cooling Cycle Tuning for Variable Environmental Conditions. *Heat Transf. Eng.* **2014**, *35*, 1035–1042. [CrossRef]
26. Nóbrega, C.E.L.; Sphaier, L.A. Desiccant-Assisted Humidity Control for Air Refrigeration Cycles. *Int. J. Refrig.* **2013**, *36*, 1183–1190. [CrossRef]
27. Nóbrega, C.E.L.; Sphaier, L.A. Modeling and Simulation of a Desiccant–Brayton Cascade Refrigeration Cycle. *Energy Build.* **2012**, *55*, 575–584. [CrossRef]
28. International Energy Agency (IEA). *The Future of Cooling Opportunities for Energy-Efficient Air Conditioning*; International Energy Agency (IEA): Paris, France, 2018.
29. ABRAVA. *Boletim Econômico ABRAVA*; ABRAVA: São Paulo, Brazil, 2022.
30. Yang, Y.; Javanroodi, K.; Nik, V.M. Climate Change and Energy Performance of European Residential Building Stocks—A Comprehensive Impact Assessment Using Climate Big Data from the Coordinated Regional Climate Downscaling Experiment. *Appl. Energy* **2021**, *298*, 117246. [CrossRef]
31. Shui, B.; Evans, M.; Somasundaram, S. *Country Report on Building Energy Codes in Australia*; U.S. Department of Energy: OakRidge, TN, USA, 2009.
32. Huang, J.; Deringer, J. *Status of Energy Efficient Building Codes in Asia*; Asia Business Council: Hong Kong, China, 2007.
33. Werner, H. *Energy Conservation Ordinance. Thermal Protection and Energy Economy in Buildings. Comments on DIN 4108-6; Energiesparverordnung. Waermeschutz und Energieeinsparung in Gebaeuden. Kommentar zu DIN V 4108-6*; Beuth: Berlin, Germany, 2001.
34. Bureau of Energy Efficiency. *Energy Conservation Building Code*; Bureau of Energy Efficiency: New Delhi, India, 2017.
35. Evans, M.; Shui, B.; Takagi, T. *Country Report on Building Energy Codes in Japan*; U.S. Department of Energy: Oak Ridge, TN, USA, 2009.
36. Youssef, A. *Climate Zone Map (Czm) Tool for Building Energy Code Compliance in Saudi Arabia*; ASHRAE: Orlando, FL, USA, 2016.
37. BRE. *SAP 2012 The Government's Standard Assessment Procedure for Energy Rating of Dwellings*; BRE: Hertfordshire, UK, 2014.
38. ASHRAE. *Standard 90.1-2019—Energy Standard for Buildings Except Low-Rise Residential Buildings*; ASHRAE: Ann Arbor, MI, USA, 2019.
39. Roriz, M.; Ghisi, E.; Lamberts, R. Bioclimatic Zoning of Brazil: A Proposal Based on the Givoni and Mahoney Methods. In Proceedings of the PLEA'99 Conference, Brisbane, Australia, 22–24 September 1999; University of Queensland: Brisbane, Australia, 1999; pp. 4–9.
40. Walsh, A.; Cóstola, D.; Labaki, L.C. Review of Methods for Climatic Zoning for Building Energy Efficiency Programs. *Build. Environ.* **2017**, *112*, 337–350. [CrossRef]

41. Walsh, A.; Cóstola, D.; Labaki, L.C. Performance-Based Validation of Climatic Zoning for Building Energy Efficiency Applications. *Appl. Energy* **2018**, *212*, 416–427. [CrossRef]
42. Cao, X.; Dai, X.; Liu, J. Building Energy-Consumption Status Worldwide and the State-of-the-Art Technologies for Zero-Energy Buildings during the Past Decade. *Energy Build.* **2016**, *128*, 198–213. [CrossRef]
43. Office of Energy Efficiency and Renewable Energy Status of State Energy Code Adoption. Available online: https://www.energycodes.gov/status (accessed on 23 September 2022).
44. International Code Council. *2021 International Energy Conservation Code (IECC)*; International Code Council: Washington, DC, USA, 2021.
45. Agence Nationale des Energies Renouvelables. *Zonage Climatique Pour La Tunisie*; Agence Nationale des Energies Renouvelables: Montplaisir, Tunis, 2004.
46. Associação Brasileira de Normas Técnicas. *ABNT NBR 15575-2: Edificações Habitacionais—Desempenho*; Associação Brasileira de Normas Técnicas: Rio de Janeiro, Brazil, 2013.
47. Fabbri, K. Energy Incidence of Historic Building: Leaving No Stone Unturned. *J. Cult. Herit.* **2013**, *14*, e25–e27. [CrossRef]
48. Martínez-Molina, A.; Tort-Ausina, I.; Cho, S.; Vivancos, J.-L. Energy Efficiency and Thermal Comfort in Historic Buildings: A Review. *Renew. Sustain. Energy Rev.* **2016**, *61*, 70–85. [CrossRef]
49. Xu, J.; Akhtar, M.; Haris, M.; Muhammad, S.; Abban, O.J.; Taghizadeh-Hesary, F. Energy Crisis, Firm Profitability, and Productivity: An Emerging Economy Perspective. *Energy Strategy Rev.* **2022**, *41*, 100849. [CrossRef]
50. EPE—Empresa de Pesquisa Energética. *Atlas of Energy Efficiency Brazil 2019*; EPE—Empresa de Pesquisa Energética: Brasília, Brazil, 2019.
51. INMETRO Programa Brasileiro de Etiquetagem. Available online: https://www.gov.br/inmetro/pt-br/assuntos/avaliacao-da-conformidade/programa-brasileiro-de-etiquetagem/conheca-o-programa (accessed on 2 September 2022).
52. Ministério de Minas e Energia. *Uso De Ar Condicionado No Setor Residencial Brasileiro: Perspectivas E Contribuições Para O Avanço Em Eficiência Energética*; Ministério de Minas e Energia: Rio de Janeiro, Brazil, 2018.
53. Dos Santos, E.R.; Salgado, M.S. BIM Na Verificação de Requisitos Do Procel Edifica. *Ii Simpósio Bras. Tecnol. Inf. Comun. Na Construção* **2019**, *2*, 1–8. [CrossRef]
54. da Luz, N.D.G. *Etiquetagem Em Edificações: Análise De Eficiência Energética De Uma Residência Unifamiliar De Toledo/PR*; Universidade Tecnológica Federal do Paraná: Curitiba, Brazil, 2021.
55. Nascimento, R.L. *Política De Eficiência Energética No Brasil*; MME: Brasília, Brazil, 2015.
56. Kavanaugh, S. Probing the Effectiveness of ANSI/ASHRAE/IES Standard 90.1. Engineered Systems. 2020. Available online: https://www.esmagazine.com/articles/100699-probing-the-effectiveness-of-ansi-asrae-ies-standard-901 (accessed on 30 September 2022).
57. Munters Munters' Desiccant Rotor—Industrial Dehumidification at Its Best. Available online: https://www.munters.com/en/about-us/history-of-munters/history-news2/munters-desiccant-rotor{-}{-}industrial-dehumidification-at-its-best/ (accessed on 11 July 2022).
58. Kent, R. Services. In *Energy Management in Plastics Processing*; Kent, R., Ed.; Elsevier: Oxford, UK, 2018; pp. 105–210. ISBN 978-0-08-102507-9.
59. Narayanan, R. Heat-Driven Cooling Technologies. In *Clean Energy for Sustainable Development*; Rasul, M.G., Azad, A.K., Sharma, S., Eds.; Academic Press: Oxford, UK, 2017; pp. 191–212. ISBN 978-0-12-805423-9.
60. Rambhad, K.S.; Walke, P.V.; Tidke, D.J. Solid Desiccant Dehumidification and Regeneration Methods—A Review. *Renew. Sustain. Energy Rev.* **2016**, *59*, 73–83. [CrossRef]
61. Guan, B.; Liu, X.; Zhang, T. Investigation of a Compact Hybrid Liquid-Desiccant Air-Conditioning System for Return Air Dehumidification. *Build. Environ.* **2021**, *187*, 107420. [CrossRef]
62. ASHRAE. *2019 ASHRAE Handbook_HVAC Applications CH35.Pdf*; ASHRAE: Orlando, FL, USA, 2019; ISBN 9781947192133.
63. Kojok, F.; Fardoun, F.; Younes, R.; Outbib, R. Hybrid Cooling Systems: A Review and an Optimized Selection Scheme. *Renew. Sustain. Energy Rev.* **2016**, *65*, 57–80. [CrossRef]
64. Munters. *Dehumidificaion Handbook*, 3rd ed.; Munters Corporation, Ed.; Munters Corporation Marketing Department: Amesbury, MA, USA, 2019; ISBN 2013206534.
65. Ling-Chin, J.; Bao, H.; Ma, Z.; Roskilly, W.T. State-of-the-Art Technologies on Low-Grade Heat Recovery and Utilization in Industry. In *Energy Conversion—Current Technologies and Future Trends*; Al-Bahadly, I.H., Ed.; IntechOpen: Rijeka, Italy, 2018; pp. 55–74. ISBN 978-1-78984-905-9.
66. Mujumdar, A.S.; Huang, L.-X.; Chen, X.D. An Overview of the Recent Advances in Spray-Drying. *Dairy Sci. Technol.* **2010**, *90*, 211–224. [CrossRef]
67. Zhang, H.; Pang, B.; Kang, S.; Fu, J.; Tang, P.; Chang, J.; Li, J.; Li, Z.; Deng, S. The Influence of Feedstock Stacking Shape on the Drying Performance of Conveyor Belt Dryer. *Heat Mass Transf.* **2022**, *58*, 157–170. [CrossRef]
68. Li, X. Analysis on the Utilization of Temperature and Humidity Independent Control Air-Conditioning System with Different Fresh-Air Handling Methods. *Procedia Eng.* **2017**, *205*, 71–78. [CrossRef]
69. Chen, T.; Norford, L. Energy Performance of Next-Generation Dedicated Outdoor Air Cooling Systems in Low-Energy Building Operations. *Energy Build.* **2020**, *209*, 109677. [CrossRef]

70. Gado, M.G.; Nasser, M.; Hassan, A.A.; Hassan, H. Adsorption-Based Atmospheric Water Harvesting Powered by Solar Energy: Comprehensive Review on Desiccant Materials and Systems. *Process. Saf. Environ. Prot.* **2022**, *160*, 166–183. [CrossRef]
71. Nizovtsev, M.I.; Borodulin, V.Y.; Letushko, V.N. Influence of Condensation on the Efficiency of Regenerative Heat Exchanger for Ventilation. *Appl. Therm. Eng.* **2017**, *111*, 997–1007. [CrossRef]
72. Calautit, J.K.; O'Connor, D.; Tien, P.W.; Wei, S.; Pantua, C.A.J.; Hughes, B. Development of a Natural Ventilation Windcatcher with Passive Heat Recovery Wheel for Mild-Cold Climates: CFD and Experimental Analysis. *Renew. Energy* **2020**, *160*, 465–482. [CrossRef]
73. Xu, Q.; Riffat, S.; Zhang, S. Review of Heat Recovery Technologies for Building Applications. *Energies* **2019**, *12*, 1285. [CrossRef]
74. Men, Y.; Liu, X.; Zhang, T. A Review of Boiler Waste Heat Recovery Technologies in the Medium-Low Temperature Range. *Energy* **2021**, *237*, 121560. [CrossRef]
75. Herath, H.M.D.P.; Wickramasinghe, M.D.A.; Polgolla, A.M.C.K.; Jayasena, A.S.; Ranasinghe, R.A.C.P.; Wijewardane, M.A. Applicability of Rotary Thermal Wheels to Hot and Humid Climates. *Energy Rep.* **2020**, *6*, 539–544. [CrossRef]
76. Liu, Z.; Li, W.; Chen, Y.; Luo, Y.; Zhang, L. Review of Energy Conservation Technologies for Fresh Air Supply in Zero Energy Buildings. *Appl. Therm. Eng.* **2019**, *148*, 544–556. [CrossRef]
77. Jani, D.B.; Mishra, M.; Sahoo, P.K. Solid Desiccant Air Conditioning—A State of the Art Review. *Renew. Sustain. Energy Rev.* **2016**, *60*, 1451–1469. [CrossRef]
78. Altomonte, S.; Schiavon, S.; Kent, M.G.; Brager, G. Indoor Environmental Quality and Occupant Satisfaction in Green-Certified Buildings. *Build. Res. Inf.* **2019**, *47*, 255–274. [CrossRef]
79. Zender–Świercz, E. A Review of Heat Recovery in Ventilation. *Energies* **2021**, *14*, 1759. [CrossRef]
80. Wang, Y.; Wang, L.; Huang, Q.; Cui, Y. Experimental and Theoretical Investigation of Cross-Flow Heat Transfer Equipment for Air Energy High Efficient Utilization. *Appl. Therm. Eng.* **2016**, *98*, 1231–1240. [CrossRef]
81. Hye-Cho, J.; Cheon, S.-Y.; Jeong, J.-W. Development of empirical models to predict latent heat exchange performance for hollow fiber membrane-based ventilation system. *Appl. Therm. Eng.* **2022**, *213*, 118686. [CrossRef]
82. Silaipillayarputhur, K.; Al-Mughanam, T. Performance of Pure Crossflow Heat Exchanger in Sensible Heat Transfer Application. *Energies* **2021**, *14*, 5489. [CrossRef]
83. Nasif, M.S. Air-to-Air Fixed Plate Energy Recovery Heat Exchangers for Building's HVAC Systems. In *Sustainable Thermal Power Resources through Future Engineering*; Sulaiman, S.A., Ed.; Springer: Singapore, 2019; pp. 63–71. ISBN 9789811329685.
84. Santos, S.M.D.; Sphaier, L.A. Transient Formulation for Evaluating Convective Coefficients in Regenerative Exchangers with Hygroscopic Channels. *Int. Commun. Heat Mass Transf.* **2020**, *116*, 104691. [CrossRef]
85. Sphaier, L.A.; Worek, W.M. Parametric Analysis of Heat and Mass Transfer Regenerators Using a Generalized Effectiveness-NTU Method. *Int. J. Heat Mass Transf.* **2009**, *52*, 2265–2272. [CrossRef]
86. Sphaier, L.A.; Worek, W.M. Numerical Solution of Periodic Heat and Mass Transfer with Adsorption in Regenerators: Analysis and Optimization. *Numer. Heat Transf. A Appl.* **2008**, *53*, 1133–1155. [CrossRef]
87. Sphaier, L.A.; Worek, W.M. The Effect of Axial Diffusion in Desiccant and Enthalpy Wheels. *Int. J. Heat Mass Transf.* **2006**, *49*, 1412–1419. [CrossRef]
88. Dong, C.; Lu, L.; Qi, R. Model development of heat/mass transfer for internally cooled dehumidifier concerning liquid film shrinkage shape and contact angles. *Build. Environ.* **2017**, *114*, 11–22. [CrossRef]
89. Sun, Q.; Dong, J.; Zhang, Y.; Tian, L.; Tu, J. Numerical study of the effect of nasopharynx airway obstruction on the transport and deposition of nanoparticles in nasal airways. *Exp. Comput. Multiph. Flow* **2022**, *4*, 399–408. [CrossRef]
90. Tian, L.; Ahmadi, G. Computational modeling of fiber transport in human respiratory airways—A review. *Exp. Comput. Multiph. Flow* **2021**, *3*, 1–20. [CrossRef]
91. Ma, Y.; Luo, Y.; Xu, H.; Du, R.; Wang, Y. Review on air and water thermal energy storage of buildings with phase change materials. *Exp. Comput. Multiph. Flow* **2021**, *3*, 77–99. [CrossRef]

processes

MDPI

Article

How Does Public Sentiment Affect the Socially Responsible Behavior of Construction Enterprises?

Xiaoxu Dang [1], Shihui Wang [1], Xiaopeng Deng [2,*], Ziming Zhang [1], Na Zhang [3] and Hongtao Mao [2]

1 School of Management, Xi'an University of Science and Technology, Xi'an 710054, China
2 School of Civil Engineering, Southeast University, Nanjing 214135, China
3 School of Civil Engineering and Architecture, Zhejiang Sci-Tech University, Hangzhou 310018, China
* Correspondence: dxp@seu.edu.cn

Abstract: The negative environmental impact of the construction sector has garnered global attention, and as the "primary force" in achieving the "double carbon" target, green development is urgent, and social responsibility practices cannot be postponed. An evolutionary game model was constructed by combining the rank-dependent expected utility (RDEU) theory and the evolutionary game theory to understand the interaction mechanism between participants' emotions and decisions, taking into account the characteristics of construction enterprises and the public regarding irrational decisions under heterogeneous emotional combinations. The study demonstrates that: (1) there is probability in the choice of rational strategies, and emotion is an irrational factor that can affect strategy choice. (2) The evolutionary trend of the strategy choice of the game subjects is altered by emotional intensity and emotional propensity. The optimism of construction enterprises inhibits their socially responsible practice, and the pessimism of the public promotes the probability of their negative strategy choice. Furthermore, moderate optimism is a safety valve for the public's positive strategy choice. (3) The interaction of emotional states leads to a heterogeneity of strategy choices exhibited under different combinations of emotions, with the emotions of construction companies having a more dominant influence on strategy. Finally, we make some feasible recommendations for improving social responsibility practices and preventing mass incidents by boosting emotional monitoring and guidance for construction businesses and the general public. Overall, this study provides important information about how to be socially responsible, maintain good relationships with the public, and protect the environment.

Keywords: corporate social responsibility; environmental corporate social responsibility; rank-dependent expected utility (RDEU) theory; evolutionary game

Citation: Dang, X.; Wang, S.; Deng, X.; Zhang, Z.; Zhang, N.; Mao, H. How Does Public Sentiment Affect the Socially Responsible Behavior of Construction Enterprises? *Processes* **2022**, *10*, 2403. https://doi.org/10.3390/pr10112403

Academic Editors: Ferdinando Salata and Virgilio Ciancio

Received: 8 October 2022
Accepted: 11 November 2022
Published: 15 November 2022

1. Introduction

Global warming and pollution have long been a source of concern. The frequency and intensity of extreme climate events caused by warming are increasing [1], and rising CO_2 emissions are regarded as the primary cause of these phenomena [2,3]. As the world's largest producer of carbon emissions, in 2020, China formally committed to strive for peak carbon emissions by 2030 and to become carbon neutral by 2060 [4].

As one of China's national pillar sectors, the construction sector has a considerable impact on the country's economic growth [5]. However, it is also the sector that uses the most energy and accounts for one-third of all global greenhouse gas (GHG) emissions [6–8]. The massive urbanization process in China will lead to a further increase in energy demand and carbon emissions [9]. In the foreseeable future, the construction industry will be the "last mile" in the shift to carbon neutrality [10,11]. Although some resource and pollution challenges have been overcome through current technological advancements [12], which show great potential for carbon emission reduction [10], the low-carbon transition is moving slowly. Corporate social responsibility (CSR) ought to be put into practice as an additional step to hasten the accomplishment of the "double carbon" goal.

CSR is defined as context-specific organizational actions and policies that take into account stakeholder expectations and the triple bottom line of economic, social, and environmental performance [13]. It restricts decision makers' single-minded pursuit of maximizing profits and aids in the creation and adoption of agendas and regulations for sustainable development [14]. With the awakening of environmental consciousness, an increasing number of businesses are shifting their CSR focus from charitable donations to environmental actions in order to "go green" [15]. Scholars have paid increased attention to CSR and environmental corporate social responsibility in accordance with the increase in environmental awareness in today's environmental evolution (e.g., climate change) [16–19]. Environmental CSR has emerged as a significant and distinct element of CSR [20,21]. It is defined as accepting responsibility for a company's operations, products, and facilities' environmental impact. It includes eliminating waste and pollution, increasing the efficiency and productivity of its resources, and minimizing practices that may affect future generations' enjoyment of national resources [19]. According to studies, implementing CSR initiatives in the construction industry helps to lessen or even eliminate the environmental impact that these activities have on the environment [22], as well as promote the sustainability of themselves and society as a whole [23].Additionally, companies' proactivity towards environmental protection can boost resource productivity and preserve their competitiveness [24], and customers are more inclined to purchase environmentally friendly products exhibiting socially responsible attributes [25–28]. In the case of construction companies, socially responsible practices not only contribute to improving the environment and promoting social development, but also bring competitive advantages and economic benefits to themselves, which is a win-win outcome [29].

Even so, construction enterprises are often accused of acting irresponsibly [30–32]. This is because its business practices consume large amounts of natural resources, are a source of pollutants and wastes (dust, harmful gases, and noise [33,34]), impose a heavy burden on the environment [35], and interfere with the daily lives of residents while posing health risks to humans [36]. The industry lags far behind other industries in terms of awareness and implementation of CSR [37,38], with its decision makers placing a greater emphasis on financial costs than on environmental habitat issues [39]. The acceleration of the construction industry due to economic growth further triggers adverse effects on environmental degradation [40], and environmental violations, as well as increased public social awareness and expectations, often lead to intensifying conflicts between construction companies and the public, thus increasing the probability of protest initiation and adverse consequences of project cancellation or postponement. Based on the above description of the necessity and utility of CSR and the negative image of construction enterprises regarding the issue of social responsibility fulfillment, it makes sense to analyze their social responsibility decisions and to consider this specific aspect of environmental CSR.

Researchers have used game theory to examine the problem of social responsibility fulfillment. Ma constructed a three-party evolutionary game model of government regulators, organic food producers, and e-commerce retailers to analyze the lack of social responsibility in the organic food supply chain [41]. Zhao provided policy recommendations for enterprises to consciously fulfill their social responsibility based on a game model involving government departments, private enterprises, and the public [42]. Liu effectively encouraged enterprises to take more social responsibility based on the analysis of a game model consisting of government, retailers, and suppliers [43]. The analysis used in the examples above assumes that decision makers are entirely or partially rational [44]. It excludes the perspective of emotions, an irrational component, which can be used to analyze the micro-influence process underlying decision making.

Emotions are often elicited by specific events [45] and play a key role in decision making by influencing the way we perceive, process, and present information [46,47]. Decision makers have diverse emotions due to differences in various factors such as their values and interests [48]. Regarding CSR implementation, the emotional state of relevant stakeholders can have a significant impact on the final degree of social responsibility practiced. It is necessary to include emotions, an irrational factor, in the analysis of the socially responsible behavior of construction enterprises. As a utility theory that focuses on players' emotional dimensions, rank-dependent expected utility (RDEU) theory overcomes the limitation exhibited by traditional game analysis approaches in that they cannot effectively explain the varied emotional states of decision makers [49]. Therefore, this paper will combine rank-dependent expected utility (RDEU) theory and evolutionary game theory to construct an emotion model with construction enterprises and the public as participants, analyze the internal logic of the role of emotions (pessimism, optimism, and rational emotions) in CSR practices, explore the micro-mechanisms of their evolution, and then enrich the existing conclusions on the influence of emotions on CSR behaviors. The main contributions of this paper are as follows:

(1) This paper extends the research boundaries of previous studies on CSR fulfillment issues in terms of content [41–43] to consider the influence of decision makers' environmental emotions on the choice of social responsibility practice strategies.
(2) This paper constructs an emotion model that combines evolutionary game theory and RDEU theory, with construction companies and the public as the main subjects, in order to enrich the application scenarios of evolutionary game theory and RDEU theory [50,51].
(3) This paper discusses the mechanisms of heterogeneous emotions and their emotional combinations regarding decision making, and relevant suggestions are made which are expected to provide effective technical and theoretical support to promote the level of social responsibility practiced by construction enterprises.

The remainder of the paper is organized as follows. Section 2 introduces the basic theory of RDEU and constructs a mathematical model to portray the mathematical performance of different game players. Section 3 analyzes the stability of strategies and the stability of strategy combinations for each game participant. Section 4 presents and discusses the results of the simulation analysis. Section 5 presents the conclusions, management implications, and directions for future research.

2. Game Model Construction

2.1. Problem Description

The fulfillment of social responsibility by construction enterprises is an inevitable choice to promote the realization of the "double carbon" goal and maintain sustainable development. Under the double pressure of maintaining growth and promoting emission reduction, it requires the efforts of multiple stakeholders, such as construction enterprises, the government, and the public, and is a process of mutual interaction and participation in the game. On the one hand, construction enterprises are the main source of energy consumption and social responsibility fulfillment, and they are also economic organizations, with natural profit-seeking motives. In the absence of certain constraints and supervision, they are inclined to save corporate costs and possess insufficient intrinsic motivation to fulfill their social responsibility. On the other hand, the relevant departments, represented by the government, have an inherent advantage in the effective management of social responsibility fulfillment. However, even a strong regulator can be opportunistic [52]. Regarding the issue of social responsibility fulfillment, the public not only assumes the role of consumers, but also the role of informal regulators, which can compensate for shortcomings in government regulation and play a significant role in promoting a virtuous cycle of social responsibility fulfillment in construction enterprises, whose environmental consumption philosophy also plays an important role in low-carbon development [27,53]. In addition, public participation has recently become an emerging theme in the issue of environmental protection [54]. Therefore, this paper chooses to include the public in the research framework of the social responsibility fulfillment of construction enterprises and constructs an evolutionary game model with construction enterprises and the public as the main subjects.

Construction enterprises have two strategic choices: active fulfillment of social responsibility and negative fulfillment of social responsibility. Active fulfillment of social responsibility shows that enterprises create profits while respecting human values, emphasizing their responsibilities to the environment, consumers, and society, and achieving a win-win situation for both enterprises and society [55]. Negative fulfillment of social responsibility means that companies continue to choose to maximize their profits and engage in a series of unethical business practices (e.g., unreasonable disposal of waste, resource waste, poor construction quality, unfair treatment of employees, etc.) [56].

The public has two strategic choices: compromise acceptance and resolute protest. Among them, "resolute protest" refers to the public voluntarily organizing themselves to disrupt the normal operation of the company by taking actions such as demonstrating and blocking traffic and causing the cancellation or postponement of the construction project [57], which has an impact on social stability. Compromising acceptance, on the other hand, refers to the public accepting the status quo and not taking additional actions.

In the issue of social responsibility fulfillment, construction enterprises are primarily responsible for the strategic deployment and practice of socially responsible behavior. The public is the victim of construction enterprises' negative fulfillment of social responsibility and will play a supervisory role in the fulfillment of CSR. When their rights and interests are infringed, they will also take action to defend their rights. Participants in the game process make their own strategic choices based on the principle of maximizing their own interests and alter their strategies based on the performance of the other side. Emotional factors can affect how construction enterprises reacts and how the public responds during this time. For example, the public, in different emotional states, holds heterogeneous beliefs and judgments about the same level of social responsibility practices. Under the effect of different emotions, construction companies and the public constantly adjust their strategies according to their own judgments, leading to diverse results in the subsequent stages of decision making. To visually describe the impact of heterogeneous emotions on the evolution of social responsibility fulfillment in construction enterprises, the conceptual diagram of the game model is shown in Figure 1.

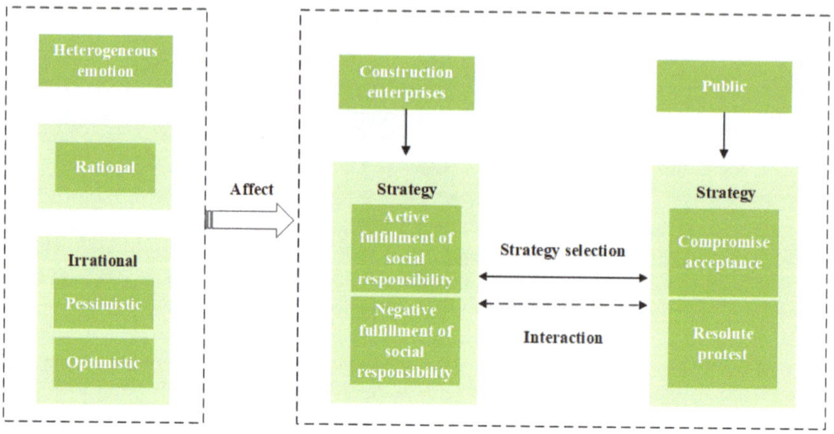

Figure 1. Conceptual diagram of the game model. Source: Created by the authors.

2.2. Model Assumptions

Hypothesis 1. *The issue of socially responsible behavior in construction firms primarily involves two game subjects: construction enterprises and the general public. In the game process, each subject is finitely rational and continuously adapts its own strategy. The game subjects have certain emotional preferences, and the emotional preferences of each game subject will affect the determination of behavioral decisions.*

Hypothesis 2. *The strategy choice of construction enterprises can be divided into the active fulfillment of social responsibility, and the negative fulfillment of social responsibility. The strategy choice of the public can be divided into compromise acceptance, and resolute protest. Construction enterprises' positive fulfillment of social responsibility will generate certain positive externalities, and construction enterprises' negative fulfillment of social responsibility will generate certain negative externalities.*

Hypothesis 3. *Construction companies incur certain costs when they actively fulfill their social responsibility. Construction enterprises will also generate certain positive externalities to benefit the public when they actively fulfill their social responsibilities. Similarly, negative fulfillment of social responsibility by construction enterprises generates certain negative externalities that harm the public. The public can choose to compromise and accept the different social responsibility behaviors of construction enterprises, and then the enterprises will give the public certain compensation; the public can also choose to resolutely protest, and the public's resolute protest will incur certain costs, and at the same time, will cause economic losses and reputation losses for construction enterprises.*

2.3. Model Construction

2.3.1. Traditional Game Model

Based on the assumptions above, the revenue perception matrix is shown in Table 1.

Table 1. Revenue perception matrix.

Participation		The Public	
		Compromise Acceptance e	**Resolute Protest** $1-e$
Construction enterprises	Active fulfillment of social responsibility p	$L_p - C_p + R$ $L_e + K_a$	$L_p - C_p$ $L_e - C_e + K_a$
	Negative fulfillment of social responsibility $1-p$	$L_p + R - W$ $L_e - K_b + W$	$L_p - D_a - D_b$ $L_e - C_e - K_b + M$

The specific parameters of the revenue perception matrix are set, as shown in Table 2.

Table 2. Parameter symbols and their meanings.

Parameter	Meaning
p	Probability of construction enterprises actively fulfilling their social responsibility
e	Probability of public compromise acceptance
C_p	Cost of actively fulfilling social responsibility in construction enterprises
C_e	Cost of resolute public protest
K_a	Positive externalities arising from the active fulfillment of social responsibility by construction enterprises
K_b	Negative externalities arising from the negative fulfillment of social responsibility by construction enterprises
L_p	Basic benefits of construction enterprises
L_e	Basic benefits of the public
D_a	Economic damage to construction enterprises as a result of resolute public protest
D_b	Reputational damage to construction enterprises as a result of resolute public protest
R	Incentives for enterprises without collective protest
W	Compensation for public compromise acceptance by construction enterprises
M	The proceeds of resolute public protest
r_1	Emotional intensity of construction enterprises
r_2	Emotional intensity of the public

The dynamic game process is shown in Figure 2:

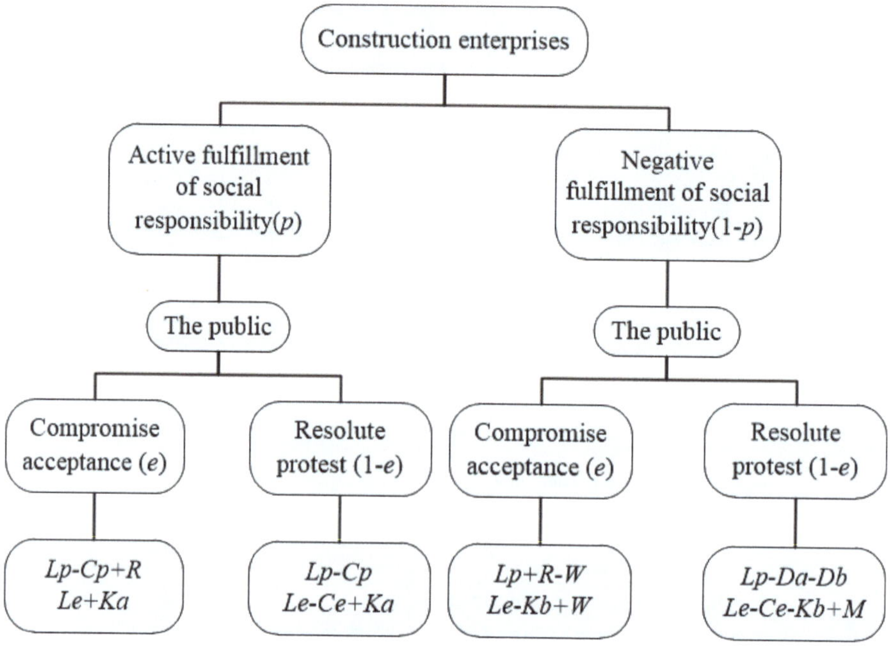

Figure 2. A dynamic game model for CSR implementation. Source: Created by the authors.

2.3.2. Rank-Dependent Expected Utility Model

The rank-dependent expected utility (RDEU) theory was first proposed by Quiggin as an extension of the expected utility function theory [58]. The utility function itself does not contain elements that reflect uncertainty, and it suffers from some shortcomings (e.g., Allais paradox, Ellsberg paradox) [59,60]. On the contrary, the rank-dependent expected utility (RDEU) theory is able to overcome the limitations of the EU theory by introducing a nonlinear function into the theory and defining decision weights using the cumulative probability of combination (inverse) rather than simple probability, which can describe the emotional attitude of decision makers as "economic agents" and their degree under uncertainty [61]. The theory has been successfully applied to analyze the influence and change process of emotions on the decision-making behavior of participants, such as insiders and the nuclear security sector [51], emitters and stakeholders [62], and expropriated farmers and local governments [50].

The core is a real-valued function V defined by a utility function $U(x)$ and a decision weight function $\pi(x)$ to represent the decision maker's preferences for different strategy choices [63], i.e., $V(x,u,\pi) = \sum_{i=1}^{n} \pi(x_i)U(x_i)$, where, for the set of strategies $X = \{x_i, i = 1,2,\cdots n\}$, obeys a probability distribution $P\{X = x_i\} = p_i$ which satisfies $p_i \geq 0, p_1 + p_2 + \cdots + p_n = 1$ Assuming that strategy x_i is ranked according to the magnitude of the utility function $U(x)$ and assigned $x_1 > x_2 > \cdots > x_n$, the utility level of strategy x_i is defined as RP_i, then the probability distribution function of the strategy is $RP_i = P(X \leq x_i) = \sum_{\tau \geq i}^{n} p_i, i = 1,2,\cdots n$. Therefore, the higher the utility rank of the benefit, the greater the probability of cumulative occurrence. At this point, the decision weight function is $\pi(x) = \omega(p_i + 1 - RP_i) - \omega(1 - RP_i)$, where $\omega(\cdot)$ denotes the participant's sentiment function, denoted by $\omega(p) = p^r (r > 0)$, and satisfies $\omega(0) = 0, \omega(1) = 1$.

The function $\omega(\cdot)$ works in three cases, as follows.

(1) When $\omega(p) < p$, $\omega(\cdot)$ is a concave function, for any $p \in [0,1]$,$\omega(\cdot)$ narrows the possibility of $X \leq x$, indicating the pessimism of the participants;

(2) When $\omega(p) > p$, $\omega(\cdot)$ is a convex function, for any $p \in [0,1]$, $\omega(\cdot)$ widens the possibility of $X \leq x$, indicating the optimistic mood of the participant;
(3) When $\omega(p) = p$, the possible is unchanged. there is no emotion affecting the participant's strategy, that is, the rational emotions among the participants.

Specifically, p is the objective probability of the decision occurring, and the subjective probability function becomes $\omega(p) = p^r (0 \leq p \leq 1)$ under the influence of emotions. When $r = 1$, the subjective probability value is the same as the objective probability value, and the game subject is in a rational state; when $r < 1$, the subjective probability value is higher than the objective probability value, and the game subject overestimates the probability of choice, showing optimism, and when $r > 1$, the subjective probability value is lower than the objective probability value, the game subject underestimates the probability of choice, showing pessimism.

Based on the above model assumptions and RDEU theory, we construct a hierarchy-dependent expected utility model for construction enterprises and the public under different strategies.

(1) strategy stability analysis of construction enterprises

For construction enterprises, the base revenue obtained when operating normally is L_p. If construction enterprises actively fulfill their social responsibility, they need to pay a series of costs C_p, including those associated with manpower, material resources, and time, but they receive additional rewards R for their contribution to maintaining social stability. If construction enterprises negatively fulfill their social responsibility and the public chooses a compromise acceptance strategy, the enterprises can still receive additional rewards R for not triggering group conflicts, but the enterprise must give the public certain compensation W. If the public chooses a resolute protest strategy, the enterprise not only faces economic losses D_a, such as compensation to the public and project delays, but also faces reputation damage D_b caused after establishing an irresponsible image, so $D_a + D_b > W$. According to the above analysis and reality, the public's resolute protest will result in large economic losses for the construction enterprises that negatively fulfill their social responsibility. When a company actively fulfills its social responsibility, public protest is ineffective; however, the public's action brings more attention to the company and increases its benefits. According to the above analysis and reality, the public's resolute protest will cause large economic losses to construction enterprises that negatively fulfill their social responsibility. On the contrary, public resolute protest will increase the attention of enterprises that actively fulfill social responsibility and bring benefits. Therefore, assuming that the benefit of construction enterprises is U_p, we can get U_p (actively fulfill social responsibility, compromise acceptance) $>U_p$ (actively fulfill social responsibility, resolute protest) $>U_p$(negatively fulfill social responsibility, compromise acceptance) $>U_p$ (negatively fulfill social responsibility, resolute protest), that is $L_p - C_p + R > L_p - C_p > L_p + R - W > L_p - D_a - D_b$.

This results in the utility, probability, rank, and decision weights corresponding to each strategy of the construction enterprises are listed in Table 3.

Table 3. RDEU of construction enterprises considering emotions.

Construction Enterprises Utility	Probability	Rank Position	Decision Weight
$L_p - C_p + R$	pe	1	$\omega A(pe)$
$L_p - C_p$	$p(1-e)$	$1-pe$	$\omega A(p) - \omega A(pe)$
$L_p + R - W$	$(1-p)e$	$1-p$	$\omega A(p+e-pe) - \omega A(p)$
$L_p - D_a - D_b$	$(1-p)(1-e)$	$1-p-e+pe$	$1 - \omega A(p+e-pe)$

The expected benefits for construction companies choosing "active fulfillment of social responsibility" and "negative fulfillment of social responsibility" are:

$$U_{1p} = [L_p - C_p + R]e^{r2} + [L_p - C_p](1 - e^{r2}) = Re^{r2} + L_p - C_p \tag{1}$$

$$U_{2p} = (L_p + R - W)e^{r2} + (L_p - D_a - D_b)(1 - e^{r2}) = (L_p - D_a - D_b) \\ +(R - W + D_a + D_b)e^{r2} \tag{2}$$

The average expected return on the strategy choice for a construction enterprise is:

$$\begin{aligned}\overline{U_p} &= (L_p - C_p + R)\omega A(pe) + (L_p - C_p)[\omega A(p) - \omega A(pe)] \\ &+(L_p + R - W)[\omega A(p + e - pe) - \omega A(p)] + (L_p - D_a - D_b) \\ &[1 - \omega A(p + e - pe)] = (W - R - C_p)p^{r1} + R(pe)^{r1} \\ &+(R - W + D_a + D_b)(p + e - pe)^{r1} + L_p - D_a - D_b\end{aligned} \tag{3}$$

The dynamic equation for construction enterprise replication is

$$F(p) = dp/dt = p^{r1}(U1p - \overline{Up}) = p^{r1}[Re^{r2} - C_p - (W - R - C_p)p^{r1} \\ -R(pe)^{r1} - (R - W + D_a + D_b)(p + e - pe)^{r1} + D_a + D_b] \tag{4}$$

Letting the dynamic equation $F(p) = 0$, we can get $p = 0$, $p = 1$ or $p = p^*$. At this time, construction companies can achieve local stability by choosing an active fulfillment of social responsibility strategy.

(2) public strategy stability analysis

For the public, the basic benefit that can be obtained from the environment is L_e. When construction companies are actively fulfilling their social responsibilities, the public can obtain benefits K_a from a good residence and living environment. When construction companies are negative fulfilling their social responsibilities, the residence and living experience become worse and can even lead to health loss K_b. If the public chooses to compromise acceptance, the enterprises will provide compensation W to them. If the public selects a resolute protest strategy, it must pay certain time and economic costs C_e to get benefits M for itself. However, as compared to construction firms, the public is in a disadvantaged position, and the benefits fought for are limited, so $K_b > M$ can also yield $K_a > M - K_b$. According to the above analysis and reality, the loss of negative fulfillment of social responsibility and the cost of resolute protest are both larger, and the rights and interests fought for by compromise acceptance are less than the benefits of resolute protest. Therefore, assuming that the public's gain is Ue, we can get Ue (active fulfillment of social responsibility, compromise acceptance) $>$ Ue (active fulfillment of social responsibility, resolute protest) $>$ Ue (negative fulfillment of social responsibility, resolute protest) $>$ Ue (negative fulfillment of social responsibility, compromise acceptance), that is $L_e + K_a > L_e - C_e + K_a > L_e - C_e - K_b + M > L_e - K_b + W$.

This results of the utility, probability, rank, and decision weights corresponding to each strategy of the public are listed in Table 4.

Table 4. RDEU of the public considering emotions.

The Public Utility	Probability	Rank Position	Decision Weight
$L_e + K_a$	pe	1	$\omega B(pe)$
$L_e - C_e + K_a$	$p(1-e)$	$1 - pe$	$\omega B(p) - \omega B(pe)$
$L_e - C_e - K_b + M$	$(1-p)(1-e)$	$1 - p$	$\omega B(1 + pe - e) - \omega B(p)$
$L_e - K_b + W$	$(1-p)e$	$e - pe$	$1 - \omega B(1 - e + pe)$

The expected benefits of "compromise acceptance" and "resolute protest" for the public are U_{1e} and U_{2e}, respectively.

$$\begin{aligned} U_{1e} &= (L_e + K_a)p^{r1} + (L_e - K_b + W)(1 - p^{r1}) \\ &= (L_e - K_b + W) + (K_a + K_b - W)p^{r1} \end{aligned} \tag{5}$$

$$\begin{aligned} U_{2e} &= (L_e - C_e + K_a)p^{r1} + (L_e - C_e - K_b + M)(1 - p^{r1}) \\ &= (L_e - C_e - K_b + M) + (K_a + K_b - M)p^{r1} \end{aligned} \tag{6}$$

The average expected return on strategy choice for the public is:

$$\begin{aligned} \overline{U_e} &= (L_e + K_a)\omega B(pe) + (L_e - C_e + K_a)[\omega B(p) - \omega B(pe)] \\ &+(L_e - C_e - K_b + M)[\omega B(1 + pe - e) - \omega B(p)] \\ &+(L_e - K_b + W)[1 - \omega B(1 - e + pe)] \\ &= (K_a - K_b + M)p^{r2} + C_e(pe)^{r2} + \\ &(M - W - C_e)(1 - e + pe)^{r2} + L_e - K_b + W \end{aligned} \tag{7}$$

The dynamic equation of construction enterprise replication is

$$\begin{aligned} F(e) &= de/dt = e^{r2}(U_{1e} - \overline{U_e}) = e^{r2} \\ &\left\{ \begin{array}{l} [(L_e + K_a)p^{r1} + (L_e - K_b + W)(1 - p^{r1}) - (K_a - K_b + M)p^{r2} \\ -C_e(pe)^{r2} - (M - W - C_e)(1 - e + pe)^{r2} - L_e + K_b - W] \end{array} \right\} \end{aligned} \tag{8}$$

Letting the dynamic equation $F(e) = 0$, we can get $e = 0$, $e = 1$ or $e = e^*$. At this time, the public can achieve local stability by choosing a compromise acceptance strategy.

3. Game Analysis

From the analysis in Section 2.3.2, it can be seen that by creating the dynamic equation $F(p) = F(e) = 0$, the five local equilibrium points of the above game model can be obtained, which are $E1(0,0)$, $E2(0,1)$, $E3(1,0)$, $E4(1,1)$, and $E5(p^*, e^*)$. According to the evolutionary game stability analysis, the stability of each game subject's strategy combination can be judged according to the Lyapunov indirect method; that is, the equilibrium point has local stability when the value of the determinant corresponding to the Jacobian matrix ($Det(J)$) is greater than zero, and the trace of the matrix ($Tr(J)$) is less than zero [64]. The Jacobian matrix of the game model is obtained from Equations (4) and (8):

$$J = \begin{bmatrix} \partial F(p)/\partial p \partial F(p)/\partial e \\ \partial F(e)/\partial p \partial F(e)/\partial e \end{bmatrix} = \begin{bmatrix} B_{11} & B_{12} \\ B_{21} & B_{22} \end{bmatrix} \tag{9}$$

where the matrix corresponds to the value of the determinant $Det(J) = B_{11}B_{22} - B_{12}B_{21}$, and the trace of the matrix $Tr(J) = B_{11} + B_{22}$.

Since the values of the Jacobian matrix are related to the values of model variables, the values of the Jacobian matrix are different under different emotional states of the game subjects, and thus, the equilibrium points obtained are different. Therefore, based on the different emotional states of the game subjects, this paper analyzes the stability of the strategy combinations of the public and the construction enterprises in four situations: (rational, rational), (emotional, emotional), (rational, emotional), and (emotional, rational), respectively.

3.1. Scenario 1: The Construction Enterprise Is Rational, and the Public Is Rational

When the construction enterprise is rational and the public is rational, the sentiment parameter at this time is $r_1 = 1, r_2 = 1$. Bringing the sentiment parameter into the dynamic equation of each replication, the stability analysis of the strategy portfolio at this time is shown in Table 5.

Table 5. Stability analysis in the strategy portfolio in the construction enterprise is rational and the public is rational scenario.

Balancing Point	$\frac{\partial F(p)}{\partial p}$	$\frac{\partial F(p)}{\partial e}$	$\frac{\partial F(e)}{\partial p}$	$\frac{\partial F(e)}{\partial e}$	**Det(J)**	**Tr(J)**	**Stability**
$E_1(0,0)$	$-C_p + D_a + D_b$	0	$2K_b - W - M$	$M - W - C_e$	-	×	Saddle point
$E_2(1,0)$	$-(-C_p + D_a + D_b)$	0	$2K_b - W - M$	$-C_e$	+	-	Stable
$E_3(0,1)$	$W - C_p$	0	$2(K_b - M)$	$M - W - C_e$	-	×	Saddle point
$E_4(1,1)$	$-(W - C_p)$	0	$2(K_b - M)$	$-C_e$	+	-	Stable
$E_5(p^*, e^*)$	A	B	C	D	×	×	Saddle point

Among which,

$$A = \left\{ \begin{array}{c} (-C_p + D_a + D_b) - 2(-C_p + D_a + D_b)p^* \\ -2(W - D_a - D_b)p^*e^* + (W - D_a - D_b)e^* \end{array} \right\}$$

$$B = p^*(W - D_a - D_b)(1 - p^*)$$

$$C = (2K_b - W - M) - (M - W)e^*$$

$$D = (M - W - C_e) - (M - W)p^*$$

As can be seen from Table 5, when both sides of the game are in a fully rational state, the values ($Det(J)$) and traces ($Tr(J)$) of the matrix determinant can be judged $E_2(1,0)$;$E_4(1,1)$ is the evolutionary stable state, that is, construction enterprises actively fulfill their social responsibility, and the public chooses to protest resolutely or compromise to accept. In reality, the Pareto optimum is achieved when construction enterprises actively fulfill their social responsibilities and the public chooses the compromise acceptance strategy. However, the above situation is a pure strategy choice when both sides of the game are emotionless. Objectively speaking, the heterogeneous combination of emotions will lead the game participants to adjust their own strategies according to the situation.

3.2. Scenario 2: The Construction Enterprise Is Emotional, and the Public Is Emotional

When the construction enterprise is emotional and the public is emotional, the sentiment parameter at this time is $r_1 \neq 1, r_2 \neq 1$. Bringing the sentiment parameter into the dynamic equation of each replication, the stability analysis of the strategy portfolio at this time is shown in Table 6.

Table 6. Stability analysis of the strategy portfolio in the construction enterprise is emotional, and the public is emotional scenario.

Balancing Point	$\frac{\partial F(p)}{\partial p}$	$\frac{\partial F(p)}{\partial e}$	$\frac{\partial F(e)}{\partial p}$	$\frac{\partial F(e)}{\partial e}$	**Det(J)**	**Tr(J)**	**Stability**
$E_1(0,0)$	0	0	0	0	0	0	Instability
$E_2(1,0)$	$r_1(C_p - D_a - D_b)$	0	0	0	0	-	Instability
$E_3(0,1)$	0	0	0	0	0	0	Instability
$E_4(1,1)$	$-r_1(W - C_p)$	$R(r_2 - r_1)$	$r_1(K_a + K_b - W)$ $-r_2(K_a - K_b + 2M - W)$	$r_2(2K_b - 2M - C_e)$	×	×	Saddle point
$E_5(p^*, e^*)$	Stability depends on specific values and emotional intensity.						

As can be seen from Table 6, when both the construction enterprise and the public have emotions, as can be judged by the values and traces of the matrix determinant, there is a saddle point $E_4(1,1)$, and the rest are unstable points. In the current situation, there is an interaction between the construction enterprise and the public holding emotions, and both have difficulty estimating each other's behavior and making their own choices. Currently, the stability of the Nash equilibrium point E_5 of the mixed strategy is difficult to judge under the joint influence of the emotion parameters r_1 and r_2 and the uncertainty of returns.

3.3. Scenario 3: The Construction Enterprise Is Rational, and the Public Is Emotional

When the construction enterprise is rational and the public is emotional, the sentiment parameter at this time is $r_1 = 1, r_2 \neq 1$. Bringing the sentiment parameter into the dynamic equation of each replication, the stability analysis of the strategy portfolio at this time is shown in Table 7.

Table 7. Stability analysis of the strategy portfolio in the construction enterprise is rational, and the public is emotional scenario.

Balancing Point	$\frac{\partial F(p)}{\partial p}$	$\frac{\partial F(p)}{\partial e}$	$\frac{\partial F(e)}{\partial p}$	$\frac{\partial F(e)}{\partial e}$	$Det(J)$	$Tr(J)$	Stability
$E_1(0,0)$	$-C_p + D_a + D_b$	0	0	0	0	+	Instability
$E_2(1,0)$	$-(-C_p + D_a + D_b)$	$-R$	0	0	0	-	Instability
$E_3(0,1)$	$W - C_p$	0	$K_a + K_b - W$	0	0	+	Instability
$E_4(1,1)$	$-(W - C_p)$	$R(r_2 - 1)$	$(K_a + K_b - W)$ $-r_2(K_a - K_b + 2M - W)$	$r_2(2K_b - 2M - C_e)$	×	×	Saddle point
$E_5(p^*, e^*)$	Stability depends on specific values and emotional intensity.						

As can be seen from Table 7, when the construction enterprise is in a rational state and the public is in a pessimistic or optimistic mood, a saddle point $E_4(1,1)$ exists as judged by the values and traces of the matrix determinant, and the rest are unstable points. The stability of the Nash equilibrium point E_5 of the hybrid strategy is difficult to judge because its sentiment index r_2 and returns are uncertain, and the results produced by different values of the variables will be different.

3.4. Scenario 4: The Construction Enterprises Is Emotional, and the Public Is Rational

When the construction enterprises is emotional and the public is emotional, the sentiment parameter at this time is $r_1 \neq 1, r_2 = 1$. Bringing the sentiment parameter into the dynamic equation of each replication, the stability analysis of the strategy portfolio at this time is shown in Table 8

Table 8. Stability analysis of the strategy portfolio in the construction enterprises emotion, the public rationality scenario.

Balancing Point	$\frac{\partial F(p)}{\partial p}$	$\frac{\partial F(p)}{\partial e}$	$\frac{\partial F(e)}{\partial p}$	$\frac{\partial F(e)}{\partial e}$	$Det(J)$	$Tr(J)$	Stability
$E_1(0,0)$	0	0	0	$-(M - W - C_e)$	0	+	Instability
$E_2(1,0)$	$C_p - D_a - D_b$	0	0	$2K_b - 2M + C_e$	×	×	Saddle point
$E_3(0,1)$	0	0	$-K_a + K_b - 2M + W$	$M - W - C_e$	0	-	Instability
$E_4(1,1)$	$-r_1(W - C_p)$	$R(1 - r_1)$	$r_1(K_a + K_b - W)$ $-K_a + K_b - 2M + W$	$2K_b - 2M - C_e$	×	×	Saddle point
$E_5(p^*, e^*)$	Stability depends on specific values and emotional intensity						

As can be seen from Table 8, in such a case, the values and traces of the determinants show that there are two saddle points, $E_2(1,0)$, $E_4(1,1)$, and the rest are unstable points. For the Nash equilibrium point E_5 of the mixed strategy, its stability is difficult to judge due to the influence of the sentiment parameter r_1, as well as the uncertainty of benefits, and different values of variables will produce different results.

4. Simulation Analysis

The preceding analysis demonstrates that the various emotions of the participants have a significant impact on the evolution of the choice of socially responsible behavior practices in construction enterprises, and the evolutionary trend of each point, as well as the evolutionary stability of the Nash equilibrium, are also affected by the size of the cost-benefit trade-off and the strength of the emotions. In this paper, numerical simulations were conducted using Matlab software to investigate evolutionary paths under various emotional combinations and emotional parameters. The specific parameters are provided in Table 9 and were determined using the gain relationship and the work of references [15,65,66].

Table 9. Parameter settings.

Parameter	p	e	C_p	C_e	K_a	K_b	L_p	L_e
Initial value	0.5	0.5	1.5	1	1	1	1.3	1
Parameter	D_a	D_b	R	W	M	r_1	r_2	
Initial value	1.6	1.6	1	1.1	1	1	1	

Section 3 of the article analyzes the stability of the strategy combinations in four contexts—(rational, rational), (emotional, emotional), (rational, emotional), and (emotional, rational)—based on the different emotional states of the construction companies and the public. Specifically, the emotional state of the game subjects can be further divided into optimistic and pessimistic emotions, and different emotional states have different effects on the mechanism of strategy choice. Based on the above parameter settings, the evolutionary stability of the system will be analyzed further under the nine specific scenarios—(rational, rational), (optimistic, optimistic), (pessimistic, pessimistic), (optimistic, pessimistic), (pessimistic, optimistic), (optimistic, rational), (pessimistic, rational), (rational, optimistic), (rational, optimistic), and (rational, pessimistic)—on the basis of the four aforementioned scenarios.

4.1. (Rational, Rational) State Analysis, That Is $r_1 = 1, r_2 = 1$

From the evolutionary trend diagram shown in Figure 3, when construction enterprises and the public are rational, construction enterprises tend to choose the negative fulfillment of social responsibility, and the public tends to choose a compromise acceptance strategy. Both have weak awareness regarding the issue of social responsibility fulfillment. Construction enterprises tend not to fulfill it, and the public also tends not to monitor the fulfilment, doing nothing, which coincides with the current phenomenon of social responsibility deficiency in the construction industry. In this game system, construction enterprises, as a powerful group, choose the most favorable strategy based on the principle of maximizing their own interests, and the public chooses to accept the status quo after a rational analysis of the gains and losses. However, the public, as a vulnerable group, will consider all the subjective and objective factors more carefully and judge things with emotions, which makes it difficult to be completely rational.

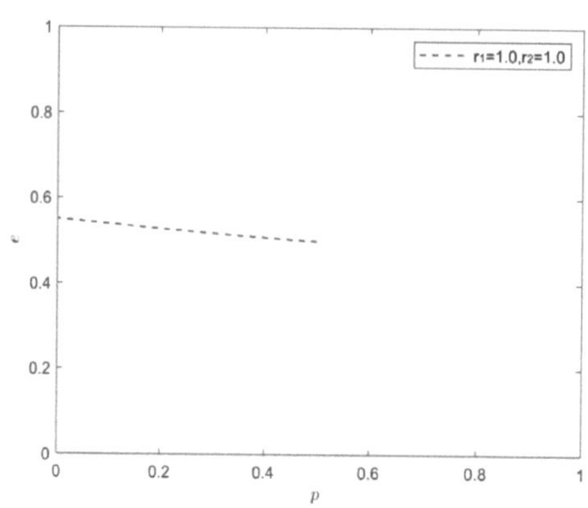

Figure 3. Evolution of game strategy in the (rational, rational) state.

4.2. (Optimistic, Rational) State Analysis, That Is $r_1 < 1, r_2 < 1$

According to the evolution trend diagram in Figure 4, when construction enterprises and the public are optimistic, the probability of the negative social responsibility fulfillment of construction enterprises increases with the depth of optimism, and the public obviously prefers the compromise acceptance strategy. Under the effect of optimism, construction enterprises have a fluke mentality, thus magnifying the effectiveness of the compensation mechanism to appease the public and neglecting the reputation loss and group conflict events caused by the negative fulfillment of social responsibility expectations. As shown in the graph, optimism contributes to a relatively harmonious situation between both enterprises and the public, while collective protests create more uncertainties, and the subsequent discontent of some groups may cause a chain reaction and bring adverse effects to society. As a result, construction enterprises must pay close attention to citizens' collective sentiments and improve their social responsibility practices.

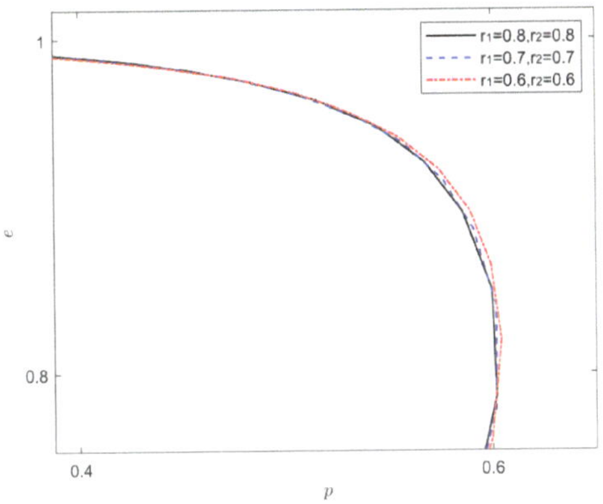

Figure 4. Evolution of game strategy in the (optimistic, optimistic) state.

4.3. (Pessimistic, Pessimistic) State Analysis, That Is $r_1 > 1, r_2 > 1$

From the evolutionary trend graph shown in Figure 5, we see that when both construction enterprises and the public are pessimistic, the probability of the public choosing a compromise acceptance strategy and the construction enterprises adopting an active social responsibility strategy is stable at 0.5. As the pessimism of both sides deepens, the lack of security under pessimism leads to a set time iteration for both sides of the game to determine their options. Unlike the situation where both construction enterprises and the public are optimistic, under the effect of pessimism, construction enterprises perceive the cost of actively fulfilling social responsibility, and the lag and uncertainty of social responsibility returns. Construction firms will maintain their present decision when all advantages and losses are considered. The public will be nervous and anxious during this time due to the effectiveness of collective protest and the time cost, further molding their pessimistic psychology, and hence their willingness to accept.

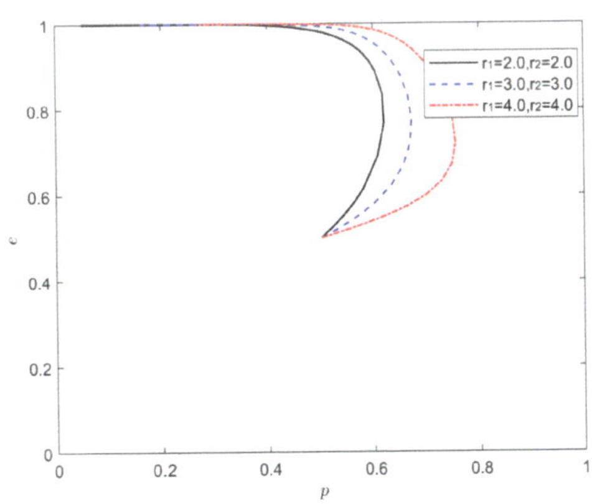

Figure 5. Evolution of game strategy in the (pessimistic, pessimistic) state.

4.4. (Optimistic, Pessimistic) State Analysis, That Is $r_1 < 1, r_2 > 1$

Figure 6 depicts the graph of the evolving trend of construction enterprise optimism and public pessimism. Figure 6 shows that when construction firms' optimism is guaranteed, the probability of the strategy choice of both the active social responsibility of construction enterprises and the public compromise acceptance reduces as public pessimism deepens. By comparing the changing trends of Figure 6a–c, it can be found that when the intensity of public pessimism is constant, with the deepening degree of optimism of construction enterprises, the trajectory of the strategy evolution of both sides of the game moves from the upper left corner to the lower left corner of the coordinate area. At this point, the probability of both the construction enterprises actively fulfilling their social responsibility and the public compromise acceptance decreases. Under the impact of growing optimism, construction enterprises, with the information advantage of their group strength, believe that they can avoid the risk of project delay, suspension, and cancellation by virtue of their brand or existing reputation, thus ignoring the negative spillover effects of social responsibility deficiency. Consistent with the behavioral strategies of construction enterprises and the public, both of which are in a state of optimism, as illustrated in Figure 4, the optimism of the construction firms inhibits them from actively fulfilling their social responsibilities.

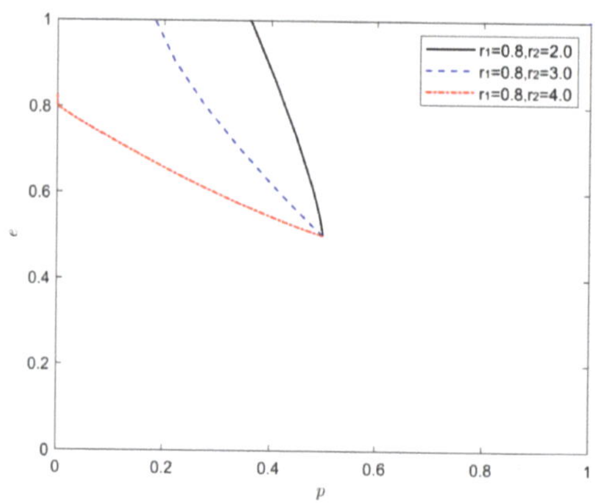

(a)

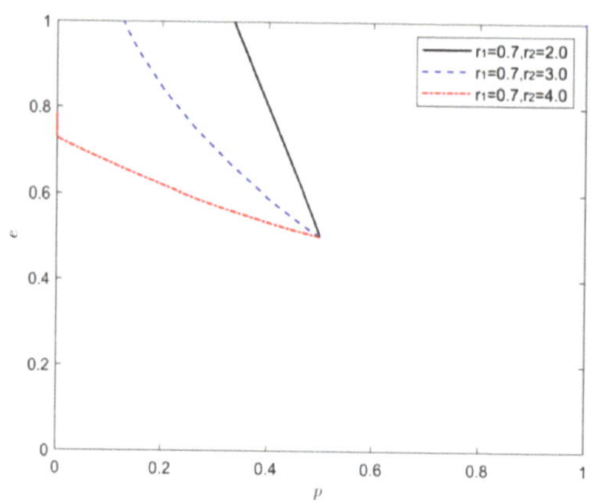

(b)

Figure 6. *Cont.*

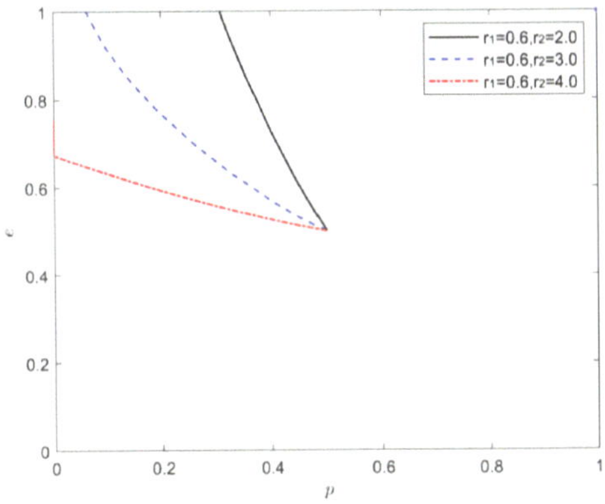

(c)

Figure 6. Evolution of game strategy in the (optimistic, pessimistic) state. (**a**) Scenario 1, (**b**) Scenario 2, (**c**) Scenario 3.

4.5. (Pessimistic, Optimistic) State Analysis, That Is $r_1 > 1, r_2 < 1$

Figure 7 reflects the evolutionary trend graph of the construction enterprises' pessimism and the public optimism. As can be seen from the figure, when public optimism is certain and as the pessimism of construction enterprises deepens, the probability of both the construction enterprises actively fulfilling their social responsibility and public compromise acceptance increases, and the system evolves toward the combined condition of (actively fulfilling social responsibility, compromise acceptance). When comparing the change trends in Figure 7a–c, it can be found that when the intensity of construction enterprises' pessimism remains constant, the mixed strategy Nash equilibrium point moves to the upper right corner of the coordinate region and then to the lower right corner of the coordinate region as public optimism grows. This suggests that the degree of public optimism in the process of influencing the evolutionary strategy has a threshold value $r_2{}^*$. When $r_2 > r_2{}^*$, i.e., the public is excessively optimistic, the probability of public compromise acceptance drops, and the system evolves steadily away from the optimal strategy combination of (actively fulfilling social responsibility, compromise acceptance).

4.6. (Optimistic, Rational) State Analysis, That Is $r_1 < 1, r_2 = 1$

Figure 8 reflects the graph of the evolutionary trend of construction enterprise optimism and public rationality. Figure 8 shows that while construction enterprises' optimism grows, the probability of the active social responsibility of construction enterprises reduces slightly, while the rate of convergence of the evolution of construction enterprises and public strategies continues to accelerate. Construction enterprises' optimism can hinder CSR practice activities, which is consistent with the findings in Figures 4 and 6.

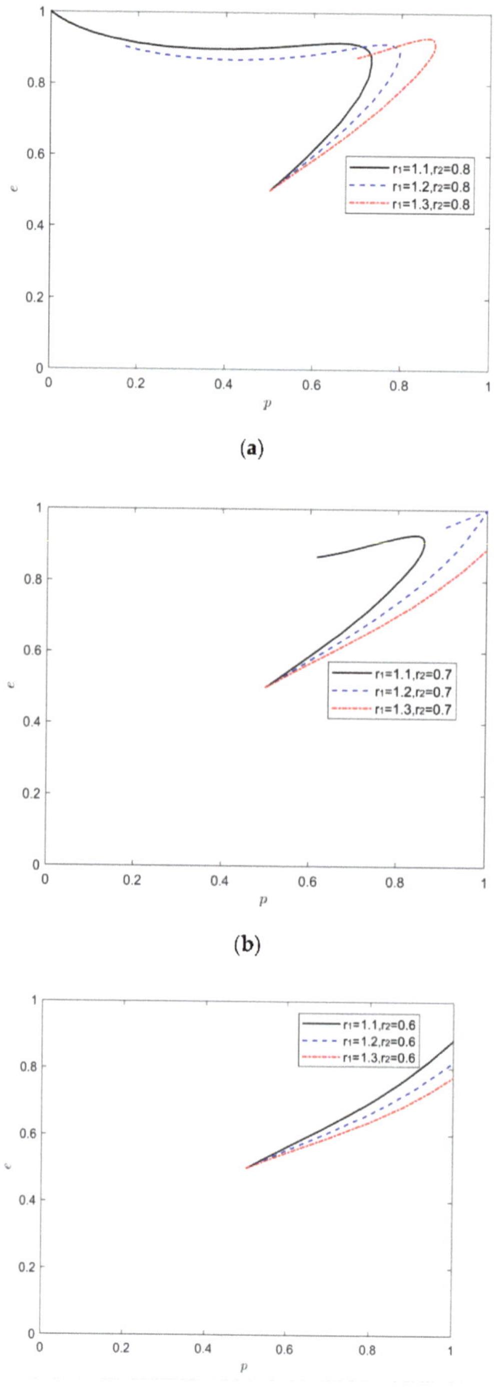

Figure 7. Evolution of game strategy in the (pessimistic, optimistic) state. (**a**) Scenario 1, (**b**) Scenario 2, (**c**) Scenario 3.

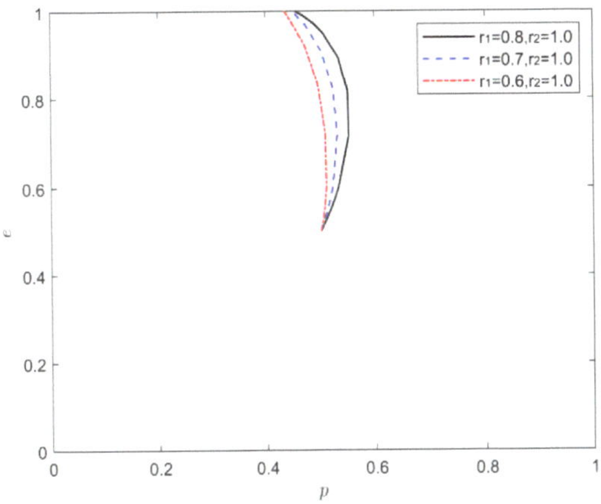

Figure 8. Evolution of game strategy in the (optimistic, rational) state.

4.7. (Pessimistic, Rational) State Analysis, That Is $r_1 > 1, r_2 = 1$

Figure 9 depicts the graph of the evolving trend of construction enterprise pessimism and public rationality. As shown in Figure 9, under current conditions, the probability of construction enterprises actively fulfilling their social responsibility rises and subsequently falls, and the probability of public compromise acceptance tends to almost 1. The rate of evolution of this system decreases as construction enterprises' pessimism grows. This indicates that construction enterprises tend to fulfill their social responsibility strategy positively at the initial stage, but will eventually embrace a negative social responsibility strategy after witnessing the public strategy selection of a rational compromise acceptance attitude. Construction enterprises are currently less active in fulfilling social responsibility, and public awareness of social responsibility is not sufficient. We should increase the publicity of the importance of social responsibility, raise public awareness of social responsibility, and create a pessimistic atmosphere for enterprises, so as not to fall into the predicament of "no supervision, no action".

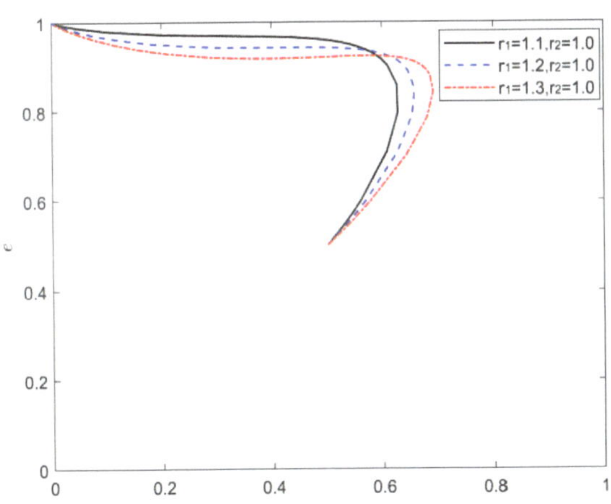

Figure 9. Evolution of game strategy in the (pessimistic, rational) state.

4.8. (Rational, Optimistic) State Analysis, That Is $r_1 = 1, r_2 < 1$

Figure 10 shows the evolution of the game when the construction enterprises are completely rational, and the public is optimistic. As can be seen from Figure 10, the game does not currently form an evolutionary stable strategy, but as public optimism deepens, the probability of construction enterprises actively fulfilling their social responsibility and the probability of public compromise acceptance both continue to increase, and the evolutionary trajectory develops in the direction of Pareto optimality. When enterprises take more social responsibility and provide green products that meet public demand, they will attract more people to choose to buy them, which will also cause enterprises to see improvement in their corporate images, while simultaneously earning profits, forming a virtuous circle. The relative optimal state of the game is also reached when the social responsibility of construction enterprises is improved.

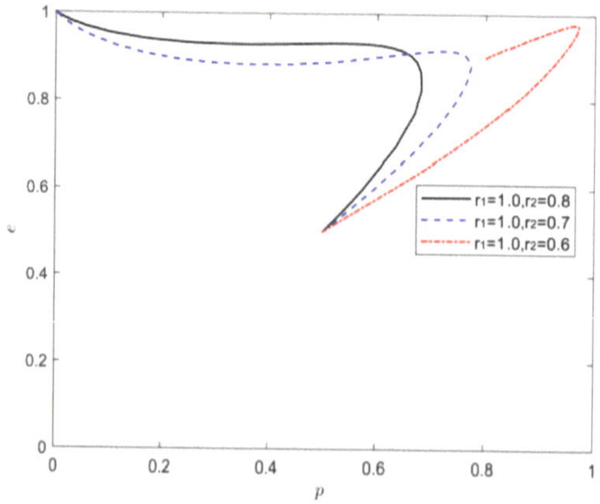

Figure 10. Evolution of game strategy in the (rational, optimistic) state.

4.9. (Rational, Pessimistic) State Analysis, That Is $r_1 = 1, r_2 > 1$

Figure 11 depicts the evolution of construction enterprises when they are rational while the public is pessimistic. As displayed in Figure 11, as public pessimism grows, the probability of public compromise acceptance rises and the rate of strategy evolution accelerates, while the probability of construction enterprises actively fulfilling their social responsibility remains at a certain level. This implies that the enterprises in the rational state have more comprehensive information, the change in emotional intensity does not cause the public to act against the interests of construction enterprises, and the enterprises do not change their strategy choices due to the public's pessimism. It is evident that there is a dominant influence of sentiment from construction enterprises on the evolutionary trend of the gaming system.

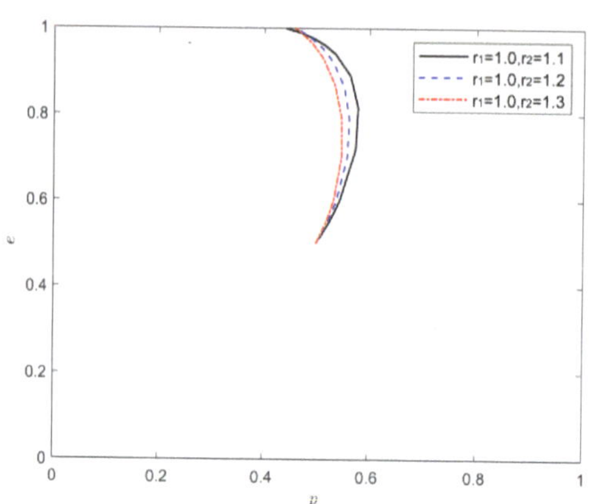

Figure 11. Evolution of game strategy in the (rational, pessimistic) state.

5. Conclusions

Based on the RDEU theory and evolutionary game theory, this paper incorporates the change in emotion into the consideration of the behavior strategy choice of construction enterprises and the public and discusses the behavior evolution law relating to construction enterprises adopting an active social responsibility strategy and the public adopting a compromise acceptance strategy under the influence of emotion by establishing the emotion model. There is a certain chance that both sides of the game, as finite rational persons, make rational equilibrium strategy choices, and the true equilibrium strategy choice will be deviated when sentiments change. The construction industry is one of the largest industries in the world, and its business activities have an impact on economic, environmental, and social spheres. While construction companies play an active role in economic development, they are often criticized and protested against for their irresponsible behaviors, such as neglecting the environment. Therefore, it is necessary to understand the intrinsic reasons for the lack of social responsibility in construction companies. This study chooses to start from emotional intensity, and through simulation analysis of the evolution process of nine emotional combination states, it explores the micro-influence mechanism of emotions on strategy choice, providing guidance for benign development in terms of social responsibility practice level improvement.

5.1. Research Findings

(1) There is probability in the choice of a rational strategy, and strategy choice is influenced by the irrational factor of emotion. When emotional fluctuations are considered during investigating the strategy selection of game subjects, rational equilibrium strategies become probabilistic events. Perceptual bias, as well as psychological fallout, are extremely likely to cause players to misjudge the status quo and their opponent's expected behavior, thus influencing the ultimate choice outcome.

(2) Emotional intensity and emotional tendency will change the evolutionary trend of game subject strategy selection. For construction enterprises, optimism will always amplify the public's understanding and tolerance of malpractices and generate perception bias, thus ignoring the possible negative impact of the current state of social responsibility fulfillment and restricting their social responsibility practice behavior. This finding differs from those in previous research, showing that "optimism leads to excellent outcomes" is conditional [50], and optimism can moderately increase the probability of decision makers choosing positive strategies [66]. This further demonstrates the uniqueness of the construction industry and the deeper reasons for the lack of social responsibility in this industry. For the public, pessimism promotes the probability of choosing their negative strategies, whereas rational and optimistic states avoid contradictory and aggravated behaviors and show a positive tendency to choose response strategies. In addition, variations in the strength of emotions can trigger changes in the behavior patterns of decision makers. For example, the public in an optimistic state possesses a sensible decision-making interval, and excessive optimism will increase the probability of their participation in protests.

(3) The interaction of emotional states leads to different emotional combinations that exhibit heterogeneity in strategy choice, and the emotional influence of construction enterprises on strategy tends to be more dominant. When both sides are in an emotional state, both sides of the game tend to choose the negative strategic combination (negative fulfillment of social responsibility, resolute protest). Whether it is an emotional combination of the constant optimism of construction enterprises and the deepening optimism of the public, or the constant pessimism of the public and the deepening optimism of construction enterprises, the strategy choices of the game subjects all evolve in a positive direction. When one side is rational and the other is emotional, it can change the rate of strategy evolution and the overall trend of the system. For example, when the construction enterprises are rational and the public pessimism is deepening, the evolution speed of the equilibrium strategy is

accelerated, and the pessimism is no longer the endogenous motivation for public resolute protest. Therefore, the emotions of construction enterprises have a stronger influence on the strategy choice of each subject. Enterprises should also be stricter with themselves and focus on their own emotional changes while strengthening the public's emotional supervision and diversion, which is the key to the sustainable development of enterprises and social stability.

5.2. Management Insights

According to the results of the evolutionary game model analysis, the following suggestions are made for the governance of the current social responsibility deficit of construction enterprises:

(1) Construction enterprises should increase the intensity and quality of social responsibility information disclosure. The establishment of a poor image regarding construction enterprises' inadequate social responsibility is mostly related to the lack of social responsibility disclosure. The improvement in the intensity and quality of information disclosure is beneficial for changing the situation regarding information asymmetry, increasing the public's emotional recognition of the behavior of construction enterprises, and establishing a bridge of trust between construction enterprises and the public. This bridge encourages the public to be optimistic about enterprise building projects and to make more scientific and rational strategic choices.

(2) There should be an increasing focus on guiding the public's awareness of social supervision and increasing the sense of responsibility of the main body by publicizing the importance of social responsibility fulfillment in a scientific manner. This strategy can mobilize public interest in CSR practices, raise public understanding of the importance of participating in CSR construction, help decision makers make rational or optimistic decisions, and avoid falling into the trap of inaction that fuels deficient social responsibility behavior. The increase in the number of supervisory subjects creates a pessimistic emotional atmosphere for construction enterprises, strengthens their perception of pressure and anxiety, and increases their awareness of the necessity and non-shrinking nature of social responsibility. When pursuing their own economic interests, more concern should be given to the green growth of the environment and the healthy and stable development of society to promote the formation of a symbiotic and co-prosperous situation among construction enterprises, the public, and society.

(3) Government departments play a vital role in improving the restraint and incentive mechanisms. Along with the establishment of SA8000, ISO26000, and other social responsibility related standards, the fulfillment of social responsibility has achieved specific results, but from an overall perspective, there is still a certain lack of social responsibility. At this time, relevant government departments are a desirable means of releasing social pressure and hostility, and they should play the role of a social safety valve. To encourage the fulfillment of social responsibility, both reward guidance and disciplinary restraint are required. The state can provide financial subsidies, tax reductions, and other preferential policies to enterprises who actively engage in social responsibility. Furthermore, for enterprises that perform poorly in regards to social responsibility, it is necessary for the government to breakdown institutional barriers by enacting special legislation and strengthening law enforcement. The expected increase in costs and losses will effectively reduce the emergence of construction enterprise behavior that is harmful to society's interests.

(4) It is necessary to set up a solid framework for the expression of interests and demands and to guarantee efficient routes for the transmission of public demands. The public in general lacks institutionalized avenues to participate and communicate their concerns concerning environmental governance policy formulation and execution, which frequently leads to the escalation of unhappiness [57]. An efficient and smooth information transmission channel can eliminate misunderstandings caused by delayed feedback of public interest demands, and it can also monitor and assess all

parties' emotional reactions and changes in real time, allowing the public's negative emotions to be channeled in a timely manner, avoiding the breeding and spreading of pessimistic emotions such as suspicion, anxiety, and nervousness. The reasonable venting of negative emotions can better ensure participants' rational tendency to solve problems and seek breakthroughs through other means rather than starting large protests.

5.3. Shortcomings and Prospects

This paper constructs an evolutionary model involving construction enterprises and the public and quantifies and integrates emotions, an irrational factor, into this game system, investigating the dynamic effects of changes in emotional intensity on the strategy choices of game participants. In fact, the stakeholders involved in the problem of social responsibility fulfillment are not only construction enterprises and the public. Future research will consider the government or specific regulatory departments as game subjects and explore their interactive behaviors with construction enterprises. Furthermore, based on the existing game system, a three-party game model can be further developed to investigate in depth the influence of emotions on the behavioral decisions of subjects related to social responsibility practices.

Author Contributions: All authors contributed to the study's conception and design. Material preparation, review, and editing were performed by S.W., X.D. (Xiaopeng Deng), Z.Z., N.Z. and H.M. The first draft of the manuscript was written by X.D. (Xiaoxu Dang); all authors commented on previous versions of the manuscript. All authors have read and agreed to the published version of the manuscript.

Funding: This research is funded by the National Natural Science Foundation of China (NSFC-72171048, NSFC-71771052, and NSFC-72101053) and the Shaanxi Provincial Department of Transportation Scientific Research Project (Grant No. 20–07R).

Conflicts of Interest: The authors declare no conflict of interest.

References

1. Seneviratne, S.I.; Donat, M.G.; Pitman, A.J.; Knutti, R.; Wilby, R.L. Allowable CO_2 emissions based on regional and impact-related climate targets. *Nature* **2016**, *529*, 477–483. [CrossRef] [PubMed]
2. Zhou, N.; Khanna, N.; Feng, W.; Ke, J.; Levine, M. Scenarios of energy efficiency and CO_2 emissions reduction potential in the buildings sector in China to year 2050. *Nat. Energy* **2018**, *3*, 978–984. [CrossRef]
3. Wang, K.; Pan, H.Y.; Zhang, T.J.; Wang, H.T. Experimental study on the radial vibration characteristics of a coal briquette in each stage of its life cycle under the action of CO2 gas explosion. *Fuel* **2022**, *320*, 123922. [CrossRef]
4. Mallapaty, S. How china could be carbon neutral by mid-century. *Nature* **2020**, *586*, 482–483. [CrossRef] [PubMed]
5. Xu, Y.; Xu, Z.; Zhou, Y.; Su, C.; Guo, L. Interactions between carbon prices and the construction industry in China: Evidence based on Network-SVAR. *Build. Environ.* **2022**, *215*, 108936. [CrossRef]
6. Lin, B.; Liu, H. CO2 mitigation potential in China's building construction industry: A comparison of energy performance. *Build. Environ.* **2015**, *94*, 239–251. [CrossRef]
7. Shi, Q.; Chen, J.; Shen, L. Driving factors of the changes in the carbon emissions in the Chinese construction industry. *J. Clean. Prod.* **2017**, *166*, 615–627. [CrossRef]
8. Huo, T.; Ren, H.; Zhang, X.; Cai, W.; Feng, W.; Zhou, N.; Wang, X. China's energy consumption in the building sector: A Statistical Yearbook-Energy Balance Sheet based splitting method. *J. Clean. Prod.* **2018**, *185*, 665–679. [CrossRef]
9. Huo, T.; Ma, Y.; Cai, W.; Liu, B.; Mu, L. Will the urbanization process influence the peak of carbon emissions in the building sector? A dynamic scenario simulation. *Energy Build.* **2021**, *232*, 110590. [CrossRef]
10. Zhang, S.; Xiang, X.; Ma, Z.; Ma, M.; Zou, C. Carbon Neutral Roadmap of Commercial Building Operations by Mid-Century: Lessons from China. *Buildings* **2021**, *11*, 510. [CrossRef]
11. Chen, M.; Ma, M.; Lin, Y.; Ma, Z.; Li, K. Carbon Kuznets curve in China's building operations: Retrospective and prospective trajectories. *Sci. Total Environ.* **2022**, *803*, 150104. [CrossRef]
12. Jiang, W.; Wong, J.K.W. Key activity areas of corporate social responsibility (CSR) in the construction industry: A study of China. *J. Clean. Prod.* **2016**, *113*, 850–860. [CrossRef]
13. Ante, A.H.G. What we know and don't know about corporate social responsibility: A review and research agenda. *J. Manag.* **2012**, *38*, 932–968. [CrossRef]

14. Cruz, J.M. Modeling the relationship of globalized supply chains and corporate social responsibility. *J. Clean. Prod.* **2013**, *56*, 73–85. [CrossRef]
15. Peng, W.; Xin, B.; Kwon, Y. Optimal Strategies of Product Price, Quality, and Corporate Environmental Responsibility. *Int. J. Environ. Res. Public Health* **2019**, *16*, 4704. [CrossRef]
16. Lee, S.-H.; Park, C.-H. Eco-Firms and the Sequential Adoption of Environmental Corporate Social Responsibility in the Managerial Delegation. *BE J. Theor. Econ.* **2017**, *19*, 20170043. [CrossRef]
17. Han, H.; Yu, J.; Kim, W. Environmental corporate social responsibility and the strategy to boost the airline's image and customer loyalty intentions. *J. Travel Tour. Mark.* **2019**, *36*, 371–383. [CrossRef]
18. Wang, L.W.X. State-enterprise relation, local economic priority, and corporate environmental responsibility. *Appl. Econ.* **2019**, *51*, 995–1009. [CrossRef]
19. Ruan, R.B.; Chen, W.; Zhu, Z.P. Research on the relationship between environmental corporate social responsibility and green innovative behavior: The moderating effect of moral identity. *Environ. Sci. Pollut. Res.* **2022**, *29*, 52189–52203. [CrossRef]
20. Jamali, D. A Stakeholder Approach to Corporate Social Responsibility: A Fresh Perspective into Theory and Practice. *J. Bus. Ethics* **2008**, *82*, 213–231. [CrossRef]
21. Rahman, N.; Post, C. Measurement Issues in Environmental Corporate Social Responsibility (ECSR): Toward a Transparent, Reliable, and Construct Valid Instrument. *J. Bus. Ethics* **2012**, *105*, 307–319. [CrossRef]
22. Xia, B.; Olanipekun, A.; Chen, Q.; Xie, L.; Liu, Y. Conceptualising the state of the art of corporate social responsibility (CSR) in the construction industry and its nexus to sustainable development. *J. Clean. Prod.* **2018**, *195*, 340–353. [CrossRef]
23. Shu, T.; Liu, Q.; Chen, S.; Wang, S.; Lai, K.K. Pricing Decisions of CSR Closed-Loop Supply Chains with Carbon Emission Constraints. *Sustainability* **2018**, *10*, 4430. [CrossRef]
24. Koseoglu, M.A.; Yick, M.Y.Y.; Parnell, J.A. The dissemination of corporate social responsibility into the intellectual structure of strategic management. *J. Clean. Prod.* **2021**, *311*, 127505. [CrossRef]
25. Auger, P.; Burke, P.; Devinney, T.M.; Louviere, J.J. What will consumers pay for social product features? *J. Bus. Ethics* **2003**, *42*, 281–304. [CrossRef]
26. Bolton, L.E.; Mattila, A.S. How Does Corporate Social Responsibility Affect Consumer Response to Service Failure in Buyer–Seller Relationships? *J. Retail.* **2015**, *91*, 140–153. [CrossRef]
27. Liu, Z.; Anderson, T.D.; Cruz, J.M. Consumer environmental awareness and competition in two-stage supply chains. *Eur. J. Oper. Res.* **2011**, *218*, 602–613. [CrossRef]
28. Bai, Q.; Chen, M.; Nikolaidis, Y.; Xu, J. Improving sustainability and social responsibility of a two-tier supply chain investing in emission reduction technology. *Appl. Math. Model.* **2021**, *95*, 688–714. [CrossRef]
29. Shen, L.-Y.; Tam, V.W.Y.; Tam, L.; Ji, Y.-B. Project feasibility study: The key to successful implementation of sustainable and socially responsible construction management practice. *J. Clean. Prod.* **2009**, *18*, 254–259. [CrossRef]
30. Xiong, B.; Lu, W.; Skitmore, M.; Chau, K.W.; Ye, M. Virtuous nexus between corporate social performance and financial performance: A study of construction enterprises in China. *J. Clean. Prod.* **2016**, *129*, 223–233. [CrossRef]
31. Lu, W.; Ye, M.; Chau, K.W.; Flanagan, R. The paradoxical nexus between corporate social responsibility and sustainable financial performance: Evidence from the international construction business. *Corp. Soc. Responsib. Environ. Manag.* **2018**, *25*, 844–852. [CrossRef]
32. Somachandra, W.; Sylva, K.K.K.; Bandara, C.S.; Dissanayake, P.B.R. Corporate social responsibility (CSR) practices in the construction industry of Sri Lanka. *Int. J. Constr. Manag.* **2022**, 1–9. [CrossRef]
33. Lien, C.-F.H.H.-C. An empirical analysis of the influences of corporate social responsibility on organizational performance of Taiwan's construction industry: Using corporate image as a mediator. *Constr. Manag. Econ.* **2012**, *30*, 263–275. [CrossRef]
34. Zhao, Z.-Y.; Zhao, X.-J.; Davidson, K.; Zuo, J. A corporate social responsibility indicator system for construction enterprises. *J. Clean. Prod.* **2012**, *29–30*, 277–289. [CrossRef]
35. Zou, P.X.W.; Couani, P. Managing risks in green building supply chain. *Archit. Eng. Des. Manag.* **2012**, *8*, 143–158. [CrossRef]
36. Bernaciak, A.; Halaburda, M.; Bernaciak, A. The Construction Industry as the Subject of Implementing Corporate Social Responsibility (the Case of Poland). *Sustainability* **2021**, *13*, 9728. [CrossRef]
37. Huang, C.F.; Lu, W.H.; Lin, T.T.; Wu, E.J. The current conditions of csr implementation in construction industry: A lesson from taiwan. *Appl. Ecol. Environ. Res.* **2017**, *15*, 67–80. [CrossRef]
38. Wang, X.P.; Zhang, C.; Deng, J.; Su, C.; Gao, Z.Z. Analysis of Factors Influencing Miners' Unsafe Behaviors in Intelligent Mines using a Novel Hybrid MCDM Model. *Int. J. Environ. Res. Public Health* **2022**, *19*, 7368. [CrossRef]
39. Chou, J.-S.; Yeh, K.-C. Life cycle carbon dioxide emissions simulation and environmental cost analysis for building construction. *J. Clean. Prod.* **2015**, *101*, 137–147. [CrossRef]
40. Spence, R.; Mulligan, H. Sustainable development and the construction industry. *Habitat Int.* **1995**, *19*, 279–292. [CrossRef]
41. Ma, Z.; Chen, J.; Tian, G.; Gong, Y.; Guo, B.; Cheng, F. Regulations on the corporate social irresponsibility in the supply chain under the multiparty game: Taking China's organic food supply chain as an example. *J. Clean. Prod.* **2021**, *317*, 128459. [CrossRef]
42. Zhao, X.; Zhao, H. Dynamic game analysis of hybrid strategy of enterprise fulfilling social responsibility. *Int. J. Electr. Eng. Educ.* **2020**, 1–9. [CrossRef]
43. Liu, Y.; Quan, B.-T.; Xu, Q.; Forrest, J.Y.-L. Corporate social responsibility and decision analysis in a supply chain through government subsidy. *J. Clean. Prod.* **2019**, *208*, 436–447. [CrossRef]

44. Zhang, Z.; Wang, X.; Su, C.; Sun, L. Evolutionary Game Analysis of Shared Manufacturing Quality Synergy under Dynamic Reward and Punishment Mechanism. *Appl. Sci.* **2022**, *12*, 6792. [CrossRef]
45. Gross, J.J. Emotion Regulation: Current Status and Future Prospects. *Psychol. Inq.* **2015**, *26*, 1–26. [CrossRef]
46. Elster, J. Emotions and Economic Theory. *J. Econ. Lit.* **1998**, *36*, 47–74. [CrossRef]
47. Bandyopadhyay, D.; Pammi, V.S.C.; Srinivasan, N. Role of affect in decision making. *Prog. Brain Res.* **2013**, *202*, 37–53. [CrossRef] [PubMed]
48. Chateauneuf, A.; Cohen, M.; Meilijson, I. More Pessimism than Greediness: A Characterization of Monotone Risk Aversion in the Rank-Dependent Expected Utility Model. *Econ. Theory* **2005**, *25*, 649–667. [CrossRef]
49. Quiggin, J. Comparative Statics for Rank-Dependent Expected Utility Theory. *J. Risk Uncertain.* **1991**, *4*, 339–350. [CrossRef]
50. Hong, K.; Zou, Y.; Zhang, Y.; Duan, K. The Weapon of the Weak: An Analysis of RDEU Game in the Conflict of Farmland Expropriation under the Influence of Emotion. *Sustainability* **2020**, *12*, 3367. [CrossRef]
51. Ni, S.; Zou, S.; Chen, J. Evolutionary Game Model of Internal Threats to Nuclear Security in Spent Fuel Reprocessing Plants Based on RDEU Theory. *Sustainability* **2022**, *14*, 2163. [CrossRef]
52. Xiaole, W.; Shuwen, X.; Qianqian, L.; Yuanwei, D. Evolutionary policy of trading of blue carbon produced by marine ranching with media participation and government supervision. *Mar. Policy* **2020**, *124*, 104302. [CrossRef]
53. Fu, J.; Geng, Y. Public participation, regulatory compliance and green development in China based on provincial panel data. *J. Clean. Prod.* **2019**, *230*, 1344–1353. [CrossRef]
54. Wu, L.; Ma, T.; Bian, Y.; Li, S.; Yi, Z. Improvement of regional environmental quality: Government environmental governance and public participation. *Sci. Total Environ.* **2020**, *717*, 137265. [CrossRef]
55. Koo, J.E.; Ki, E.S. Corporate Social Responsibility and Employee Safety: Evidence from Korea. *Sustainability* **2020**, *12*, 2649. [CrossRef]
56. Nunes, M.F.; Park, C.L.; Shin, H. Corporate social and environmental irresponsibilities in supply chains, contamination, and damage of intangible resources: A behavioural approach. *Int. J. Prod. Econ.* **2021**, *241*, 108275. [CrossRef]
57. He, G.; Mol, A.P.J.; Lu, Y. Public protests against the Beijing-Shenyang high-speed railway in China. *Transp. Res. Part D-Transp. Environ.* **2016**, *43*, 1–16. [CrossRef]
58. John, Q. A theory of anticipated utility. *J. Econ. Behav. Organ.* **1982**, *3*, 323–343. [CrossRef]
59. Machina, M.J. Choice Under Uncertainty: Problems Solved and Unsolved. *J. Econ. Perspect.* **1987**, *1*, 121–154. [CrossRef]
60. Diecidue, E.; Wakker, P.P. On the Intuition of Rank-Dependent Utility. *J. Risk Uncertain.* **2001**, *23*, 281–298. [CrossRef]
61. Guo, D.; Chen, H.; Long, R. How to involve individuals in personal carbon trading? A game model taking into account the heterogeneous emotions of government and individuals. *Nat. Hazards* **2019**, *95*, 419–435. [CrossRef]
62. Liu, J.; Lyu, Y.; Zhao, H.; Chen, J. Game analysis of nuclear wastewater discharge under different attitudes: Seeking a potential equilibrium solution. *Sci. Total Environ.* **2021**, *801*, 149762. [CrossRef] [PubMed]
63. Starmer, C. Developments in Non-Expected Utility Theory: The Hunt for a Descriptive Theory of Choice under Risk. *J. Econ. Lit.* **2000**, *38*, 332–382. [CrossRef]
64. Lyapunov, A.M. The general problem of the stability of motion. *Int. J. Control.* **1992**, *55*, 531–534. [CrossRef]
65. Hartl, B.; Sabitzer, T.; Hofmann, E.; Penz, E. Sustainability is a nice bonus the role of sustainability in carsharing from a consumer perspective. *J. Clean. Prod.* **2018**, *202*, 88–100. [CrossRef]
66. Zhang, Z.; Wang, X.; Su, C.; Sun, L. Evolutionary Game Analysis of Shared Manufacturing Quality Innovation Synergetic Behavior Considering a Subject's Heterogeneous Emotions. *Processes* **2022**, *10*, 1233. [CrossRef]

processes

Review

Existing Stature and Possible Outlook of Renewable Power in Comprehensive Electricity Market

Shreya Shree Das [1], Jayendra Kumar [1], Subhojit Dawn [2] and Ferdinando Salata [3,*]

1 School of Electronics Engineering, VIT-AP Campus, Amaravati 522237, India; shreyashree36@gmail.com (S.S.D.); jayendra854330@gmail.com (J.K.)
2 Department of Electrical and Electronics Engineering, Velagapudi Ramakrishna Siddhartha Engineering College, Vijayawada 520007, India; subhojit.dawn@gmail.com
3 DIAEE Faculty of Civil and Industrial Engineering, "Sapienza", University of Rome, 00185 Roma, Italy
* Correspondence: ferdinando.salata@uniroma1.it

Abstract: The massive growth in power demand and the sharp decay in the availability of conventional energy sources forces society to move toward renewable power consumption. The rise in renewable energy utilization is one of the greatest strategies involved in making the environment clean, green, and emission-free. The investment cost is slightly high for this kind of resource, but their running costsare very minimal, which encourages power producers to invest in renewable power plants. Power producers always focus on their economic profit and possible feasibilities before the investment in a new power plant setup. Wind, small hydro, and solar photovoltaics have been considered the foremost efficient and feasible renewable sources. In 2021, the worldwide renewable power capacity had grown to 17%, despite supply chain disturbances and surges in prices. The renewable power market set a target of achieving a 95% increment in global power capacity by 2026. Hence, humankind should consider wind, hydro, and solar photovoltaics as the main energy sources, and there is a need to increase use of such sources. This paper highlights the latest developments, enormous technologies, and upcoming aspects of wind power, hydropower, and solar photovoltaics in India, as well as around the globe. The present status of renewable energy can motivate global power producers to spend their money on the installation of a new renewable power plant to obtain more economic benefits and give societal economic and environmental payback to humans.

Keywords: global power scenario; India power scenario; grid-interactive power; tariff; renewable energy sources; environment

Citation: Das, S.S.; Kumar, J.; Dawn, S.; Salata, F. Existing Stature and Possible Outlook of Renewable Power in Comprehensive Electricity Market. *Processes* **2023**, *11*, 1849. https://doi.org/10.3390/pr11061849

Academic Editor: Jingyu Ran

Received: 18 May 2023
Revised: 7 June 2023
Accepted: 17 June 2023
Published: 19 June 2023

1. Introduction

In this era, the electricity demand is increasing enormously due to rapid expansion in technologies. Day by day, human beings are looking for a maintained and comfortable lifestyle. Nowadays, the maximum electricity demand is fulfilled by conventional energy sources. As the availability of conventional sources is very limited, governments should switch toward renewable sources to fulfill the increasing demand. Over the past few decades, every country has faced enormous problems regarding environmental pollution, and one of its main causes is the power sector as it consumes a huge number of conventional resources. To mitigate this issue, the power network has introduced a vast area for the renewable power sector. The main attention of the power sector is grabbed by renewable energy as it causes much fewer environmental issues compared to conventional energy. The growing concern of society regarding nature and the harsh effect of emissions in the power sector has pushed strategies to increase the scope of renewable energy in the electricity market. Power generated from renewable energy sources causes minor pollutants from decarbonization-generating technologies [1]. According to the United States Environmental Protection Agency, greenhouse gases are one of the major air-polluting agents. The electrical power sector was responsible for 25% of total greenhouse gas emissions, which was the second largest share of greenhouse gas emissions in the USA in 2020 [2]. Although the figures are specificallyfrom the USA, further studies reveal that almost similar trends are followed throughout the world. To counter environmental pollution, the engagement of renewable power generation has become a necessity. Among the different renewable sources, wind, solar, and hydropower deserve special mention due to their clean, reliable, and versatile nature for long-term usage. There are many other advantages of using a renewable source, such as reduced tariffs, reduced pollution levels, etc. In 2020, around 29% of power was generated from renewable sources, and 256 GW capacity of renewable power has been newly added worldwide. Despite having huge potential for renewable generation, India has exploited a very small amount of the total capacity [3]. The government of India (GOI) set an aim to generate 175 GW of power from renewable sources by 2022, of which 100 GW from solar, 60 GW from wind, and the rest from other renewable sources [4]. To encourage the citizens of India to use renewable energy, the GOI has launched several subsidy policies for renewable power installation.

In the past few decades, researchers have mainly focused on wind, hydro, and solar-power-generation techniques. Ming et al. [5] focus on the policy barrier, status, and tools for hydro energy in China. The benefits and scenarios of small hydropower plants were displayed in [6] for the year 2014. Mishra et al. [7] examined the present status and plans of small hydro in the Indian power sector. In recent times, many countries' governments have focused on small hydropower plants for electricity production. The present and past scenarios of hydropower were exploited in [8–11] for Pakistan, Malaysia, and Thailand, respectively. The status of the hydropower plant in the western Himalayas region (Jammu and Kashmir) was displayed by Sharma et al. in [12,13]. The author of [14] illustrated a review of ultra-modern computational optimization techniques in sustainable and renewable energy. This offered a clear image of the newest research achievements in the arena of sustainable energy. Ashwani et al. [15] focused on the outline of India's renewable energy sources and their opportunities, as well as their current status, availability, and notable achievements. The authors also discussed governmental initiatives for eliminating complications and enhancing the placement of renewable energy in the future.

The rise in wind power plant installation has been seen in the last two decades around the globe and the average lifespan of a wind power plant is considered to be around 20 years. The author of [16] proposed some strategies thatneed to be implemented by the government and private power plant owners to maximize the benefits of renewable sources. The optimal location of a wind power plant and the related factors that limit its installation was discussed by Rediske et al. [17]. Nowadays, offshore wind plants are located far from the shore and deeper into the sea. As wind energy prices are reducing, the wind power market is becoming more competitive [18]. Over the past few decades, sustainability has entered policy debate. Government bodies are imposing several regulations to mitigate the adverse effects of CO_2 on the environment [19]. Kunz et al. [20] displayed that the most evolution in a sustainable energy system depends on wind as lynchpin technology.

- After completing comprehensive and careful analysis, itwas revealed that a few pieces of literature have already been completed by some researchers in the arena of wind, hydro, and solar power, but as per the authors' best knowledge, no one has displayed all the probable features related to wind, solar, and hydropower in the world, as well as in the Indian power region.
- All the recent strategies, objectives, government skills, and achievements that are crucial for the development of renewable power in India, as well as global power systems, are stated in this paper.
- The adaptation of wind, solar, and hydropower plants, as well as the present power status around the globe, are also focused on in this paper with real-time data.
- This paper also presents comparative studies among the different renewable energy sources, which provide a clear idea as to how global power producerscanfind the best options to invest their money in new renewable power plant installation.

2. Renewable Power Plant Scenarios

This section presents the basic generating process of different types of renewable sources, along with the world's largest renewable power plant details.

2.1. Hydroelectric Power Plant

In a hydropower plant, a reservoir and a dam are constructed to store the water and release it when required. When water falls from a height, then potential energy is converted into kinetic energy, which gives a great thrust to the turbine present at the power plant. Dams are mainly constructed in mountain areas where they can properly achieve the head difference between the upper and lower basin. The turbine rotates the shafts and in turn, electricity is produced. The catchment area is the entire space behind the dam, draining into a waterway through which the dam has been constructed at an appropriate place. The reservoir is a huge gathering of water behind a hydropower dam and delivers water uniformly to the turbine over the year to generate electricity. A surge tank is placed in the lower channel end area of a feeder, barrage pipe, and closed conduit to absorb unexpected rises in pressure and speedily supply necessary water [21].

Recently, a new kind of hydropower generator has been introduced. This type of generator uses the vortex present in the water to turn the turbine and generate electricity [22]. This generator can be located even at lesser natural altitudes, which is an additional advantage of this kind of generator. Conventional generators need ample water heads to produce the required thrust. However, most of the rivers in India do not have enough water throughout the year, apart from during the monsoon period. If instances of turbulence are to be assumed, then this type of generator might receive massive demand in the future.

The hydropower plant is not only considered a resource to produce electricity, but they also do not contaminate the air, soil, and water like other power plants. Hydroelectric power plays a ruling part in the development of the world's power sector. Both the small and large hydropower industries remained contributors to the power industry. The world's first hydropower plant started its commercial operation in the year 1882, on the Fox River in Wisconsin, USA [23].

Table 1 presents the largest hydropower plant in the world as of December 2022 [24]. Three Gorges is the leading hydropower plant in the globe. Its construction started in 1993, and it took almost 20 years to complete, with the largest capacity of 22.5 GW. It is 181 m tall and 2335 m wide. The yearly estimated power production of this hydropower plant is 85 TWh. It is maintained and functioned by China Three Gorges Corporation (CTGC). For operation, it uses the water of the Yangtze River. Figure 1 depicts the dam's contribution to the country's electricity capacity for Paraguay, Venezuela, Brazil, and China.

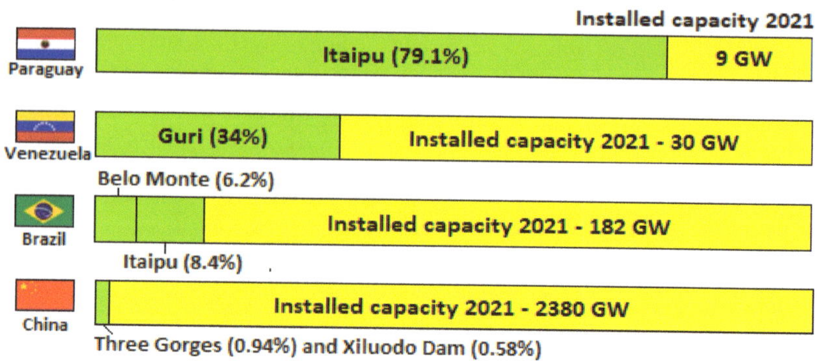

Figure 1. Dam's Contribution to Country's Electricity Capacity [24].

Table 1. Biggest Hydroelectric Power Plants in World (as of December 2022) [24].

World Ranking	Name	Country	Capacity (GW)	Completion Year	Source River
1	Three Gorges Dam	China	22.5	2003	Yangtze
2	Itaipu Dam	Brazil and Paraguay	14	1984	Parana
3	Xiluodu Dam	China	13.9	2013	Jinsha
4	Belo Monte Dam	Brazil	11.2	2016	Xingu
5	Guri Dam	Venezuela	10.2	1978	Caroni
6	Tucurui Dam	Brazil	8.37	1984	Tocantins
7	Grand Coulee Dam	USA	6.8	1980	Columbia
8	Xiangjiaba Dam	China	6.44	2012	Jinsha
9	Longtan Dam	China	6.42	2008	Hongshui
10	Sayano–Shushenskaya Dam	Russia	6.4	2010	Yenisei

India's first hydropower plant was built in 1897 in Darjeeling, West Bengal. The overall expected potential of India in the hydro sector is 145,000 MW with a load factor of 60%. The hydropower sector has exploited about 26% of the potential and around 85,000 MW demand so far in India [25].

2.2. Wind Power Plant

Wind turbines are a recent version of windmills, which avail the power of the wind to produce electrical energy. In a wind farm, the observable parts are towering turbines. The turbines consist of sets of blades, shafts, and a control box beside them that contains a bearing, hub, hydraulics, gearbox, inverter, rotor, and generator, which is called a nacelle [26]. The spinning of blades due to wind flow creates kinetic energy. The rotating blades allow the shaft, which is attached to the nacelle, to turn, and the generators in the nacelle convert kinetic energy into electrical energy. The wind turbine can be operated at any wind speed, from very low to very high wind. Around 80% of the time, it can generate electricity but not always at full capacity [27].

Wind farms need to be situated in the breeziest place to opt for extreme energy, which will let them produce the best output. They are more oftenseen at hilltop or coastal regions. Wind farms that are located in coastal regions or the seaside are called offshore wind farms, and wind farms that are situated in drylands are called onshore wind farms [28]. The Haliade-X wind turbine is a new-generation wind turbine, declared to be the furthermost powerful offshore wind turbine [29]. The world's first known wind turbine was created by Prof. James Blyth of Anderson's College, Scotland to produce electricity. The cloth-sailed wind turbine had a height of 10 m. It was used to charge accumulators for lighting cottages [30]. The details of the world's ten largest wind farms are depicted in Table 2.

2.3. Solar Power Plant

Solar is considered an infinite energy source. Solar power converts sunlight into electrical power either through mirrors or through photovoltaic (PV) panels. This energy can be stored in batteries or used to generate electricity. Solar panels are mainly made up of silicon or germanium, which is usually mounted in a metal panel with a glass covering. When the small particles of light or photons hit the fine silicon or germanium surface of the solar panel, they strike electrons off the silicon atoms. Solar power is considered to be clean energy as it does not create any harmful greenhouse gas.

Table 2. Biggest Wind Farms in World (as of August 2022) [31].

World Ranking	Name	Country	Capacity	Completion Year
1	Jiuquan Wind Power Base	China	20 GW	2010
2	Jaisalmer Wind Urja Park	India	1600 MW	2010
3	Alta Wind Energy Centre	US	1548 MW	2011
4	Muppandal Wind Farm	India	1500 MW	1986
5	Shepherds Flat Wind Farm	US	845 MW	2012
6	Roscoe Wind Farm	US	781.5 MW	2009
7	Horse Hollow Wind Energy Centre	Texas, US	735.5 MW	2006
8	Capricorn Ridge Wind Farm	Texas, US	662.5 MW	2008
9	Walney Extension Offshore Wind Farm	UK	659 MW	2011
10	London Array Offshore Wind Farm	UK	630 MW	2012

It has great efficiency with operating times of more than 25 years. The carbon footprint of a solar panel is very small. The material used in a solar panel is recyclable, so the carbon footprint continues to shrink [32]. NTPC has commissioned India's largest floating solar power project in Ramagundam with a capacity of 100 MW. It has an eco-friendly nature and advanced technology.

In 1839, a French physicist, Edmond Becquerel, discovered a PV effect during the conduction of an experiment with a metal cell's electrodes in a conducting compound solution. The cell generated more energy when kept direct under sunlight [33]. Tengger solar park is known as the 'the great wall of Solar' and is located in Zhongwei, Ningxia, China. This power plant is owned by China National Grid and the Zhongwei Power Supply Company. The project started in 2012 and was completed in 2017. The biggest solar power plant details in the world are displayed in Table 3 [34].

Table 3. Biggest Solar Power Plants in World [34].

World Ranking	Name	Country	Capacity (MW)
1	Tengger Desert Solar Park	China	1547
2	Sweihan Photovoltaic Independent Power Project	UAE	1177
3	Yanchi Ningxia Solar Park	China	1000
4	Datong Solar Power Top Runner Base	China	1070
5	Kurnool Ultra Mega Solar Park	India	1000
6	Longyangxia Dam Solar Park	China	850
7	Enel Villanueva PV Plant	Mexico	828
8	Kamuthi Solar Power Station	India	648
9	Solar Star Projects	US	579
10	Topaz Solar Farm	US	550

3. Global Power Scenario

Throwing light on the past 15 years'worth of data on the consumption of energy, a very sharp rise grabs the attention, as is evident from Figure 2. To keep up with this growing demand, a steeper rise in the production curve is evident.

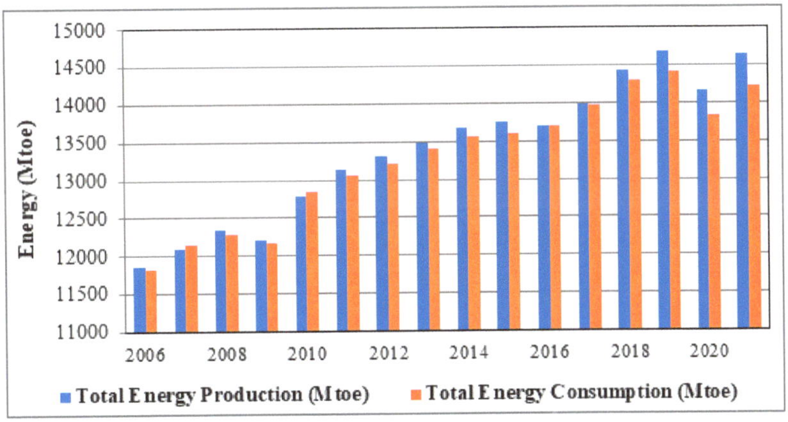

Figure 2. Comparison of Energy Production and Consumption (Mtoe).

In the race of keeping the production and demand line in synchronization, the rising levels of pollution went unnoticed, which resulted in a rise in greenhouse gas emissions and environmental pollution. Figure 3 displays the rising trend in the level of CO_2 emission for the past 15 years [35]. These data are for emissions from power plants only. It is evident from the graph that CO_2 emissions reached a whopping level of 32,969 metric tons in 2021, compared to 27,464 metric tons in 2006.

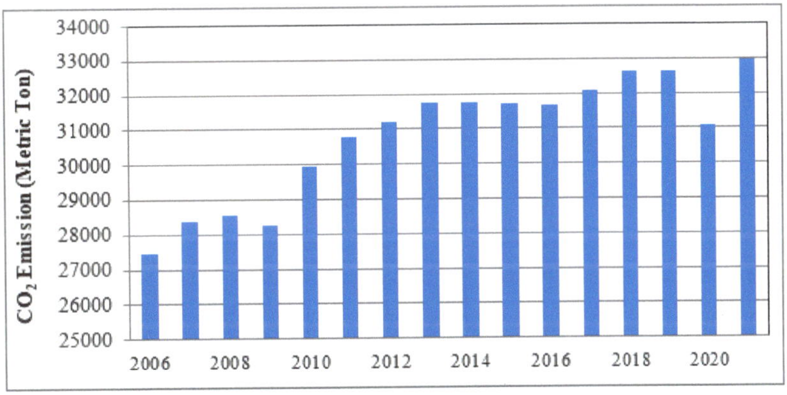

Figure 3. Global CO_2 Emission in Metric Tons [35].

However, a growing consciousness is now being witnessed in all parts of the world. Consequently, the usage of renewable energy for electricity production has gained momentum. In particular, 2017 was a record-breaking year for renewable energy. This year witnessed the largest-ever increase in renewable power capacity, as well as increases in investment and advances in enabling technologies. Figure 4 shows the percentage of energy shared by different sources throughout the world. Fossil fuels remain one of the major sources of energy and is a matter of concern [36]. However, with a large investment inflow in renewable energy sources, this situation can be changed. A comparison of energy production between 2011 and 2021 indicates a rise in the use of modern renewable sources throughout the world. Almost 28.3% of electricity produced in 2021 came from renewable energy sources, which is larger than the renewable energy used in 2011 with only 20.4% total energy conversion. Among the renewable energy sources, hydropower is one of the most consistent and efficient electricity-generation sources (shown in Figure 4). Approximately 15% of global electricity was produced by hydropower plants in 2021 [36].

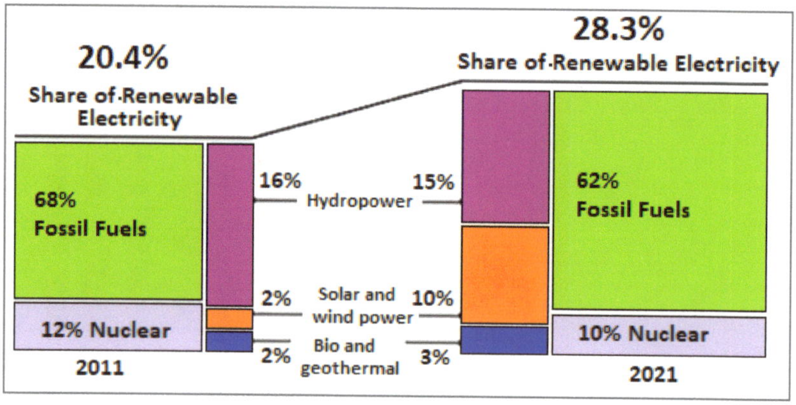

Figure 4. Renewable Share of Total Energy Consumption, 2011 and 2021 [36].

Figure 5 shows the details about global electricity production by source and the share of renewable energy from 2010 to 2020. In this figure, renewable energy is divided into two parts, i.e., renewable with and without hydropower. Starting from 2010, electricity production increased day by day with all types of sources, but the most important thing is that the electricity-generation rate from renewable energy sources increased at the very firstrate [36]. This happened due to the large involvement of governments regarding the use of renewable energy sources. Table 4 depicts the renewable energy indicators at the end of 2020. Approximately USD five billion in new annual investment was observed in 2020 from the immediate previous year (i.e., 2019). The additional investment in renewable energy directly indicates the motive of several countries toward the renewable dominating power system [36].

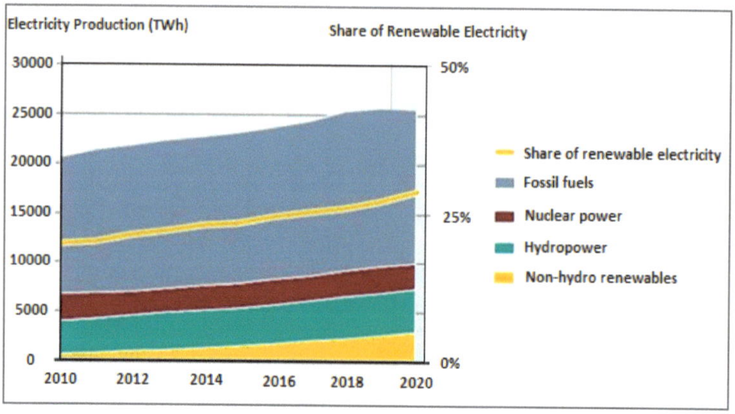

Figure 5. Global Electricity Production by Source and Share of Renewable, 2010–2020 [36].

As per the report of 'Our World in Data', approximately 11% of the world's total power demand has been fulfilled by renewable energy sources. Figure 6 displays the global renewable-energy-generation status from 1965 to 2019 [37]. From 1965 to date, the maximum power has been generated by hydropower plants. Due to their dependable nature and high production capacity, hydropower plantsare the best energy production option among non-conventional energy sources. Starting with approx. 1000 TWh in 1965, renewable energy generation reached the maximum capacity of 7625 Twh at the end of 2021.

Table 4. Renewable Energy Indicators, 2020 and 2021.

Details	2020	2021
New annual investment in renewable power generation along with fuels (billion USD)	342.7	365.9
Renewable power generation including hydropower (GW)	2840	3146
Renewable power generation excluding hydropower (GW)	1672	1945
Hydropower generation (GW)	1168	1195
Solar PV generation (GW)	767	942
Wind power generation (GW)	745	845
Bio-power generation (GW)	133	143
Geothermal power generation (GW)	14.2	14.5

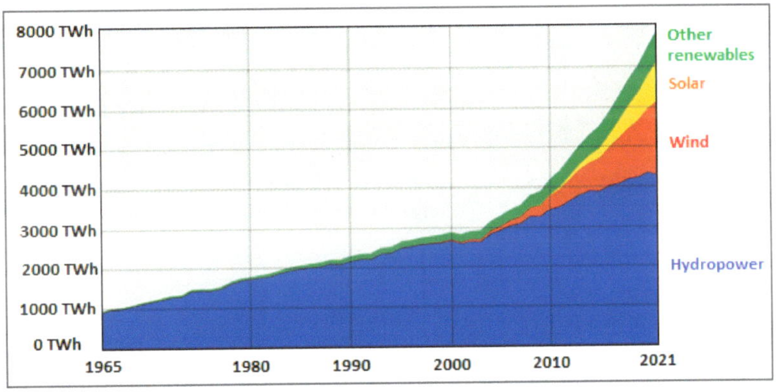

Figure 6. Renewable Energy Generation in World [37].

Annual additions of renewable power by technology from 2014 to 2021 are depicted in Figure 7. There is a clear indication of growing power generation from all renewable sources. In 2020, the addition of power from solar PV sources reached a high point of 138 GW (approx.), whereas this capacity was only 40 GW in 2014 [36]. Due to the new technological accomplishments in wind power generation and solar PV generation, their production ratesare also very high throughout the world.

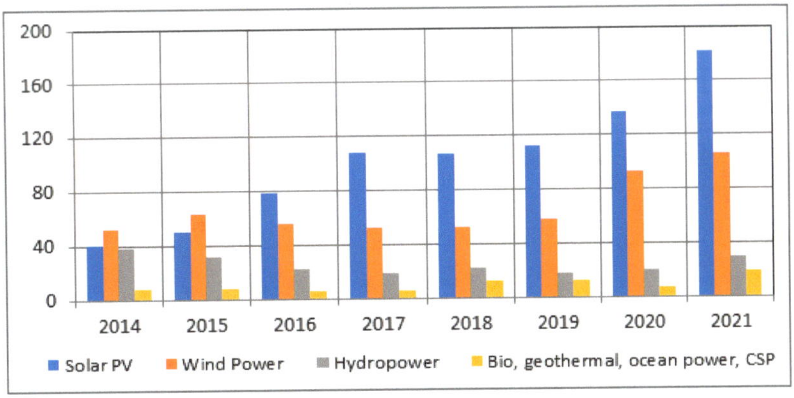

Figure 7. Annual Additions of Renewable Power by Technology, 2014–2021.

The world's top five countries for renewable power generation in 2021 are shown in Table 5. In the category of annual investment and total capacity addition, China secured the top position in terms of hydro, solar PV, and wind power generation [36]. In terms of total capacity addition, India was in the top four ranks in the world for solar PV, hydro, and wind power. The United States also maintained its progress in all respects of renewable energy sources. These leading countries have encouraged other countries to develop renewable power plants.

Table 5. Top Five Countries for Renewable Power Generation in 2021.

Details Related to Renewable Energy	World Ranking				
	1	2	3	4	5
Annual Investment/Production/Net Capacity Addition in 2021					
Solar PV Capacity	China	United States	India	Japan	Brazil
Wind power capacity	China	United States	Brazil	Vietnam	United Kingdom
Hydropower Capacity	China	Canada	India	Nepal	Lao PDR
Total Capacity of Generation at the End of 2021					
Hydropower capacity	China	United States	Brazil	India	Germany
Solar PV capacity	China	United States	Japan	India	Germany
Wind power capacity	China	United States	Germany	India	Spain

4. Global Renewable Power Scenario

In recent years, the power requirement has increased at a high rate, whereas the availability of thermal power plants is very limited, which forces power producers to think about power generation from non-conventional energy sources. This section provides the detailed status of the global hydro, wind, and solar power scenario.

4.1. Global Hydropower Scenario

Hydropower is one of the oldest and leading sources of environmentally friendly energy. Carbon production is much lower in hydropower plants. Figure 8 shows the status of hydropower generation by region from 1965 to 2019 [37]. The entire world is divided into seven regions, i.e.,the Middle East, Africa, CIS, Europe, North America, South and Central America, and Asia-Pacific. In Eastern Africa, the hydropower generation increment rate in 2019 was more than 2.5 times that in 1985. The hydropower generation was 19.68 TWh in 1985, reaching 68.85 TWh in 2019 [37].

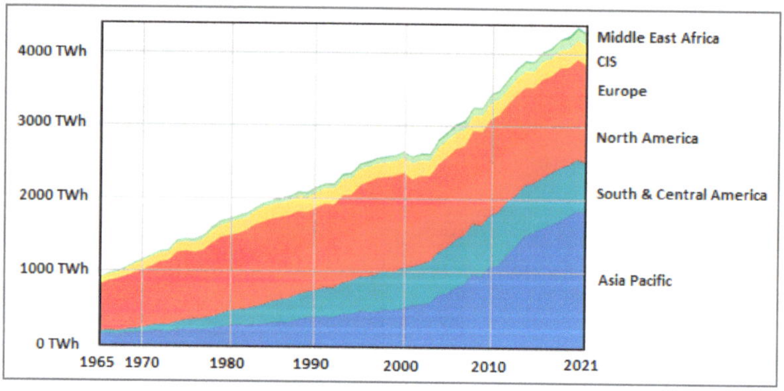

Figure 8. Hydropower Generation by Region [37].

The same tradition has also been followed by other regions. The increment in the electricity generation from hydropower plants increased approximately 1.68 times, 4.57 times, and 1.84 times in 2019 compared to1985 for Africa, Asia Pacific, and Central America, respectively. In Europe, the hydropower-generation capacity was 632.5383 TWh in 2019, and in North America, the capacity was 676.9256 TWh. According to the 'Our World in Data' report, the hydropower-generation capacity is still more than 60% of entirely renewable energy sources excluding biomass (traditional).

At the end of 2020, China secured the chief position in hydropower generation with 29% of total hydropower generation capacity, followed by Brazil, Canada, the United States, and India with 9%, 7%, 7%, and 4%, respectively. Figure 9 shows the global hydropower capacity and the shares of the top countries, as well as the rest of the world, in 2020. Other countries in the sixth to tenth position in hydropower generation are the Russian Federation, Norway, Turkey, Japan, and France, respectively [36].

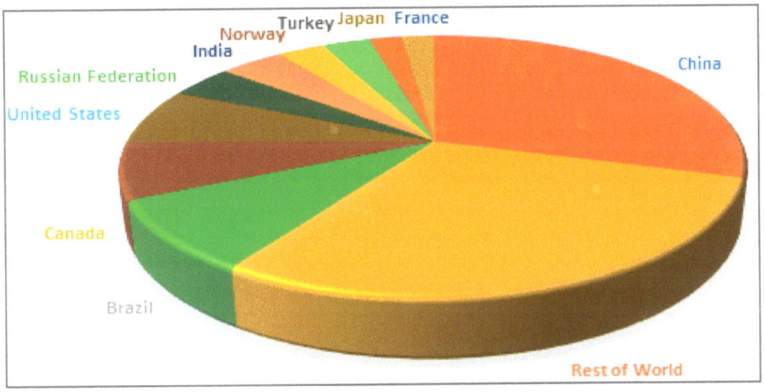

Figure 9. Hydropower Global Capacity of Top 10 Countries, 2021.

Figures 10 and 11 display the hydropower capacity and addition of the world's top countries [36,37]. The hydropower-generation capacity is increasing for maximum countries day by day. When looking at the case of China, an increment of 20.6% in hydropower generation was observed in 2021 from2020. This scenario was followed by some other countries, such as Canada, India, Nepal, Lao PDR, Turkey, Indonesia, Norway, Zambia, Kazakhstan, etc.

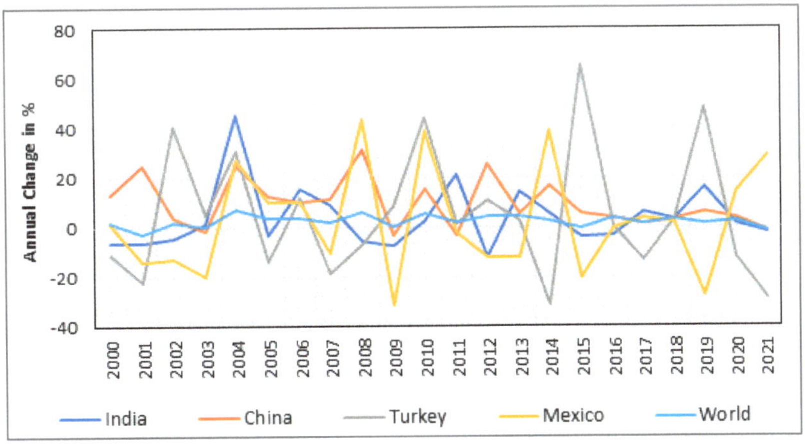

Figure 10. Annual Percentage Change in Hydropower Generation [38].

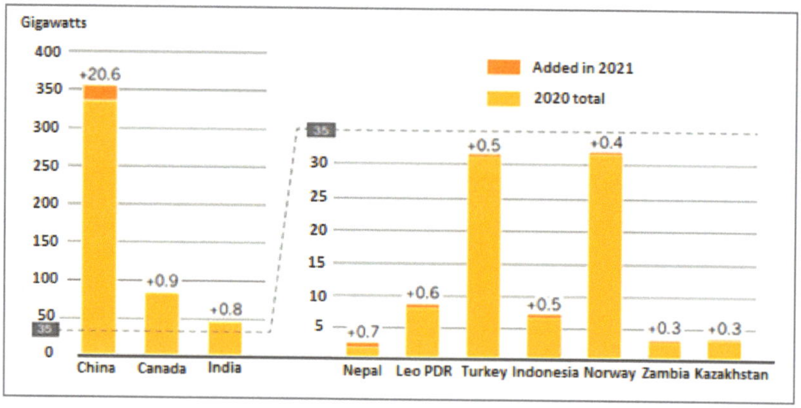

Figure 11. Hydropower Capacity and Additions for Top 10 Countries, 2021 [36].

Figure 12 displays the annual change in hydropower generation between 2014 and 2021 [37]. It can be concluded from the graphs that there is a huge change in the capacity addition in hydropower throughout the world. Hydropower generates electricity at a lesser rate than conventional sources, as well as at a lesser rate than almost all other renewable energy sources. Dams built for hydropower generation also help in checking flood conditions and canals help in irrigation. To sum, hydropower sourcesare not only a source of power generation, but a gold mine in disguise.

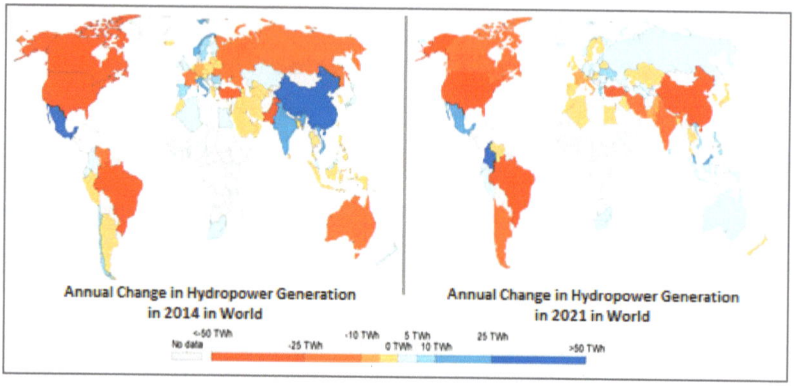

Figure 12. Annual Change in Hydropower Generation in 2014 and 2021 in World [37].

4.2. Global Wind Power Scenario

The usage of wind power is rising around the globe as its cost drops. In the past two decades, a spike of 75% has been reported in offshore and onshore wind-generation capacity, rising from 7.5 GW in 1997 to 564 GW in 2018. Wind electricity production has doubled between 2009 and 2013 [37]. Wind turbines were first introduced a century ago, succeeding the invention of the electric generator in 1830. In 1887 and 1888, wind-power-generation installation took place in the United Kingdom and the United States, respectively. The first modern wind power generatorwas developed in Denmark. The volume of power gathered from the wind depends on the blade's length and the turbine's size. Production is directly related to the rotor's dimensions and the cube of the wind speed. The capacity of wind turbines has increased over this period. Previously, the turbine-rated capacity was 0.5 MW, and the diameter of the rotor was 15 m. Currently, the wind turbine capacity has reached 8 MW, and the rotor diameter is up to 164 m.

Figure 13 shows the status of wind power generation by region from 1965 to 2021. In 1978, Europe was the first region where wind energy generation was observed. Then in 1989, Europe and North America were the regions that generated around 3 TWh of wind energy. Gradually, in 2000, the Middle East, Africa, Europe, North America, South and Central America, and the Asia Pacific generated around 31 TWh. All of the sevenregions started to generate wind energy in 2010, with a capacity of 346 TWh. In 2021, wind energy generation increased to 1862 TWh [37].

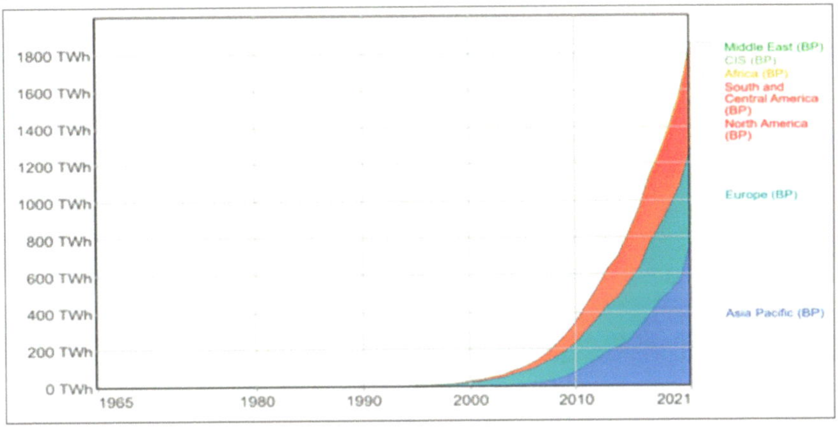

Figure 13. Wind Power Generation by Region [37].

Figure 14 shows the new offshore and onshore wind power installation in 2021. China, the United Kingdom, Vietnam, Denmark, and the Netherlands are the world's top five countries in terms of offshore wind power markets. Wind power of 21.1 GW was installed in 2021, of which 99.5% was been installed by these five countries. China made up over 80%, the United Kingdom made up around 10.99%, Denmark made up around 2.9%, and the Netherlands made up around 1.9%, while the rest of the world installed 0.5%. The top ten new onshore wind-installed countries are also shown in Figure 14. Total new onshore wind power of 72.5 GW was installed in 2021. Around 42.34% was installed by China, making it the highest generating country, followed by the USA at around 17.62%. Brazil generated 5.31%, Vietnam made up 3.7%, Sweden installed 2.9%, Germany installed 2.7%, Australia installed about 2.4%, India made up 2%, Turkey installed 1.9%, 1.6% was installed by France, and the rest of world installed 17.52% [37].

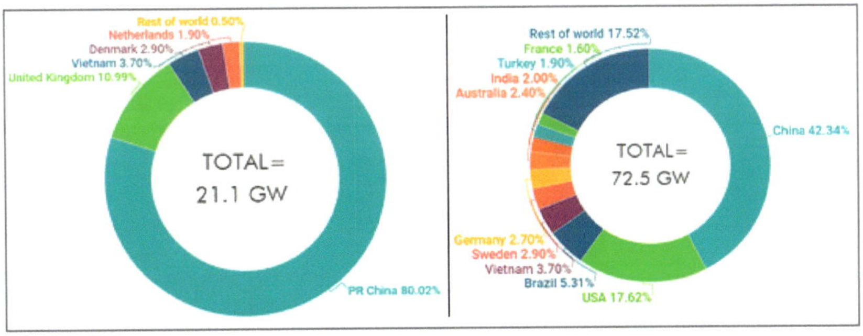

Figure 14. New Offshore (21.1 GW) and Onshore (72.5 GW) Global Installation of Wind Power: Shares of Top 10 Countries and Rest of World, 2021.

The annual percentage change in wind power generation is demonstrated in Figure 15. The annual changing status of wind generation in comparison to the previous years is delivered in this figure. China, India, US, Brazil, and UK are the main contributors to the generation of wind power, whereas Sweden's contribution is adverse, so they cannot achieve the generating demand and are instead lagging behind [37]. From the last consecutive 13th year, the prime regional market of wind power was in Asia. China added 55.9 GW of wind power capacity in 2021 from the previous year, whereas the rest of the world only added 15 GW of wind power capacity in 2020 (shown in Figure 16).

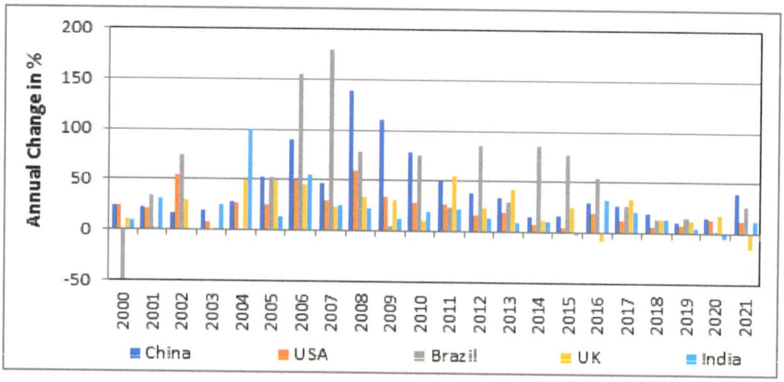

Figure 15. Annual Percentage Change in Wind Power Generation.

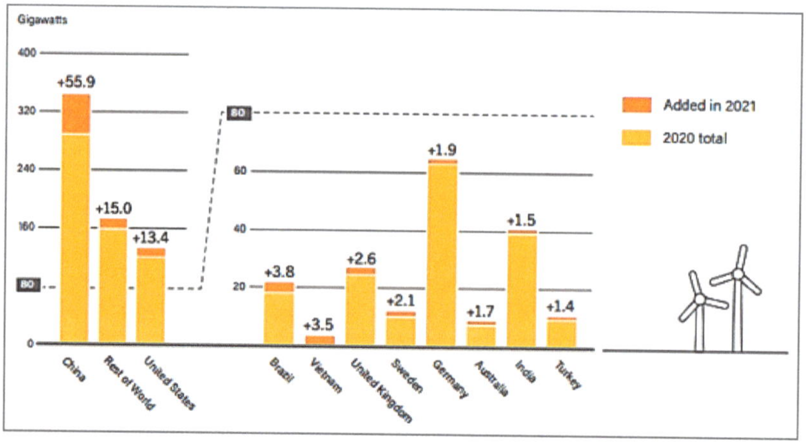

Figure 16. Wind Power Capacity and Additions: Top 10 Countries for Capacity Added, 2021 [36].

Figure 17 demonstrates the annual modification in wind power generation between 2014 and 2021 [37]. A massive change in the capacity addition of wind power throughout the globe can be noticed.

4.3. Global Solar Power Scenario

Solar energy has been used worldwide to generate electricity, purifywater, or for heating purposes. Solar photovoltaics was invented in 1954 in the United States. They convert direct sunlight into electricity. Presently, they are the fastest-growing renewable technology and are prepared to take a key role in the future comprehensive electricity market. Solar PV materials can be installed to provide configuration for small-scale projects, personal use, mini-grids, or commercial use. This initiative can also help people who cannot easilyaccess power transmission lines. Solar panels have become costeffective in the last few decades, which makes them affordable for the consumer to use and the cheapest form of electricity [37].

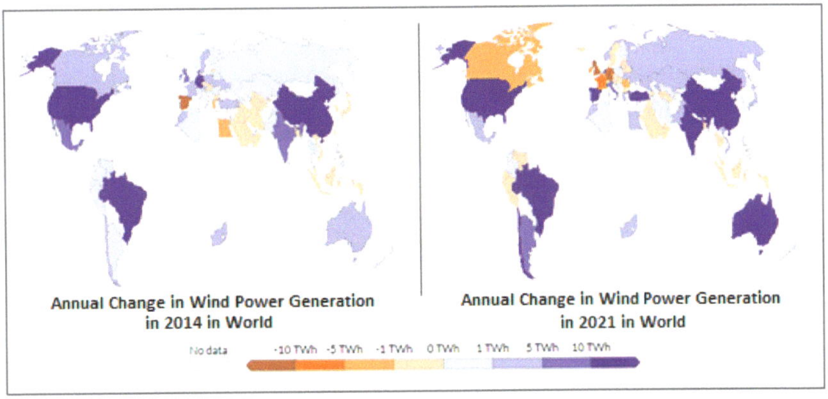

Figure 17. Annual Change in Wind Power Generation in 2014 and 2021 in World.

The stature of solar power generation by region from 1965 to 2021 is illustrated in Figure 18 [36,37]. The growth in solar power generation was observed to be than 1 TWh in the year 1983 in North America. In 2005, South and Central America, North America, Europe, Asia Pacific, and Africa had a slight growth of around 4 TWh. An amount of 34 TWh generation was noted in 2010. In 2021, a jump of 1033 TWh of solar generation transformed the scenario of the global solar industry. Figures 18–20 show the performance of the top countries, where the indication is clear regarding the performance reduction in 2021 (61%) compared to 2020 (66%). China, India, Japan, Brazil, and the United States together cover about 61% of recently installed solar power capacity. More players have entered the market in response to the falling operational and capital costs of solar PV materials. France, Germany, Spain, the Republic of Korea, and Australia were the next five countries in 2021. The annual performance of the top ten countries spiked from 3 GW in 2020 to 3.4 GW in 2021.

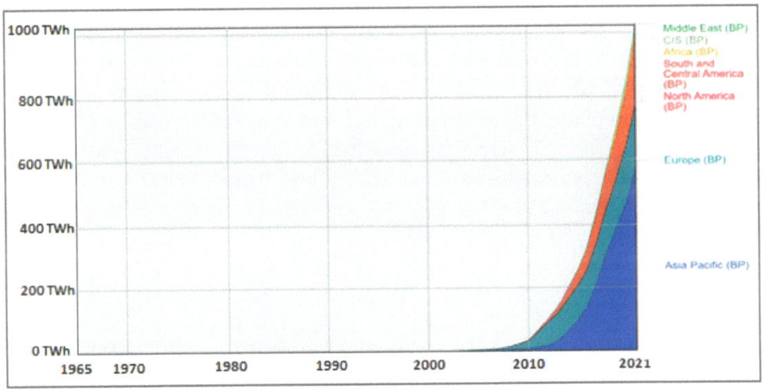

Figure 18. Solar Power Generation by Region.

The top countries according to cumulative solar PV capacity are China, India, United States, Germany, and Japan, whereas Australia, Germany, and the Netherlands are the leading countries per capita capacity in solar power. These figures give a clear view of the solar power capacity; shares of the top ten countries and the rest of the world; annual percentage change in the solar power sector; and capacity and additions in the top ten countries for capacity addition around the globe [36].

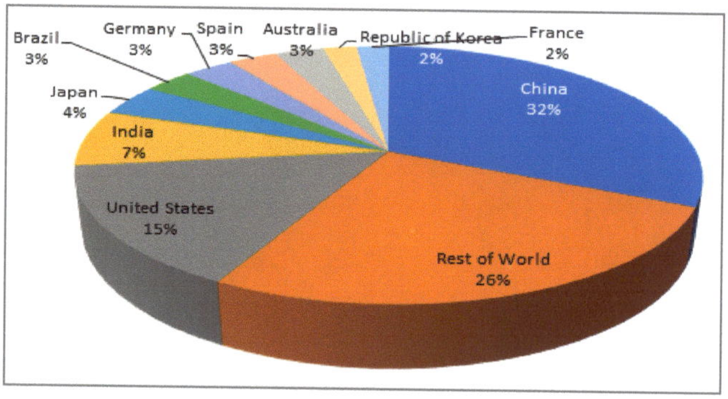

Figure 19. Solar Power Global Capacity of Top 10 Countries, 2021.

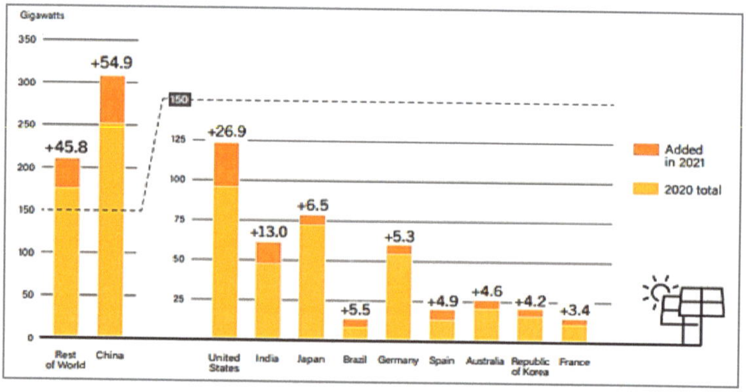

Figure 20. Solar Power Capacity and Additions, Top 10 Countries for Capacity Added, 2021.

Figure 21 illustrates the annual change in solar power generation in 2014 and 2021 [37]. The figure shows the vast changes in the addition of solar power. It can be concluded that there has been a huge change in the capacity addition of solar power throughout the globe. Solar energy is limitless. It is a clean source of energy. Its maintenance costs are very minimal as solar panels last for 20–25 years. The technology in this industry is advancing very drastically. As such, power producers can easily attract the installation of new solar power plants.

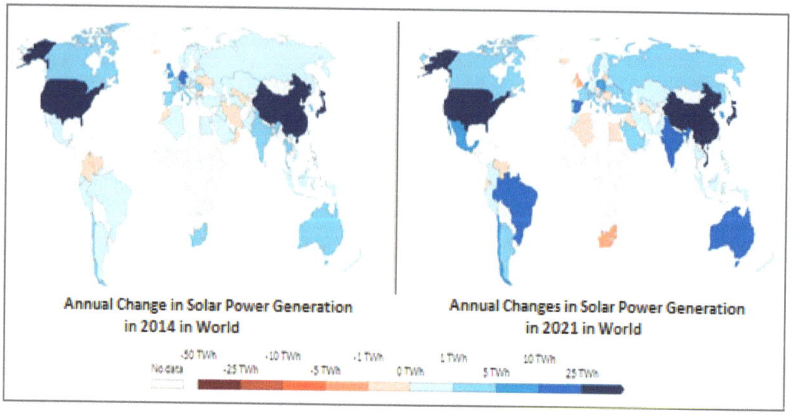

Figure 21. Annual Change in Solar Power Generation in 2014 and 2021 in World.

5. Renewable Power Scenario in India

Energy generation capacity is one of the main elements for the growth of a country's economy. It is a huge challenge for a country such as India, where the population is about 1.46 billion as of 1 July 2022 [38]. The urban population is considered to be the major consumer of electricity, and the government is expecting urban growth to contribute to 73% of the total population by 2036 [39]. As conventional sources are expected to be exhausted shortly, the renewable sector needs to receive more attention. The Indian power sector has begun to set up innovative solutions. India is bounded by a coastline from three edges, which has good projections for harnessing offshore wind energy. India has been gifted with huge solar energy potential, which is being used for the installation of solar panels. According to the speech of the Hon'ble Prime Minister of India at the UN Conference on climate change (2015), the GoI has engaged in numerous initiatives to reduce emission levels by 33% to 35% from the year 2005 to 2030. It has similarly set up the target to generate around 40% of complete power from non-conventional sources by 2030. In India, the maximum utilization of energy per capita is in the northern region. All India region-wise energy and power scenarios are shown in Tables 6 and 7 for July 2021 and July 2022, respectively. As per the Ministry of Power figures, at the end of July 2021, the energy demand was 124,167 MU, while availability remained at 123,720 MU with a shortage rate of 0.4%.

Table 6. All India Region-Wise Energy Scenarios in July 2021and 2022 (as of 31 July 2022).

Region	Energy (MU)				Deficit (%)	
	Demand		Availability			
	July 2021	July 2022	July 2021	July 2022	July 2021	July 2022
Northern	45,180	45,049	44,793	44,860	−0.9	−0.4
Western	33,799	35,707	33,796	35,700	0.0	0.0
Southern	28,072	28,777	28,065	28,775	0.0	0.0
Eastern	15,374	16,800	15,325	16,610	−0.3	−1.1
Northeastern	1742	1904	1741	1896	−0.1	−0.4
All India	**124,167**	**128,237**	**123,720**	**127,841**	**−0.4**	**−0.3**

Table 7. All India Region-Wise Power Scenarios in July 2021 and 2022 (as of 31 July 2022).

Region	Power (MW)				Deficit (%)	
	Peak Demand		Peak Availability			
	July 2021	July 2022	July 2021	July 2022	July 2021	July 2022
Northern	73,305	72,935	72,935	741,413	−0.5	−0.8
Western	59,036	58,995	60,966	53,550	−0.1	0.0
Southern	50,121	50,121	58,430	48,810	0.0	−0.1
Eastern	26,019	25,145	25,145	26,330	−0.34	−1.8
Northeastern	3117	3053	3053	3675	−2.1	−0.6
All India	**203,014**	**200,539**	**200,539**	**190,386**	**−1.2**	**−0.6**

At the end of July 2022, the energy demand increased by 128,237 MU, whereas 127,841 MU was available with a shortage rate of 0.3% [40,41]. As per the statistics, it can be determined that there is more or less ascarcity of power in the Indian power sector. Considering this status, many renewable energy projects have been introduced by the GoIto reduce the power deficit rate. Similar to the energy scenarios, power also experiences the same scenario in India. At the end of July 2021, the peak power demand was 203,014 MW, while peak power availability was 200,539 MW with a deficit rate of 1.2%. At the end of July 2022, the peak power demand increased to 200539 MW, while availability was 190,386 MW with a shortage rate of 0.6%.

Indian resource-wise, region-wise, and sector-wise power scenarios are shown in Tables 8 and 9. India has one of the largest power sectors according to number of consumers and geographical area. The overall installed power capacity in India is 404,132.95 MW (as of 31 July 2022).The central government power sector has installed 99,004.93 MW, the state government power sector has installed 104,969.33 MW, and the private power sector has installed 200,158.70 MW [41]. A huge quota of total installed power capacity is fulfilled by the private sector. The central sector generates 24%, the state sector generates 26%, and maximum power is generated by the private sector at 50%. Overall power generation in the Indian power sector and its development is shown in Figure 22 from 2009 to 2020. As the power demand is increasing continuously, this enlargement has triggered economic development. India is rising as a developing country, and its frequency of power consumption is drastically growing in every area. To satisfy demand, generation is increasing progressively. The percentage of growth differs every year as per the sources. The government of India takes numerous initiatives to move toward renewable power plants quickly to keep the system stable and enhance grid security due to massive power shortages in the Indian power sector. For hydropower plant investors, the state government offers subsidies and make the NOC process considerably easier. The scheme-wise physical growth in power generation in 2020–2021and the cumulative status up to June 2020 are shown in Table 10 [42]. Nowadays, people are aware of the importance of renewable power, and they want to shift toward it to reap the benefitsof this kind of power.

The modern renewable-energy-generation scenario by sources in India is shown in Figure 23. Hydro, solar, and wind are the leading renewable energy sources that make India a renewable-commanding country. These statistics illustrate simply the increase in renewable generation year by year.

Table 8. Region-Wise and Resource-Wise Power Scenarios in India (MW) (as of 31 July 2022).

Region	Thermal	Nuclear	Hydro	Renewable Energy	Total
Northern	63,702.13	1620.00	20,751.77	29,230.41	115,304.31
Western	86,779.63	1840.00	7562.50	34,760.24	130,942.37
Southern	55,870.76	3320.00	11,827.48	48,137.35	119,155.59
Eastern	27,329.70	0.00	4764.42	1768.28	33,862.40
Northeastern	2343.16	0.00	1944.00	502.66	4789.81
Islands	40.05	0.00	0.00	38.43	78.48
All India	**236,065.42**	**6780.00**	**46,850.17**	**114,437.37**	**404,132.95**

Table 9. Sector-Wise and Resource-Wise Power Scenarios in India (MW) (as of 31 July 2022).

Sector	Thermal	Nuclear	Hydro	Renewable Energy	Total
State	75,261.56	0.00	27,254.45	2453.31	104,969.33
Private	85,875.95	0.00	3931.00	110,351.75	200,158.70
Central	74,927.91	6780.00	15,664.72	1632.30	99,004.93
All India	**236,065.42**	**6780.00**	**46,850.17**	**95,012.59**	**404,132.95**

Information regarding the power supply position in India for different regions is deliberated in the presented statistics. In 2020–2021, the southern region faced a huge energy deficit, while the total Indian power sector gained some surplus. This was possible due to the endless promotion activities conducted by the government of India regarding renewable energy usage.

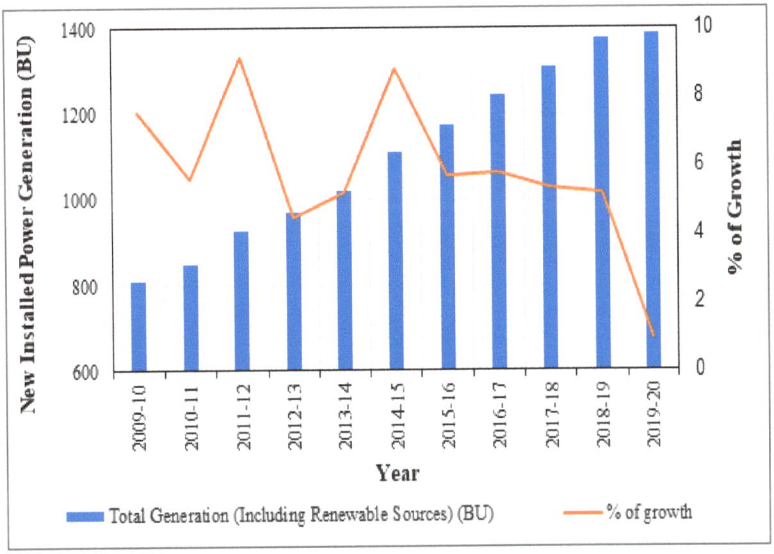

Figure 22. Power Generation and Its Growth Rate in the Indian Power Sector Over the Previous Year.

Table 10. Scheme-wise Physical Progress in 2022–2023and Cumulative Status up to July 2022.

Renewable Energy Scheme	Target of Power Generation during 2022–2023	Achievement in Power Generation during (April–July 2022)	Cumulative Achievements (as of July 2022)
Grid-Interactive Power (Capacities in MW)			
Wind Power	3000.00	535.75	40,893.33
Solar Power	9000.00	3977.26	57,973.80
Small Hydropower	100.00	39.00	4887.90
Total	14,380.00	4552.01	114,437.39
Off-Grid/Captive Power (Capacities in MWEQ)			
Waste to Energy	10.00	0.00	253.61
Waste to Power	500.00	0.00	223.14
Total	510.00	191.87	1368.44

5.1. Hydropower Scenario in India

Depending on size, hydropower plants are mainly classified into two types, i.e., small hydro and large hydro plants. Every country has a different capacity range for the classification of a hydropower plant. In India, a capacity of 25 MW or below is considered for small hydro plants, which is further classified into micro, mini, and small hydro plants. The range of capacity in micro hydropower plants is between 100 kW or below; mini hydropower plants are between 101 kW to 2 MW; and small hydropower plants are between 2–25 MW [43]. In the beginning, hydropower was looked after by the Ministry of Power with the support of the state electricity boards. India's oldest hydropower plant was built in 1897 in Darjeeling with an installed capacity of 130 KW [44]. The GoI is playing an active role in encouraging the development of small hydropower plants in a planned manner and increasing the quality and reliability of the projects. In India, the public sector has produced 92.5% of total hydropower. National Hydroelectric Power Corporation (NHPC), NTPC-Hydro, Northeast Electric Power Company (NEEPCO), THDC, and Satluj Jal Vidyut Nigam (SJVNL) are the leading public sector companies thatgenerate hydropower in India.

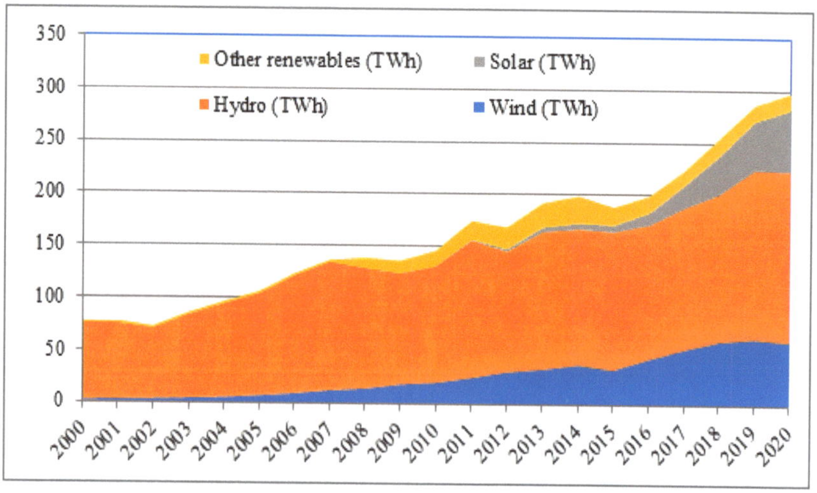

Figure 23. Modern Renewable Energy Generation by Sources in India [37].

The details regarding the ongoing major hydropower plants in India are as follows [45]:

- Tehri Dam (three stages) with a capacity of 2400 MW is located in Uttarakhand and operated by THDC Limited, New Tehri, Uttarakhand, India.
- Koyna Hydroelectric Project (four stages) with a capacity of 1960 MW is located inMaharastra and operated by MAHAGENCO, Maharashtra State Power Generation Co., Ltd., Mumbai, India.
- Srisailam with a capacity of 1670 MW is located in Andhra Pradesh and operated by APGENCO, Vijayawada, India.
- NathpaJhakri with a capacity of 1500 MW is located in Himachal Pradesh and operated by Satluj Jal Vidyut Nigam, Shimla, India.
- Sardar Sarovar Dam with a capacity of 1450 MW is located in Gujarat and operated by Sardar Sarovar Narmada Nigam Ltd., Gandhinagar, India.
- Bhakra Nangal Dam with a capacity of 1325 MW is located in Himachal Pradesh and operated by Bhakra Beas Management Board, Chandigadh, India.
- Indira Sagar Dam with a capacity of 1000 MW is located in Madhya Pradesh and operated by Narmada Valley Development Authority, Bhopal, India.

Some of the other hydropower plants located in several states of India, with a minimum generation capacity of 500 MW, include Belimela Dam, Dehar (Pandoh) Power Project, KarchamWangtoo Hydroelectric Plant, Omkareshwar Dam, Purulia Pass Dam, Nagarjuna Sagar Dam, Ranjit Sagar Dam Salal I and II, Upper Indravati Dam, and Teesta Dam. Figure 24 displays the overall target and installed capacity of hydropower in India in the years 2021 and 2020. The installed capacity was 4.8 GW in 2021 and 4.7 GW in 2020, and the overall target was 5 GW [46]. From the figure, it can be seenthat the installed capacity of the hydropower plant increases year by year and is reaching the target.

5.2. Wind Power Scenario in India

Wind power is considered to be a site-specific and discontinuous type of energy resource. As such, a particular site needs to be selected where extensive wind resource assessment is possible. India has a coastal boundary on its three sides of around 7600 km, which is a great scenario for harnessing wind energy. The Indian wind energy sector has been showing consistent progress as it is led by wind-power-manufacturing industries. Increasing the number of wind power plants has led to a great ecosystem, development procedure abilities, and manufacturing. India is ranked 4th according to its wind installed capacity with a total installed capacity of 39.25 GW. In 2020–2021, India generated around 60.149 billion units [47].

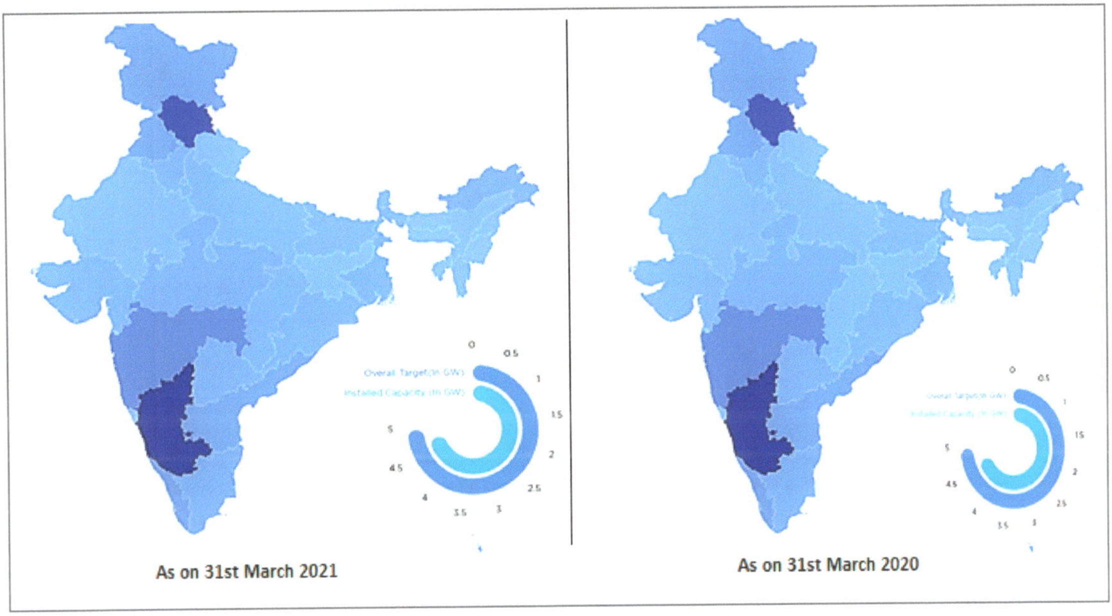

Figure 24. Status of Hydropower Generation in India.

The government of India is also promoting wind power projects across the country. The private sector is also providing investment for various financial incentives to encourage customers. Some of the incentives introduced to encourage wind projects across the country [48] are as follows:

- A guideline has been introduced to permit distribution licensees to acquire wind power costeffectively.
- To ease interstate wind power transmission and distribution, transmission charges and losses have been waived for solar and wind projects.
- Proper site-specific identification and technical support are handled bythe National Institute of Wind Energy.

The wind potential for India's top seven states at heights of 100 m and 120 m is displayed in Figure 25. India's first offshore wind power project with a capacity of 1 GW was planned in Gujarat to bring the localization of essential ecology and low-cost scaling for the offshore wind power sector. The details regarding some of the major wind power plants in India are as follows [49]:

- The Muppandal Wind Farm with a capacity of 1500 MW is located in Tamil Nadu.
- The JaisalmerWind Park with a capacity of 1064 MW is located in Rajasthan.
- The BrahmanvelWind Farm with a capacity of 528 MW is located in Maharashtra.
- The Dhalgaon Wind Farm with a capacity of 278 MW is located in Maharashtra.
- The VankusawadeWind Park with a capacity of 259 MW is located in Maharashtra.

Figure 26 shows the overall target and installed capacity of wind power generation in India in 2021 and 2020. The installed capacity was 39.2 GW in 2021, whereas 37.7 GW found in 2020, and the overall target was 60 GW. The figure shows the improvement in the solar energy sector and display the trend toward reaching its goal.

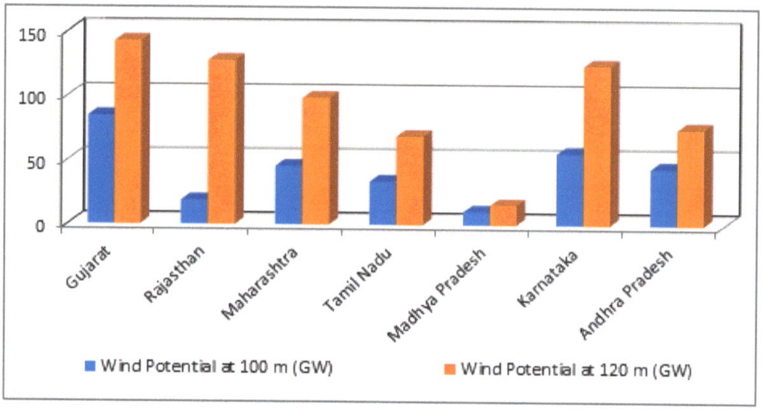

Figure 25. Wind Potential for 7 Highest States at 100 m and 120 m [47].

5.3. Solar Power Scenario in India

Around five thousand trillion kWh per year of energy is directly sent across the country and major parts of it have achieved rates of 4–7 kWh per sq. m per day. Solar photovoltaics have a vast scope in India. It is chosen as the utmost secure source compared to others as it is richly available. Over the last few decades, solar energy has shown a visible rise that is evident in the Indian energy scenario. The distributed and decentralized applications based on solar energy have benefittedmany people in an eco-friendly manner. The solar energy sector in India has developed over the years and become a significant player in generating capacity in the grid-connected system. It has become a vital part that meets the energy needs of the nation, holds up the government schema of sustainable development, and plays an important role in energy safety.

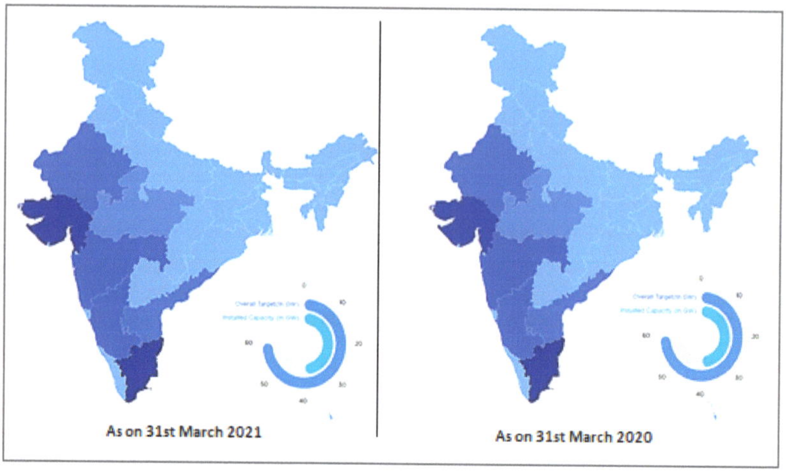

Figure 26. Status of Wind Power Generation in India [47].

The GoI has introduced several schemes to boost the generation of solar power. Solar rooftop schemes, bundling schemes, canal bank and canal top schemes, grid-connected system, defense schemes, VGF schemes, etc., are a few of them. India's first commercial solar power plant was inaugurated in the year 2009 with a capacity of 2 MW in Amritsar [50]. Figure 27 depicts the overall target and installed capacity of solar power generation in India for 2021 and 2020. The installed capacity was 40.1% of the overall target in 2021, whereas it was 34.6% in 2020.

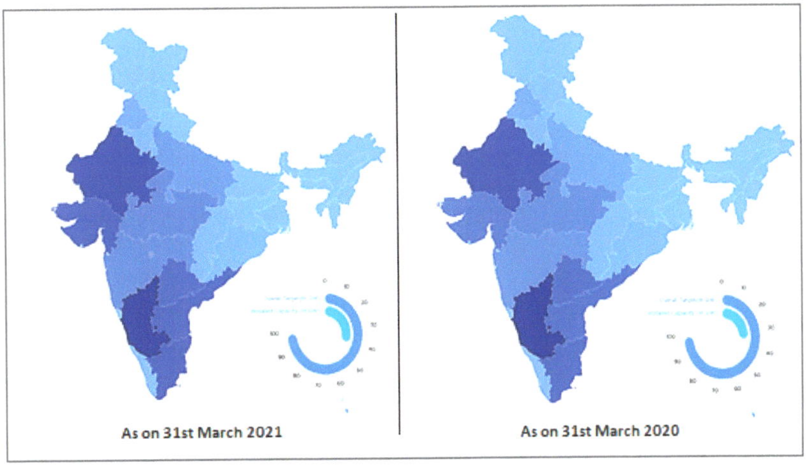

Figure 27. Status of Solar Power Generation in India.

India holds the fifth position globally in solar power deployment. In the past five years, solar power capacity has augmented by more than 11 times from 2.6 GW in 2014 to 30 GW in 2019 [51]. India's power sector is currently adopting mixed electricity sources, where the nation can achieve the biggest profit by releasingcarbon emissions. Some of the biggest public sector companies have also dived into a mix of energy source without supporting the nation'sconversion, which is hardhitting. The details regarding the ongoing major solar power plants in India are as follows [52].

- Bhadla Solar Park (10,000 acres) with a capacity of 2245 MW is located in Rajasthan.
- Pavagada Solar Park, covering an area of 13,000 acres and with a capacity of 2050 MW, is located in Karnataka.
- Kurnool Ultra Mega Solar Park, covering an area of 5932 acres with a capacity of 10,000 MW, is located in Andhra Pradesh.
- NP Kunta Ultra Mega Solar Park, covering an area of 7924 acres with a capacity of 978 MW, is located in Andhra Pradesh.
- Rewa Ultra Mega Solar Plant, covering an area of 1590 acres with a capacity of 750 MW, is located in Madhya Pradesh.

6. Application of Renewable Sources in the Power Sector: Glimpses from the Review

In the recent past, several researchers have performed their work in the field of renewable energy sources and their application in the power sector. Ref. [53] illustrated that the adaptation of huge renewable energy technologies needs vast transmission system extension. The authors compared two market models to observe social welfare while facing in wind variability. In a deregulated power market, it is very crucial to pay special attention to the expansion of renewable energy sources due to socio-economic and environmental issues of conventional sources. An ideal design for handling congestion in a competitive electricity market for both renewable and conventional sources was discussed in [54]. Rong-Gang Cong [55] identified and reviewed numerous important aspects influencing the progress of renewable energy generation in the present research. After a detailed investigation, an innovative optimization technique to use renewable energy resources optimally for generation purposes in the future was introduced.

The author of [56] discussed the current scenario of electricity generation using clean technology with its unpredictable nature and the importance of storage devices in the electricity network. The joint operation of renewable sources and energy-storage devices, such as wind farms, solar photovoltaic materials, pump storage, and energy-storage devices, is deliberated on in terms of ancillary service and the energy market. The uncertain nature of solar photovoltaic materials and wind farms and their adverse effects on the power market was discussed in [57]. Renewable energy sources and energy-coherent models have numerous benefits, such as mitigating climate change, increasing safety, and increasing sustainability in the power system [58]. Piyasak [59] illustrated the benefits of renewable sources with proper storage planning that can decrease instabilities in solar photovoltaic and wind energy generation. The wind energy sector was broadly discussed in [60]. The author reviewed various facts about designing wind energy systems, i.e., policies, investments, and socio-economic benefits, for merging into the competitive power market.

The advantages of using variable renewable resources, such as solar photovoltaic materials and wind power, were discussedby Simon Sinsel et al. [61]. These renewable sourcesare considered to be the key techniques to achieve the decarbonization of the electricity market and have significantly different scenarios from conventional sources. Dalia et al. [62] analyzed several approaches to enriching the performance of power systems with a huge expanse of power generation from renewable sources. They mainly pointed to the methodology applied to improve solar PV and wind power with low-voltage ride-through skills to please grid necessities. The authors of [63] reviewed the hybrid integration of solar and wind energy systems. The increasing electricity demand is also making the power market switch toward renewable as they are eco-friendly. Many countries are installing renewable energy into their existing projects as they natively exist in their nation. They offer numerous additional benefits compared to other energy resources. The cost of power generated from renewable energy sources is a vital factor as it is very low, which subsidizes the penetration of renewables in the deregulated market [64]. Because the marginal cost of renewable sources is very low or even zero, they negatively affect the market clearing price by eliminating more expensive generating sources from the dispatch program [65].

Anuj et al. [66] displayed the energy and operational reserve procurement from renewable power plants on an actual time basis in a short-term power market with a sequential dispatching technique. A deregulated power market aims to deliver excellent quality power to its consumers, which should be economical and environmentally friendly. Renewable sources are integrated into the distribution system to complete the inconstant load demand, and the decarbonizing effects are covered in depth by B. Muruganantham et al. [67]. Yin et al. [68] analyzed the current electricity market by integrating solar power and the global industrial scenario. By using the viable solution model, a detailed investigation was conductedon solar uncertainty in the short-term electricity market.

Hydropower is not only sustainable energy, but it also has great storage capability and flexibility, which make it improve the stability of the grid and help other alternating renewable energy sources, such as wind and solar placement, was further discussed by the author [69]. The authors of [70] proposed a framework for the ideal generation approach of a micro grid that integrates solar photovoltaic materials, wind farms, tidal steam turbines, fuel cells, boilers, and energy-storage devices. The uncertain nature of solar photovoltaic radiation, wind speed, and tidal steam generation affects the quality and quantity of power and market pricing. Farzana et al. [71] deliberated an outline to reschedule power generation with and without integrating renewable energy sources to maintain power flows in transmission lines within specific limits in a deregulated power market. The authors of [72] presented a novel framework for combining wind–thermal plants for optimal investment strategies. Wind forecasting was evaluated using the data mining technique, and the Monte Carlo technique was used to determine the fuel price and electricity demand in a deregulated power market. The authors of [73] provided information to researchers, policymakers, and designers regarding renewable energy production through the potential of a biomass, wind, and hybrid system. Dey et al. [74] elaborated on the relationship between renewable sources and sustainable development with their future scopes. The authors of [75] described the complete structure, lifecycle, and costing model of wind power from a large-scale point and checked the imbalance cost models and structure between a variety of wind farms using different economic analysis approaches. An ANN-based method was presented in [76] to study the optimal conditions using the load frequency control method for a distributed power network. A bacterial foraging algorithm was implemented in [77] to augment learning of the neural network model for automatic generation.Al-Majidi et al. [78] presented the ANFIS MPPT controller design for a remote PV system.

The technical efficiency of renewable sources is improving with the modernization of energy-harnessing methods, as well as the use of efficient energy collectors. The advancement of the efficiency of renewable energy sources encourages power plant installers, policymakers, and utility services to go towards renewable sources throughout the world. The authors of [79] presented a process for cleansing the forecasting of renewable sources by its intraday tuning and explored the measures for keeping renewable sources with unbalanced production. Table 11 displays a comprehensive analysis of the literature in the field of renewable energy sources.

After the comprehensive literature review, it can be concluded that hydropower, wind power, and solar power are now the leading backup energy sources throughout the world. However, the day is very close when these sources must act as the foremost and superlative energy sources. As such, these studies will give a clear-cut idea about the present status and future perspective to industrialists for their new renewable power plants.

Table 11. Comprehensive Analysis of State-of-the-Art Reviews in Renewable Energy Sources.

Paper ID	Renewable Energy Sources						Remarks
	Generalized	Wind	Solar	Hydro	Biomass	Geothermal	
14		√	√	√	√	√	Introduced ultra-modern computational optimization techniques in the field of renewable energy
15		√	√		√	√	Renewable energy scenario of India
16		√					Analysis and plan of the end-of-life of wind power plants
17		√					Optimal location for placing a wind power plant
18		√					Large-scale offshore wind power plant operation and project status
19	√						Several regulations are imposed by government bodies to mitigate the adverse of CO_2 on the environment.
20		√					Evolution of sustainable energy
54		√					Two market models are proposed to observe the social welfare of the customer while facing wind variability
55	√						Importance of expanding renewable energy sources in the deregulated market
56		√	√		√		Important aspects influencing the progress of renewable energy generation
57		√					Current scenario of electricity generation using clean technology
58		√	√		√		Renewable energy sources and energy-storage devices affect the deregulated power market
59	√	√	√				Benefits of renewable energy sources and energy-coherent models in the power system
60	√	√	√				Renewable energy sources with proper storage planning can decrease instabilities in the system
61	√	√					The wind energy sector has been broadly discussed
62	√	√					Advantages of using variable renewable resources
63		√	√				Several approaches are suggested for enriching the performance of power systems with a huge expanse of power generation from renewable sources
64		√	√				Hybrid integration of solar and wind renewable energy systems
65		√	√		√		The benefit of installing renewable energy sources compared to the other type of sources

Table 11. *Cont.*

Paper ID	Renewable Energy Sources						Remarks
	Generalized	Wind	Solar	Hydro	Biomass	Geothermal	
66	√		√				MCP of renewable energy sources compared to other energy sources and its benefit
67		√	√	√			The energy and operational reserve procurement from renewable power has been discussed
68	√		√				Economic and environmental benefits of the renewable source
69			√				The current electricity market by integrating solar
70		√	√				Pumped hydro energy storage technology has been thoroughly deliberated
71		√	√				The ideal generation approach of a micro grid, integrating solar photovoltaic materials, wind farms, tidal steam turbines, combined heat and power units, fuel cells, boilers, and energy-storage devices is discussed
72	√						Reschedule power generation with and without integrating renewable energy sources in a deregulated power market
73		√					Combining wind–thermal plants for optimal investment strategies

7. Technical Advancement in Renewable Energy

In the recent era, renewable energy has developed new and innovative technologies. Some of them are discussed as follows:

- *Solar Power*

The demand for solar energy has increased over the past few years due to its several benefits. The solar photovoltaic effect is the technique of a solar cell that can convert sunlight into electricity. Solar power has acted an important part in the conversion to renewable energy so far and there is no chance of it declining. Here are some technologies related to solar power.

Floating Solar Panels

Floating solar panels are panels that are made up of solar cells and are placed upon a floating body, such as ponds, lakes, water reservoirs, oceans, etc. They are mounted with the help of an anchor reaching the bottom of the reservoir bed. The generated power is transferred to the shore with the help of underwater cables. This technology has achieved outstanding grip due to its several benefits. They do not take up much space on land, unlike traditional solar panels. As such, they avoid interrupting the view and densely populated areas where land is scarce. The placement of the floating solar panel helps in the reduction in evaporation from the reservoirs.

Perovskite Solar Cells

Perovskite solar cellsare another innovation in solar power. This solar cell uses a specific material known as perovskite. They have unique crystal structures and are considered to have low cost and high efficiency compared to traditional silicon ones. These cells are very flexible and light in weight, which makes them perfect for transportable and wearable devices.

Solar Windows

This unique innovative idea was designed so that the house windows can supply electricity. In this, the solar PV cell is integrated into the glass of the window. These cells can be transparent or semi-transparent, permitting the sun's rays to pass for generating power. This idea leads the house owner to avail electricity easily and effectively.

- *Wind Power*

Wind energy is a renewable energy source that has the potential to achieve the increasing power demand. It uses the eternal power of the wind to produce electricity without harming the environment. This technology uses wind turbines to convert kinetic energy into mechanical energy and then convert it into electricity.

Vertical-Axis Wind Turbines

Vertical-axis wind turbines have a specific structure that helps to catch the wind from any direction and convert it into wind energy. They are compact in design, which allows them to be installed in residential as well as commercial buildings. The blades are placed vertically, and they rotate around the central axis. This makes them convenient to use in cities and areas where the direction of the wind is unpredictable. They create less noise compared to traditional horizontal-axis wind turbines.

Offshore Wind Turbines

Offshore wind turbines are placed in water bodies, such as lakes or oceans, where a consistent flow of wind is available to generate electricity. The primary advantage of them is that they can avail stronger and more consistent wind flow near the shore. They have a unique potential to produce huge amounts of energy.

Airborne Wind Energy

Airborne wind energy is a unique innovation of renewable energy technology in which electricity is generated using wind turbines thatare placed upon flying devices. This technology uses stronger and more consistent winds thatare found at higher altitudes. They use airborne devices, such as kites, drones, etc., to use wind energy at elevated places with robust and consistent winds. The primary advantage of them is that they can be deployed in such areas where traditional wind turbines are not workable.

- *Wave Energy*

Wave energy is a power-generating technology that uses kinetic energy from ocean waves. It uses floating devices that move upward and downward depending on the waves, which are connected to a generator that converts the energy into electricity. It is a predictable source that can be used for large-scale power generation.

8. Energy Efficiency in Renewable Sources

Energy efficiency means minimizing the waste of energy. It states that this occurs by performing the same task by utilizing less amounts of energy. It has a variety of benefits, such as minimizing the emission of greenhouse gas, decreasing the energy import demand, and bringing costs to domestic and economic standards. The technologies thatare related to renewable energy sources help to achieve these objectives, providing efficient energy economically and immediately. Different scopes exist for efficiency enhancements in each economic sector, whether it is buildings, transport, commerce, or power generation.

- *Buildings*

The designers of buildings focus on optimizing building efficiency. They can integrate renewable-energy-related technologies thatcreate zero-energy buildings. Variations can be made to the present buildings to decrease the usage of energy and make them economical. Small efforts can be considered, such as selecting LED light bulbs and energy-efficient appliances, or huge steps can be taken, such as upgrading the insulation.

- *Energy Generation and Distribution*

The 'waste' heat energy available from the power plant can be used for heating /cooling purposes or hot water for nearby facilities. This can increase the energy efficiency of power generation. The smart grid also improves the efficiency of energy generation and distribution.

- *Community Design*

Societies are designed for both development purposes and safety.Having available choices for hiking, biking, and public transport helps to reduce the requirement for private automobiles for travel. Energy-efficient automobiles need a small amount of fuel to complete a certain distance. This makes the operation comparatively less expensive and causes less pollution.

- *Freight*

Freight movement can be more efficient by improving truck and rail communication.

9. Conclusions

In recent years, the power sector has faced huge issues all over the globe. A major reason for this is the use of conventional-based power plants as a primary source, which hurts the environment and causes economic burden due to its scarcity. This inspires the power industries to shift towards renewable-energy-based plants. It is important to move over to renewable energy sources as environmental degradation and the lack of availability of fossil fuels are increasing day by day. In the present scenario, the government has increased its interest in wind, solar, and hydropower plants as they all have their benefits with respect to power generation. Wind power plants are considered to be more efficient when it comes to clean energy. They have the potential to meetincreasing power demands. They do not depend on daylight or the sun to generate power. They are beneficial on a utilityscale. Solar is considered to be the most common residential renewable source. It is a more realistic solution for residential homes looking to tap into greener energy alternatives. It is a sustainable source of energy and has zero carbon footprint, which alleviates climate change. Hydropower plants have huge potential around the world; policymakers are shifting from other energy sources to them. They can be used to produce electricity on a large scale. They are a type of technology that has remained for decades but keeps evolving. In this work, the present status of wind, solar, and hydropower plants with background study has been discussed. The researchers are motivated to discover more by observing the huge potential of renewable power and will perhaps provide an enhanced structure to the renewable power sector. It will inspire researchers to discover more by witnessing the huge potential of wind, solar, and hydropower, possibly giving better shape to renewable power.

Author Contributions: Conceptualization, S.S.D., J.K. and S.D.; methodology, S.S.D. and J.K.; software, S.S.D.; validation, J.K., S.D. and F.S.; formal analysis, S.D. and F.S.; investigation, S.S.D.; resources, J.K., S.D. and F.S.; data curation, F.S.; writing—original draft preparation, S.S.D.; writing—review and editing, J.K. and S.D.; visualization, S.D. and F.S.; supervision, J.K.; project administration, S.D. and F.S.; funding acquisition, F.S. All authors have read and agreed to the published version of the manuscript.

Funding: This research received no external funding.

Data Availability Statement: Not available.

Conflicts of Interest: The authors declare no conflict of interest.

References

1. Rissman, J.; Helseth, J. Technologies and policies to decarbonize global industry: Review and assessment of mitigation drivers through 2070. *Appl. Energy* **2020**, *266*, 114–848. [CrossRef]
2. Clean Power Plan. United States Environmental Protection Agency. 2016. Available online: https://19january2017snapshot.epa.gov/cleanpowerplan/learn-about-carbon-pollution-power-plants_.html (accessed on 10 August 2022).
3. Renewable Source. Centre for Climate and Energy Solution. Available online: https://www.c2es.org/content/renewable-energy/ (accessed on 10 August 2022).
4. Sustainable Development. Department of Economic and Social Affairs. Available online: https://sdgs.un.org/partnerships (accessed on 10 August 2022).
5. Ming, Z.; Kun, Z.; Daoxin, L. Overall review of pumped-hydro energy storage in China: Status quo, operation mechanism and policy barriers. *Renew. Sustain. Energy Rev.* **2013**, *17*, 35–43. [CrossRef]

6. Ardizzon, G.; Cavazzini, G.; Pavesi, G. A new generation of small hydro and pumped-hydro power plants: Advances and future challenges. *Renew. Sustain. Energy Rev.* **2014**, *31*, 746–761. [CrossRef]
7. Mishra, M.K.; Khare, N.; Agrawal, A.B. Small hydropower in India: Current status and future perspectives. *Renew. Sustain. Energy Rev.* **2015**, *51*, 101–115. [CrossRef]
8. Uddin, W.; Zeb, K.; Haider, A.; Khan, B.; ul Islam, S.; Ishfaq, M.; Khan, I.; Adil, M.; Kim, H.J. Current and future prospects of small hydropower in Pakistan: A survey. *Energy Strategy Rev.* **2019**, *24*, 166–177. [CrossRef]
9. Kadier, A.; Kalil, M.S.; Pudukudy, M.; Hasan, H.A.; Mohamed, A.; Hamid, A.A. Pico hydropower (PHP) development in Malaysia: Potential, present status, barriers and future perspectives. *Renew. Sustain. Energy Rev.* **2018**, *81*, 2796–2805. [CrossRef]
10. Yah, N.F.; Oumer, A.N.; Idris, M.S. Small scale hydro-power as a source of renewable energy in Malaysia: A review. *Renew. Sustain. Energy Rev.* **2017**, *72*, 228–239. [CrossRef]
11. Aroonrat, K.; Wongwises, S. Current status and potential of hydro energy in Thailand: A review. *Renew. Sustain. Energy Rev.* **2015**, *46*, 70–78. [CrossRef]
12. Sharma, A.K.; Thakur, N.S. Resource potential and development of small hydro power projects in Jammu and Kashmir in the western Himalayan region: India. *Renew. Sustain. Energy Rev.* **2015**, *52*, 1354–1368. [CrossRef]
13. Sharma, A.K.; Thakur, N.S. Energy situation, current status, and resource potential of run of the river (RoR) large hydro power projects in Jammu and Kashmir: India. *Renew. Sustain. Energy Rev.* **2017**, *78*, 233–251. [CrossRef]
14. Baños, R.; Manzano-Agugliaro, F.; Montoya, F.; Gil, C.; Alcayde, A.; Gómez, J. Optimization methods applied to renewable and sustainable energy: A review. *Renew. Sustain. Energy Rev.* **2011**, *15*, 1753–1766. [CrossRef]
15. Kumar, A.; Kumar, K.; Kaushik, N.; Sharma, S.; Mishra, S. Renewable energy in India: Current status and future potentials. *Renew. Sustain. Energy Rev.* **2010**, *14*, 2434–2442. [CrossRef]
16. Ramírez, F.J.; Villena-Ruiz, R.; Honrubia-Escribano, A.; Pérez-Barroso, A.; Gómez-Lázaro, E. Assessment of different end-of-life strategies for wind power plants under uncertainty. *Energy Convers. Manag. J.* **2022**, *270*, 116–158. [CrossRef]
17. Rediske, G.; Burin, H.; Rigo, P.; Rosa, C.; Michels, L.; Siluk, J. Wind power plant site selection: A systematic review. *Renew. Sustain. Energy Rev. J.* **2021**, *148*, 111–293. [CrossRef]
18. Soares-Ramos, E.P.; de Oliveira-Assis, L.; Sarrias-Mena, R.; Fernández-Ramírez, L.M. Current status and future trends of offshore wind power in Europe. *Energy J.* **2020**, *202*, 117–787. [CrossRef]
19. A Policy Framework for Climate and Energy in the Period from 2020 to 2030. Available online: http://eur-lex.europa.eu/legal-content/EN/TXT/?uri=CELEX:52014DC0015.2014 (accessed on 10 August 2022).
20. Kunz, F. Improving congestion management: How to facilitate the integration of renewable generation in Germany. *Energy J.* **2013**, *34*, 55–78. [CrossRef]
21. Energy Efficiency & Renewable Energy, US Department of Energy. Available online: https://www.energy.gov/eere/ (accessed on 12 August 2022).
22. The Most Profitable Micro Hydropower Plant: Turbulent. Available online: https://www.turbulent.be/ (accessed on 12 August 2022).
23. America's Story from America's Library. Available online: https://www.americaslibrary.gov/jb/gilded/jb_gilded_hydro_1.html (accessed on 12 August 2022).
24. These Are the World's Largest Hydroelectric Dams, World Economic Forum. Available online: www.weforum.org/agenda/2022/12/worlds-largest-hydroelectric-dams-renewable-energy/ (accessed on 15 December 2022).
25. Ministry of Power, Government of India. Available online: https://powermin.nic.in/en/content/faqs-hydropower (accessed on 15 August 2022).
26. Wind Power Engineering & Development. Available online: https://www.windpowerengineering.com/ (accessed on 15 August 2022).
27. National Grid UK. Available online: https://www.nationalgrid.com/ (accessed on 15 August 2022).
28. The Renewable Energy Hub UK. Available online: https://www.renewableenergyhub.co.uk/ (accessed on 15 August 2022).
29. GE. Available online: https://www.ge.com/ (accessed on 15 August 2022).
30. Renewable Energy World. Available online: https://www.renewableenergyworld.com/ (accessed on 16 August 2022).
31. World's Biggest Wind Power Plants, Power Technology. Available online: https://www.power-technology.com/ (accessed on 16 August 2022).
32. Renewable Energy, National Thermal Power Corporation (NTPC). Available online: https://www.ntpc.co.in/en/power-generation/renewable-energy (accessed on 16 August 2022).
33. Edmond Becquerel: The Man Behind Solar Panels, Solenergy Systems Inc. on Facts. Available online: https://solenergy.com.ph/solar-panel-philippines-edmond-becquerel/ (accessed on 16 August 2022).
34. The World's Biggest Solar Power Plant, Power Technology. Available online: https://www.power-technology.com/analysis/the-worlds-biggest-solar-power-plants/ (accessed on 20 August 2022).
35. Global Energy Statistical Yearbook 2022. Enerdata. Available online: https://yearbook.enerdata.net/ (accessed on 20 August 2022).
36. Renewables 2022 Global Status Report, REN21. Available online: https://www.ren21.net/reports/global-status-report/ (accessed on 20 August 2022).

37. How Much of Our Primary Energy Comes from Renewables? Our World in Data. Available online: https://ourworldindata.org/renewable-energy (accessed on 20 August 2022).
38. Indian Population 2022, World Population Review. Available online: https://worldpopulationreview.com/countries/india-population (accessed on 20 August 2022).
39. REFORMS IN URBAN PLANNING CAPACITY IN INDIA–Final Report (September 2021). Available online: https://www.niti.gov.in/sites/default/files/2021-09/UrbanPlanningCapacity-in-India-16092021.pdf (accessed on 20 August 2022).
40. Executive Summary Report, July-2021. Central Electricity Authority, Government of India. Available online: https://cea.nic.in/executive-summary-report/?lang=en (accessed on 20 August 2022).
41. Executive Summary Report, July-2022. Central Electricity Authority, Government of India. Available online: https://cea.nic.in/executive-summary-report/?lang=en (accessed on 20 August 2022).
42. Ministry of Power, Government of India. Available online: https://powermin.nic.in/en/content/power-sector-glance-all-india (accessed on 20 August 2022).
43. Small Hydro Plant. Available online: https://mnre.gov.in/small-hydro/current-status (accessed on 24 August 2022).
44. Oldest Hydropower Plant in India. Available online: https://powermin.gov.in/en/content/faqs-hydropower (accessed on 24 August 2022).
45. Major Hydropower Plants in India. Available online: https://www.mapsofindia.com/maps/india/hydropowerproject.html (accessed on 24 August 2022).
46. Hydropower Generation in India (2021). Available online: https://www.ibef.org/download/Renewable-Energy-July-2021.pdf (accessed on 24 August 2022).
47. Wind Energy. Available online: https://mnre.gov.in/wind/current-status/ (accessed on 24 August 2022).
48. Generation Based Incentive (GBI) Scheme. Available online: https://pib.gov.in/newsite/PrintRelease.aspx (accessed on 24 August 2022).
49. Top Wind Power Plants in India. Available online: https://www.nsenergybusiness.com/features/top-wind-power-farms-india/ (accessed on 24 August 2022).
50. India's First Solar Power Plant Opens in Punjab. 2009. Available online: https://www.indiatoday.in/india/story/indias-first-solar-power-plant-opens-in-punjab (accessed on 24 August 2022).
51. Solar Energy. Available online: https://mnre.gov.in/solar/current-status/ (accessed on 24 August 2022).
52. List of Solar Power Plants in India. Available online: https://www.jagranjosh.com/general-knowledge/list-of-solar-power-plants-in- (accessed on 26 August 2022).
53. Maurovich-Horvat, L.; Boomsma, T.K.; Fleten, S.-E.; Siddiqui, A.S. Transmission and Wind Investment in a Deregulated Electricity Industry. *IEEE Trans. Power Syst.* **2015**, *30*, 1633–1643. [CrossRef]
54. Sood, Y.R.; Singh, R. Optimal model of congestion management in deregulated environment of power sector with promotion of renewable energy sources. *Renew. Energy* **2010**, *35*, 1828–1836. [CrossRef]
55. Cong, R.-G. An optimization model for renewable energy generation and its application in China: A perspective of maximum utilization. *Renew. Sustain. Energy Rev.* **2013**, *17*, 94–103. [CrossRef]
56. Ferreira, H.L.; Garde, R.; Fulli, G.; Kling, W.; Lopes, J.P. Characterisation of electrical energy storage technologies. *Energy* **2013**, *53*, 288–298. [CrossRef]
57. Parastegari, M.; Hooshmand, R.A.; Khodabakhshian, A.; Zare, A.H. Joint operation of wind farm, photovoltaic, pump-storage, and energy storage devices in energy and reserve markets. *Int. J. Electr. Power Energy Syst.* **2015**, *64*, 275–284. [CrossRef]
58. Zahedi, A. A review of drivers, benefits, and challenges in integrating renewable energy sources into the electricity grid. *Renew. Sustain. Energy Rev.* **2011**, *15*, 4775–4779. [CrossRef]
59. Poonpun, P.; Jewell, W.T. Analysis of the Cost per Kilowatt Hour to Store Electricity. *IEEE Trans. Energy Convers.* **2008**, *23*, 529–534. [CrossRef]
60. Chinmoy, L.; Iniyana, S.; Goic, R. Modeling wind power investments, policies and social benefits for deregulated electricity market–A review. *Appl. Energy* **2019**, *242*, 364–377. [CrossRef]
61. Sinsel, S.R.; Riemke, R.L.; Hoffmann, V.H. Challenges and solution technologies for the integration of variable renewable energy sources—A review. *Renew. Energy* **2020**, *145*, 2271–2285. [CrossRef]
62. Eltigani, D.; Masri, S. Challenges of integrating renewable energy sources to smart grids: A review. *Renew. Sustain. Energy Rev.* **2015**, *52*, 770–780. [CrossRef]
63. Khare, V.; Nema, S.; Baredar, P. Solar–wind hybrid renewable energy system: A review. *Renew. Sustain. Energy Rev.* **2016**, *58*, 23–33. [CrossRef]
64. Banshwar, A.; Sharma, N.K.; Sood, Y.R.; Shrivastava, R. Market Based Procurement of Energy and Ancillary Services from Renewable Energy Sources in Deregulated Environment. *Renew. Energy* **2017**, *101*, 1390–1400. [CrossRef]
65. Lund, D.; Lindgren, J.; Mikkola, J.; Salpakari, J. Review of energy system flexibility measures to enable high levels of variable renewable electricity. *Renew. Sustain. Energy Rev.* **2015**, *45*, 785–807. [CrossRef]
66. Banshwar, A.; Sharma, N.K.; Sood, Y.R.; Shrivastava, R. Real-time procurement of energy and operating reserve from Renewable Energy Sources in deregulated environment considering imbalance penalties. *Renew. Energy* **2017**, *113*, 855–866. [CrossRef]
67. Murugananthama, B.; Gnanadassa, R.; Padhy, N. Challenges with renewable energy sources and storage in practical distribution systems. *Renew. Sustain. Energy Rev.* **2017**, *73*, 125–134. [CrossRef]

68. Yin, S.; Wang, J.; Li, Z.; Fang, X. State-of-the-art short-term electricity market operation with solar generation: A review. *Renew. Sustain. Energy Rev.* **2021**, *138*, 110–647. [CrossRef]
69. Rehman, S.; Al-Hadhrami, L.M.; Alam, M.M. Pumped hydro energy storage system: A technological review. *Renew. Sustain. Energy Rev.* **2015**, *44*, 586–598. [CrossRef]
70. Jafari, E.; Soleymani, S.; Mozafari, B.; Amraee, T. Scenario-based stochastic optimal operation of wind/PV/FC/CHP/boiler/tidal/energy storage system considering DR programs and uncertainties. *Energy Sustain. Soc.* **2018**, *8*, 2. [CrossRef]
71. Farzana, D.F.; Mahadevan, K. Performance comparison using firefly and PSO algorithms on congestion management of deregulated power market involving renewable energy sources. *Soft Comput.* **2020**, *24*, 1473–1482. [CrossRef]
72. Askari, M.T.; Kadir, M.Z.A.A.; Tahmasebi, M.; Bolandifar, E. Modeling optimal long-term investment strategies of hybrid wind-thermal companies in restructured power market. *J. Mod. Power Syst. Clean Energy* **2019**, *7*, 1267–1279. [CrossRef]
73. Amjith, L.R.; Bavanish, B. A review on biomass and wind as renewable energy for sustainable environment. *Chemosphere* **2022**, *293*, 133579. [CrossRef] [PubMed]
74. Dey, S.; Sreenivasulu, A.; Veerendra, G.T.N.; Rao, K.V.; Babu, P.S.S.A. Renewable energy present status and future potentials in India: An overview. *Chemosphere* **2022**, *293*, 100006. [CrossRef]
75. Liu, J.; Song, D.; Li, Q.; Yang, J.; Hu, Y.; Fang, F.; Joo, Y.H. Life cycle cost modelling and economic analysis of wind power: A state of art review. *Energy Convers. Manag.* **2023**, *277*, 116628. [CrossRef]
76. Al-Majidi, S.D.; AL-Nussairi, M.K.; Mohammed, A.J.; Dakhil, A.M.; Abbod, M.F.; Al-Raweshidy, H.S. Design of a Load Frequency Controller Based on an Optimal Neural Network. *Energies* **2022**, *15*, 6223. [CrossRef]
77. Al-Majidi, S.D.; Altai, H.D.S.; Lazim, M.H.; Al-Nussairi, M.K.; Abbod, M.F.; Al-Raweshidy, H.S. Bacterial Foraging Algorithm for a Neural Network Learning Improvement in an Automatic Generation Controller. *Energies* **2023**, *16*, 2802. [CrossRef]
78. Al-Majidi, S.D.; Abbod, M.F.; Al-Raweshidy, H.S. Maximum Power Point Tracking Technique based on a Neural-Fuzzy Approach for Stand-alone Photovoltaic System. In Proceedings of the 2020 55th International Universities Power Engineering Conference (UPEC), Turin, Italy, 1–4 September 2020. [CrossRef]
79. Smolarz, A.; Lezhniuk, P.; Kudrya, S.; Komar, V.; Lysiak, V.; Hunko, I.; Amirgaliyeva, S.; Smailova, S.; Orazbekov, Z. Increasing Technical Efficiency of Renewable Energy Sources in Power Systems. *Energies* **2023**, *16*, 2828. [CrossRef]

processes

Article

Life Cycle Assessment and Cumulative Energy Demand Analyses of a Photovoltaic/Thermal System with MWCNT/Water and GNP/Water Nanofluids

Gülşah Karaca Dolgun [1], Meltem Koşan [2], Muhammet Kayfeci [3], Aleksandar G. Georgiev [4,*] and Ali Keçebaş [1]

1 Department of Energy Systems Engineering, Technology Faculty, Muğla Sıtkı Koçman University, 48000 Muğla, Turkey
2 Department of Energy Systems Engineering, Elbistan Engineering Faculty, Kahramanmaras Istiklal University, 46100 Kahramanmaras, Turkey
3 Department of Energy Systems Engineering, Technology Faculty, Karabük University, 78100 Karabük, Turkey
4 Department of General engineering, University of Telecommunications and Posts, 1 Akad. Stefan Mladenov str., 1700 Sofia, Bulgaria
* Correspondence: ageorgiev@gmx.de

Abstract: The global climate crisis has led society toward cleaner energy sources. Another reason is the limited reserves of fossil energy resources. Efforts to increase the efficiency of photovoltaic modules (PVs) have gained momentum. The high temperature is the biggest factor causing a decrease in the efficiency of PVs. In this study, a commercial PV was cooled with distilled water, a multiwalled carbon nanotubes (MWCNT)/water mixture, and a graphene nanoplatelets (GNP)/water mixture. The environmental impact of electricity, total energetic efficiency, energy payback time, energy return on investment, and embodied energy of the PV/thermal (PV/T) system were compared using life cycle assessment and cumulative energy demand. The electrical efficiency of the PV/T changed between 13.5% and 14.4%. The total efficiency of PV/T changed between 39.5% and 45.7%. The energy returns on investment were 1.76, 1.80, and 1.85 for PV/T-distilled water, the PV/T-MWCNT/water mixture, and the PV/T-GNP/water mixture, respectively. Moreover, the embodied energy evaluation values were 3975.88 MJ for PV/T-distilled water, 4081.06 MJ for the PV/T-MWCNT/water mixture, and 4077.86 MJ for the PV/T-GNP/water mixture. The main objective of this research was to study the energy and environmental performances of PVs cooled with different nanofluids and draw general conclusions about the applicability of these systems.

Keywords: cooling PV module; life cycle assessment; cumulative energy demand; total energy requirement; energy payback time

Citation: Dolgun, G.K.; Koşan, M.; Kayfeci, M.; Georgiev, A.G.; Keçebaş, A. Life Cycle Assessment and Cumulative Energy Demand Analyses of a Photovoltaic/Thermal System with MWCNT/Water and GNP/Water Nanofluids. *Processes* **2023**, *11*, 832. https://doi.org/10.3390/pr11030832

Academic Editors: Ferdinando Salata and Virgilio Ciancio

Received: 3 February 2023
Revised: 2 March 2023
Accepted: 7 March 2023
Published: 10 March 2023

1. Introduction

In recent years, the use of alternative energy sources has become more important than using fossil-based fuels due to environmental problems, such as climate change and global warming. With the advancement of technology, electricity consumption is also increasing rapidly. Solar photovoltaic (PV) technologies are considered the cleanest alternative source for electricity generation because their greenhouse gas emissions into the atmosphere are lower than those of other sources [1,2]. However, a certain amount of energy is consumed in the production, installation, processing, and maintenance stages of PV technologies. Life cycle assessment (LCA) is considered an index by which the environmental impacts of PVs can be calculated throughout the entire life cycle, starting from the acquisition of raw materials to processing, production, use, end-of-life, and disposal. The environmental sustainability of PV and its systems can be determined using LCA methodology [3,4]. LCA study is applied in terms of the environment and energy profiles. While different LCA methodologies are used for environmental profiling, energy profiling is performed using the cumulative energy demand (CED). Likewise, embedded energy evaluation is essential to understanding this energy profile assessment and is determined by the CED methodology [5].

In the literature, the emission reduction and environmental impact have been examined for the environmental profiles of PVs, concentrated PVs, PV/thermal (PV/T) systems, and conventional solar systems. Carnevale et al. [6] compared the environmental profile of PV modules with solar hot water systems using LCA analysis. The energy payback times of silicon-based PV modules, thin-film PV modules, and hot water systems are 2.6, 1.0, and 1.2 years, respectively. Ehtiwesh et al. [7] performed an exergetic LCA analysis of a concentrated solar power plant (50 MW_e capacity) from cradle to grave using with the CED and Eco-indicator 99 methods. The percentages of the Human Health, Resource, and Ecosystem Quality damage categories were 69% (14.4 MPoints), 24% (5 MPoints), and 7% (1.4 MPoints), respectively. The highest impact between materials was seen in steel, at 9.77 MPoints (46.9%), followed by molten salt at 5.19 MPoints (24.9%) and synthetic oil at 4.27 MPoints (21%). Santoyo-Castelazo et al. [8] carried out an LCA study of a PV system with an installed power of 3 kW_p connected to the grid. The carbon footprint was calculated to be 47.156 g CO_2.eq/kWh and the estimated normalized greenhouse gas emissions were between 20 and 90 g CO_2.eq/kWh. It was observed that most of the environmental loads arose from the manufacture of the materials required for the PV module. Krebs-Moberg et al. [9] analyzed the effects on the lifetimes of multi-crystalline silicon, organic thin-film, and perovskite thin-film PV modules. They reported that the manufacture and use of multi-crystalline silicon PV modules had the greatest impacts across all hazard categories. The recycling process reduced the environmental impacts of all module types. Li et al. [10] designed and tested a semi-transparent PV window. An LCA analysis was performed by investigating the energy and environmental benefits of the system. The energy payback period and greenhouse gas payback period of this system were 13.8 years and 10.4 years, respectively. Rao et al. [11] investigated multi-crystalline PV/T systems installed at open-field and rooftop locations. The energy payback time was between 6.53 and 11.38 years, with a positive energy yield. For a 2 m^2 single-crystalline PV/T system at 1200 W/m^2, the energy payback times were 6.93 and 7.59 years for the rooftop and open field locations, respectively. The embodied energy values for a single-crystalline PV/T system were 3178 kWh/m^2 on a rooftop and 3478 kWh/m^2 in an open field for a 25-year lifetime period with a 5-year battery-replacement cycle. Li et al. [12] used a multi-index life cycle assessment (LCA) technique that included economic, energy, and environmental variables for bifacial photovoltaic (BPV) modules installed on buildings. Compared with the mono-facial PV modules, the BPV module increased the power generation performance by 10.7–12.7%. The EPBT of the BPV solar house and inclined roof decreased by 5.7% and 7.4%, respectively, and the GPBT of the BPV solar house and inclined roof decreased (2.3 and 1.7 years) according to the mono-facial PV module. BPV modules produced more power and had a higher return on investment (ROI) of more than 10.7%, despite costing 5% more than a mono-facial PV module. Goel et al. [13] installed a 3.4 kW rooftop stand-alone photovoltaic system (SAPV) in India and performed life cycle cost and energy analysis of the system by the present value method and embodied energy basis, respectively. The internal rate of return (IRR), life cycle cost of energy (LCOE), benefit–cost ratio (BCR), and energy payback time (EPBT) of the SAPV were 2.02, INR 5.40/kWh, 0.57, and 4.61 years, respectively. The IRR, LCOE, BCR, and EPBT of the grid-connected PV system were 13.42%, INR 3.17 per kWh, 1.11, and 3.78 years, respectively. Jurcevic et al. [14] investigated the performance of a photovoltaic-thermal (PVT) collector. Water was used to cool the PV, and hot water was stored in a phase-change material (PCM) container, where pork fat was used as the PCM. In August, cooling increased the PV's electrical efficiency from 11.7–12.0% to 12.0–12.4%. The increase in the PVT's electrical efficiency was negligible. The total energy generated by the PVT collector would have a levelized cost of between EUR 0.056 and 0.083 per kWh.

The embodied energy of the solar thermal collectors and PVs and/or PV/T systems for energy profiling has been investigated in the literature. Lamnatou et al. [15] performed the LCA analysis of building-integrated solar thermal collectors according to the embodied energy and embodied carbon methodologies. By recycling, the embodied energy values of the collectors were reduced to around 0.4–0.5 GJ/m^2 from 3 GJ/m^2 (without recycling). With recycling, the embodied carbon values were reduced to around 0.02–0.03 t CO_2.eq/m^2 from 0.16 t CO_2.eq/m^2. It was seen that recycling led to remarkable reductions in all parameters. Hassani et al. [16] modeled a PV/T system, theoretically calculated the life cycle exergy of the nanofluid-based system, and compared it with standard PV and PV/T systems. The embodied emissions of the PV/T system were between 691 and 896 kg CO_2.eq/m^2, and the amount of CO_2 emissions prevented was 448 kg CO_2.eq/m^2/year in a nanofluid-based PV/T system. The exergy payback time of the PV/T system changed between 2.0 and 2.58 years. Ren et al. [17] assessed the dynamic life cycle cost and environmental impact of residential PV systems in a real prototype house in Boston. Two systems, such as grid-connected (GC) and standalone (SA) solar PV systems, were compared according to their life cycle cost, LCA, and CED. The LCC savings for the SA were $754.9 in 2018 with 18.5 years of investment payback time when 40 panels and 40 batteries were used, and the LCC savings for the GC were $1739.4 with 16.8 years of investment payback time. The life cycle reductions in the SA and GC were 2.1 TJ and 2.3 TJ of CED, respectively. Bahlawan et al. [18] performed the LCA analysis of energy systems for residential applications. The CEDs of the PV, solar thermal collector, and hot water storage (100 L capacity) were calculated. Their CED values were 1.53 GJ/m^2, 4.69 GJ/m^2, and 1.27 GJ, respectively. The above studies indicate that, despite the advanced development of PV technology, there has not been much focus on the energy and environmental profiles of PV and PV-related systems. Herrando et al. [19] performed a life cycle assessment (LCA) of a solar combined cooling, heating, and power (S-CCHP) system that provides electricity, domestic hot water, space heating, and cooling. An S-CCHP system, PV system, and grid-based system were compared. The environmental impact of the S-CCHP system was 4.48 kPts and 82.4 tons of CO_2.eq, which is equal to half that of the grid-based system, according to the ReCiPe 2016 Endpoint (H/A) and the IPCC GWP 100a methods. The environmental impact of the PV system was 30% lower than that of the grid-based system. Morini et al. [20] investigated the carbon footprint (CF) and embodied energy (EE) of photovoltaic and wind power plants. The CF and EE of the photovoltaic plant (monocrystalline silicon cell) were 16.21 gCO_2/kWh and 0.0638 kWh/kWh, respectively.

The electrical performance of PV systems can be improved by using nanofluids with high thermal conductivity for PV module cooling [21]. The nanoparticles dispersed in the base fluid increase the contact surface area and thermal conductivity of the fluid [22]. According to the literature studies, it can be seen that the use of carbon- and metal-based nanoparticles in solar collectors and the cooling of PV modules and/or PV/T systems is common. The effects of cooling on the electrical and thermal efficiencies of PV with various working fluids have been sought. Fayaz et al. [23] numerically and experimentally investigated the performance of a PV/T panel using multiwalled carbon nanotubes (MWCNT)/water nanofluid to increase its overall efficiency. An improvement in the electrical efficiency of 10.72% was observed in the PV/T panel using a 0.75% MWCNT/water mixture with a flow rate of 120 L/h. The thermal efficiency was also calculated to be 79.1%. In another similar study, Abdallah et al. [24] conducted an experiment in a PV/T system using MWCNT/water nanofluid. The best system efficiency was obtained at a 0.075% volume concentration, and a temperature drop of 12 °C was achieved in the PV module. As a result, the total system efficiency was found to be 83.26%. Alous et al. [25] experimentally investigated the effects of using MWCNT/water and graphene nanoplatelets (GNP)/water at a 0.5 wt.% concentration on the performance of a PV/T system. They reported that the overall energy efficiency of PV/T systems increased by 53.4% for distilled water, 57.2% for MWCNT/water, and 63.1% for GNP/water. Sangeetha et al. [26] experimentally tested the effects of three different nanofluids, MWCNT/water, Al_2O_3/water, and TiO_2/water, on the PV/T system. It was found that the MWCNT/water nanofluid exhibited superior thermal conductivity and better physical properties than the Al_2O_3/water and TiO_2/water nanofluids. In the PV/T systems using MWCNT/water, Al_2O_3/water, and TiO_2/water nanofluids, 47%, 33%, and 27% increases in electrical efficiency and 48%, 37%, and 36% decreases in PV temperature, respectively, were obtained. In addition to the positive physical, thermal, and flow properties of nanofluids, negative effects, such as precipitation, surface erosion, clogging in narrow passages, and increased pressure drop, make them important in the design of a PV/T system. Therefore, many studies have been carried out for each nanoparticle and its concentration.

This study aimed to submit a detailed experimental and environmental study of the entire lifetime of an energy system consisting of PV modules, the cooling unit behind the PV module (PV/T), the nanofluid and water storage tanks, and the pump. To the authors' knowledge, this is the first study to combine experimental work with energy and environmental profiles for LCA and CED methodologies for a PV/T system using nanofluids, such as MWCNT/water and GNP/water. Thus, the gaps in the literature are filled with the following scientific research innovations:

- A comprehensive LCA analysis from cradle to grave of the PV and PV/T systems was conducted;
- PV and PV/T experiments were conducted under the same environmental conditions to understand the differences in total energy efficiency;
- The use of three different working fluids: distilled water, MWCNT/water mixture, and GNP/water mixture;
- Important indicators, such as the efficiency, energy return factor, energy payback time, and the environmental impact of electricity generation were presented;
- How energy systems whose environmental impact is thought to be close to zero increase their environmental impact was demonstrated.

2. Experimental Setup and Experimental Procedure

To achieve the purpose of this study, an experimental setup was developed in the energy labs of Karabük University, Karabük, Turkey. A schematic representation of the experimental setup is shown in Figure 1. As can be seen in these figures, the considered experimental setup consisted of a PV module (reference), a PV/T system, a storage tank with a spiral-coil heat exchanger (30 L of water), a nanofluid tank (6 L), a pump, a support system, and other parts.

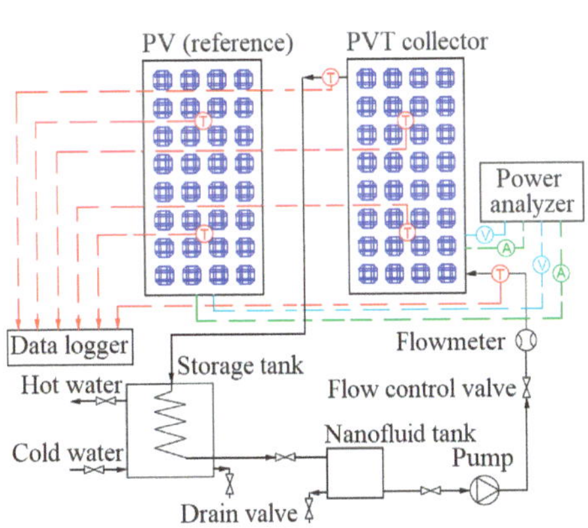

Figure 1. Flow chart of the experimental setup for the PV module and PV/T collector.

The PV/T system consisted of a mono-crystalline silicon module with 41.2 × 33.6 cm dimensions and a heat extraction unit laminated with the back surface of the PV module to provide thermal contact; two types of experimental setup were investigated accordingly: a PV/T system (see Figure 1 right) and an only PV module (same features as in the PV/T system; see Figure 1 left). The flat plate PV modules had the following technical features: a maximum power of 40 W, an open circuit voltage of 22.1 V, a maximum power voltage of 18 V, a short circuit current of 2.58 A, a maximum power current of 2.22 A, and a thermal absorptivity of 0.7. In the PV/T system, the heat extraction unit consisted of a serpentine heat exchanger, which was produced by soldering a serpentine copper tube of 1.0 cm and 0.8 cm outer and inner diameter, respectively, to a copper absorber plate of $67.5 \times 38 \times 0.07$ cm. The total length of the copper tube needed was about 302 cm. To investigate the PV's electrical performance, the heat extraction unit was assembled on the back surface of the PV panel.

In the spiral-coil heat exchanger and the heat extraction unit, distilled water, a multi-walled carbon nanotubes (MWCNT)/water mixture, and a graphene nanoplatelets (GNP)/water mixture were used as carrying fluids. The selected carrier fluids could work with a closed circuit flow in forced (pumped) operation, as illustrated in Figure 1. For this reason, a pump (model: RS25/4G-130, Nova Company) was used for the circulation of the fluids from the 6 L nanofluid tank to the heat extraction unit of the PV/T system and then the spiral-coil heat exchanger of the storage tank. The experimental setup requires approximately 3.5 L of fluid to operate.

GNP and MWCNT nanoparticles were purchased from Nanografi Co., Ltd. and prepared at 0.5% wt. concentrations. The aqueous dispersion of the single-layer graphene nanoplatelets had more than 99.3% wt. purity, 500–1200 m^2/g specific surface area, 1–12 μm diameter, and 0.55–1.2 nm thickness. MWCNTs' aqueous dispersion had more than 96% purity and was 8–35 μm in length and 18–28 nm in outside diameter. Transmission electron microscopy (TEM) images of the GNP and MWCNTs are displayed in Figure 2. No sedimentation was visually observed throughout and after the experimentation.

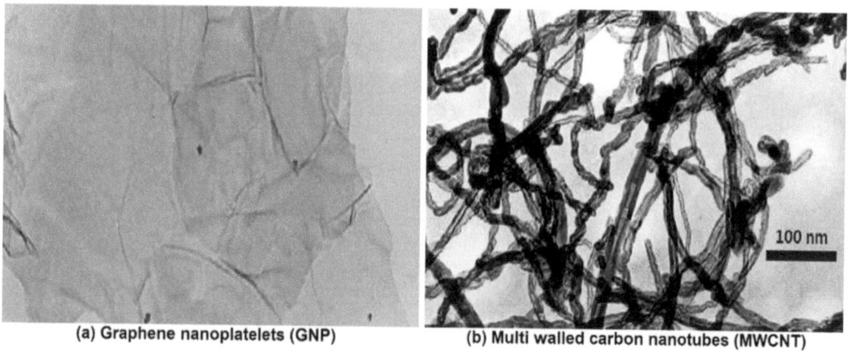

Figure 2. TEM images of nanoparticles and the nanofluids: (**a**) GNP and (**b**) MWCNT [25].

The experimental setup, consisting of a PV module and PV/T system, was equipped with measuring devices and was prepared for the separate use of water and nanofluids to cool the PV/T system. The experiments were conducted at Karabük University, Karabük, Turkey (41.19° N latitude and 32.62° E longitude) in August and September, daily, from 9:00 a.m. to 5:00 p.m. For the fixed-position PV module and PV/T system, the tilt angle was determined to be 30° in the south direction. Preliminary studies were carried out using water for both the stable regime behaviors of the system and the calibration/accuracy of the measurements. The mass flow rates of water as a carrying fluid were 0.5, 1.0, and 2.0 L/min. In the experimental procedure, the experimental days were also extended as both water and nanofluids could not be run at the same time. To replace the carrying fluid, the heat extraction unit of the PV/T system, the nanofluid tank, spiral-coil heat exchanger, and pipeline were cleaned using water and air. In the process of cleaning with water, the color of the water was observed. The experiments under similar environmental conditions were taken care of. As experiments were repeated, the average values were used for this study.

3. Analysis Methodology

3.1. Thermal Analysis

The incident current, I in A, and voltage, V in V, values were measured on the system; multiplying these values equaled the electrical power, P in W, thus, the electrical powers of the PV module and PV/T system were calculated with Equation (1).

$$\dot{P} = I \times V \quad (1)$$

The thermal cooling power, $\dot{Q}$ in W, taken from the PV/T system was determined by:

$$\dot{Q} = \dot{m}C_p(T_o - T_i) \quad (2)$$

where C_p is the coolant-specific heat in J/kgK, $\dot{m}$ is the mass flow rate of the coolant in kg/s, and T_o and T_i are outlet and inlet temperatures of the coolant, respectively.

A PV/T's energy efficiency indicates the amount of thermal and electrical energy that the PV/T takes from solar radiation. Thus, the electrical (η_{el}) and thermal (η_{th}) efficiencies are calculated as:

$$\eta_{el} = \frac{\dot{P}}{I_R \times A_{PV}} \quad (3)$$

$$\eta_{th} = \frac{\dot{Q}}{I_R \times A_{th}} \quad (4)$$

where A_{PV} and A_{th} are the surface areas of the PV module and PV/T module in m^2, respectively, and I_R denotes the total incident solar radiation in W/m^2. The total energy efficiency of the PV/T system can be calculated by Equation (5) [27].

$$\eta_{tot} = \eta_{th} + r\eta_{el} \quad (5)$$

where r is the packing factor and can be obtained by dividing the surface area of the PV module by the surface area of the PV/T system as A_{PV}/A_{th}. In this study, r was assumed to be 1 (actually 0.999) as $A_{PV} = A_{th}$.

The methodology proposed by Kline and McClintock [28] was used to evaluate the accuracy of the data collected in the experimental study. The uncertainties in the measured parameters over the range of experiments were calculated by using the uncertainty data listed in Table 1. The maximum uncertainties in the thermal and electrical efficiencies were 1.3% and 1.0%, respectively.

Table 1. Experimental uncertainty values of the measurement instruments [25].

Equipment	Parameter	Maximum Experimental Uncertainty
Pyranometer (MS-602, EKO instruments)	Solar irradiance	±5.2 W/m^2
K-types thermocouple	Temperature	±0.6 °C
Flowmeter (YF-S201, Sea company)	Volumetric flow rate	±0.003 L/min
Data collecting board (USB TC-08, Pico)	Voltage Current	±0.06 V ±0.02 A

3.2. Life Cycle Assessment

The main purpose of this study was to obtain a detailed energy and environmental profile of the solar-focused PV/T system and the use of nanofluids. Therefore, life cycle assessment (LCA) and cumulative energy demand (CED) analyses were performed to quantify the energy consumption and environmental impact from the production, use, disposal, and operational phases of various components of the PV/T system with and without nanofluid. The system boundaries used for the analyses are illustrated in Figure 3. This study took the "cradle-to-grave" approach by limiting the analyses to the sub-processes of construction materials, their weight, and disposal processing. All data sources used in the analysis were selected according to their reliability. Cutting-edge processes/technologies were considered for all intermediate production stages of the components of the experimental setup presented in Figure 1, as well as for the raw materials used.

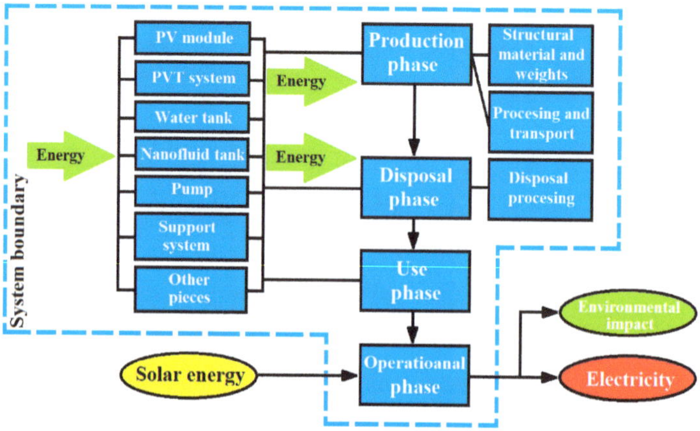

Figure 3. Flow chart of the experimental setup for the PV module and PV/T collector.

The environmental impact of a product (production, processing, maintenance, disposal, etc.) is considered in LCA analysis, which is a useful tool for evaluating the environmental impact of a system [29,30]. Different LCA methodologies can be used for environmental impact assessment. For the environmental profile, in this study, Eco-indicator 99 was used because it was open to the literature. The data were cross-checked with the data accompanying the SimaPro software in addition to the tables of the Eco-indicator 99. For the energy profile, the energy return factor (ERF) and energy payback period (EPT) with the CED analysis were used.

LCA analysis consists of an inventory analysis in parts, including the scope, target, assessment of impacts, and interpretation. The collection of inventories of the main flows was carried out according to the international standard approach [31,32]. It included the collection of input–output data on material and energy streams about the experimental setup. It was necessary to collect information about the sizes, weights, core material, manufacturing process of the system and its components, and scrap output of all of the components needed to assemble the experimental setup. Thus, the experimental setup, consisting of a PV module and PV/T system, was basically divided into six components (e.g., the PV/T system, a water tank, a nanofluid tank, a pump, a support system, and other parts), as shown in Table 2. Detailed information about the structural materials and weights of six components is listed in Table 2. In this study, it was assumed that there was no chemical reaction between the components. For environmental impact evaluation, Eco-indicator 99 was chosen as a quantitative indicator. The indicator's results were reported as eco-indicator points (pts). The points per unit and energies per unit values used in the LCA and CED analyses for the material, process, and disposal phases, respectively, are given in Table 3. Using Tables 2 and 3, the environmental impact of the kth component, $\dot{Y}_k$, can be calculated by:

$$\dot{Y}_k = \dot{Y}_k^{CO} + \dot{Y}_k^{OM} + \dot{Y}_k^{DI} \quad (6)$$

where the subscripts CO, OM, and DI denote construction, operation/maintenance, and disposal/dismantling, respectively.

Table 2. Components and their material weights of the whole PV/T system.

PV/T System (0.23146 m^2)		Water (Storage) Tank (30 L)		Nanofluid Tank (6 L)		Pump (2.55 kg)		Support System		Other Parts	
Material	**Mass (kg)**	**Material**	**Mass (kg)**	**Material**	**Mass (kg)**	**Material**	**Mass (kg)**	**Material**	**Mass (kg)**	**Material**	**Mass (kg)**
Solar glass	2.93	Main frame (galvanized iron)	2.35	Main frame (galvanized iron)	2.76	Copper	0.20	Main frame (galvanized iron)	7.15	Copper	1.25
PV cell	0.02	Insulation (PUR)	1.03	Insulation (PUR)	1.23	Steel	0.68			Plastic tube	2.55
Back sheet	0.13	Water	30.17	Water	5.982	Cast iron	1.63			Insulation (PUR)	0.97
Only PV module	3.08					Aluminum	0.03				
PV frame (aluminum)	4.32					Plastic cover	0.01				
Tube (copper)	3.525										
Sheet (copper)	3.525										
Main frame (galvanized iron)	2.33										
Insulation (PUR)	1.73										
Water	1.85										
Only thermal system	17.28										
Total weight	**20.36**		**33.55**		**9.97**		**2.55**		**7.15**		**4.77**

Table 3. The points and energies per unit in kg used in the LCA and CED analyses.

Components	Indicators for the LCA Analysis			Energy for the CED Analysis	Processing Details
	Material (mPts/kg)	**Process (mPts/kg)**	**Disposal (mPts/kg)**	**Energy per Unit (MJ/kg)**	
PV module					
Solar glass (40.4%)	58.00	0.00	2.20	15.00	-
PV cell (2%)	58.00	10.00	2.20	39714.50	Electricity roof, packing glass
Backsheet (Polyethylene) (1.2%)	360.00	35.90	−1.00	32.88	Blow foil extrusion, injection molding, milling/turning/drilling, and pressure forming
PV/T system					
Tube (Copper) (23%)	1400.00	72.00	0.00	57.00	Extruction
Water (6%)	0.03	0.00	0.00	9.20	-
Insulation (PUR) (5.6%)	420.00	35.90	3.10	101.50	Blow foil extrusion, injection molding, milling/turning/drilling, and pressure forming
Frame (aluminum) (14.1%)	780.00	74.70	−23.00	218.00	Extruction, bending, shearing/stamping, spot welding

Table 3. *Cont.*

Components	Indicators for the LCA Analysis			Energy for the CED Analysis	Processing Details
	Material (mPts/kg)	Process (mPts/kg)	Disposal (mPts/kg)	Energy per Unit (MJ/kg)	
Galvanized iron (%0.5 zinc coating steel) (7.6%)	241.70	243.50	−5.61	34.80	Sheet production, shearing/stamping, bending, and band zinc coating
Water tank					
Galvanized iron (%0.5 zinc coating steel)	241.70	243.50	−5.61	34.80	Sheet production, shearing/stamping, bending, and band zinc coating
Insulation (PUR)	420.00	35.90	3.10	101.50	Blow foil extrusion, injection molding, milling/turning/drilling, and pressure forming
Water	0.03	0.00	0.00	9.20	
Nanofluid tank					
Galvanized iron (%0.5 zinc coating steel)	241.70	243.50	−5.61	34.80	Sheet production, shearing/stamping, bending, and band zinc coating
Insulation (PUR)	420.00	35.90	3.10	101.50	Blow foil extrusion, injection molding, milling/turning/drilling, and pressure forming
Water	0.03	0.00	0.00	9.20	
Pump					
Copper (8%)	1400.00	72.00	−23.00	57.00	Extruction
Steel (26.5%)	86.00	30.00	0.00	35.40	Sheet production, shearing/stamping, and bending
Cast iron (64%)	240.00	5.30	0.00	34.80	Heat gas (industrial furnace)
Aluminum (1%)	780.00	74.70	−23.00	218.00	Extruction, bending, shearing/stamping, and spot welding
Plastic cover (0.5%)	380.00	35.90	3.10	77.70	Blow foil extrusion, injection molding, milling/turning/drilling, and pressure forming
Support structure					
Galvanized iron (%0.5 zinc coating steel)	241.70	243.50	−5.61	34.80	Sheet production, shearing/stamping, bending, and band zinc coating
Other pieces					
Plastic tube (53.5%)	240.00	35.90	3.10	101.50	Blow foil extrusion, injection molding, milling/turning/drilling, and pressure forming
Copper (26.2%)	1400.00	72.00	0.00	57.00	Extruction
Insulation (PUR) (20.3%)	420.00	35.90	3.10	101.50	Blow foil extrusion, injection molding, milling/turning/drilling, and pressure forming

Note: The energies per unit for the graphene nanoplatelets and MCWNT are 260.41 and 295 MJ/kg, respectively.

The most important parts in the manufacture of the PV/T system are the production of the monocrystalline silicon module and the heat extraction unit. In the production of PV modules, silicon, glass, and polyethylene materials are mainly used. In addition, processes such as blow foil extrusion, injection molding, milling/turning/drilling, and pressure forming, are performed for silicon production, mono-Si wafer, cell production, and module assembly operations. The processes performed for other components are described in detail in Table 3.

3.3. Cumulative Energy Demand

The energy profile was evaluated using the cumulative energy demand (CED) methodology for the installation phase. Two indicators were used for energy profile assessment throughout the entire system's life cycle: the energy return factor (ERF) and the energy payback time (EPT). CED is a methodology for measuring indirect and direct energy use in MJ units throughout the life cycle of a process or product, including the energy consumption during the disposal, production, and extraction of the materials. To evaluate the life cycle energy of a product, the sum of three energy terms is considered: embodied energy (consisting of the energy consumed during the production phase and energy used for maintenance and healing during the operation phase), operational energy, and destruction energy. The energies per unit for the material, process, and disposal phases in the CED analysis are provided in Table 3. The CED value for each component was determined by multiplying the weight of the component material presented in Table 2 and the relevant energy per unit listed in Table 3. The CED value of the overall system was then found by summing the CED values obtained for each component. The CED values obtained for the PV module and the PV/T system were compared with the results obtained from the SimaPro software. As a result, determining the CED for the whole life cycle energy assessment of a given system enabled the calculation of the ERF and EPT.

The ERF is defined as the ratio of the energy produced by a system during its entire life to the CED of the system (see Equation (7)). In other words, it shows how many times the energy required to produce a system is generated. In the case of electricity generation technologies, the ERF requires the comparison of the electricity produced with the amount of primary energy used at different life cycle stages. The ERF (dimensionless) is calculated as the ratio of supplied energy to the energy costs and is given as follows:

$$\mathrm{ERF} = \frac{\mathrm{E_{Global}}}{\mathrm{CED}} \tag{7}$$

where $\mathrm{E_{Global}}$ and CED denote the total amount of energy produced over the life of a system and the energy input (cumulative energy demand) of a system/total energy demand, respectively. If the ERF value is less than 1, it means that there will be no net energy output during the life of the system. Generally, it should be greater than 1.

The EPT is described as the duration necessary for a system to produce the same amount of energy as that used to produce the system itself. It is obtained by Equation (8), and its unit is years.

$$\mathrm{EPT} = \frac{\mathrm{N}}{\mathrm{ERF}} \tag{8}$$

where N denotes the lifetime in years. It was assumed to be 25 years at all stages of the analyses.

4. Results and Discussion

In this study, the performance, energy, and environmental profiles of a PV module and a PV/T system cooled by nanofluids were evaluated. Therefore, real-time data were collected from the experimental setup for heat carrier fluids, such as distilled water, multi-walled carbon nanotubes (MWCNT)/water mixture, and graphene nanoplatelets (GNP)/water mixture in the PV/T system. Thermal, LCA, and CED analyses were performed using validated data to compare performance from energetic and environmental perspectives.

The temperature variations in the environment, PV, and PV/T according to different flow rates of distilled water are displayed in Figure 4. As shown in Figure 4, the temperature of the PV/T system with different flow rates was below the PV module temperature throughout the day. While the ambient temperature was 29.1 °C on a daily average, the temperature in the PV reached a maximum of 55.8 °C and an average of 51 °C. The difference between the PV and ambient temperatures was 21.9 °C on average. The lowest PV temperature in the PV/T system occurred at a flow rate of 2 L/min for an average of 37.9 °C. It was followed by the temperature at the flow rate of 1 L/min, which was almost as high as its temperature (about 38.1 °C). However, the highest PV temperature occurred at a flow rate of 0.5 L/min, with an average value of 39.2 °C. It can be seen from Figure 4 that passing the heat-carrying fluid (here, water) through the heat-extraction unit of the PV/T system caused the temperature of the PV module to decrease. In addition, Pantzali et al. [33] reported that the improvement in heat transfer with nanofluids was clearer at lower Reynolds numbers. Therefore, this article presents the results obtained at a flow rate of 0.5 L/min.

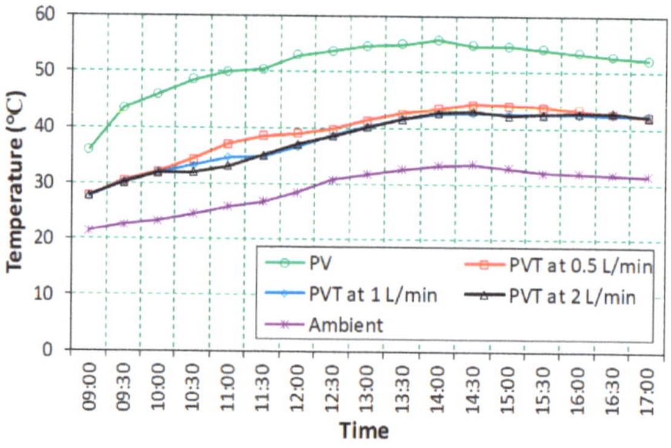

Figure 4. Changes in temperature at different flow rates for distilled water.

The experiment period was extended because different heat-carrying fluids were used in the experimental setup. In the experiments, the PV module and the PV/T system were placed side-by-side and at the same angle to conduct comparisons under the same conditions. To observe the performance of the PV/T system, it was very important that both were exposed to solar irradiance, especially at the same angle. Therefore, the instantaneous changes in ambient temperature and solar radiation for the most suitable days when different carrying fluids were used are presented in Figure 5. The average solar radiation and ambient temperature were, respectively, 727.35 W/m^2 and 29.1 °C for the PV/T system with distilled water, 702.06 W/m^2 and 28.6 °C for the system with the MWCNT/water mixture, and 693.82 W/m^2 and 28.4 °C for the system with the graphene nanoplatelets/water mixture. According to Figure 5, all three experiments were carried out under similar environmental conditions, as their values were close to each other. The experiment day was selected when both values of the PV/T system with distilled water were higher compared with the other two systems. In addition, its radiation value showed a slight increase after 13:00 pm. In addition, the days when both the ambient temperature and solar radiation of the PV/T system with the graphene nanoplatelets/water mixture were lower than those of the system with the MWCNT/water mixture were selected.

The power rate changes generated by the PV module and the PV/T system throughout the experiments are presented in Figure 6. The average power rates for the PV and PV/T modules were, in the same order, 22.11 W and 24.33 W for the PV/T system with distilled water, 22.62 W and 24.84 W for the system with the MWCNT/water mixture, and 23.14 W and 25.32 W for the system with the graphene nanoplatelets/water mixture. As shown in Figure 6, the highest power generation rate occurred in the PV/T system with a graphene nanoplatelets/water mixture. It was followed by the system with the MWCNT/water mixture and the system with distilled water. In Figure 6, can be seen that the PV/T systems with solar radiation from the highest to the lowest were the systems with distilled water, with the MWCNT/water mixture, and with the graphene nanoplatelets/water mixture. However, it can be seen from Figure 6 that the reverse order of this situation occurred for the amount of power produced. The highest power generation rate occurred in the PV/T system with a graphene nanoplatelet/water mixture.

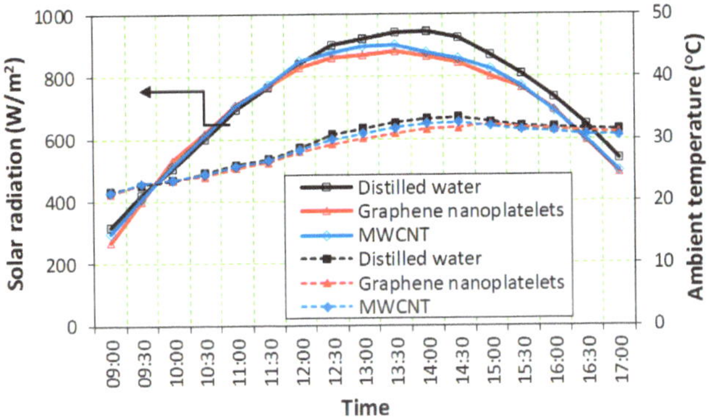

Figure 5. Changes in solar radiation and ambient temperature in the experiments selected with the same conditions at a flow rate of 0.5 L/min.

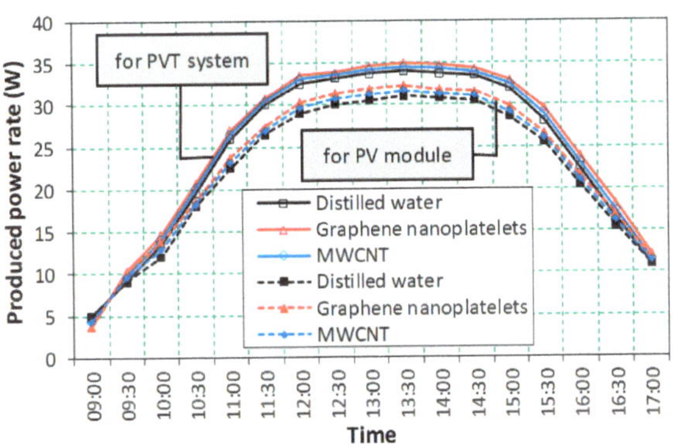

Figure 6. Change in the electrical power rate produced by PV as a module and a system.

Figure 7 illustrates the efficiency changes in the PV module and PV/T system for different nanofluids. As seen in Figure 7, the highest electrical efficiency of a PV module according to Equation (8) was obtained in the system with a graphene nanoplatelets/water mixture (14.6% in the daily average). For the PV/T systems with a MWCNT/water mixture and distilled water, the average electrical efficiencies of PV were 13.6% and 13.4%, respectively. However, until 10:20 a.m., the PV electrical efficiency of the PV/T system with distilled water was higher than that of the others. This was a result of the use of different carrier fluids in the experimental setup, prolonging the experiment day. Regarding Figure 7, the carrier fluid order did not change with the total efficiency changes in the PV/T system according to Equation (5), and it was similar to the change in the electrical efficiency of the PV module. For example, the overall efficiency values were 45.7%, 42.6%, and 39.5% as daily averages for the PV/T systems with the graphene nanoplatelets/water mixture, with the MWCNT/water mixture, and with distilled water, respectively. It can be reported that the electrical efficiency of the PV module was increased by 2–9% with the use of PV/T systems and especially nanofluids. In addition, small decreases in electrical efficiency were observed due to increases in the temperature of the PV module during the periods of maximum solar radiation (between 11:40 and 15:40). Consequently, the best performance was obtained by cooling the PV module with a graphene nanoplatelets/water mixture (Figure 7), despite the special choices made for the environmental conditions of the experimental days (see Figures 4 and 5).

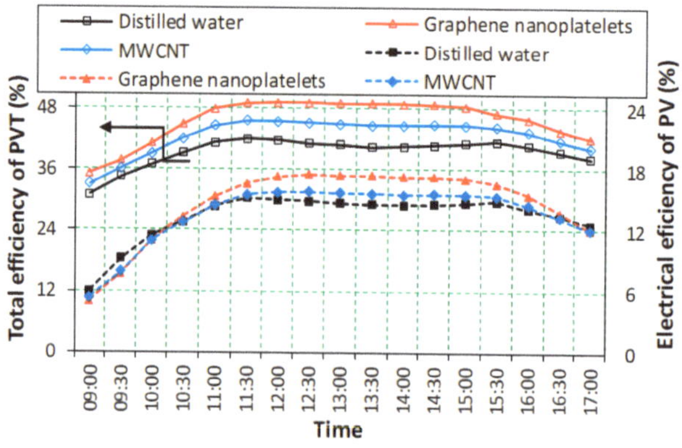

Figure 7. Changes in efficiency for the PV module and PV/T systems with different nanofluids.

In a study conducted by Shalaby et al. [34], the electrical efficiency of the PV panel was found to be 13.8% on average by cooling it with water. Additionally, in another similar study, Zilli et al. [35] obtained an electrical efficiency of 9.09% for a PV/T. Nasrin et al. [36] attained an increase in electrical efficiency of 9.2% for MWCNT nanofluid. Naghdbishi et al. [37] reported that MWCNT nanoparticles in PV/T systems increased the electrical efficiency by 4.21% compared with pure water. Gundala et al. [38] used 0.05% by-weight graphene nanoplatelets to cool the PV/T and observed that the electrical performance of the PV/T increased by 8.5%. Hassan et al. [39] obtained a maximum electrical efficiency of 14% by using graphene nanoparticles and phase change material together in the PV/T system. The results of this study showed that it was comparable with similar studies in the literature.

The performance evaluation of the above-mentioned PV/T system and nanofluid usage were discussed in detail. However, there were two major aspects to consider when evaluating a PV/T system: environmental and energy. The LCA and CED methods were applied to the experimental setup, which consisted of a PV module and a PV/T system, to determine the environmental impact of the electricity produced. The analyses were performed using the Eco-indicator 99 and SimaPro software according to the international standard approach [31,32]. Thus, Figure 8 depicts the changes in the environmental impact and embodied energy based on the system's components for the experimental setup. The total environmental impact of the experimental setup was determined to be 178.7 mPts/h (31.3 Pts for lifetime). As can be seen in Figure 8, the component with the highest environmental impact was the PV/T system, accounting for 51.5% of the total, followed by the water tank (19.5%), the support system (11%), and the other components (9.6%). The reason why the PV/T system had such a high environmental impact percentage was the presence of a copper tube and absorber plate in the heat-extraction unit (shown as "only thermal" in Figure 8). Similarly, the spiral-coil heat exchanger in the water tank was made of copper. The PV module accounted for 0.7% of the total (1.3 mPts/h). The largest percentage of it belonged to solar glass as a sub-component. The most striking issue in Figure 8 was that the environmental effect of nanofluids could not be observed. In the experimental setup, there was approximately 7.85 L of water in the nanofluid tank and its closed-loop line. Approximately 0.5% (by wt.) of water contained nanoparticles. Therefore, even if the production, use, transport, and disposal phases were taken into account for very small amounts in the system, the environmental impact could not be shown here.

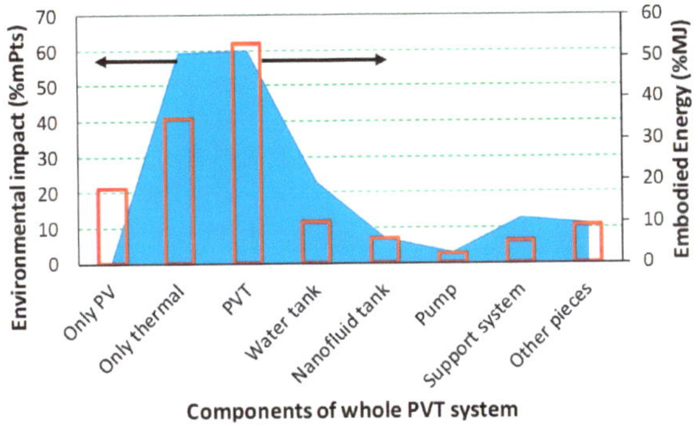

Figure 8. Percentage changes in the environmental impact vs. embodied energy in the components of the experimental setup.

However, in terms of embodied energy, the embodied energies of the experimental setup were calculated to be 3975.9 MJ, 4077.9 MJ, and 4081.1 MJ, respectively, for the PV/T systems with distilled water, the graphene nanoplatelets/water mixture, and the MWCNT/water mixture. In Figure 8, for the PV/T system with distilled water, the highest embodied energy occurred in the heat extraction unit, with 40.7% of 3975.9 MJ. It was followed by the PV module (21.2%), water tank (11.7%), other components (10.5%), and nanofluid tank (6.9%). The high embodied energy of the PV/T system was due to the following reason: the use of an aluminum frame in the heat extraction unit resulted in increased embodied energy. In addition, the embodied energy of the PV module had a very high value of 794.3 MJ. The lowest embodied energy was in the pump (about 2.5%).

The embodied energy percentage of the system components for different nanoparticles is indicated in Figure 9. The PV/T system with the highest embodied energy was that with a MWCNT/water mixture. After that came the PV/T systems with distilled water and a graphene nanoplatelets/water mixture. Similar to Figure 9, for different nanoparticle/water mixtures, the PV/T system was the most energy-intensive component of the experimental setup and was followed only by the thermal part (heat-extraction unit), PV module, water tank, other pieces, nanofluid tank, and support system (see Figure 9). As shown in Figure 9, there was a difference in the percentage of embodied energy between different heat-carrying fluids in the nanofluid tank, only the thermal part (heat-extraction unit) and water tank (spiral-coil heat exchanger), as the heat-carrying fluid of the experimental setup was circulated in a closed loop between the nanofluid tank, pump, heat extraction unit, and spiral-coil heat exchanger components. From Figure 9, the lowest and highest percentages of embodied energy in the nanofluid tank occurred in the distilled water, with 6.9%, and the MWCNT/water mixture, with 8.9%. In the water tank, the percentage of embodied energy from smallest to largest was ranked as follows: MWCNT/water mixture (11.3%), graphene nanoplatelets/water mixture (11.4%), and distilled water (11.7%). For the heat-extraction unit (only thermal), they were ranked following the same order as that for the water tank, as 40.2%, 40.3%, and 40.7%.

The general results of the overall analyses conducted for the PV module and PV/T system are summarized in Table 4. As seen in Table 4, the superiority of the PV/T system with the graphene nanoplatelets/water mixture for performance, energy, and environmental profiles was reached in all parameters compared with the other two systems. The average daily electricity production of the PV module was determined to be about 19.7 W. As a result of the experimental study, it was observed that the daily instantaneous power and the electrical efficiency increased up to 23.88 W and 14.4%, respectively, when the PV modules were cooled with a graphene nanoplatelets/water mixture. Similarly, the total efficiency of the PV/T system scaled up (to 45.7%). For example, the daily average overall efficiency values were 45.7%, 42.6%, and 39.5% for the PV/T systems with a graphene nanoplatelets/water mixture, a MWCNT/water mixture, and distilled water, respectively.

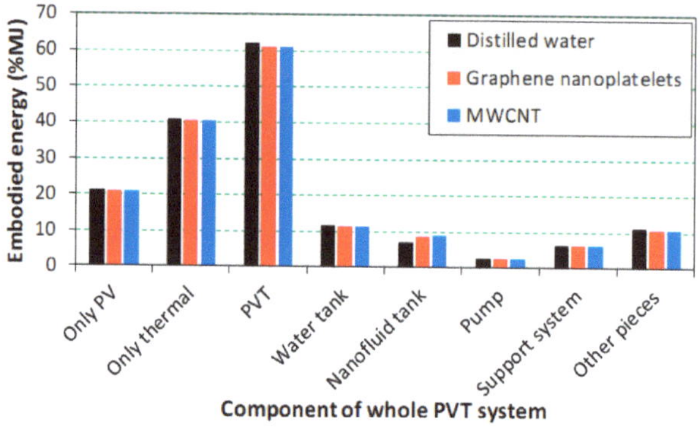

Figure 9. Percentage changes in environmental impact vs. embodied energy in the components of the experimental setup.

Table 4. General results of the thermal, LCA, and CED analyses for the experimental setup.

Parameters\Nanofluids	Distilled Water	MWCNT/Water Mixture	Graphene Nanoplatelets/ Water Mixture
Electrical efficiency of PV, %	13.5	13.6	14.4
Total efficiency of PV/T, %	39.5	42.6	45.7
Daily instant power, W	22.13	23.25	23.88
Total embodied energy, MJ	3975.88	4081.06	4077.86
Energy return factor (ERF)	1.76	1.80	1.85
Energy payback time (EPT), year	11.39	11.13	10.83
Environmental impact of electricity produced (EIE), Pts/kWh	0.046	0.044	0.043

As shown in Table 4, the total embodied energies of the PV/T systems were calculated to be 3975.88 MJ, 4081.06 MJ, and 4077.86 MJ for distilled water, the graphene nanoplatelets/water mixture, and the MWCNT/water mixture, respectively. The PV/T system with graphene nanoplatelets/water mixture needed the most energy during installation. As described in Equations (7) and (8), the energy return factor (ERF) and the energy payback time (EPT) were 1.76 and 11.39 years for a PV/T system with distilled water, 1.80 and 11.13 years for a system with a MWCNT/water mixture, and 1.85 and 10.83 years for a system with a graphene nanoplatelets/water mixture, respectively. The highest ERF value was 1.85 for the PV/T system with a graphene nanoplatelets/water mixture. This means that 1.85 times the energy (electricity generation) can be produced from the energy required in the installation phase. In addition, the PV/T system with a graphene nanoplatelets/water mixture had the lowest EPT value of 10.8 years. This means that approximately 10.8 years are necessary to recover the energy used to produce the system. While the embodied energy of the PV/T system with a graphene nanoplatelets/water mixture during the assembly phase was higher than that of other PV/T systems, its EPT value was lower. This was due to the higher energy output produced by the PV/T system with a graphene nanoplatelets/water mixture, which compensated for the embedded energy invested in the assembly phase.

In addition, the environmental impact per unit of electricity generated was calculated for different PV/T systems, and given in Table 4. From the table, the environmental impact of electricity produced (EIE) was 0.046, 0.044, and 0.043 Pts/kWh for the PV/T systems with distilled water, with a MWCNT/water mixture, and with a graphene nanoplatelets/water mixture, respectively. According to the Eco-Indicator 99 [40], the environmental impact on electricity generation of the experimental setup in this study was 4.7 times higher (for the system with distilled water) compared with the environmental impact of electricity generation at 0.0097 and 0.0072 Pts/kWh for small installations (3 kW_p) with mono-crystalline cells on a building facade and on a building roof, respectively. Using nanofluid instead of distilled water, especially a graphene nanoplatelets/water mixture, reduced the EIE value. Consequently, the results of this study show that the use of nanofluids could both increase the electrical efficiency of PV/T systems and decrease their environmental effects.

In Table 5, the results of some studies conducted in the literature are compared for the LCA analysis of the PV module and PV/T systems. The results of this study appear to be consistent with the results of the studies listed in Table 5.

Table 5. A comparison of the LCA analysis results of the presented study with those of studies in the literature.

Authors	Location	PV Type—System	Efficiency %	System Life-time (Year)	Embodied Energy CED	Energy Payback Time (Years)	Other Results
Tiwari et al. [41]	New Delhi and Leh, India	Monocrystalline silicon PV–PV/T system with and without glazed	11	-	4968–7480.8 MJ/m^2	11.4–14.33	Energy return on investment (dimensionless) 2.4–6.6
Chow and Ji [42]	Hong Kong	building-integrated PV/T system	13	-	4690.8–6220.8 MJ/m^2	2.8- 3.8	Energy return on investment (dimensionless) 5.9–8 Embodied emissions 297 g CO_2.eq/m^2
Kim et al. [43]	South Korea	Multi-crystalline silicon PV	14.91	25	0.44 MJ/kWh	3.68	Global warming potential 31.5 g CO_2.eq/kWh
		Monocrystalline silicon PV	15.96	30	0.56 MJ/kWh	4.65	Global warming potential 41.8 g CO_2.eq/kWh
Chen et al. [44]	China	Monocrystalline silicon PV	15.7	25	-	0.42–0.91	GHG 5.60–12.07 g CO_2/kWh
Hou et al. [45]	Northwest China and East China	Monocrystalline silicon PV	15	25	1123–1186 MJ/m^2	1.7–2.3	GHG 65.2–87.3 g CO_2/kWh
		Multi-crystalline silicon PV	17.5	25	1034–1094 MJ/m^2	1.6–2.1	GHG 60.1–80.5 g CO_2/kWh
Sagani et al. [46]	Athens, Greece	Polycrystalline silicon PV	14.4	25	-	1.8–4.1	CO_2 payback time 1.5–3.5 years Environmental benefits of PV are between 3.85–19.55 tons CO_2.eq/year compared with the conventional technology
This study	Turkey	Monocrystalline silicon PV–PV/T system	13.5–14.4	25	3975 MJ, 4077 MJ, and 4081 MJ for water, graphene, and MWCNT, respectively	11.39, 10.83, and 11.13 for water, graphene, and MWCNT, respectively.	Energy return on investment values were 1.76, 1.85, and 1.80; environmental impacts of electricity produced were 0.046, 0.043, and 0.044 Pts/kWh; and total efficiencies were 39.5%, 45.7%, and 42.6%, for water, graphene, and MWCNT, respectively

5. Conclusions

This paper presents a detailed experimental study and a complete environmental assessment using LCA and CED analyses for the entire lifetime of a PV/T system cooled with nanofluids, such as a MWCNT/water mixture and a GNP/water mixture. The LCA and CED analyses enabled detailed evaluations of the energy and environmental profiles of the PV/T system and its components throughout the manufacturing, installation, and operation phases. The results of the study yield the general conclusions listed below:

- According to the PV module, electrical efficiency increased in the PV/T system. The electrical and total efficiencies changed between 13.5–14.4% and 39.5–45.7%, respectively, in the PV/T system. Among the three working fluids, the PV/T system with the graphene nanoplatelets/water mixture had the highest efficiency values.
- The embodied energy values were 3975.08 MJ for the PV/T with distilled water, 4081.06 MJ for the PV/T with a MWCNT/water mixture, and 4077.86 MJ for the PV/T system with a graphene nanoplatelets/water mixture. This result relates to the large amount of silicon, galvanized iron, and aluminum used in the manufacture of the PV module and PV/T system, respectively.
- The energy payback time was changed between 10.83 and 11.39 years and had the lowest value in the PV/T system with graphene nanoplatelets/water mixture. The use of nanoparticles could shorten the energy return time.
- The environmental impact of the produced electricity was 0.046, 0.044, and 0.043 Pts/kWh for the PV/T systems with distilled water, MWCNT/water, and graphene nanoplatelets/water, respectively. These values were 0.0097 and 0.0072 Pts/kWh for small installations with monocrystalline cells on a façade and on a building roof, respectively. Thus, although the values of the PV/T system were very high, the use of nanoparticles decreased the environmental impact somewhat compared with the use of distilled water.
- Based on the LCA results, it was evident that the PV/T systems were less environmentally friendly compared with PV modules due to the increased amount of copper and aluminum used during their fabrication.
- The installation and operation of the PV/T systems require less energy-intensive materials (nanoparticles), resulting in lower environmental impacts during their processing compared with the PV/T system with distilled water.
- The energy consumed and environmental impacts of all manufacturing, installation, and operation processes to be followed were increased when the electrical efficiency of the PV module was increased. Therefore, for even the smallest improvement in the PV module and/or PV/T system, its energy and environmental impact should be considered.
- Nanofluids should be used in the cooling of PV/T systems, increasing their electrical efficiency and decreasing their environmental impact.

Author Contributions: Conceptualization, G.K.D.; methodology, A.K.; validation, A.G.G., M.K. (Muhammet Kayfeci) and A.K.; investigation, M.K. (Meltem Koşan); data curation, M.K. (Meltem Koşan) and A.G.G.; writing—original draft preparation, G.K.D. and A.K.; writing—review and editing, G.K.D., M.K. (Meltem Koşan), M.K. (Muhammet Kayfeci), A.G.G., and A.K.; visualization, A.K.; supervision, A.G.G. and M.K. (Muhammet Kayfeci); project administration, M.K. (Muhammet Kayfeci). All authors have read and agreed to the published version of the manuscript.

Funding: This research was funded by the Karabük University Scientific Research Projects Coordination Unit, grant number KBÜ-BAP-17-DR-262.

Data Availability Statement: Not applicable.

Acknowledgments: The authors would like to kindly express their thankfulness and gratitude to the Karabük University Scientific Research Projects Coordination Unit for the financial support provided under the project number of KBÜ-BAP-17-DR-262.

Conflicts of Interest: The authors declare no conflict of interest.

References

1. Hossain, F.; Karim, M.R.; Bhuiyan, A.A. A review on recent advancements of the usage of nano fluid in hybrid photovoltaic/thermal (PV/T) solar systems. *Renew. Energy* **2022**, *188*, 114–131. [CrossRef]
2. Todde, G.; Murgia, L.; Carrelo, I.; Hogan, R.; Pazzona, A.; Ledda, L.; Narvarte, L. Embodied energy and environmental impact of large-power stand-alone photovoltaic irrigation systems. *Energies* **2018**, *11*, 2110. [CrossRef]
3. Arnaoutakis, N.; Souliotis, M.; Papaefthimiou, S. Comparative experimental Life Cycle Assessment of two commercial solar thermal devices for domestic applications. *Renew. Energy* **2017**, *111*, 187–200. [CrossRef]
4. Souliotis, M.; Arnaoutakis, N.; Panaras, G.; Kavga, A.; Papaefthimiou, S. Experimental study and Life Cycle Assessment (LCA) of Hybrid Photovoltaic/Thermal (PV/T) solar systems for domestic applications. *Renew. Energy* **2018**, *126*, 708–723. [CrossRef]
5. Wong, J.H.; Royapoor, M.; Chan, C.W. Review of life cycle analyses and embodied energy requirements of single-crystalline and multi-crystalline silicon photovoltaic systems. *Renew. Sustain. Energy Rev.* **2016**, *58*, 608–618. [CrossRef]
6. Carnevale, E.; Lombardi, L.; Zanchi, L. Life cycle assessment of solar energy systems: Comparison of photovoltaic and water thermal heater at domestic scale. *Energy* **2014**, *77*, 434–446. [CrossRef]
7. Ehtiwesh, I.A.S.; Coelho, M.C.; Sousa, A.C.M. Exergetic and environmental life cycle assessment analysis of concentrated solar power plants. *Renew. Sustain. Energy Rev.* **2016**, *56*, 145–155. [CrossRef]
8. Santoyo-Castelazo, E.; Solano-Olivares, K.; Martínez, E.; García, E.O.; Santoyo, E. Life cycle assessment for a grid-connected multi-crystalline silicon photovoltaic system of 3 kW_p: A case study for Mexico. *J. Clean. Prod.* **2021**, *316*, 128314. [CrossRef]
9. Krebs-Moberg, M.; Pitz, M.; Dorsette, T.L.; Gheewala, S.H. Third generation of photovoltaic panels: A life cycle assessment. *Renew. Energy* **2021**, *164*, 556–565. [CrossRef]
10. Li, Z.; Zhang, W.; Xie, L.; Wang, W.; Tian, H.; Chen, M.; Li, J. Life cycle assessment of semi-transparent photovoltaic window applied on building. *J. Clean. Prod.* **2021**, *295*, 126403. [CrossRef]
11. Tirupati Rao, V.; Sekhar, Y.R. Comparative analysis on embodied energy and CO_2 emissions for stand-alone crystalline silicon photovoltaic thermal (PVT) systems for tropical climatic regions of India. *Sustain. Cities Soc.* **2022**, *78*, 103650. [CrossRef]
12. Li, Z.; Zhang, W.; He, B.; Xie, L.; Chen, M.; Li, J.; Zhao, O.; Wu, X. A Comprehensive life cycle assessment study of innovative bifacial photovoltaic applied on building. *Energy* **2022**, *245*, 123212. [CrossRef]
13. Goel, S.; Sharma, R.; Jena, B. Life cycle cost and energy assessment of a 3.4 kW_p rooftop solar photovoltaic system in India. *Int. J. Amb. Energy* **2022**, *43*, 1913221. [CrossRef]
14. Jurčević, M.; Nižetić, S.; Čoko, D.; Arıcı, M.; Hoang, A.T.; Giama, E.; Papadopoulos, A. Techno-economic and environmental evaluation of photovoltaic-thermal collector design with pork fat as phase change material. *Energy* **2022**, *254*, 124284. [CrossRef]
15. Lamnatou, C.; Notton, G.; Chemisana, D.; Cristofari, C. Life cycle analysis of a building-integrated solar thermal collector, based on embodied energy and embodied carbon methodologies. *Energy Build.* **2014**, *84*, 378–387. [CrossRef]
16. Hassani, S.; Saidur, R.; Mekhilef, S.; Taylor, R.A. Environmental and exergy benefit of nanofluid-based hybrid PV/T systems. *Energy Convers. Manag.* **2016**, *123*, 431–444. [CrossRef]
17. Ren, M.; Mitchell, C.R.; Mo, W. Dynamic life cycle economic and environmental assessment of residential solar photovoltaic systems. *Sci. Total Environ.* **2020**, *722*, 137932. [CrossRef]
18. Bahlawan, H.; Poganietz, W.R.; Spina, P.R.; Venturini, M. Cradle-to-gate life cycle assessment of energy systems for residential applications by accounting for scaling effects. *Appl. Therm. Eng.* **2020**, *171*, 115062. [CrossRef]
19. Herrando, M.; Elduque, D.; Javierre, C.; Fueyo, N. Life cycle assessment of solar energy systems for the provision of heating, cooling and electricity in buildings: A comparative analysis. *Energy Convers. Manag.* **2022**, *257*, 115402. [CrossRef]
20. Morini, A.A.; Hotza, D.; Ribeiro, M.J. Embodied energy and carbon footprint comparison in wind and photovoltaic power plants. *Int. J. Energy Environ. Eng.* **2022**, *13*, 457–467. [CrossRef]
21. Khanjari, Y.; Pourfayaz, F.; Kasaeian, A.B. Numerical investigation on using of nanofluid in a water-cooled photovoltaic thermal system. *Energy Convers. Manag.* **2016**, *122*, 263–278. [CrossRef]
22. Yazdanifard, F.; Ameri, M.; Ebrahimnia-Bajestan, E. Performance of nanofluid-based photovoltaic/thermal systems: A review. *Renew. Sustain. Energy Rev.* **2017**, *76*, 323–352. [CrossRef]
23. Fayaz, H.; Nasrin, R.; Rahim, N.A.; Hasanuzzaman, M. Energy and exergy analysis of the PVT system: Effect of nanofluid flow rate. *Sol. Energy* **2018**, *169*, 217–230. [CrossRef]
24. Abdallah, S.R.; Saidani-Scott, H.; Abdellatif, O.E. Performance analysis for hybrid PV/T system using low concentration MWCNT (water-based) nanofluid. *Sol. Energy* **2019**, *181*, 108–115. [CrossRef]
25. Alous, S.; Kayfeci, M.; Uysal, A. Experimental investigations of using MWCNTs and graphene nanoplatelets water-based nanofluids as coolants in PVT systems. *Appl. Therm. Eng.* **2019**, *162*, 114265. [CrossRef]
26. Sangeetha, M.; Manigandan, S.; Ashok, B.; Brindhadevi, K.; Pugazhendhi, A. Experimental investigation of nanofluid based photovoltaic thermal (PV/T) system for superior electrical efficiency and hydrogen production. *Fuel* **2021**, *286*, 119422. [CrossRef]
27. Chow, T.T.; Pei, G.; Fong, K.F.; Lin, Z.; Chan, A.L.S.; Ji, J. Energy and exergy analysis of photovoltaic-thermal collector with and without glass cover. *Appl. Energy* **2009**, *86*, 310–316. [CrossRef]
28. Kline, S.J.; McClintock, F.A. Describing uncertainties in single-sample experiments. *Mech. Eng.* **1953**, *75*, 3–8.
29. SimaPro. *User's Manual*; Pre Consultants BV: Amersfoort, The Netherlands, 2007.

30. Meyer, L.; Tsatsaronis, G.; Buchgeister, J.; Schebek, L. Exergoenvironmental analysis for evaluation of the environmental impact of energy conversion systems. *Energy* **2009**, *34*, 75–89. [CrossRef]
31. *ISO 14040*; Environmental Management-Life Cycle Assessment-Principles and Framework. ISO (International Organization for Standardization): Geneva, Switzerland, 2006.
32. *ISO 14044*; Environmental Management-Life Cycle Assessment-Requirements and Guidelines. ISO (International Organization for Standardization): Geneva, Switzerland, 2006.
33. Pantzali, M.N.; Kanaris, A.G.; Antoniadis, K.D.; Mouza, A.A.; Paras, S.V. Effect of nanofluids on the performance of a miniature plate heat exchanger with modulated surface. *Int. J. Heat Fluid Flow* **2009**, *30*, 691–699. [CrossRef]
34. Shalaby, S.M.; Elfakharany, M.K.; Moharram, B.M.; Abosheiasha, H.F. Experimental study on the performance of PV with water cooling. *Energy Rep.* **2022**, *8*, 957–961. [CrossRef]
35. Zilli, B.M.; Lenz, A.M.; de Souza, S.N.M.; Secco, D.; Nogueira, C.E.C.; Junior, O.H.A.; Nadaleti, W.C.; Siqueira, J.A.C.; Gurgacz, F. Performance and effect of water-cooling on a microgeneration system of photovoltaic solar energy in Paraná Brazil. *J. Clean. Prod.* **2018**, *192*, 477–485. [CrossRef]
36. Nasrin, R.; Rahim, N.A.; Fayaz, H.; Hasanuzzaman, M. Water/MWCNT nanofluid based cooling system of PVT: Experimental and numerical research. *Renew. Energy* **2018**, *121*, 286–300. [CrossRef]
37. Naghdbishi, A.; Yazdi, M.E.; Akbari, G. Experimental investigation of the effect of multi-wall carbon nanotube–Water/glycol based nanofluids on a PVT system integrated with PCM-covered collector. *Appl. Therm. Eng.* **2020**, *178*, 115556. [CrossRef]
38. Gundala, S.; Basha, M.M.; Madhurima, V.; Praveena, N.; Kumar, S.V. An experimental performance on solar photovoltaic thermal collector with nanofluids for sustainable development. *J. Nanomater.* **2021**, *2021*, 6946540. [CrossRef]
39. Hassan, A.; Wahab, A.; Qasim, M.A.; Janjua, M.M.; Ali, M.A.; Ali, H.M.; Jadoon, T.R.; Ali, E.; Raza, A.; Javaid, N. Thermal management and uniform temperature regulation of photovoltaic modules using hybrid phase change materials-nanofluids system. *Renew. Energy* **2020**, *145*, 282–293. [CrossRef]
40. Goedkoop, M.; Spriensma, R. The Eco-Indicator 99: A Damage Oriented Method for Life Cycle Impact Assessment. 2000. Available online: www.pre.nl (accessed on 31 October 2022).
41. Tiwari, A.; Raman, V.; Tiwari, G.N. Embodied energy analysis of hybrid photovoltaic thermal (PV/T) water collector. *Int. J. Ambien. Energy* **2007**, *28*, 181–188. [CrossRef]
42. Chow, T.T.; Ji, J. Environmental life-cycle analysis of hybrid solar photovoltaic/thermal systems for use in Hong Kong. *Int. J. Photoenergy* **2012**, *2012*, 101968. [CrossRef]
43. Kim, B.J.; Lee, J.Y.; Kim, K.H.; Hur, T. Evaluation of the environmental performance of sc-Si and mc-Si PV systems in Korea. *Sol. Energy* **2014**, *99*, 100–114. [CrossRef]
44. Chen, W.; Hong, J.; Yuan, X.; Liu, J. Environmental impact assessment of monocrystalline silicon solar photovoltaic cell production: A case study in China. *J. Clean. Prod.* **2016**, *112*, 1025–1032. [CrossRef]
45. Hou, G.; Sun, H.; Jiang, Z.; Pan, Z.; Wang, Y.; Zhang, X.; Zhao, Y.; Yao, Q. Life cycle assessment of grid-connected photovoltaic power generation from crystalline silicon solar modules in China. *Appl. Energy* **2016**, *164*, 882–890. [CrossRef]
46. Sagani, A.; Mihelis, J.; Dedoussis, V. Techno-economic analysis and life-cycle environmental impacts of small-scale building-integrated PV systems in Greece. *Energy Build.* **2017**, *139*, 277–290. [CrossRef]

processes

Article

Modeling and Multi-Stage Planning of Cement-IIES Considering Carbon-Green Certificate Trading

Zhaochu Guo and Suyang Zhou *

School of Electrical Engineering, Southeast University, Nanjing 210096, China; 213203614@seu.edu.cn
* Correspondence: suyang.zhou@seu.edu.cn

Abstract: The cement industry is an important industrial entity responsible for implementing carbon emission reduction targets. Considering the carbon trading and green certificate trading mechanisms, this paper presents a multi-stage planning approach for the constructed Cement-Industrial Integrated Energy System (Cement-IIES). Carbon reduction technologies represented by low-temperature waste heat recovery, as well as phased changes in economic and technical parameters, are considered in the model. The case study shows that the proposed method not only optimizes the design economy of the Cement-IIES but also achieves a substantial carbon emission reduction in the cement production process and energy supply system. Compared with the traditional single-stage planning, the proposed method improves the system's economic efficiency by 13.88% and flexibly adapts to changes in policies such as "coal reform", green certificate trading and carbon quotas. The low-temperature waste heat recovery technology helps the system energy utilization efficiency in the two stages increase by 0.45% and 0.86%, respectively, whilst oxygen-enriched combustion and carbon capture technologies can reduce the total carbon emissions by about 83%. In addition, the negative carbon emission effect of biomass gives the system access to annual benefits of CNY 3.10×10^7 and CNY 7.89×10^7 in the two stages, respectively.

Keywords: cement industry; carbon reduction; carbon trading; green certificate trading; integrated energy system; multi-stage planning

Citation: Guo, Z.; Zhou, S. Modeling and Multi-Stage Planning of Cement-IIES Considering Carbon-Green Certificate Trading. *Processes* **2023**, *11*, 1219. https://doi.org/10.3390/pr11041219

Academic Editors: Ferdinando Salata and Virgilio Ciancio

Received: 13 March 2023
Revised: 10 April 2023
Accepted: 12 April 2023
Published: 15 April 2023

1. Introduction

1.1. Motivation

Electricity and heat, industry and transport sectors are important entities responsible for global carbon emissions. According to the International Energy Agency (IEA), the carbon emissions of the global industry sector in 2022 reach 9.15 Gton, accounting for 24.86% of the total [1]. To this end, China [2], the United States [3], the United Kingdom [4], Germany [5] and other countries have formulated industrial low-carbon action plans, including energy efficiency improvement plans, decarbonization roadmaps and environmental protection laws.

Meanwhile, low-carbon energy technologies that can improve the sustainability, autonomy and security of energy systems have been extensively studied, thus providing a theoretical and technical basis for the action plans. Ref. [6] systematically sorted out the research directions and achievements supporting the sustainable development of energy systems from the two dimensions of technological and social. Specifically, new forms of energy utilization (typically hydrogen energy) [7], carbon abatement technology [8], distributed energy autonomous methods [9] and top-to-bottom energy policies [10] are all hot topics. In addition, refs. [11–13] and other studies tried to explore the market regulation potential in guiding renewable energy utilization and low-carbon energy consumption behavior.

It should be noted that carbon emissions in industry sectors come from energy use and industrial processes. Therefore, it is necessary to explore specialized and customized carbon reduction paths in combination with the production processes and energy demand of various industrial sectors.

1.2. Literature Review

The current research on the Industrial Integrated Energy System (IIES) focuses on exploring energy efficiency improvement means and low-carbon economic operation mode from the perspectives of technical and economic analysis, planning, simulation and operation optimization [14,15].

In terms of technical and economic analysis, scholars have attempted to evaluate the performance of IIES under new forms of energy utilization and energy system structures. To satisfy the economy and emission reduction needs of typical iron and steel parks, ref. [16] analyzed the benefits of building a "green power-hydrogen-industry" coupled energy system under different hydrogen energy substitution speed scenarios. Ref. [17] considered the interaction between the multi-energy system and the production process of process industry users and summarized the means of flexible control of the system from two aspects: direct regulation of cogeneration units and electrical equipment, and indirect regulation of non-electrical equipment. Authors of [18] proposed a national-level tool capable of modeling complex energy systems in the power, heat, transport and industry sectors, which could calculate the hourly carbon emission information of each sector. Based on the input of energy material information, the possibility of Kazakhstan in terms of transitioning to a 100% renewable energy system by 2050 was validated.

The planning and simulation research on IIES formulates planning and design schemes for industrial parks or enterprises and uses methods including simulation tools to evaluate the optimality of the schemes. Ref. [19] considered the environmental externality costs of coal, wind and solar power in the capacity allocation and operation optimization of the IES in industrial parks. The impact of environmental externality costs on the planning results was validated through a case study. In the hybrid energy system proposed in [20], hydrogen produced from nuclear/renewable energy is combined with coal to produce fuel gas and chemical raw materials. For coal-intensive countries such as China and the United States, compared with coal gasification to produce syngas, the proposed hybrid energy system can significantly reduce carbon emissions. Ref. [21] combined two energy analysis tools, discrete event simulation (DES) and energy system optimization (ESO), to provide the best operation mode for an iron foundry. To help an IES project in Zhejiang, China, realize the optimization of system planning and design, an evaluation method based on modified fuzzy integrals and prospect theory was proposed in [22].

To better coordinate the management of energy equipment in the IIES, many scholars have carried out operation optimization research to tap the system potential of flexible adjustment [23]. Authors of [24] incorporated electricity, heating and cooling demands into generalized demand-side resources, and established a demand response strategy and optimization model for the IIES. Power companies, CCHP and industrial users could interact through demand-side resources, and their economic costs of energy consumption were all significantly reduced. Ref. [25] built a demand response model for IIES considering the ladder carbon trading mechanism, which utilized the carbon reduction and demand response potential of industrial users in smart parks. To realize the flexible and economic dispatch of the industrial park-type regional IES, ref. [26] proposed an energy management strategy for energy cascade utilization, in which the partial load ratio is considered in the segmental linear model of the gas turbine, and the thermal energy is divided into high-, middle- and low-grade. The reinforcement learning method was applied to the multi-energy flow scheduling of the IIES in [27], and the collaborative management of multiple energy storage, renewable energy generation, energy conversion and energy trading was realized.

It can be seen that the existing research has made great attempts to tap the energy-saving and emission-reduction potential of IIES with high energy demand and tightly coupled multi-energy utilization. However, the actual production process of the industry to which a certain IIES belongs is seldom considered, and what differentiates IIES from residential and commercial IES is the interaction between multi-energy systems and production processes in IIES. Therefore, it is necessary to carry out targeted research in combination with specific industrial processes. In this paper, the cement industry with high energy consumption and high carbon emissions is selected as the research object.

As the third largest source of greenhouse gas emissions after power and the steel industry, the cement industry should undertake the task of reducing carbon emissions from now on. Specifically, 561 to 622 kg of carbon dioxide is emitted per ton of cement produced [28]. In 2022, the world's total cement production reaches 4.1 billion tons, and China's cement production accounts for 51.66% of the total, ranking first in the world [29]. At present, coal is still the main energy consumption of cement production in China. Coal combustion and limestone calcination produce 95% of the total carbon emissions [30].

Nowadays, the mainstream recognized carbon emission reduction methods for the cement industry include [31–33]: replacing cement with other additives (reducing cement demand in construction, water conservancy and other scenarios); improving the thermal efficiency of cement production; oxygen-enriched combustion; replacing fossil fuels; electrification of cement kilns; carbon capture and storage (CCS) technology. In [34], a cement kiln waste heat utilization poly generation system with integrated carbon capture was proposed, and the expensive carbon capture technology was subsidized with the system's electricity, heat and cold energy sales revenue. Ref. [35] carried out a coupling study on oxygen-enriched combustion technology and XDL energy-saving calcination technology, and the experiment confirmed that oxygen-enriched combustion technology can promote coal saving and flue gas emission reduction in cement clinker production.

The emergence of the carbon market and green certificate transactions have enriched the connotation of energy and production raw material consumption in the cement industry. By the end of 2021, a total of 33 carbon emission trading markets have been put into operation in the world, including countries or regions such as the European Union, North America, South Korea and China. Regarding carbon emission trading and carbon price formulation, each carbon market determines the carbon market participation mechanism, taxation scope and pricing method based on the needs of the country [36,37]. To solve the problem of renewable energy consumption, countries including the United States, Australia and the United Kingdom have successively launched quota-based green certificate trading systems to convert clean-energy subsidies from government behavior to market behavior [38]. In addition, some scholars have gradually paid attention to the research on the electricity-carbon-green certificate joint market [39]. In the context of carbon and green certificate trading, the cement industry is bound to pursue efficient and clean energy system structures, as well as planning and operation methods.

1.3. Contribution

In the above context, this paper analyzes the production process of the cement industry, constructs the basic structure of Cement-IIES and proposes a multi-stage planning method. The specific contributions are as follows.

(1) Combined with the flow of multi-energy and raw materials of the actual cement production line, a Cement-IIES model that comprehensively considers multiple carbon reduction technologies is built. Carbon reduction technologies include low-temperature waste heat recovery, oxygen-enriched combustion, carbon capture, hydrogen/oxygen production, energy storage and fuel/technology substitution.

(2) Considering carbon and green certificate trading, a multi-stage flexible planning method for Cement-IIES is proposed. The phased changes in technical and economic parameters are taken into account, such as energy equipment cost and efficiency and external energy/carbon/green certificate prices.

(3) The effectiveness of the proposed method is verified based on a cement industrial park on the southeast coast of China. Comparative analysis reveals the adaptability of the multi-stage planning method to policies such as "coal reform", green certificate trading and carbon quotas, as well as the comprehensive benefits of various carbon reduction technologies.

1.4. Organization

The rest of the paper is organized as follows. Section 2 builds the mathematical model of equipment and technology in Cement-IIES; Section 3 proposes a multi-stage planning method, including specific expressions of objective functions and constraints. Section 4 is a case study, and the discussion is given in Section 5. Section 6 presents the conclusion.

2. Cement-IIES Modeling

The power requirement of a cement factory varies depending on the heat treatment process in use, which can be divided into dry production and wet production according to the method to prepare the raw material [40]. This paper focuses on using the dry process with rapid heat transfer and high heating efficiency for cement production. Taking into account the energy demand and carbon emission during the cement production process, a Cement Industry Integrated Energy System (Cement-IIES) is proposed and its structure is shown in Figure 1.

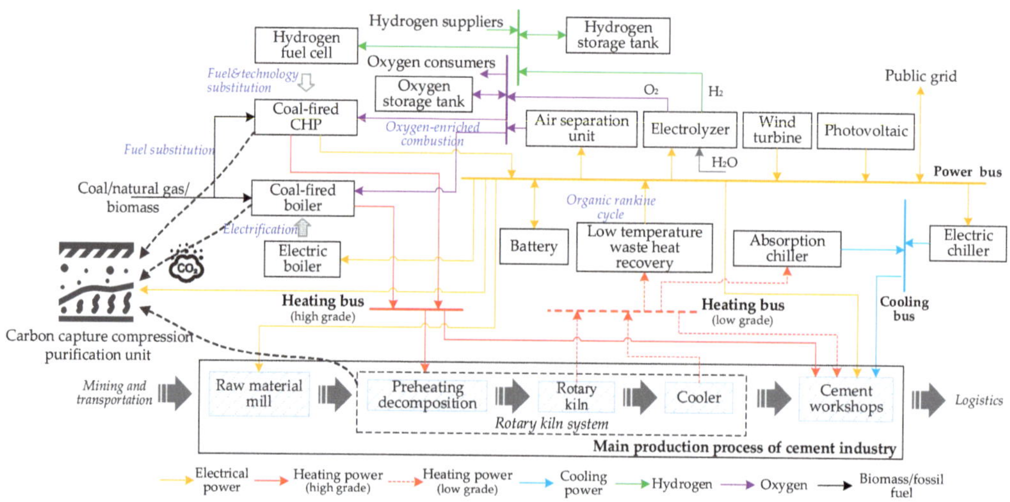

Figure 1. Structure of Cement Industry Integrated Energy System (Cement-IIES).

As shown in Figure 1, the cement production line mainly composes of raw material collection and transportation, raw material preparation and homogenization, clinker calcination, clinker cooling and cement grinding and packaging [28]. The carbon dioxide emitted during the cement production process mainly comes from the limestone calcination process. More specifically, the limestone calcination for producing quicklime contributes to 55–70% of the total carbon emission. It is worth noting that the high-temperature calcination process needs to burn fuel, which can emit 25–40% of carbon dioxide. The calcined clinker needs to be rapidly cooled by a cooler before entering the cement workshop to complete the grinding and packaging processes. Eventually, the product is transported to storage facilities or downstream entities.

In this paper, the load demand mainly comes from the raw material mill, rotary kiln system and cement workshops. The rotary kiln system composes of three sessions including preheating decomposition, rotary kiln and cooler. The internal structure and material energy flow will be described in Section 2.8. The load of the cement unit is mainly attributed to the circulating water pump room, the conveying machinery of the joint storage and the dust collector of the packaging machine in the packaging unit. In addition, a Cement-IIES also needs to provide the energy for lighting, heating/hot water and cooling of each working unit, living area and office area.

In the proposed Cement-IIES, the energy production equipment includes Combined Heating and Power (CHP) plant, Fired Boiler (FB) and renewable generation units such as Wind Turbine (WT) and Photovoltaic (PV). For the energy conversion equipment, the Electrolyzer (ELZ), Hydrogen Fuel Cell (HFC), Electric Boiler (EB), Electric Chiller (EC), Absorption Chiller (AC) and Air Separation Unit (ASU) are used. To provide energy storage capability, equipment including Battery (BT), Hydrogen Storage Tank (HST) and Oxygen Storage Tank (OST) are deployed. To promote tiered energy utilization and reduce carbon emissions, various technologies such as low-temperature waste heat power generation, oxyfuel combustion and carbon capture are also considered in the proposed Cement-IIES.

The detailed mathematical models of the technology and equipment employed in the Cement-IIES are presented in the following sections.

2.1. Wind Turbine

The output of a Wind Turbine (WT) is affected by factors such as blade radius, wind speed and energy conversion efficiency [41]. In stage w of a typical day d at time t, the predicted WT output can be calculated using (1).

$$P_{wt.fore}^{wdt} = 0.5 \times 10^{-6} \times \pi \cdot \eta_{wt} \rho_{air} \sum_{k \leq w} N_{wt}^{k} R_{wt}^{2} \cdot \left(u^{wdt}\right)^{3} \tag{1}$$

where N_{wt}^{k} represents the number of WTs installed in stage k ($k \leq w$); η_{wt} is the energy conversion efficiency of the WT whilst ρ_{air} is the air density (kg/m^3); R_{wt} denotes the radius (m) of the WT blades and u^{wdt} indicates the wind speed at stage w of a typical day d at time t.

2.2. Photovoltaic

The output of Photovoltaic (PV) is affected by factors including radiation intensity, panel area and inclination and energy conversion efficiency. Equation (2) illustrates how the predicted output is derived [41].

$$P_{pv.fore}^{wdt} = \Re^{wdt} \cdot \cos\theta_{pv} \cdot \eta_{MPPT} \cdot \sum_{k \leq w} A_{pv}^{k} \cdot \eta_{pv} \times 10^{-3} \tag{2}$$

where A_{pv}^{k} is the installed PV area (m^2) in stage k ($k \leq w$) and $\Re^{wdt}$ indicates the radiation intensity (kW/m^2) in stage w of a typical day d at time t; θ_{pv} denotes the inclination angle (°) of the PV panel whilst η_{MPPT} and η_{pv} represents the energy conversion efficiency of MPPT and PV component, respectively.

2.3. Electrolyzer

An Electrolyzer (ELZ) uses water and electricity to produce hydrogen and oxygen [42]. The relationship between the hydrogen generated by the ELZ and the input power is shown in (3) and the ratio of the flow rate between hydrogen and oxygen is presented in (4).

$$\eta_{elz} \cdot P_{elz}^{wdt} = LHV_{h} \cdot V_{elz,h}^{wdt} \times 10^{-3} \tag{3}$$

$$V_{elz,oxy}^{wdt} = \lambda_{elz} \cdot V_{elz,h}^{wdt} \tag{4}$$

where $V_{elz,o}^{wdt}$ and $V_{elz,h}^{wdt}$ represent the flow rate (m^3/h) of the oxygen and hydrogen produced by the ELZ, respectively; LHV_h indicates the calorific value of hydrogen (kWh/m^3); λ_{elz} is the ratio coefficient of the hydrogen-oxygen production. In theory, electrolyzing 1 L water can produce 1234.7 L (111 g) hydrogen and 622 L (889 g) oxygen under standard atmospheric pressure.

2.4. Combined Heating and Power Unit

Combined Heating and Power (CHP) units in the cement industry mainly use coal as fuel, which is called coal-fired CHP. Technologies applying alternative fuels to reduce the use of coal or increase the efficiency of coal combustion are gaining attention in various industries including the cement industry. The use of alternative fuels can save the requirement of primary fossil resources and reduce carbon emissions whilst maintaining the same level of energy consumption for clinker production. In the Cement-IIES proposed in this paper, natural gas and biomass can be used as alternative fuels and the associated system is referred to as natural gas-Fired CHP (ng-fired CHP) and biomass-Fired CHP (bio-fired CHP). Assuming that all three types of CHP units are operating in the heat-constant power mode, there will be differences in the thermoelectric efficiency [43]. The electric power and heating power output can be obtained using (5) and (6), respectively.

$$P_{chp}^{wdt} = P_{chp,ng}^{wdt} + P_{chp,coal}^{wdt} + P_{chp,bio}^{wdt}$$
$$\begin{cases} P_{chp,ng}^{wdt} = \eta_{chp,ng} LHV_{ng} V_{chp,ng}^{wdt} \times 10^{-3} \\ P_{chp,coal}^{wdt} = \eta_{chp,coal} LHV_{coal} m_{chp,coal}^{wdt} \\ P_{chp,bio}^{wdt} = \eta_{chp,bio} LHV_{bio} m_{chp,bio}^{wdt} \end{cases} \tag{5}$$

$$Q_{chp}^{wdt} = Q_{chp,ng}^{wdt} + Q_{chp,coal}^{wdt} + Q_{chp,bio}^{wdt}$$
$$\begin{cases} Q_{chp,ng}^{wdt} = r_{chp,ng} \cdot P_{chp,ng}^{wdt} \\ Q_{chp,coal}^{wdt} = r_{chp,coal} \cdot P_{chp,coal}^{wdt} \\ Q_{chp,bio}^{wdt} = r_{chp,bio} \cdot P_{chp,bio}^{wdt} \end{cases} \tag{6}$$

where $P_{chp,ng}^{wdt}$, $P_{chp,coal}^{wdt}$ and $P_{chp,bio}^{wdt}$ represent the output power (MW) of the three types of CHP units in stage w of a typical day d at time t; $\eta_{chp,ng}$, $\eta_{chp,coal}$ and $\eta_{chp,coal}$ show the power generation efficiency of the three types of CHP units; LHV_{ng} indicates the calorific value (kWh/m^3) of the natural gas whilst the calorific value (MWh/ton) of coal and biomass fuel is denoted by LHV_{coal} and LHV_{bio}, respectively; $V_{chp,ng}^{wdt}$ is the input flow rate (m^3/h) of the natural gas to the ng-fired CHP; $m_{chp,coal}^{wdt}$ and $m_{chp,bio}^{wdt}$ represent the fuel input volume (ton/h) of coal and biomass to the coal-fired CHP and bio-fired CHP, respectively; the output heat power (MW) of the three types of CHP units is indicated by $Q_{chp,ng}^{wdt}$, $Q_{chp,coal}^{wdt}$ and $Q_{chp,bio}^{wdt}$, respectively. Last but not least, $r_{chp,ng}$, $r_{chp,coal}$ and $r_{chp,bio}$ is the heat-to-power ratios of the three types of CHP units, respectively.

The electric power output of the CHP unit is also constrained by the ramping capability as specified in (7).

$$\begin{cases} -\pi_{chp,ng,\max} \sum\limits_{k \le w} Cap_{chp,ng}^{k} \le P_{chp,ng}^{wdt} - P_{chp,ng}^{wd(t-1)} \le \pi_{chp,ng,\max} \sum\limits_{k \le w} Cap_{chp,ng}^{k} \\ -\pi_{chp,coal,\max} \sum\limits_{k \le w} Cap_{chp,coal}^{k} \le P_{chp,coal}^{wdt} - P_{chp,coal}^{wd(t-1)} \le \pi_{chp,coal,\max} \sum\limits_{k \le w} Cap_{chp,coal}^{k} \\ -\pi_{chp,bio,\max} \sum\limits_{k \le w} Cap_{chp,bio}^{k} \le P_{chp,bio}^{wdt} - P_{chp,bio}^{wd(t-1)} \le \pi_{chp,bio,\max} \sum\limits_{k \le w} Cap_{chp,bio}^{k} \end{cases} \tag{7}$$

where $Cap_{chp,ng}^{k}$, $Cap_{chp,coal}^{k}$ and $Cap_{chp,bio}^{k}$ represent the installed capacity (MW) of the three types of CHP units in stage k ($k \le w$); $\pi_{chp,ng,\max}$, $\pi_{chp,coal,\max}$ and $\pi_{chp,bio,\max}$ is the maximum ramping capability (MW/h) of the three types of CHP units, respectively.

2.5. Fired Boiler

Similar to the CHP unit, the Fired Boiler (FB) can also use a variety of fuels including coal, natural gas and biomass and these can be categorized as coal-fired FB, ng-fired FB and bio-fired FB [44]. The output heating power of these FBs can be calculated using Equation (8).

$$\begin{cases} Q_{fb}^{wdt} = Q_{fb,ng}^{wdt} + Q_{fb,coal}^{wdt} + Q_{fb,bio}^{wdt} \\ Q_{fb,ng}^{wdt} = \eta_{fb,ng} LHV_{ng} V_{fb,ng}^{wdt} \times 10^{-3} \\ Q_{fb,coal}^{wdt} = \eta_{fb,coal} LHV_{coal} m_{fb,coal}^{wdt} \\ Q_{fb,bio}^{wdt} = \eta_{fb,bio} LHV_{bio} m_{fb,bio}^{wdt} \end{cases} \tag{8}$$

where $Q_{fb,ng}^{wdt}$, $Q_{fb,coal}^{wdt}$ and $Q_{fb,bio}^{wdt}$ denote the output electric power (MW) of the three types of FBs in stage w of a typical day d at time t, respectively; $\eta_{fb,ng}$, $\eta_{fb,coal}$ and $\eta_{fb,bio}$ are the energy conversion efficiency of the three types of FBs, respectively. The input flow rate (m^3/h) of the natural gas to the ng-fired CHP is represented by $V_{fb,ng}^{wdt}$ whilst the input volume (ton/h) of coal and biomass fuel to the coal-fired FB and bio-fired FB is shown by $m_{fb,coal}^{wdt}$ and $m_{fb,bio}^{wdt}$, respectively.

2.6. Hydrogen Fuel Cell

The Hydrogen Fuel Cell (HFC) uses hydrogen as fuel to achieve combined heat and power generation and its power generation efficiency is higher than that of the aforementioned three types of CHP units [45]. The model of the electric heating power output of an HFC is similar to that of a CHP unit, which can be defined using Equations (9)–(11).

$$P_{hfc}^{wdt} = LHV_h \cdot \eta_{hfc} V_{hfc}^{wdt} \times 10^{-3} \tag{9}$$

$$Q_{hfc}^{wdt} = r_{hfc} \cdot P_{hfc}^{wdt} \tag{10}$$

$$-\pi_{hfc,\max} \sum_{k \le w} Cap_{hfc}^{k} \le P_{hfc}^{wdt} - P_{hfc}^{wd(t-1)} \le \pi_{hfc,\max} \sum_{k \le w} Cap_{hfc}^{k} \tag{11}$$

where P_{hfc}^{wdt} and Q_{hfc}^{wdt} is the output electric power and heating power (MW) of the HFC in stage w of a typical day d at time t whilst η_{hfc} and r_{hfc} represent the power generation efficiency and heat-to-electricity ratio of the HFC, respectively; V_{hfc}^{wdt} is the input flow rate (m^3/h) of hydrogen to the HFC.

2.7. Energy Storage Equipment

The energy storage equipment deployed in Cement-IIES includes the Oxygen Storage Tank (OST), Hydrogen Storage Tank (HST) and Battery (BT). The mathematical models of the three types of energy storage are similar and the main differences are the storage medium and charging/discharging capabilities. Model details of the OST are illustrated in (12)–(14) [46].

2.7.1. Constraint for Energy Storage State

The gas storage state of the OST at each time interval needs to satisfy the upper and lower limit of the gas tank, which is generally expressed as the product of the limit coefficient and the installed capacity. This can be shown in (12).

$$\mu_{ost,\min} \sum_{k \le w} Cap_{ost}^{k} \le S_{ost}^{wdt} \le \mu_{ost,\max} \sum_{k \le w} Cap_{ost}^{k} \tag{12}$$

where $\mu_{ost,\min}$ and $\mu_{ost,\max}$ indicate the minimum and maximum coefficients, respectively; the installed capacity (m^3) of the OST in stage k ($k \le w$) is denoted by Cap_{ost}^{k}.

It can be observed from (13) that the value of S_{ost}^{wdt} is related to the gas storage state at the previous time interval $S_{ost}^{wd(t-1)}$ and the charging and discharging state at the current moment in stage w of a typical day d at time t. To ensure that the scheduling instructions can be executed periodically, the value of S_{ost}^{wdt} is fixed at the initial time and end time of each typical day.

$$\begin{cases} S_{ost}^{wdt} = 0.5 \cdot \sum_{k \leq w} Cap_{ost}^{k} \quad t = 1, T \\ S_{ost}^{wdt} = S_{ost}^{wd(t-1)}(1 - \sigma_{ost}) + \left(V_{ost,ch}^{wdt} \cdot \eta_{ost,ch} - V_{ost,dis}^{wdt} / \eta_{ost,dis}\right)\Delta t \quad 2 \leq t \leq T-1 \end{cases} \tag{13}$$

where σ_{ost} indicates the self-loss coefficient of the OST gas storage; $V_{ost,ch}^{wdt}$ and $V_{ost,dis}^{wdt}$ represents the gas inflating and deflating rate (m^3/h) in stage w of a typical day d at time t, respectively; the charge and discharge efficiency of the OST is denoted by $\eta_{ost,ch}$ and $\eta_{ost,dis}$, respectively.

2.7.2. Constraints for Inflation and Deflation Flow and Electric Power

The inflation and deflation behavior of the OST also needs to satisfy the constraints specified in (14).

$$\begin{cases} 0 \leq V_{ost,ch}^{wdt} \leq \gamma_{ost,ch} x_{ost,ch}^{wdt} \sum_{k \leq w} Cap_{ost}^{k} \\ 0 \leq V_{ost,dis}^{wdt} \leq \gamma_{ost,dis} x_{ost,dis}^{wdt} \sum_{k \leq w} Cap_{ost}^{k} \\ x_{ost,ch}^{wdt} + x_{ost,dis}^{wdt} \leq 1 \end{cases} \tag{14}$$

where $\gamma_{ost,ch}$ and $\gamma_{ost,dis}$ represent the inflation and deflation ratio of the OST, reflecting the relationship between the inflation and deflation flow rate and the OST capacity; $x_{ost,ch}^{wdt}$ and $x_{ost,dis}^{wdt}$ are binary variables which place a restriction that the OST cannot inflate and deflate at the same time.

2.8. *Power Generation Device Using Low-Temperature Waste Heat*

The cement rotary kiln system is the core of cement production and its internal material balance and heat balance relationship is illustrated in Figure 2 [47].

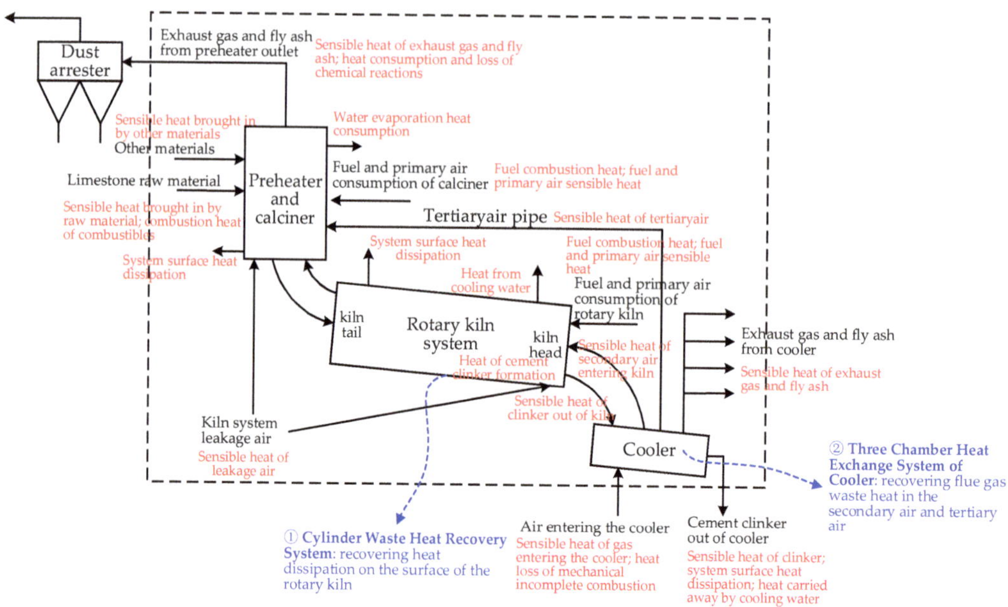

Figure 2. Internal Material Balance and Heat Balance of Cement Rotary Kiln System.

During the calcination process of the clinker within the rotary kiln, the surface temperature of the cylinder body can reach approximately 240–350 °C and this amount of the low-temperature waste heat will be dissipated in the air if not utilized.

Meanwhile, the temperature of the cement clinker produced from the kiln head can reach 1450 °C and it needs to enter the cooler to be cooled down to about 100 °C by cold air. The waste heat from the clinker can heat the cold air to 500–900 °C with considerable recycling potential. The existing three-chamber heat exchange system of the cooler already utilizes the waste heat of the clinker to heat the secondary air entering the kiln and the tertiary air entering the furnace, which minimizes the fuel consumption of the rotary kiln system. However, there is still 200–250 °C low-temperature flue gas waste heat in the secondary air and tertiary air that can be reused.

As a result, the waste heat utilization of the cement rotary kiln considered in this paper mainly focuses on the waste heat recovery from the Cylinder Waste Heat Recovery System (CWHC) and the Three Chamber Heat Exchange System of Cooler (TCHE).

2.8.1. Cylinder Waste Heat Recovery System

Taking the cement production line with a daily output of 5000 tons in the existing cement factory as an example [48], the amount of reusable heating power can be calculated using Equation (15) when the CWHC adopts convective heat transfer and radiation heat transfer.

$$Q_{cwhc}^{wdt} = h_{cht} \cdot \Delta T_{cwhc}^{wdt} \cdot S_{cwhc}^{w} \tag{15}$$

where h_{cht} represents the convective heat transfer coefficient in $W/(m^2 \cdot K)$ whilst ΔT_{cwhc}^{wdt} indicates the difference between the average surface temperature of the cylinder and the ambient temperature K. The surface area (m^2) of the cylinder is illustrated by S_{cwhc}^{w}.

2.8.2. Three-Chamber Heat Exchange System of Cooler

The amount of heating power that can be recovered from the secondary air and tertiary air by TCHE is calculated using (16).

$$Q_{tche}^{wdt} = c_{fume,p} \cdot \rho_{fume} \cdot V_{tche}^{wdt} \cdot \Delta T_{tche}^{wdt} \tag{16}$$

where $c_{fume,p}$ shows the constant pressure-heat ratio ($kJ/(kg \cdot K)$) of the flue gas whilst ρ_{fume} represents the density (kg/m^3) of the flue gas. V_{tche}^{wdt} is the flow rate (m^3/h) of flue gas during the waste heat recovery and ΔT_{tche}^{wdt} is the temperature difference (K) before and after the flue gas recovery.

2.8.3. Heat Balance during Low-Temperature Waste Heat Recovery

Part of the low-temperature waste heat recovered by CWHC and TCHE can be used for power generation when applying the Organic Rankine Cycle (ORC) technology. Similarly, part of the input heat for the Absorption Chiller (AC) can also be used to supply the heating and hot water demand of cement units, office buildings and dormitory areas. The deficit in demand for heating and domestic hot water can be covered by high-grade heating energy. Equation (17) indicates the heat balance relationship during the low-temperature waste heat recovery, which can also be referred to as the low-grade heating bus balance.

$$Q_{cwhc}^{wdt} + Q_{tche}^{wdt} = P_{orc}^{wdt} / \eta_{orc} + Q_{ac}^{wdt} / COP_{ac} + L_{heat,low}^{wdt} \tag{17}$$

where P_{orc}^{wdt} is the output electric power (MW) of the ORC system and η_{orc} is the power generation efficiency of the ORC system; Q_{ac}^{wdt} denotes the output cooling power (MW) of the AC whilst COP_{ac} is the energy efficiency coefficient of the AC; $L_{heat,low}^{wdt}$ indicates the low-grade heat demand (MW) that needs to be met during the production and living activities (referred to "non-core activities") other than the limestone calcination.

2.9. Oxygen-Enriched Combustion and Carbon Capture Technologies

Major carbon capture technologies include pre-combustion capture technology, post-combustion capture technology and oxygen-enriched combustion technology. The oxygen-enriched combustion technology has gained popularity because of its high carbon capture rate and low investment cost [35]. Its working principle is to replace the air within the combustion chamber with a mixture of oxygen and circulating CO_2 flow to generate a highly concentrated CO_2 flow (unchanged total amount with improved purity), which is easier to be captured by the carbon capture Compression Purification Unit (CPU) [49].

Carbon capturing is a costly process and cement kilns typically cannot provide all the required energy. It is a common practice to provide the energy for carbon capture in cement factories from coal-fired power plants or by burning natural gas after adding waste heat boilers [34,50]. In this paper, there are multiple approaches to meet the energy demand of the CPU including the utilization of three types of CHP units and FB, as well as the low-temperature waste heat recovery of the rotary kiln system, which can effectively reduce the CPU's dependence on expensive external energy.

2.9.1. Power Demand of CPU

The power demand of the CPU varies depending on the source of the carbon emission. $P_{cpu,cal}^{wdt}$, $P_{cpu,chp}^{wdt}$ and $P_{cpu,fb}^{wdt}$ represent the power demand of CPU for capturing carbon from limestone calcination, CHP unit and FB and can be derived using Equation (18).

$$P_{cpu}^{wdt} = P_{cpu,cal}^{wdt} + P_{cpu,chp}^{wdt} + P_{cpu,fb}^{wdt}$$
$$\begin{cases} P_{cpu,cal}^{wdt} = \kappa_{cpu} \alpha_{cal}^{wdt} \varepsilon_{cal}^{carb} m_{cal}^{wdt} \\ P_{cpu,chp}^{wdt} = \kappa_{cpu} \left(\varepsilon_{chp,ng}^{carb} \alpha_{chp,ng}^{wdt} P_{chp,ng}^{wdt} + \varepsilon_{chp,coal}^{carb} \alpha_{chp,coal}^{wdt} P_{chp,coal}^{wdt} + \varepsilon_{chp,bio}^{carb} \alpha_{chp,bio}^{wdt} P_{chp,bio}^{wdt} \right) \\ P_{cpu,fb}^{wdt} = \kappa_{cpu} \left(\varepsilon_{fb,ng}^{carb} \alpha_{fb,ng}^{wdt} Q_{fb,ng}^{wdt} + \varepsilon_{fb,coal}^{carb} \alpha_{fb,coal}^{wdt} Q_{fb,coal}^{wdt} + \varepsilon_{fb,bio}^{carb} \alpha_{fb,bio}^{wdt} Q_{fb,bio}^{wdt} \right) \end{cases} \tag{18}$$

where κ_{cpu} represents the energy (MWh/ton) required by the CPU to capture one unit of CO_2 whilst α_{cal}^{wdt} reflects the carbon capture level of the CPU for limestone calcination; $\alpha_{chp,ng}^{wdt}$, $\alpha_{chp,coal}^{wdt}$ and $\alpha_{chp,bio}^{wdt}$ show the carbon capture levels of the CPU for three types of CHP units and $\alpha_{fb,ng}^{wdt}$, $\alpha_{fb,coal}^{wdt}$ and $\alpha_{fb,bio}^{wdt}$ indicate that of the CPU for three types of FB; ε_{cal}^{carb} denotes the carbon emission intensity (ton/ton) during the limestone calcination. The carbon emission intensity (ton/MWh) of the three types of CHP units are denoted as $\varepsilon_{chp,ng}^{carb}$, $\varepsilon_{chp,coal}^{carb}$ and $\varepsilon_{chp,bio}^{carb}$ whilst the intensity (ton/MWh) of the three types of FB are represented by $\varepsilon_{fb,ng}^{carb}$, $\varepsilon_{fb,coal}^{carb}$ and $\varepsilon_{fb,bio}^{carb}$.

It should be noted that the carbon dioxide produced from biomass fuel does not need to specifically proceed. When considering carbon capture in this paper, it is equivalent that the bio-fired CHP and bio-fired FB will produce "negative carbon emission" and can obtain a reward for carbon reduction.

2.9.2. Constraints for CPU Carbon Capture Level

In the CPU model established in this paper, it is assumed that the upper limit of the CPU's carbon capture level for the CHP unit and GB relies on whether the CHP unit and GB use oxygen-enriched combustion. Binary variables ζ_{chp}^{wdt} and ζ_{gb}^{wdt} are introduced to indicate whether the CHP unit and GB are in the oxygen-enriched combustion state. The expression (19) defines the relevant constraints.

$$\begin{cases} 0 \le \alpha_{chp}^{wdt} \le \zeta_{chp}^{wdt} \alpha_{chp,\max}^{oxy} + \left(1 - \zeta_{chp}^{wdt}\right) \alpha_{chp,\max} \\ 0 \le \alpha_{fb}^{wdt} \le \zeta_{fb}^{wdt} \alpha_{fb,\max}^{oxy} + \left(1 - \zeta_{fb}^{wdt}\right) \alpha_{fb,\max} \\ 0 \le \alpha_{cal}^{wdt} \le \alpha_{cal,\max} \end{cases} \tag{19}$$

where $\alpha_{cal,\max}$ is the maximum carbon capture level of the CPU for limestone calcination; $\alpha_{chp,\max}^{oxy}$ and $\alpha_{chp,\max}$ represent the maximum carbon capture level of the CPU for the

CHP unit with and without oxygen-enriched combustion. Similarly, $\alpha_{fb,\max}^{oxy}$ and $\alpha_{fb,\max}$ indicate the maximum carbon capture level of the CPU for the FB with and without oxygen-enriched combustion.

2.9.3. Net Carbon Emission of Limestone Calcination, CHP Units and FB

When the oxygen-enriched combustion and carbon capture are considered, the net carbon emissions of the limestone calcination, CHP units and FB can be calculated using (20).

$$\begin{cases} E_{cal,carb}^{wdt} = \left(1-\alpha_{cal}^{wdt}\right)\varepsilon_{cal,carb} m_{cal}^{wdt} \\ E_{chp,carb}^{wdt} = \left(1-\alpha_{chp}^{wdt}\right)\left(\varepsilon_{chp,ng}^{carb} P_{chp,ng}^{wdt} + \varepsilon_{chp,coal}^{carb} P_{chp,coal}^{wdt}\right) \\ E_{fb,carb}^{wdt} = \left(1-\alpha_{fb}^{wdt}\right)\left(\varepsilon_{fb,ng}^{carb} Q_{fb,ng}^{wdt} + \varepsilon_{fb,coal}^{carb} Q_{fb,coal}^{wdt}\right) \\ E_{bio,carb}^{wdt} = \alpha_{chp}^{wdt}\varepsilon_{chp,bio}^{carb} P_{chp,bio}^{wdt} + \alpha_{fb}^{wdt}\varepsilon_{fb,bio}^{carb} Q_{fb,bio}^{wdt} \end{cases} \quad (20)$$

where $E_{cal,carb}^{wdt}$, $E_{chp,carb}^{wdt}$ and $E_{fb,carb}^{wdt}$ denote the net carbon emissions of the limestone calcination, CHP units and FB; $E_{bio,carb}^{wdt}$ is the "negative carbon emission" of biomass fuel because of carbon capture.

2.9.4. Oxygen Requirement for Oxygen-Enriched Combustion

The relationship between the amount of oxygen consumed by the CHP unit and FB when they are in an oxygen-enriched combustion state and the associated output power is shown in (21) [51].

$$\begin{cases} V_{chp,oxy}^{wdt} = \zeta_{chp}^{wdt}\chi_{chp,oxy} P_{chp}^{wdt} \\ V_{fb,oxy}^{wdt} = \zeta_{fb}^{wdt}\chi_{fb,oxy} Q_{fb}^{wdt} \end{cases} \quad (21)$$

where $\chi_{chp,oxy}$ and $\chi_{fb,oxy}$ represent the amount of oxygen required (m^3/MWh) to output unit power when the CHP unit and FB are in the oxygen-enriched combustion state.

3. Problem Formulation

After establishing the model of Cement-IIES in Section 2, this section builds a multi-stage planning model for the Cement-IIES including the objective function and constraints.

3.1. Objective Function

The system planning aims to minimize the expenditure in the entire planning cycle. As shown in (22), the expenditure includes initial equipment technology investment cost at each stage, power grid electricity exchange cost, fuel purchase cost, oxygen selling income, system equipment operation and maintenance cost, renewable energy curtailment and heat reduction penalty cost, carbon trading cost/gain and green certificate cost/gain.

$$\min T_{tot} = C_{inv} + \sum_{w\in N_w}\sum_{j\in N_y}\frac{1}{(1+r)^{(w-1)N_y+j}}\sum_{d\in N_d} n(d)\left(\begin{array}{c} C_{grid}^{wd} + C_{fuel}^{wd} - C_{oxy}^{wd} + C_{om}^{wd} + \\ C_{res,curt}^{wd} + C_{heat,curt}^{wd} + C_{carb}^{wd} + C_{gc}^{wd} \end{array}\right) \quad (22)$$

where N_w indicates the number of planning stages in the planning cycle and N_y represents the number of years involved in each stage; N_d is the number of typical days being considered for each year whilst $n(d)$ is the number of days for each typical day group; r is the discount rate.

In what follows, the mathematical expressions for each cost/income in Equation (22) are described in detail in Sections 3.1.1–3.1.9.

3.1.1. Investment Cost

The investment cost C_{inv} can be further divided according to the type of equipment, as shown in (23).

$$C_{inv} = \begin{pmatrix} C_{inv,wt} + C_{inv,pv} + C_{inv,elz} + C_{inv,chp} + C_{inv,fb} + C_{inv,hfc} + \\ C_{inv,eb} + C_{inv,ec} + C_{inv,ac} + C_{inv,ost} + C_{inv,hst} + C_{inv,bt} + \\ C_{inv,cwhc} + C_{inv,tche} + C_{inv,orc} + C_{inv,cpu} + C_{inv,asu} \end{pmatrix} \tag{23}$$

where the investment cost of WT and PV is indicated by $C_{inv,wt}$ and $C_{inv,pv}$ and they can be calculated by (24) and (25), respectively. The investment cost of other equipment can be derived using (26).

$$C_{inv,wt} = \sum_{w \in N_w} k_{inv,wt}^{w} N_{wt}^{lw} \frac{1}{(1+r)^{(w-1)N_y}} \tag{24}$$

$$C_{inv,pv} = \sum_{w \in N_w} k_{inv,pv}^{w} A_{pv}^{w} \frac{1}{(1+r)^{(w-1)N_y}} \tag{25}$$

$$C_{inv,\vartheta} = \sum_{w \in N_w} k_{inv,\vartheta}^{w} Cap_{\vartheta}^{lw} \frac{1}{(1+r)^{(w-1)N_y}} \tag{26}$$

where $k_{inv,\vartheta}^{w}$ denotes the unit investment cost of the equipment in stage w and the unit can be CNY/MW, CNY/MWh or CNY/m^3 depending on the equipment type.

Except for C_{inv}, the other costs are related to a typical day which will be described in the following sections.

3.1.2. Power Grid Electricity Exchange Cost

The power grid electricity exchange cost C_{grid}^{wd} consists of two parts including the electricity purchase fee and electricity sale revenue.

$$C_{grid}^{wd} = \sum_{t \in N_t} \left(k_{grid,buy}^{wdt} P_{grid,buy}^{wdt} - k_{grid,sell}^{wdt} P_{grid,sell}^{wdt} \right) \Delta t \tag{27}$$

where $P_{grid,buy}^{wdt}$ and $P_{grid,sell}^{wdt}$ represent the power (MW) purchased and sold in stage w of a typical day d at time t. The price (CNY/MWh) of purchasing and selling electricity is represented by $k_{grid,buy}^{wdt}$ and $k_{grid,sell}^{wdt}$, respectively.

3.1.3. Cost of Purchasing Various Types of Fuel

The purchase cost of fuel includes that for natural gas, coal, biomass fuel and hydrogen and can be calculated using Equation (28).

$$C_{fuel}^{wd} = \sum_{t \in N_t} \left(k_{ng,buy}^{wdt} V_{ng,buy}^{wdt} + k_{coal,buy}^{wdt} m_{coal,buy}^{wdt} + k_{bio,buy}^{wdt} m_{bio,buy}^{wdt} + k_{h,buy}^{wdt} V_{h,buy}^{wdt} \right) \Delta t \tag{28}$$

where $V_{h,buy}^{wdt}$ is the amount (m^3/h) of hydrogen purchased for the system; the purchase price (CNY/m^3) of natural gas and hydrogen is denoted by $k_{ng,buy}^{wdt}$ and $k_{h,buy}^{wdt}$. For the purchase price (MWh/ton) of coal and biomass fuel, they are represented by $k_{coal,buy}^{wdt}$ and $k_{bio,buy}^{wdt}$.

3.1.4. Cost of Selling Oxygen

The income of oxygen selling can be calculated using Equation (29).

$$C_{oxy}^{wd} = \sum_{t \in N_t} \left(k_{oxy,sell}^{wdt} V_{oxy,sell}^{wdt} \right) \Delta t \tag{29}$$

where $V^{wdt}_{oxy,sell}$ is the amount (m^3/h) of selling oxygen from the system; the selling price (CNY/m^3) of oxygen is denoted by $k^{wdt}_{oxy,sell}$.

3.1.5. Operation and Maintenance Cost of Equipment

The operation and maintenance cost of the system C^{wd}_{om} is shown in (30).

$$C^{wd}_{om} = \sum_{t \in N_t} \begin{bmatrix} k_{om,wt}P^{wdt}_{wt} + k_{om,pv}P^{wdt}_{pv} + k_{om,elz}P^{wdt}_{elz} + k_{om,chp}P^{wdt}_{chp} + k_{om,fb}Q^{wdt}_{fb} + \\ k_{om,hfc}P^{wdt}_{hfc} + k_{om,eb}Q^{wdt}_{eb} + k_{om,ec}Q^{wdt}_{ec} + k_{om,ac}Q^{wdt}_{ac} + k_{om,ost}\left(V^{lwdt}_{ost,ch} + V^{lwdt}_{ost,dis}\right) + \\ k_{om,hst}\left(V^{lwdt}_{hst,ch} + V^{lwdt}_{hst,dis}\right) + k_{om,bt}\left(P^{wdt}_{bt,ch} + P^{wdt}_{bt,dis}\right) + k_{om,cwhc}Q^{wdt}_{cwhc} + \\ k_{om,tche}Q^{wdt}_{tche} + k_{om,orc}P^{wdt}_{orc} + k_{om,cpu}P^{wdt}_{cpu} + k_{om,asu}P^{wdt}_{asu} \end{bmatrix} \Delta t \quad (30)$$

where $k_{inv,\vartheta}$ is the unit operation and maintenance cost of the equipment and the unit can be CNY/MWh or CNY/m^3 depending on the type of equipment.

3.1.6. Penalty Fee for Renewable Energy Curtailment

There will be a cost for curtailing wind and solar energy as calculated in (31).

$$C^{wd}_{res,curt} = \sum_{t \in N_t} \left[k^{wdt}_{wt,curt}\left(P^{wdt}_{wt,fore} - P^{wdt}_{wt}\right) + k^{wdt}_{pv,curt}\left(P^{wdt}_{pv,fore} - P^{wdt}_{pv}\right)\right]\Delta t \quad (31)$$

where $k^{wdt}_{wt,curt}$ and $k^{wdt}_{pv,curt}$ indicate the unit penalty fee (CNY/MWh) for wind and solar curtailment.

3.1.7. Penalty Fee for Heating Supply Reduction

If the total heat supply exceeds the heat demand of the system, a penalty fee needs to be paid to reduce the heat supply, as shown in (32):

$$C^{wd}_{heat,curt} = \sum_{t \in N_t} k^{wdt}_{heat,curt}\left(L^{wdt}_{heat,low} + L^{wdt}_{heat,high} - L^{wdt}_{heat}\right)\Delta t \quad (32)$$

where $k^{wdt}_{heat,curt}$ indicates the unit penalty fee (CNY/MWh) to reduce the heat supply; L^{wdt}_{heat} is the total heat demand (MW) of non-core activities; the high-grade heat demand (MW) of non-core activities is denoted by $L^{wdt}_{heat,high}$.

3.1.8. Carbon Trading Cost/Gain

The operating mechanism of the carbon market can be summarized into four parts: total amount setting, allowance allocation, trading and penalty for breach of contract [52]. Among them, the setting of the total amount, the allocation of carbon allowance and the filing of certified emission reductions are mainly completed in the primary market, and the transactions of carbon allowance and certified emission reductions are mainly carried out in the secondary market.

In this paper, the baseline method is adopted to determine the carbon allowance of the generator set in Cement-IIES. When the amount of carbon emission exceeds the free carbon allowance, expense needs to be paid and it can be calculated by (33) [53].

$$C^{wd}_{carb} = \sum_{t \in N_t} k^{w}_{carb,tax}\begin{pmatrix} E^{wdt}_{cal,carb} + E^{wdt}_{chp,carb} + E^{wdt}_{fb,carb} - E^{wdt}_{bio,carb} + \\ \varepsilon^{w}_{grid,carb}P^{wdt}_{grid,buy} - EA^{w}_{chp,ng} - EA^{w}_{chp,coal} \end{pmatrix}\Delta t \quad (33)$$

where $\varepsilon^{w}_{grid,carb}$ is the carbon emission factor (ton/MWh) of the external power grid power in stage w whilst $EA^{w}_{chp,ng}$ and $EA^{w}_{chp,coal}$ represents the carbon allowance (ton/h) of the ng-fired CHP and coal-fired CHP, respectively.

The carbon allowance is calculated using (34).

$$\begin{cases} EA_{chp,ng}^{wdt} = P_{chp,ng}^{wdt} b_{e,ng}^{w} f_{r,ng} + Q_{chp,ng}^{wdt} b_{h,ng}^{w} \times 3600 \times 10^{-3} \\ EA_{chp,coal}^{wdt} = P_{chp,coal}^{wdt} b_{e,coal}^{w} f_{l,coal} f_{r,coal} f_{f,coal} + Q_{chp,coal}^{wdt} b_{h,coal}^{w} \times 3600 \times 10^{-3} \end{cases} \quad (34)$$

where $b_{e,ng}^{w}$ is the power supply reference value (ton/MWh) of the gas-fired unit and $f_{r,ng}$ ($f_{r,ng} = 1 - 0.6r_{chp,ng}$) is the heat supply correction coefficient of the gas-fired unit; $b_{h,ng}^{w}$ represents the heat supply reference value (ton/GJ) of the gas-fired unit whilst $b_{e,coal}^{w}$ indicates the power supply reference value (ton/MWh) of the coal-fired unit. The correction coefficient of the cooling mode of the coal-fired unit can be denoted as $f_{l,coal}$ and will take the value 1; $f_{r,coal}$($f_{r,coal} = 1 - 0.22r_{chp,coal}$) is the correction coefficient of the heat supply of the coal-fired unit and $f_{f,coal}$ is the correction coefficient of the output of the coal-fired unit whose value is 1; $b_{h,coal}^{w}$ indicates the heat supply reference value (ton /GJ) of the coal-fired unit.

It is worth noting that, in China, there is no carbon allowance for pure heating facilities without power generation capability such as FB. In addition, the planning work for Cement-IIES in this paper does not consider dynamic carbon trading, but only sets the carbon price according to the current situation of China's carbon trading market.

3.1.9. Green Certificate Cost/Gain

Green certificate is a certificate issued by the national energy authority to renewable energy power generators, which proves that the generator has generated a portion of its electricity from renewable sources [54]. The green certificate mechanism is a supporting measure to ensure the effective implementation of the Renewable Portfolio Standard (RPS), and the purpose of implementing the RPS and green certificate trading mechanisms is to gradually transition renewable energy generation from government subsidies to market-based subsidies.

When the weight of the renewable energy power consumption is lower than the required value, the entity which is responsible to consume renewables can purchase green certificates from the market to compensate for the deficit. By contrast, the green certificate can be sold to earn a profit if the weight of the renewable consumption is higher than the target value [39]. Equation (35) shows how the green certificate cost and gain are derived.

$$C_{gc}^{wd} = \sum_{t \in N_t} k_{gc}^{w} \left[\left(P_{chp}^{wdt} + \xi_{chp} Q_{chp}^{wdt} \right) \beta_{ge}^{w} - \left(P_{wt}^{wdt} + P_{pv}^{wdt} \right) \left(1 - \beta_{ge}^{w} \right) \right] \Delta t \quad (35)$$

where ξ_{chp} is the conversion coefficient of power-heat supply of CHP; β_{ge}^{w} represents the required value for the weight of renewable energy consumption in stage w whilst k_{gc}^{w} denotes the price (CNY/MWh) of the green certificate.

3.2. Constraints

3.2.1. Balance of Multi-Energy Flow Buses

There are power, heat, cooling, hydrogen and oxygen buses in the system.

(1) Power Bus

The power bus balance is shown in (36).

$$\left(\begin{array}{l} P_{grid,buy}^{wdt} + P_{wt}^{wdt} + P_{pv}^{wdt} + P_{chp}^{wdt} + P_{hfc}^{wdt} + P_{bt,dis}^{wdt} + P_{orc}^{wdt} = \\ P_{grid,sell}^{wdt} + P_{elz}^{wdt} + Q_{eb}^{wdt}/\eta_{eb} + Q_{ec}^{wdt}/COP_{ec} + P_{bt,ch}^{wdt} + P_{cpu}^{wdt} + V_{asu}^{wdt}\kappa_{asu} + P_{raw}^{wdt} + L_{elec}^{wdt} \end{array} \right) \quad (36)$$

where Q_{eb}^{wdt} represents the output heating power (MW) of the EB with η_{eb} indicating the energy conversion efficiency of the EB. Q_{ec}^{wdt} is the output cooling power (MW) of the EC with COP_{ec} denoting the energy efficiency coefficient of EC. The amount of oxygen (m^3/h) produced by the ASU is represented by V_{asu}^{wdt} whilst κ_{asu} is the energy (MWh/m^3) required for ASU oxygen production. L_{elec}^{wdt} indicates the power demand (MW) of non-core activities.

(2) Heating Bus

In addition to the low-grade heating bus balance defined in (17), the system also needs to satisfy the high-grade heating bus balance constraints, as shown in (37).

$$\begin{array}{l} Q_{chp}^{wdt} + Q_{fb}^{wdt} + Q_{hfc}^{wdt} + Q_{eb}^{wdt} = Q_{cal}^{wdt} + L_{heat,high}^{wdt} \\ Q_{cal}^{wdt} = \kappa_{cal} m_{lime}^{wdt} \theta_{cal} \end{array} \quad (37)$$

where Q_{cal}^{wdt} indicates the heat demand (MW) during the limestone calcination process; κ_{cal} is the heat demand (MWh/ton) to calcine limestone; m_{lime}^{wdt} denotes the amount (ton/h) of cement produced; θ_{cal} represents the proportional coefficient between the amount of limestone calcined and the amount of cement produced and the typical value is 1.31.

The high-grade and low-grade heat demands of non-core activities also need to satisfy the conditions specified in the expression (38).

$$\begin{array}{l} L_{heat,low}^{wdt} + L_{heat,high}^{wdt} \geq L_{heat}^{wdt} \\ L_{heat,high}^{wdt} \geq \gamma_{heat} \cdot L_{heat}^{wdt} \end{array} \quad (38)$$

where γ_{heat} shows the requirement for the minimum proportion of high-grade heat demand within the total heat demand.

(3) Cooling Bus

The cold energy demand of the system is provided by both the EC and AC and the constraint for cooling bus balance is illustrated by (39).

$$Q_{ac}^{wdt} + Q_{ec}^{wdt} \geq L_{cool}^{wdt} \quad (39)$$

where L_{cool}^{wdt} is the cold energy demand (MW) of non-core activities.

(4) Oxygen Bus

From (40), the oxygen is supplied by the ELZ and ASU and consumed by the CHP unit, the FB where oxygen-enriched combustion occurs and the oxygen trading with external systems. It is also worth noting that the OST participates in the peak-shaving and valley-filling of the oxygen bus.

$$V_{elz,oxy}^{wdt} + V_{ost,dis}^{wdt} + V_{asu}^{wdt} = V_{ost,ch}^{wdt} + V_{chp,oxy}^{wdt} + V_{fb,oxy}^{wdt} + V_{oxy,sell}^{wdt} \quad (40)$$

where $V_{oxy,sell}^{wdt}$ is the volume (m^3/h) of oxygen being sold.

(5) Hydrogen Bus

The balance of the hydrogen bus is depicted by (41).

$$V_{elz,h}^{wdt} + V_{hst,dis}^{wdt} + V_{h,buy}^{wdt} = V_{hst,ch}^{wdt} + V_{hfc}^{wdt} \quad (41)$$

where $V_{h,buy}^{wdt}$ is the amount (m^3/h) of hydrogen purchased from the hydrogen suppliers.

3.2.2. Constraint for Equipment Installed Capacity

The installed capacity of the WT and PV in each stage also needs to meet the upper and lower capacity limit, as shown in (42) and (43), whilst other equipment needs to satisfy the constraints defined in (44).

$$0 \leq \pi \cdot (5R_{wt})^2 \cdot N_{wt}^w \leq A_{wt,\max}^w \quad (42)$$

$$0 \leq A_{pv}^w \leq A_{pv,\max}^w \quad (43)$$

$$0 \leq Cap_{\vartheta}^{w} \leq Cap_{\vartheta,\max}^{w} \tag{44}$$

where $A_{wt,\max}^{w}$ and $A_{pv,\max}^{w}$ represent the maximum construction area (m^2) of the WT and PV in stage w, respectively. $Cap_{\vartheta,\max}^{w}$ is the maximum installed capacity of the equipment in stage w and the unit can be CNY/MW, CNY/MWh or CNY/m^3 depending on the equipment type.

3.2.3. Constraint for Equipment Operating State

The actual electric power used by the WT and PV should be less than the associated predicted values, as shown in expression (45) and (46). The electrical output or heating output of other equipment needs to meet the constraints defined in (47).

$$0 \leq P_{wt}^{wdt} \leq P_{wt.fore}^{wdt} \tag{45}$$

$$0 \leq P_{pv}^{wdt} \leq P_{pv.fore}^{wdt} \tag{46}$$

$$\begin{cases} \mu_{\chi,\min} \sum\limits_{k \leq w} Cap_{\chi}^{k} \leq P_{\chi}^{wdt} \leq \mu_{\chi,\max} \sum\limits_{k \leq w} Cap_{\chi}^{k} \\ \mu_{\chi,\min} \sum\limits_{k \leq w} Cap_{\chi}^{k} \leq Q_{\chi}^{wdt} \leq \mu_{\chi,\max} \sum\limits_{k \leq w} Cap_{\chi}^{k} \end{cases} \tag{47}$$

where $\mu_{\chi,\min}$ and $\mu_{\chi,\max}$ represent the minimum and maximum values of the proportional coefficients between the equipment output and capacity.

3.2.4. Constraint for Grid Power Exchange

Electricity will be purchased from the external grid when there is an insufficient power supply in the system. By contrast, the system will sell electricity when there is excess power. The electricity exchange needs to satisfy the constraints shown in (48).

$$\begin{cases} x_{grid,buy}^{wdt} \cdot P_{grid,buy,\min}^{w} \leq P_{grid,buy}^{wdt} \leq x_{grid,buy}^{wdt} \cdot P_{grid,buy,\max}^{w} \\ x_{grid,sell}^{wdt} \cdot P_{grid,sell,\min}^{w} \leq P_{grid,sell}^{wdt} \leq x_{grid,sell}^{wdt} \cdot P_{grid,sell,\max}^{w} \\ x_{grid,buy}^{wdt} + x_{grid,sell}^{wdt} \leq 1 \end{cases} \tag{48}$$

where $P_{grid,buy,\min}^{w}$ and $P_{grid,buy,\max}^{w}$ denote the minimum and maximum amount (MW) of purchased electricity in stage w. $P_{grid,sell,\min}^{w}$ and $P_{grid,sell,\max}^{w}$ represent the minimum and maximum amount (MW) of sold electricity. $x_{grid,buy}^{wdt}$ and $x_{grid,sell}^{wdt}$ are binary variables which are used to indicate that the system cannot simultaneously purchase and sell electricity.

3.2.5. Constraint for Fuel Purchase and Selling

The amount of purchased and sold natural gas, coal, biomass fuel, hydrogen and oxygen need to satisfy the constraints specified in (49).

$$\begin{cases} V_{ng,buy}^{wdt} = V_{chp,ng}^{wdt} + V_{fb,ng}^{wdt}, V_{ng,buy,\min}^{w} \leq V_{ng,buy}^{wdt} \leq V_{ng,buy,\max}^{w} \\ m_{coal,buy}^{wdt} = m_{chp,coal}^{wdt} + m_{fb,coal}^{wdt}, m_{coal,buy;\min}^{w} \leq m_{coal,buy}^{wdt} \leq m_{coal,buy,\max}^{w} \\ m_{bio,buy}^{wdt} = m_{chp,bio}^{wdt} + m_{fb,bio}^{wdt}, m_{bio,buy,\min}^{w} \leq m_{bio,buy}^{wdt} \leq m_{bio,buy,\max}^{w} \\ V_{h,buy,\min}^{w} \leq V_{h,buy}^{wdt} \leq V_{h,buy,\max}^{w} \\ V_{oxy,sell,\min}^{w} \leq V_{oxy,sell}^{wdt} \leq V_{oxy,sell,\max}^{w} \end{cases} \tag{49}$$

where $V_{ng,buy,\min}^{w}$ and $V_{ng,buy,\max}^{w}$ show the minimum and maximum amount (m^3/h) of natural gas being purchased in stage w. The minimum and maximum purchased amount (ton/h) of coal in stage w is represented by $m_{coal,buy,\min}^{w}$ and $m_{coal,buy,\max}^{w}$. $m_{bio,buy,\min}^{w}$ and $m_{bio,buy,\max}^{w}$ depict the minimum and maximum purchased amount (ton/h) of biomass fuel in stage w whilst $V^{w}_{h,buy,\min}$ and $V^{w}_{h,buy,\max}$ denote the minimum and maximum volume (m^3/h) of hydrogen purchased in stage w. The minimum and maximum volume (m^3/h) of oxygen sold in stage w is $V^{w}_{oxy,sell,\min}$ and $V^{w}_{oxy,sell,\max}$.

4. Case Study

Considering the current production situation of the cement factory production line, the planning of Cement-IIES is carried out in an industrial park in the southeast coastal area of China. Figure 3a shows the production characteristics of typical days in the spring-autumn transition season, summer and winter. In addition to the core production activities, the characteristics of the electric heating and cooling demand of the cement workshops, office buildings and dormitory areas are presented in Figure 3b–d. On the summer typical day in Figure 3b, the electric load is dominant while the heating load is at its trough throughout the year. When it comes to the winter typical day in Figure 3c, the system's heating load other than the lime production line exceeds the cooling load because of low temperature. The load characteristics of spring/autumn typical days in Figure 3d are between summer and winter typical days.

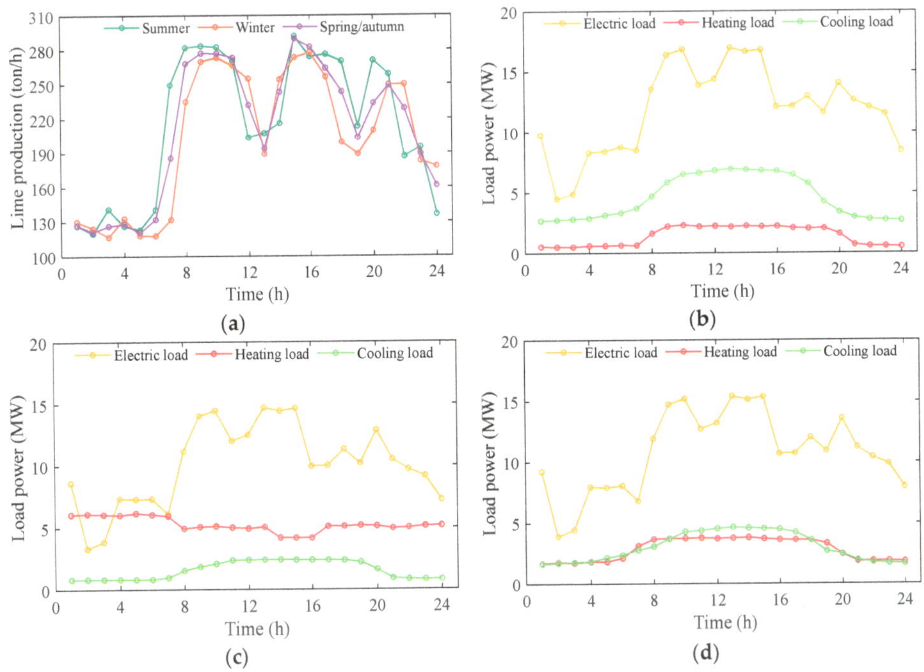

Figure 3. Characteristics of Cement Production and Seasonal Load of Cement Factory: (**a**) Cement Production; (**b**) Summer Load; (**c**) Winter Load; (**d**) Transition Season Load (Spring–Autumn).

4.1. Case Design

The planning cycle of the cement factory energy system is 20 years and it is intended to expand the production line to increase cement production in the later stage of the planning cycle. The multi-stage planning model proposed in this paper is applied to conduct multi-stage flexible planning research on the cement factory with a planning cycle divided into two stages and each stage is 10 years long. In the planning model, the influence of various technical, economic, political and environmental factors is considered when setting the parameters. More specifically, these factors include the resource characteristics of the area, the policy environment for carbon trading and green certificate trading, the efficiency of energy production, cost changes, price trends of external energy market transactions and constraints such as available construction area.

To evaluate the performance of the proposed planning method, four sub-cases are compared. Case 1 is configured as the base case and uses the phased implementation method which is the modeling and planning method proposed in this paper. For Case 2, the traditional single-stage planning method (i.e., one-step-reach) is applied without considering the phased changes in the load and complex parameters. Different from Case 1, Case 3 does not consider the recycling of low-grade waste heat from the rotary kiln system whilst Case 4 does not involve the use of oxygen-enriched combustion and carbon capture technologies. The parameter configuration of the test system can be seen in Appendix A.

4.2. Comparison of Case Results

The case study is run on a desktop computer with an i7-9700KF processor and 32 GB memory. The program is written in Matlab R2022a and solved by Yalmip + Gurobi (Version 9.5.1). Table 1 shows the results of Cases 1–4.

Table 1. Comparison of Case Results: Cost and Benefit.

	Case 1		Case 2	Case 3		Case 4	
	Stage I	Stage II		Stage I	Stage II	Stage I	Stage II
Total cost (CNY 10^{10})	1.11		1.29	1.13		1.33	
Total cost by stages (CNY 10^9)	7.29	3.78	-	7.43	3.85	8.36	4.97
Initial devices investment cost (CNY 10^8)	12.42	3.61	12.30	12.51	3.58	10.92	3.30
Annual green certificate trading cost (CNY 10^7)	−1.36	−2.65	−1.26	−1.36	−2.65	−1.40	−2.71
Annual natural gas purchasing cost (CNY 10^8)	2.53	3.34	2.90	2.56	3.39	2.18	2.73
Annual coal purchasing cost (CNY 10^8)	3.17	4.73	4.73	3.17	4.74	3.15	4.67
Annual biomass purchasing cost (CNY 10^8)	1.22	2.28	1.90	1.22	2.28	1.22	2.28
Annual hydrogen purchasing cost (CNY 10^6)	0.00	0.00	0.00	0.00	0.00	5.25	0.00
Annual electricity trading cost (CNY 10^7)	11.68	9.58	12.83	12.35	10.87	5.11	0.57
Annual oxygen trading cost (CNY 10^7)	2.99	3.58	3.58	2.99	3.58	2.99	3.58
Annual maintenance cost (CNY 10^7)	4.34	5.41	5.07	4.21	5.28	4.04	5.01
Fuel carbon cost (CNY 10^7)	2.59	1.48	2.69	3.21	1.60	20.03	32.17
Electricity carbon cost (CNY 10^7)	1.40	1.28	1.38	1.47	1.48	0.59	0.32
Limestone calcination carbon cost (CNY 10^7)	1.70	0.52	2.10	1.90	0.52	8.89	14.94
Biomass carbon income (CNY 10^7)	3.10	7.89	4.91	3.05	7.92	0.00	0.00
Annual fuel CO_2 emission (10^5 ton)	3.59	2.23	3.96	4.11	2.30	18.07	20.46
Annual electricity CO_2 emission (10^5 ton)	1.17	0.76	1.15	1.22	0.88	0.49	0.19
Annual limestone calcination CO_2 emission (10^5 ton)	1.42	0.31	1.75	1.58	0.31	7.41	8.89
Penalty fee for wind/solar power curtailment (CNY)	0.00	0.00	0.00	0.00	0.00	0.00	0.00
Penalty fee for heating power curtailment (CNY)	0.00	0.00	0.00	0.00	0.00	0.00	0.00
Energy utilization efficiency (%)	68.24	70.18	69.52	67.79	69.32	69.89	71.91

In terms of economy, the total cost of Case 1 in the entire planning cycle is lower than that of the other three cases and accounts for 86.12% of the cost of Case 2, 98.16% of the cost of Case 3 and 83.02% of the cost of Case 4. Case 2 does not consider the development of energy equipment technology, the available area for renewable energy construction and the energy demand of cement factories in various stages, resulting in excess requirements for traditional energy equipment in the early stage of planning and insufficient utilization of renewable resources including WT and PV in the later stage. Meanwhile, carbon capture reduces the cost of the carbon emission and increases income via the "negative carbon emission" of biomass fuel. Furthermore, the reuse of low-temperature waste heat reduces the system's reliance on the external energy source. All of the aforementioned factors improve the economic performance of the system.

From the perspective of environmental protection, Case 1 has the lowest amount of total carbon emission from limestone calcination, fossil fuel use and electricity consumption because the carbon emitted from the combustion of coal and natural gas is captured and stored by the CPU. Compared with Case 4, the carbon emission of Case 1 is approximately 17% of that in Case 4. In addition, there are gains of CNY 3.10×10^7 and CNY 7.89×10^7 from the "negative carbon emission" of biomass fuel after using oxygen-enriched combustion and carbon capture.

Concerning energy efficiency, the energy utilization efficiency of Case 1, Case 3 and Case 4 in the second stage is higher than that in the first stage due to the improvement in the energy conversion efficiency of the WT, PV, ELZ, HFC and other equipment. It can be observed that the energy efficiency of Case 2 where single-stage planning is applied is not the lowest and the energy efficiency ratio in both stages of Case 4 is higher than that of Case 1. This is because the cost of oxygen storage and energy loss during carbon capture can slightly lower the system's energy efficiency even though the oxygen-enriched combustion and carbon capture technologies deployed in Case 1 and Case 3 can significantly improve the environmental performance of the system. Moreover, the energy efficiency ratios of the two stages in Case 1 are higher by 0.45% and 0.86% compared to Case 3 because of the reuse and utilization of low-grade waste heat from the rotary kiln system.

4.3. Capacity and Investment Costs of Equipment in Various Stages

The results of Case 1 will be comprehensively analyzed in the following sections. The installed equipment capacity is shown in Figure 4 whilst the investment cost proportion for the two stages is presented in Figure 5. It should be noted that the meanings of the upper and lower case abbreviations for the equipment or technologies in Figures 4 and 5 are given in the Nomenclature.

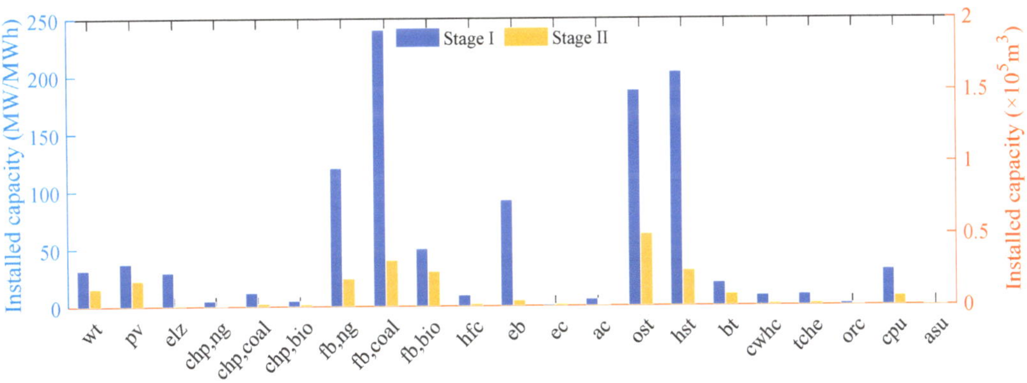

Figure 4. Installed Capacity of Equipment in Two Stages of Case 1.

Since cement production requires a massive amount of heat energy, the installed capacity of the FB and EB used to provide the heat load of calcination is significantly higher than other energy-producing and consuming equipment. In the first stage, the eight types of equipment with the highest investment cost proportion are the WT, PV, ELZ, FB, HFC, CHP, EB and CPU. In the second stage, the ranking slightly changes but the renewable units, FB and CHP still account for the major portion of the cost.

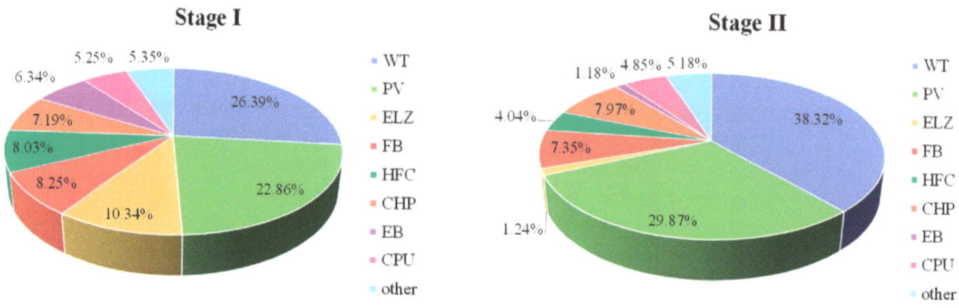

Figure 5. Proportion of Equipment Investment Cost in Two Stages of Case 1.

The total installed capacity of WT and PV is 69.5 MW and 38.5 MW. Due to the high construction and maintenance cost of new energy units, WT and PV account for 49.25% of the total investment in the first phase and reaches 68.19% in the second stage.

CHP has installed a capacity of 22 MW and 5.4 MW in two stages, accounting for 7.19% and 7.97% of the total investment. Although the installed capacity of the bio-fired CHP is not large, it accounted for 6.02% of the total investment in the second stage due to its relatively high investment cost (around 6.3 times coal-fired CHP and 3.7 times ng-fired CHP).

The total investment in FB accounted for 8.25% and 7.35% in the first and second stages, respectively. The coal-fired FB in the first stage has 240 MW capacity and accounts for 4.06% of the investment. Limited by the "coal reform" policy, the newly installed capacity of the bio-fired FB in the second stage exceeds that of the coal-fired FB, which accounts for 2.78% of the total investment.

The EB can perform as an alternative source of heat and its installed capacity is 91.99 MW and 4.71 MW in the first and second stages, respectively. The corresponding investment cost accounts for 6.34% and 1.18%, and the EB achieves partial electrification replacement of FB.

The CPU can significantly reduce carbon emission during limestone calcination and fuel combustion and its investment cost accounts for 5.25% of the total cost in the first stage and 4.85% in the second stage.

Compared with the heat demand for limestone calcination, the demand for heat and cold energy in the non-core activities of the industrial park is relatively low. Therefore, the installed capacity and investment proportion of the AC and EC equipment are relatively small.

4.4. System Operation Analysis

In the previous subsections, the overall configuration results of planning schemes for Cement-IIES have been presented and compared. This section focuses on the detailed operation status of Cement-IIES in Case 4, which adopts the proposed modeling and planning method. To be specific, multi-energy flow balance, key equipment operation output and combustion and carbon capture state are analyzed in sequence.

4.4.1. Multi-Energy Flow Balance

The system energy demand at terminals and the energy flow balance are shown in Figure 6. The energy flow relationship of electricity, high-grade heating, low-grade heating and cooling is given in Figure 6a–d, respectively.

As shown in Figure 6a, the power of the system is mainly provided by the renewable units, purchased electricity and CHP unit whilst most of the power is consumed by the EB, ELZ and CPU. It can be observed that the system will have two power demand peaks on different typical days, which is mainly affected by the output state of the EB. To fill the peak power demand, the BT will switch to the discharging state and the HFC will generate full power during the peak hours. The peak of electrical load for non-core activities occurs between 8:00 and 16:00, which coincides with the peak output of WT and PV. Hence, there is no large amount of renewable energy available to be used by the ELZ to produce hydrogen and oxygen in this period.

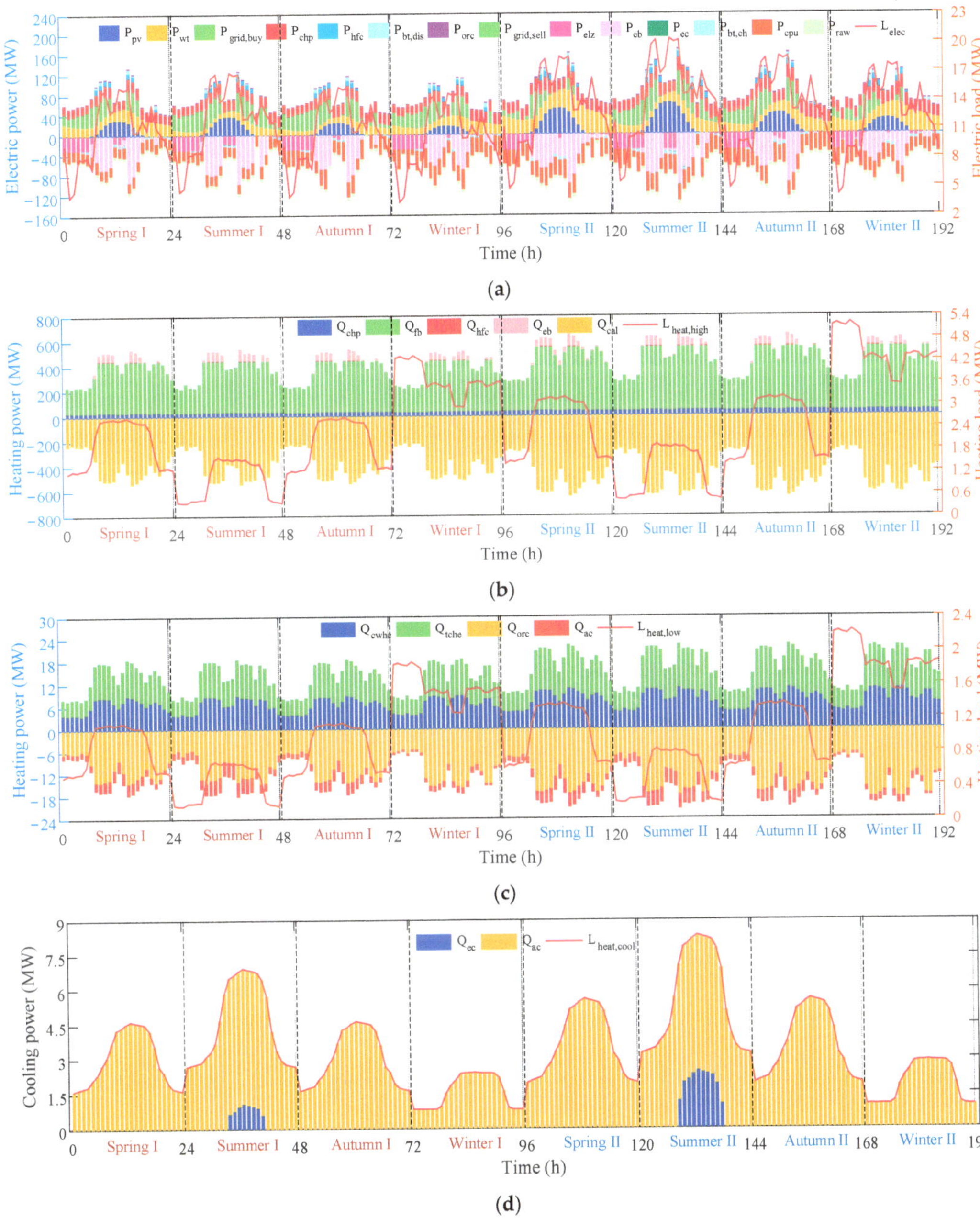

Figure 6. Multi-Energy Flow Bus Balance in Two Stages of Case 1: (**a**) Electrical Power; (**b**) High-Grade Heat Demand; (**c**) Low-Grade Heat Demand; (**d**) Cooling Energy.

The high-grade heating energy in the cement industrial park is mainly used for limestone calcination and simultaneously satisfies the high-grade heating demand of non-core activities. The low-grade heating energy is mainly used for low-temperature waste heat power generation and meets the low-grade heating demand of non-core activities. In the case study, the high-grade heat demand of non-core activities accounts for at least 70% of the total heat demand. Based on Figure 6b, most of the high-grade heating energy is consumed by limestone calcination, accounting for approximately 99.45% of the total high-grade heating energy demand. This high-grade heating energy is mainly supplied by the FB with the CHP, EB and HFC used as the supplementary heating providers.

By contrast, the low-grade heating demand of the system is covered by the low-temperature waste heat recovered from the rotary kiln system, as shown in Figure 6c. It can be seen that the surplus low-temperature waste heat is used to drive ORC power generation and supply AC refrigeration.

In Figure 6d, the cooling energy demand of the system is mainly provided by the AC and EC acts as the supplementary source for the cooling demand peak on a typical summer day.

4.4.2. Operation Output of Key Equipment

Based on the energy flow balance analysis in Section 4.4.1, Figures 7 and 8 further illustrate the output of the CHP and FB. More specifically, the coal-fired CHP and bio-fired CHP keep generating at full power and this is because the CHP discussed in this paper operates in the heat-constant power mode which means the heat-to-power ratio of the coal-fired CHP ($r_{chp,coal}$ in Equation (6)) and bio-fired CHP ($r_{chp,bio}$ in Equation (6)) is higher than that of the ng-fired CHP ($r_{chp,ng}$ in Equation (6)). To satisfy the high heat demand, the former two CHPs have a higher priority to generate than the ng-fired CHP.

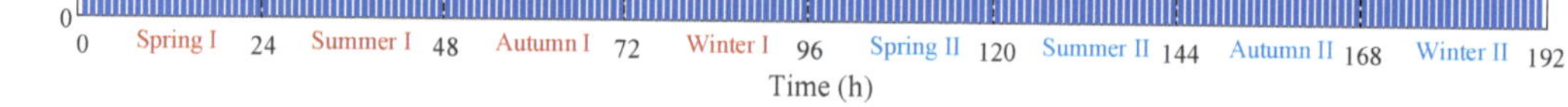

Figure 7. CHP Output in Two Stages of Case 1.

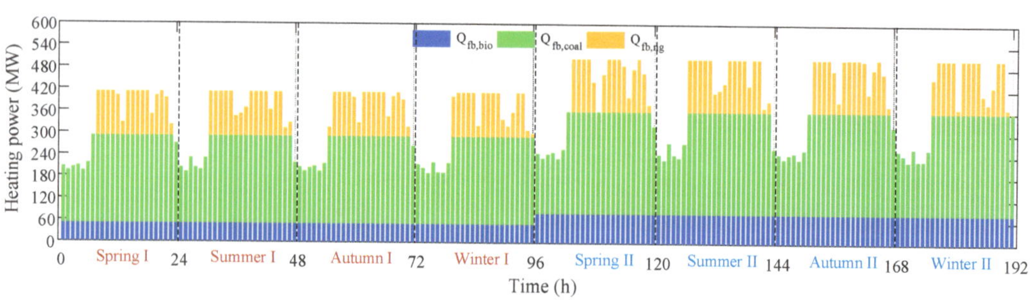

Figure 8. FB Output in Two Stages of Case 1.

Regarding FB, the heating efficiency of bio-fired FB ($\eta_{fb,bio}$ in Equation (8)) can reach 0.9 which is higher than 0.6 for coal-fired FB ($\eta_{fb,coal}$ in Equation (8)). In addition, biomass fuel can be used in conjunction with CPU to achieve "negative carbon emission" so that carbon trading subsidies can be obtained. Consequently, the bio-fired FB has the highest output priority and it typically keeps generating at full power.

Figure 9 depicts the relationship between the supply and demand of hydrogen and oxygen relationship, as well as the operating status of the HST and OST.

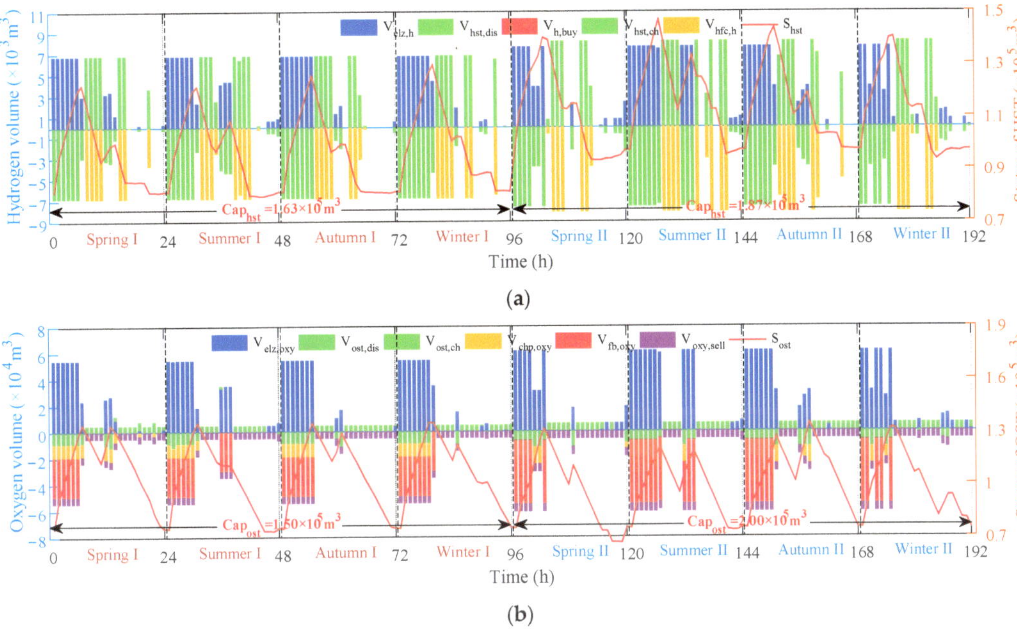

Figure 9. Relationship between Oxygen/hydrogen Input and Output in Two Stages of Case 1: (**a**) Hydrogen; (**b**) Oxygen.

Figure 9a presents the relationship between oxygen input and output in two stages of Case 1. It can be seen that the HST installed capacity in the first stage is 1.63×10^5 m^3 and it is expanded to 1.87×10^5 m^3 in the second stage. The HST is mainly used to store the hydrogen produced by the ELZ during the low power demand period and supply hydrogen when the HFC is dispatched to shave the peak demand. In addition, the hydrogen supply in the system is sufficient, so there is no need to purchase additional hydrogen from external suppliers.

Concerning the OST, the installed capacity in the first stage is 1.50×10^5 m^3, which is expanded to 2.00×10^5 m^3 in the second stage, as shown in Figure 9b. The oxygen discussed in this paper is the by-product when the ELZ electrolyzes water to produce hydrogen and can satisfy the oxygen demand of the CHP and FB to achieve oxygen-enriched combustion. The relationship between the oxygen input and output in Figure 9b is more complex than that of hydrogen, where most of the oxygen is injected into the FB and CHP. Due to the high market price of oxygen, the system keeps selling 5×10^3 m^3/h and 6×10^3 m^3/h of oxygen to external parties in the first and second stages, respectively. It is worth noting that the OST has a relatively weak capability for peak-shaving and valley-filling because of limited construction capacity.

4.4.3. State of Oxygen-Enriched Combustion and Carbon Capture

Since the oxygen-enriched combustion and carbon capture technologies have a significant influence on the system's environmental performance, this subsection analyzes their operation status in detail. The hourly carbon emission and capture during the limestone calcination process on each typical day are shown in Figure 10. The dark blue part of the stacked histogram represents the amount of carbon emission from the limestone calcination process after applying the carbon capture technology whilst the light blue session indicates the amount of reduced carbon emissions by using carbon capture technology. A large portion of CO_2 is produced during the limestone calcination process with the carbon emission peaking at 118.11 ton/h and 136.81 ton/h in the first and second stages, respectively. In certain cement production peak periods in the first stage, the capture level of CPU is 0 but it typically can reach the highest value of 98% during other peak periods. In the second stage, the installed capacity of the CPU is further expanded and 98% of the carbon emissions are captured most of the time.

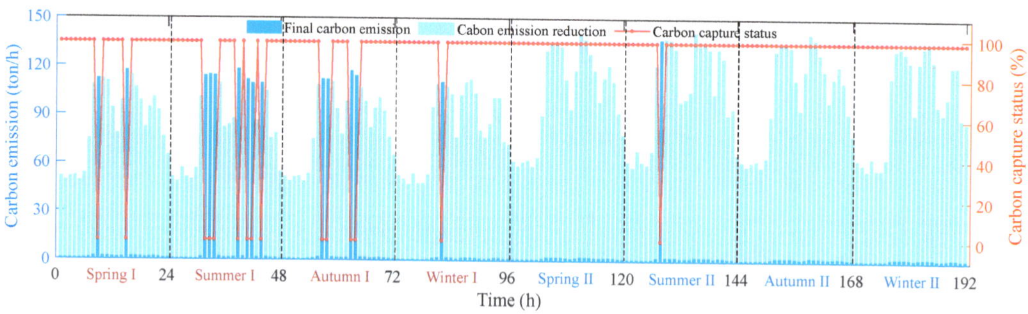

Figure 10. Carbon Emission and Capture during Limestone Calcination in Two Stages of Case 1.

The application of oxygen-enriched combustion in the CHP and FB can affect the upper limit for the carbon capture level of the CPU. More specifically, the carbon capture level can reach 98% if oxygen-enriched combustion is employed. Otherwise, the upper limit of the carbon capture level will be 90%. The state of oxygen-enriched combustion and carbon capture of the three types of CHP and FB under the aforementioned condition is shown in Figure 11. When the carbon capture level is higher than the red dotted line (90%), it means the CHP/FB has implemented oxygen-enriched combustion.

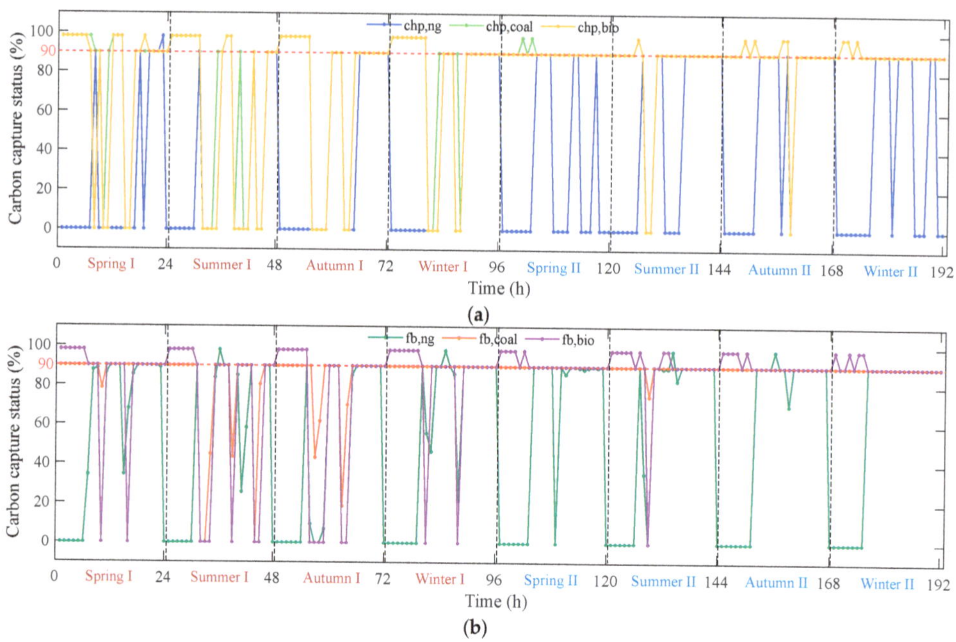

Figure 11. State of Oxygen-Enriched Combustion and Carbon Capture of CHP and FB in Two Stages of Case 1: (**a**) CHP; (**b**) FB.

The CHP in Figure 11a and FB in Figure 11b operate in the oxygen-rich combustion state mostly at night, that is, 0:00–6:00. In this period, the green power is abundant while the power load is low, so the surplus renewable generation can be used to drive the ELZ and produce hydrogen and oxygen. The oxygen is then supplied to the CHP and FB to support oxygen-rich combustion. This finding is consistent with the trends shown in Figures 6a and 9b.

Comparing Figure 10 with Figure 11, it can be observed that the moment when CHP and FB are in the state of oxygen-enriched combustion misalign with the moment when the carbon capture level of limestone calcination reaches the upper limit. This is because of the upper limit on the total amount of carbon that can be captured by the CPU.

This paper also considers the "negative carbon emission" effect of biomass fuel, and the carbon emission of bio-fired CHP and bio-fired FB after carbon capture is illustrated in Figure 12.

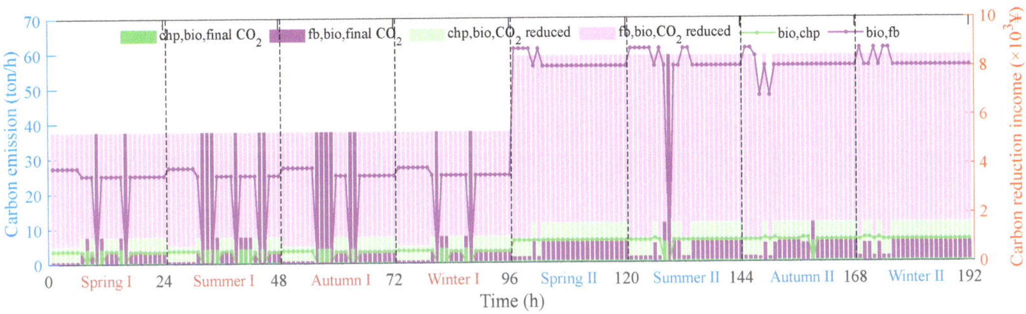

Figure 12. "Negative Carbon Emission" Effect of Biomass Fuel Equipment in Two Stages of Case 1.

In Figure 12, the dark green and dark purple parts of the stacked histogram represent the carbon emission from the bio-fired CHP and bio-fired FB when carbon capture technology is applied whilst the light green and light purple session indicates the amount of reduced carbon emission by using carbon capture technology. It can be seen that the maximum income of 4.41 × 10^3 CNY/h and 9.66 × 10^3 CNY/h can be obtained in the first and second stages if carbon capture is implemented on biomass fuel equipment. With the potential economic benefits of "negative carbon emission" from biomass fuels, among the three types of CHP and FB, bio-fired CHP and bio-fired FB are optimized to operate more in the oxygen-rich combustion state, so as to achieve a higher level of carbon capture, which can also be found in Figure 11.

5. Discussion

This paper effectively addressed the lack of IIES planning methods that combine specific industrial processes. Taking the cement industry as an example, we focus on the massive carbon emissions caused by lime calcination and skillfully couple the production process with the multi-energy system to convert low-temperature waste heat dissipated in the air into useful energy for production and living needs. Especially, the integration of oxygen-enriched combustion and carbon capture technology can help cement plants improve their environmental performance at the source of carbon emissions.

In addition to the comprehensive utilization of carbon reduction technologies, the multi-stage planning method proposed in this paper can overcome the inflexibility issue of traditional single-stage planning, and help the cement industry achieve optimal performance in terms of environmental protection, economic efficiency and energy efficiency, which have been validated in the above case study.

Although a series of valuable qualitative and quantitative findings have been obtained, there are still some shortcomings or areas for further research.

From a technical perspective, the impact of oxygen-enriched combustion on generation set operating conditions outside of carbon emissions, especially the emissions characteristics of SO_2 and NOx, has not been considered. In addition, the next step in dealing with CO_2 after the compression and purification by the CPU has not been fully explored. If underground storage is considered, issues such as requirements on geological conditions and transportation, as well as impacts on the surrounding environment need to be analyzed.

The planning approach and method proposed in this paper can be applied to other energy-intensive and carbon-intensive industries. In China, although pilot carbon markets in some regions have covered some industrial and transportation sectors, the national carbon market currently only includes the power generation sector, and the carbon allowance in this paper only considers power generation units. If heating systems, cement and other industries' production processes can be included in the national carbon market in the future, the planning model in this paper can be further improved.

Due to the long-term nature of planning issues, the carbon trading and green certificate trading mechanisms and prices are relatively simple in this paper, with fixed carbon prices and green certificate prices at specific stages. It will be meaningful to further conduct research on real-time or intra-day optimal operation strategies for Cement-IIES and analyze the system's potential of flexibly adjusting to the dynamic carbon price and green certificate price. In particular, it is meaningful to analyze the sensitivity of low-carbon energy technologies to incentive signals such as carbon price, green certificate price and external system flexibility requirements.

6. Conclusions

This paper focuses on the modeling and multi-stage planning of the Cement-IIES with high energy consumption and carbon emission. Combined with the energy flow and material flow of the actual cement production line, a Cement-IIES considering multiple carbon reduction technologies including low-temperature waste heat recovery, oxygen-enriched combustion, carbon capture, hydrogen and oxygen production using electrolysis, energy storage and fuel/technology substitution is proposed. Furthermore, a multi-stage flexible planning method for the Cement-IIES with consideration of green certificate trading and carbon trading is proposed by examining the periodic changes in technical and economic parameters. The case study demonstrates that the economic efficiency of the proposed planning method improves by 13.88% when compared with the traditional single-stage planning method. Meanwhile, it can flexibly adapt to changes in various policies including the "coal reform", green certificate trading and carbon allowance and thus allocate reasonable investment in equipment in different stages. The application of low-temperature waste heat recovery technology can increase the system energy efficiency ratio by 0.45% and 0.86% in the first and second stages whilst the oxygen-enriched combustion and carbon capture technology can reduce the total carbon emission of the system by approximately 83%. In addition, the "negative carbon emission" effect of the biomass fuel equipment can help the system earn 3.10×10^7 CNY/year and 7.89×10^7 CNY/year in the first and second stages, respectively. In addition, the limitations of this work and our future research directions are discussed.

Author Contributions: Conceptualization, Z.G. and S.Z.; methodology, Z.G. and S.Z.; software, Z.G.; validation, S.Z.; writing—original draft preparation, Z.G. All authors have read and agreed to the published version of the manuscript.

Funding: This research was funded by the National Natural Science Foundation of China with grant number 52177076.

Institutional Review Board Statement: Not applicable.

Informed Consent Statement: Not applicable.

Data Availability Statement: The data presented in this study are available on request from the corresponding author. The data are not publicly available due to privacy.

Conflicts of Interest: The authors declare no conflict of interest.

Nomenclature

WT, wt	Wind turbine
PV, pv	Photovoltaic
ELZ, elz	Electrolyzer
CHP, chp	Combined heat and power
FB, fb	Fired boiler
HFC, hfc	Hydrogen fuel cell
EB, eb	Electric boiler
EC, ec	Electric chiller
AC, ac	Absorption chiller
OST, ost	Oxygen storage tank
HST, hst	Hydrogen storage tank
BT, bt	Battery
CWHC, cwhc	Cylinder waste heat recovery system
TCHE, tche	Three chamber heat exchange system of cooler
ORC, orc	Organic rankine cycle
CPU, cpu	Compression purification unit
ASU, asu	Air separation unit

Appendix A

Table A1. Values of fixed technical and economical parameters in Case study.

No.	Name	Value (Stage I/ Stage II)	Unit	No.	Name	Value (Stage I/ Stage II)	Unit
1	η_{wt}	0.55/0.572	-	45	ρ_{fume}	0.748	kg/m^3
2	ρ_{air}	1.23	kg/m^3	46	V_{tche}	387,000/464,400	m^3/h
3	R_{wt}	40	m	47	η_{orc}	0.12	-
4	θ_{pv}	23	°	48	COP_{ac}	1.5	-
5	η_{MPPT}	0.75	-	49	κ_{cpu}	0.0893	MWh/ton
6	η_{pv}	0.16/0.1664	-	50	ε_{cal}^{carb}	0.4043	ton/ton
7	η_{elz}	0.64/0.672	-	51	$\varepsilon_{chp,ng}^{carb}$	0.3904	ton/MWh
8	λ_{elz}	8.01	-	52	$\varepsilon_{chp,coal}^{carb}$	1.08	ton/MWh
9	LHV_h	2.792	kWh/m^3	53	$\varepsilon_{chp,bio}^{carb}$	0.864	ton/MWh
10	LHV_{ng}	9.886	kWh/m^3	54	$\varepsilon_{fb,ng}^{carb}$	0.234	ton/MWh
11	LHV_{coal}	8.141	MWh/ton	55	$\varepsilon_{fb,coal}^{carb}$	0.83	ton/MWh
12	LHV_{bio}	4.652	MWh/ton	56	$\varepsilon_{fb,bio}^{carb}$	0.664	ton/MWh
13	$\eta_{chp,ng}$	0.45	-	57	$\alpha_{cal,max}$	0.98	-
14	$\eta_{chp,coal}$	0.3	-	58	$\alpha_{chp,max}^{oxy}$	0.98	-
15	$\eta_{chp,bio}$	0.3	-	59	$\alpha_{chp,max}$	0.9	-
16	$r_{chp,ng}$	0.8	-	60	$\alpha_{fb,max}^{oxy}$	0.98	-
17	$r_{chp,coal}$	1.9	-	61	$\alpha_{fb,max}$	0.98	-
18	$r_{chp,bio}$	1.7	-	62	$\chi_{chp,oxy}$	600	m^3/MWh
19	$\pi_{chp,ng,max}$	100%	-	63	$\chi_{fb,oxy}$	600	m^3/MWh
20	$\pi_{chp,coal,max}$	100%	-	64	$b_{e,ng}$	0.3901/0.3121	ton/MWh
21	$\pi_{chp,bio,max}$	100%	-	65	$b_{h,ng}$	0.0557/0.0446	ton/GJ
22	$\eta_{fb,ng}$	0.9	-	66	$b_{e,coal}$	0.8729/0.6983	ton/MWh
23	$\eta_{fb,coal}$	0.6	-	67	$b_{h,coal}$	0.1104/0.0883	ton/GJ
24	$\eta_{fb,bio}$	0.9	-	68	$f_{l,coal}$	1	-
25	η_{hfc}	0.5/0.52	-	69	$f_{f,coal}$	1	-
26	r_{hfc}	0.65	-	70	η_{eb}	0.95	
27	$\pi_{hfc,max}$	100%	-	71	COP_{ec}	5	

Table A1. *Cont.*

No.	Name	Value (Stage I/ Stage II)	Unit	No.	Name	Value (Stage I/ Stage II)	Unit
28	$\mu_{ost,min}$	0.075	-	72	κ_{asu}	3.03×10^{-4}	MWh/m^3
29	$\mu_{ost,max}$	0.9	-	73	$\varepsilon_{grid,carb}$	0.5810/0.5229	ton/MWh
30	σ_{ost}	0.001	-	74	ξ_{chp}	0.3	
31	$\eta_{ost,ch}$	0.95	-	75	β_{ge}	0.18/0.252	
32	$\eta_{ost,dis}$	0.95	-	76	k_{gc}	100/140	CNY/MWh
33	$\gamma_{ost,ch}$	0.08	h^{-1}	77	κ_{cal}	1.4162	MWh/ton
34	$\gamma_{ost,dis}$	0.08	h^{-1}	78	θ_{cal}	1.31	
35	$\mu_{hst,min}$	0.075	-	79	γ_{heat}	0.7	
36	$\mu_{hst,max}$	0.9	-	80	$k_{grid,sell}$	300/360	CNY/MWh
37	σ_{hst}	0.001	-	81	$k_{ng,buy}$	3.5/4.2	CNY/m^3
38	$\eta_{hst,ch}$	0.95	-	82	$k_{h,buy}$	6.5/5.2	CNY/m^3
39	$\eta_{hst,dis}$	0.95	-	83	$k_{coal,buy}$	734/954	MWh/ton
40	$\gamma_{hst,ch}$	0.042	h^{-1}	84	$k_{bio,buy}$	900/1080	MWh/ton
41	$\gamma_{hst,dis}$	0.042	h^{-1}	85	$k_{oxy,sell}$	0.682	CNY/m^3
42	h_{cht}	23.44	$W/(m^2 \cdot K)$	86	$k_{wt,curt}$	40	CNY/MWh
43	S_{cwhc}	1006/1207.2	m^2	87	$k_{pv,curt}$	40	CNY/MWh
44	$c_{fume,p}$	1.097	$kJ/(kg \cdot K)$	88	$k_{heat,curt}$	50	CNY/MWh

Table A2. Values of time-varying electricity purchasing price in Case study.

Time	$P_{grid,sell}$ (CNY/MWh)
0:00–1:00	0.0259
1:00–2:00	0.0259
2:00–3:00	0.0259
3:00–4:00	0.0259
4:00–5:00	0.0259
5:00–6:00	0.0259
6:00–7:00	0.0259
7:00–8:00	0.0259
8:00–9:00	0.1035
9:00–10:00	0.1035
10:00–11:00	0.1035
11:00–12:00	0.1035
12:00–13:00	0.0607
13:00–14:00	0.0607
14:00–15:00	0.0607
15:00–16:00	0.0607
16:00–17:00	0.0607
17:00–18:00	0.1035
18:00–19:00	0.1035
19:00–20:00	0.1035
20:00–21:00	0.1035
21:00–22:00	0.0607
22:00–23:00	0.0607
23:00–24:00	0.0607

Table A3. Values of time-varying wind speed and radiation intensity in Case study.

Time	u (m/s)				$\Re$ (kW/m^2)			
	Spring	Summer	Autumn	Winter	Spring	Summer	Autumn	Winter
0:00–1:00	5.54	4.50	4.58	5.17	0.000	0.000	0.000	0.000
1:00–2:00	5.38	4.33	4.42	5.00	0.000	0.000	0.000	0.000

Table A3. *Cont.*

Time	u (m/s)				$\Re$ (kW/m^2)			
	Spring	Summer	Autumn	Winter	Spring	Summer	Autumn	Winter
2:00–3:00	5.23	4.17	4.25	4.83	0.000	0.000	0.000	0.000
3:00–4:00	5.15	4.08	4.17	4.75	0.000	0.000	0.000	0.000
4:00–5:00	5.08	4.00	4.08	4.67	0.000	0.006	0.000	0.000
5:00–6:00	5.00	3.92	4.00	4.58	0.015	0.018	0.012	0.000
6:00–7:00	4.96	3.88	3.96	4.54	0.045	0.054	0.036	0.026
7:00–8:00	4.92	3.83	3.92	4.50	0.113	0.135	0.091	0.065
8:00–9:00	4.92	3.83	3.92	4.50	0.279	0.334	0.224	0.160
9:00–10:00	4.92	3.83	3.92	4.50	0.430	0.515	0.345	0.246
10:00–11:00	5.08	4.00	4.08	4.67	0.528	0.632	0.424	0.302
11:00–12:00	5.23	4.17	4.25	4.83	0.559	0.668	0.448	0.320
12:00–13:00	5.46	4.42	4.50	5.08	0.566	0.677	0.454	0.324
13:00–14:00	5.69	4.67	4.75	5.33	0.551	0.659	0.442	0.315
14:00–15:00	5.85	4.83	4.92	5.50	0.476	0.569	0.381	0.272
15:00–16:00	6.00	5.00	5.08	5.67	0.340	0.406	0.272	0.194
16:00–17:00	6.08	5.08	5.17	5.75	0.189	0.226	0.151	0.108
17:00–18:00	6.15	5.17	5.25	5.83	0.068	0.081	0.054	0.000
18:00–19:00	6.08	5.08	5.17	5.75	0.000	0.018	0.000	0.000
19:00–20:00	6.00	5.00	5.08	5.67	0.000	0.000	0.000	0.000
20:00–21:00	5.85	4.83	4.92	5.50	0.000	0.000	0.000	0.000
21:00–22:00	5.77	4.75	4.83	5.42	0.000	0.000	0.000	0.000
22:00–23:00	5.62	4.58	4.67	5.25	0.000	0.000	0.000	0.000
23:00–24:00	5.54	4.50	4.58	5.17	0.000	0.000	0.000	0.000

Table A4. Values of time-varying parameters related to low-temperature waste heat in Case study.

Time	ΔT_{cwhc} (°C)				ΔT_{tche} (°C)			
	Spring	Summer	Autumn	Winter	Spring	Summer	Autumn	Winter
0:00–1:00	166.79	164.81	166.79	178.16	48.28	47.71	48.28	51.57
1:00–2:00	159.01	156.13	159.01	170.36	46.03	45.19	46.03	49.31
2:00–3:00	165.77	183.42	165.77	160.04	47.99	53.10	47.99	46.33
3:00–4:00	168.33	164.44	168.33	181.98	48.73	47.60	48.73	52.68
4:00–5:00	158.40	160.19	158.40	161.67	45.85	46.37	45.85	46.80
5:00–6:00	173.23	182.68	173.23	161.45	50.15	52.88	50.15	46.74
6:00–7:00	243.41	324.56	243.41	180.42	70.46	93.95	70.46	52.23
7:00–8:00	351.45	366.82	351.45	321.39	101.74	106.19	101.74	93.03
8:00–9:00	363.29	368.73	363.29	369.29	105.16	106.74	105.16	106.90
9:00–10:00	363.02	367.44	363.02	373.95	105.08	106.36	105.08	108.25
10:00–11:00	358.35	352.11	358.35	364.56	103.73	101.93	103.73	105.53
11:00–12:00	304.05	264.79	304.05	348.73	88.01	76.65	88.01	100.95
12:00–13:00	254.10	269.66	254.10	258.86	73.55	78.06	73.55	74.93
13:00–14:00	318.94	281.13	318.94	347.69	92.32	81.38	92.32	100.65
14:00–15:00	380.00	380.00	380.00	374.23	110.00	110.00	110.00	108.33
15:00–16:00	370.63	356.78	370.63	380.00	107.29	103.28	107.29	110.00
16:00–17:00	346.41	359.37	346.41	350.92	100.28	104.03	100.28	101.58
17:00–18:00	319.27	351.46	319.27	272.66	92.42	101.74	92.42	78.93
18:00–19:00	266.60	277.46	266.60	258.64	77.17	80.32	77.17	74.87
19:00–20:00	306.11	352.27	306.11	286.71	88.61	101.97	88.61	83.00
20:00–21:00	326.79	336.70	326.79	341.46	94.60	97.47	94.60	98.84
21:00–22:00	300.21	242.98	300.21	341.82	86.90	70.34	86.90	98.95
22:00–23:00	247.84	253.51	247.84	250.19	71.74	73.38	71.74	72.42
23:00–24:00	211.29	177.25	211.29	243.42	61.16	51.31	61.16	70.47

References

1. International Energy Agency. *CO_2 Emissions in 2022*; International Energy Agency: Paris, France, 2023.
2. Ministry of Industry and Information Technology. Notice of the Ministry of Industry and Information Technology and Other Six Departments on the Issuance of Industrial Energy Efficiency Improvement Action Plan. 2022. Available online: http://www.gov.cn/zhengce/zhengceku/2022-06/29/content_5698410.htm (accessed on 23 June 2022).
3. Office of Energy Efficiency & Renewable Energy. DOE Industrial Decarbonization Roadmap. 2022. Available online: https://www.energy.gov/eere/doe-industrial-decarbonization-roadmap (accessed on 23 February 2023).
4. Department for Business, Energy and Industrial Energy. Industrial Decarbonisation Strategy. 2021. Available online: https://www.gov.uk/government/publications/industrial-decarbonisation-strategy (accessed on 1 April 2021).
5. Engel, H.; Helmcke, S.; Heuss, R.; Hieronimus, S. Net-Zero Germany: Chances and Challenges on the Path to Climate Neutrality by 2045. 2021. Available online: https://www.mckinsey.com/capabilities/sustainability/our-insights/net-zero-germany-chances-and-challenges-on-the-path-to-climate-neutrality-by-2045 (accessed on 19 October 2021).
6. Kyriakopoulos, G.L. *Should Low Carbon Energy Technologies Be Envisaged in the Context of Sustainable Energy Systems?* Academic Press: Cambridge, MA, USA, 2021; pp. 357–389.
7. Kovač, A.; Paranos, M.; Marciuš, D. Hydrogen in energy transition: A review. *Int. J. Hydrogen Energy* **2021**, *46*, 10016–10035. [CrossRef]
8. Mikulčić, H.; Skov, I.R.; Dominković, D.F.; Alwi, S.R.W.; Manan, Z.A.; Tan, R.; Duić, N.; Mohamad, S.N.H.; Wang, X. Flexible Carbon Capture and Utilization technologies in future energy systems and the utilization pathways of captured CO_2. *Renew. Sustain. Energy Rev.* **2019**, *114*, 109338. [CrossRef]
9. Weinand, J.M.; Scheller, F.; McKenna, R. Reviewing energy system modelling of decentralized energy autonomy. *Energy* **2020**, *203*, 117817. [CrossRef]
10. Kabeyi, M.J.B.; Olanrewaju, O.A. Sustainable Energy Transition for Renewable and Low Carbon Grid Electricity Generation and Supply. *Front. Energy Res.* **2022**, *9*, 1032. [CrossRef]
11. Drosos, D.; Kyriakopoulos, G.L.; Arabatzis, G.; Tsotsolas, N. Evaluating Customer Satisfaction in Energy Markets Using a Multicriteria Method: The Case of Electricity Market in Greece. *Sustainability* **2020**, *12*, 3862. [CrossRef]
12. Van den Bergh, J.; Botzen, W. Low-carbon transition is improbable without carbon pricing. *Proc. Natl. Acad. Sci. USA* **2020**, *117*, 23219–23220. [CrossRef]
13. Guo, H.; Chen, Q.; Xia, Q.; Kang, C. Modeling Strategic Behaviors of Renewable Energy with Joint Consideration on Energy and Tradable Green Certificate Markets. *IEEE Trans. Power Syst.* **2020**, *35*, 1898–1910. [CrossRef]
14. Liang, Z.; Zhangyi, L.; Wei, P.; Hao, X.; Rui, S.; Jun, H.; Xuefeng, L. Tie-line power control method for an industrial park based on integrated demand response. *Electr. Power Eng. Technol.* **2021**, *40*, 106–113.
15. Yue, Q.; Shuai, L.; Hai, L.; Enbo, L.; Wei, G.; Wennan, Z. Flexibility of Integrated Energy System: Basic Connotation, Mathematical Model and Research Framework. *Autom. Electr. Power Syst.* **2022**, *46*, 16–43.
16. Pengcheng, Z.; Jian, X.; Yuanzhang, S.; Deping, K.; Siyang, L. Low-carbon development mode for energy system of iron and steel park driven by hydrogen energy. *Autom. Electr. Power Syst.* **2022**, *46*, 10–20.
17. Xiandong, X.; Wenliang, S.; Yunfan, C.; Xiaodan, Y.; Hongjie, J. A perspective on electricity flexibility of integrated energy systems in processing industries. *J. Tianjin Univ. Sci. Technol.* **2021**, *54*, 1212–1220.
18. Bogdanov, D.; Gulagi, A.; Fasihi, M.; Breyer, C. Full energy sector transition towards 100% renewable energy supply: Integrating power, heat, transport and industry sectors including desalination. *Appl. Energy* **2021**, *283*, 116273. [CrossRef]
19. Sun, J.; Ruze, N.; Zhang, J.; Shi, J.; Shen, B. Capacity planning and optimization for integrated energy system in industrial park considering environmental externalities. *Renew. Energy* **2021**, *167*, 56–65. [CrossRef]
20. Chen, Q.; Lv, M.; Gu, Y.; Yang, X.; Tang, Z.; Sun, Y.; Jiang, M. Hybrid Energy System for a Coal-Based Chemical Industry. *Joule* **2018**, *2*, 607–620. [CrossRef]
21. Mardan, N.; Klahr, R. Combining optimisation and simulation in an energy systems analysis of a Swedish iron foundry. *Energy* **2012**, *44*, 410–419. [CrossRef]
22. Yuan, J.; Li, Y.; Luo, X.; Zhang, Z.; Ruan, Y.; Zhou, Q. A new hybrid multi-criteria decision-making approach for developing integrated energy systems in industrial parks. *J. Clean. Prod.* **2020**, *270*, 122119. [CrossRef]
23. Siyun, Y.; Chen, W.; Dengji, Z. Optimization of integrated electricity and gas system considering hydrogen-natural-gas mixture transportation. *Electr. Power Eng. Technol.* **2021**, *40*, 10–16+49.
24. Jiang, Z.; Hao, R.; Ai, Q.; Yu, Z.; Xiao, F. Extended multi-energy demand response scheme for industrial integrated energy system. *IET Gener. Transm. Distrib.* **2018**, *12*, 3186–3192. [CrossRef]
25. Aidong, Z.; Yuhang, Z.; Sipeng, H.; Jia, N.; Lianghua, N. Comprehensive demand response strategy of industrial users in the park considering the stepped carbon trading mechanism. *High Volt. Eng.* **2022**, *48*, 4352–4363.
26. Zhu, X.; Yang, J.; Pan, X.; Li, G.; Rao, Y. Regional integrated energy system energy management in an industrial park considering energy stepped utilization. *Energy* **2020**, *201*, 117589. [CrossRef]
27. Xu, Z.; Han, G.; Liu, L.; Martinez-Garcia, M.; Wang, Z. Multi-Energy Scheduling of an Industrial Integrated Energy System by Reinforcement Learning-Based Differential Evolution. *IEEE Trans. Green Commun. Netw.* **2021**, *5*, 1077–1090. [CrossRef]
28. Fennell, P.S.; Davis, S.J.; Mohammed, A. Decarbonizing cement production. *Joule* **2021**, *5*, 1305–1311. [CrossRef]

29. Statista. Cement Production Worldwide from 1995 to 2022. 2023. Available online: https://www.statista.com/statistics/1087115/global-cement-production-volume/#:~:text=Feb%203%2C%202023,industry%20has%20grown%20since%20then (accessed on 23 January 2023).
30. Woetzel, J.; Xu, H.; Wang, X.; Liao, X. *China Accelerates towards Carbon Neutrality: Pathways to Carbon Reduction in the Cement Industry*; McKinsey&Company: Atlanta, GA, USA, 2021.
31. Rissman, J.; Bataille, C.; Masanet, E.; Aden, N.; Morrow, W.R.; Zhou, N.; Elliott, N.; Dell, R.; Heeren, N.; Huckestein, B.; et al. Technologies and policies to decarbonize global industry: Review and assessment of mitigation drivers through 2070. *Appl. Energy* **2020**, *266*, 114848. [CrossRef]
32. Ali, M.B.; Saidur, R.; Hossain, M.S. A review on emission analysis in cement industries. *Renew. Sustain. Energy Rev.* **2011**, *15*, 2252–2261. [CrossRef]
33. Madlool, N.A.; Saidur, R.; Rahim, N.A.; Kamalisarvestani, M. An overview of energy savings measures for cement industries. *Renew. Sustain. Energy Rev.* **2013**, *19*, 18–29. [CrossRef]
34. Decao, X.; Xiantao, L.; Nan, L.; Yihan, W.; Heng, C.; Peiyuan, P. Performance analysis of waste heat deep utilization of cement plant based on carbon dioxide capture. *Clean Coal Technol.* **2023**, 1–18.
35. Yonghua, D. *Theoretical and Experimental Study on Oxygen Enriched Combustion Technology in Cement Industry*; Xi'an University of Architecture and Technology: Xi'an, China, 2015.
36. Teixidó, J.; Verde, S.F.; Nicolli, F. The impact of the EU Emissions Trading System on low-carbon technological change: The empirical evidence. *Ecol. Econ.* **2019**, *164*, 106347. [CrossRef]
37. Wang, H.; Chen, Z.; Wu, X.; Nie, X. Can a carbon trading system promote the transformation of a low-carbon economy under the framework of the porter hypothesis?—Empirical analysis based on the PSM-DID method. *Energy Policy* **2019**, *129*, 930–938. [CrossRef]
38. Li, X.; Wang, W.; Wang, H.; Wu, J.; Fan, X.; Xu, Q. Dynamic environmental economic dispatch of hybrid renewable energy systems based on tradable green certificates. *Energy* **2020**, *193*, 116699. [CrossRef]
39. Yu, X.; Dong, Z.; Zhou, D.; Sang, X.; Chang, C.T.; Huang, X. Integration of tradable green certificates trading and carbon emissions trading: How will Chinese power industry do? *J. Clean. Prod.* **2021**, *279*, 123485. [CrossRef]
40. Moya, J.A.; Boulamanti, A. *Production Costs from Energy-Intensive Industries in the EU and Third Countries*; The European Commission: Brussels, Belgium, 2016.
41. Qiu, Y.; Zhou, S.; Gu, W.; Ding, S.; Han, G.; Zhang, K.; Lv, H. Multi-stage flexible planning of regional electricity-HCNG-integrated energy system considering gas pipeline retrofit and expansion. *IET Renew. Power Gener.* **2022**, *16*, 3339–3367. [CrossRef]
42. Marocco, P.; Ferrero, D.; Martelli, E.; Santarelli, M.; Lanzini, A. An MILP approach for the optimal design of renewable battery-hydrogen energy systems for off-grid insular communities. *Energy Convers. Manag.* **2021**, *245*, 114564. [CrossRef]
43. He, T.; Cao, Y.; Si, F. Thermodynamic analysis and optimization of a compressed carbon dioxide energy storage system coupled with a combined heating and power unit. *Energy Convers. Manag.* **2023**, *277*, 116618. [CrossRef]
44. Zhang, Y.; Hao, J.; Ge, Z.; Zhang, F.; Du, X. Optimal clean heating mode of the integrated electricity and heat energy system considering the comprehensive energy-carbon price. *Energy* **2021**, *231*, 120919. [CrossRef]
45. Samy, M.M.; Mosaad, M.I.; Barakat, S. Optimal economic study of hybrid PV-wind-fuel cell system integrated to unreliable electric utility using hybrid search optimization technique. *Int. J. Hydrogen Energy* **2021**, *46*, 11217–11231. [CrossRef]
46. Hemmati, M.; Mirzaei, M.A.; Abapour, M.; Zare, K.; Mohammadi-Ivatloo, B.; Mehrjerdi, H.; Marzband, M. Economic-environmental analysis of combined heat and power-based reconfigurable microgrid integrated with multiple energy storage and demand response program. *Sustain. Cities Soc.* **2021**, *69*, 102790. [CrossRef]
47. *GB/T 26281-2021*; Methods for the Calculation of Heat Balance, Heat Efficiency and Comprehensive Energy Consumption of Cement Rotary Kiln. State Administration for Market Regulation; Standardization Administration of the People's Republic of China: Beijing, China, 2021.
48. Fan, Y.; Liu, E.; Pan, J.; Qiao, J.; Li, W.; Zhang, W.; Liu, C. Waste heat recovery and utilization system of cement clinker production line and energy saving analysis. *Energy Conserv.* **2020**, *39*, 101–103.
49. Laribi, S.; Dubois, L.; De Weireld, G.; Thomas, D. Study of the post-combustion CO_2 capture process by absorption-regeneration using amine solvents applied to cement plant flue gases with high CO_2 contents. *Int. J. Greenh. Gas Control* **2019**, *90*, 102799. [CrossRef]
50. Yihan, W. *Integration and Optimization of Deep Waste Heat Utilization Systems in Cement Plants Based on System Coupling and Carbon Dioxide Capture*; North China Electric Power University: Beijing, China, 2022.
51. Yang, C.; Peng, Z.; Wuzhi, Z.; Zheng, W.; Peng, Z.; Yuting, Z. Low-carbon economic dispatch of electro-gas-thermal integrated energy system based on oxy-combustion technology. *Proc. CSEE* **2021**, *41*, 592–608.
52. Liu, X.; Zhou, X.; Zhu, B.; He, K.; Wang, P. Measuring the maturity of carbon market in China: An entropy-based TOPSIS approach. *J. Clean. Prod.* **2019**, *229*, 94–103. [CrossRef]

53. Chun, Q.; Zhanfeng, Y.; Yi, F.; Jianhu, Y. Optimal operation of hybrid energy storage integrated micro-energy network considering carbon quote. *Electr. Power Eng. Technol.* **2022**, *41*, 119–127.
54. Zhao, L.; Hongzhi, L.; Weijie, Z.; Xin, S.; Xin, L.; Yunrui, L. Collaborative operation optimization of cross-border integrated energy system considering joint trading of carbon and green certificates. *Electr. Power Autom. Equip.* **2023**.

processes MDPI

Article

Energy and Exergy Analysis of a Modified Absorption Heat Pump (MAHP) to Produce Electrical Energy and Revaluated Heat

Javier Alejandro Hernández-Magallanes [1,*], L. A. Domínguez-Inzunza [2], Shadai Lugo-Loredo [1], K. C. Sanal [1], Andrea Cerdán-Pasarán [1], Salvador Tututi-Avila [1] and L. I. Morales [3]

1 Facultad de Ciencias Químicas, Universidad Autónoma de Nuevo León, San Nicolas de los Garza 66455, Mexico
2 Tecnológico Nacional de México/I.T. Culiacán, Culiacán Rosales 80220, Mexico
3 Facultad de Ciencias Químicas e Ingeniería, Universidad Autónoma del Estado de Morelos, Cuernavaca 62209, Mexico
* Correspondence: javier.hernandezmg@uanl.edu.mx

Abstract: The novel modified absorption heat pump (MAHP) with the H_2O-LiBr working mixture for cogeneration applications is introduced. The MAHP can simultaneously produce electric energy and heat revaluation. The proposed system has the particularity that it can be powered by alternative thermal sources (such as solar energy, biomass, geothermal) or industrial waste heat, thus promoting the production and efficient use of clean energy. The effects of pressure ratio (RP), source or supply temperature (T_{GH}), and the energy revaluation gradient (GTL) are analyzed. The critical parameters of the proposed system are evaluated, including thermal efficiency (η_{Th}), exergetic efficiency (η_{Ex}), revaluated heat ($\dot{Q}_A$), as well as net power produced ($\dot{W}_{net}$). For the MAHP analysis, RP and T_{GH} operating ranges were chosen at 1.1–15.0 and 100–160 °C, respectively. The results show that η_{Ex} of 87% can be obtained, having the maximum performance in T_{GH} of 120 °C, RP of 1.1, and GTL of 35 °C. The η_{Th} varies between 51% and 55%, having a maximum GTL of 45 °C. On the other hand, $\dot{W}_{net}$ achieves values between 260 and 582 kW, depending on the defined operating conditions.

Keywords: cogeneration; absorption; heat pump; power; revaluated heat; H_2O-LiBr

Citation: Hernández-Magallanes, J.A.; Domínguez-Inzunza, L.A.; Lugo-Loredo, S.; Sanal, K.C.; Cerdán-Pasarán, A.; Tututi-Avila, S.; Morales, L.I. Energy and Exergy Analysis of a Modified Absorption Heat Pump (MAHP) to Produce Electrical Energy and Revaluated Heat. *Processes* **2022**, *10*, 1567. https://doi.org/10.3390/pr10081567

Academic Editors: Ferdinando Salata and Virgilio Ciancio

Received: 7 July 2022
Accepted: 9 August 2022
Published: 10 August 2022

1. Introduction

Industrial development worldwide has generated a growing energy demand, bringing an energy crisis and damage to the environment, having fossil fuels as the principal sources [1]. It is estimated that world energy consumption will continue increasing in the coming decades due to the forecast projected economic growth. In 2018, the economic growth rate was 3.7%, exceeding the 3.5% average annual growth since 2010. However, this also reflected a 2.3% increase in global energy consumption, nearly twice the average growth rate since 2010, driven by more significant heating and cooling needs in some parts of the world [2]. In this context, electricity plays a vital role in human life due to the gradual demand for energy, the accelerated increase in population, and new lifestyle trends [3].

Regarding the 2015 Paris Agreement, the urgency of the climate change challenge is highlighted [4], trying to keep the global temperature increase below 2 °C by reducing CO_2 production [5]. Therefore, several countries worldwide have adopted ambitious targets to reduce their carbon footprint [4]. An alternative is using a sustainable energy system, which implies achieving adequate energy efficiency and the potential use of renewable energies; both can achieve industrial development while maintaining care for the environment and social welfare [6]. In this background, more than 50% of the energy used worldwide is wasted as heat [7]. Currently, technological developments are based on promoting sustainable energy systems and where various thermal supply sources are used. Renewable energies and industrial waste heat have enormous potential for action to address the problems caused by climate change.

Solar thermal energy can significantly boost renewable energy development. In turn, it can largely offset the adverse effects of climate change and improve the world's energy supply. For example, solar thermal energy is used in industrial, commercial, and residential applications through different technologies, including steam, heating, cooling, and even electricity generation. Temperatures can be produced from 45 °C to over 300 °C, making it potentially helpful energy for many sectors. By 2021, the annual solar thermal energy yield amounted to 425 TWh, which correlates to savings of 45.7 million tons of oil and 147.5 million tons of CO_2 avoided [8].

Fan et al. [9] presented a bibliographic review of the state of the art of solar sorption systems (absorption and adsorption). The analysis shows that solar-powered sorption technologies are attractive alternatives to demand energy conservation and environmental protection. Ayou et al. [10] and López-Villada et al. [11] presented various systems for simultaneous electrical energy and cooling production. They obtained high first and second law efficiencies due to the greater internal use of the thermal resource, guaranteeing that the absorption systems have a broad spectrum of applications. Arabkoohsar et al. [12,13] analyze the effect of employing parabolic through solar collectors-Powered absorption chiller for co-supply in heating and cooling systems. The results demonstrate that the system feeds district heating with a heat supply rate of over 1 MW for about three months, and the internal rate of return of the system is 5.7% for 8 years of operation. Rahman et al. [14] implemented a simulation model of a solar absorption cooling system for air conditioning in Pakistan. The evacuated glass tube collectors use R-410A as heat transfer fluid. The system has successfully provided the desired room temperature of 26 °C with total reliance on solar thermal energy. Düzcan et al. [15] analyze the usage potential of the evacuated tube collectors (ETCs) and the flat plate collectors (FPCs) for assisting an absorption chiller with the H_2O-LiBr pair. Through the tests carried out, the evacuated tube collector obtained the best results. Ali et al. [16] show the design and performance of an absorption system using a non-tracking semi-circular trough (SCT) solar collector. The experimental results show that a chilled water output temperature of around 10–15 °C, with a record 3.3 °C, was achieved when the chiller reached an operating temperature of 120 °C. Behzadi et al. [17] propose a novel hybrid renewable-based cold production system consisting of evacuated solar collectors integrated with a biomass heater, thermal storage tanks, and an absorption machine. The system's performance is simulated and assessed for a study case hospital in India. Nami et al. [18] propose a modeling and analysis of biomass- and solar-driven combined cooling, heating, and power system. Results showed that the designed system could feed 1241 kW of space heating, 101.5 kW of domestic hot water, and 55.35 kW of chilled water as space cooling, besides delivering 1 MWe power. Nasir et al. [19] present an assessment of a trigeneration system powered by biomass. The system analyzed was designated to provide 250 kW of electricity, cooling values between 19 and 22 kW, and heating values ranging from 879 to 1255 kW. Furthermore, the second law efficiency of the system was found to be approximately 56%. Siddiqui et al. [20] present a new solar and geothermal-based integrated system developed for electricity, fresh water, hydrogen, and cooling. The system's overall energy efficiency obtained was 42.3%. Chen et al. [21] analyze a system for the combined production of heat and power (CHP), taking advantage of a geothermal energy source. The results demonstrate that the increased allocation ratio of mixed water for the heat pumps raised the output and coefficient of performance (COP).

On the other hand, many industries worldwide reject this energy into the atmosphere at temperatures close to 230 °C. In addition, residual heat can produce helpful energy products, reduce CO_2 emissions, and create sustainable systems and efficiency. Waste heat recovery systems are a promising group of technologies that can take energy wasted from a process and transform it into usable thermal, electrical, or mechanical energy. Today, conventional waste heat recovery technologies operate mainly with different systems based on thermodynamic cycles, which can work with cleaner/sustainable sources such as solar thermal energy [22]. For the latter, the most suitable are the organic Rankine cycle (ORC), the Kalina cycle (KC), the Goswami cycle (GC), absorption heat pump (AHP), and cogeneration cycles to obtain energy, calefaction, and refrigeration [23]. KC is another thermodynamic cycle for low and medium-temperature heat recovery. Unlike a pure organic fluid, such as the one used in the ORC, it has the main characteristic of a non-constant temperature during vaporization and condensation. Compared to its competitors, it translated into better performance under the same operating conditions [24]. The GC combines the power cycle with the absorption refrigeration cycle and thus provides various operating conditions where the system offers both power and sensible cooling [25].

Regarding the implementation of sorption systems in cogeneration applications, a detailed bibliographic review carried out by Kumar et al. [26] establishes that one of the main advantages of using these systems is the low temperatures that are needed for their operation, which can range between 50 and 95 °C, with applications in refrigeration, desalination, and air conditioning. Köse et al. [27] conducted a theoretical comparison through simulations between the KC and ORC, considered the most critical low-temperature energy conversion systems in industrial waste heat utilization. The most attractive design for using residual heat was ORC, obtaining a thermal efficiency of 25.95%, operating with the working fluid n-pentane. Nevertheless, KC was more economically viable, with an estimated payback period of 3.93 years. Akimoto et al. [28] propose an analysis of the combined performance of power generation and economic evaluation for integrating a system composed of AHP coupled to KC, operating with a low-temperature heat source of approximately 100 °C. The conclusions showed that the system comprising KC + AHP + KC generally delivers the best results, with a power generation between 10 and 25 kW, a thermal efficiency of 1.2–2.9%, and an exergetic efficiency of around 10%. Aksar et al. [29] designed, analyzed, and compared the KC, the Rankine cycle, and the simple cycle of pure ammonia with residual gases at 350 °C. The results showed that the KC with cogeneration reported a maximum thermal and exergetic efficiency of 72.13% and 78.60%, respectively. Marshall et al. [30] performed a techno-economic evaluation of multi-cycle organic Rankine-assisted heat pumps (AHP-ORC). The maximum values of COP and efficiency were obtained with R161/n-pentane, with values of 13.2% and 4.1%, respectively. Liu et al. [31] proposed the theoretical evaluation of a system composed of KC coupled to an H_2O-LiBr absorption refrigeration system. The power generation was improved by 45% compared to the conventional KC, and the COP increased from 0.1158 to 0.1678. Cao et al. [32] used a thermodynamic analysis establishing a mathematical model for a simultaneous cooling and power generation cycle based on an absorption refrigeration cycle coupled to KC, driven by a low-temperature heat source between 100 and 135 °C. The results obtained were that the exergetic efficiency increases with the expander inlet temperature, as well as with the concentration of the basic NH_3-H_2O solution. Wang et al. [33] suggested a new combined refrigeration and energy production at a maximum temperature of 200 °C. The output power of the turbine, the cooling power, and the thermal and exergetic efficiency of the system had their maximum values of 37.63 kW, 75.12 kW, 13.72%, and 5.40%, respectively. Padilla et al. [34] present a parametric analysis of the GC using the binary mixture NH_3-H_2O. The heat source temperature was selected between 90 and 170 °C, finding the maximum theoretical energy and exergy efficiencies of 21% and 92%, respectively. To improve the performance of the Goswami cycle, multiple researchers such as Zare et al. [35], Xu et al. [36], and Demirkaya et al. [37] proposed the use of internal rectification, which consists of dividing the flow at the outlet of the pump in such a way that one current goes to the economizer. In contrast, the other goes directly to the rectifier.

As seen in the theoretical framework, alternative energy sources have a vast application, coupled with a wide variety of systems that can take advantage of these resources. For example, solar thermal energy and biomass can meet urban and rural needs. The waste heat recovery systems are mainly used to meet industrial needs, and geothermal energy can offer a wide range of possible applications in geographical areas where the resource is available [38].

Based on the bibliographic review, a novel modified absorption heat pump (MAHP) study is presented for cogeneration applications. The system can simultaneously produce electrical energy and revalued heat. The proposed method has the peculiarity that it can be powered by alternative thermal sources (such as solar energy, biomass, or geothermal energy) or industrial waste heat. In both cases, it is intended to produce clean energy to promote the use of sustainable processes that are friendlier to the environment. For this reason, it is essential to analyze the operating conditions, the amounts of energy, and the energy performance achieved with the MAHP.

2. Description of the MAHP

The MAHP cycle is composed of the union of a type I heat pump, a type II heat pump, and a turbine to take advantage of the pressure gradient in the system [39,40] (see Figure 1). In the case of MAHP, the maximum pressure and temperature of the system are used to provide mechanical energy for generating electricity. The expansion in the turbine is carried out using the maximum and minimum pressure of the system, P_H and P_L, respectively, to ensure higher power output. In addition, the revalued thermal energy or $\dot{Q}_A$ from the absorption process can be used in secondary subsystems [39], offering greater versatility in operations and a higher performance system.

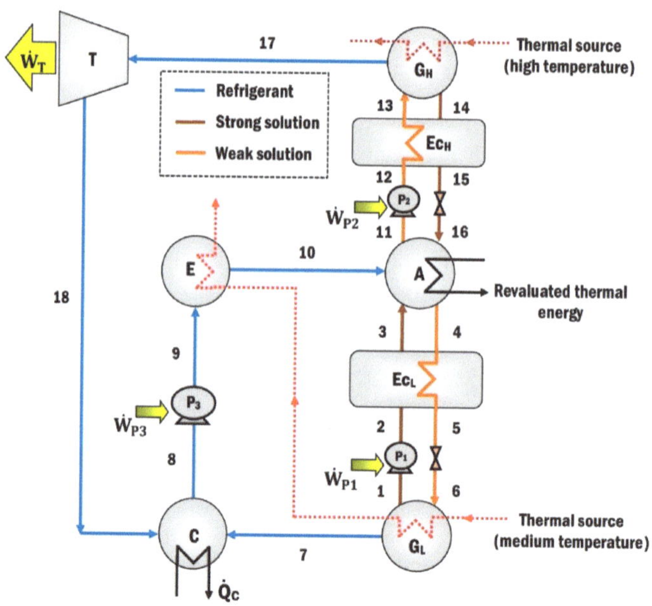

Figure 1. The modified absorption heat pump (MAHP).

The system operates on three pressure and four temperature levels (two temperatures from a thermal source, a revalued energy temperature, and a thermal sink). The external source ($\dot{Q}_{GH}$) exchanges heat in the high-pressure generator (G_H) to produce refrigerant at a high temperature and pressure (state 17). The water vapor is expanded in the turbine (T), producing mechanical energy ($\dot{W}_T$). The refrigerant leaves the turbine at low pressure and temperature (state 18) and goes toward the condenser (C), where the water vapor is condensed, extracting a quantity of heat ($\dot{Q}_C$).

A second heat source is supplied at an intermediate temperature to the evaporator (E) and the low-pressure generator (G_L). The heat provided to the G_L is used to heat the dilute solution (weak solution state 6) that arrives from the absorber to have a second production of the refrigerant vapor (state 7). The concentrated solution (strong solution) leaving the G_L

is pumped (state 2) and preheated (state 3) in the heat exchanger (Ec_L) before entering the absorber (A). The steam produced in the G_L (state 7) passes to the condenser, which joins with the current that leaves the turbine (state 18) and condenses. The liquid refrigerant that exits the condenser (state 8) is pumped to the evaporator (state 9), where it receives heat from the second heat source ($\dot{Q}_E$) to evaporate the refrigerant again. Subsequently, the vapor refrigerant is sent to the absorber (state 10), where it is absorbed by the concentrated solution that comes from the G_L (state 3) and G_H (state 16). As a result of the absorption process, the available heat ($\dot{Q}_A$) is released, which is considered revalued because it is at higher energy levels than the thermal loads supplied in the G_L and E. The weak solution, a product of the refrigerant absorption process, is divided into two streams (states 4 and 11). Finally, flow 4 is sent to the G_L and flow 11 to the G_H to restart the thermodynamic cycle of refrigerant production.

As mentioned above, the MAHP needs four energy levels to function correctly. First, the high-temperature thermal source (between 100 and 160 °C [39,40]) has the enormous potential of using various power sources, such as solar energy with evacuated tube collectors or concentration systems [13–16], biomass [17–19], geothermal energy [20,21], or industrial waste heat [1,3,4]. The medium-temperature thermal source (60–100 °C [39,40]) has excellent versatility in using different energy sources due to its low and easily attainable temperature range [12]. Therefore, in both cases, it is preponderant to promote renewable energies.

3. H_2O-LiBr Work Solution Mixture

Absorption is a mass transfer process in which a substance can capture and hold another substance within itself at low pressures and temperatures. In absorption cooling, the property of a liquid substance (absorbent) is used to absorb into itself a gas vapor (refrigerant) and thus form a solution (binary mixture) [41]. The gas vapor previously absorbed can be separated by supplying thermal energy to the solution by increasing its temperature and pressure [42]. Absorption systems and conventional ones use the latent heat from the liquid-vapor phase change of the refrigerant to remove or transfer heat to the environment. However, absorption systems use two working fluids: the refrigerant responsible for removing and releasing heat and the absorbent that helps the movement of the refrigerant within the cycle [43]. There are many binary mixtures for absorption cycles. However, two of the most common mixtures are NH_3-H_2O (where NH_3 is the refrigerant and H_2O is the absorbent) and H_2O-LiBr (where H_2O is the refrigerant and LiBr is the absorbent) [44]. In the latter, the concentration of the LiBr solution is a ratio between the constituent's mass and the mixture´s mass [44]. H_2O-LiBr is the only binary mixture in which water is the refrigerant. The formation of solids due to their melting point at 0 °C limits their application in conventional compression systems. LiBr is a chemical compound of lithium and bromine that is hygroscopic, has the appearance of a white crystalline powder, and its solubility in water is 177 g/100 mL. In absorption refrigeration systems, there are two concentrations of LiBr, the concentrated (strong solution) that comes out of the generator and the diluted (weak solution) that comes out of the absorber. Strong concentrations of the solution range from 50% to 70% for normal operating conditions, as can be seen in Figure 2.

As shown in Figure 2, the most significant limitation of the H_2O-LiBr working solution is the crystallization zone, an area where the absorption system will not operate. The formation of solids begins when the solubility limit is exceeded, which would cause a pipe to be blocked entirely or decrease the heat exchange power.

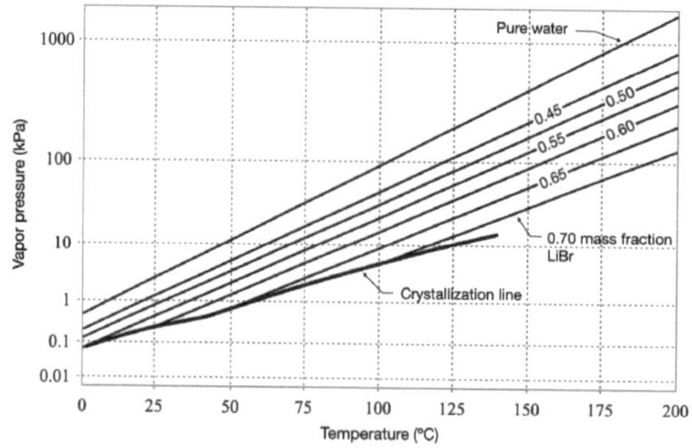

Figure 2. Pressure-temperature diagram for water-lithium bromide [45].

4. MAHP Performance Parameters

This section presents the performance parameters used to carry out the thermodynamic analysis of MAHP operating with the H_2O-LiBr working mixture.

4.1. Thermal Efficiency (η_{Th})

Thermal efficiency can be expressed as the ratio of helpful energy produced to the energy supplied from the process. The thermal efficiency parameter can be applied to an absorption machine and is calculated based on the primary or secondary fluids. For the present study, the primary fluids or "internal flows" are considered, which are those used internally by the absorption machine (in this case, the solution H_2O-LiBr) [40].

$$\eta_{Th} = \frac{\text{Useful energy output}}{\text{Total energy input}} = \frac{\dot{W}_{net} + \dot{Q}_A}{\dot{Q}_{supply}}, \tag{1}$$

where $\dot{Q}_A$ is the revalued heat (useful heat) by the MAHP, the $\dot{W}_{net}$ represents the net power produced (power produced by the turbine minus the power consumed by the pumps). Both terms describe the helpful energy obtained by the MAHP. On the other hand, $\dot{Q}_{supply}$ represent the thermal energy supplied to the MAHP to get the desired products. In this case, the thermal loads supplied in the generators and evaporator ($\dot{Q}_{GL}$, $\dot{Q}_{GH}$ and $\dot{Q}_E$).

4.2. Exergetic Efficiency (η_{Ex})

Exergy refers to the maximum work achievable from a particular form of energy transformed from one thermodynamic state in equilibrium to another. Hence, it is possible to infer the true potential of different types of energies. However, due to thermodynamic irreversibility, the exergetic efficiency indicates the degradation of the quality of different energies [46].

$$\eta_{Ex} = \frac{\dot{W}_{net} + \dot{Q}_A\left(1 - \frac{T_0}{T_A}\right)}{\sum \dot{Q}_{in}\left(1 - \frac{T_0}{T_{in}}\right)}, \tag{2}$$

Using the thermodynamic states of Figure 1, Equation (2) can be expressed as:

$$\eta_{ex} = \frac{\dot{W}_{net} + \dot{Q}_A\left(1 - \frac{T_0}{T_A}\right)}{\dot{Q}_{GL}\left(1 - \frac{T_0}{T_{GL}}\right) + \dot{Q}_E\left(1 - \frac{T_0}{T_E}\right) + Q_{GH}\left(1 - \frac{T_0}{T_{GH}}\right)}, \tag{3}$$

The ambient temperature (T_0) is equivalent to the sink temperature used for condensing the refrigerant (state 8). In this way, the four thermal levels of the MAHP are contemplated.

4.3. Gross Temperature Lift (GTL)

This parameter illustrates the increase or gains in temperature depending on the supply source. It represents the revaluation of energy obtained through the absorption of refrigerant in the solution, starting from the established operating conditions. GTL is defined as the temperature difference between the absorption temperature (revaluated thermal energy) and the evaporation temperature (thermal source at medium temperature) of MAHP [47].

$$GTL = T_A - T_E, \tag{4}$$

Therefore, the higher the GTL, the higher the T_A, which means that the MAHP can produce a better revalued thermal load ($\dot{Q}_A$). The higher the $\dot{Q}_A$, the MAHP is expected to perform better. In addition, there is more significant potential to use the thermal load in secondary processes.

5. Mathematical Modeling

The thermodynamic simulation of the MAHP was carried out using the Engineering Equation Solver (EES) software [48]. Different relevant assumptions were established for the simulation:

- The system is in thermodynamic equilibrium and a steady state [31–33,49,50];
- Thermal losses to the surroundings are not considered, just as friction losses are not [32,33,51];
- The H_2O-LiBr solution at the generator/absorber outlet is saturated liquid [49];
- The coolant is water, and there is no carry-over of lithium bromide during the process;
- The processes in expansion valves are isenthalpic [32,33,49];
- The pumps and turbine's isentropic efficiencies of 0.80 [10] and 0.85 [11] were considered, respectively;
- The thermal efficiency for the economizer is 0.7 [42].

The mass and energy balances for each component of the MAHP, supported by the thermodynamic states of Figure 1, are shown below.

High-pressure generator (G_H):

$$\dot{m}_{13} = \dot{m}_{14} + \dot{m}_{17}, \tag{5}$$

$$\dot{m}_{13}x_{13} = \dot{m}_{14}x_{14} + \dot{m}_{17}x_{17}, \tag{6}$$

$$\dot{Q}_{GH} = \dot{m}_{14}h_{14} + \dot{m}_{17}h_{17} - \dot{m}_{13}h_{13}, \tag{7}$$

Absorber (A):

$$\dot{m}_3 + \dot{m}_{16} + \dot{m}_{10} = \dot{m}_{11} + \dot{m}_4, \tag{8}$$

$$\dot{m}_3x_3 + \dot{m}_{16}x_{16} + \dot{m}_{10}x_{10} = \dot{m}_{11}x_{11} + \dot{m}_4x_4, \tag{9}$$

$$\dot{Q}_A = \dot{m}_3h_3 + \dot{m}_{16}h_{16} + \dot{m}_{10}h_{10} - \dot{m}_{11}h_{11} - \dot{m}_4h_4, \tag{10}$$

Low-pressure generator (G_L):

$$\dot{m}_6 = \dot{m}_1 + \dot{m}_7, \tag{11}$$

$$\dot{m}_6x_6 = \dot{m}_1x_1 + \dot{m}_7x_7, \tag{12}$$

$$\dot{Q}_{GL} = \dot{m}_1h_1 + \dot{m}_7h_7 - \dot{m}_6h_6, \tag{13}$$

Turbine (T):

$$\dot{m}_{17} = \dot{m}_{18}, \tag{14}$$

$$\dot{m}_{17}x_{17} = \dot{m}_{18}x_{18}, \tag{15}$$

$$\dot{W}_T = \dot{m}_{17}(h_{17} - h_{18}), \tag{16}$$

$$\eta_T = \frac{h_{17} - h_{18}}{h_{17} - h_{18s}}, \tag{17}$$

Evaporator (E):

$$\dot{m}_9 = \dot{m}_{10}, \tag{18}$$

$$\dot{m}_9 x_9 = \dot{m}_{10} x_{10}, \tag{19}$$

$$\dot{Q}_E = \dot{m}_{10}(h_{10} - h_9), \tag{20}$$

Condenser (C):

$$\dot{m}_8 = \dot{m}_7 + \dot{m}_{18}, \tag{21}$$

$$\dot{m}_8 x_8 = \dot{m}_7 x_7 + \dot{m}_{18} x_{18}, \tag{22}$$

$$\dot{Q}_c = \dot{m}_{18} h_{18} + \dot{m}_7 h_7 - \dot{m}_8 h_8, \tag{23}$$

Pump 1 (P_1):

$$\dot{m}_1 = \dot{m}_2, \tag{24}$$

$$W_{P1} = V_1(P_2 - P_1), \tag{25}$$

$$\dot{W}_{P1} = (\dot{m}_1 W_{P1})/\eta_p, \tag{26}$$

Pump 2 (P_2):

$$\dot{m}_{11} = \dot{m}_{12}, \tag{27}$$

$$W_{P2} = V_{11}(P_{12} - P_{11}), \tag{28}$$

$$\dot{W}_{P2} = (\dot{m}_{11} W_{P2})/\eta_p, \tag{29}$$

Pump 3 (P_3):

$$\dot{m}_8 = \dot{m}_9, \tag{30}$$

$$W_{P3} = V_8(P_9 - P_8), \tag{31}$$

$$\dot{W}_{P3} = (\dot{m}_8 W_{P3})/\eta_p, \tag{32}$$

Net power ($\dot{W}_{net}$):

$$\dot{W}_{net} = \dot{W}_T - \dot{W}_{P1} - \dot{W}_{P2} - \dot{W}_{P3}. \tag{33}$$

6. MAHP Operating Variables

As mentioned in Section 3, solution concentration performs a fundamental role in the design and operation of heat pumps. Primarily due to the operating variables range (T and P) that can move concentrations into no-go zones (zones where the system cannot work due to crystallization). The MAHP consists of two solution circuits, see Figure 1 (loop 1: states 1–2–3–4–5–6, loop 2: states 11–12–13–14–15–16). These conditions determine two concentration gradients: $\Delta X_1 = X_4 - X_1$ (between the absorber and low-pressure generator) and $\Delta X_2 = X_{11} - X_{14}$ (between the absorber and high-pressure generator). It is essential to have a concentration gradient greater than zero so that the system has the potential to produce refrigerant and can obtain the desired products ($\dot{W}_{net}$ y $\dot{Q}_A$). Figure 3a shows the variation of the concentration gradients and mass flows of the MAHP for a particular study case (RP = 1.3, T_C = 30 °C, T_{GL} = 60 °C, T_{GH} = 140 °C). In Figure 3a, as the GTL increases (higher GTL equals higher energy revalued by the MAHP), the concentration gradients gradually decrease until ΔX_1 and ΔX_2 tend to 0 and 0.17, respectively (at a GTL of 34 °C). The system operation stops at 34 °C because the ΔX_1 = 0. This condition means no longer a concentration gradient that allows the continuous production of refrigerant. For the defined conditions, ΔX_2 has concentration gradients far from zero. This condition allows the solution mass flow of the upper sub-cycle (states 11 –12 –13 –14 –15 –16) to be low and stable. For example, $\dot{m}_{11}$ reaches values below 5 kg/s, which guarantees low energy consumption for pumping and, therefore, a better system's net power. For its part, when the GTL exceeds 24 °C, ΔX_1 has concentration gradients close to zero. Moreover, as the GTL continues to increase, ΔX_1 tends to zero. This scenario makes the mass flows of the lower sub-cycle (states 1 –2 –3 –4 –5 –6) increasingly larger, showing an exponential behavior. The mass flow $\dot{m}_4$ reaches values of up to 844 kg/s for a GTL of 34 °C. These results show that it is essential to establish an adequate balance between obtaining revalued energy (higher GTL) and high pumping consumption without sacrificing the overall efficiency of the MAHP.

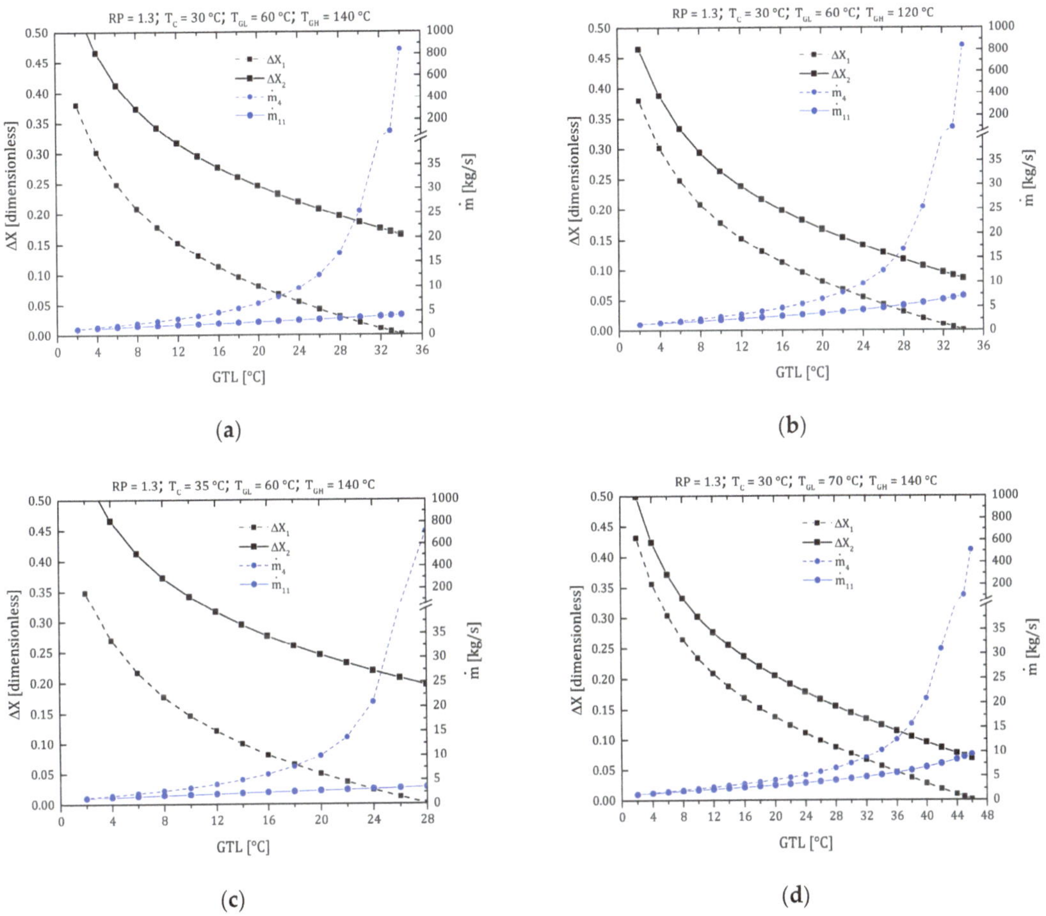

Figure 3. *Cont.*

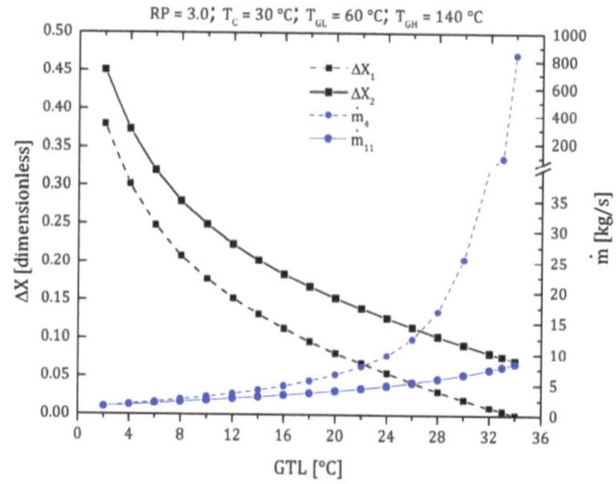

(e)

Figure 3. Concentration gradients and mass flow ratio of the MAHP. (**a**) Case study: RP = 1.3, T_C = 30 °C, T_{GL} = 60 °C, T_{GH} = 140 °C; (**b**) T_{GH} effect; (**c**) T_C effect; (**d**) T_{GL} effect; (**e**) RPT_{GH} effect.

Figure 3b shows the effect of the T_{GH} variation compared to the case study. As can be seen, T_{GH} directly affects the conditions of the upper sub-cycle (ΔX_1 and $\dot{m}_4$ remain constant). By reducing the supply power from 140 to 120 °C, the high-pressure generator has fewer resources to produce refrigerant. For this reason, ΔX_2 is smaller, slightly raising $\dot{m}_{11}$ up to values of 7.5 kg/s. Figure 3c shows the effect of the T_C variation compared to the case study (going from a T_C of 30 to 35 °C). In this case, ΔX_1 starts from 0.35, a value 9% lower than in the case study. This reduction is mainly because the T_C is closely linked with the P_L of the system. Increasing the T_C, P_L also increases, modifying the ΔX_1 and limiting the cycle operation. In other words, the system loses its ability to absorb the refrigerant. The higher the temperature in the refrigerant, the less efficient the absorption process is. Therefore, the energy revaluation capacity is reduced (GTL is reduced from 34 to 28 °C).

Figure 3d shows the effect of the T_{GL} variation compared to the case study (the TGL is increased from 60 to 70 °C). In this scenario, a combined behavior is presented because a medium-temperature thermal source is supplied both in the low-pressure generator and in the evaporator. In the first part, the increase in T_{GH} allows the system to have a greater refrigerant production capacity and to operate with better versatility since a higher GTL range can be achieved (from 34 to 46 °C of revalued thermal energy). Secondly, increasing T_E also increases the system's P_M, gradually reducing ΔX_2. Under these conditions, ΔX_1 continues to be the limiting gradient in the operation of the MAHP, reaching mass flows of $\dot{m}_4$ of up to 513 kg/s.

Finally, Figure 3e shows the effect of the RP variation compared to the study case (the RP is increased from 1.3 to 3.0). Although the P_L and P_M are defined by the phase change in the condenser and evaporator, P_H is determined by the solution concentrations outside the crystallization range. As expected, changes in P_H directly affect ΔX_2, reducing its value and thus the operating range of the cycle (the lower sub-cycle remains intact). In addition to the above, it is expected that by increasing the system's P_H, there will be a greater capacity to produce $\dot{W}_{net}$. However, this same increase reduces the operating range. In conclusion, the thermodynamic conditions of operation of the MAHP establish concentration gradients that depend on the desired applications, the levels of the thermal sources, and the environment.

7. Thermodynamic Simulation and Results

To evaluate the results of the proposed model, different operating conditions where the MAHP can work were varied. The operating ranges are based on the choice of pressures and temperatures outside the crystallization zone of the H_2O-LiBr working solution. Therefore, the RP (to define the concentrations of the solution), T_{GH} (sources temperatures that feeds the generators), and the T_A were varied (this temperature is related to the GTL or revaluated thermal gain). The ranges of operating conditions are defined in Table 1. In addition, the T_{GH} and T_{GL} temperature ranges are achievable with renewable energy or industrial waste heat. On the other hand, $T_E = T_{GL}$; this consideration is relevant to improving the absorber's energy recovery. Therefore, it is convenient to supply the medium-temperature thermal source in the evaporator and low-pressure generator in parallel [39,40]. Finally, the condensation temperature (T_C) was established as constant, which depends on the requirements of the environment where the system will be installed.

Table 1. Operating conditions of the MAHP.

Operating Parameters	Range	Unit
Condenser temperature (T_C)	30	°C
Evaporator and low-pressure generator temperature (T_E–T_{GL})	70	°C
High-pressure generator temperature (T_{GH})	100–160	°C
Low pressure (P_L)	4.3	kPa
Medium pressure (P_M)	31.2	kPa
Pressure ratio ($RP = P_H/P_M$)	1.1–15	-
Refrigerant mass flow ($\dot{m}_7$, $\dot{m}_{17}$)	1	kg/s

Table 2 shows a MAHP case study where the system operates at 140 °C (40 °C as energy revaluation) and with an RP of 1.1. For these conditions, the MAHP reaches the highest exergetic performance. That is, the operating conditions of the case study offer the best relationship between the products obtained concerning the sources supplied in the MAHP. As seen in Table 2, the MAHP can produce 4762 kW of $\dot{Q}_A$ and 265 kW of $\dot{W}_{net}$, getting 0.509 and 0.865 energy and exergetic performance in the system, respectively. The RP variable is vital in the operating conditions of the cycle. Although expansion work is expected to be high at high RP, it is a variable that cannot be deliberately increased, as seen below.

Table 2. Thermodynamic simulation of the MAHP. Case study: T_C = 30 °C, T_{GL} = 70 °C, T_{GH} = 120 °C, RP = 1.1.

Thermodynamic State	T_i [°C]	P_i [kPa]	X_i [-]	h_i [kJ/kg]	m_i [kg/s]	$\dot{W}_{net}$ [kW]	$\dot{Q}_A$ [kW]	η_{Th} [-]	η_{Ex} [-]	GTL [°C]
1	70.0	4.2	0.583	167.4	19.9	265	4762	0.509	0.865	40
2	70.0	31.2	0.583	167.5	19.9					
3	100.9	31.2	0.583	229.6	19.9					
4	110.0	31.2	0.556	241.7	20.9					
5	82.0	31.2	0.556	182.6	20.9					
6	64.2	4.2	0.556	182.6	20.9					
7	70.0	4.2	0	2631.0	1					
8	30.0	4.2	0	125.7	2					
9	30.0	31.2	0	125.7	2					
10	70.0	31.2	0	2626.0	2					
11	110.0	31.2	0.556	241.7	17.9					
12	110.0	34.3	0.556	241.7	17.9					
13	116.3	34.3	0.556	255.2	17.9					
14	120.0	34.3	0.589	270.0	16.9					
15	113.0	34.3	0.589	255.7	16.9					
16	117.4	31.2	0.589	255.7	16.9					
17	120.0	34.3	0	2723.0	1					
18	30.0	4.2	0	2457.0	1					

7.1. MAHP Products

The main products obtained from MAHP are shown in Figure 4. In each section of Figure 4, the evolution of net power $\left(\dot{W}_{net}\right)$ and revaluated heat $\left(\dot{Q}_A\right)$ is revealed concerning the effect of T_{GH} (in decreasing order from 160 to 100 °C). The system's net power is directly proportional to the T_{GH} and RP, the latter variable affecting the most. This behavior is expected because, at higher T_{GH}, a refrigerant with a higher degree of overheating is supplied at the turbine inlet. In addition, at higher RP, the net power increases due to the pressure gradient between the high and low-pressure zone of the MAHP, obtaining higher expansion power. As expected, by setting the T_C and T_{GH} (the sink temperature and the source temperature, respectively), the pressure gradients remain fixed, which causes the powers produced by the turbine to be constant throughout its range of operation. Regarding the revalued thermal load, as the MAHP obtains a higher GTL, the system shows a slight increase during the first 35 °C. However, as mentioned in Section 5, concentration gradients limit the functioning of the cycle. For this reason, above 35 °C of GTL, $\dot{Q}_A$ decreases abruptly, reducing its operating performance and reaching a maximum GTL of 45 °C. Figure 4a shows variations in net power and revalued heat between 283–582 kW and 2500–5100 kW, respectively. Furthermore, with a T_{GH} of 160 °C, the system can operate with a maximum RP of 15. As expected, increasing RP offers better net power output. However, RP cannot be deliberately increased because it limits the operation of the thermodynamic cycle (as explained in Figure 3e). For example, with RP of 1.1, 45 °C of GTL is achieved, while with RP of 15, 7 °C is barely reached (having little operating margin). Regarding Figure 4b–d, the source temperature gradually reduces from 160 to 100 °C. As expected, having the less thermal resource, the MAHP has less capacity to produce $\dot{W}_{net}$ and $\dot{Q}_A$. The RP operating range is also reduced (to operate in areas far from solution crystallization). For a T_{GH} of 140, 120, and 100 °C, maximum RPs of 9, 5, and 3 are reached, with net powers of 515, 441, and 375 kW, respectively.

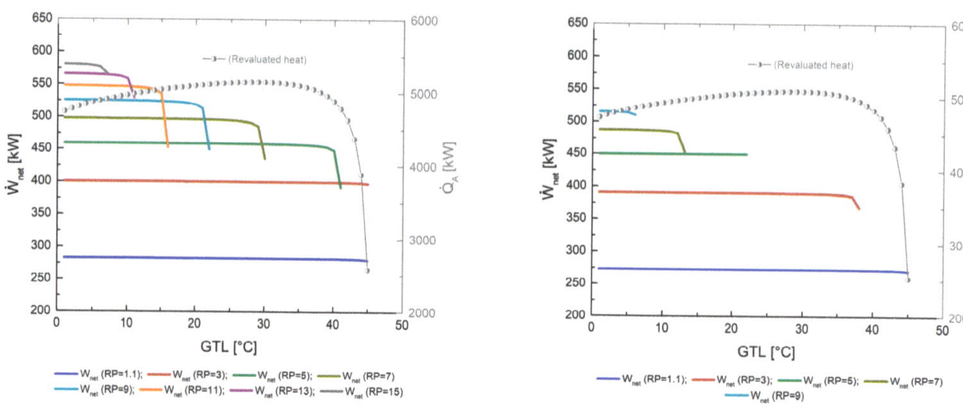

(**a**) (**b**)

Figure 4. *Cont.*

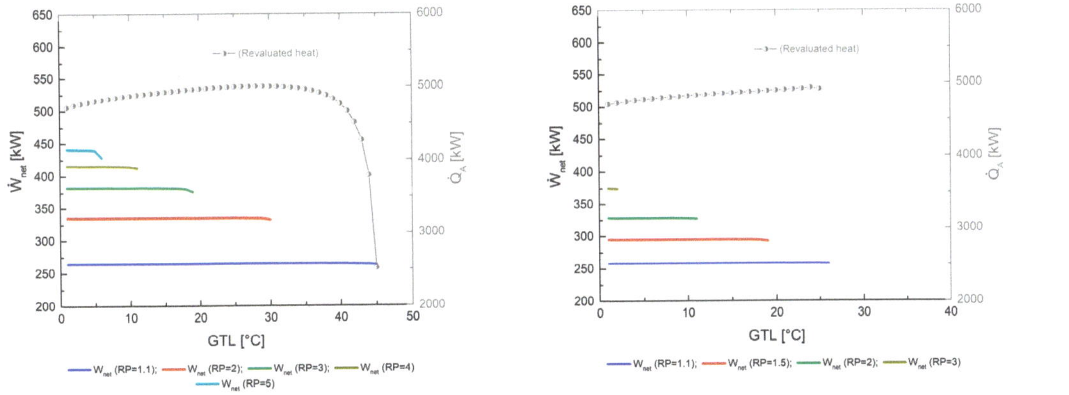

(**c**) (**d**)

Figure 4. Net power ($\dot{W}_{net}$) and revaluated heat ($\dot{Q}_A$) as a function of GTL. (**a**) T_{GH} = 160 °C; (**b**) T_{GH} = 140 °C; (**c**) T_{GH} = 120 °C; (**d**) T_{GH} = 100 °C.

7.2. Thermal Efficiency (η_{Th})

The thermal efficiency of the MAHP, as observed in Figure 5, is directly related to the pressure ratio (RP) and the changes in the temperature of the high-pressure generator (T_{GH}). As previously observed in Figure 4, reducing the source temperature can considerably reduce the net power produced, not the revalued heat. These factors cause the thermal efficiency of the system to show slight changes. However, the most noticeable effect on the MAHP is the reduction in the RP's operating ranges, mainly shortening the maximum achievable GTL. For example, in Figure 5a, heat is supplied at 160 °C; with the energy level of 160 °C, the MAHP reaches values of η_{Th} between 0.51 and 0.55 (for the first 35 °C of GTL, the thermal efficiency varies very little). In addition, the MAHP can operate in a greater range of operation between the RP of 1.1 and 3, reaching 45 °C of GTL. In addition, for RP of 5, 7, 9, 11, 13, and 15, the GTL went reduced to 34, 22, 15, 11, 7, and 4 °C, respectively. The maximum values of GTL are compared with poor thermal performances (tending to zero). A better GTL point would be around 30–35 °C without sacrificing the η_{Th} and without the need to operate outside the thermodynamic limit of the system. A more significant pressure gradient serves to produce greater power in the turbine. Nevertheless, the system's thermal efficiency does not change significantly, mainly because the thermal supplies in both generators and evaporators also increase, thus compensating for the output produced concerning the required input of the MAHP. For example, in Figure 5b–d, the thermal efficiencies achieved remain almost constant. Only the RP range and GTL reduction are sacrificed with less thermal supplied. Having a greater thermal source, the MAHP operates with greater fluidity.

7.3. Exergetic Efficiency (η_{Ex})

The exergetic efficiency, as already mentioned in Section 4, is related to the amount of thermal energy that will be usable (it is a better way of equating the revalued thermal load with the power produced). Finally, in Figure 6, the exergetic efficiency of MAHP is observed as a function of GTL for different temperatures of the supply source.

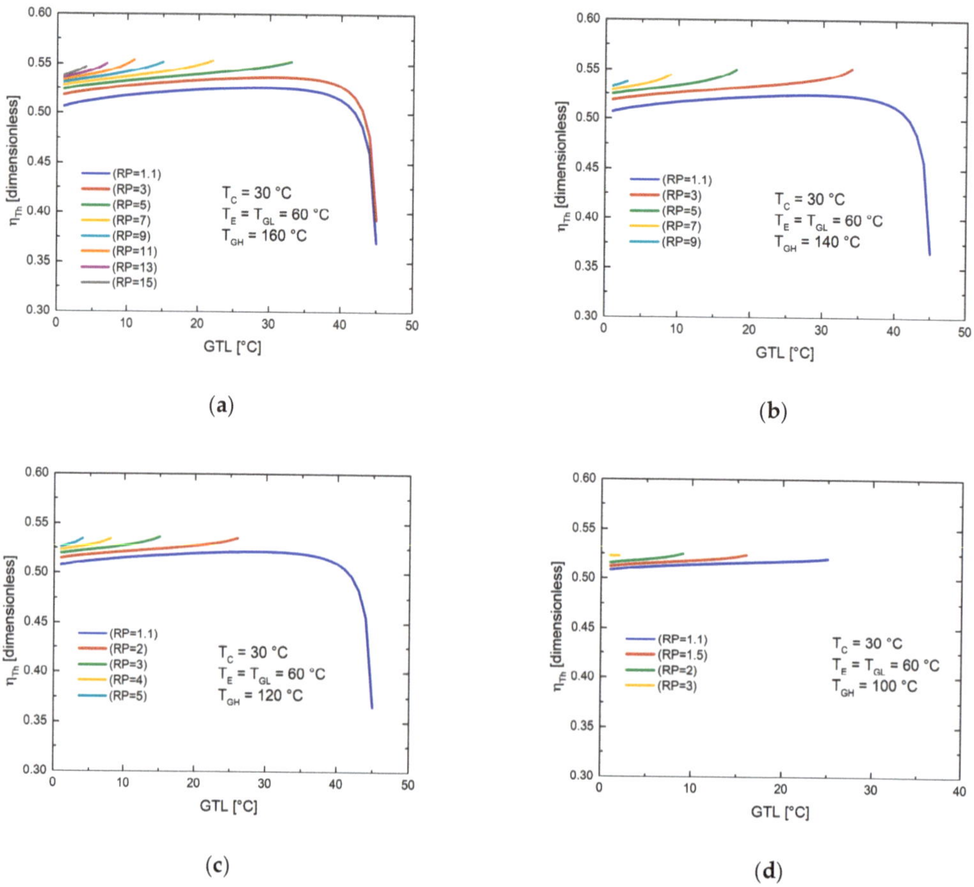

Figure 5. Thermal efficiency (η_{Th}) as a function of GTL at different pressure ratios (RP). (**a**) T_{GH} = 160 °C; (**b**) T_{GH} = 140 °C; (**c**) T_{GH} = 120 °C; (**d**) T_{GH} = 100 °C.

For example, in Figure 6, unlike Figure 5, more marked behaviors are shown for what has been explained above. For a fixed source temperature and RP, increasing T_A favors exergetic efficiency as $\dot{Q}_A$ continues to increase ($\dot{W}_{net}$ remains constant during operation). The thermodynamic limit operative abruptly reduces the revalued heat of the system (Figure 4), which directly affects the exergetic efficiency of the cycle (Equation (3)). For this reason, efficiencies are constantly growing as GTL increases. Subsequently, a maximum value is reached, and finally, there is an abrupt reduction in the efficiencies in the maximum values of GTL. By increasing the RP, the pressure gradient of the MAHP is improved, which benefits the $\dot{W}_{net}$ production of the system. Through the exergetic efficiency, it is deduced that the RP cannot be increased without any control. For each condition, there is an appropriate relationship between $\dot{W}_{net}$ and $\dot{Q}_A$ that offers the best use of the energy supplied concerning that produced. For each section of Figure 6, at the same RP (for example, RP = 1.1), the exergetic efficiencies increase as there is a minor thermal supply source (or at least for low RP). This behavior is to be expected because the exergy supplied in the high-pressure generator has a more significant influence than that achieved with the desired products. Secondly, operating at higher supply temperatures means higher environmental thermal losses. Figure 6a shows that maximum η_{Ex} (between 32 and 40 °C of GTL) of 76%, 83%, 84%, and 80% are reached, for the RP of 1.1, 3, 5, and 7, respectively, further reducing its exergetic efficiency for higher RP (0.72 with RP of 15). Figure 6b achieves its best performances for RP between 1.1 and 3, GTL between 30 and 35 °C, and maximum η_{Ex} of 0.86. Figure 6c achieves its best performances for RP between 1.1 and 2, GTL between 30 and 39 °C, and ultimate η_{Ex} of 0.87. Finally, Figure 6d is considered the worst operating condition since the MAHP works with the least thermal resource. In this case, it is only possible to operate between 1.1 and 2 RP and 27 °C of GTL. However, it is essential to mention that despite the extreme conditions of the last scenario, the MAHP continues to function and produce the desired outputs, greatly favoring its application in various sustainable processes.

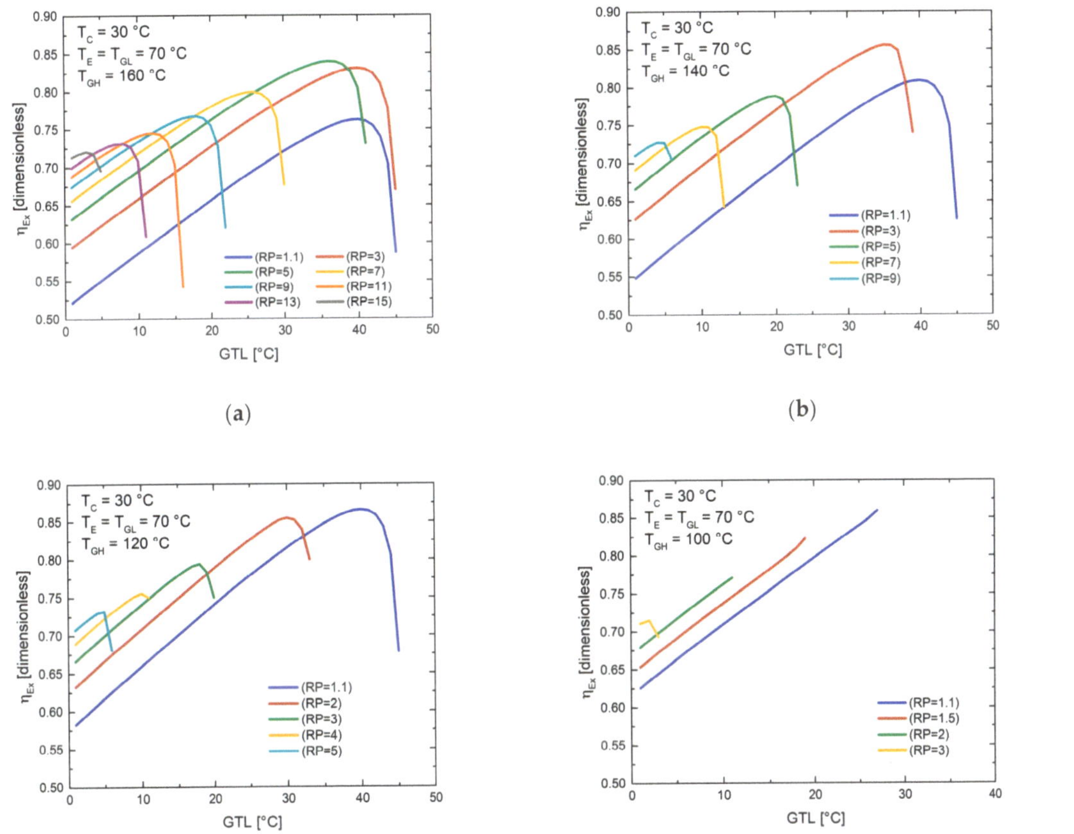

Figure 6. Exergetic efficiency (η_{Ex}) as a function of GTL at different pressure ratios (RP). (**a**) T_{GH} = 160 °C; (**b**) T_{GH} = 140 °C; (**c**) T_{GH} = 120 °C; (**d**) T_{GH} = 100 °C.

As seen in Section 7, the MAHP is a novel system with high efficiencies compared to the systems presented in the introduction. The proposed configuration of nested heat pumps coupled with a turbine provides enormous versatility of operation. On the one hand, MAHP can operate with various sources of supply; on the other, working conditions can be sought to produce high revalued thermal loads or better expansion power. Otherwise, these characteristics allow the MAHP to have an important field of development for applications in the industrial and renewable energy sectors.

8. Conclusions

The analysis of a modified absorption heat pump (MAHP) for cogeneration applications has been made. The MAHP works with the binary solution H_2O-LiBr and simultaneously produces expansion work ($\dot{W}_{net}$) and the revaluation of a thermal source ($\dot{Q}_A$). The working mixture presents enormous potential for development in these new cogeneration systems. However, operating the cycle in regions far from the crystallization zones is essential. In this way, the thermodynamic operation will not be affected. Instead, it will have a more significant margin of process and obtain the expected products in a versatile way. In addition to the problem of crystallization, it is essential to have operating conditions that guarantee gradients greater than zero in the nested sub-cycles of the MAHP. This condition can be fulfilled when there are three levels of pressure and four temperatures. For the simulation conditions, the T_{GH} (100 to 160 °C), the T_A (to improve the degree of revalued heat, GTL), and RP (1.1 to 15) were varied. The results showed that $\dot{W}_{net}$ increases with T_{GH} and RP and decreases with absorber temperature when maximum GTLs are reached. Regarding the revalued thermal load, as the MAHP produces higher T_A, $\dot{Q}_A$ presents a sustained increase grade by grade until the most restricted zone of operation is reached. Therefore, above 35 °C GTL, $\dot{Q}_A$ decreases sharply, reaching a maximum of 45 °C GTL. The maximum $\dot{W}_{net}$ and $\dot{Q}_A$ produced were 582 and 5100 kW, respectively. These significant potencies were obtained for a T_{GH} of 160 °C, RP of 15, and a GTL of 8 °C. The η_{Th} of the MAHP is not significantly affected by the T_{GH} and RP variables. However, for moderate GTL gradients, the thermal efficiencies differ very little, reaching values between 51% and 55%. Secondly, rises above 35 °C cause a drop considerably in the thermal efficiency to 36% (this applies to T_{GH} of 160, 140, and 120 °C). Finally, the maximum η_{Ex} of 87% was reached for a T_{GH} of 120 °C and an RP of 1.1. Therefore, it is essential to mention that for each T_{GH} supply temperature, an optimal RP favors the relationship between the desired products and energy consumption. The η_{Ex} presents a behavior composed mainly of $\dot{W}_{net}$ and $\dot{Q}_A$. Accordingly, low increases in GTL are accompanied by constant increases in η_{Ex} And for high increases in GTL, η_{Ex} reaches a maximum value and subsequently has an abrupt decay. The MAHP is presented as a versatile system with high performance to improve the efficient use of energy, promote the use of alternative energy sources, and use industrial waste heat.

Author Contributions: Conceptualization, J.A.H.-M. and K.C.S.; Data curation, A.C.-P.; Formal analysis, J.A.H.-M., L.A.D.-I., A.C.-P., S.T.-A. and L.I.M.; Investigation, L.A.D.-I., S.L.-L. and S.T.-A.; Methodology, K.C.S.; Software, J.A.H.-M. and S.T.-A.; Validation, J.A.H.-M., S.L.-L., K.C.S., S.T.-A. and L.I.M.; Writing—review and editing, L.A.D.-I., S.L.-L. and A.C.-P. All authors have read and agreed to the published version of the manuscript.

Funding: This research received external funding from PAICYT-UANL (592-IT-2022).

Institutional Review Board Statement: Not applicable.

Informed Consent Statement: Not applicable.

Data Availability Statement: Not applicable.

Conflicts of Interest: The authors declare no conflict of interest.

Nomenclatures

1,2, ..., 18	thermodynamic state points
A	absorber
AHP	absorption heat pump
AHT	absorption heat transformer
C	condenser
CHP	combine heat and power
COP	coefficient of performance
E	evaporator
EC	economizer
EES	engineering equation solver
ETC	evacuated tube collectors
FPC	flat plate collectors
G	generator
GC	Goswami cycle
GTL	gross temperature lift [°C]
h	enthalpy [kJ/kg]
KC	Kalina cycle
LiBr	lithium bromide
H_2O	water
MAHP	modified absorption heat pump
$\dot{m}$	mass flow rate [kg/s]
ORC	organic Rankine cycle
P	pressure [kPa]
$\dot{Q}$	thermal capacity [kW]
RP	pressure ratio [PH/PM]
SHE	solution heat exchanger
SRC	steam Rankine cycle
T	temperature [°C]
$\dot{W}$	mechanical power [kW]
X	concentration of the solution [-]
Subscripts	
A	absorption
C	condensation
E	evaporation
Ex	exergetic
G	generation
H	high
L	low
M	medium
net	net
0	ambient temperature
P	pump
T	turbine
Th	thermic
Greek Symbols	
η	efficiency [-]
v	specific volume [m^3/kg]

References

1. Su, Z.; Zhang, M.; Xu, P.; Zhao, Z.; Wang, Z.; Huang, H.; Ouyang, T. Opportunities and strategies for multigrade waste heat utilization in various industries: A recent review. *Energy Convers. Manag.* **2021**, *229*, 113769. [CrossRef]
2. International Energy Agency (IEA). *Global Energy and CO_2 Status Report 2018*; International Energy Agency: Paris, France, 2019; p. 562.
3. Loni, R.; Najafi, G.; Bellos, E.; Rajaee, F.; Said, Z.; Mazlan, M. A review of industrial waste heat recovery system for power generation with Organic Rankine Cycle: Recent challenges and future outlook. *J. Clean. Prod.* **2021**, *287*, 125070. [CrossRef]

4. Arnaudo, M.; Dalgren, J.; Topel, M.; Laumert, B. Waste heat recovery in low temperature networks versus domestic heat pumps—A techno-economic and environmental analysis. *Energy* **2021**, *219*, 119675. [CrossRef]
5. Turnbull, R.; Muneer, T. A Two Year Comparison of Energy and CO2 Emissions of an Industrial Refrigeration Plant after the Installation of a Waste Heat Recovery System. *Energy Procedia* **2019**, *161*, 251–258. [CrossRef]
6. Holzleitner, M.; Moser, S.; Puschnigg, S. Evaluation of the impact of the new Renewable Energy Directive 2018/2001 on third-party access to district heating networks to enforce the feed-in of industrial waste heat. *Util. Policy* **2020**, *66*, 101088. [CrossRef]
7. Mahmoudi, A.; Fazli, M.; Morad, M. A recent review of waste heat recovery by Organic Rankine Cycle. *Appl. Therm. Eng.* **2018**, *143*, 660–675. [CrossRef]
8. Weiss, W.; Spörk-Dür, M. Global Market Development and Trends 2021 Detailed Market Figures 2020. In *Solar Heat Worldwide Edition 2022*; AEE—Institute for Sustainable Technologi: Gleisdor, Austria, 2022.
9. Fan, Y.; Luo, L.; Souyri, B. Review of solar sorption refrigeration technologies: Development and applications. *Renew. Sustain. Energy Rev.* **2007**, *11*, 1758–1775. [CrossRef]
10. Ayou, D.S.; Bruno, J.C.; Coronas, A. Combined absorption power and refrigeration cycles using low- and mid-grade heat sources. *Sci. Technol. Built Environ.* **2015**, *21*, 934–943. [CrossRef]
11. López-Villada, J.; Ayou, D.S.; Bruno, J.C.; Coronas, A. Modelling, simulation and analysis of solar absorption power-cooling systems. *Int. J. Refrig.* **2014**, *39*, 125–136. [CrossRef]
12. Arabkoohsar, A.; Andresen, G. A smart combination of a solar assisted absorption chiller and a power productive gas expansion unit for cogeneration of power and cooling. *Renew. Energy* **2018**, *115*, 489–500. [CrossRef]
13. Arabkoohsar, A.; Sadi, M. A solar PTC powered absorption chiller design for Co-supply of district heating and cooling systems in Denmark. *Energy* **2020**, *193*, 116789. [CrossRef]
14. Rahman, A.; Abas, N.; Dilshad, S.; Saleem, M.S. A case study of thermal analysis of a solar assisted absorption air-conditioning system using R-410A for domestic applications. *Case Stud. Therm. Eng.* **2021**, *26*, 101008. [CrossRef]
15. Düzcan, A.; Kara, Y.A. Investigation of the usage potential of the evacuated tube and the flat plate collectors to assist an absorption chiller. *Sustain. Energy Technol. Assess.* **2021**, *47*, 101437. [CrossRef]
16. Ali, D.; Ratismith, W. A semicircular trough solar collector for air-conditioning system using a single effect NH_3–H_2O absorption chiller. *Energy* **2020**, *215*, 119073. [CrossRef]
17. Behzadi, A.; Arabkoohsar, A.; Sadi, M.; Chakravarty, K.H. A novel hybrid solar-biomass design for green off-grid cold production, techno-economic analysis and optimization. *Sol. Energy* **2021**, *218*, 639–651. [CrossRef]
18. Nami, H.; Anvari-Moghaddam, A.; Nemati, A. Modeling and analysis of a solar boosted biomass-driven combined cooling, heating and power plant for domestic applications. *Sustain. Energy Technol. Assess.* **2021**, *47*, 101326. [CrossRef]
19. Nasir, M.; Ekwonu, M.; Park, Y.; Esfahani, J.; Kim, K. Assessment of a District Trigeneration Biomass Powered Double Organic Rankine Cycle as Primed Mover and Supported Cooling. *Energies* **2021**, *14*, 1030. [CrossRef]
20. Siddiqui, O.; Dincer, I. A new solar and geothermal based integrated ammonia fuel cell system for multigeneration. *Int. J. Hydrogen Energy* **2020**, *45*, 34637–34653. [CrossRef]
21. Chen, Y.; Wang, J.; Lund, P.D. Thermodynamic performance analysis and multi-criteria optimization of a hybrid combined heat and power system coupled with geothermal energy. *Energy Convers. Manag.* **2020**, *210*, 112741. [CrossRef]
22. Ochieng, A.O.; Megahed, T.F.; Ookawara, S.; Hassan, H. Comprehensive review in waste heat recovery in different thermal energy-consuming processes using thermoelectric generators for electrical power generation. *Process Saf. Environ. Prot.* **2022**, *162*, 134–154. [CrossRef]
23. Wang, Y.; Yang, H.; Xu, K. Comparative environmental impacts and emission reductions of introducing the novel organic Rankine & Kalina cycles to recover waste heat for a roller kiln. *Appl. Therm. Eng.* **2021**, *190*, 116821. [CrossRef]
24. Kalan, A.S.; Ghiasirad, H.; Saray, R.K.; Mirmasoumi, S. Thermo-economic evaluation and multi-objective optimization of a waste heat driven combined cooling and power system based on a modified Kalina cycle. *Energy Convers. Manag.* **2021**, *247*, 114723. [CrossRef]
25. Singh, A.; Das, R. A novel combined power and cooling cycle design and a modified conditional exergy destruction approach. *Energy Convers. Manag.* **2021**, *233*, 113943. [CrossRef]
26. Kumar, A.; Rakshit, D. A critical review on waste heat recovery utilization with special focus on Organic Rankine Cycle applications. *Clean. Eng. Technol.* **2021**, *5*, 100292. [CrossRef]
27. Köse, Ö.; Koç, Y.; Yağlı, H. Is Kalina cycle or organic Rankine cycle for industrial waste heat recovery applications? A detailed performance, economic and environment based comprehensive analysis. *Process Saf. Environ. Prot.* **2022**, *163*, 421–437. [CrossRef]
28. Akimoto, R.; Yamaki, T.; Nakaiwa, M.; Matsuda, K. Evaluation of a power generation system that integrates multiple Kalina cycles and absorption heat pumps. *Case Stud. Therm. Eng.* **2021**, *28*, 101363. [CrossRef]
29. Aksar, M.; Yağlı, H.; Koç, Y.; Koç, A.; Sohani, A.; Yumrutaş, R. Why Kalina (Ammonia-Water) cycle rather than steam Rankine cycle and pure ammonia cycle: A comparative and comprehensive case study for a cogeneration system. *Energy Convers. Manag.* **2022**, *265*, 115739. [CrossRef]
30. Marshall, Z.; Duquette, J. A techno-economic evaluation of low global warming potential heat pump assisted organic Rankine cycle systems for data center waste heat recovery. *Energy* **2022**, *242*, 122528. [CrossRef]
31. Liu, Z.; Xie, N.; Yang, S. Thermodynamic and parametric analysis of a coupled LiBr/H2O absorption chiller/Kalina cycle for cascade utilization of low-grade waste heat. *Energy Convers. Manag.* **2020**, *205*, 112370. [CrossRef]

32. Cao, L.; Wang, J.; Wang, H.; Zhao, P.; Dai, Y. Thermodynamic analysis of a Kalina-based combined cooling and power cycle driven by low-grade heat source. *Appl. Therm. Eng.* **2017**, *111*, 8–19. [CrossRef]
33. Wang, J.; Wang, J.; Zhao, P.; Dai, Y. Thermodynamic analysis of a new combined cooling and power system using ammonia–water mixture. *Energy Convers. Manag.* **2016**, *117*, 335–342. [CrossRef]
34. Padilla, R.V.; Demirkaya, G.; Goswami, D.Y.; Stefanakos, E.; Rahman, M.M. Analysis of power and cooling cogeneration using ammonia-water mixture. *Energy* **2010**, *35*, 4649–4657. [CrossRef]
35. Zare, V.; Mahmoudi, S.; Yari, M.; Amidpour, M. Thermoeconomic analysis and optimization of an ammonia–water pow-er/cooling cogeneration cycle. *Energy* **2012**, *47*, 271–283. [CrossRef]
36. Xu, F.; Goswami, D.Y.; Bhagwat, S.S. A combined power/cooling cycle. *Energy* **2000**, *25*, 233–246. [CrossRef]
37. Demirkaya, G.; Padilla, R.V.; Goswami, D.Y.; Stefanakos, E.; Rahman, M.M. Analysis of a combined power and cooling cycle for low-grade heat sources. *Int. J. Energy Res.* **2010**, *35*, 1145–1157. [CrossRef]
38. Chakravarty, K.H.; Sadi, M.; Chakravarty, H.; Alsagri, A.S.; Howard, T.J.; Arabkoohsar, A. A review on integration of renewable energy processes in vapor absorption chiller for sustainable cooling. *Sustain. Energy Technol. Assess.* **2022**, *50*, 101822. [CrossRef]
39. Magallanes, J.A.H.; Heard, C.; Best, R.; Rivera, W. Modeling of a new absorption heat pump-transformer used to produce heat and power simultaneously. *Energy* **2018**, *165*, 112–133. [CrossRef]
40. Hernández-Magallanes, J.A.; Tututi-Avila, S.; Cerdán-Pasarán, A.; Morales, L.; Rivera, W. Thermodynamic simulation of an absorption heat pump-transformer-power cycle operating with the ammonia-water mixture. *Appl. Therm. Eng.* **2021**, *182*, 116174. [CrossRef]
41. Allouhi, A.; Kousksou, T.; Jamil, A.; Bruel, P.; Mourad, Y.; Zeraouli, Y. Solar driven cooling systems: An updated review. *Renew. Sustain. Energy Rev.* **2015**, *44*, 159–181. [CrossRef]
42. Hernández-Magallanes, J.A. Desarrollo y Evaluación de un Sistema de Enfriamiento Solar Tipo Vertical Operando con la Mezcla Nitrato de Litio-Amoniaco. Ph.D. Thesis, Universidad Nacional Autónoma de México, Mexico City, Mexico, 2017.
43. Bahador, B. Process Integration of Absorption Heat Pumps. Ph.D. Thesis, Universidad de Montreal, Montréal, QC, Canada, 2009.
44. Espinosa, F.C. Diseño de un enfriador con el sistema de absorción Agua/Bromuro de litio operando con energía solar. Master's Thesis, Instituto Tecnológico y de Estudios Superiores de Monterrey, Monterrey, Mexico, 2007.
45. Herold, K.E.; Radermacher, R.; Klein, S.A. *Absorption Chillers and Heat Pumps*, 2nd ed.; CRC Press: Boca Raton, FL, USA, 2016.
46. Panahizadeh, F.; Bozorgan, N. The energy and exergy analysis of single effect absorption chiller. *Int. J. Adv. Des. Manuf. Technol.* **2011**, *4*, 19–26.
47. Magallanes, J.A.H.; Rivera, W.; Coronas, A. Comparison of single and double stage absorption and resorption heat transformers operating with the ammonia-lithium nitrate mixture. *Appl. Therm. Eng.* **2017**, *125*, 53–68. [CrossRef]
48. Nellis, G.F.; Klein, S.A. *Engineering Equation Solver (EES)*; F-Chart Software: Madison, WI, USA, 2020.
49. Jing, X.; Zheng, D. Effect of cycle coupling-configuration on energy cascade utilization for a new power and cooling cogeneration cycle. *Energy Convers. Manag.* **2014**, *78*, 58–64. [CrossRef]
50. Rashidi, J.; Ifaei, P.; Esfahani, I.J.; Ataei, A.; Yoo, C.K. Thermodynamic and economic studies of two new high efficient power-cooling cogeneration systems based on Kalina and absorption refrigeration cycles. *Energy Convers. Manag.* **2016**, *127*, 170–186. [CrossRef]
51. He, J.; Liu, C.; Xu, X.; Li, Y.; Wu, S.; Xu, J. Performance research on modified KCS (Kalina cycle system) 11 without throttle valve. *Energy* **2014**, *64*, 389–397. [CrossRef]

processes

MDPI

Article

Impact of Electric Vehicles on Energy Efficiency with Energy Boosters in Coordination for Sustainable Energy in Smart Cities

Pawan Kumar [1,*], Srete Nikolovski [2,*], Ikbal Ali [3], Mini S. Thomas [3] and Hemant Ahuja [4]

1 Electrical and Instrumentation Engineering Department, Thapar Institute of Engineering and Technology, Patiala 147004, India
2 Power Engineering Department, Faculty of Electrical Engineering, Computer Science and Information Technology, J. J. Strossmayer University of Osijek, K. Trpimira 2B, HR-31000 Osijek, Croatia
3 Department of Electrical Engineering, Faculty of Engineering & Technology, Jamia Millia Islamia, New Delhi 110025, India
4 Department of Electrical and Electronics Engineering, Ajay Kumar Garg Engineering College, Ghaziabad 201009, India
* Correspondence: pawanror@gmail.com (P.K.); srete.nikolovski@ferit.hr (S.N.)

Abstract: The use of electric vehicles (EVs) has recently increased in a smart city environment. With this, the optimal location of the charging station is a great challenge and, hence, the energy efficiency performance (EEP) of an electrical system is important. Ideally, the EEP is realized through passive energy boosters (PEBs) and active energy boosters (AEBs). PEBs require no external resources, and EEP is achieved through altering the network topology and loading patterns, whereas, in AEBs, integrating external energy resources is a must. The EEP has also become dynamic with the integration of an energy storage system (ESS) in a deregulated environment. Customer energy requirement varies daily, weekly, and seasonally. In this scenario, the frequent change in network topology requires modifying the size and location of AEBs. It alters the customers' voltage profile, loadability margin, and supply reliability when the EV works differently as a load or source. Therefore, a comprehensive EEP analysis with different probabilistic loading patterns, including ESS, must be performed at the planning stage. This work uses a harmony search algorithm to evaluate EEP for AEBs and PEBs, in coordination, when ESS works as a load or source, at four locations, for customers' and utilities' benefits.

Keywords: probabilistic loading patterns; energy efficiency performance; distribution systems; passive energy boosters; active energy boosters; energy storage system; harmony search algorithm; sustainable energy and society

Citation: Kumar, P.; Nikolovski, S.; Ali, I.; Thomas, M.S.; Ahuja, H. Impact of Electric Vehicles on Energy Efficiency with Energy Boosters in Coordination for Sustainable Energy in Smart Cities. *Processes* **2022**, *10*, 1593. https://doi.org/10.3390/pr10081593

Academic Editors: Ferdinando Salata and Virgilio Ciancio

Received: 21 July 2022
Accepted: 9 August 2022
Published: 12 August 2022

1. Introduction

Technological development paves the way to modernization with significant enhancements in energy consumption, where electricity plays a crucial role. In this scenario, sustainable energy is the primary requirement for a sustainable society. Though energy resources are depleting day by day, it is necessary to explore the ever-best approach for energy efficiency while generating, distributing, and utilizing electricity. In the current scenario, societal development mainly depends upon the availability of energy resources. Conversely, the consumers' energy demand has grown unexpectedly, and agencies are working tirelessly to explore various energy resources. The availability of the resources is limited. Therefore, the smart systems for energy efficiency in generation, distribution, and utilization can be considered alternatives to the resources for sustainable energy and society.

A smart energy system has two aspects: namely, (a) realization of technical parameters for energy efficiency and (b) bidirectional communication with power components for their control and monitoring. The first part is more critical to developing a smart infrastructure for operation and control. In the present scenario, the loading pattern varies with the state of the economy and system voltage profile. The involvement of smart devices encompasses power electronic devices and converters; the loading pattern has become highly non-linear and needs to be managed on an hourly basis [1]. It further alters the loadability margin and reliability of power supply, at respective load points, during power delivery. However, in recent years, the technological evolution in electric vehicles has introduced a new scope of research for energy efficiency in smart energy systems.

Further, in a smart city environment, the EEP of a distribution system depends upon several parameters. It changes with the representation of loads, network topology, integration of DGs and shunt capacitors, and the implementation of various DSM approaches during power delivery. In addition, in recent years, the integration of EVs has evolved the new research challenges for energy efficiency. In this work, EEP is realized with and without external energy resources to benefit utility and consumers. The approach for EEP without external energy resources is defined as a passive energy booster, whereas the approach for EEP with external energy resources is the active energy booster. In the past, the researchers have presented several approaches, with slight modifications, based on PEBs and AEBs, exclusively. From consumers' perspectives, in this work, the EEP is realized by PEBs and AEBs in coordination with ESS, and a comprehensive evaluation is presented based on several operating parameters.

Several approaches exist for the EEP of the distribution system, and using single and multi-objective functions, with line losses, are a significant concern. Ideally, it is beneficial for utilities, whereas it has little significance in consumers' interests. Ali et al. [2] have presented several indices-based approaches for the comprehensive evaluation of EEP of the distribution system. Here, the authors considered different combinations of the load models, and EEP is realized with and without the integration of external energy resources exclusively. A heuristic approach is developed, and the weightage of different indices is varied to show their effects on EEP. However, in this approach, the loads are represented with fixed weightage to the particular load class or type, as well as under fixed weightage to the energy parameters. As an extension to the method presented in [2], it is required to evaluate the system performance for a possible number of customers, as well as the reliability of power supplies under different probabilistic loading patterns, including ESS.

Authors in [3,4] have developed a heuristic approach, for improving a system's voltage profile and EEP, by considering a different objective in optimization that allows the more significant number of customers to be supplied at a particular time instant. However, authors in [5] have developed the DSM approach under different loads, such as electric vehicles, mobile charge, and energy storage devices, for smart grid systems with fixed network topology. Habeeb et al. [6] have presented the energy management system, with local generation and energy storage, for domestic applications. Conversely, in [7], authors have developed schemes for the operation of different power equipment throughout the day. The real-time implementation of these approaches tends to alter the loading patterns hourly. In this scenario, the representation of loads in power system operation and stability studies plays an important role. Bin et al. [8] have presented a multi-objective optimization problem in a hybrid microgrid environment, where power generation from the PV system, windmills, and ESS is simulated using fuzzy techniques. Here, the authors have also considered the static loading patterns, and no emphasis has been laid on the voltage dependency of loading patterns during a change in network topology. In [9], the integration of these power generating systems has been simulated with DSM, for an agriculture farm, to reduce the energy demand of the power grid.

With the implementation of the DSM approach, the load demand was found to be varying significantly in [10], where the energy demand of the commercial building is evaluated for the intensification of energy efficiency. However, Asrari et al. [11] and Dorostkar et al. [12] have presented the impact of DER on the electrical system during network reconfiguration for EEP. Similarly, in [13,14], the reconfiguration is performed to improve the reliability of supplying power, with a vehicle, to grid and DG allocation in an active distribution system, respectively. Conversely, Fu et al. [15] revealed that the reconfiguration is limited due to switching loss. Therefore, it requires development of a timeframe for a particular configuration for the capacity enhancement of existing networks. Ali et al. [16] have also presented the reconfiguration, in the presence of ESS, considering static loads. In [17], reconfiguration is performed under an unbalanced distribution system to coordinate voltage regulators and renewable resources. Song et al. [18] considered the voltage volatility due to DER, and reconfiguration is performed to minimize it, but in [19], DG allocation is presented in the presence of ESS for energy performance.

Naguib et al. [20] conferred that environmental uncertainty can affect the system performance during reconfiguration and DG allocation. Cattani et al. [21] addressed the distribution system planning by using reconfiguration, and in [22], an AC-DC hybrid distribution system is developed for the customer's energy management. Takenobu et al. [23] presented the annual energy consumption, based on the reconfiguration of the distribution system, under different operating conditions. This allows system operation flexibility, and it benefits both the customer and the utility. Conversely, Arsari et al. [24] presented a new framework of DG allocation, based on time management, hourly—ahead of the system's decision making—to reduce the necessity of reconfiguration. In [25], the authors emphasize the short-term improvement in the voltage profile when using network reconfiguration. However, in [26,27], the process of reconfiguration is highlighted, and heuristic approaches are developed for single or multilevel switching operations, whereas in [28–30], network reconfiguration is performed for loss reduction under a fixed loading scenario.

The above literature has revealed that network reconfiguration and DG allocation play crucial roles in the EEP of an electrical system. However, in these approaches, the impact of ESS on EEP has not been considered; instead, they focused on the conventional approach of network reconfiguration and DG allocation. These approaches can be realized in two ways: namely, (i) the approaches in which no external energy resource is required to improve the EEP and (ii) the approach in which external source is a must to improve the EEP. The former is called a passive energy booster, and later, is an active energy booster. Further, in recent years, the new type of load, featured with ESS, can be recognized as a dynamic load since it can work as a load or source, while charging or discharging, respectively.

The performance of the electrical power system with ESS has been presented in [31–33]. However, in these studies, the representation of physical loads and network reconfiguration have not been considered; instead, results are demonstrated under constant power loads only. However, in smart energy infrastructure, the voltage dependency of these loads cannot be ignored. An extensive review presented in [34] has emphasized the need of ESS for a sustainable microgrid. Considering the above facts, this work intended to develop a probabilistic loading pattern for the comprehensive evaluation of EEP of the electrical systems in a smart energy system's environment. Moreover, in this work, DG allocation and network reconfiguration are performed, in coordination, under different loading patterns, including ESS, which is beneficial for the customer and utilities in a sustainable city and society.

The organization of the paper is as follows: The extensive literature review is presented under the introduction in Section 1. The mathematical formulations, with operating constraints, load modeling, and EEP parameters, are described in Section 2. However, Section 3 presents a detailed description of the harmony search algorithm, whereas Section 4 presents the procedural steps involved in the proposed algorithm. Section 5 describes the test system and the case study cases. Section 6 presents the test results and discussion, and finally, the results and findings are concluded in Section 7.

2. Mathematical Formulations for Energy Efficiency

2.1. Problem Formulation

With the involvement of smart devices in energy systems, the operation and control has become fast. As a result, it has improved the decision-making capacity under the frequent change in operating conditions for energy efficiency, which involves the reduction in power loss, improvement in voltage profile, and loadability margin—exclusively or in coordination—with certain weightage of each parameter. However, the change in working conditions may compromise one parameter over another. In this scenario, the primary function of PEBs and AEBs is to deal with the constraints, such as active and reactive power demand, voltage limits, line capacity, and network radiality. Therefore, the optimization problem is formulated as follows:

$$\text{Minimize} \rightarrow \text{f(x)} \tag{1}$$

The above problem is subject to the following constraints:

a. *Active and reactive power limit:* The algebraic sum of all the generations, power demands, and power losses should be zero, and it is defined as:

$$\sum_{i=1}^{n} P_{Gi} - \sum_{i=1}^{n} P_{Di} - \sum_{i=1}^{n} P_{Li} = 0 \tag{2}$$

$$\sum_{i=1}^{n} Q_{Gi} - \sum_{i=1}^{n} Q_{Di} - \sum_{i=1}^{n} Q_{Li} = 0 \tag{3}$$

Here, in (2) and (3), the P_{Gi} and Q_{Gi} are the generated power at the ith node, P_{Di} and Q_{Di} are the demands at the ith node, and P_{Li} and Q_{Li} is the power losses beyond ith node.

b. *Voltage limit:* The PEBs and AEBs are responsible for improving the system voltage profile. However, PEBs can improve up to a certain extent, whereas AEBs can improve the voltage profile even beyond 1.0 pu. Therefore, voltage limit restrains the overcompensation due to AEBs, and it is defined in (4):

$$V_{i,min} \leq \; V_i \leq \; V_{i,max} \tag{4}$$

c. *Capacity limit:* In power distribution, the lines have their capacity to transfer power and, hence, the loadability. The line capacity is usually evaluated with current limits, and the maximum load that can be connected at a particular node is defined with kVA loading. These limits are described in (5).

$$\left.\begin{array}{c} I_i < I_{i,\;max} \\ kVA_i < kVA_{i,\;max} \end{array}\right] \tag{5}$$

d. *Network radiality limit:* A distribution system is generally operated as radial to coordinate protecting devices. The branches are arranged in a downward stream. In a set of branches, the index value k_{ij} is equal to '1' for all connected branches between the 'ith' and 'jth' nodes; otherwise, it is zero if that branch is considered as a tie-line in the configuration. In a radial network,

$$kij = 1 \rightarrow \rightarrow \text{for all connected branches}$$

$$kij = 0 \rightarrow \rightarrow \text{for tie-lines}$$

Therefore, for network radiality the following is to be satisfied.

$$\sum k_{ij} = n - 1 \tag{6}$$

Here, in (6), 'n' is the total number of nodes in a radial path, including the substation.

e. *State of charge limit:* An energy storage system (ESS) in electrical power distribution can work as a load, as well as an energy resource, depending on its state of charge during operation. ESS below its lower limit of SOC tends to work as load, whereas it can meet load demand under peak loading conditions when SOC is above the prescribed limit.

$$SOC_{b,min} \le SOC_b < SOC_{b,max} \tag{7}$$

f. *AEB's limit:* The external DGs are the source of AEBs. If AEBs are operated below their minimum prescribed limit, they become uneconomical. The lower limit is taken as 0.3 times of its full capacity in this work. Therefore, the following limits are imposed for AEBs for the energy-efficient operation of the power system.

$$0.3 \times AEB_{i,max} \le AEB_i < AEB_{i,max} \tag{8}$$

2.2. Energy Efficiency Analysis

In the deregulated competitive energy market, the energy efficiency is not limited to a mere single parameter, such as the reduction in power loss. Rather, it depends upon several other parameters, such as time of operation, loading patterns, state of the economy, environmental conditions, and the number of customers to be supplied, i.e., loadability limit, voltage profile at the customer end, reliability of supplying power, as well as the quantity and quality of supplying power. To improve the EEP, utilities usually implement energy boosters during operation. These energy boosters can be classified as PEBs and AEBs. The effect of various PEBs and AEBs on the EEP is summarized in Table 1.

Table 1. Parameters for energy efficiency performance.

S. No.	Parameters	Significance in EEP
1	kW demand (Ps_i)	The integration of PEBs and AEBs improves the voltage profile, altering the loading pattern.
2	kVAr demand (Qs_i)	The reactive power flow in the system directly affects the voltage profile. The higher the reactive power lower will be the voltage and hence the poor energy efficiency.
3	kVA demand (Ss_i)	The number of customers to be supplied depends upon the kVA rating of transformer at load point.
4	Node voltage profile (V_i)	The loads are voltage-dependent, and hence, the voltage profile can alter the loading pattern and the operating efficiency of the connected loads.
5	Loadability index (f_{si})	Loadability margin limits the maximum number of customers supplied at a particular node.
6	Power loss (P_L)	PEBs and AEBs reduce the power loss by improving voltage profile and local generation.
7	Margin of Reliability (MR)	The reliability of supplying power is influenced by the possible number of consumers, voltage limit, and thermal limit. The PEBs and AEBs improve these aspects and hence the reliability.

2.3. Load Representation and Modeling

Practical loads are the combination of several types of loads having different characteristics [35]. The operating characteristics of these loads may change with time and the system voltage profile. Depending on their voltage profile sensitivity, loads are classified as residential, commercial, and industrial loads. Conversely, the operating characteristics can be represented as constant power, constant current, and constant impedance for individual loads. In practice, these loads act together, and therefore, it is required to consider the various combinations of different load classes and types, as shown in Table 2.

Table 2. Load type and load class with their voltage exponents.

Load Type	Voltage Exponents		Load Class	Voltage Exponents	
	α	β		α	β
Constant power	0	0	Industrial	0.18	6.0
Constant current	1	1	Commercial	0.99	3.5
Constant impedance	2	2	Residential	1.2	2.9

In this work, based on the above representation, the loads are modeled as follows:

a. *Load model-1 (LM-1):* In LM-1, the load at each node is considered to be a constant power type and represented as

$$P_i = a * P_{i0}\left[\frac{V_i}{V_{i0}}\right]^{\alpha} \quad Q_i = a * Q_{i0}\left[\frac{V_i}{V_{i0}}\right]^{\beta} \tag{9}$$

b. *Load model-2 (LM-2):* In LM-2, the loads are considered as the combinations of various load types, such as constant power, constant current, and constant impedance loads, and it is represented as

$$P_i = P_{i0}\left(a\left[\frac{V_i}{V_{i0}}\right]^{\alpha_1} + b\left[\frac{V_i}{V_{i0}}\right]^{\alpha_2} + c\left[\frac{V_i}{V_{i0}}\right]^{\alpha_3}\right) \quad Q_i = Q_{i0}\left(a\left[\frac{V_i}{V_{i0}}\right]^{\beta_1} + b\left[\frac{V_i}{V_{i0}}\right]^{\beta_2} + c\left[\frac{V_i}{V_{i0}}\right]^{\beta_3}\right) \tag{10}$$

c. *Load model-3 (LM-3):* In LM-3, the loads are considered as the combinations of various load classes, such as residential, commercial, and industrial loads, and it is represented as

$$P_i = P_{i0}\left(x\left[\frac{V_i}{V_{i0}}\right]^{\alpha_1} + y\left[\frac{V_i}{V_{i0}}\right]^{\alpha_2} + z\left[\frac{V_i}{V_{i0}}\right]^{\alpha_3}\right) \quad Q_i = Q_{i0}\left(x\left[\frac{V_i}{V_{i0}}\right]^{\beta_1} + y\left[\frac{V_i}{V_{i0}}\right]^{\beta_2} + z\left[\frac{V_i}{V_{i0}}\right]^{\beta_3}\right) \tag{11}$$

In (10), voltage exponents $\alpha_1 = \beta_1 = 0$, $\alpha_2 = \beta_2 = 1$, and $\alpha_3 = \beta_3 = 2$, as well as 'a', 'b', and 'c' are the fractional contributions in the LM-2. Conversely, in (11), the voltage exponents $\alpha_1 = 0.18$, $\beta_1 = 6.0$, $\alpha_2 = 0.99$, $\beta_2 = 3.5$, and $\alpha_3 = 1.2$, $\beta_3 = 2.9$, as well as 'x', 'y', and 'z' are the fractional contributions in LM-3. In various LMs, the fraction 'a', 'b', 'c' and 'x', 'y', 'z' are obtained using (21), where it varies at random as a function of probabilistic constant factor (PCF).

Considering the above formulation, the node voltage and loadability are calculated using equivalent radial network between two nodes, as shown in Figure 1:

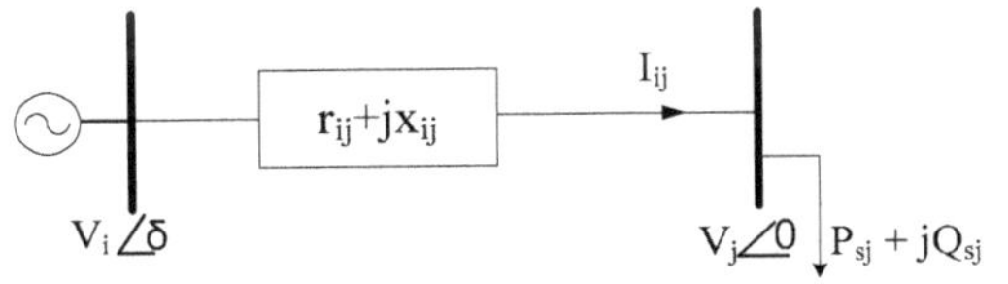

Figure 1. Representation of radial network between two nodes.

Let the receiving-end power factor angle (θ), sending-end voltage angle (δ), and $\Psi = \theta + \delta$, so the following equations can be written:

$$V_i cos\,\Psi = V_j cos\,\theta + I_{ij}r_{ij}\,; \tag{12}$$

$$V_i sin\,\Psi = V_j sin\,\theta + I_{ij}x_{ij}\,; \tag{13}$$

$$V_i^2 = (V_j cos\theta + I_{ij} r_{ij})^2 + (V_j sin\theta + I_{ij} x_{ij})^2 \quad (14)$$

Equation (14) can be written as (15), in terms of load demand, as follows:

$$V_i^2 = V_j^2 + 2r_{ij}P_{sj} + 2x_{ij}Q_{sj} + \left(r_{ij}^2 + x_{ij}^2\right)\left[\frac{P_{sj}^2 + Q_{ij}^2}{V_j^2}\right] \quad (15)$$

Further re-arranging (15) gives receiving-end voltage, represented, as follows, by (16):

$$V_j = \left[\frac{V_i^2}{2} - \left(r_{ij}P_{sj} + x_{ij}Q_{sj}\right) \pm \left[\left\{\frac{V_i^2}{2} - \left(r_{ij}P_{sj} + x_{ij}Q_{sj}\right)\right\}^2 - \left(r_{ij}^2 + x_{ij}^2\right)\left(P_{sj}^2 + Q_{sj}^2\right)\right]^{\frac{1}{2}}\right]^{\frac{1}{2}}; \quad (16)$$

The loadability limit is the maximum volt-ampere capacity of a radial feeder. Equation (16), for receiving-end node voltage, is used to find the loadability limit. The following inequality constraint is to be satisfied for the feasible solution of (16)

$$\left\{\frac{V_i^2}{2} - (r_{ij}P_{sj} + x_{ij}Q_{sj})\right\}^2 - \left(r_{ij}^2 + x_{ij}^2\right)\left(P_{sj}^2 + Q_{sj}^2\right) \geq 0. \quad (17)$$

In order to define the maximum load that can be connected at the respective nodes, the existing load is replaced by $f_{sj} \times (P_{sj} + jQ_{sj})$, and (17) is further modified as a quadratic form of 'f_{sj}'. On solving, it yields

$$f_{sj} = \frac{V_i^2\left[-(r_{ij}P_{sj} + x_{ij}Q_{sj}) + \sqrt{\left(r_{ij}^2 + x_{ij}^2\right)\left(P_{sj}^2 + Q_{sj}^2\right)}\right]}{2(x_{ij}P_{sj} - r_{ij}Q_{sj})^2} \quad (18)$$

Here, 'f_{sj}' is the factor that describes the possible additional load that can be connected at respective nodes before voltage collapse. The value of 'f_{sj}' may vary differently at each node in different configurations. To show the effect of improvement in the voltage profile at respective nodes, in a reconfigured network, the sending-end node voltage is assumed constant for the calculation of loadability factor at any subsequent node. The total power losses in the system are calculated as follows in (19) and (20):

$$P_{LT} = \sum \frac{\left(P_{sj}^2 + Q_{sj}^2\right)}{V_j^2} r_{ij}\ ; \quad (19)$$

$$Q_{LT} = \sum \frac{\left(P_{sj}^2 + Q_{sj}^2\right)}{V_j^2} x_{ij}\ ; \quad (20)$$

Equations (19) and (20) show the voltage profile dependence of power losses in the system.

2.4. Probabilistic Loading Patterns (λ) and Cases of Study

In practice, the loading pattern, at a particular node or load point, varies at random, depending on the contribution of a specific load type or class. Equations (21) and (22) describe different loading patterns.

$$\lambda_1 = 1 - PCF \times rand(\,) \quad (21)$$

$$\lambda_2 = \frac{1 - \lambda_1}{2} \quad (22)$$

The value of PCF is operator-dependent, and it can lie between 0 and 1. Here, in (21), the '0' value of PCF means the contribution of 'λ_1' is 100%. It is independent of random function, whereas if PCF > 0, then the loading pattern 'λ_1' will depend upon the random function at that moment, e.g., if PCF is 0.6 and *rand*() is obtained as '1' at any moment, then 'λ_1' will be 40%, which is its minimum value. Similarly, at any other moment, if *rand*() is '0', then the loading pattern (λ_1) will be 100% again. This allows the selection of 'a', 'b', 'c' and/or 'x', 'y', 'z', as well as the corresponding study cases for EEP, of a power distribution system under consideration.

Considering the above formulation, the three different cases of probabilistic loading patterns (PLP), under LM-2 and LM-3, are studied and described in Table 3.

Table 3. PLP under LM-2 and LM-3.

LP	LM-2	LM-3
PLP-1	$a = \lambda_1$ and $b = c = \lambda_2$	$x = \lambda_1$ and $y = z = \lambda_2$
PLP-2	$b = \lambda_1$ and $a = c = \lambda_2$	$y = \lambda_1$ and $x = z = \lambda_2$
PLP-3	$c = \lambda_1$ and $b = a = \lambda_2$	$z = \lambda_1$ and $y = x = \lambda_2$

The selection of 'a', 'b', 'c' and/or 'x', 'y', 'z' is such that

$$a + b + c = 1 \tag{23}$$

$$x + y + z = 1 \tag{24}$$

Therefore, with one component known, the others are equally divided using (22). In LM-2 and LM-3, the fractional values 'x', 'y', and 'z' are the same as 'a', 'b', and 'c'. However, in the analysis, they may be taken differently for the comprehensive evaluation of EEP during power delivery.

2.5. Representation of Energy Storage System

The state of charging and discharging the battery defines the behavior of the energy storage system (ESS) to be worked as a load or source. However, all the ESS may not be fully charged or discharged simultaneously. In this scenario, the representation of the ESS can be modeled with the help of a random distribution function. The ESS, under charging state, works as a load, whereas it will work as a source under discharge state. The modeling of ESS is represented as follows

$$ESS = \pm\, ESS_{in}(1 - PCF \times rand(\,)) \tag{25}$$

In (25), plus and minus represent the behavior of ESS as a load or source, respectively. ESS_{in} is the initial state of ESS that may be fully charged or discharged, as per ESS operation, of load or source. PCF is the probabilistic constant function that is taken as 0, 0.25, and 0.5 for constant power loads. In contrast, for voltage-dependent loads, PCF is taken as 0.4 under different cases of PLPs shown in Table 3. It means, if the value of *rand*() is '1', the minimum or maximum state of charge or discharge of ESS depends upon the PCF.

2.6. Margin of Reliability

During operation, the reliability of supplying power depends upon the connected loads and time of operation. In this scenario, the loading patterns play an important role since, under different circumstances, the behavior of voltage-dependent loads are found to be different, contributing to the reliability of supplying power and energy efficiency at that instant. From [2], it can be observed that the relative value of reliability can exceed the unity since it is the indexed value. Considering the indexed value of reliability of supplying power, the variation in the margin of reliability (MR) can be evaluated as follows:

$$\%MR = \frac{RI_{base} - RI_{Reconfiguration}}{RI_{base}} \times 100 \tag{26}$$

In (26), the margin of reliability can be plus or minus, depending upon the indexed value of the reliability index (RI), in reconfigured networks. Therefore, if MR is positive, the reliability of supplying power in the reconfigured network will improve and vice-versa, as compared to the base or original configuration.

3. Harmony Search Algorithm

The improvisation of musical harmony inspires the harmony search algorithm. HSA gives a set of solutions where the best solution is a near-optimum solution. The important feature of HSA is that it can independently consider each component variable in a single solution vector while it generates a new vector. This feature can accelerate the convergence rate of the HS algorithm by imposing constraints in the process of harmony improvisation (i.e., pitch adjustment, memory consideration, and random selection) for selecting a new decision variable. In HSA, discrete and continuous variables are managed together [3,4,35–38].

In the optimization process, HSA performs three functions described as:

a. Initialization of harmony memory
b. Improvisation of harmony memory
c. Updating the harmony memory

The harmony memory (HM) is a collection of possible solutions, depending upon the nature of the problem. Initially, HM is selected at random and/or based on prior information. The size of HM is operator-dependent, which further varies with the number of solution vectors and the complexity of the problem. Conversely, 'HM's improvisation can be achieved by selecting a new solution vector, at random, from either the initial solution space or the HM. The selection of a new solution vector from HM is decided based on the harmony memory considering rate (HMCR). Later, the pitch and bandwidth of such a selection is adjusted with a certain probability. Finally, the new harmony is evaluated, and the HM is updated based upon the audience response, i.e., improvement in the objective function corresponding to that selection.

A detailed explanation of HSA parameter selection is given in [3,4,35–38]. In Table 4, the effect of HSA parameters is summarized in the optimization function. Here, it can be noticed that the minimum power loss is obtained when HMCR is 0.85, PAR is 0.35, and HMS is 12 for the system under consideration. Therefore, with the change in HMCR, there is a significant change in the objective function.

Table 4. Results based on various HSA parameters for a 33-node radial distribution system.

Cases	HAS Parameter			Objective Function as Power Loss (kW)
	HMCR	PAR	HMS	
1	0.90	0.25	12	147.63
	0.70	0.25	12	143.44
	0.45	0.25	12	144.35
	0.35	0.25	12	155.73
2	0.85	0.35	12	139.55
	0.85	0.45	12	140.02
	0.85	0.55	12	139.97
	0.85	0.65	12	140.65

Table 4. *Cont.*

Cases	HAS Parameter			Objective Function as Power Loss (kW)
	HMCR	PAR	HMS	
3	0.80	0.40	5	159.81
	0.80	0.40	20	147.90
	0.80	0.40	25	162.68
	0.80	0.40	35	163.82

3.1. Harmony Improvisations

In HSA, the critical aspect is harmony improvisation. In harmony improvisation, there are three rules: namely, (a) random selection, (b) harmony memory consideration, and (c) pitch adjustment.

In the case of random selection, the new solution vector is selected from the global solution space with a probability of '1-HMCR'. Initially, with lower HMCR, the possibility of selection from global space is high. This is to explore the possible solution and to generate a solution vector in HM.

$$HMCR\ (t) = HMCR_{min} + \frac{(HMCR_{max} - HMCR_{min}) \times t}{T_{max}} \quad (27)$$

From (27), it can be observed that, with the increase in iteration, the HMCR increases, which, in turn, allows the higher probability of selection of a new solution vector from the HM. Further, the solution vector selected from HM has to be adjusted with the neighbor solution vector with PAR probability. The PAR increases with iteration and finally reaches '1', which means every solution vector needs to be pitch adjusted. The new vector, at this stage, is further modified with bandwidth. The PAR and bandwidth are defined using the following:

$$PAR\ (t) = PAR_{min} + \frac{(PAR_{max} - PAR_{min}) \times t}{T_{max}} \quad (28)$$

$$BW(t) = BW * e^{-\ln\left(\frac{BW_{min}}{BW_{max}}\right)\frac{t}{T_{max}}} \quad (29)$$

In (28) and (29), the PAR increases linearly, whereas BW decreases exponentially with an increase in iteration. This allows for the increase in probability for pitch adjustment with every increase in iteration, but it reduces the requirement of change in bandwidth at the same solution.

3.2. Solution Vector for PEBs

In the case of PEBs, the solution vector stored in the HM is limited to the tie-lines only. Therefore, in this case, the HM can be defined as

$$HSV^1 = \left[TL_1^1\ TL_2^1\ TL_3^1\ TL_4^1\ TL_5^1 \ldots\ldots\ldots\ldots\right] \quad (30)$$

In (30), each solution vector is defined corresponding to the respective tie-lines, and there may be any number of tie-lines in the sample system. Therefore, for a large network, it is required to generate a single solution vector in HSA.

3.3. Solution Vector for AEBs

In the case of AEBs, the solution vector stored in the HM is limited to twice the number of AEBs. Therefore, in this case, the HM can be defined as

$$HSV^1 = \left[DG_1^1\ DG_2^1\ DG_3^1 \ldots\ldots\ldots ND_4^1\ ND_5^1\ ND_6^1 \ldots\ldots\ldots\ldots\right] \quad (31)$$

In (31), $DG_1^1\ DG_2^1\ DG_3^1 \ldots$ represents the solution vector, corresponding to the size of AEBs, and $ND_4^1\ ND_5^1\ ND_6^1 \ldots$ represents the solution vector, corresponding to the locations of AEBs.

3.4. Solution Vector for PEBs and AEBs in Coordination

In the case of coordinated operation, the solution vector stored in the HM is the tie-lines, DG size, and its location at a particular node. Therefore, in this case, the HM can be defined as

$$HSV^1 = \left[TL_1^1\ TL_2^1\ TL_3^1\ TL_4^1\ TL_5^1 \ldots\ldots\ DG_6^1\ DG_7^1\ DG_8^1 \ldots\ldots ND_9^1\ ND_{10}^1\ ND_{11}^1 \ldots\ldots\right] \quad (32)$$

In (32), the solution vectors are represented, in coordination, for PEBs and AEBs, which were defined exclusively in (30) and (31), respectively.

4. Proposed Algorithm and Flowchart

Figure 2 shows the flow chart of the proposed algorithm. The steps involved in the proposed algorithm are described as follows:

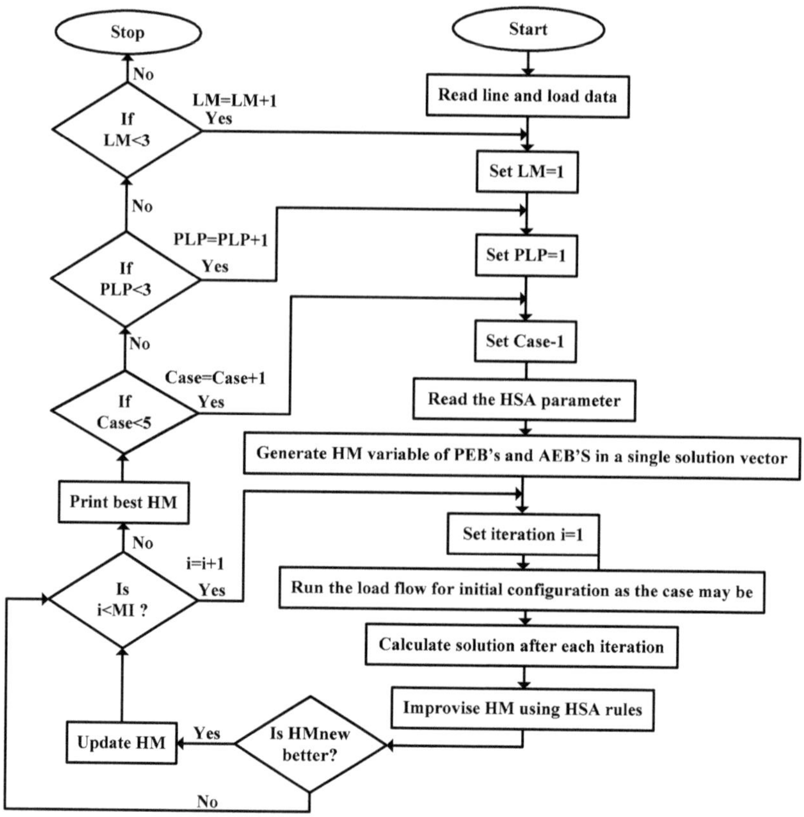

Figure 2. Flowchart of the proposed algorithm.

Step-1: Read the line and load data.
Step-2: Set LMs, PLPs, and cases of study
Step-3: Run the load flow for initial configuration and save the result as reference.
Step-4: Read the HSA parameters and generate HM, as per the solution vector described in Section 3.

Step-5: Set iteration counts = 1.
Step-6: Calculate the solution after each iteration and apply HSA rules for HM improvisation.
Step-7: If the new HM is better than the old, update HM, or else set counts = count + 1.
Step-8: In case-1, consider the following:

a. *run the load flow for LM-1, LM-2, and LM-3 under light, normal, and overloading scenario and find optimal configuration.*
b. *run the load flow for LM-1, LM-2, and LM-3 with 1DG, 2DG, and 3DG allocation in base configuration.*
c. *run the load flow for LM-1, LM-2, and LM-3 with 1DG, 2DG, and 3DG allocation in optimal configuration, which is obtained in (a).*

Step-9: In case-2, consider the following:

a. *run the load flow for LM-1 under normal loading scenario for 0%, 25%, and 50% SOC, with 1EV, 2EV, 3EV, and 4EVs as a load and find optimal configuration.*
b. *run the load flow individually for LM-2 and 3 under normal loading scenario for PLP-1, PLP-2, and PLP-3, with 1EV, 2EV, 3EV, and 4EVs as a load and find optimal configuration.*

Step-10: In case-3, consider the following;

a. *run the load flow for LM-1 under normal loading scenario for 0%, 25%, and 50% SOC, with 1EV, 2EV, 3EV, and 4EVs as a source and find optimal configuration.*
b. *run the load flow individually for LM-2 and 3 under normal loading scenario for PLP-1, PLP-2, and PLP-3, with 1EV, 2EV, 3EV, and 4EVs as a source and find optimal configuration.*

Step-11: In case-4, consider the following;

a. *run the load flow for LM-1 under normal loading scenario for 0%, 25%, and 50% SOC, with 1EV, 2EV, 3EV, and 4EVs as a load, at the most occurred location obtained in case 2 and 3, and find optimal configuration and DG allocation in coordination.*
b. *run the load flow individually for LM-2 and 3 under normal loading scenario for PLP-1, PLP-2, and PLP-3, with 1EV, 2EV, 3EV, and 4EVs as a load, at the most occurred location obtained in case 2 and 3, and find optimal configuration and DG allocation in coordination.*

Step-12: In case-5, consider the following;

a. *run the load flow for LM-1 under normal loading scenario for 0%, 25%, and 50% SOC, with 1EV, 2EV, 3EV, and 4EVs as a source, at the most occurred location obtained in case 2 and 3, and find optimal configuration and DG allocation in coordination.*
b. *run the load flow individually for LM-2 and 3 under normal loading scenario for PLP-1, PLP-2, and PLP-3, with 1EV, 2EV, 3EV, and 4EVs as a source, at the most occurred location obtained in case 2 and 3, and find optimal configuration and DG allocation in coordination.*

Step-13: Print the EEP under different LMs for different SOC, PLP, EVs, and DG allocations, as the case may be, of energy efficient configuration.

5. Test System, Assumptions and Cases of Study

5.1. Test System

In this work, a sample 33-node radial distribution system, as shown in Figure 3, is considered, and the proposed algorithm is demonstrated in the MATLAB environment. In this network, the branches or lines are divided into two parts, such as tie-lines and connected branches. The tie-lines are the branches made OFF from the network during operation, whereas the connected branches are the branches that remain in operation. Here, the base configuration is represented when tie-lines 33–37 are opened, and the other configurations are obtained by changing the status of the switch from ON to OFF in other branches during optimization. The first node is the substation, and the voltage at that node is maintained as 1.0 pu.

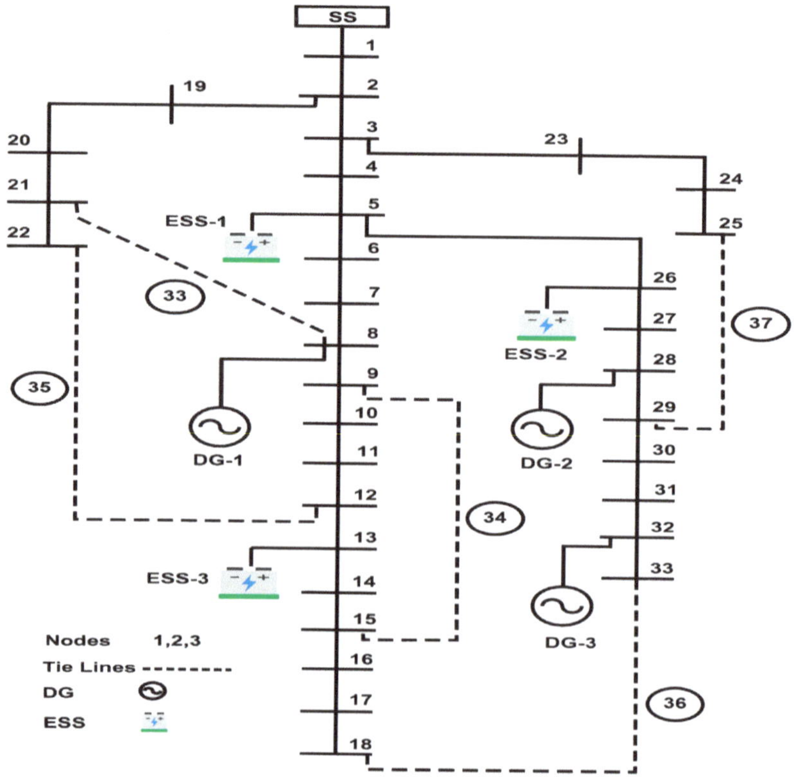

Figure 3. A sample 33-node radial distribution system.

5.2. Assumptions

The following are the assumptions while performing optimization in this work:

(a) The substation can meet the power demand of the system
(b) The maximum voltage at the substation is 1.0 pu.
(c) The minimum and maximum voltage, at respective nodes, are 0.90 pu and 1.05 pu.
(d) The maximum capacity of a single AEB is 1 MW at one location.
(e) The switching loss is negligible.
(f) The appropriate size of AEBs is available at the optimal location and can operate at their maximum capacity.

5.3. Energy Efficiency Analysis under Different Cases of Study

Table 5 shows the five different study cases for EEP analysis of the distribution system. However, these cases can be redefined as EEP evaluation when ESS works as a load or source. In the case-1, the EEP is analyzed without ESS. EEP realization of the network under consideration is presented in the following sections, from 6.1 to 6.5, exclusively, for all cases.

Table 5. The cases of study for EEP evaluation with PLP under different LMs.

Cases	Description	Remarks
Case-1	EEP evaluation of new configuration with PEBs and AEBs exclusively	This allows the EEP evaluation of reconfigured network and AEBs allocation in the original network and reconfigured network under different LMs **without ESS**.
Case-2	EEP evaluation with PEBs	This allows the change in a network topology for single/multi objectives and the EEP evaluation of the resulting configuration under different LMs **with ESS as a LOAD**.
Case-3	EEP evaluation with PEBs	This allows the change in a network topology for single/multi objectives and the EEP evaluation of the resulting configuration under different LMs **with ESS as a SOURCE**.
Case-4	EEP evaluation with PEBs and AEBs in coordination	This allows EEP evaluation in the coordination of PEBs and AEBs for energy-efficient operation under different LMs **with ESS as a LOAD**.
Case-5	EEP evaluation with PEBs and AEBs in coordination	This allows EEP evaluation in the coordination of PEBs and AEBs for energy-efficient operation under different LMs **with ESS as a SOURCE**.

6. Test Results and Discussions

6.1. Case-1: EEP Realization in Original and Reconfigured Topology (without ESS)

Table 6 shows the test result of the 33-node electrical system under different LMs for case-1. Here, the tie-lines in the original configuration are 33, 34, 35, 36, and 37, fixed, and the EEP is evaluated under light, normal, and overloading scenarios. Part-A of Table 6 shows that the optimal configuration obtained under the overloading scenario is different from the light and normal loading scenario in LM-1. However, in LM-2, the optimal configuration under light loading is different from the normal and overloading scenario. Conversely, in LM-3, the optimal configuration is the same under light, normal, and overloading scenarios.

The power loss under LM-2 and 3 is less than the LM-1 because the load profile in LM-1 is fixed, whereas it is voltage-dependent under LM-2 and 3, which draws less power if the voltage profile is lower than their nominal voltages at the load bus. As a result, the loadability margin has improved in the case of LM-2 and 3. Here, the %MR has been reduced over the various loading scenario, as shown in Part A of Table 6.

Similarly, Part B of Table 6 shows the EEP analysis of the original configuration when 1DG, 2DG, and 3DGs are placed at optimal locations under different LMs. All EEP parameters vary with the DG placement, but %MR in LM-1 is found unchanged. This is because the system's reliability depends upon the network configuration and loading patterns, which, in this case, are fixed. However, MR has increased in LM-2 and 3 with DG placements. Part-C of Table 6 shows the EEP analysis of reconfigured networks, obtained in Part-A, with optimal 1DG, 2DG, and 3DG placement. Here, it can be noticed that the power losses have reduced, and MR has increased under all LMs. When DGs are placed in the optimal configuration, obtained in Part-A, the reduced DG size can significantly improve the EEP compared to the original configuration's DG allocation.

6.2. Case-2: EEP Realization with PEBs (ESS Works as a Load)

Table 7 shows the test result of the 33-node electrical system under different LMs for case-2. The original configuration, with tie lines as 33, 34, 35, 36, and 37, is fixed, and the EEP is evaluated under different LMs. In LM-1, three different cases are developed where PCF is taken as 0.0, 0.25, and 0.50 when ESS works as a load. The EEP of an electrical system is also evaluated for single, two, three, and four ESS.

Table 6. EEP for Case-1 (reconfiguration and DG allocation exclusively).

A. EEP analysis of reconfigured network									
	EEP evaluation under LM-1			EEP evaluation under LM-2			EEP evaluation under LM-3		
Loading scenario	Light	Normal	Over	Light	Normal	Over	Light	Normal	Over
Optimal Conf	7, 14, 9, 32, 37	7, 14, 9, 32, 37	7, 14, 9, 32, 28	7, 14, 9, 32, 28	7, 14, 9, 32, 37	7, 14, 9, 32, 37	7, 14, 9, 32, 37	7, 14, 9, 32, 37	7, 14, 9, 32, 37
V_{min}	0.9508	0.9378	0.9287	0.9553	0.9410	0.9292	0.9512	0.9384	0.9254
$F_{si,min}$	18.53	14.73	12.20	19.11	15.32	12.80	18.63	14.84	12.31
P_L, (kW)	87.59	139.55	205.62	81.86	127.47	183.59	86.43	137.20	200.80
%MR	1.2812	1.2812	2.3501	2.2355	1.1569	1.1302	1.2622	1.2566	1.2505
B. EEP analysis of original network with DG allocations									
	EEP evaluation under LM-1			EEP evaluation under LM-2			EEP evaluation under LM-3		
AEBs scenario	1DG	2DG	3DG	1DG	2DG	3DG	1DG	2DG	3DG
OriginalConf	33, 34, 35, 36, 37	33, 34, 35, 36, 37	33, 34, 35, 36, 37	33, 34, 35, 36, 37	33, 34, 35, 36, 37	33, 34, 35, 36, 37	33, 34, 35, 36, 37	33, 34, 35, 36, 37	33, 34, 35, 36, 37
DG size (Node)	1975 (7)	1078 (11) 1032 (30)	945 (30) 796 (25) 716 (14)	1976 (7)	1304 (30) 780 (13)	882 (10) 943 (30) 868 (24)	2190 (7)	1138 (29) 779 (14)	765 (25) 726 (14) 952 (30)
V_{min}	0.9450	0.9673	0.9629	0.9472	0.9708	0.9620	0.9515	0.9659	0.9658
$F_{si,min}$	22.13	29.29	25.26	22.83	29.36	27.03	23.35	29.08	26.87
P_L, (kW)	108.31	86.24	73.49	102.50	85.51	72.87	97.88	81.11	68.50
%MR	–	–	–	−0.0420	−0.0302	−0.0178	−0.0765	−0.1250	−0.1327
C. EEP analysis of reconfigured network with DG allocations									
	EEP evaluation under LM-1			EEP evaluation under LM-2			EEP evaluation under LM-3		
AEBs scenario	1DG	2DG	3DG	1DG	2DG	3DG	1DG	2DG	3DG
Optimal Conf	7, 14, 9, 32, 37	7, 14, 9, 32, 37	7, 14, 9, 32, 28	7, 14, 9, 32, 28	7, 14, 9, 32, 37	7, 14, 9, 32, 37	7, 14, 9, 32, 37	7, 14, 9, 32, 37	7, 14, 9, 32, 37
DG size (Node)	1030 (30)	881 (9) 1080 (29)	944 (30) 963 (8) 855 (24)	1571(25)	1693 (25) 865 (8)	861 (9) 792 (30) 879 (24)	1570 (25)	907 (8) 1372 (25)	903 (24) 759 (30) 811 (8)
V_{min}	0.9478	0.9707	0.9732	0.9489	0.9725	0.9732	0.9507	0.9694	0.9720
$F_{si,min}$	14.75	30.08	32.53	14.94	30.13	29.64	15.42	32.24	29.43
P_L, (kW)	98.19	73.54	59.46	85.70	61.86	59.28	81.58	60.06	56.45
%MR	−0.8123	−0.8123	−0.8123	−0.7195	−0.7468	−0.7103	−0.7716	−0.8404	−0.8292

Table 7. EEP for Case-2 (when ESS works as a load).

	A. EEP evaluation under LM-1											
With SOC	With 1EV Station			With 2EV Station			With 3EV Station			With 4EV Station		
	0%	25%	50%	0%	25%	50%	0%	25%	50%	0%	25%	50%
Optimal EVS (Location)	24	2	5	5, 2	27, 2	10, 19	23, 5, 2	30, 6, 2	24, 2, 19	18, 5, 19, 2	19, 4, 28, 13	13, 23, 2, 28
Optimal Conf	7, 14, 10, 32, 37	7, 13, 9, 36, 28	7, 14, 9, 32, 37	7, 14, 10, 32, 28	7, 13, 9, 32, 28	7, 14, 9, 32, 28	6, 14, 10, 36, 37	7, 14, 10, 36,28	7, 14, 9, 36, 37	7, 34, 10, 17, 28	7, 14, 10, 36, 28	7, 34, 10, 36, 28
V_{min}	0.9370	0.9377	0.9369	0.9403	0.9402	0.9411	0.9336	0.9327	0.9330	0.9226	0.9354	0.9360
$F_{si,min}$	14.72	15.58	14.73	14.71	14.68	13.83	13.19	15.26	15.62	14.24	13.80	14.54
P_L, (kW)	148.10	145.29	143.01	147.77	149.81	147.01	162.97	164.78	148.31	183.94	174.35	163.09
%MR	−1.0526	−0.6973	−0.8123	−0.9409	−0.7283	−0.6982	−1.5227	−0.9105	−0.7816	−0.5477	−0.9105	−0.7326
	B. EEP evaluation under LM-2											
With SOC (60–100%)	With 1EV Station			With 2EV Station			With 3EV Station			With 4EV Station		
	PLP = 1	PLP = 2	PLP = 3	PLP = 1	PLP = 2	PLP = 3	PLP = 1	PLP = 2	PLP = 3	PLP = 1	PLP = 2	PLP = 3
Optimal EVS (Location)	18	33	32	18, 32	32, 15	15, 31	10, 33, 15	31, 15, 18	16, 15, 18	15, 30, 32, 31	16, 32, 18, 14	32, 30, 18, 14
Optimal Conf	7, 14, 9, 32, 37	7, 14, 10, 32, 28	7, 14, 9, 31, 28	7, 13, 9, 31, 28	7, 14, 9, 31, 37	7, 14, 9, 30, 37	6, 14, 9, 31, 37	6, 14, 9, 31, 37	7, 13, 10, 30, 37	7, 34, 10, 31, 27	7, 14, 10, 30, 37	7, 13, 9, 30, 37
V_{min}	0.9396	0.9442	0.9460	0.9526	0.9524	0.9446	0.9559	0.9500	0.9440	0.9512	0.9601	0.9553
$F_{si,min}$	16.48	17.05	17.35	17.35	16.92	15.82	17.07	17.16	17.74	17.41	20.51	18.96
P_L, (kW)	124.17	118.33	104.40	103.07	100.24	92.91	87.19	91.43	87.19	82.25	80.21	74.88
%MR	−0.8491	−1.0015	−0.7938	−0.7784	−0.9003	−0.6252	−1.4383	−1.4025	−0.9168	−0.7571	−0.8689	−0.6699
	C. EEP evaluation under LM-3											
With SOC (60–100%)	With 1EV Station			With 2EV Station			With 3EV Station			With 4EV Station		
	PLP = 1	PLP = 2	PLP = 3	PLP = 1	PLP = 2	PLP = 3	PLP = 1	PLP = 2	PLP = 3	PLP = 1	PLP = 2	PLP = 3
Optimal EVS (Location)	31	16	16	30,32	29,18	15,32	18,29,15	8,31,33	32,30,14	30,29, 17,31	32,31, 33,25	32,30, 16,18
Optimal Conf	7, 14, 9, 36, 37	7, 14, 9, 31, 37	7, 14, 9, 31, 37	7, 14, 9, 31, 28	7, 14, 10, 31, 37	7, 14, 10, 31, 28	7, 14, 9, 31, 37	7, 14, 10, 31, 28	7, 13, 9, 31, 37	7, 14, 9, 30, 28	7, 14, 10, 31, 28	6, 13, 9, 31, 37
V_{min}	0.9441	0.9388	0.9337	0.9437	0.9446	0.9507	0.9542	0.9568	0.9469	0.9479	0.9639	0.9528
$F_{si,min}$	16.27	14.64	13.75	14.82	15.27	16.36	17.61	18.28	17.27	15.86	18.27	16.06
P_L, (kW)	116.19	114.33	124.65	103.44	102.09	105.97	85.70	83.18	89.70	80.11	74.13	84.53
%MR	−0.8572	−0.8825	−0.8515	−0.7700	−1.1233	−0.9918	−0.9064	−1.0288	−0.9881	−0.4787	−1.0305	−1.3883

The location of ESS, in these cases, is found to be different under different LMs, as shown in Table 7. The 0% SOC means ESS is wholly discharged and acts as a load of its total capacity, similar to 25% and 50%. Since ESS works as a load, the overall loading patterns of the system will increase, and as a result, the power loss may increase under different LMs. Here, the optimization is performed for ESS locations under different LMs. Part A of Table 7 shows the optimal location of ESS when it works as a load, found at node numbers 24, 2, and 5 in LM-1, with SOC as 0, 25, and 50%, respectively.

The optimal location of ESS varies with LMs and is found at 18, 33, and 32 in LM-2, as well as 31, 16, and 16 in LM-3 for one EV station, as shown in Part B and C of Table 7. Here, it can also be noticed that the optimum location of ESS and the resulting configurations are found to be different for 2EV, 3EV, and 4EVs under different LMs. However, the MR is increased compared to the original configuration in these configurations.

Here, it can also be observed that the optimal configuration, under different loading patterns, is found to be different. The variation in the resulting configuration and the ESS location may further alter the %MR for 1EV, 2EV, 3EV, and 4EVs. Though ESS works as a load, the power loss in the resulting optimal configuration is slightly greater compared to the optimal configurations obtained in case-1, as shown in Table 6.

6.3. Case-3: EEP Realization with PEBs (ESS Works as a Source)

Table 8 shows the test result for EEP realization with PEBs for case-3 when ESS works as a source. In this case, the EVs location was found to be different in comparison with the previous case. Additionally, the optimal configuration and other EEP parameters vary significantly. The optimal configuration, with tie-lines 7, 9, 14, 23, and 27, is obtained for LM-1 with 1EV under 0% SOC. Similarly, the optimal configuration is the same for 25% and 50% SOC, whereas the other EEP parameters are different.

The optimal configuration under the same EVs may be different. It can be noticed that the optimal configuration with 2EV, 3EV, and 4EVs under 50% SOC are found to be 7, 9, 14, 32, 37, 7, 9, 14, 32, 28, as well as 7, 9, 14, 32, and 37. The difference in optimal configuration can also be observed under 0% and 25% SOC. EVs work as a source to alter the load profile and relieve the demand at that node. Here, the system losses have been reduced, and the loadability margin has improved because of the improvement in voltage profile. From the results, it can be observed that the margin of reliability has also been improved.

However, under LM-1, the power loss across 1EV, 2EV 3EV, and 4EVs varies slightly, whereas the reduction in loss, in the case of LM-2 and LM-3, is found to be very significant. These losses are 135.47, 134.25, 132.47, and 126.48, in the case on LM-2, and 126.26, 123.84, 121.60, and 119.77 kW in the case of LM-3, respectively. The reduction in power loss and improvement in the voltage profile further help improve the loadability margin and MR. For example, with 1EV, the f_{si} for 0% SOC in LM-1 is 14.74, whereas, for LM-2 and LM-3, they are 15.05 and 15.46, respectively.

6.4. Case-4: EEP Realization with PEBs and AEBs in Coordination (ESS Works as a Load)

Unlike the above two cases, case-4 considers the PEBs and AEBs in coordination, where optimization is performed when ESS works as a load. The coordinated operation of PEBs and AEBs allows for finding the optimal configuration and DGs allocation with 1EV, 2EV, 3EV, and 4EVs under different loading patterns, as shown in Table 9. Part A of Table 9 offers the EEP evaluation under LM-1. Here, three different cases are considered, such as 0%, 25%, and 50% SOC in the case of LM-1, as well as PLP-1, PLP-2, and PLP-3 in the case of LM-2 and LM-3. The optimal location of EVs is decided based on their locations in case-2 and 3, as shown in Tables 7 and 8. Here, the most occurred site of EVs is identified, and later, with their fixed location, the optimization is performed for coordinated operation. In LM-1, with 1EV, the most occurred site is at node 18, and for 2EVs, it is at node 2 and 19. A similar check is performed for 3EVs and 4EVs across all cases under LM-2 and LM-3.

Table 8. EEP for Case-3 (when ESS works as a source).

A. EEP evaluation under LM-1												
With SOC	With 1EV Station			With 2EV Station			With 3EV Station			With 4EV Station		
	0%	25%	50%	0%	25%	50%	0%	25%	50%	0%	25%	50%
Optimal EVS(Location)	31	32	33	31, 17	9, 18	29, 32	8, 18, 31	16, 30, 31	17, 31, 32	16, 24, 31, 32	8, 15, 27, 33	17, 18, 28, 33
Optimal Conf	7, 9, 14, 23, 37	7, 9, 14, 23, 37	7, 9, 14, 23, 37	7, 9, 14, 23, 37	7, 9, 14, 23, 37	7, 9, 14, 32, 37	7, 14, 10, 32, 37	7, 9, 14, 32, 37	7, 9, 14, 28, 32	7, 9, 14, 32, 37	7, 9, 14, 32, 37	7, 9, 14, 32, 37
V_{min}	0.9386	0.9386	0.9378	0.9387	0.9378	0.9390	0.9391	0.9398	0.9429	0.9405	0.9384	0.9385
$F_{si,min}$	14.73	14.73	14.89	15.04	15.21	14.77	15.33	15.02	15.02	14.98	15.46	15.40
P_L, (kW)	137.87	138.03	138.24	135.55	136.19	136.97	133.44	133.24	133.97	131.70	133.11	133.02
%MR	−0.8123	−0.8123	−0.8123	−0.8123	−0.8123	−0.8123	−1.0526	−0.8123	−0.6982	−0.8123	−0.8123	−0.8123

B. EEP evaluation under LM-2												
With SOC (60–100%)	With 1EV Station			With 2EV Station			With 3EV Station			With 4EV Station		
	PLP = 1	PLP = 2	PLP = 3	PLP = 1	PLP = 2	PLP = 3	PLP = 1	PLP = 2	PLP = 3	PLP = 1	PLP = 2	PLP = 3
Optimal EVS (Location)	12	27	9	15, 31	8, 33	9, 32	15, 17, 32	15, 16, 33	14, 30, 31	8, 10, 29, 32	7, 17, 32, 15	14, 12, 15, 32
Optimal Conf	7, 9, 14, 32, 37	7, 9, 14, 32, 37	7, 9, 14, 32, 37	7, 9, 14, 32, 37	7, 9, 14, 28, 32	7, 9, 14, 32, 37	7, 9, 14, 32, 37	7, 9, 14, 28, 32	7, 10, 14, 32, 37	7, 9, 14, 32, 37	7, 14, 10, 32, 37	7, 9, 14, 28, 32
V_{min}	0.9386	0.9413	0.9433	0.9394	0.9441	0.9429	0.9396	0.9442	0.9454	0.9416	0.9428	0.9470
$F_{si,min}$	15.06	15.32	15.91	14.99	15.76	15.75	15.20	16.16	15.95	15.48	15.72	16.53
P_L, (kW)	135.47	126.77	118.73	134.25	125.25	120.93	132.47	122.71	115.13	126.48	121.93	114.41
%MR	−0.8274	−0.8707	−0.9091	−0.8226	−0.7482	−0.8912	−0.8217	−0.7523	−1.1480	−0.8474	−1.1124	−0.7822

C. EEP evaluation under LM-3												
With SOC (60–100%)	With 1EV Station			With 2EV Station			With 3EV Station			With 4EV Station		
	PLP = 1	PLP = 2	PLP = 3	PLP = 1	PLP = 2	PLP = 3	PLP = 1	PLP = 2	PLP = 3	PLP = 1	PLP = 2	PLP = 3
Optimal EVS(Location)	18	32	33	18,30	24,31	15,31	12,28,29	18,30, 33	14,31,33	16,17, 30, 33	30,31, 27,18	18,31, 32,33
Optimal Conf	7, 9, 14, 32, 37	7, 9, 14, 32, 37	7, 9, 14, 32, 37	7, 9, 14, 32, 37	7, 9, 14, 28, 32	7, 9, 14, 28, 32	7, 9, 14, 32, 37	7, 14, 10, 32, 37	7, 9, 14, 32, 37	7, 9, 14, 32, 37	7, 9, 14, 32, 37	7, 14, 10, 32, 37
V_{min}	0.9410	0.9413	0.9393	0.9417	0.9449	0.9427	0.9431	0.9415	0.9397	0.9424	0.9433	0.9405
$F_{si,min}$	15.46	15.26	15.19	15.64	15.26	15.13	15.54	15.72	15.34	16.06	15.46	15.54
P_L, (kW)	126.65	127.53	132.50	123.84	126.60	133.29	121.60	124.39	131.37	119.77	121.85	128.37
%MR	−0.8722	−0.8660	−0.8409	−0.8781	−0.7413	−0.7141	−0.8767	−1.1081	−0.8270	−0.8865	−0.8723	−1.0800

Table 9. EEP for Case-4 (ESS works as a load).

A. EEP evaluation under LM-1												
With SOC	With 1EV Station			With 2EV Station			With 3EV Station			With 4EV Station		
	0%	25%	50%	0%	25%	50%	0%	25%	50%	0%	25%	50%
Optimal Conf	7, 14, 10, 28, 31	9, 28, 31, 33, 34	7, 9, 13, 28, 31	9, 26, 30, 33, 34	7, 10, 28, 30, 34	7, 9, 13, 28, 36	7, 9, 14, 28, 30	7, 12, 28, 31, 35	7, 13, 28, 32, 35	6, 9, 13, 27, 30	10, 28, 31, 33, 34	10, 14, 28, 31, 33
Optimal EVS (Location)	18	18	18	2, 19	2, 19	2, 19	2, 19, 23	2, 19, 23	2, 19, 23	2, 19, 23, 30	2, 19, 23, 30	2, 19, 23, 30
DG	691 (33) 428 (12) 877 (25)	933 (15) 885 (7) 990 (30)	952 (29) 548 (8) 615 (17)	890 (17) 568 (7) 923 (25)	820 (21) 693 (17) 932 (25)	773 (32) 836 (25) 709 (15)	612 (13) 795 (33) 867 (25)	696 (18) 996 (29) 568 (8)	662 (29) 977 (25) 775 (16)	778 (18) 975 (29) 797 (7)	755 (33) 954 (25) 721 (26)	837 (30) 688 (15) 784 (7)
V_{min}	0.9684	0.9656	0.9710	0.9599	0.9553	0.9785	0.9732	0.9721	0.9728	0.9681	0.9703	0.9712
$F_{si,min}$	27.16	31.73	28.19	26.07	24.58	28.47	28.04	30.81	27.31	24.46	30.78	28.96
P_L, (kW)	59.71	55.39	59.04	57.86	58.97	59.77	57.64	59.87	61.60	59.53	56.42	61.55
%MR	−0.9553	0.0017	−0.7455	0.2208	−0.4815	−0.6964	−0.4128	−0.4004	−0.6657	−0.9202	−0.2524	−0.4389
B. EEP evaluation under LM-2												
With SOC (60–100%)	With 1EV Station			With 2EV Station			With 3EV Station			With 4EV Station		
	PLP = 1	PLP = 2	PLP = 3	PLP = 1	PLP = 2	PLP = 3	PLP = 1	PLP = 2	PLP = 3	PLP = 1	PLP = 2	PLP = 3
Optimal Conf	7, 10, 13, 28, 32	7, 9, 28, 32, 34	9, 14, 28, 30, 33	10, 14, 27, 30, 33	6, 13, 16, 35,37	10, 14, 28, 30, 33	9, 14, 28, 30, 33	9, 28, 31, 33, 34	7, 10, 13, 28, 31	7, 9, 13, 32, 37	10, 13, 28, 30, 33	6, 9, 26, 31, 34
Optimal EVS (Location)	18	18	18	2, 19	2, 19	2, 19	2, 19, 23	2, 19, 23	2, 19, 23	2, 19, 23, 30	2, 19, 23, 30	2, 19, 23, 30
DG	785 (31) 746 (14) 996 (24)	662 (31) 816 (24) 826 (14)	977 (25) 699 (17) 930 (7)	670 (25) 655 (33) 931 (8)	963 (30) 996 (8) 955 (24)	610 (25) 921 (7) 626 (32)	946 (7) 812 (29) 753 (33)	890 (4) 914 (30) 787 (17)	683 (24) 771 (16) 693 (30)	863 (25) 852 (30) 691 (15)	625 (18) 925 (29) 601 (8)	947 (29) 612 (8) 793 (15)
V_{min}	0.9757	0.9745	0.9739	0.9714	0.9660	0.9728	0.9728	0.9722	9632	0.9699	0.9631	0.9633
$F_{si,min}$	26.71	29.64	29.46	29.58	35.42	31.21	28.58	30.79	22.60	25.72	34.77	28.27
P_L, (kW)	58.39	56.11	53.37	58.68	57.02	53.96	57.57	56.57	55.16	58.09	57.52	55.30
%MR	−0.9806	−0.6383	−0.0501	−0.1342	−1.0598	−0.3125	0.1254	−0.1120	−1.1570	−0.8885	−0.2734	−1.3336
C. EEP evaluation under LM-3												
With SOC (60–100%)	With 1EV Station			With 2EV Station			With 3EV Station			With 4EV Station		
	PLP = 1	PLP = 2	PLP = 3	PLP = 1	PLP = 2	PLP = 3	PLP = 1	PLP = 2	PLP = 3	PLP = 1	PLP = 2	PLP = 3
Optimal Conf	6, 10, 13, 28, 30	9, 27, 30, 33, 34	7, 13, 26, 32, 35	7, 9, 14, 28,30	7, 9, 13, 27,30	7, 9, 13, 28, 32	7,8,28, 32, 34	7, 9, 13, 28, 32	7, 10, 14, 28,30	7,9,12, 28, 31	10, 27, 30, 33, 34	7, 12, 28, 32, 35
Optimal EVS (Location)	18	18	18	2, 19	2, 19	2, 19	2, 19, 23	2, 19, 23	2, 19, 23	2, 19, 23, 30	2, 19, 23, 30	2, 19, 23, 30
DG	632 (9) 947 (25) 610 (33)	957 (25) 920 (8) 741 (18)	831 (29) 856 (15) 860 (25)	667 (11) 827 (30) 641 (31)	735 (12) 797 (32) 984 (25)	666 (25) 733 (15) 979 (29)	664 (14) 922 (24) 750 (29)	745 (9) 848 (24) 807 (29)	934 (29) 942 (8) 482 (31)	985 (25) 613 (23) 834 (18)	984 (18) 749 (25) 889 (7)	935 (30) 886 (24) 883 (15)
V_{min}	0.9614	0.9579	0.9741	0.9680	0.9717	0.9760	0.9692	0.9687	0.9701	0.9702	0.9732	0.9760
$F_{si,min}$	21.05	24.21	30.09	27.42	25.60	26.51	25.80	27.51	28.95	22.80	31.06	30.57
P_L (kW)	55.32	54.99	58.58	57.33	56.81	59.58	56.29	57.65	57.23	57.56	54.64	56.64
%MR	−1.3101	0.2732	−0.8319	−0.5304	−0.5192	−0.7529	−0.6434	−0.8319	−0.7244	−0.5743	−0.0033	−0.4092

As shown in Part-A of Table 9, the optimal configuration with 1EV for 0% SOC is 7, 14, 10, 28, 31, for 25% SOC is 9, 28, 31, 33, 34, and for 50% SOC is 7, 9, 13, 28, 31. The location of EVs is the same across PLPs, whereas the optimal configuration is found to be different. The DG allocations vary differently, and they are 632 kW at 9th node, 947 kW at node 25, and 610 kW at node 33 for PLP-1, whereas, for PLP-2, the DGs are 957 kW at node 25, 920 kW at node 8, and 741 kW at node 18 in the case of LM-3, as shown in Part C of Table 9. In these cases, the reduction in power loss is found significantly improved compared to the original configuration, as shown in Table 6. Here, it can also be observed that improvement in node voltage increases the loading profile at respective nodes with the DG placement. Compared to the original configuration, some configurations can reduce the MR.

6.5. Case-5: EEP Realization with PEBs and AEBs in Coordination (ESS Works as Source)

Table 10 shows the EEP realization of the sample distribution system, with PEBs and AEBs in coordination, when ESS works as a source. Here, three different cases are also considered for EEP evaluation.

Part-A of Table 10 shows the EEP analysis with LM-1 under different cases of SOC. Similar to case-4, the most occurred location of EVs is identified, and later, with their fixed location, the optimization is performed for a coordinated operation. In LM-1, with 1EV, the most occurred location is at node 32, and for 2EVs, it is at node 18 and 32. A similar check is performed for 3EVs and 4EVs across all cases under LM-2 and LM-3.

As shown in Part-A of Table 10, the optimal configuration with 1EV for 0% SOC is 10, 27, 30, 33, 34, for 25% SOC is 7, 8, 12, 26, 31, and for 50% SOC is 7, 10, 14, 28, and 30. In these configurations, the EEP is found to be different.

6.6. Comparative Analysis of EEP with PEBs and AEBs across All the Cases

In a smart city environment, the EEP varies differently in the presence of ESS. In recent years, the use of electric vehicles has increased manifold, which demands smart operation, control, and monitoring. The smart process does require realizing the EEP from different perspectives. Sections 6.1–6.5 present the EEP realization in five separate cases. As described in Section 5, ESS is involved in all the cases except case-1. Therefore, case-1 represents the generalized EEP under different loading scenarios, whereas, in other cases, the ESS either works as a load or source depending upon their charging or discharging states. Table 6 presents the EEP of distribution with three loading scenarios: light, normal, and overload. With LM-1, under normal loading, as shown in Part-A of Table 6, the optimal configuration is 7, 9, 14, 32, 37, whereas, for overloading, it is 7, 9, 14, 28, 32. It shows that the optimal configuration is different when loading patterns change, even in the absence of ESS.

Conversely, integrating ESS with the existing system, as a load or source, can significantly affect the loading patterns. Therefore, EEP of the distribution network needs to be realized at the planning stage to identify the most economical location may not be optimal for EV stations. This aspect has been realized, in this work, through four different cases where EEP is evaluated with PEBs and AEBs when ESS works exclusively as a load or source. However, case-2 and 3 present the EEP with PEBs and ESS. In these cases, the optimal configuration under LM-1 (with 50% SOC) is found to be 7, 9, 14, 32, 37 and 7, 9, 14, 23, 37, respectively. Here, it can be noticed that, in case-2, the ESS works as a load, whereas case-3 works as a source. On comparing case-2 and 3 with case-1, the EEP is found to vary differently. The third aspect of the proposed approach, i.e., case-4 and 5, is EEP realization with PEBs, AEBs, and ESS in coordination. EEP, in cases 2 and 3, varies with ESS as a load or source; therefore, in cases 4 and 5, the coordinated operation is realized accordingly. In these cases, the optimal location of DGs and the ESS differ. Here, some locations are found repeatedly across different LMs in case-4 and 5. From across four cases, the optimal location of ESS is identified: node 18 for 1EV, nodes 2 and 19 for 2EVs, nodes 2, 19, and 23 for 3EVs, and nodes 2, 19, 23, and 30 for 4EVs when ESS works as a load. On the other hand, the optimal location of EV stations when ESS works as a source is node 32 for 1EV, nodes 18 and 32 for 2EVs, nodes 17, 18, and 32 for 3EVs, and nodes 17, 18, 30, and 32 for 4EVs.

Table 10. EEP for Case-5 (ESS works as a source).

A. EEP evaluation under LM-1												
With SOC	With 1EV Station			With 2EV Station			With 3EV Station			With 4EV Station		
	0%	25%	50%	0%	25%	50%	0%	25%	50%	0%	25%	50%
Optimal Conf	10, 27, 30, 33, 34	7, 8, 12, 26, 31	7, 10, 14, 28, 30	10, 28, 31, 33, 34	10, 13, 28, 30, 33	6, 9, 13, 28, 30	10, 14, 28, 30, 33	10, 14, 28, 30, 33	6, 9, 12, 17, 28	10, 27, 31, 33, 34	10, 13, 26, 30, 33	7, 12, 10, 27, 32
Optimal EVS(Location)	32	32	32	18, 32	18, 32	18, 32	18, 32, 17	18, 32, 17	18, 32, 17	18, 32, 17, 30	18, 32, 17, 30	18, 32, 17, 30
DG	904 (18) 767 (7) 833 (29)	857 (16) 838 (24) 942 (29)	964 (25) 549 (13) 883 (33)	630 (15) 734 (25) 908 (7)	782 (33) 834 (7) 889 (25)	766 (18) 925 (29) 832 (8)	816 (7) 650 (32) 902 (30)	915 (26) 808 (25) 764 (31)	574 (25) 862 (9) 755 (32)	800 (26) 824 (16) 713 (29)	842 (8) 731 (33) 944 (25)	906 (29) 804 (15) 882 (24)
V_{min}	0.9692	0.9593	0.9765	0.9582	0.9752	0.9703	0.9754	0.9754	0.9721	0.9687	0.9748	0.9728
$F_{si,min}$	29.01	22.50	28.78	27.84	34.39	25.27	31.39	31.39	27.97	26.74	32.45	29.58
P_L, (kW)	54.94	59.78	56.97	58.66	55.49	57.58	56.77	57.69	61.11	58.56	57.07	56.25
%MR	−0.1137	−0.5635	−0.6639	−0.2555	−0.1613	−0.9775	−0.1322	−0.1320	−0.8058	−0.2126	−0.2690	−0.6587
B. EEP evaluation under LM-2												
With SOC (60–100%)	With 1EV Station			With 2EV Station			With 3EV Station			With 4EV Station		
	PLP = 1	PLP = 2	PLP = 3	PLP = 1	PLP = 2	PLP = 3	PLP = 1	PLP = 2	PLP = 3	PLP = 1	PLP = 2	PLP = 3
Optimal Conf	11, 14, 28, 30, 33	7, 10, 14, 28, 30	7, 12, 10, 32, 37	7, 8, 32, 34, 37	7, 13, 28, 30, 35	9, 13, 26, 30, 33	7, 9, 14, 28, 30	9, 26, 31, 33, 34	6, 9, 13, 31, 37	7, 9, 14, 28, 30	7, 9, 13, 28, 30	7, 10, 27, 31, 34
Optimal EVS (Location)	32	32	32	18, 32	18, 32	18, 32	18, 32, 17	18, 32, 17	18, 32, 17	18, 32, 17, 30	18, 32, 17, 30	18, 32, 17, 30
DG	880 (25) 877 (5) 790 (32)	594 (32) 831 (22) 985 (25)	790 (30) 856 (15) 755 (25)	806 (25) 874 (12) 965 (28)	628 (33) 886 (25) 736 (8)	830 (29) 561 (18) 812 (6)	858 (18) 756 (7) 967 (25)	966 (30) 557 (8) 871 (15)	531 (29) 876 (15) 810 (24)	915 (25) 556 (32) 437 (22)	613 (31) 966 (24) 778 (20)	472 (18) 555 (8) 950 (29)
V_{min}	0.9755	0.9668	0.9714	0.9662	0.9659	0.9664	0.9672	0.9706	0.9625	0.9603	0.9643	0.9664
$F_{si,min}$	34.26	24.37	31.39	30.87	28.84	32.12	19.94	29.57	20.58	21.88	22.95	29.62
P_L, (kW)	57.81	53.02	53.14	60.50	55.49	54.80	57.38	55.35	55.63	58.36	57.58	53.67
%MR	−0.0852	−0.7734	−1.0477	−0.6443	−0.4915	−0.1927	−0.4482	−0.2100	−1.5506	−0.4392	−0.5456	−0.9368
C. EEP evaluation under LM-3												
With SOC (60–100%)	With 1EV Station			With 2EV Station			With 3EV Station			With 4EV Station		
	PLP = 1	PLP = 2	PLP = 3	PLP = 1	PLP = 2	PLP = 3	PLP = 1	PLP = 2	PLP = 3	PLP = 1	PLP = 2	PLP = 3
Optimal Conf	10, 28, 31, 33, 34	7, 8, 32, 34, 37	10, 14, 28, 31, 33	7, 11, 14, 31, 37	7, 8, 28, 32, 34	7, 9, 12, 27, 30	7, 12, 27, 31, 35	10, 13, 28, 31, 33	7, 9, 25, 32, 34	7, 9, 26, 30, 34	6, 8, 13, 27, 31	6, 9, 12, 16, 28
Optimal EVS (Location)	32	32	32	18, 32	18.32	18, 32	18, 32, 17	18, 32, 17	18, 32, 17	18, 32, 17, 30	18, 32, 17, 30	18, 32, 17, 30
DG	878 (16) 979 (6) 714 (31)	848 (25) 678 (12) 889 (30)	610 (26) 653 (18) 946 (29)	854 (16) 834 (24) 941 (29)	793 (15) 663 (31) 820 (25)	851 (25) 839 (17) 673 (22)	908 (24) 866 (15) 918 (30)	912 (25) 815 (8) 607 (33)	680 (15) 813 (24) 977 (29)	807 (33) 648 (8) 914 (29)	690 (8) 670 (17) 870 (25)	843 (24) 871 (32) 893 (8)
V_{min}	0.9738	0.9610	0.9740	0.9724	0.9788	0.9644	0.9605	0.9712	0.9708	0.9719	0.9672	0.9748
$F_{si,min}$	28.55	26.02	31.41	23.79	29.41	24.11	24.89	31.48	26.77	28.59	27.79	31.51
P_L, (kW)	55.88	58.51	57.06	56.58	56.30	58.77	56.99	55.51	57.33	57.18	57.64	57.01
%MR	−0.3905	−0.7449	−0.4915	−1.1597	−0.6402	−0.1447	−0.4603	−0.5888	−0.6973	−0.4448	−1.3562	−0.7979

Table 11 shows the comparative statement of various techniques available in the literature for network reconfiguration and DG allocation. The test results show that the optimal configuration and DG allocation are found to be different using different techniques with the marginal difference between the power losses in the resulting configuration. Conversely, the proposed approach is demonstrated for the coordinated operation of DG allocation and reconfiguration in the presence of ESS, as a load and source, exclusively. Here, it can be noticed that the existing approaches are tested for constant power load, i.e., LM-1, across all cases. In contrast, the proposed approach is tested for LM-1, LM-2, and LM-3, which are formulated using several load combinations in a practical scenario. On comparing the DG allocation and network reconfiguration without ESS, and for constant power load, the power loss in [28], using the TLBO approach, is found to be 58.08 kW with DG allocation as 1329 (8), 1172 (24), and 726 (31), which offers the total size of 3227 kW. On the other hand, under identical operating conditions, the power loss in the proposed approach is 58.10 kW with DG allocation as 890 (25), 856 (9), and 792 (31), which is in a total size of 2538 kW. Here, it can be noticed that the optimization, performed using existing approaches, may not yield energy-efficient operation during power delivery in the present competitive energy market scenario.

The results show that the integration of electric vehicles in the existing network affects the EEP when ESS works as a load and or source differently. Therefore, it further requires investigating the ideal location of EV stations before their installation because the location of ESS, as a load, may not offer the optimal results when ESS works as a source. Further, the proposed approach performs better with enhanced EEP in terms of voltage profile, loadability, power loss, DG size, and margin of reliability, which is beneficial for both utilities and customers.

Table 11. Comparison of different approaches for 3-DG allocation and power loss reduction.

Case/Methods	Configuration	EV Location	Power Loss (kW)	DG Size	DG Location
Base Configuration	33, 34, 35, 36, 37	–	202.67	–	–
Optimal configuration	7, 9, 14, 32, 37	–	139.5	–	–
DG allocation in base Configuration under,		–			
LM-1	33, 34, 35, 36, 37		78.25	652, 999, 656	14, 26, 32
LM-2			76.78	781, 991, 407	32, 7, 18
LM-3			69.72	970, 542, 505	29, 9, 14
DG allocation in optimal configuration under,					
LM-1	7, 9, 14, 32, 37	–	72.19	968, 780, 997	23, 16, 27
LM-2			75.05	419, 408, 653	6, 18, 28
LM-3			65.07	680, 828, 622	12, 7, 30
DG allocation and Reconfiguration under,					
LM-1	7, 13, 11, 36, 27	–	58.10	890, 856, 792	25, 9, 31
LM-2	7, 12, 10, 32, 25		56.39	984, 811, 754	25, 15, 31
LM-3	7, 12, 10, 31, 27		52.18	892, 967, 780	29, 9, 16
DG allocation and Reconfiguration with EVs as a load					
LM-1	9, 28, 31, 33, 34	18	55.64	933, 885, 990	15, 7, 30
LM-2	9, 14, 28, 30, 33	18	53.37	977, 699, 930	25, 17, 7
LM-3	9, 27, 30, 33, 34	18	54.99	957, 920, 741	25, 8, 18

Table 11. *Cont.*

Case/Methods	Configuration	EV Location	Power Loss (kW)	DG Size	DG Location
DG allocation and Reconfiguration with EVs as a source					
LM-1	10, 27, 30, 33, 34	32	54.94	904, 767, 833	18, 7, 29
LM-2	7, 10, 14, 28, 30	32	53.02	594, 831, 985	32, 22, 25
LM-3	10, 28, 31, 33, 34	32	55.88	878, 979, 714	16, 6, 31
Ref. [28]					
IPSO	33, 34, 9, 32, 28	–	59.63	557, 922, 931	18, 7, 30
TLBO	6,14, 10, 32, 37		58.08	1329, 1172, 726	8, 24, 31
PSO	7, 13, 11, 32, 27		59.37	1732, 809, 550	29, 16, 7
Jaya	33, 13, 9, 28, 30		58.49	801, 1215, 745	18, 25, 9
Ref. [29]	7, 14, 10, 31, 28	–	73.05	526, 559, 584	28, 31, 33
Ref. [30]	7, 9, 14, 17, 37	–	92.98	55, 151, 103	18, 31, 32

7. Conclusions

In this paper, the energy efficiency performance (EEP) of the distribution system is realized through passive energy boosters (PEBs) and active energy boosters (AEBs), exclusively and in coordination, during power delivery in the presence of energy storage devices (ESS). The electric vehicles are the movable ESS that can work as load or source and are considered a significant constituent of ESS. For load representation, several combinations of voltage-dependent loads are modeled using a random distribution function to show the contribution of different loading patterns under various operating conditions. Here, the optimization results are obtained using the harmony search algorithm for the 33-node radial distribution system. From the test results, it has been observed that the EEP of the distribution system varies differently under different load models in the base and optimal configurations. However, the variation in power loss is slight, using various optimization techniques, but other parameters, such as the voltage profile, loadability, and DG size, are found to be significantly varying. The test results show that the optimal location of EVs, as a load and source, is not the same and can affect the EEP differently.

Considering the present scenarios of the integration of EVs at different locations, this work performs optimization. The EEP of the power distribution system under consideration has improved significantly, and it is beneficial for both utilities and customers. This work emphasizes the technical aspects of a smart energy system; however, in the future, the impact of renewable energy resources and ESS on EEP needs to be considered with hourly loading patterns by developing a bidirectional communication framework for control and monitoring. The optimal location of EVs is realized when ESS works as a load or source independently, whereas the common location requires further investigations.

Author Contributions: Conceptualization, P.K. and S.N.; data curation, P.K., S.N. and I.A.; formal analysis, P.K. and S.N.; funding acquisition, P.K. and H.A.; methodology, P.K., S.N. and H.A.; project administration, I.A. and M.S.T.; resources, P.K. and H.A.; software, P.K. and H.A.; supervision, I.A. and M.S.T.; validation, I.A. and M.S.T.; visualization, I.A. and M.S.T.; writing—original draft, P.K. and S.N.; writing—review and editing, I.A. and M.S.T. and H.A. All authors have read and agreed to the published version of the manuscript.

Funding: This research received no external funding.

Institutional Review Board Statement: Not applicable.

Informed Consent Statement: Not applicable.

Data Availability Statement: Not applicable.

Conflicts of Interest: The authors declare no conflict of interest.

References

1. Kumar, P.; Thanki, D.V.; Singh, S.; Nikolovski, S. A new framework for intensification of energy efficiency in commercial and residential use by imposing social, technical and environmental constraints. *Sustain. Cities Soc.* **2020**, *62*, 102400. [CrossRef]
2. Ali, I.; Thomas, M.S.; Kumar, P. Energy efficient reconfiguration for practical load combinations in distribution systems. *IET Gener. Transm. Distrib.* **2015**, *9*, 1051–1060. [CrossRef]
3. Kumar, P.; Ali, I.; Thomas, M.S.; Singh, S. Imposing voltage security and network radiality for reconfiguration of distribution systems using efficient heuristic and meta-heuristic approach. *IET Gener. Transm. Distrib.* **2017**, *11*, 2457–2467. [CrossRef]
4. Kumar, P.; Ali, I.; Thomas, M.S.; Singh, S. A coordinated framework of dg allocation and operating strategy in distribution system for configuration management under varying loading patterns. *Electr. Power Compon. Syst.* **2020**, *48*, 12–29. [CrossRef]
5. Wang, K.; Li, H.; Maharjan, S.; Zhang, Y.; Guo, S. Green energy scheduling for demand side management in the smart grid. *IEEE Trans. Green Commun. Netw.* **2018**, *2*, 596–611. [CrossRef]
6. Haseeb, M.; Kazmi, S.A.; Malik, M.M.; Ali, S.; Bukhari, S.B.; Shin, D.R. Multi objective based framework for energy management of smart micro-grid. *IEEE Access* **2020**, *8*, 220302–220319. [CrossRef]
7. Kumar, P.; Singh, S.; Ali, I.; Ustun, T.S. Handbook of Research on Power and Energy System Optimization. In *2018 IGI Global, Engineering Science Reference*; IGI Global: Hershey, PA, USA, 2018.
8. Cao, B.; Dong, W.; Lv, Z.; Gu, Y.; Singh, S.; Kumar, P. Hybrid microgrid many-objective sizing optimization with fuzzy decision. *IEEE Trans. Fuzzy Syst.* **2020**, *28*, 2702–2710. [CrossRef]
9. Zhang, K.; Yu, J.; Ren, Y. Demand side management of energy consumption in a photovoltaic integrated greenhouse. *Int. J. Electr. Power Energy Syst.* **2022**, *134*, 107433. [CrossRef]
10. Kumar, P.; Brar, G.S.; Singh, S.; Nikolovski, S.; Baghaee, H.R.; Balkić, Z. Perspectives and intensification of energy efficiency in commercial and residential buildings using strategic auditing and demand-side management. *Energies* **2019**, *12*, 4539. [CrossRef]
11. Asrari, A.; Wu, T.; Lotfifard, S. The impacts of distributed energy sources on distribution network reconfiguration. *IEEE Trans. Energy Convers.* **2016**, *31*, 606–613. [CrossRef]
12. Dorostkar, G.M.R.; Fotuhi, F.M.; Lehtonen, M.; Safdarian, A. Value of distribution network reconfiguration in presence of renewable energy resources. *IEEE Trans. Power Syst.* **2015**, *31*, 1879–1888. [CrossRef]
13. Kavousi, F.A.; Rostami, M.A.; Niknam, T. Reliability-oriented reconfiguration of vehicle-to-grid networks. *IEEE Trans. Ind. Inform.* **2015**, *11*, 682–691. [CrossRef]
14. Xing, H.; Sun, X. Distributed generation locating and sizing in active distribution network considering network reconfiguration. *IEEE Access* **2017**, *5*, 14768–14774. [CrossRef]
15. Fu, Y.Y.; Chiang, H.D. Toward optimal multiperiod network reconfiguration for increasing the hosting capacity of distribution networks. *IEEE Trans Power Deliv.* **2018**, *33*, 2294–2304. [CrossRef]
16. Azizivahed, A.; Arefi, A.; Ghavidel, S.; Shafie, K.M.; Li, L.; Zhang, J.; Catalão, J.P. Energy management strategy in dynamic distribution network reconfiguration considering renewable energy resources and storage. *IEEE Trans. Sustain. Energy* **2019**, *11*, 662–673. [CrossRef]
17. Liu, Y.; Li, J.; Wu, L. Coordinated optimal network reconfiguration and voltage regulator/DER control for unbalanced distribution systems. *IEEE Trans. Smart Grid* **2018**, *10*, 2912–2922. [CrossRef]
18. Song, Y.; Zheng, Y.; Liu, T.; Lei, S.; Hill, D.J. A new formulation of distribution network reconfiguration for reducing the voltage volatility induced by distributed generation. *IEEE Trans. Power Syst.* **2019**, *35*, 496–507. [CrossRef]
19. Nick, M.; Cherkaoui, R.; Paolone, M. Optimal planning of distributed energy storage systems in active distribution networks embedding grid reconfiguration. *IEEE Trans. Power Syst.* **2017**, *33*, 1577–1590. [CrossRef]
20. Naguib, M.; Omran, W.A.; Talaat, H.E. Performance Enhancement of Distribution Systems via Distribution Network Reconfiguration and Distributed Generator Allocation Considering Uncertain Environment. *J. Mod. Power Syst. Clean Energy* **2021**, *10*, 647–655. [CrossRef]
21. Cattani, I.B.; Chaparro, E.; Barán, B. Distribution system operation and expansion planning using network reconfiguration. *IEEE Lat. Am. Trans.* **2020**, *18*, 845–852. [CrossRef]
22. Ahmed, H.M.; Salama, M.M. Energy management of AC–DC hybrid distribution systems considering network reconfiguration. *IEEE Trans. Power Syst.* **2019**, *34*, 4583–4594. [CrossRef]
23. Takenobu, Y.; Yasuda, N.; Kawano, S.; Minato, S.I.; Hayashi, Y. Evaluation of annual energy loss reduction based on reconfiguration scheduling. *IEEE Trans. Smart Grid* **2016**, *9*, 1986–1996. [CrossRef]
24. Asrari, A.; Ansari, M.; Khazaei, J.; Fajri, P.; Amini, M.H.; Ramos, B. The impacts of a decision making framework on distribution network reconfiguration. *IEEE Trans. Sustain. Energy* **2020**, *12*, 634–645. [CrossRef]
25. Huang, W.; Zheng, W.; Hill, D.J. Distribution network reconfiguration for short-term voltage stability enhancement: An efficient deep learning approach. *IEEE Trans. Smart Grid* **2021**, *12*, 5385–5395. [CrossRef]
26. Zhan, J.; Liu, W.; Chung, C.Y.; Yang, J. Switch opening and exchange method for stochastic distribution network reconfiguration. *IEEE Trans. Smart Grid* **2020**, *11*, 2995–3007. [CrossRef]
27. Gao, H.; Ma, W.; Xiang, Y.; Tang, Z.; Xu, X.; Pan, H.; Zhang, F.; Liu, J. Multi-Objective Dynamic Reconfiguration for Urban Distribution Network Considering Multi-Level Switching Modes. *J. Mod. Power Syst. Clean Energy* **2021**, 1–14. [CrossRef]

28. Rawat, M.S.; Vadhera, S. Heuristic optimization techniques for voltage stability enhancement of radial distribution network with simultaneous consideration of network reconfiguration and DG sizing and allocations. *Turk. J. Electr. Eng. Comput. Sci.* **2019**, *27*, 330–345. [CrossRef]
29. Rao, R.S.; Ravindra, K.; Satish, K.; Narasimham, S.V. Power loss minimization in distribution system using network reconfiguration in the presence of distributed generation. *IEEE Trans. Power Syst.* **2012**, *28*, 317–325. [CrossRef]
30. Esmaeilian, H.R.; Fadaeinedjad, R. Energy loss minimization in distribution systems utilizing an enhanced reconfiguration method integrating distributed generation. *IEEE Syst. J.* **2014**, *9*, 1430–1439. [CrossRef]
31. Wang, B.; Zhang, C.; Dong, Z.Y. Interval optimization based coordination of demand response and battery energy storage system considering SoC management in a microgrid. *IEEE Trans. Sustain. Energy* **2020**, *11*, 2922–2931. [CrossRef]
32. Diab, A.A.; Sultan, H.M.; Mohamed, I.S.; Kuznetsov, O.N.; Do, T.D. Application of different optimization algorithms for optimal sizing of PV/wind/diesel/battery storage stand-alone hybrid microgrid. *IEEE Access* **2019**, *7*, 119223–119245. [CrossRef]
33. Islam, M.M.; Nagrial, M.; Rizk, J.; Hellany, A. Dual stage microgrid energy resource optimization strategy considering renewable and battery storage systems. *Int. J. Energy Res.* **2021**, *45*, 21340–21364. [CrossRef]
34. Faisal, M.; Hannan, M.A.; Ker, P.J.; Hussain, A.; Mansor, M.B.; Blaabjerg, F. Review of energy storage system technologies in microgrid applications: Issues and challenges. *IEEE Access* **2018**, *6*, 35143–35164. [CrossRef]
35. Kumar, P.; Ali, I.; Thomas, M.S. Energy efficiency analysis of reconfigured distribution system for practical loads. *Perspect. Sci.* **2016**, *8*, 498–501. [CrossRef]
36. Geem, Z.W.; Kim, J.H.; Loganathan, G.V. A new heuristic optimization algorithm: Harmony search. *Simulation* **2001**, *76*, 60–68. [CrossRef]
37. Dubey, M.; Kumar, V.; Kaur, M.; Dao, T.P. A systematic review on harmony search algorithm: Theory, literature, and applications. *Math. Probl. Eng.* **2021**, *2021*, 5594267. [CrossRef]
38. Ala'a, A.; Alsewari, A.A.; Alamri, H.S.; Zamli, K.Z. Comprehensive review of the development of the harmony search algorithm and its applications. *IEEE Access* **2019**, *7*, 14233–14245.

processes

Article

Dynamic Performance Assessment of PMSG and DFIG-Based WECS with the Support of Manta Ray Foraging Optimizer Considering MPPT, Pitch Control, and FRT Capability Issues

Mohamed Metwally Mahmoud [1,*], Basiony Shehata Atia [2], Almoataz Y. Abdelaziz [3] and Noura A. Nour Aldin [4,*]

1 Electrical Engineering Department, Faculty of Energy Engineering, Aswan University, Aswan 81528, Egypt
2 Field Service Engineer at Rapiscan Systems, 2805 Columbia St., Torrance, CA 90503, USA
3 Faculty of Engineering and Technology, Future University in Egypt, Cairo 11835, Egypt
4 Electrical Department, Faculty of Technology and Education, Suez University, Suez 43533, Egypt
* Correspondence: metwally_m@aswu.edu.eg (M.M.M.); noura.nouraldin@ind.suezuni.edu.eg (N.A.N.A.)

Abstract: Wind generators have attracted a lot of attention in the realm of renewable energy systems, but they are vulnerable to harsh environmental conditions and grid faults. The influence of the manta ray foraging optimizer (MRFO) on the dynamic performance of the two commonly used variable speed wind generators (VSWGs), called the permanent magnet synchronous generator (PMSG) and doubly-fed induction generator (DFIG), is investigated in this research article. The PMSG and DFIG were exposed to identical wind speed changes depending on their wind turbine characteristics, as well as a dangerous three-phase fault, to evaluate the durability of MRFO-based wind side controllers. To protect VSWGs from hazardous gusts and obtain the optimum power from incoming wind speeds, we utilized a pitch angle controller and optimal torque controller, respectively, in our study. During faults, the commonly utilized industrial approach (crowbar system) was exclusively employed to aid the studied VSWGs in achieving fault ride-through (FRT) capability and control of the DC link voltage. Furthermore, an MRFO-based PI controller was used to develop a crowbar system. The modeling of PMSG, DFIG, and MRFO was performed using the MATLAB/Simulink toolbox. We compared performances of PMSG and DFIG in reference tracking and resilience against changes in system parameters under regular and irregular circumstances. The effectiveness and reliability of the optimized controllers in mitigating the adverse impacts of faults and wind gusts were demonstrated by the simulation results. Without considering the exterior circuit of VSWGs or modifying the original architecture, MRFO-PI controllers in the presence of a crowbar system may help cost-effectively alleviate FRT concerns for both studied VSWGs.

Keywords: crowbar system; DFIG; FRT; PMSG; manta ray foraging optimizer (MRFO); wind energy

Citation: Mahmoud, M.M.; Atia, B.S.; Abdelaziz, A.Y.; Aldin, N.A.N. Dynamic Performance Assessment of PMSG and DFIG-Based WECS with the Support of Manta Ray Foraging Optimizer Considering MPPT, Pitch Control, and FRT Capability Issues. *Processes* **2022**, *10*, 2723. https://doi.org/10.3390/pr10122723

Academic Editors: Ferdinando Salata and Virgilio Ciancio

Received: 14 November 2022
Accepted: 4 December 2022
Published: 16 December 2022

1. Introduction

The International Energy Agency (IEA) report released in 2020 to address electrical power systems (EPSs) security concerns states that, until 2040, the average yearly contribution of renewable energy sources (RESs) will reach 45% of all generations. The security of EPSs is jeopardized by the changeable nature of RESs [1]. EPS security procurement is now much more essential in light of recent pandemics, such as COVID-19 [2]. To make EPSs resilient to grid faults and the fluctuating power output of RESs, security considerations must be incorporated into their operation [3]. One of the RESs that has been rapidly expanding as a source of power in recent years is wind energy (WE), which is used to contribute to the demand side of the electricity supply chain. This RES is used to address issues such as global CO_2 emissions, inadequate load demands, and fossil fuel shortages [4]. Global installed WE capacity will continue to increase due to the decreased costs and high reliability of these systems [5]. WE generation from on- and off-shore wind turbines (WTs) has gained pace, and is currently the cheapest kind of energy in many major markets, signaling that WE is a viable option [5,6]. During electricity transition, the most critical challenges are cost, efficiency, supply security/timing, and wind integration [7–9].

WGs in WE markets these days are based on fixed or variable speed concepts in the market [9]. In the past, WGs were based on fixed concepts due to their features such as simplicity and low cost; however, the main drawback was its need for reactive power (Q) to assist voltage support. The latest standard for installed WE is variable-speed wind generators (VSWGs). VSWGs efficiently capture energy and have good voltage control [9,10]. Widely used VSWGs include the doubly-fed induction wind generator (DFIWG) and permanent magnet synchronous wind generator (PMSWG), with a back-to-back (BTB) power converter strategy [8,11,12]. The DFIWG contains a gearbox, and only 20–30% of its BTB converter rating is needed for its working speed range of 0.7–1.3 pu; however, the PMSWG has a high initial cost due to its use of full-rated BTB power converters [13]. Due to features such as gearless design, minimal maintenance, decreased losses, strong controllability, realized MPPT, new grid code requirements, and high efficiency, the PMSWG is recommended [14].

Utilizing the greatest amount of WE possible is crucial given the growing uptake of WE in the power grid. To do this, the WE system must monitor the maximum power point. There is a respectable range of publication reports on maximum power point tracking (MPPT) algorithms for WE systems. However, selecting the precise MPPT algorithm for a given situation requires considerable expertise because each method has its own advantages and disadvantages. In [15], various MPPT algorithms that could be used to extract the most power were discussed. These techniques were categorized based on whether they use a direct or indirect power controller to monitor power. The benefits, drawbacks, and a thorough comparison of the various MPPT algorithms were also described in terms of their complexity, required wind speed, prior training, speed responses, etc., as well as their capacity to obtain maximum energy production. The presented study in [16] used a control method called Kalman MPPT for the extraction of maximum power from grid-connected wind systems under speed variations; moreover, the pitch angle control (PAC) effect was negligible. To achieve faster convergence and less oscillation when used with variable power sources, the golden section search (GSS), perturb and observe (P&O), and incremental conductance (INC) approaches for MPPT were combined in [17]. The results demonstrate the viability and efficacy of the suggested MPPT technique, but harmonic and fault analyses were not conducted. The work in [18] used a real-time fuzzy-based MPPT controller to provide the extremely efficient operation and step-up power conversion of a standalone PV system under low voltage penetration. Grid-connected PV using FLC has major limitations that are not demonstrated. In order to gain quick and maximum PV power with no oscillation tracking, Ref. [19] offered an adaptive neuro-fuzzy inference system–particle swarm optimization (ANFIS–PSO)-based hybrid MPPT method. However, the suggested ANFIS–PSO continues to introduce greater power oscillations over a longer length of time.

In the DFIWG, the BTB power converter lies between the rotor and the grid side. The DFIWG may run at various speeds depending on the incoming wind speeds, allowing for improved WE harvesting [20]. Due to the PAC and dynamic slip management methods of the DFIWG, rebuilding the terminal voltage after a grid disruption is considerably simpler [21]. Furthermore, with the DFIWG, controlling active power (P) and Q using decoupling principles is significantly easier. When the power converters of the DFIWG are exposed to lower voltages, they fall into standby mode [22]. However, during grid faults over threshold voltages, the DFIWG quickly synchronizes with the electricity grid [22]. In comparison with the DFIWG, the PMSWG provides more flexibility [23,24]. As a result, while employing the PMSWG, P and Q regulation is more successful.

Under fault conditions, WTs may be disconnected from the grid, necessitating the use of new grid codes to improve fault ride-through (FRT) capabilities. FRT has two requirements: WTs must remain connected to the grid even if the voltage is above or below the rated value, and Q injections must occur under abnormal situations such as faults [25,26]. DFIWG stator terminals are directly connected to the grid and their rotor terminals are connected to the grid via a BTB converter. Its smaller size converter leads to reduced power losses and is cost-efficient, although Q supply capability is small due to its size. FACTS devices are equipped with the DFIWG to increase Q supply. Grid faults lead to oscillations in rotor speed and electromagnetic torque, which is transferred to the grid voltage. The rotor side converter (RSC) is deactivated with only a conventional crowbar, and this leads to a reduction of injected Q [27,28]. The PMSWG is directly connected to the power grid through a full-scale BTB converter. This converter decouples the PMSWG from the grid, making it less sensitive to grid faults compared with the DFIWG. It is capable of injecting the rated Q to fully meet grid code requirements for voltage support [29].

For improving the transient stability of the DFIWG and PMSWG, different FRT control strategies have been presented in the literature, such as fault current limiters (FCLs), a crowbar switch, DC chopper circuitry, a parallel capacitor, energy storage systems, FACTS, and sliding mode controls (SMC) [30–37]. Different studies evaluated the DFIWG utilizing various control techniques [38], with a focus on the usage of MPPT and PAC using different algorithms [39], whereas peak current limiting and MPPT were used by [40,41], respectively. For augmentation of the FRT capability of the DFIWG, a series of FCL was combined with a metal oxide varistor [42]. The use of a multistep bridge-type FCL for the PMSWG was reported by [43] to increase its FRT performance. The FRT capability of a wind farm (WF) constituting of DFIWGs was improved using a neuro-fuzzy-logic-controlled (FLC) parallel-resonance-type FCL scheme by [44], whereas complete power systems utilizing an FLC capacitive-bridge-type FCL scheme were examined by [45]. An SMC based on the bridge-type FCL was employed in [46] for the FRT-improved DFIWG performance, whereas another way of employing a dynamic multi-cell FCL was reported to enhance the FRT performance of the WF, based on DFIWG control [47]. FRT capability enhancement methods for WGs are summarized in the literature [30].

By providing a comparative analysis between the proposed work for the PMSWG and DFIWG with recently published methods based on FLC, model predictive controller (MPC), SMC, and optimization concepts, we demonstrate the role of MRFO. Table 1 presents and summarizes a comparison of the results between the proposed and previously published techniques in the PMSWG. Furthermore, this comparison is performed for the DFIWG in Table 2.

To resolve thought-provoking engineering challenges in geometry, a wide variety of algorithms motivated by societal, cosmological, and animal behavior have been suggested. The manta ray foraging optimizer (MRFO) technique is applied to optimal controller design (PI) to enhance the dynamic performance of the PMSWG and DFIWG under regular and irregular conditions. As PMSWGs and DFIWGs are widely used and account for the biggest proportion of WE markets, we were interested in investigating their impact on power systems. In light of the earlier discussion, this comparative study presents the effects of wind gusts and grid faults on the dynamic performance of the PMSWG and DFIWG, considering FRT capability, MPPT operation, and PAC issues. Previous research investigations did not include the application of the MRFO technique for the operation of optimum controllers based on the PMSWG and DFIWG. Furthermore, gathering the three issues for the two investigated renewable generators has not appeared in any single research article. As a result, our research aims to fill the gap in the literature. The DFIWG and PMSWG are discussed in terms of their properties, modeling, and control systems. The PMSWG and DFIWG that were tested had the same capacity and were subject to identical wind gusts and severe faults. Furthermore, both VSWTs operated at their rated speed under the assumed fault state (85% voltage dip), depending on their MPPT characteristics. The main benefit of this control technique is its high effectiveness, small overshoot, quick dynamic response, and successful handling of three critical issues in dominant VSWGs.

Table 1. Comparison of the proposed work with previously published works on PMSWG.

Refs.	Publisher	Year	Developed Controllers		Contribution of Study			Remarks
			MSC	GSC	MPPT	PAC	FRT	
[13]	Elsevier	2017	✓	×	✓	×	✓	Three different control systems (PI, ISMC, and FCS-MPC) were provided, and their effectiveness was evaluated. Outcome: FCS-MPC was the fastest controller, whereas ISMC had the best performance.
[23]	Springer	2020	✓	×	✓	×	✓	A PSO, WOA, and GWO-based PI controller was given. GWO performed more smoothly and quickly than the other approaches that were being compared.
[48]	IEEE	2018	✓	✓	×	×	✓	The system was improved by figuring out the PI controller optimum gain values using the GWO, GA, and simplex methods. The GWO method was reported to have the best convergence to the minimal value, as well as the finest reaction to faults.
[49]	MDPI	2021	×	✓	×	×	✓	The variable switching frequency issue in the traditional FCS-MPC was fixed by a unique MMPC, which also resulted in decreased THD in stator current and shorter simulation times. A coordinated LVRT was used under faults, and the MMPC operated smoothly and with a quick dynamic response.
[50]	MDPI	2022	✓	✓	✓	×	×	Adjusted the machine and GSC to follow the MPPT-established standard for wind speed and to address the chattering issue brought on by the traditional SMC. It was superior to five modern controllers under wind variations.
[51]	Springer	2022	✓	×	✓	✓	×	OTC along with an FLC was implemented to decrease installation costs and improve the system's overall efficiency. FLC was superior to PI under two wind profiles.
[52]	Taylor & Francis	2022	✓	×	✓	×	✓	Optimized PI controller with WHO was presented. WHO was superior to the Ziglar–Nicolas method in terms of fast transient response and smooth operation.
[53]	SAGE	2021	×	✓	×	×	✓	In addition to backing choppers, FLC was employed to enhance system performance in the event of three-phase faults. In terms of quick transient reaction and lag-free operation, FLC outperformed PI.
Current study			✓	×	✓	✓	✓	With the MRFO-PI control of MSC and BC, the three issues are realized. This is the first study that considers the three issues. The optimized controller has a fast response and smooth operation

Table 2. Comparison of the proposed work with previously published works on DFIWG.

Refs.	Publisher	Year	Developed Controllers		Contribution of Study			Remarks
			RSC	GSC	MPPT	PAC	FRT	
[54]	IEEE	2018	✓	×	×	×	✓	Applied feed-forward current control that reduced the transient current in the rotor circuit when a fault quickly occurred.
[55]	IET-Wiely	2018	×	✓	×	×	✓	The overcurrents in the stator and rotor were decreased, and Q was quickly injected during voltage dips, using a combined vector and direct power controller.
[56]	IEEE	2021	✓	✓	✓	×	✓	In order to mitigate (malfunctioning of sensors and parameter fluctuations) and assure acceptable performance during faults or wind speed conditions, a modified adaptive control architecture was added to the existing conventional vector control and was effective.
[47]	Hindawi	2020	×	×	×	×	✓	To improve system performance, a nonlinear SMC-based FCL was linked at the POCC. The results showed that SMC performs well with nonlinear dynamics and unanticipated voltage dip levels.
[46]	MDPI	2020	×	×	×	×	✓	The dynamic adaptive multi-cell FCL topology was coupled at the POCC, which significantly improved system performance and offered an adaptable voltage dip compensation mechanism depending on the level of voltage. Comparison assessment with the single-cell FCL verified the suggested scheme's efficacy.
[57]	Elsevier	2021	✓	✓	✓	✓	×	Performance comparisons between the FLC, H infinity (H∞), and PI controllers were conducted, and H∞ was shown to be the best. Mixed controllers, calledthe FL-H∞, and PI- andPID-filter derivative (Fd)-H∞ gave better performance and resulted in decreasing harmonics.
[58]	MDPI	2022	✓	×	✓	✓	×	The precision of the three controllers (SM, PI, and advanced backstepping (AB)), which provided the lowest tracking error, was studied and assessed. The ABC's benefits included target monitoring, current waveform compatibility, quick response times, and robustness.
[59]	MDPI	2022	✓	×	✓	✓	×	The MPC system was utilized to maximize the amount of wind energy extracted, even when the wind speed was erratic or the WT was uncertain, and it was more efficient than the PI type.
	Current study		✓	×	✓	✓	✓	With the MRFO-PI control of RSC and crowbar, the three issues were realized. This is the first study that considers the three vital issues. The optimized controller had a fast response and smooth operation.

This paper is prepared as follows: The introduction, relevant literature, and purpose are described in Section 1 of this work, which is divided into six parts. Modeling, development, and the failure ratio of VSWGs and basic concepts of the investigated WE systems are presented in Section 2. In Section 3, the application of the MRFO method is discussed in detail. In Section 4, control of the crowbar system with MRFO-PI is studied for solving FRT issues in the investigated VSWGs. A discussion of simulated results is described in Section 5. Finally, concluding remarks are found in Section 6.

2. Modeling of the Studied WE Systems

Kinetic energy is converted into mechanical energy with a WT. The modeling of WT is discussed in detail [30,34].

$$P_M = 0.5C_p(\lambda.\beta)\ \rho A {v_W}^3 \tag{1}$$

$$C_p\ (\lambda.\beta) = 0.5176\left(\frac{116}{\lambda i} - 0.4\beta - 5\right) \exp^{-\frac{21}{\lambda i}} + 0.0068\lambda \tag{2}$$

$$\frac{1}{\lambda i} = \frac{1}{\lambda + 0.08\beta} - \frac{0.035}{\beta^3 + 1} \tag{3}$$

$$\lambda = \frac{\omega_r\, R}{V_W} \tag{4}$$

From Equation (4), we can obtain the value of ω_r at optimal λ and operated V_W:

$$T_m = \frac{P_M}{\omega_r} \tag{5}$$

$$T_m = J_{eq}\frac{d\omega_r}{dt} + B_{eq}\omega_r + T_e \tag{6}$$

where the variables T_m, J_{eq}, B_{eq}, and T_e are the turbine torque, total equivalent inertia of turbine, generator, damping coefficient, and electromagnetic torque of the generator, respectively.

Modeling of WT was performed according to Equations (1)–(6), as shown in Figure 1. The WE power capture is maximized at different wind speeds; the VSWG has the capability to do this for a wide speed range. PAC, one of the software solutions, assists in FRT by keeping the generator operating at rated wind speeds. When a WT is exposed to wind gusts, PAC increases to reduce Cp by controlling the yaw mechanism, and therefore, output power decreases where PA equals zero at normal wind speeds are shown in Figure 2. A variety of MPPT control strategies have been developed, thence MPPT with optimal torque control (OTC) is proposed to be applied due to its merits, such as its power smoothing capability. Figure 3 shows the OTC-MPPT algorithm of WECS.

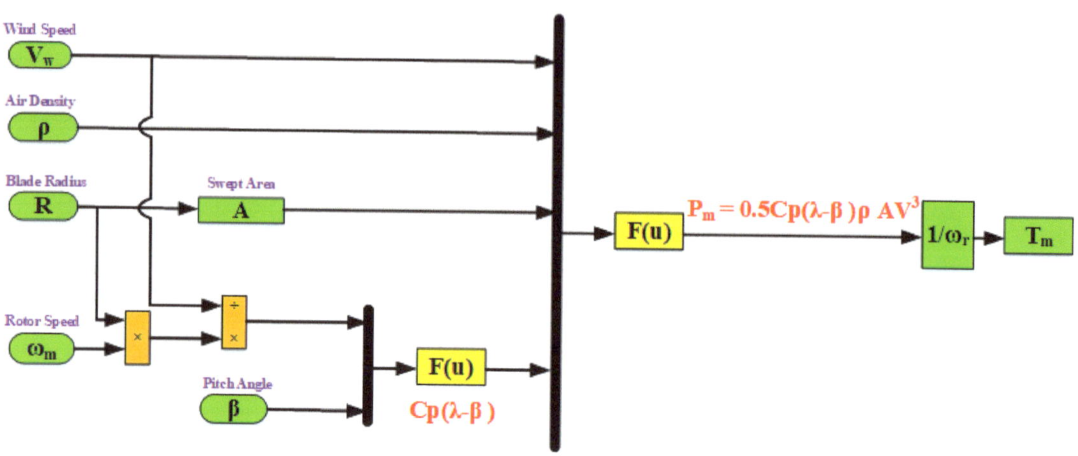

Figure 1. Modeling of a WT.

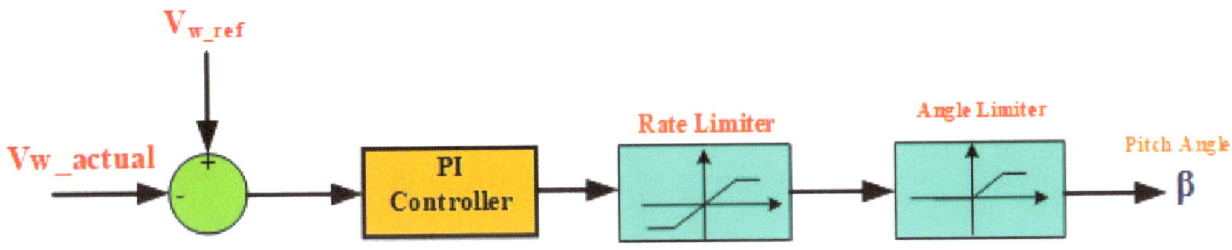

Figure 2. PAC system.

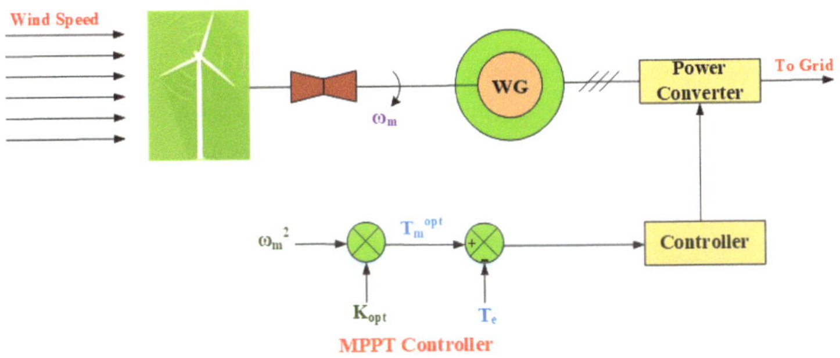

Figure 3. OTC method for MPPT operation.

2.1. Modeling of the PMSWG

The PMSWG model was run using a d-q equivalent electrical circuit. A full description of the proposed PMSWG with its control system is shown in Figure 4. The parameter definitions are found in [60]. Dynamic equations of the mathematical model are as follows [13,52]:

$$V_{ds} = R_s I_d + \lambda^{\cdot}{}_d - \omega_e \psi_q \tag{7}$$

$$V_{qs} = R_s I_q + \lambda^{\cdot}{}_q - \omega_e \psi_d \tag{8}$$

$$\psi_d = L_d I_d + \psi_{pm} \tag{9}$$

$$\psi_q = L_q I_q \tag{10}$$

$$\lambda_d = L_d I_d + \psi_{pm} \tag{11}$$

$$T_e = \frac{3}{2} n_p \left(\psi_{pm} I_q\right) \tag{12}$$

$$C\frac{dV_{dc}}{dt} = \frac{P_{MSC}}{V_{dc}} - \frac{P_{GSC}}{V_{dc}} \tag{13}$$

$$V_{gd}{}^{*} = V_{id} - R_g\, I_{gd} - L_g \frac{d}{dt} I_{gd} - L_g \omega_e\, I_{gq} \tag{14}$$

$$V_{gq}{}^{*} = V_{iq} - R_g\, I_{gq} - L_g \frac{d}{dt} I_{gq} - L_g \omega_e\, I_{gq} \tag{15}$$

$$P_g = \frac{3}{2} V_{gd}\, I_{gd} \tag{16}$$

$$Q_g = \frac{3}{2} V_{gd}\, I_{gq} \tag{17}$$

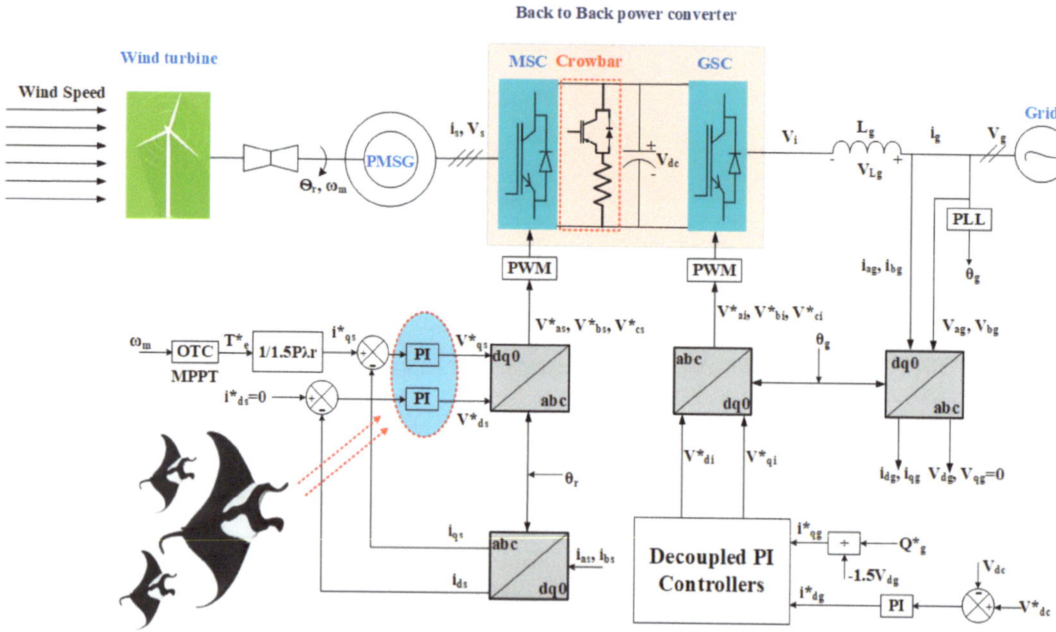

Figure 4. PMSWG with its proposed control system.

Equations (16) and (17) indicate that P_g and Q_g are controlled by controlling I_{gd} and I_{gq} currents, respectively. To transfer all of the P_g generated from the wind turbine, DC-bus voltage must be constant, according to Equation (13). P_{GSC} and P_{MSC} are the active power to the grid and from the WG, respectively.

2.2. Modeling of the DFIWG

The DFIWG is represented through a fifth-order model [30,61]. This model consists of four electrical differential equations (two equations for both the stator and rotor voltages). The electrical equations, expressed in the direct-quadrature (dq) reference frame rotating at synchronous speed (ω_S), are given by Equations (18)–(26). The DFIWG with its proposed control configuration is depicted in Figure 5.

$$V_{ds} = R_S\, I_{ds} - \frac{d\psi_{ds}}{dt} - \omega_S\, \psi_{qs} \tag{18}$$

$$V_{qs} = R_S\, I_{qs} + \frac{d\psi_{qs}}{dt} + \omega_S\, \psi_{ds} \tag{19}$$

$$V_{dr} = R_r\, I_{dr} + \frac{d\psi_{dr}}{dt} - (\omega_S - \omega_r)\psi_{qr} \tag{20}$$

$$V_{qr} = R_r\, I_{qr} + \frac{d\psi_{qr}}{dt} + (\omega_S - \omega_r)\, \psi_{dr} \tag{21}$$

$$\psi_{ds} = L_s\, I_{ds} + L_m\, I_{dr} \tag{22}$$

$$\psi_{qs} = L_s\, I_{qs} + L_m\, I_{qs} \tag{23}$$

$$\psi_{dr} = L_r\, I_{dr} + L_m\, I_{dr} \tag{24}$$

$$\psi_{qr} = L_r\, I_{qr} + L_m\, I_{qs} \tag{25}$$

$$T_e = \frac{3}{2}\, P\, (\psi_{ds}\, I_{qs} - \psi_{qs}\, I_{ds}) \tag{26}$$

where V denotes voltage, ψ represents magnetic flux, R denotes resistance, I denotes current, L denotes inductance, the index m denotes magnetization, T_e is the electromagnetic torque of the generator, P is number of pole pairs, and indexes s and r refer to stator and rotor, respectively.

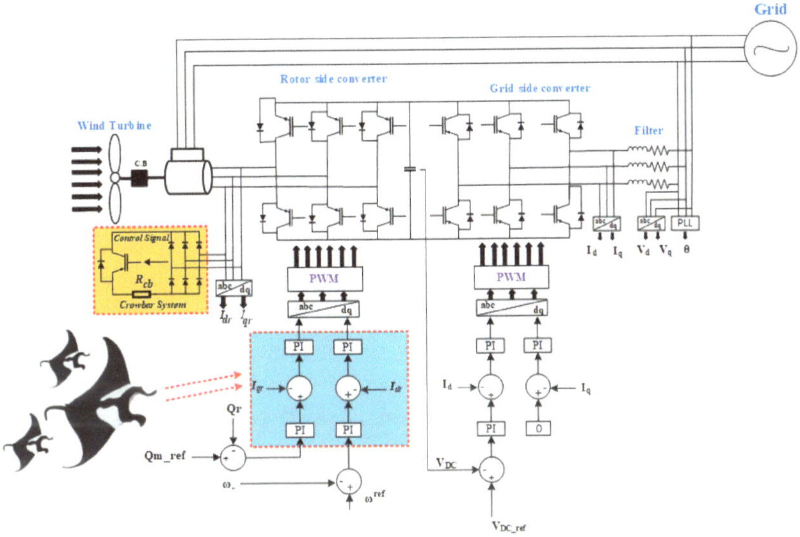

Figure 5. DFIWG with its proposed control system.

2.3. *Development and Failure Ratio of Wind-Driven Power Generators*

The failure ratio in WG components is shown in Figure 6 [62–64]. PAC systems and power converters represent a high ratio of failure for these components, and this is due to the mechanical stress that happens due to the nature of wind speed and faults, where faults lead to an increase in the speed of WG. Thus, protection topologies have great importance, and aid in decreasing the failure ratio [60,65,66]. Manufacturers and technologies for the top five WTs are shown in Table 3. From this table, it can be deduced that the PMSWG and DFIWG are produced by dominant manufacturing companies. With the fastest growth rate in WE between 2009–2018, full-scale power converter WGs, such as the PMSWG, have become dominant in the WE market. Table 4 indicates the top ten biggest turbines [8,9].

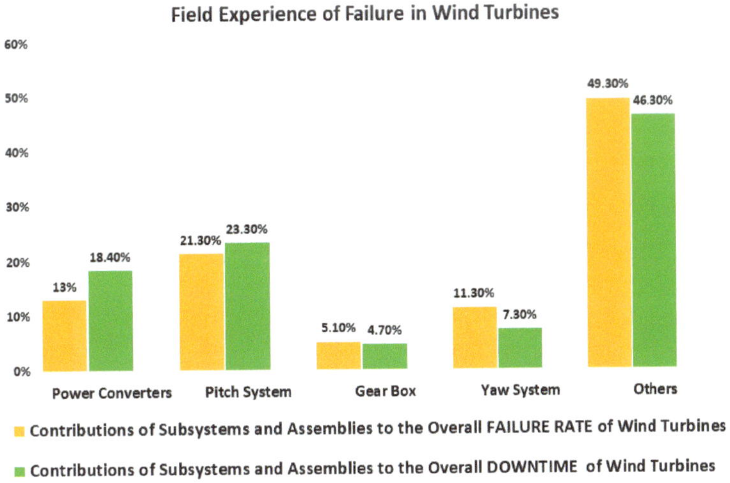

Figure 6. Failure in WE system components.

Table 3. Top 5 WT manufacturers and technologies.

Manufacturer	Concept	Rotor Diameter (m)	Power Range (MW)
Vestas (Denmark)	DFIG	90–120	2.0–2.2
	PMSG	105–162	3.4–9.5
Siemens Gamesa (Spain)	SCIG	154–167	6.0–8.0
	DFIG	120–142	3.5–4.3
	PMSG	114–145	2.1–4.5
Gold wind (China)	PMSG	-	2.0–6.0
GE (USA)	DFIG	116–158	2.0–5.0
	PMSG	150	6.0
Enercon (Germany)	WRSG	82–138	2.0–4.2

Table 4. Top 10 biggest WTs.

Manufacturer	Power Rating (MW)	Rotor Diameter (m)	Drive Train	IEC Class
MHI Vestas	9.5	164	Medium-speed geared	S
Siemens Gamesa	8	167	Direct drive	S (IB)
Gold wind	6.7	154	PM direct drive	I
Senvion	6.15	152	High-speed geared	S
GE	6	150	Direct drive	IB
Ming Yang	6	140	Medium-speed geared	IIB
Doosan	5.5	140	High-speed geared	I
Hitachi	5.2	126–136	Medium-speed geared	S
Nbjj	5	151	High-speed geared	IIB
Adwen	5	135	Low-speed geared	IA

3. MRFO Algorithm

(a) MRFO mathematical model

A novel meta-heuristic technique called MRFO is motivated by the smart and strategic behavior of manta rays (MRs) when they are looking for prey. It has already been demonstrated that using this approach to solve engineering challenges yields remarkably positive outcomes. Chain foraging, cyclone foraging, and somersault foraging are the three processes that the MRFO mimics in the MR eating technique, as depicted in Figure 7 [67–69].

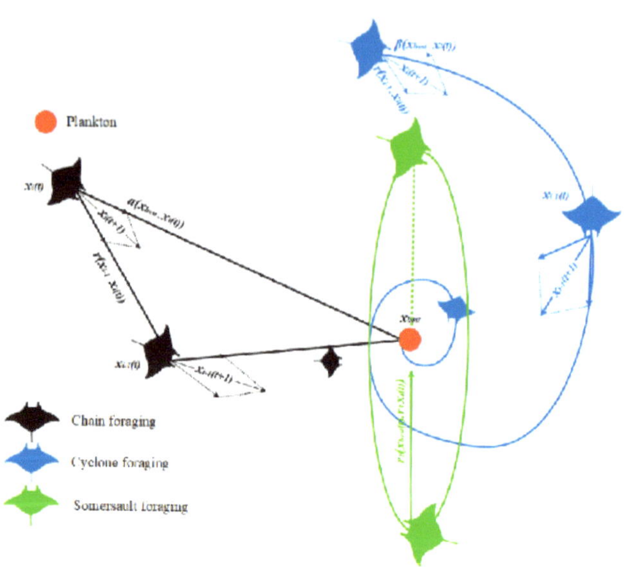

Figure 7. MRFO strategy.

Step 1: chain foraging

MRs move in a foraging chain by swimming together to an area with more plankton. The other MRs track the first as it goes toward the meal, each moving in the same direction as the first. At every time point, an MR adjusts its place with the best option that is open to both it and the one in front of it. Equation (27) provides the mathematical formulation of this circumstance [68,69].

$$X_i^{d}(t+1) = \begin{Bmatrix} X_i^{d}(t) + (X_{best}{}^{d}(t) - X_i^{d}(t))\ (r+\alpha)\ if\ i = 1 \\ X_i^{d}(t) + r\left(X_{i-1}{}^{d}(t) - X_i^{d}(t)\right) + \alpha(X_{best}{}^{d}(t) - X_i^{d}(t))\ else \end{Bmatrix} \tag{27}$$

where $X_i^{d}(t)$ and $X_{best}{}^{d}(t)$ are the place of the ith individual and the finest solution, respectively, in the t iteration. The random vector (r) ranges from [0–1]. The weighting coefficient (α) is presented in Equation (28).

$$\alpha = 2r|\log(r)|^{0.5} \tag{28}$$

Step 2: cyclone foraging

Each MR in the second step tracks the one next to it while also pursuing the meal by spinning its body around. The scenario's model is provided in Equation (29) [67].

$$\begin{aligned} X_i(t+1) &= X_{best} + r(X_{i-1}(t) - X_i(t)) + e^{bw}\cos(2\pi\omega)(X_{best} - X_i(t)) \\ Y_i(t+1) &= Y_{best} + r(Y_{i-1}(t) - Y_i(t)) + e^{bw}\sin(2\pi\omega)(Y_{best} - Y_i(t)) \end{aligned} \tag{29}$$

where ω is a random number ranging from [0,1]. If the motions made by MRs are enlarged in the d space, its model can be written as in Equation (30).

$$X_i^{d}(t+1) = \begin{Bmatrix} (X_{best}{}^{d}(t) + (X_{best}{}^{d}(t) - X_i^{d}(t))\ (r+\beta)\ if\ i = 1 \\ (X_{best}{}^{d}(t) + r\left(X_{i-1}{}^{d}(t) - X_i^{d}(t)\right) + \beta(X_{best}{}^{d}(t) - X_i^{d}(t))\ else \end{Bmatrix} \tag{30}$$

where β denotes the weighting factor, as seen in Equation (31).

$$\beta = 2e^{\frac{r_1(T-t+1)}{T}}\ sin(2\pi r_1) \tag{31}$$

In this case, T stands for the total number of iterations, and r_1 is a random number between [0, 1]. At this point, the MRs leave the area and assume new places, aiding in the search method. Consequently, a thorough global investigation is carried out. The following is the plot's mathematical formula [67,69].

$$X_{rand}{}^{d} = Lb^{d} + r\left(Ub^{d} - Lb^{d}\right) \tag{32}$$

$$X_i^{d}(t+1) = \begin{Bmatrix} (X_{rand}{}^{d}(t) + (X_{rand}{}^{d}(t) - X_i^{d}(t))\ (r+\beta)\ if\ i = 1 \\ (X_{rand}{}^{d}(t) + r\left(X_{i-1}{}^{d}(t) - X_i^{d}(t)\right) + \beta(X_{rand}{}^{d}(t) - X_i^{d}(t))\ else \end{Bmatrix} \tag{33}$$

where $X_{rand}{}^{d}$ is an arbitrary number in the search planetary. The Lb^{d} and Ub^{d} are the upper and lower limits of the dth dimension, respectively.

Step 3: somersault foraging

The third step is where the meal is noticed as the key element. Each MR now usually swims back and forth to the center, changing positions. As a result, each MR's location is constantly changed to be at the ideal location. This group's model is described in Equation (34).

$$X_i^{d}(t+1) = X_i^{d}(t) + S\left(r_2\ X_{best}{}^{d}(t) - r_3 X_i^{d}(t)\right) \tag{34}$$

where S refers to the MR somersault factor value, and r_2 and r_3 are two randomly selected values between [0, 1]. Each MR can therefore go to any position between their current location and the search space. As a result, the change in individuals' present positions becomes increasingly smaller as they eventually get closer to the ideal answer. The actions taken are shown in Figure 8.

(b) Application of MRFO

MRFO is applied on the machine side controllers of the PMSWG and DFIWG to fine-tune the PI controllers' gains. This tuning is performed to improve the dynamic performance of the investigated systems during normal and abnormal conditions. The optimization of the systems under study (taking control cost (CC) into account) can be formally represented in Equation (35). Tables 5 and 6 list the findings of controller gain calculations based on the MRFO approach. Furthermore, the other controller system gains used in grid-side controllers are presented in these tables. The objective function for CC that is employed is written as follows: Minimize F(x)

$$= \int_0^T W_1|P - P^*| + W_2|\omega_m - \omega^*{}_m| + W_3|Q_s - Q^*{}_s| + W_4|V_{dc} - V^*{}_{dc}| + W_5|C_P - C^*{}_P| \tag{35}$$

where W_1, W_2, W_3, W_4, and W_5 are constants used for the estimation of the CS function, which is 4×10^5 here. T denotes the average time and 100 and 6 are the number of iterations and agents, respectively.

Table 5. Data of converter controller for PMSWG.

Technique	Optimized MSC Controller Gains		GSC Controller Gains	
	Gain	Value	Gain	Value
MRFO	K_{p1}	2.8971	K_{p3}	0.83
	K_{i1}	199.7842	K_{i3}	5
	K_{p2}	2.8971	K_{p4}	8
	K_{i2}	199.7842	K_{i4}	400
	-	-	K_{p5}	0.83
	-	-	K_{i5}	5

Table 6. Converter controller data for DFIWG.

Technique	Optimized RSC Controller Gains						GSC Gains					
	Voltage Regulator		Torque Regulator		Voltage Regulators		Voltage Regulator		Torque Regulator		Voltage Regulators	
MRFO	$K_p = 7.9712$	$K_i = 0.0319$	$K_p = 2.7839$	$K_i = 0.0937$	$K_p = 0.2981$	$K_i = 97.278$	$K_p = 3$	$K_i = 0.02$	$K_p = 8$	$K_i = 500$	$K_p = 1.2$	$K_i = 5$

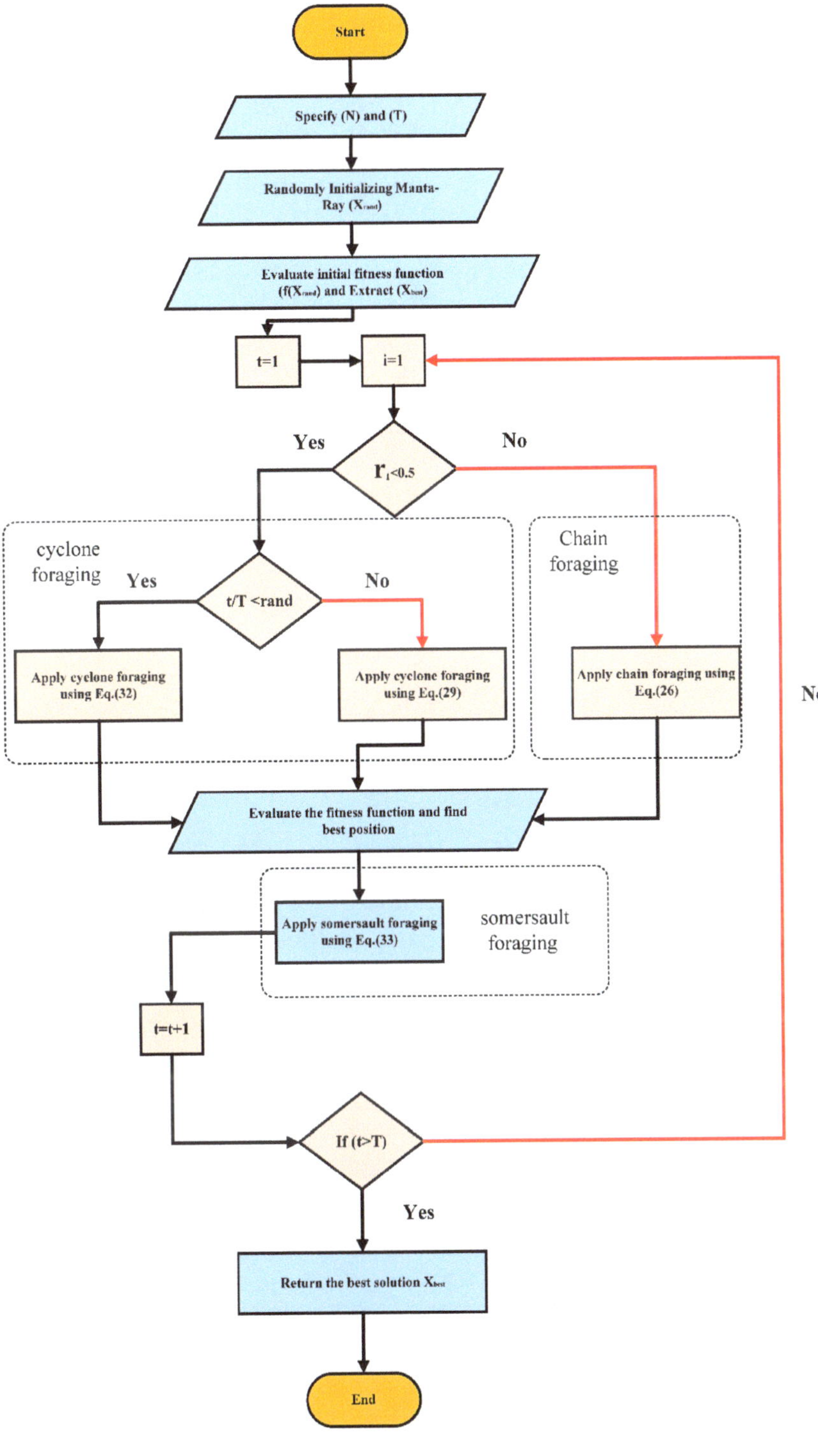

Figure 8. MRFO flowchart.

4. Crowbar Control System for Improving FRT Capability

Faults on the grid have an adverse effect on the dynamic performance of WGs. These faults lead to generator speed-ups, oscillations in electromagnetic torque, overcurrents, overvoltages at the DC link, and reduction of the output P from the VSWG [25,30]. Power electronic converters are exposed to damage due to current limitations of the converters, and these converters are the highest cost of the system; thus, hardware and software solutions have been implemented by researchers to protect the VSWGs from all mentioned bad results. Reliability/security of supply, efficiency, cost, volume, protection, control of P and Q power electronics-enabling technology, and ride-through operation are the important issues for the converters used in WECS [70].

A braking chopper/crowbar is chosen as a hardware solution to successfully protect the DC capacitor from overvoltages by dissipating the surplus energy during abnormal conditions. It has been inserted with the PMSWG to enhance its dynamic performance during grid faults, as seen in Figure 4. It is only inserted during voltage sag [31]. An active crowbar is used to protect the RSC of the DFIWG and improve its dynamic behavior during grid faults, as depicted in Figure 5. Utilizing a crowbar allows the DFIWG to ride through the fault and continue the power supply, even during grid faults. The optimized values of K_p and K_i are 0.06993 for the PMSWG. The optimized values of K_p and K_i are 0.05417 for the DFIWG. The monitored and referenced values are the controller's inputs, and the controller's output determines how to operate the system while taking into account the sawtooth signal. The proposed control strategy of the crowbar for both investigated VSWGs is depicted in Figure 9. Hence, a comparison between different hardware protection apparatuses used for FRT enhancement based on cost is shown in Table 7 [29]. A comparison of FRT strategies of different crowbar protection circuits is shown in Table 8 [27].

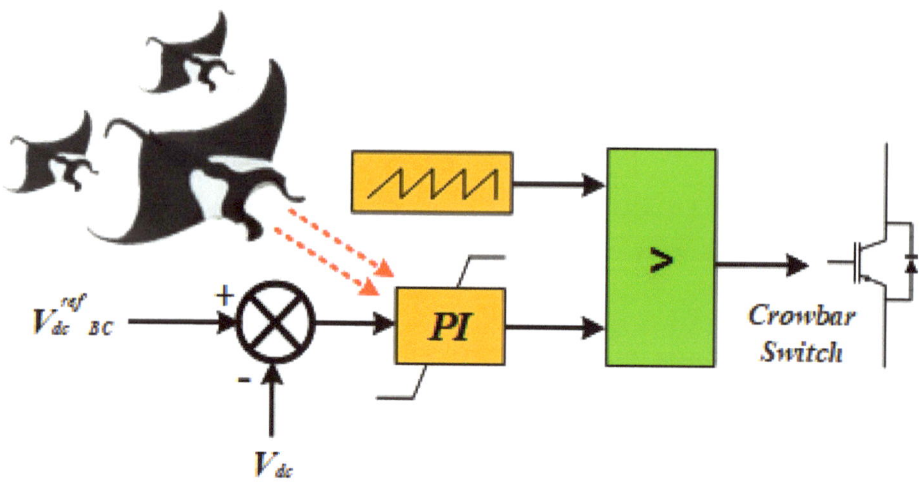

Figure 9. Control of crowbar system based on MRFO.

Table 7. Comparison of different hardware systems used for FRT enhancement based on cost.

Hardware Protection Device	Price (US$)
Classical DVR	67,229.99
Low-cost DVR	36,778.79
STATCOM	200,000.00
Conventional crowbar	5.00–75.00
Active crowbar	85.00

Table 8. Comparison of FRT strategies of different protection circuits for DFIWG.

Protection Scheme	Rotor Current Limit (pu)	Status of RSC	V_{DC} Limit (pu)	Remarks
Crowbar circuit with resistances only	<2.4	Blocked	<1.25	Useful for symmetrical faults only
Crowbar with DVR	<2.0	Partly maintained	<1.25	Useful for all fault types
Crowbar with chopper	<2.4	Blocked	<1.25	Useful for all fault types
Crowbar with R–L	<2.4	Partly maintained	<1.25	Useful for all fault types
ACB_P	<2.0	Partly maintained	<1.08	Useful for all fault types

5. Simulation Results and Discussion

Simulation for the investigated WGs is carried out by using MATLAB/Simulink to verify the aforementioned analysis and the effectiveness of proposed schemes which are PAC and OTC. Models' solutions are tested on a detailed model and in MW class-based WECS. The point of common coupling (POCC) is a significant point in which a bolted fault occurs to evaluate FRT capability. To validate the simulation models, the crowbar parameters are listed in Table 9 [29] and the PMSWG and DFIWG data are listed in Table 10 [13,27]. Furthermore, the impact of the inertia of the dynamic systems is deliberated on in the Appendix A.

Table 9. Crowbar resistance parameters.

Resistance	1.5 Ω
Rated power	12 kW
Maximum temperature	150 °C
Thermal time constant	4 min
Weight	30 kg
Dimensions	(750.330.150) mm

Table 10. Simulated WG data.

PMSWG Parameters	Value	DFIWG Parameters	Value
Rated power	1.5 MW	Rated power	1.5 MW
Rated stator voltage	575 V	Rated stator voltage	575 V
Rated frequency	60 Hz	Rated frequency	60 Hz
DC-link voltage	1150 V	DC-link voltage	1150 V
Pole pairs	40	Pole pairs	3
Generator inductance in the d frame	0.7 pu	Stator resistance	0.023 pu
Generator inductance in the q frame	0.7 pu	Rotor leakage inductance	0.16 pu
Generator stator resistance	0.01 pu	Mutual inductance	2.9 pu
A flux of the permanent magnets	0.9 pu	Stator leakage inductance	0.18 pu
Line inductance	0.3 pu	Rotor resistance	0.016 pu
Line resistance	0.003 pu	Inertia constant	0.685 pu

5.1. Impact of Wind Speed Variation under Regular Grid Conditions

The studied system is illustrated in Figure 4, where the wind speed profile is applied to two worst case scenarios. In the former, a lower wind speed is applied to verify the system in extracting MPPT, and in the other case, the system is exposed to wind gusts to verify the effectiveness of PAC, as depicted in Figure 10a. If the PA equals zero, this signifies MPPT realization; if it is bigger than zero, this means the WG does not operate at MPPT to prevent the WT from wind gusts that may cause a failure in the system components. Figure 10b,c show both λ and C_p, where they quickly track the wind speed profile and achieve optimal values, respectively. C_p is affected by the step change in wind speed that is due to changes in λ and the earliest operator change, according to Equation (4). Figure 10d depicts the PAC response where the PA increases to reach 2.974^0. Figure 10e,f display the changes in both T_e and ω_r as a result of wind speed changes, respectively. Figure 10g depicts the injected P and Q to the grid as a result of wind speed changes. The Q is kept at zero because the system operates at unity power factor (UPF). Equations (1), (13), (14), and (15) give a strong explanation for T_e, ω_r, and the P and Q responses, respectively. Figure 10h depicts that the V_{DC} is constant, indicating that all generated power is transferred into the grid. Oscillations that occur in the simulated parameters are very small and indicate the success of PAC and OTC in the presence of MRFO-PI controllers.

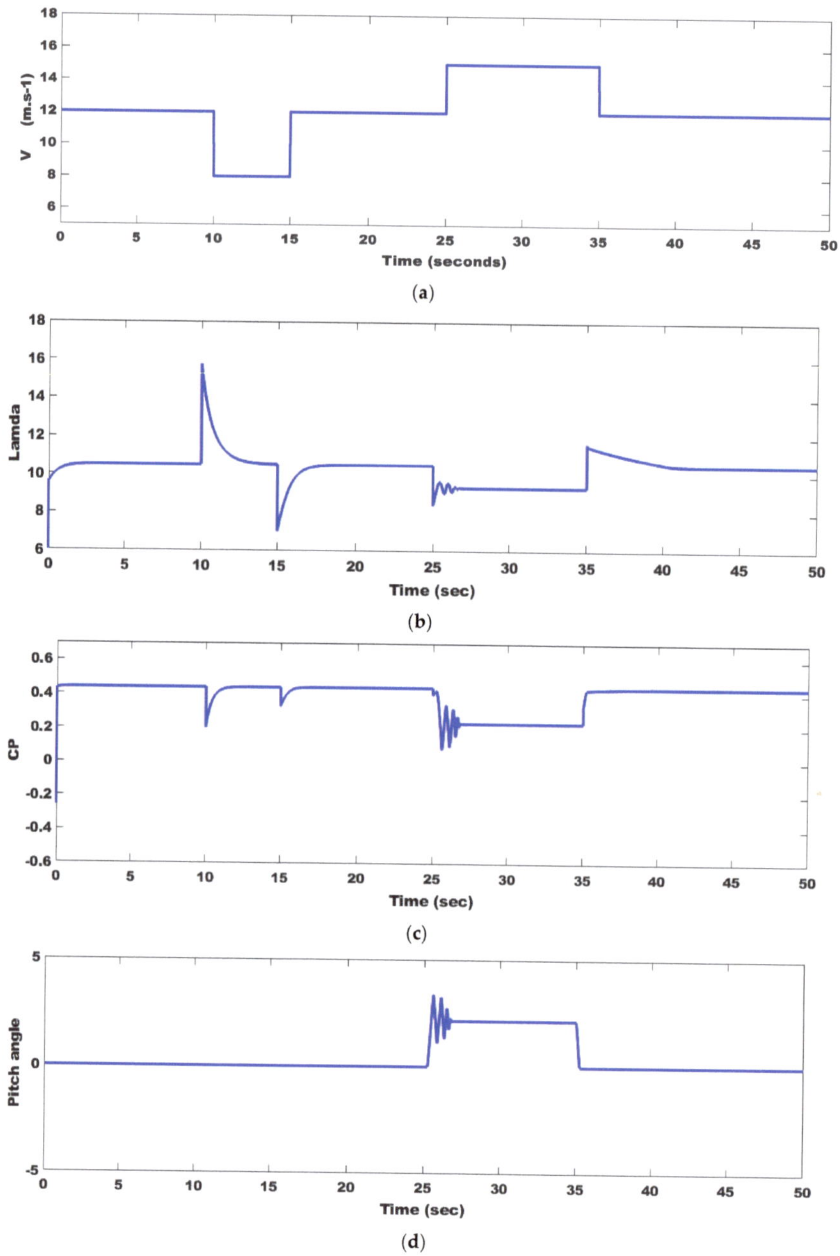

Figure 10. *Cont.*

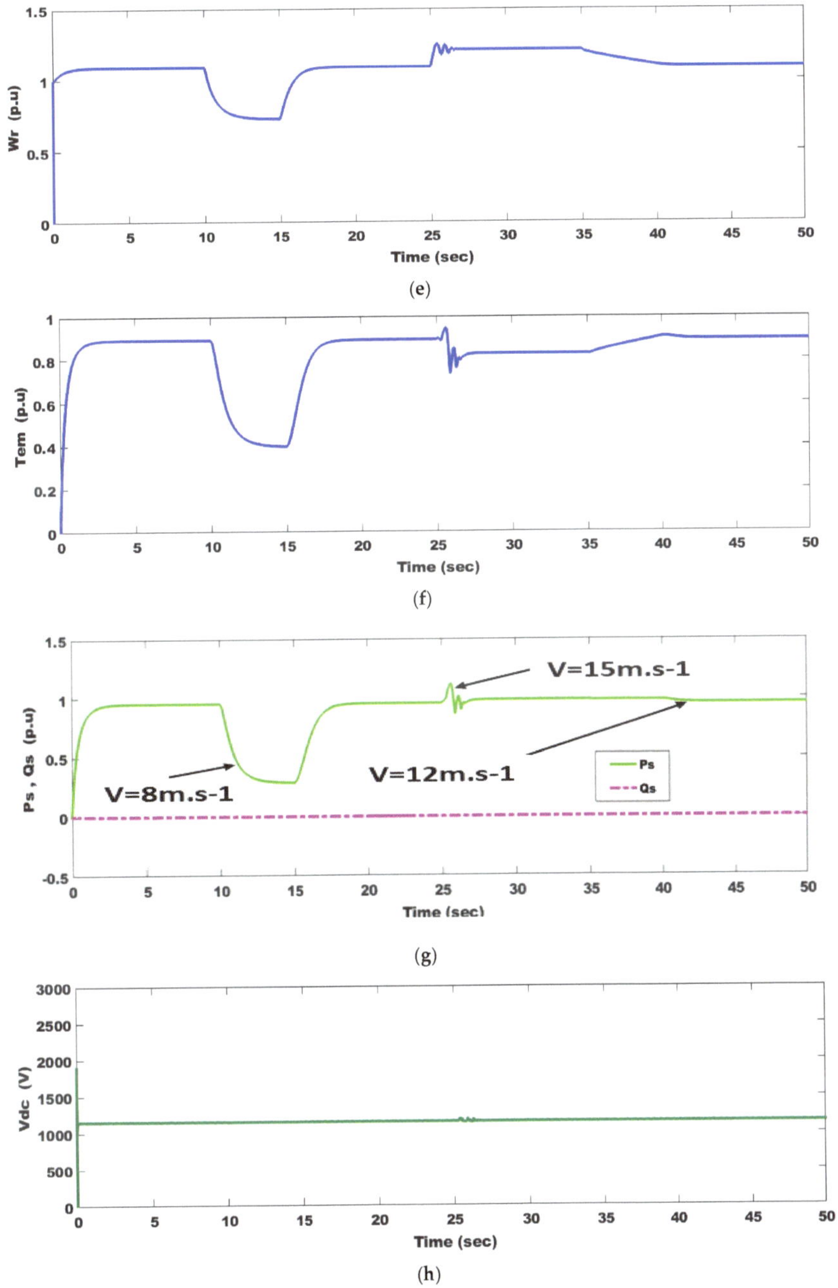

Figure 10. PMSWG system parameter responses as a result of wind speed changes: (**a**) wind speed profile; (**b**) tip speed ratio; (**c**) power coefficient; (**d**) pitch angle; (**e**) angular speed; (**f**) electromagnetic torque; (**g**) injected active and reactive power to the grid; and (**h**) DC-link capacitor voltage.

5.2. Realization of FRT under 85% Voltage Dip

Transient response enhancement during and after clearing the fault becomes a crucial requirement for new grid codes. The fault is assumed to occur at 3 s and cleared at 3.15 s in the grid voltage, as seen in Figure 11a, and wind speed is constant at 12 m·s^{-1}. During the fault period, both P and T_e decrease, as seen in Figure 11c,f, respectively, but an increase occurs in I, Q, V_{DC}, and ω_r, as seen in Figure 11b,d,e,g, respectively, because of this voltage dip. The braking chopper gets rid of surplus power by dissipating it in the form of thermal power to keep V_{DC} in the whole rated range. The proposed technique is successful in this issue where the overshoot of P, oscillations in ω_r, and all parameters are damped. Injection of Q after clearing the fault and the PMSWG maintaining grid connection shows FRT capability realization.

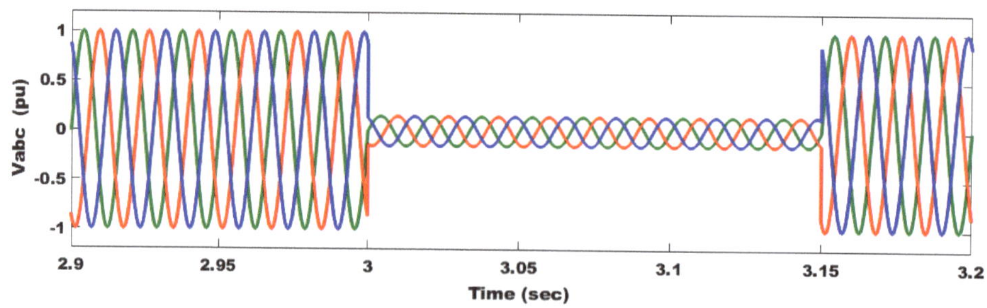

(a)

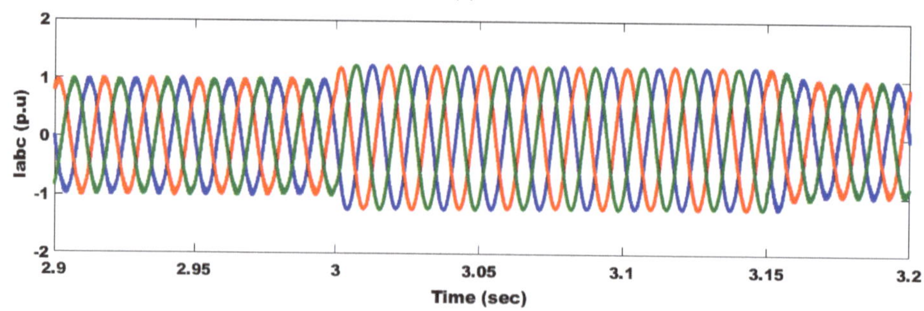

(b)

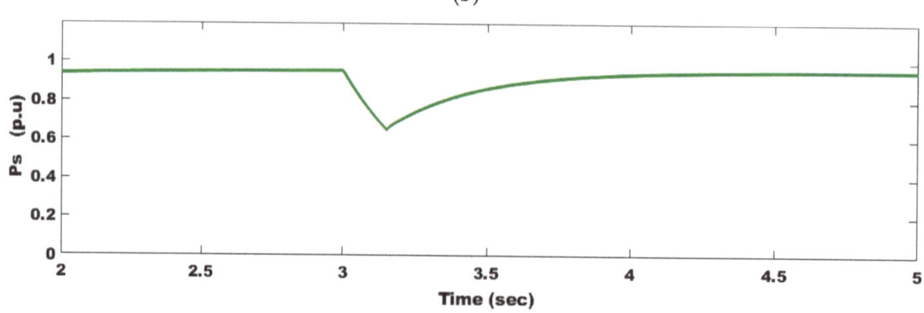

(c)

Figure 11. *Cont.*

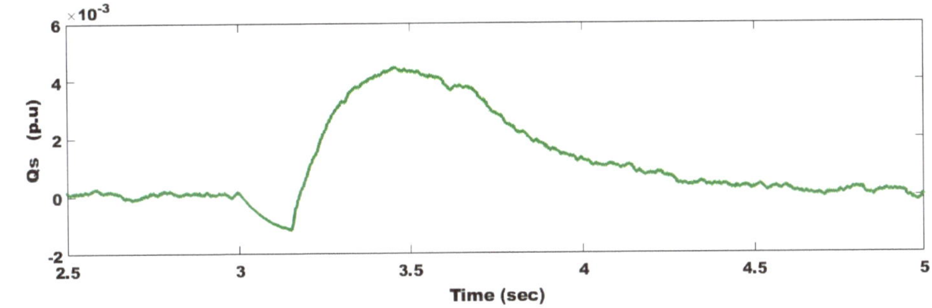

(**d**)

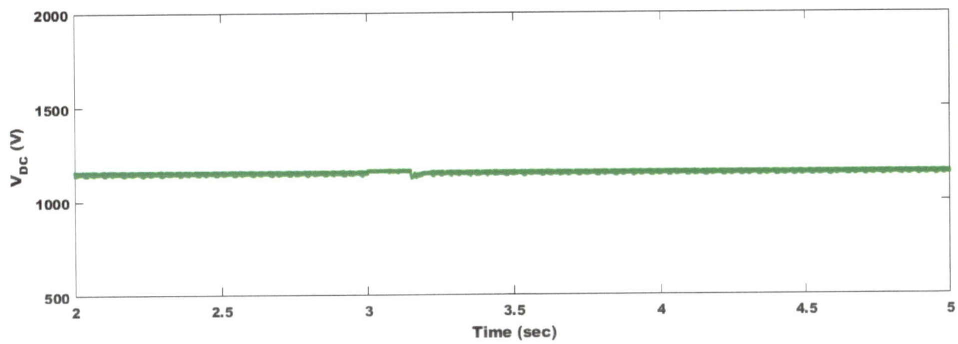

(**e**)

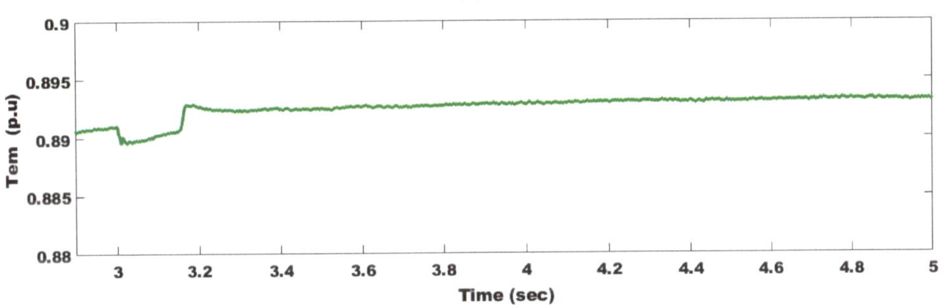

(**f**)

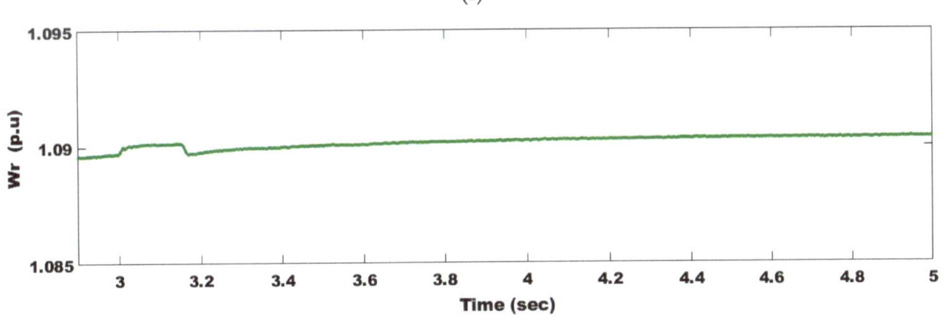

(**g**)

Figure 11. PMSG system parameter responses as a result of an 85% voltage dip: (**a**) system voltage; (**b**) system currents; (**c**) injected active power to the grid; (**d**) injected reactive power to the grid; (**e**) DC-link voltage; (**f**) electromagnetic torque; and (**g**) angular speed.

5.3. Impact of Wind Speed Variation under Regular Grid Conditions

The investigated wind system is shown in Figure 5. Figure 12a shows the variations in the studied wind speed profile. Both λ and C_p are depicted in Figure 12b,c, respectively, and they quickly track the wind speed profile and achieve their desired values. In response to a step change in wind speed, C_p is impacted by λ. Figure 12d depicts the PAC response where the PA increases to reach 9.874^0. The PAC successfully prevents the DFIWG from overwinding speeds by increasing PA from when PA is larger than zero. The parameters λ and C_p decrease to reduce output power until this gust disappears to keep the WT working. Figure 12e,f display the changes in both T_e and ω_r as a result of wind speed changes, respectively. P and Q are shown in Figure 12g as a function of changing wind speeds. Due to the UPF operation, Q is maintained at zero. The unchanging V_{DC} in Figure 12h implies that all produced electricity is transmitted to the power grid. Very minor oscillations in the simulated parameters show the effectiveness of PAC and OTC in the presence of MRFO-PI controllers. The findings of the simulation indicate that the DFIWG has more variability than the PMSWG.

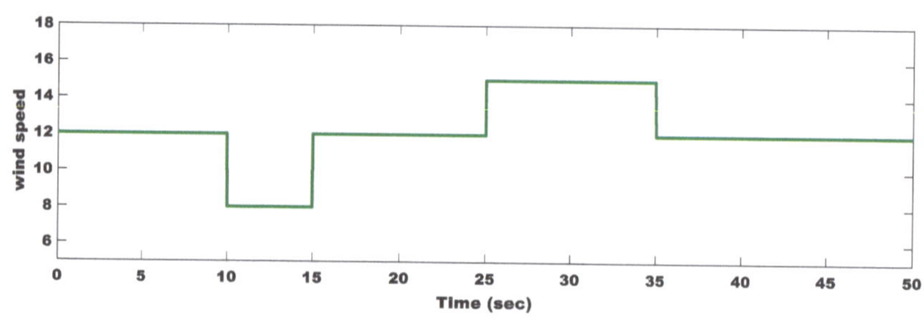

(a)

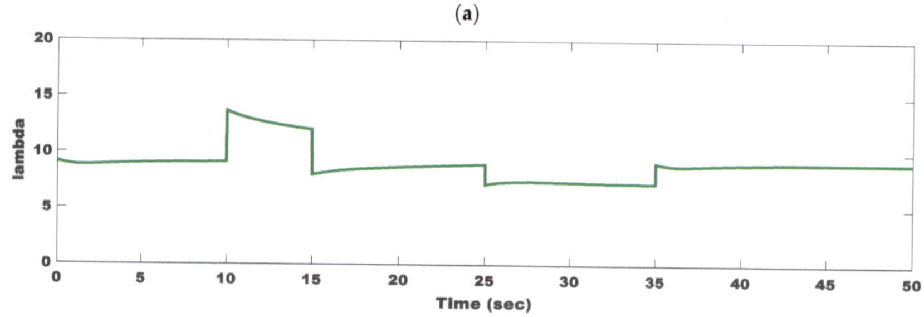

(b)

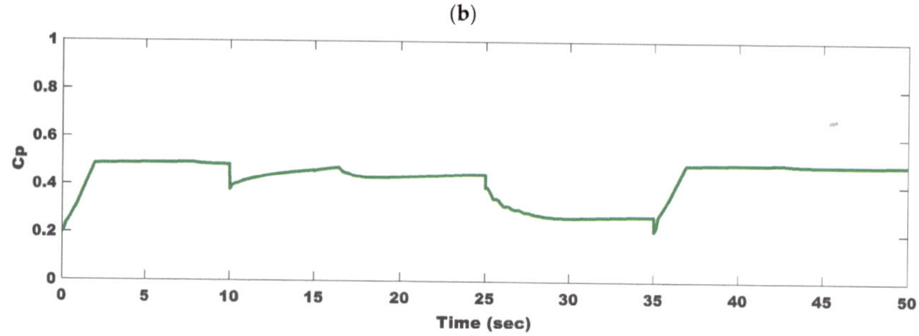

(c)

Figure 12. *Cont.*

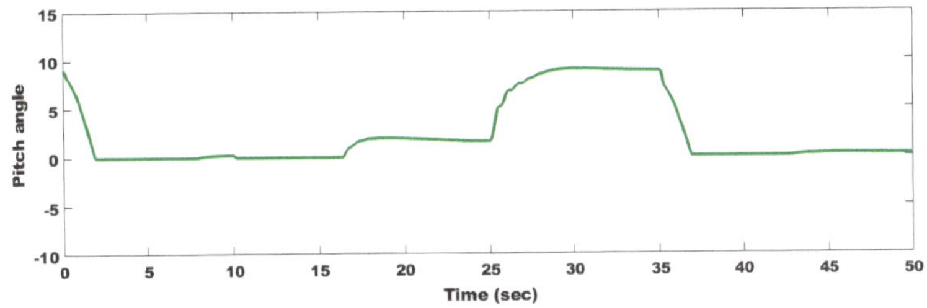

(d)

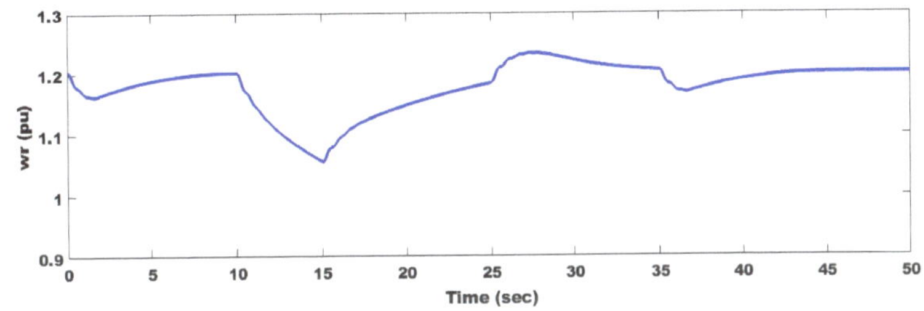

(e)

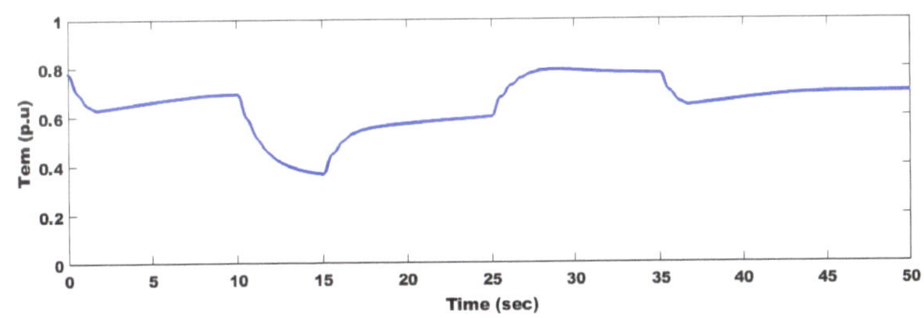

(f)

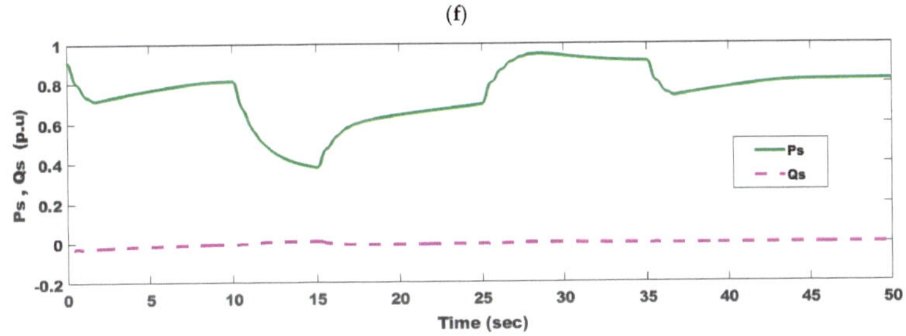

(g)

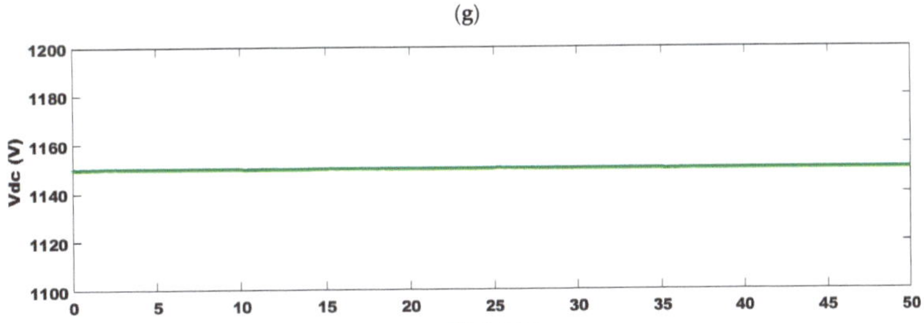

(h)

Figure 12. DFIWG system parameter responses as a result of wind speed changes: (**a**) wind speed profile; (**b**) tip speed ratio; (**c**) power coefficient; (**d**) pitch angle; (**e**) angular speed; (**f**) electromagnetic torque; (**g**) delivered active and reactive powers to the grid; and (**h**) DC-link voltage.

5.4. Realization of FRT at 85% Voltage Dip

The performance of the DFIWG with a proposed active crowbar and optimized controllers under an 85% voltage dip on the grid, as seen in Figure 13a, is evaluated in this section. The duration of the fault period is assumed to be 150 ms, according to the worst case in new grid codes, to test the efficacy of the proposed schemes. During the fault, both P and T_e decrease, as seen in Figure 13c,f, but an increase occurs in I, Q, V_{DC}, and ω_r, as seen in Figure 13b,d,e,g, respectively, because of the drop in the grid voltage. In order to keep V_{DC} within the specified range, the braking chopper dissipates excess power as thermal energy, as seen in Figure 13e. The suggested method works well in this case, where all parameters, including oscillations in ω_r and overshoot of P, are damped. When Q is injected after a fault has been repaired and the DFIWG is still tied to the grid, FRT capability is achieved. The observed simulated results show that the DFIWG continues to operate appropriately, even in the face of serious failures. All the studied systems' parameter fluctuations for the cases under study are summarized and listed in Table 11.

Table 11. All investigated systems' parameter changes for the scenarios under study.

Studied Cases	Parameters	
	DFIWG	PMSWG
Wind speed scenario	C_p change (0.2 → 0.48)	C_p change (0.18 → 0.44)
	V_{DC} ripple (±14 V)	V_{DC} ripple (±9 V)
	λ change (8.7 → 12.9)	λ change (7 → 16)
	P change (0.3 → 0.97)	P change (0.27 → 1.03)
	T_e change (0.3 → 0.8)	T_e change (0.28 → 0.93)
	ω_r change (1.03 → 1.24)	ω_r change (0.72 → 1.19)
Voltage dip scenario	Overvoltage at V_{DC} 1.0783 pu	Overvoltage at V_{DC} 1.0584 pu
	P change ≈ (0.17 → 1.28)	P change ≈ (0.64 → 0.965)
	T_e change ≈ (−0.53 → 1.63)	T_e change ≈ (0.889 → 0.894)
	ω_r change ≈ (1.15 → 1.22)	ω_r change ≈ (1.087 → 1.095)
	Q change ≈ (−0.54 → 0.61)	Q change ≈ (−0.001 → 0.005)
	I change (−2.27 → 2.27)	I change (−1.2→ 1.2)

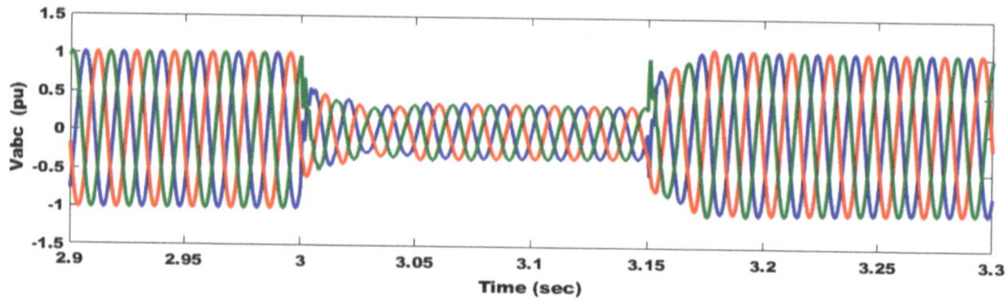

(a)

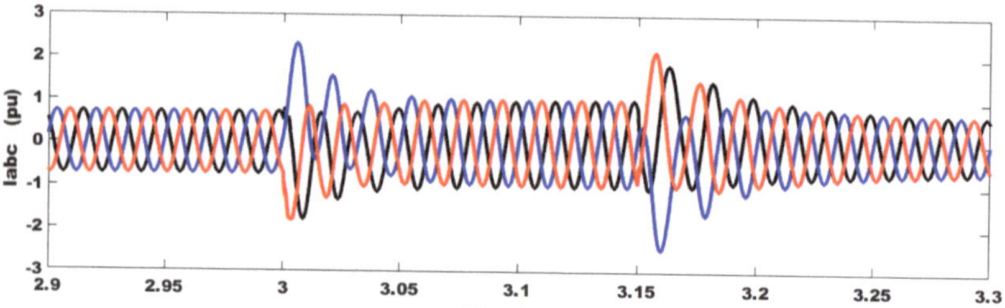

(b)

Figure 13. *Cont.*

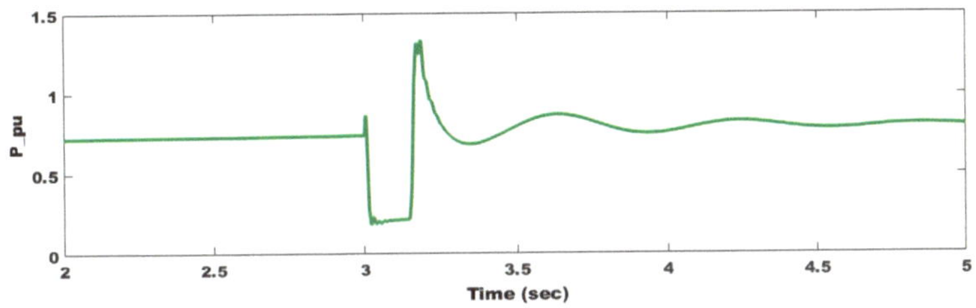

(c)

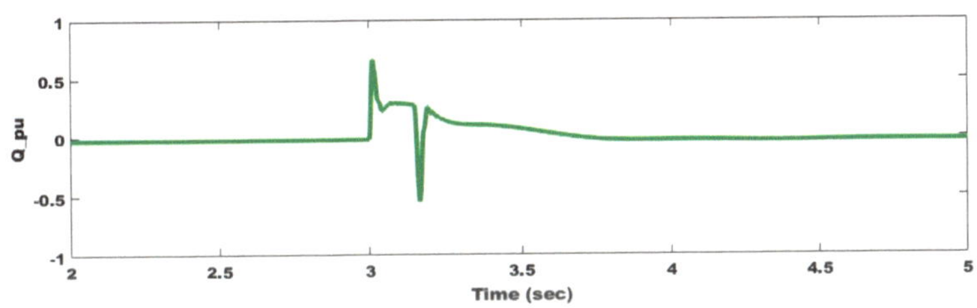

(d)

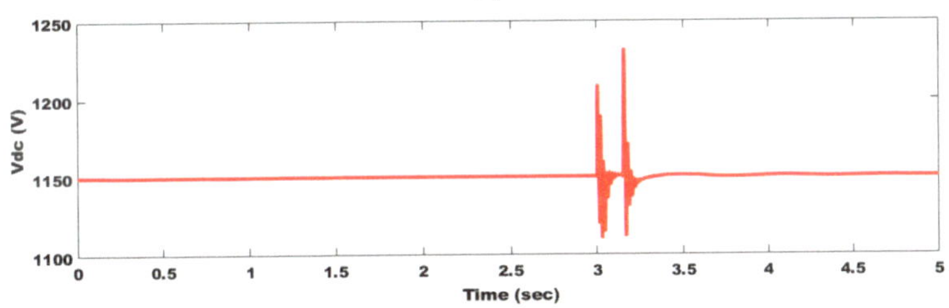

(e)

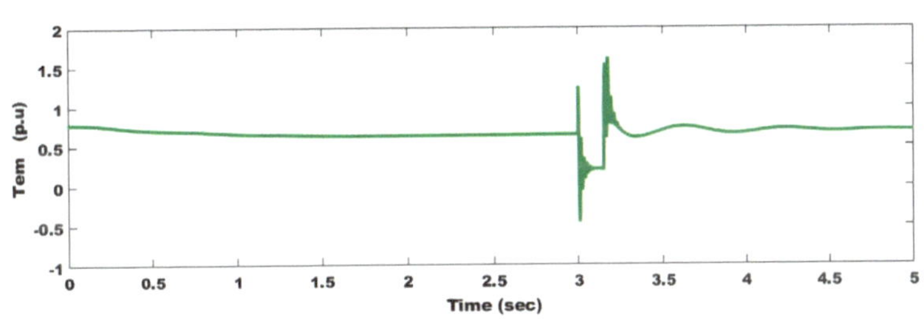

(f)

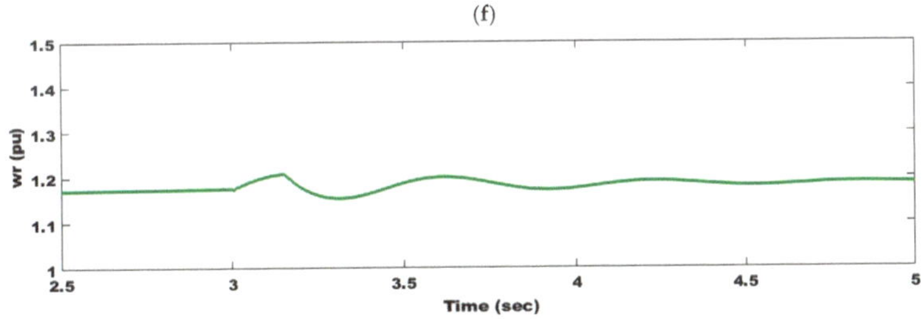

(g)

Figure 13. DFIWG system parameter responses as a result of 85% voltage dip: (**a**) system voltage; (**b**) system current; (**c**) supplied active power to the grid; (**d**) supplied reactive power to the grid; (**e**) DC-link voltage; (**f**) electromagnetic torque; and (**g**) angular speed.

6. Conclusions

We investigated the dynamic performances of the DFIWG and PMSWG with MRFO support during wind speed fluctuations (8–15 ↑↓) and an 85% grid voltage decrease while taking advantage of MPPT, PAC, and FRT capability. MRFO-based wind side controllers and a chopper controller was designed and implemented for the optimum performance of both the PMSWG and DFIWG. The DFIWG's performance during regular grid and transient operation was significantly enhanced and the V_{DC} was maintained below the permitted limits when using optimized controllers. Simulated results of the DFIWG's system parameters showed that the system successfully operated at MPPT and PAC regions and realized an enhanced FRT capability. When the PMSWG operated with optimized controllers, its performance during normal grid and transient operation was greatly improved and the V_{DC} was kept below its allowable limits. Simulated results of the PMSWG's system parameters showed that the system successfully operated at MPPT and PAC regions and realized an enhanced FRT capability. With the proposed control schemes, the obtained results indicate that:

- Both WGs are able to function in the PAC zone, have FRT capabilities, and have optimized controllers, all of which have a significant impact on how well they perform in the instances under study.
- The FRT issues may be made easier with an appropriate choice of controller gains based on the WT design. Compared with the majority of current FRT methods for WTs, this may be a more affordable method without taking external circuitry into account.
- Blocking of converters for the DFIWG was eliminated with the proposed technique, which is the main problem for DFIWGs.
- The change in the parameters of the studied wind systems was evident due to the violent change in wind speed and three-phase fault. The change was smaller in the PMSWG in the case of wind speed because it contained more poles; the change was smaller in the case of the fault due to the direct connection of the DFIWG to the network. Table 11 summarizes all the events and changes in parameters.
- The simulation results showed that the PMSWG was able to track the reference wind speed faster than the DFIWG, where the settling time for C_P was found to be 0.784 s with the PMSWG compared with 1.248 s with the DFIWG.
- The results showed that, with the prosed schemes, the V_{DC} was below limits (1.02 under regular conditions and below 1.1 pu under faults).
- A small oscillation in the PMSWG, compared with the DFIWG, reveals that it has more power-smoothing capability.
- The simulation results showed the superiority of the PMSWG over the DFIWG, especially in the event of large disturbances due to the latter's direct connection to the grid.

Author Contributions: Conceptualization, M.M.M.; methodology, M.M.M. and B.S.A.; software, M.M.M. and B.S.A.; validation, A.Y.A. and M.M.M.; formal analysis, B.S.A.; investigation, M.M.M. and N.A.N.A.; writing—original draft preparation, M.M.M.; writing—review and editing, M.M.M. and B.S.A.; visualization, N.A.N.A. and M.M.M.; supervision, A.Y.A. and M.M.M. All authors have read and agreed to the published version of the manuscript.

Funding: This research received no external funding.

Data Availability Statement: Not applicable.

Conflicts of Interest: The authors declare no conflict of interest.

List of Abbreviations

EPSs: Electrical power systems
RESs: Renewable energy sources
WE: Wind energy
VSWGs: Variable-speed wind generators
IEA: International energy agency
PWM: Pulse-width modulation
WTs: Wind turbines
FRT: Fault ride-through

PMSWG: Permanent magnet synchronous wind generator
DFIWG: Doubly-fed induction wind generator
PIC: PI controller
MPPT: Maximum power point tracking
OTC: Optimal torque control
Q: Reactive power
MRFO: Manta ray foraging optimizer
FLC: Fuzzy logic control
CC: Control cost
FACTS: Flexible AC transmission systems
ANFIS: Adaptive neuro-fuzzy inference system
POCC: Point of common coupling
INC: Incremental conductance
dq: Direct quadrature
BTB: Back-to-back
PAC: Pitch angle control
RSC: Rotor side converter
WF: Wind farm
P: Active power
FCLs: Fault current limiters
SMC: Sliding mode control
MRs: Manta rays
PSO: Particle swarm optimization
P&O: Perturb and observe
GSS: Golden section search
MPC: Model predictive controller

Appendix A

Influence of inertia on electromagnetic torque [51].

$$T_e \pm T_m = f\omega_m + J\omega^{\cdot}{}_m \tag{A1}$$

(−) and (+) signs represent acceleration and deceleration modes, respectively.
Acceleration mode ($\Delta\omega/\Delta t > 0$)

$$T_e - T_m = f\omega_m + J\omega^{\cdot}{}_m \tag{A2}$$

In step change, $\Delta t \rightarrow 0$ is a very small value, so that

$$T_e \uparrow\uparrow \alpha J \uparrow \frac{d\omega_m}{dt \downarrow\downarrow} \alpha J \frac{\Delta\omega_m}{\Delta t} \tag{A3}$$

Deceleration mode ($\Delta\omega/\Delta t < 0$)

$$T_e + T_m = f\omega_m + J\omega^{\cdot}{}_m \tag{A4}$$

Moreover,

$$T_e \downarrow\downarrow \alpha J \uparrow \frac{d\omega_m}{dt \downarrow\downarrow} \alpha J \frac{\Delta\omega_m}{\Delta t} \tag{A5}$$

where $\Delta\omega = \omega_{new} - \omega_{old}$

References

1. Ahmadi, A.; Nezhad, A.E.; Hredzak, B. Security-Constrained Unit Commitment in Presence of Lithium-Ion Battery Storage Units Using Information-Gap Decision Theory. *IEEE Trans. Ind. Inform.* **2019**, *15*, 148–157. [CrossRef]
2. Chen, C.; Li, C.; Reniers, G.; Yang, F. Safety and security of oil and gas pipeline transportation: A systematic analysis of research trends and future needs using WoS. *J. Clean. Prod.* **2021**, *279*, 123583. [CrossRef]
3. Groissböck, M.; Gusmão, A. Impact of renewable resource quality on security of supply with high shares of renewable energies. *Appl. Energy* **2020**, *277*, 115567. [CrossRef]
4. Radzka, E.; Rymuza, K.; Michalak, A. Wind power as a renewable energy source. *J. Ecol. Eng.* **2019**, *20*, 167–171. [CrossRef]
5. ANEEL. Annual Wind Energy Report 2020. 2020. Available online: http://abeeolica.org.br/wp-content/uploads/2021/06/PT_Boletim-Anual-de-Geracao_2020.pdf (accessed on 1 January 2020).
6. Jensen, C.U.; Panduro, T.E.; Lundhede, T.H.; Nielsen, A.S.E.; Dalsgaard, M.; Thorsen, B.J. The impact of on-shore and off-shore wind turbine farms on property prices. *Energy Policy* **2018**, *116*, 50–59. [CrossRef]
7. Pilpola, S.; Lund, P.D. Different flexibility options for better system integration of wind power. *Energy Strateg. Rev.* **2019**, *26*, 100368. [CrossRef]
8. Holttinen, H. Wind integration: Experience, issues, and challenges. *Wiley Interdiscip. Rev. Energy Environ.* **2012**, *1*, 243–255. [CrossRef]
9. Ahmed, S.D.; Al-Ismail, F.S.M.; Shafiullah, M.; Al-Sulaiman, F.A.; El-Amin, I.M. Grid Integration Challenges of Wind Energy: A Review. *IEEE Access* **2020**, *8*, 10857–10878. [CrossRef]
10. Mahmoud, M.M.; Aly, M.M.; Salama, H.S.; Abdel-Rahim, A.M.M. Dynamic evaluation of optimization techniques–based proportional–integral controller for wind-driven permanent magnet synchronous generator. *Wind Eng.* **2021**, *45*, 696–709. [CrossRef]

11. Xing, F.; Xu, Z.; Zhang, Z.; Dan, Y.; Zhu, Y. Resonance stability analysis of large-scale wind power bases with type-IV wind generators. *Energies* **2020**, *13*, 5220. [CrossRef]
12. Majout, B.; El Alami, H.; Salime, H.; Laabidine, N.Z.; El Mourabit, Y.; Motahhir, S.; Bouderbala, M.; Karim, M.; Bossoufi, B. A Review on Popular Control Applications in Wind Energy Generator PMSG. *Energy* **2022**, *15*, 6238.
13. Shehata, E.G. A comparative study of current control schemes for a direct-driven PMSG wind energy generation system. *Electr. Power Syst. Res.* **2017**, *143*, 197–205. [CrossRef]
14. Xu, Y.; An, L.T.; Jia, X.P.; Jia, B.Z.; Maki, N. Optimization study of the main parameters of different types of wind turbine generators. *Supercond. Sci. Technol.* **2022**, *35*, 035007. [CrossRef]
15. Kumar, D.; Chatterjee, K. A review of conventional and advanced MPPT algorithms for wind energy systems. *Renew. Sustain. Energy Rev.* **2016**, *55*, 957–970. [CrossRef]
16. Buduma, P.; Vulisi, N.K.; Panda, G. Robust Control and Kalman MPPT for Grid-Assimilated Wind Energy Conversion System. *IEEE Trans. Ind. Appl.* **2021**, *57*, 1274–1284. [CrossRef]
17. Xu, S.; Shao, R.; Cao, B.; Chang, L. Single-phase grid-connected PV system with golden section search-based MPPT algorithm. *Chin. J. Electr. Eng.* **2021**, *7*, 25–36. [CrossRef]
18. Bayrak, G.; Ghaderi, D. An improved step-up converter with a developed real-time fuzzy-based MPPT controller for PV-based residential applications. *Int. Trans. Electr. Energy Syst.* **2019**, *29*, 12140. [CrossRef]
19. Priyadarshi, N.; Padmanaban, S.; Holm-Nielsen, J.B.; Blaabjerg, F.; Bhaskar, M.S. An Experimental Estimation of Hybrid ANFIS-PSO-Based MPPT for PV Grid Integration under Fluctuating Sun Irradiance. *IEEE Syst. J.* **2020**, *14*, 1218–1229. [CrossRef]
20. Dong, Z.; Li, Z.; Dong, Y.; Jiang, S.; Ding, Z. Fully-distributed deloading operation of DFIG-based wind farm for load sharing. *IEEE Trans. Sustain. Energy* **2021**, *12*, 430–440. [CrossRef]
21. Okedu, K.E.; Barghash, H.F.A. Investigating Variable Speed Wind Turbine Transient Performance Considering Different Inverter Schemes and SDBR. *Front. Energy Res.* **2021**, *8*, 604338. [CrossRef]
22. Marques, G.; Silva, F.A. Doubly Fed Induction Machine ? Modeling and Control for Wind Energy Generation [Book News]. *IEEE Ind. Electron. Mag.* **2015**, *9*, 54–55. [CrossRef]
23. Mahmoud, M.M.; Aly, M.M.; Abdel-Rahim, A.M.M. Enhancing the dynamic performance of a wind-driven PMSG implementing different optimization techniques. *SN Appl. Sci.* **2020**, *2*, 684. [CrossRef]
24. Mahmoud, M.M.; Ratib, M.K.; Aly, M.M.; Moamen, A.; Rahim, M.A. Effect of Grid Faults on Dominant Wind Generators for Electric Power System Integration: A Comparison and Assessment. *Energy Syst. Res.* **2021**, *4*, 70–78.
25. Mahela, O.P.; Gupta, N.; Khosravy, M.; Patel, N. Comprehensive overview of low voltage ride through methods of grid integrated wind generator. *IEEE Access* **2019**, *7*, 99299–99326. [CrossRef]
26. Mahmoud, M.M.; Hemeida, A.M.; Senjy, T.; Ewais, A.M. Fault Ride-Through Capability Enhancement for Grid-Connected Permanent Magnet Synchronous Generator Driven by Wind Turbines. In Proceedings of the IEEE Conference on Power Electronics and Renewable Energy, CPERE 2019, Aswan City, Egypt, 23–25 October 2019; pp. 567–572.
27. Swain, S.; Ray, P.K. Short circuit fault analysis in a grid connected DFIG based wind energy system with active crowbar protection circuit for ridethrough capability and power quality improvement. *Int. J. Electr. Power Energy Syst.* **2017**, *84*, 64–75. [CrossRef]
28. Mossa, M.A.; Abdelhamid, M.K.; Hassan, A.A.; Bianchi, N. Improving the Dynamic Performance of a Variable Speed DFIG for Energy Conversion Purposes Using an Effective Control System. *Processes* **2022**, *10*, 456. [CrossRef]
29. El Moursi, M.S.; Zeineldin, H.H. A parallel capacitor control strategy for enhanced FRT capability of DFIG. *IEEE Trans. Sustain. Energy* **2014**, *6*, 303–312.
30. Hiremath, R.; Moger, T. Comprehensive review on low voltage ride through capability of wind turbine generators. *Int. Trans. Electr. Energy Syst.* **2020**, *30*, 12524. [CrossRef]
31. Mahmoud, M.M.; Mohamed Hemeida, A.; Abdel-Rahim, A.M.M. Behavior of PMSG Wind Turbines with Active Crowbar Protection under Faults. In Proceedings of the 2019 Innovations in Power and Advanced Computing Technologies, i-PACT 2019, Vellore, India, 22–23 March 2019; pp. 1–6.
32. Mohammadi, J.; Afsharnia, S.; Vaez-Zadeh, S.; Farhangi, S. Improved fault ride through strategy for doubly fed induction generator based wind turbines under both symmetrical and asymmetrical grid faults. *IET Renew. Power Gener.* **2016**, *10*, 1114–1122. [CrossRef]
33. Mahmoud, M.M.; Aly, M.M.; Salama, H.S.; Abdel-Rahim, A.M.M. An internal parallel capacitor control strategy for DC-link voltage stabilization of PMSG-based wind turbine under various fault conditions. *Wind Eng.* **2021**, *46*, 983–992. [CrossRef]
34. Tohidi, S.; Behnam, M.I. A comprehensive review of low voltage ride through of doubly fed induction wind generators. *Renew. Sustain. Energy Rev.* **2016**, *57*, 412–419. [CrossRef]
35. Angala Parameswari, G.; Habeebullah Sait, H. A comprehensive review of fault ride-through capability of wind turbines with grid-connected doubly fed induction generator. *Int. Trans. Electr. Energy Syst.* **2020**, *30*, 12395. [CrossRef]
36. Zheng, Z.; Xie, Q.; Xiao, X.; Zhou, Y. Review of SMES and SFCL Based FRT Strategy for PMSG-Based Wind Turbine Generator. In Proceedings of the 2020 IEEE International Conference on Applied Superconductivity and Electromagnetic Devices, ASEMD 2020, Tianjin, China, 16–18 October 2020.
37. Mahmoud, M.M.; Esmail, Y.M.; Atia, B.S.; Kamel, O.M.; Aboras, K.M.; Bajaj, M.; Sabir, S.; Bukhari, H.; Eutyche, D.; Wapet, M. Voltage Quality Enhancement of Low-Voltage Smart Distribution System Using Robust and Optimized DVR Controllers: Application of the Harris Hawks Algorithm. *Int. Trans. Electr. Energy Syst.* **2022**, *2022*, 18. [CrossRef]

38. Tourou, P.; Sourkounis, C. Review of control strategies for DFIG-based wind turbines under unsymmetrical grid faults. In Proceedings of the 2014 9th International Conference on Ecological Vehicles and Renewable Energies, EVER 2014, Monte-Carlo, Monaco, 25–27 March 2014.
39. Adetokun, B.B.; Muriithi, C.M. Impact of integrating large-scale DFIG-based wind energy conversion system on the voltage stability of weak national grids: A case study of the Nigerian power grid. *Energy Rep.* **2021**, *7*, 654–666. [CrossRef]
40. Nasiri, M.; Mohammadi, R. Peak Current Limitation for Grid Side Inverter by Limited Active Power in PMSG-Based Wind Turbines During Different Grid Faults. *IEEE Trans. Sustain. Energy* **2017**, *8*, 3–12. [CrossRef]
41. Gencer, A. Analysis and control of fault ride through capability improvement pmsg based on WECS using active crowbar system during different fault conditions. *Elektron. Ir Elektrotechnika* **2018**, *24*, 63–69. [CrossRef]
42. Moghimian, M.M.; Radmehr, M.; Firouzi, M. Series Resonance Fault Current Limiter (SRFCL) with MOV for LVRT Enhancement in DFIG-Based Wind Farms. *Electr. Power Compon. Syst.* **2019**, *47*, 1814–1825. [CrossRef]
43. Firouzi, M.; Nasiri, M.; Benbouzid, M.; Gharehpetian, G.B. Application of multi-step bridge-type fault current limiter for fault ride-through capability enhancement of permanent magnet synchronous generator-based wind turbines. *Int. Trans. Electr. Energy Syst.* **2020**, *30*, 12611. [CrossRef]
44. Islam, M.R.; Hasan, J.; Shipon, M.R.R.; Sadi, M.A.H.; Abuhussein, A.; Roy, T.K. Neuro Fuzzy Logic Controlled Parallel Resonance Type Fault Current Limiter to Improve the Fault Ride through Capability of DFIG Based Wind Farm. *IEEE Access* **2020**, *8*, 115314–115334. [CrossRef]
45. Sadi, M.A.H.; Abuhussein, A.; Shoeb, M.A. Transient Performance Improvement of Power Systems Using Fuzzy Logic Controlled Capacitive-Bridge Type Fault Current Limiter. *IEEE Trans. Power Syst.* **2021**, *36*, 323–335. [CrossRef]
46. Shafiee, M.R.; Kartijkolaie, H.S.; Firouzi, M.; Mobayen, S.; Fekih, A. A Dynamic Multi-Cell FCL to Improve the Fault Ride through Capability of DFIG-Based Wind Farms. *Energies* **2020**, *13*, 6071. [CrossRef]
47. Firouzi, M.; Nasiri, M.; Mobayen, S.; Gharehpetian, G.B. Sliding Mode Controller-Based BFCL for Fault Ride-Through Performance Enhancement of DFIG-Based Wind Turbines. *Complexity* **2020**, *2020*, 1259539. [CrossRef]
48. Qais, M.H.; Hasanien, H.M.; Alghuwainem, S. A Grey Wolf Optimizer for Optimum Parameters of Multiple PI Controllers of a Grid-Connected PMSG Driven by Variable Speed Wind Turbine. *IEEE Access* **2018**, *6*, 44120–44128. [CrossRef]
49. Ghany, A.A.; Shehata, E.G.; Elsayed, A.H.M.; Mohamed, Y.S.; Alhelou, H.H.; Siano, P.; Diab, A.A.Z. Novel switching frequency fcs-mpc of pmsg for grid-connected wind energy conversion system with coordinated low voltage ride through. *Electron.* **2021**, *10*, 492. [CrossRef]
50. Majout, B.; Bossoufi, B.; Bouderbala, M.; Masud, M.; Al-Amri, J.F.; Taoussi, M.; El Mahfoud, M.; Motahhir, S.; Karim, M. Improvement of PMSG-Based Wind Energy Conversion System Using Developed Sliding Mode Control. *Energies* **2022**, *15*, 1625. [CrossRef]
51. Mahmoud, M.M.; Aly, M.M.; Salama, H.S.; Abdel-Rahim, A.M.M. A combination of an OTC based MPPT and fuzzy logic current control for a wind-driven PMSG under variability of wind speed. *Energy Syst.* **2022**, *13*, 1075–1098. [CrossRef]
52. Mahmoud, M.M. Improved current control loops in wind side converter with the support of wild horse optimizer for enhancing the dynamic performance of PMSG-based wind generation system. *Int. J. Model. Simul.* **2022**, 1–15. [CrossRef]
53. Metwally Mahmoud, M.; Salama, H.S.; Aly, M.M.; Abdel-Rahim, A.M.M. Design and implementation of FLC system for fault ride-through capability enhancement in PMSG-wind systems. *Wind Eng.* **2021**, *45*, 1361–1373. [CrossRef]
54. Zhu, D.; Zou, X.; Zhou, S.; Dong, W.; Kang, Y.; Hu, J. Feedforward Current References Control for DFIG-Based Wind Turbine to Improve Transient Control Performance During Grid Faults. *IEEE Trans. Energy Convers.* **2018**, *33*, 670–681. [CrossRef]
55. Mohammadi, J.; Vaez-Zadeh, S.; Ebrahimzadeh, E.; Blaabjerg, F. Combined control method for grid-side converter of doubly fed induction generatorbased wind energy conversion systems. *IET Renew. Power Gener.* **2018**, *12*, 943–952. [CrossRef]
56. Nair, A.R.; Bhattarai, R.; Smith, M.; Kamalasadan, S. Parametrically Robust Identification Based Sensorless Control Approach for Doubly Fed Induction Generator. *IEEE Trans. Ind. Appl.* **2021**, *57*, 1024–1034. [CrossRef]
57. Wadawa, B.; Errami, Y.; Obbadi, A.; Sahnoun, S. Robustification of the H∞ controller combined with fuzzy logic and PI&PID-Fd for hybrid control of Wind Energy Conversion System Connected to the Power Grid Based on DFIG. *Energy Rep.* **2021**, *7*, 7539–7571. [CrossRef]
58. Chojaa, H.; Derouich, A.; Taoussi, M.; Chehaidia, S.E.; Zamzoum, O.; Mosaad, M.I.; Alhejji, A.; Yessef, M. Nonlinear Control Strategies for Enhancing the Performance of DFIG-Based WECS under a Real Wind Profile. *Energies* **2022**, *15*, 6650. [CrossRef]
59. Shiravani, F.; Cortajarena, J.A.; Alkorta, P.; Barambones, O. Generalized Predictive Control Scheme for a Wind Turbine System. *Sustainability* **2022**, *14*, 8865. [CrossRef]
60. Krause, P.C.; Wasynczuk, O.; Sudhoff, S.D. *Analysis of Electric Machinery and Drive Systems*; John Wiley & Sons: New York, NY, USA, 2002. [CrossRef]
61. Nkosi, N.R.; Bansal, R.C.; Adefarati, T.; Naidoo, R.M.; Bansal, S.K. A review of small-signal stability analysis of DFIG-based wind power system. *Int. J. Model. Simul.* **2022**, 1–18. [CrossRef]
62. IRENA. Renewable Capacity Statistics 2020. 2020. Available online: https://www.irena.org/-/media/Files/IRENA/Agency/Publication/2020/Mar/IRENA_RE_Capacity_Statistics_2020.pdf (accessed on 1 January 2020).
63. Chao, H.; Hu, B.; Xie, K.; Tai, H.M.; Yan, J.; Li, Y. A Sequential MCMC Model for Reliability Evaluation of Offshore Wind Farms Considering Severe Weather Conditions. *IEEE Access* **2019**, *7*, 132552–132562. [CrossRef]

64. Qiu, Y.; Feng, Y.; Sun, J.; Zhang, W.; Infield, D. Applying thermophysics for wind turbine drivetrain fault diagnosis using SCADA data. *IET Renew. Power Gener.* **2016**, *10*, 661–668. [CrossRef]
65. Xie, S.; Wang, X.; Qu, C.; Wang, X.; Guo, J. Impacts of different wind speed simulation methods on conditional reliability indices. *Int. Trans. Electr. Energy Syst.* **2013**, *20*, 1–6. [CrossRef]
66. Li, H.; Chen, Z. Overview of different wind generator systems and their comparisons. *IET Renew. Power Gener.* **2008**, *2*, 123–138. [CrossRef]
67. Kahraman, H.T.; Akbel, M.; Duman, S. Optimization of optimal power flow problem using multi-objective manta ray foraging optimizer. *Appl. Soft Comput.* **2022**, *116*, 108334. [CrossRef]
68. Selem, S.I.; Hasanien, H.M.; El-Fergany, A.A. Parameters extraction of PEMFC's model using manta rays foraging optimizer. *Int. J. Energy Res.* **2020**, *44*, 4629–4640. [CrossRef]
69. Micev, M.; Ćalasan, M.; Ali, Z.M.; Hasanien, H.M.; Abdel Aleem, S.H.E. Optimal design of automatic voltage regulation controller using hybrid simulated annealing—Manta ray foraging optimization algorithm. *Ain Shams Eng. J.* **2021**, *12*, 641–657. [CrossRef]
70. Yuan, L.; Meng, K.; Huang, J.; Yang Dong, Z.; Zhang, W.; Xie, X. Development of HVRT and LVRT control strategy for pmsg-based wind turbine generators y. *Energies* **2020**, *13*, 5442. [CrossRef]

processes

MDPI

Article

A Different Approach to Develop a District Heating Grid Based on the Optimization of Building Clusters

Laura Pompei [1,*], Jacob Mannhardt [2], Fabio Nardecchia [1], Lorenzo M. Pastore [1] and Livio de Santoli [1]

1 Department of Astronautical Electrical and Energy Engineering (DIAEE), Sapienza University of Rome, Via Eudossiana 18, 00184 Rome, Italy
2 RWTH Aachen University, 52062 Aachen, Germany
* Correspondence: laura.pompei@uniroma1.it

Abstract: Currently, reducing energy consumption and fossil fuel emissions are key factors placed in the first position on the European agenda. District heating technology is an attractive solution, able to satisfy the energy and environmental goals of policymakers and designers. In line with this, a different approach to planning a district heating grid based on the optimization of building clusters is presented. The case study is Wilhelmsburg, a district of Hamburg city. This approach also investigates the usage of industrial waste heat as the grid's heat source, which is CO2-neutral. First, the data acquisition regarding the buildings' location and heat demand are described in detail. Based on the derived data and the source of the industrial waste heat, the district heating grid is created by clustering the buildings and connecting the obtained nodes. Furthermore, the grid's efficiency is improved by eliminating nodes, which are too distant from the heat source, or have lower heat demand. Finally, a single building is simulated in Matlab/Simulink, showing the energy-savings and ecological results. The usage of the district heating grid saves 97.32 GWh annually, which results in financial savings of €5.83 million, and avoided CO2 emissions of 19,585 *tCO2*.

Keywords: district heating; industrial waste heat; optimization; building clustering; GIS

Citation: Pompei, L.; Mannhardt, J.; Nardecchia, F.; Pastore, L.M.; de Santoli, L. A Different Approach to Develop a District Heating Grid Based on the Optimization of Building Clusters. *Processes* **2022**, *10*, 1575. https://doi.org/10.3390/pr10081575

Academic Editor: Matti Lehtonen

Received: 18 July 2022
Accepted: 7 August 2022
Published: 11 August 2022

1. Introduction

The building sector is one of the largest energy consumers in Europe today, therefore the identification of appropriate measures to reach the European reduction target is a current challenge for policymakers and designers [1]. As a result, many interventions and measures are now forecasted using planning tools and methods [2,3]. Local energy planning has therefore gained popularity in recent years, as geographic information system (GIS) applications spread and computing power has increased [4,5]. Recently, there have been numerous strategies proposed for conserving energy in the building sector, including thermal insulation, double and triple glazing, solar shadings [6,7], the efficient usage of HVAC equipment [8], hybrid energy system [9] and using renewable energy sources [10,11]. As well as these technologies, district heating systems are another viable solution for improving energy efficiency [12] and sustainable assessment [13] in buildings. Consequently, this technology is ideal for urban areas with high thermal demand densities. Heat sources, users, and distribution networks constitute the general components of DHSs. The complexity of a DHS depends on various factors [14,15]. Starting from the beginning of this technology, it was not easy to obtain proper and coherent results in line with the real problems of a district network [14].

Many advantages of this energy system are well-known, starting from the renewable energy source inclusion [16] to positive environmental impacts, also for improving the outdoor comfort for citizens [17,18]. Due to its relevant benefits, different works were developed during the last years to improve its capabilities, such as optimizing the operation of the thermal plants [19] and the supply temperature [20] or the settings of the pumping system [21]. Those approaches are often focused on specific components or issue of the DH, missing a global approach to plan the entire network. Therefore, there is still an urgent call to develop and investigated affordable approach to design the district heating in different urban areas [22–24]. The two most famous planning methods available in literature are the Danish method [25] and the German method [26]. In both methods, smaller branches are merged into bigger branches and the heat loads of the removed branches are incorporated into the preserved branches. Among this, Guelpa et al. [27] proposed a method to increase the performance of district heating minimizing the thermal peaks. A clustering approach is also involved to optimize the pipe grid. Results demonstrate an average reduction of the thermal peak load up to 14% thanks to this optimization. Another work [28] is focused on optimizing the multi-source energy plant sizing using different climatic scenarios, using Matlab for modelling the environment understudy. As well, Widen and Aberg [29] developed a fixed model structure (FMS) that requires only general information about DH systems for cost-optimization studies. The results demonstrate the usefulness of the FMS for DH system optimisation studies, and that increased building energy efficiency leads to a reduction in fossil fuel and biomass consumption.

A few papers have been published on multi-source production plants or distribution hubs and how to expand or redesign a distribution hub to connect new consumers to it. Among this, biomass-based district-heating networks, which represents the basis of Italian industry's successful production, are designed using a system optimisation approach [30]. Instead, Burer et al. [31] proposed an optimized district heating and cooling plant providing heat and power to a small number of residential buildings, to minimize cost and CO2 emissions. According to Corrado et al. [32], heat pumps, wood boilers, condensing boilers, and solar energy (thermal and photovoltaic) can be combined to generate energy. Cogeneration with solar energy and wind energy was combined by Sontag et al. [33]. An interesting study developed by Trillat et al. [34], integrated CHP with an absorption chiller and desiccant cooling; while Lee et al. [35] studied the integration of solar water heating, solar photovoltaic, ground source heat pumps, electric chillers, and gas boilers into an integrated renewable energy system.

In this framework, this study aims to fulfill this field of research proposing a different approach to connect existing district heating with new one, optimizing the building clustering. In fact, few papers explore the issue of expanding or redesigning a DHS to connect new consumers to multi-source production plants. The developed method is applied to a district of Hamburg, named Wilhelmsburg. Industrial waste heat is used as heat source, which is CO2-neutral. Based on the data acquisition, the district heating grid is created by clustering the buildings and connecting the obtained nodes, also using GIS technology. The grid of the network is improved, cleaning and reducing the branches and nodes wherein the heat demand is not so relevant. To show the advantages of this approach, a single building is simulated with Matlab/Simulink. Results underlined the financial and ecological benefits not only for the building case study, but for the entire district heating network.

2. Materials and Methods

2.1. Building Location and Heat Demand Data

Firstly, the data acquisition and allocation regarding the buildings' location and heat demand are presented in detail. It is necessary to acquire detailed information on the buildings' location, area, and heat demand in the district, to model a realistic heating grid. The averaged and assumed building properties are not sufficient, since the location and heat demand determine the characteristics of the heating grid. Furthermore, yearly averaged heat demand data cannot take the essential seasonal differences in the heating behavior into account. The buildings' location and area are acquired from the open-source project OpenStreetMap [36], where a polygon spanning the district, or an arbitrary area can be selected and exported. The data file contains information on buildings, land use, natural objects, places, points, railways, roads, and waterways in the selected area. The land use highlights the usage of the area within the polygon. This land use is of varying types, e.g., residential, commercial, but also grass or farmland. For this project, only the buildings and the land use are considered. The data is imported into the software QGIS, where the geoinformation is stored and viewed, and new variables are calculated.

Table 1 shows exemplary fictional building data, here for a house and a commercial building. Each building has a unique ID, a type, an area, and a longitudinal and latitudinal coordinate. Here, the variable x_value denotes the longitude of the building's center, and y_value the latitude. In Wilhelmsburg, the longitude ranges from x_{min} = 9.9707° to x_{max} = 10.0610°, and the latitude ranges from y_{min} = 53.4615° to y_{max} = 53.5239°. The type of a building is often not declared, and hence the land use is considered. For that, the buildings' information is intersected with the land use data and hence each building is assigned its land use type, which is used whenever the building's type is unknown. Subsequently, the types are matched to their corresponding categories, i.e., the category accommodation contains, amongst others, the types of houses, apartments, and residential. Thereby, the buildings can be summarized more broadly. Figure 1 shows real exemplary building data from Wilhelmsburg. Here, accommodational buildings are indicated in red and commercial buildings in blue. The extracted data contains the information on 6.129 buildings in Wilhelmsburg. In this manuscript, only accommodation and commercial buildings are considered, because the other categories, such as civic or religious, are miscellaneous and can hence not be easily considered regarding their heat demand.

Table 1. Exemplary, fictional building data extracted in QGIS.

Id [-]	Type [-]	Area [m^2]	X_Value [°]	Y_Value [°]
1	house	100.123	10.0100	53.4900
2	commercial	200.456	10.0200	53.5100

Figure 1. Exemplary building data for an apartment (red) and a commercial (blue) building.

The heat demand for the accommodational and commercial buildings is obtained from the Wärmekataster Hamburg (heat registry Hamburg) [37]. The registry provides the heat demand and geoinformation of accommodational and commercial buildings in Hamburg.

For data security reasons, the buildings are allocated in clusters of minimum 5 buildings. Hence, the heat demand cannot be matched to a single, unique building, though applying the same intersection as used for the land use matches the heat demand to each building. Some buildings are not listed in the registry and are thus outside any cluster, so the broader block of buildings is used. The available information contains:

- The specific annual heat demand for accommodational buildings in the cluster [kWh/(y m^2)]
- The absolute annual heat demand for all buildings in the cluster [kWh/y]
- The area of accommodational buildings in the cluster [m^2]
- The area of commercial buildings in the cluster [m^2]

To calculate the absolute heat demand of each building, the specific heat demand for accommodational and commercial buildings is needed. The specific heat demand for accommodational buildings $q_{spec,acc}$ is already listed in the registry's data, whereas the specific heat for commercial buildings $q_{spec,com}$ can be calculated using $q_{spec,acc}$, the total absolute heat demand $q_{abs,tot}$, the area of accommodational buildings A_{acc} and the area of commercial buildings A_{com}:

$$q_{abs,tot} = q_{spec,acc} A_{acc} + q_{spec,com} A_{com} \rightarrow q_{spec,com} = \frac{1}{A_{com}} \left(q_{abs,tot} - q_{spec,acc} A_{acc} \right) \tag{1}$$

If the specific heat demand cannot be retrieved for a building, the heat demand of the building from the same category with the closest area is used. Since the information on the heat demand is only available for the building categories accommodation and commercial, the number of buildings is reduced to those buildings that can be unambiguously matched to these categories. Furthermore, outliers are ignored by allowing only those buildings with an area smaller than 3000 m^2 and a specific annual heat demand lower than 250 kWh/(a m^2). Thereby, the number of considered buildings is reduced to 4884 buildings, which will be considered in the district heating grid.

So far, only the information on the annual heat demand is available, but for a detailed analysis and the subsequent dimensioning of the heating grid, a more detailed allocation is desired. The BDEW [38] propose generic heat load profiles as a function of the ambient temperature $\vartheta = T_{amb}$. Each building type is matched to a sigmoid equation $h = f(\vartheta)$ with the building type specific parameters A, B, C, D and ϑ_0:

$$h = \frac{A}{1 + \left(\frac{B}{\vartheta - \vartheta_0} \right)^C} + D \tag{2}$$

The parameters for the building types of single-family household, multi-family household, commercial, and industrial are listed in Table 2, whose evolution over the ambient temperature are presented in Figure 2. Note that the accommodational buildings have a flatter curve than the commercial buildings and are hence not as sensitive to a change in temperature. It is assumed that the h-values of accommodational buildings do not depend on the day of the week, whereas the commercial buildings have a higher heat demand on weekdays and a lower demand on weekends. The correction factors F_{WD} for the h-values of commercial buildings are listed in Table 3. Since the h-values of accommodational buildings are independent of the day, $F_{WD} = 1$ is set for each day and building type [38].

Table 2. Parameters for sigmoid equations for different building types [38].

Building Type	A	B	C	D	ϑ_0
Single-family household	3.0722215	−37.1842844	5.6975234	0.0904188	40
Multi-family household	2.4207684	−34.7277917	5.7668252	0.1082275	40
Commercial	3.5811214	−36.9650065	7.2256947	0.0448416	40
Industrial	3.7882424	−34.8806130	6.5951899	0.0540329	40

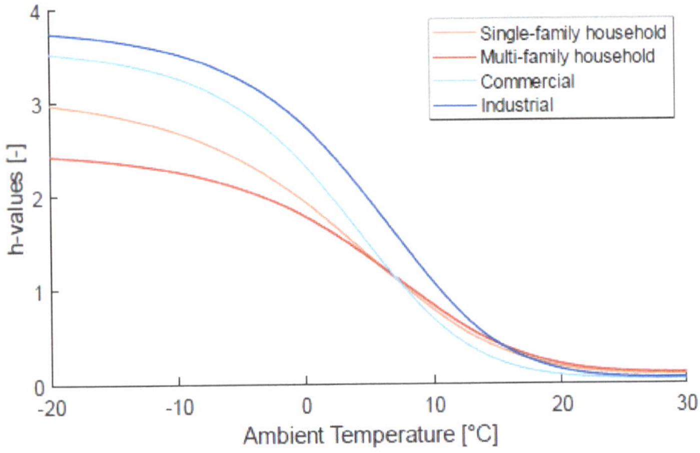

Figure 2. Sigmoid equations for different building types over ambient temperature.

Table 3. Daily correction factors F_{WD} for h-values of commercial buildings [38].

Building Type	F_{Mon}	F_{Tue}	F_{Wed}	F_{Thu}	F_{Fri}	F_{Sat}	F_{Sun}
Commercial	1.0358	1.0232	1.0252	1.0295	1.0253	0.9675	0.8935
Industrial	1.0699	1.0365	0.9933	0.9948	1.0659	0.9362	0.9034

The ambient temperature data T_{amb} for the year 2019 is obtained from the DWD (German Meteorological Service), which provides the data in an hourly resolution [39]. Hence, the hourly heat demand q_{hourly} can be calculated for each of the 8760 h of the year:

$$q_{hourly}(t) = \alpha\, h(\vartheta(t))F_{WD}(t) \tag{3}$$

The multiplication factor α is different for each building and is calculated with the annual heat demand q_{annual}, derived from the heat registry Hamburg:

$$q_{annual} = \sum_{t=1}^{8760h} q_{hourly}(t) = \sum_{t=1}^{8760h} \alpha h(\theta(t))F_{WD}(t) \tag{4}$$

Figure 3 shows the load duration curve for total heat demand in all considered buildings in Wilhelmsburg. The peak load is 64 MW, which corresponds to an ambient temperature of −7.8 °C. The higher the ambient temperature, the lower is the heat demand, whose minimum value is 1.9 MW. Here, the ambient temperature is 35.5 °C. In conclusion, utilizing the building data from OpenStreetMaps, the heat data from the heat registry Hamburg [36], the sigmoid equations from the BDEW [26] and the temperature data from the DWD [37], the hourly heat demand for each of the 4884 considered buildings in Wilhelmsburg is calculated.

2.2. Building Clustering and Grid Improvement

Based on the derived data and the source of the industrial waste heat, the district heating grid is created by clustering the buildings and connecting the obtained allocated nodes. Furthermore, the grid's efficiency is improved by eliminating nodes, which are too distant from the heat source or have too little heat demand.

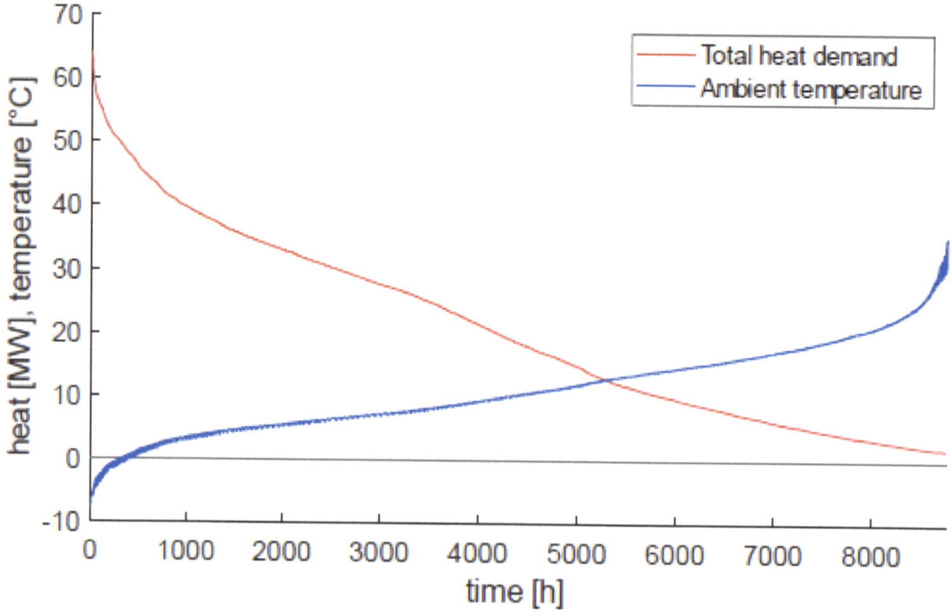

Figure 3. Heat load duration curve for total heat demand, and corresponding ambient temperature.

To reduce the computational load, the buildings are allocated into clusters, which then represent the buildings' heat demand. The general idea is to create 15 bins in x-direction and 10 bins in y-direction, resulting to 150 clusters, which are equidistantly distributed. Subsequently, the size of the clusters is adjusted in order to level the heat demand between clusters, i.e., seeking an equal heat demand in each cluster. The clusters with no buildings within are dropped and not considered any further. Hence, the number of clusters is reduced from 150 to 123. Afterwards, the cluster centers are calculated as the heat weighted mean coordinates of the buildings, which are located within the cluster. The calculation is heat weighted, so that the center of the cluster is closer to the buildings with the highest heat demand. Subsequently, the buildings are allocated to the closest cluster center. Hence, the clusters lose their rectangular shape for a polygonal boundary, and buildings change clusters. The center coordinates are not reevaluated.

Figure 4 shows the final clusters with their cluster centers and boundaries. The size of the buildings' markers correlates with their heat demand, i.e., larger circles mean higher heat demands. The clusters vary in shape and size, and as desired the clusters in denser areas are smaller than the clusters in the outskirts. Each cluster with all its buildings and the center node is transformed into a graph object using the graph function in Matlab, which creates a connection between each pair of buildings. Subsequently, the minspantree function is applied to the graph to find the shortest connection of all buildings, starting from the center node. The length of the connections is saved as the internal pipe length of the cluster.

The connection of the cluster nodes depends on the root node, i.e., the location where the heat is supplied to the grid. In this model it is assumed that only one heat source is present. Most district heating grids are supplied by a heating plant or a combined heat and power (CHP) plant, which burn fossil fuels to generate heat—and in the case of CHP additionally power. The usage of fossil fuels emits CO2, and it must hence be critically discussed whether that can provide a more sustainable alternative for domestic gas boilers. Alternatively, biomass can be used to power the heating plant, which is then carbon-neutral, but the amount of bioenergy necessary to supply an entire district with heat exceeds any feasible scale. On the other hand, the largest share of Hamburg's energy-intensive industry is located in the port and is hence close to Wilhelmsburg.

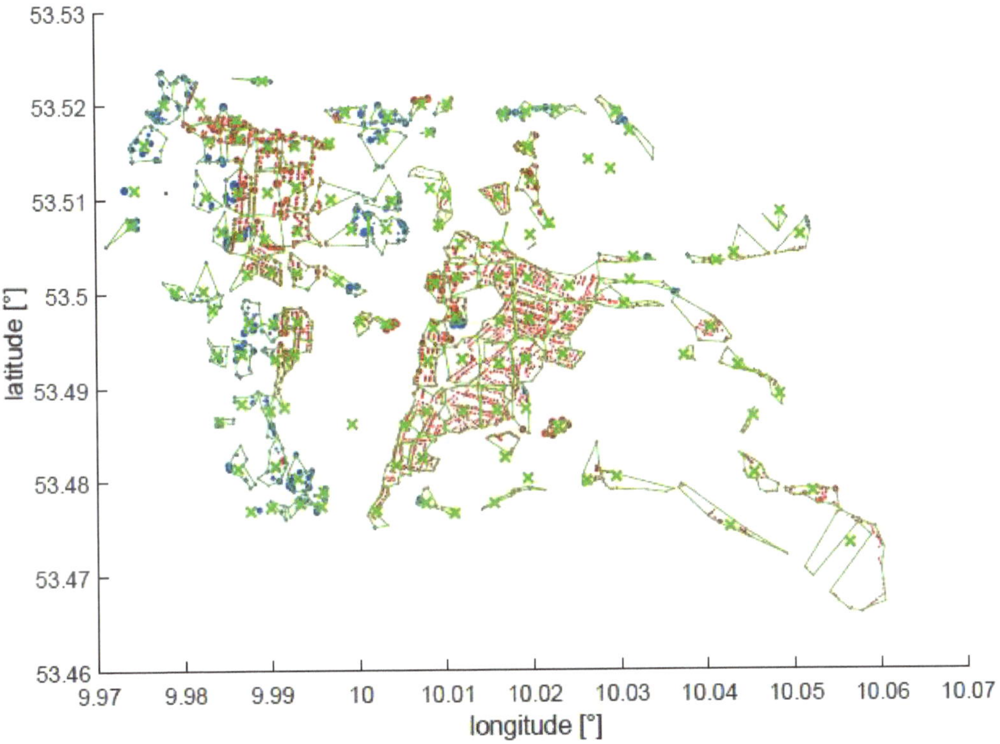

Figure 4. Clustering of accommodational (red) and commercial (blue) buildings. Centers (x) and boundaries of clusters in green.

As is common for heating systems, the base load provider is only designed for the base load and not the peak load, because otherwise the system would be oversized for the few moments in which the peak load is demanded. Hence, the 18 MW are an appropriate base load. Furthermore, it is assumed that the temperatures at the root node are constantly 90/60 °C, and the maximum supplied heat power $\dot{Q}_{sup,max}$ is constant and amounts to 18 MW.

Analogous to the connections in each cluster, the general heating grid is derived by applying Matlab's minspantree function to the nodes, transformed into a graph object. The 123 nodes are hence connected by 122 connections, because each node needs exactly one connection leading to it. The root node however does not have a connection, because the grid starts there. The initial heating grid is depicted in the upper plot in Figure 4. The nodes are numbered from 1 to 123, beginning in the bottom left corner and ending in the upper right corner. The root node is node 111. Note that the heating grid is finely branched, with long distances to the end nodes, e.g., node 1 in the bottom left corner or node 10 in the bottom right corner. Longer distances evidently result in higher heat losses, which lower the efficiency of the heating grid. It is hence desired to balance the delivered and lost heat. Consequently, the connectivity density $r_{con,node}$ is calculated as the ratio between the supplied heat $Q_{sup,node}$ and the distance to the node S_{node} [40]:

$$r_{con,node} = \frac{Q_{sup,node}}{S_{node}} = \frac{Q_{sup,node}}{2(S_{root,node} + S_{in,node})} \quad (5)$$

The distance to the node S_{node} is composed of the distance from the root node $S_{root,node}$ and the distance of the grid within the cluster $S_{in,node}$, multiplied by 2 to account for the return flow. Hence, to discard the least efficient nodes, the nodes with the worst connectivity density are iteratively eliminated from the graph. Afterwards, the minspantree is reevaluated for the new, smaller grid. Thereby, less heat is supplied, but on the other hand, the heat losses are simultaneously reduced. This elimination is continued until the supplied heat falls below 85% of the initially demanded heat.

2.3. Building Location and Heat Demand Data

In the next step, the properties of the heating grid, including especially the diameter of the pipes, are calculated. This determines the heat loss along the edges of the grid, as shown below. Consequently, the lost heat reduces the feed temperature and raises the return temperature. It is still assumed that the first node has the supplied feed and return temperatures of 90 °C and 60 °C.

The cumulated heat flow on each edge in hourly resolution $\dot{Q}_{cum,edge}$, i.e., the total heat flow on this edge which supplies all subsequent nodes, is the product of the mass flow on the edge $\dot{m}_{edge}$, the heat capacity of water c_w, and the difference of the feed and the return temperature on this edge $T_{in,edge}$ and $T_{out,edge}$:

$$\dot{Q}_{cum,edge} = c_w \dot{m}_{edge}\left(T_{in,edge} - T_{out,edge}\right) \rightarrow \dot{m}_{edge} = \frac{\dot{Q}_{cum,edge}}{c_w\left(T_{in,edge} - T_{out,edge}\right)} \quad (6)$$

It is assumed that the temperatures remain constant on each edge, and are equal to the temperatures of the previous node. Thus, for example, the two edges leading away from the root node have the same temperature as the root node.

The diameter of an edge D_{edge} is a function of the maximum mass flow on the edge $\dot{m}_{edge,max}$. With a fixed specific pressure loss of $R = \Delta p/l = 300$ Pa/m, the flow coefficient λ, the density of water ρ_w and $\dot{m}_{edge,max}$, D_{edge} is calculated as [40]:

$$D_{edge} = \sqrt[5]{\frac{8\lambda \dot{m}^2_{edge,max}}{\rho_w \pi^2 R}} \quad (7)$$

For high enough Reynold numbers, λ depends on D_{edge} and the roughness height k, which is set to 0.01 mm [28]:

$$\lambda = \frac{0.25}{\left[log_{10}\left(\frac{3.715 D_{edge}}{k}\right)\right]^2} \quad (8)$$

The two equations for D_{edge} and λ are iteratively repeated until the relative change in λ is $\Delta\lambda_{rel} < 10^{-3}$. The calculation converges after 4 iterations. With D_{edge}, the specific heat loss coefficient U_{edge} is calculated. The Planungshandbuch Fernwärme (planning manual district heating) [40] proposes empirical values for U_{edge}, which are then logarithmically approximated. The heat loss in feed direction $\dot{Q}_{loss,in,edge}$ and in the return flow $\dot{Q}_{loss,out,edge}$ are subsequently calculated with the edge length l_{edge} as:

$$\dot{Q}_{loss,in,edge} = l_{edge} U_{edge}\left(T_{in,edge} - T_{amb}\right) \quad (9)$$

$$\dot{Q}_{loss,out,edge} = l_{edge} U_{edge}\left(T_{out,edge} - T_{amb}\right) \quad (10)$$

The ambient temperature T_{amb} is the ground temperature in Wilhelmsburg for the year 2019 at a depth of 1 m [39]. Hence, the flow into a node or from a node loses heat, which can be translated in a lost temperature difference $\Delta T_{in,edge}$ and $\Delta T_{out,edge}$:

$$\Delta T_{in,edge} = \frac{\dot{Q}_{loss,in,edge}}{\dot{m}_{edge} c_w} \tag{11}$$

$$\Delta T_{in,edge} = \frac{\dot{Q}_{loss,out,edge}}{\dot{m}_{edge} c_w} \tag{12}$$

As before, the edge corresponding to a node is the edge leading to the root node. The node temperatures $T_{in,node}$ and $T_{out,node}$ are then calculated as:

$$T_{in,node} = T_{in,prev} - \Delta T_{in,edge} \tag{13}$$

$$T_{out,node} = T_{out,prev} - \Delta T_{out,edge} \tag{14}$$

The index *prev* denotes here the previous node connected by the edge, which is assumed to have the same temperature as the edge.

As seen above, the heat loss determines the temperatures, which in return determine the heat loss. Thus, the calculation is started at the root node where the temperatures are set as a boundary, and subsequently the heat loss on the adjacent edges is calculated. The temperatures on the then adjacent nodes are calculated next. This procedure is continued until the end nodes. The problem with this procedure is that the physicality of the solution cannot be guaranteed because the feed and return flows are calculated separately. Hence, at the furthest nodes, the temperatures can change places. The reason for this behavior is that far nodes with relatively little heat demand have a higher heat loss than heat demand, which is not physical. Hence in this project, the nodes' heat demand is iteratively increased. The new cumulated heat demand $\dot{Q}_{cum,node,new}$ is the sum of the previous heat demand $\dot{Q}_{cum,node,prev}$ and the heat losses which occur in both flows until the node $\dot{Q}_{loss,node}$:

$$\dot{Q}_{cum,node,new} = \dot{Q}_{cum,node,prev} + \dot{Q}_{loss,node} \tag{15}$$

But a higher heat demand leads to a higher mass flow, which leads to a larger diameter, which itself leads to a higher heat loss coefficient. Hence, the heat losses themselves are higher as well. Thus, the recalculation of the heat demand is iteratively repeated until the change in heat demand is lower than 0.1%. The result, which converges after three iterations, can be seen in the bottom plot of Figure 5.

2.4. Numerical Model

This chapter described the heating circuit of one building as an example for the reader. The Simulink model is split into three parts: first the calculation of the necessary feed and return temperatures of the heating circuit, second the partial supply of heat by the district heating, and third the peak load boiler.

Figure 6 depicts the general structure of the heating in the building. The cold heating water temperature T_{return} is increased by the provided heat from the district heating $\dot{Q}_{prov}$ to a medium temperature T_{DH}, which varies dependent on the demanded feed temperature T_{feed} and the provided heat. If the district heating can supply the entire heat demand, the boiler is not used and $T_{DH} = T_{feed}$. The peak load boiler then burns gas represented by the chemical power $\dot{Q}_{gas}$ to supply the rest of the demanded heat $\dot{Q}_{boiler}$:

$$\dot{Q}_{boiler} = \dot{Q}_{demand} - \dot{Q}_{prov} \tag{16}$$

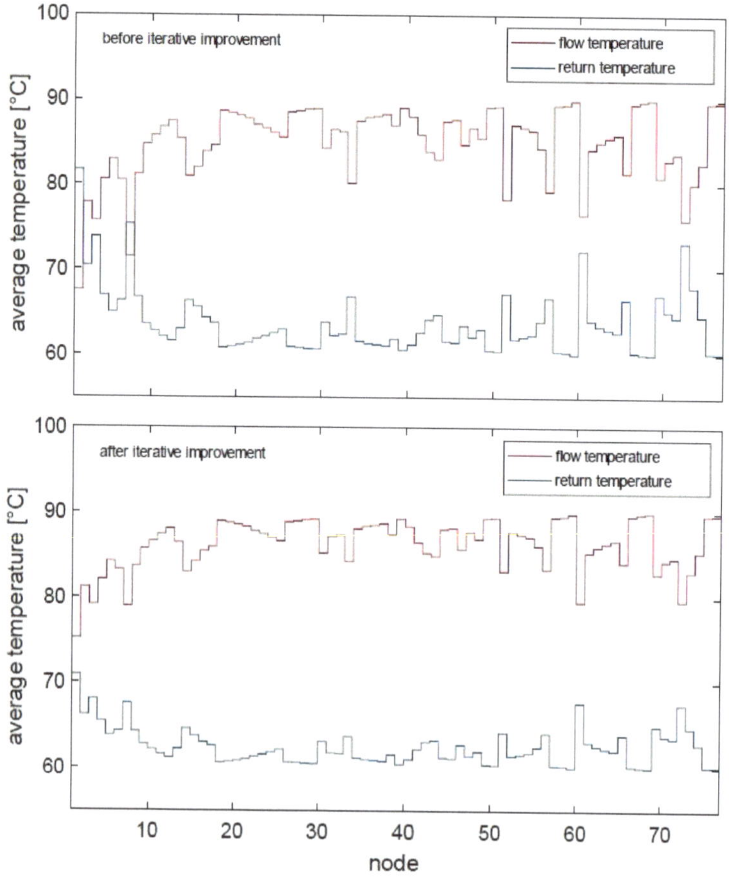

Figure 5. Mean feed and return temperature of nodes before and after iterative improvement.

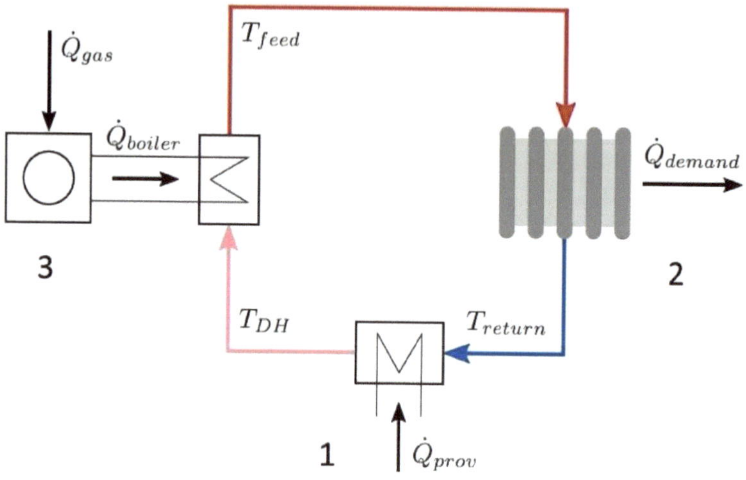

Figure 6. Schematic heating circuit with district heating and boiler.

First, the heating temperatures T_{feed} and T_{return} are calculated. Assuming a constant mass flow in the heating circuit $\dot{m}_{w,heat}$, the load ratio m_{demand} between current demand $\dot{Q}_{demand}$ and nominal, i.e., maximum possible demand $\dot{Q}_{demand,0}$ is:

$$m_{demand} = \frac{\dot{Q}_{demand}}{\dot{Q}_{demand,0}} = \frac{\dot{m}_{w,heat} c_w \left(T_{feed} - T_{return}\right)}{\dot{m}_{w,heat} c_w \left(T_{feed} - T_{return}\right)_0} = \frac{\left(T_{feed} - T_{return}\right)}{\left(T_{feed} - T_{return}\right)_0} \tag{17}$$

The nominal temperatures are $T_{feed,0}$ = 70 °C and $T_{return,0}$ = 55 °C. Derived from the thermodynamic characteristics of radiators, the mean water temperature $T_w = 1/2(T_{feed} + T_{return})$ is [41]:

$$T_w = T_l + (T_w - T_l)_0 m_{demand}^{1/n} \tag{18}$$

with the room temperature T_l = 20 °C and the exponent n = 1.3. T_{feed} and T_{return} are then calculated as [41]:

$$T_{feed} = T_w + 1/2\left(T_{feed} + T_{return}\right)_0 m_{demand} \tag{19}$$

$$T_{return} = T_w - 1/2\left(T_{feed} + T_{return}\right)_0 m_{demand} \tag{20}$$

Note that the temperature difference increases for higher loads. Figure 7 shows the Simulink subsystem to calculate the feed and return temperatures. Most of the inputs are constants, whereas $\dot{Q}_{demand}$ is variable and hence passed to the model as a timeseries object through a "From Workspace" block. Subsequently, the provided heat $\dot{Q}_{prov}$ is translated into a temperature difference ΔT_{DH} in the heating circuit with the $\dot{m}_{w,heat}$ and c_w (21):

$$\Delta T_{DH} = \frac{\dot{Q}_{prov}}{\dot{m}_{w,heat} c_w} \tag{21}$$

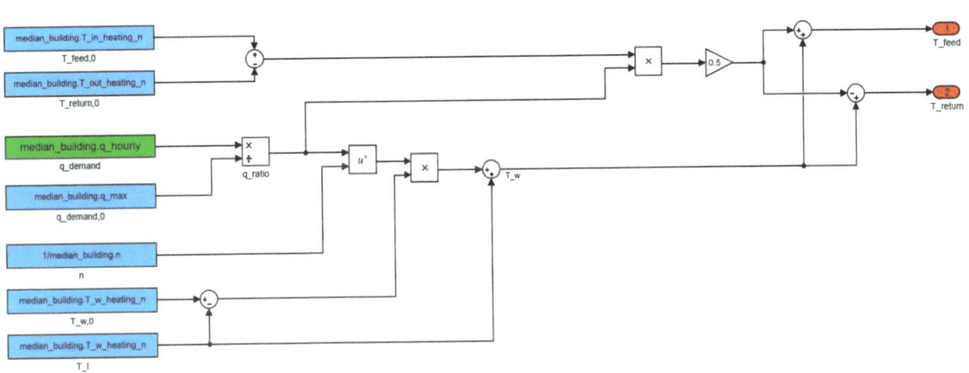

Figure 7. Simulink subsystem of feed and return temperature calculation. Constants (blue), outputs (red) and From Workspace blocks (green).

Thus, the temperature after the district heating T_{DH} is calculated, as seen in Figure 8. $\dot{Q}_{prov}$ is variable and hence passed as a timeseries object. The heat transfer from the district heating grid to the domestic heating circuit is not modeled in detail. Furthermore, the Boolean $b_{enableDH}$ sets whether the district heating is used in the simulation. To calculate $\dot{Q}_{boiler}$ and the boiler efficiency η_{boiler}, T_{feed} and T_{DH} are passed to the subsystem "Boiler".

In the subsystem "Boiler" (Figure 9), the toolbox Carnot is used to simulate the boiler operation with the Condensing Boiler block, as seen in Figure 8. Furthermore, a PID controller is used to control T_{feed}. The controller manipulates the supplied chemical energy to the boiler. Inherently, the integral behavior of the PID controller ensures no steady-state control error in T_{feed}. The fuel block is set to gas, and the water temperature entering the boiler is set to T_{DH}. The efficiency η_{boiler}, $\dot{Q}_{boiler}$ and the load factor $\beta_{boiler} = \dot{Q}_{boiler} / \dot{Q}_{demand,0}$ are passed back to the workspace.

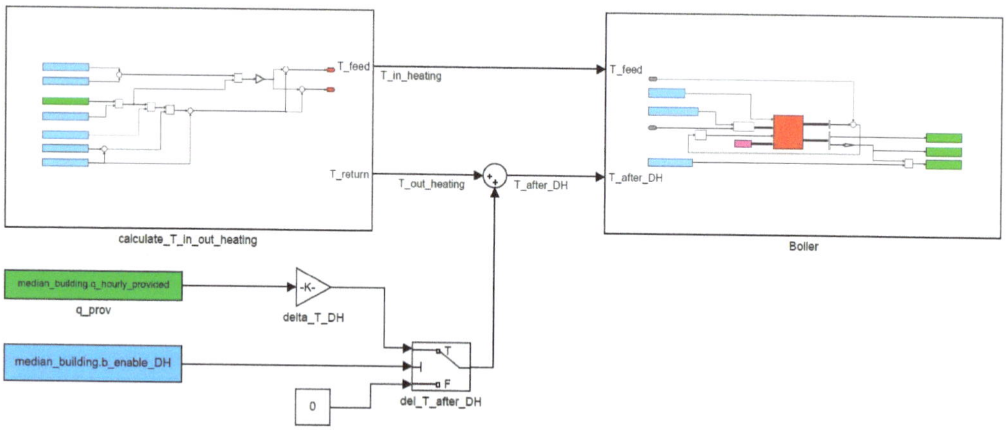

Figure 8. Simulink connection of the two subsystems calculation of heating temperatures and boiler, and calculation of temperature after district heating. Constants (blue) and From Workspace blocks (green).

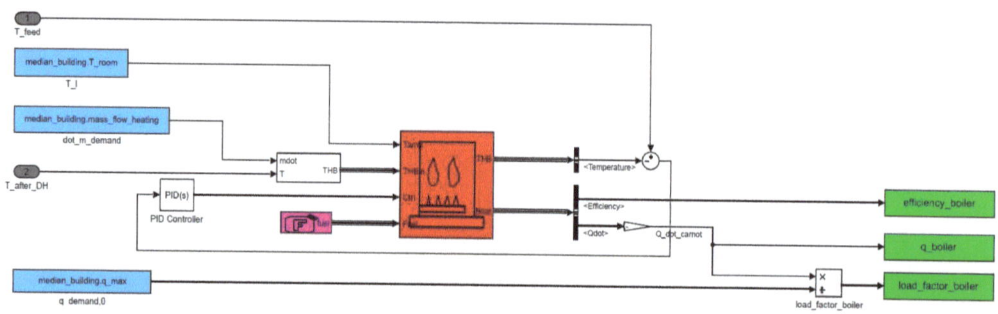

Figure 9. Simulink subsystem to calculate the boiler power and efficiency. Usage of toolbox Carnot and PID controller. Constants (blue), inputs (grey), Condensing Boiler block (red) and To Workspace blocks (green). The block to calculate THD concatenate the mass flow and temperature information to be passed to the boiler block.

η_{boiler} is calculated as (22):

$$\eta_{boiler} = \frac{\dot{Q}_{boiler}}{\dot{Q}_{gas}} \tag{22}$$

The Carnot toolbox works on a seconds-based time scale, and the variable input signals $\dot{Q}_{prov}$ and $\dot{Q}_{demand}$ are therefore interpolated from an hours-based to a seconds-based time scale. Furthermore, a simulation interval of two days is selected where the district heating can sometimes provide the entire heat, and sometimes cannot. Figure 10 shows the results of the two simulated days. The district heating provides the base load, while the boiler provides the peak load. In the afternoon of both days, the district heating covers the entire heat demand, and the boiler is off. The extracted variables η_{boiler} and β_{boiler} are used to calculate a generic fit of η_{boiler} as a function of β_{boiler}. Since it is assumed that this characteristic is the same for each building and boiler, the load factor and subsequently the efficiency are calculated for each building.

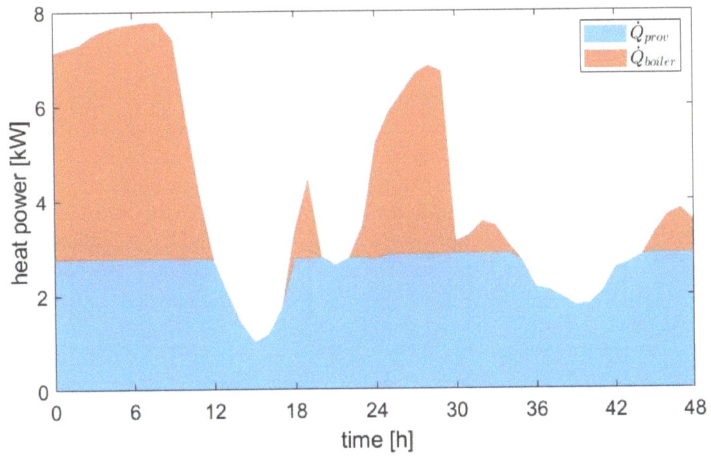

Figure 10. Power results from simulation of median building, shared supply between district heating (blue) and boiler (orange).

2.5. Building Location and Heat Demand Data

Wilhelmsburg is located south of the northern main river branch on the largest river isle, as depicted in Figure 11. It is Hamburg's largest district by area (35.4 km^2) and its fifth largest district by population (53,519), according to the Statistikamt Nord (census bureau north) [42]. Historically, the southern districts were closely interlinked with the port, the second largest port in Europe. Due to its location close to the port and south of the Elbe, Wilhelmsburg traditionally offered low priced housing and hosted low-income workers who were employed at the port or the adjacent industry. As in all of Germany, after the second World War, many immigrant families came to Hamburg to work in the industry and hence lived predominantly in districts like Wilhelmsburg [43,44].

Hamburg has a widespread district heating grid, which satisfies circa 22% of the city's heat demand, and is hence Germany's second largest district heating grid. As of 1 December 2020, more than 500,000 households are connected to the heating grid [45]. Though, this heating grid is only placed north of the Elbe, i.e., in the districts adjacent to the city center. Wilhelmsburg is not part of this heating grid, even though it is close to the heating plants supplying the grid.

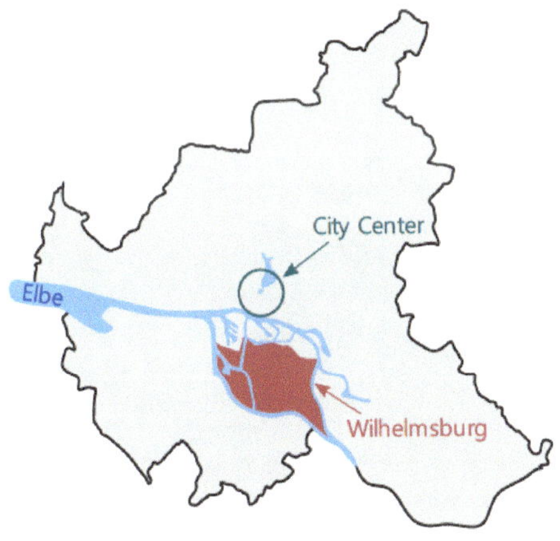

Figure 11. Schematic map of Hamburg, the river Elbe and Wilhelmsburg.

3. Results

The goal of district heating is to substitute the usage of conventional heating partially or entirely, which are mainly domestic gas boilers in each building. Hence, the previous results of the district heating grid are used to calculate the saved gas in Wilhelmsburg. The boiler efficiency is derived by implementing the district heating connection and the boiler in Simulink for the median building of the district. Afterwards, the calculated efficiency characteristics are applied to all buildings to calculate the saved gas. Figure 12 depicts the elimination of nodes, beginning with the upper plot with the initial nodes, to the lower plot, where the remaining 77 nodes after the elimination are shown. Note that the nodes and connections close to the root node remain the same, since their distance to the root node is small and thus their connectivity density is high.

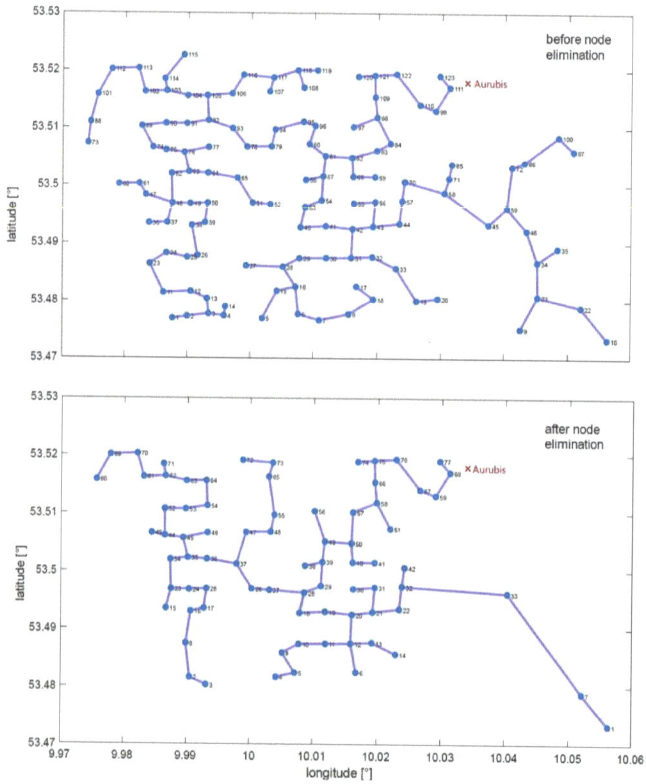

Figure 12. District heating grid with nodes and connection before (**top**) and after (**bottom**) node elimination. Reduction from 123 to 77 nodes.

Figure 13 shows the resulting heat demand at each node and hence the necessary heat flow along the connections of the heating grid. The heat is supplied at the root node 68 and distributed across the grid to each node. In summary, by allocating the buildings to clusters, the computational load can be reduced. Hence, the 4884 buildings are condensed to 123 clusters, which then incorporate the heat demand of the buildings in each cluster. The clusters are connected to a heating grid by utilizing the graph theory functions in Matlab, especially minspantree, to calculate the shortest connection between the nodes. Furthermore, to increase the overall connectivity density and hence efficiency of the heating grid, the 46 nodes with the lowest connectivity density are eliminated from the grid.

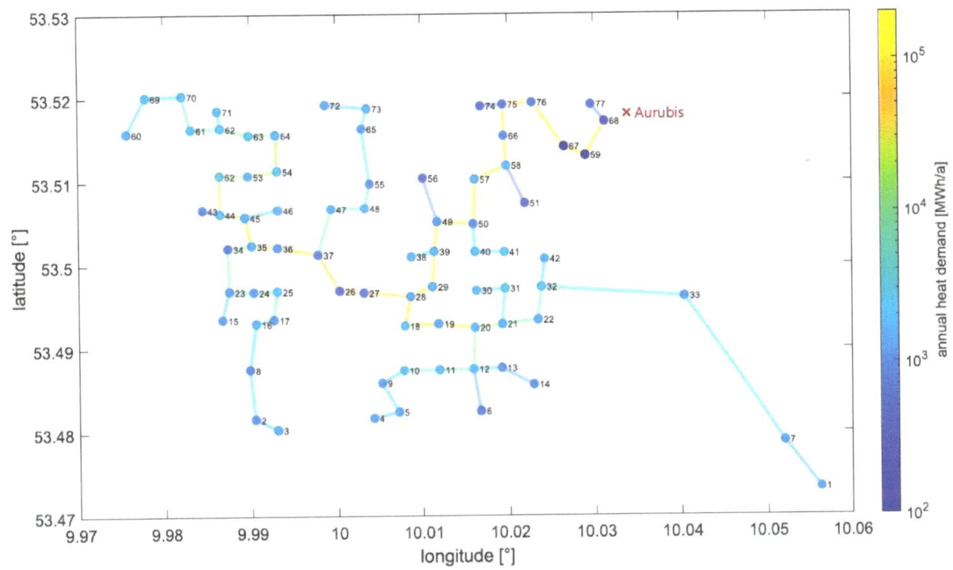

Figure 13. Annual heat demand in district heating grid at each node and connection (logarithmic scale).

All feed temperatures are higher than the return temperatures, even though the furthest nodes still have the lowest temperature difference. In practice, the temperature difference at node 1 might be too small to be technically feasible, which is neglected in this article. Figure 14 depicts the load duration curves of the heat demand before node elimination (compare Figure 3), the heat demand after node elimination, the provided heat at the buildings $\dot{Q}_{prov,buildings}$, and the provided heat with losses, which is equivalent to the effectively supplied heat from the energy provider at the root node $\dot{Q}_{sup,root}$. There is an almost constant offset of roughly 2.5 MW between $\dot{Q}_{sup,root}$ and $\dot{Q}_{prov,buildings}$, which indicates the overall heat loss. This results in an average efficiency of the district heating grid η_{grid} of (23):

$$\eta_{grid} = \frac{\sum_{t=1}^{8760h} \dot{Q}_{prov,buildings}}{\sum_{t=1}^{8760h} \dot{Q}_{sup,root}} = 83.54\% \rightarrow \Delta Q_{loss,rel} = 1 - \eta_{grid} = 16.46\% \quad (23)$$

Thus, the overall relative heat loss amounts to $\Delta Q_{loss,rel}$ = 16.46%, which is undistinguishably similar to the benchmark of $\Delta Q_{loss,rel,bm}$ = 16.5% of a real district heating grid presented by the Planungshandbuch Fernwärme [40]. Aurubis provides a maximum heat power of 18 MW and an annual amount of energy of 120 GWh [46]. Applying the same maximum heat power to the model in this article as seen in Figure 14, the provided annual energy amounts to 119.63 GWh. That shows that the assumptions made in this project hold and thus the model represents the behavior of real heating grids well. In summary, in this project a district heating grid is developed based on openly accessible data for the buildings in Wilhelmsburg and their heat demand, which are then allocated in clusters. Subsequently, the clusters are connected to a heating grid starting from the root node. In the last step, the least efficient nodes are eliminated and the heat loss along the edges is calculated. The heat loss is then iteratively added to the supplied heat.

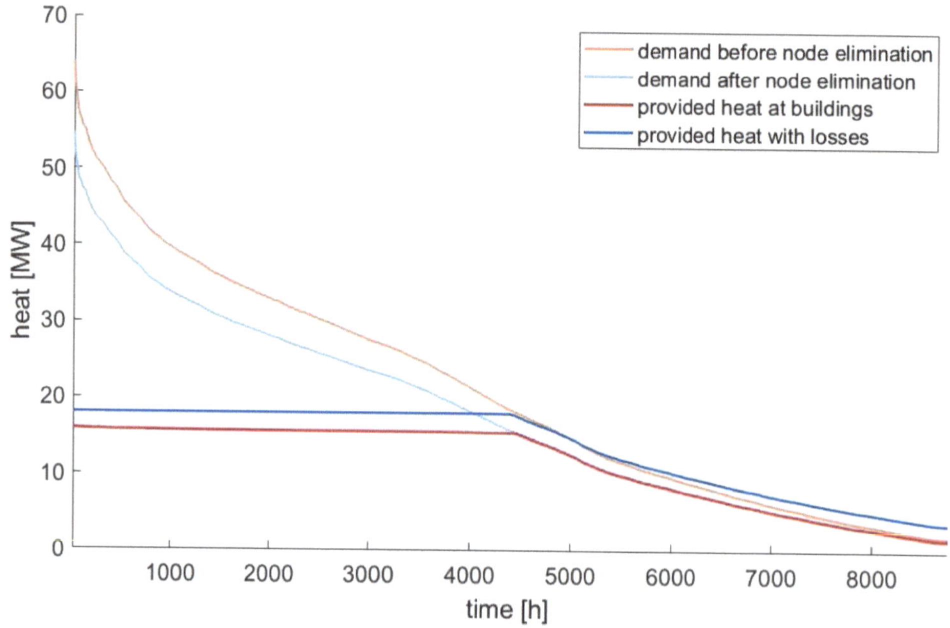

Figure 14. Load duration curves of heat demand before node elimination (orange), heat demand after node elimination (light blue), provided heat at buildings (red) and provided heat with losses (blue).

As seen in Figure 14, the supplied heat is lower than the overall heat demand, first due to the elimination of nodes in the grid and second due to the maximum heat power of 18 MW. The remaining heat is thus still supplied by domestic gas boilers, which are then used as peak load boilers. Hence, the cold heating water is first heated by the district heating and then—if necessary—heated further by the boiler to the desired feed temperature. It is assumed that each building shows the same boiler properties, i.e., the boiler efficiency characteristics are the same for every building. Thus, a single building is selected, and the boiler efficiency is calculated. To select a building which shows an appropriate characteristic, the median heat demand of all accommodational buildings is selected and matched to its building. Then, the building's heat demand $\dot{Q}_{demand}$ and the provided heat $\dot{Q}_{prov}$ are extracted and fed into the Simulink model.

The amount of consumed gas is characterized by the consumed chemical energy $Q_{gas} = Q_{boiler}/\eta_{boiler}(\beta_{boiler})$. With the previous calculations, the chemical energy is derived for each building and summed up for all 4884 buildings. Due to the different load factors—and hence efficiencies—for when the district heating is used and when it is not, Q_{gas} is evaluated for both cases $Q_{gas,noDH}$ and $Q_{gas,DH}$. The entire consumption is Q_{demand} = 182.59 GWh, while the provided heat by the district heating grid is Q_{prov} = 99.91 GWh. Without district heating the consumed amount of gas is $Q_{gas,noDH}$ = 186.49 GWh, while this is reduced to $Q_{gas,DH}$ = 89.17 GWh when the district heating is used. Hence, by utilizing the district heating, a total amount of gas of ΔQ_{gas} = 97.32 GWh per year is saved. The Bundesnetzagentur (Federal Network Agency of Germany) lists an average domestic gas price in Hamburg for the year 2019 of k_{gas} = 0.0599 €/kWh [47]. The CO2 emissions of domestic gas are e_{gas} = 0.2012 kg_{CO2}/kWh according to the Umweltbundesamt (German Environment Agency) [48,49]. Hence, by saving ΔQ_{gas} = 97.32 GWh of gas annually, CO2 emissions of Δm_{CO2} = 19,585 t_{CO2} are avoided, and the customers save ΔK_{gas} = €5.83 mil annually. In the existing project by Aurubis and Enercity, the benchmark use of the district heating grid avoids $\Delta m_{CO2,bm}$ = 20,000 t_{CO2} annually [46]. As before, the model presented in this article tracks the benchmark project well in terms of CO2 avoidance.

In summary, the simulation of the heating circuit in an exemplary median building yields the boiler efficiency as a function of the boiler's load, which is then used to calculate the saved gas of every building. The usage of the district heating grid saves ΔQ_{gas} = 97.32 GWh annually, which results in financial savings of ΔK_{gas} = €5.83 mil and avoided CO2 emissions of Δm_{CO2} = 19,585 t_{CO2}.

4. Discussions and Conclusions

In this paper, a different approach is presented to plan a district heating grid based on the optimization of building clusters with the aim of improving the nodes of the grid. GIS technology was also employed to collect and group data of the buildings, from geometrical to heat demand. In the literature, few works described in detail those approaches, therefore this research wants to fill this gap. A case study is chosen to validate the method proposed, in the district of Hamburg city, Wilhelmsburg. The scope is to connect this district with the existing thermal network of the city. Moreover, this approach investigates the usage of industrial waste heat as the grid's heat source, which is CO2-neutral. Most of Hamburg's energy-intensive and high-temperature industry is located at the port, which spans across the islands in the river. To simulate the model, Matlab/Simulink is used. After the data collection, the district heating grid is created by clustering the buildings and connecting the obtained nodes. A single building is modelled to investigate the benefit in terms of energy saving due to the replacement of gas boilers with DH systems. As a result, the usage of the district heating grid saves 97.32 GWh annually, which provides financial savings of €5.83 million, and avoided CO2 emissions of 19,585 t_{CO2}. Last but not least, the proposed method allows other designers and practitioners to adapt it to different urban settlements; this issue is also the core of this research. Future developments will be involved for the investigation of the substations of the buildings connected to the grid, being an essential key for the economic and environmental assessment of users.

Author Contributions: Conceptualization, L.P., J.M. and F.N.; methodology, J.M.; software, J.M.; validation, L.P. and F.N.; data curation, L.M.P.; writing—original draft preparation L.P., J.M. and F.N.; writing—review and editing L.P., J.M. and F.N.; visualization, L.P. and F.N.; supervision L.d.S.; funding acquisition L.M.P. All authors have read and agreed to the published version of the manuscript.

Funding: The research activities leading to this study have been partially supported by the project "Progetti per Avvio alla Ricerca" n. AR12117A8A8E6664 funded by Sapienza University of Rome.

Conflicts of Interest: The authors declare no conflict of interest.

Nomenclature

q	annual heat demand [kWh/yr]	e	Emission factor [kgCO$_2$/kWh]
A	Area [m^2]	K	Gas expenditure [€]
ϑ	Temperature [°C]	*Subscripts*	
F	Correction factor [-]	spec	specific
Q	Heat Power [MW]	abs	absolute
r	connectivity density []	acc	accommodational buildings
$\dot{Q}$	Heat flow [kW/m^2]	tot	total
c	Heat capacity [kWh/kg K]	com	commercial buildings
$\dot{m}$	Mass flow [kg/s]	amb	ambient
D	Diameter [m]	w	water
R	specific pressure loss [Pa/m]	prov	provided
λ	flow coefficient [-]	*Abbreviations*	
ρ	Density [kg/m^3]	GIS	Geographic information systems
k	roughness height [mm]	DH	District Heating
U	specific heat loss coefficient [W/m K]	DWD	German Meteorological Service
β	load factor [-]	CHP	Combined Heat and Power
η	Efficiency [-]		

References

1. Eurostat. *Sustainable Development in the European Union—Monitoring Report on Progress towards the SDGs in an EU Context*, 2017th ed.; Publications Office of the European Union: Luxembourg, 2017.
2. Battaglia, N.; Massarotti, L.; Vanoli, L. Urban regeneration plans: Bridging the gap between planning and design energy districts. *Energy* **2022**, *254*, 124239. [CrossRef]
3. Nardecchia, F.; Minniti, S.; Bisegna, F.; Gugliermetti, L.; Puglisi, G. An alternative tool for the energy evaluation and the management of thermal networks: The exergy analysis. In Proceedings of the EEEIC 2016—International Conference on Environment and Electrical Engineering, Florence, Italy, 7–10 June 2016; p. 7555645.
4. Delmastro, C.; Mutani, G.; Schranz, L. The evaluation of buildings energy consumption and the optimization of district heating networks: A GIS-based model. *Int. J. Energy Environ. Eng.* **2016**, *7*, 343–351. [CrossRef]
5. Lavagno, E. *Advanced Local Energy Planning (ALEP): A Guidebook*; International Energy Agency: Paris, France, 2000.
6. Pompei, L.; Blaso, L.; Fumagalli, S.; Bisegna, F. The impact of key parameters on the energy requirements for artificial lighting in Italian buildings based on standard EN 15193-1:2017. *Energy Build.* **2022**, *263*, 112025. [CrossRef]
7. Mattoni, B.; Mangione, A.; Pompei, L.; Bisegna, F.; Iatauro, D.; Spinelli, F.; Zinzi, M.; Casaccia, E.C.R. An alternative method for the assessment of the typical lighting energy numeric indicator for different outdoor illuminance conditions. *Build. Simul. Conf. Proc.* **2019**, *2*, 1224–1230.
8. Nardecchia, F.; Pompei, L.; Bisegna, F. Environmental parameters assessment of a new diffuser for air cooling/heating system: Measurements and numerical validation. *Build. Simul.* **2022**, *15*, 1111–1132. [CrossRef]
9. Pastore, L.M.; Sforzini, M.; Lo Basso, G.; de Santoli, L. H2NG environmental-energy-economic effects in hybrid energy systems for building refurbishment in future National Power to Gas scenarios. *Int. J. Hydrogen Energy* **2022**, *47*, 11289–11301. [CrossRef]
10. Pastore, L.M.; Lo Basso, G.; Sforzini, M.; de Santoli, L. Technical, economic and environmental issues related to electrolysers capacity targets according to the Italian Hydrogen Strategy: A critical analysis. *Renew. Sustain. Energy Rev.* **2022**, *166*, 112685. [CrossRef]
11. Pastore, L.M.; Lo Basso, G.; de Santoli, L. Can the renewable energy share increase in electricity and gas grids takes out the competitiveness of gas-driven CHP plants for distributed generation? *Energy* **2022**, *256*, 124659. [CrossRef]
12. Vesterlund, M.; Dahl, J. A method for the simulation and optimization of district heating systems with meshed networks. *Energy Convers. Manag.* **2015**, *89*, 555–567. [CrossRef]
13. Asdrubali, F.; Guattari, C.; Roncone, M.; Baldinelli, G.; Gul, E.; Piselli, C.; Pisello, A.L.; Presciutti, A.; Bianchi, F.; Pompei, L.; et al. A Round Robin Test on the dynamic simulation and the LEED protocol evaluation of a green building. *Sustain. Cities Soc.* **2022**, *78*, 103654. [CrossRef]
14. Sakawa, M.; Kato, K.; Ushiro, S. Operational planning of district heating and cooling plants through genetic algorithms for mixed 0–1 linear program-ming. *Eur. J. Oper. Res.* **2002**, *137*, 677–687. [CrossRef]
15. Weber, C.; Maréchal, F.; Favrat, D. Design and optimization of district energy systems. In Proceedings of the 10th International Symposiumon District Heating and Cooling, Reykjavik, Iceland, 3–5 September 2006; Valentin, P., Serban, A.P., Eds.; Elsevier: Hanover, Germany, 2007; pp. 1127–1132.
16. Pompei, L.; Nardecchia, F.; Mattoni, B.; Gugliermetti, L.; Bisegna, F. Combining the exergy and energy analysis for the assessment of district heating powered by renewable sources. In Proceedings of the 2019 IEEE International Conference on Environment and Electrical Engineering and 2019 IEEE Industrial and Commercial Power Systems Europe, EEEIC/I and CPS Europe, Genova, Italy, 11–14 June 2019; p. 8783426.
17. Salata, F.; Golasi, I.; Ciancio, V.; Rosso, F. Dressed for the season: Clothing and outdoor thermal comfort in the Mediterranean population. *Build. Environ.* **2018**, *146*, 50–63. [CrossRef]
18. Salata, F.; Golasi, I.; Verrusio, W.; de Lieto Vollaro, E.; Cacciafesta, M.; de Lieto Vollaro, A. On the necessities to analyse the thermohygrometric perception in aged people. A review about indoor thermal comfort, health and energetic aspects and a perspective for future studies. *Sustain. Cities Soc.* **2018**, *41*, 469–480. [CrossRef]
19. Wang, H.; Yin, W.; Abdollahi, E.; Lahdelma, R.; Jiao, W. Modelling and optimization of CHP based district heating system with renewable energy production and energy storage. *Appl. Energy* **2015**, *159*, 401–421. [CrossRef]
20. Gadd, H.; Werner, S. Achieving low return temperatures from district heating substations. *Appl. Energy* **2014**, *136*, 59–67. [CrossRef]
21. Sciacovelli, A.; Guelpa, E.; Verda, V. Pumping cost minimization in an existing district heating network. In Proceedings of the ASME 2013 International Mechanical Engineering Congress and Exposition, San Diego, CA, USA, 15–21 November 2013; American Society of Mechanical Engineers: New York, NY, USA, 2013; p. V06AT07A066.
22. Caputo, P.; Costa, G.; Ferrari, S. A supporting method for defining energy strategies in the building sector at urban scale. *Energy Policy* **2013**, *55*, 261–270. [CrossRef]
23. Galante, A.; Torri, M. A methodology for the energy performance classification of residential building stock on an urban scale. *Energy Build.* **2012**, *48*, 211–219.
24. Söderman, J. Optimisation of structure and operation of district cooling networks in urban regions. *Appl. Therm. Eng.* **2007**, *27*, 2665–2676. [CrossRef]
25. Larsen, H. A comparison of aggregated models for simulation and operational optimisation of district heating networks. *Energy Convers. Manag.* **2004**, *45*, 1119–1139. [CrossRef]

26. Larsen, H. Aggregated dynamic simulation model of district heating networks. *Energy Convers. Manag.* **2002**, *43*, 995–1019. [CrossRef]
27. Guelpa, E.; Deputato, S.; Verda, V. Thermal request optimization in district heating networks using a clustering approach. *Appl. Energy* **2018**, *228*, 608–617. [CrossRef]
28. Barbieri, E.S.; Dai, Y.J.; Morini, M.; Pinelli, M.; Spina, P.R.; Sun, P.; Wang, R.Z. Optimal sizing of a multi-source energy plant for power heat and cooling generation. *Appl. Therm. Eng.* **2014**, *71*, 736–750. [CrossRef]
29. Åberg, M.; Widén, J. Development, validation and application of a fixed district heating model tructure that requires small amounts of input data. *Energy Convers. Manag.* **2013**, *75*, 74–85. [CrossRef]
30. Chinese, D.; Meneghetti, A. Optimisation models for decision support in the development of biomass-based industrial district-heating networks in Italy. *Appl. Energy* **2005**, *82*, 228–254. [CrossRef]
31. Burer, M.; Tanaka, K.; Favrat, D.; Yamada, K. Multi-criteria optimization of a district cogeneration plant integrating a solid oxide fuel cell–gas turbine combined cycle, heat pumps and chillers. *Energy* **2003**, *28*, 497–518. [CrossRef]
32. Fabrizio, E.; Corrado, V.; Filippi, M. A model to design and optimize multienergy systems in buildings at the design concept stage. *Renew. Energy* **2010**, *35*, 644–665. [CrossRef]
33. Sontag, R.; Lange, A. Cost effectiveness of decentralized energy supply system taking solar and wind utilization plants into account. *Renew. Energy* **2003**, *28*, 1865–1880. [CrossRef]
34. Trillat-Berdal, V.; Souyri, B.; Fraisse, G. Experimental study of ground-coupled heat pump combined with thermal solar collectors. *Energy Build.* **2006**, *38*, 1477–1484. [CrossRef]
35. Lee, K.H.; Lee, D.W.; Baek, N.C.; Kwon, H.M.; Lee, C.J. Preliminary determination of optimal size for renewable energy resources in building using RETScreen. *Energy* **2012**, *47*, 83–96. [CrossRef]
36. Available online: openstreetmap.org (accessed on 17 December 2020).
37. Behörde für Umwelt, Klima, Energie und Agrarwirtsc. Wärmekataster für die Freie und Hansestadt Hamburg. Available online: https://www.hamburg.de/energiewende/waermekataster/8342506/waermekataster-fuer-die-fhh/ (accessed on 17 December 2020).
38. GEODE—Groupement Européen des entreprises et Organismes de Distribution d'Énergie; EWIV. *Bundesverband der Energie- und Wasserwirtschaft e. V., V. k*; Abwicklung von Standardlastprofilen Gas: Berlin, Germany, 2018.
39. Deutscher Wetterdienst. Climate Data Center Suche. Available online: https://cdc.dwd.de/portal/202007291339/searchview (accessed on 17 December 2020).
40. Arbeitsgemeinschaft QM Fernwärme. *Planungshandbuch Fernwärme*; EnergieSchweiz: Bern, Switzerland, 2017.
41. Lucas, U.-P.K. *Vorlesung Energiesystemtechnik Vorlesungsumdruck*; Lehrstuhl für Technische Thermodynamik: Aachen, Germany, 2013.
42. Statistikamt Nord. *Hamburger Stadtteil-Profile Berichtsjahr 2019*; Statistikamt Nord: Hamburg, Germany, 2019.
43. Statistikamt Nord. *Hamburger Stadtteil-Profile 2016*; Statistikamt Nord: Hamburg, Germany, 2016.
44. Bezirksamt Hamburg-Mitte. *Sozialraumbeschreibung Wilhelmsburg*; Bezirksamt Hamburg-Mitte: Hamburg, Germany, 2015.
45. Beckereit, M. *Anschluss der 500.000. Wohneinheit*; Wärme Hamburg: Hamburg, Germany, 2020.
46. Aurubis and Enercity. *Industriewärme—Ein Klimabündnis von Aurubis und Enercity*; Aurubis and Enercity: Hamburg, Germany, 2018.
47. Bundesnetzagentur. *Verbraucher-Kennzahlen Monitorbericht 2019*; Bundesnetzagentur: Bonn, Germany, 2019.
48. Umweltbundesamt. *CO2-Emissionsfaktoren für Fossile Brennstoffe*; Umweltbundesamt: Dessau-Roßlau, Germany, 2016.
49. Umweltbundesamt. *Grenzwerte für Schadstoffemissionen von PKW*; Umweltbundesamt: Dessau-Roßlau, Germany, 2016.

processes

MDPI

Article

Fast-Frequency-Response Control Method for Electrode Boilers Supporting New Energy Accommodation

Tao Shi [1,2,*], Zhiqiang Chen [1], Shufeng Guo [3] and Dan Li [3]

[1] College of Automation, Nanjing University of Posts and Telecommunications, Nanjing 210023, China; 1022051525@njupt.edu.cn
[2] Institute of Advanced Technology for Carbon Neutrality, Nanjing University of Posts and Telecommunications, Nanjing 210023, China
[3] Qinghai Green Energy Data Co., Ltd., Xining 810008, China; guoshufeng123@126.com (S.G.); haiou8081@163.com (D.L.)
* Correspondence: shitao@njupt.edu.cn

Abstract: With the large-scale integration of new energy generation, represented by wind and photovoltaic power, into the power grid, the intermittency, randomness, and fluctuations of their output pose significant challenges to the safe and stable operation of the power system. Therefore, this paper proposes a control method for electrode boiler systems participating in rapid grid frequency response based on a fuzzy control strategy. This method improves the traditional electrode boiler control strategy, giving it characteristics similar to those of synchronous generators in terms of active power–frequency droop, allowing it to actively adjust active power based on system frequency disturbances. Furthermore, it optimizes its control performance indicators using fuzzy algorithms. The simulation results on the Matlab/Simulink platform demonstrate that the modified electrode boiler control system, when applying this method, can effectively address power disturbances in the system, reduce system frequency deviations, and contribute to enhancing the grid frequency regulation capability and system stability.

Keywords: primary frequency control; electrode boiler; fuzzy control; clean heating; frequency stability

Citation: Shi, T.; Chen, Z.; Guo, S.; Li, D. Fast-Frequency-Response Control Method for Electrode Boilers Supporting New Energy Accommodation. *Processes* **2023**, *11*, 3098. https://doi.org/10.3390/pr11113098

Academic Editors: Ferdinando Salata, Davide Papurello and Virgilio Ciancio

Received: 6 October 2023
Revised: 26 October 2023
Accepted: 27 October 2023
Published: 28 October 2023

1. Introduction

In recent years, new energy generation technologies represented by wind power and photovoltaics have rapidly developed, and the installed capacity has been increasing year by year. As of 2022, China's newly added installed capacity of wind power and photovoltaic power generation reached 125 million kilowatts, breaking through 100 million kilowatts for three consecutive years, reaching a historical high. The electricity generation from wind power and photovoltaics reached 11.9 trillion kilowatt-hours, an increase of 207.3 billion kilowatt-hours compared to 2021, representing a year-on-year growth of 21%. This accounted for 13.8% of the total electricity consumption in the entire country, an increase of two percentage points compared to the previous year.

However, due to the random, intermittent, and fluctuating nature of renewable energy generation, there has been a significant increase in the need for flexible adjustments in the power system. This has led to a substantial occurrence of the curtailment of wind and solar power in some regions. At the same time, with the large-scale development of renewable energy, a significant number of conventional power generation units have been replaced by renewable energy units. However, these renewable energy units, when connected to the grid through power electronic components, cannot provide effective rotational inertia. This results in a decrease in the overall system inertia, leading to a deterioration in the system's frequency and voltage regulation capabilities. This situation presents a serious challenge to the safe and stable operation of the power system [1].

To enhance the integration capacity of high proportions of renewable energy, clean heating has been vigorously promoted in northern China, implementing electricity substitution to increase the space for accommodating renewable energy generation. Researchers have conducted extensive studies on this topic. In [2], a dynamic optimization scheduling method was proposed for an electric–thermal combined heat and power system that considers the optimal control of combined heat and power units along with electric boilers. This method aims to reduce the thermal–electric coupling and enhance the system's regulation capability. By establishing models for the electric and thermal output of combined heat and power units that consider the optimal control process, the impact of load variations on the thermal–electric characteristics and unit pressure safety was analyzed. Based on this analysis, optimization scheduling was carried out with the constraint of minimizing the system's operational cost. The article demonstrated that considering the optimal control of combined heat and power units and the optimal scheduling of electric boilers can promote wind power consumption and improve the overall economic performance of the system. In [3], an optimal economic scheduling model for a cogeneration energy system was proposed to minimize the total operating cost and the amount of abandoned air of an electric boiler. The proposed model considers the thermal characteristics of the building and the constraints of electrothermal balance, cogeneration units, heat storage tanks, electric boiler power, and wind power generation. Based on this model, a linear optimization model was established to analyze the different scheduling strategies under three working conditions: cogeneration; cogeneration and heat storage; and cogeneration, a heat storage unit, and an electric boiler. Taking a six-node cogeneration system as an example, the effectiveness of the model and the solution method was verified, and the influence of the heat storage tank and the electric boiler on wind abandonment and coal consumption was illustrated. In [4], a multisource system unit combination model considering high wind power penetration with thermal storage electric boilers was presented. This model harnesses the energy storage capabilities of thermal storage electric boilers to perform load shifting, thereby improving the wind power integration capacity and the economic operation of the system. In [5], a scheme for accommodating excess wind power using energy coupling devices in a comprehensive energy system was studied. The use of thermal storage electric boilers as energy coupling devices decouples the heating network from the electrical grid. Models for energy storage electric boilers and control strategies were established to support combined heat and power plants in meeting their heat demand while reducing their electrical output, thus increasing the utilization of wind power. In [6], to address the issue of challenging wind power integration in China's Three-North region, a multiobjective optimization model for wind power regulation was established, based on a wind power–thermal storage electric boiler heating system. The objectives of this model were to maximize wind power regulation capacity while minimizing the number of electric boiler regulation events and operational costs. An optimized solution was designed using an improved multiobjective particle swarm algorithm. This proposed approach was validated based on real engineering data and demonstrated its ability to reduce operational costs and enhance the capacity for wind power integration. In [7], in order to improve the acceptability of curtailment, a scheduling model of indirect heating coordination between a regenerative electric boiler and energy storage battery was established. The selection process of the electric boiler electrode was optimized according to the characteristics of curtailment, and the optimal operation strategy of the hybrid energy storage system was proposed. Based on the operation data of a real wind farm, different control methods were compared and analyzed. The results show that the ability of the power system to accept wind power can be further improved under the requirement of meeting heating demand and reducing the amount of electric boiler electrode regulation.

The abovementioned studies primarily focus on using electrode boilers for heating to increase the integration capacity of renewable energy generation. However, they do not address the active participation of electrode boilers in power system frequency response and control. At the same time, in view of the frequency stability of the power system under the high proportion of new energy access, researchers try to transform the power electronic converter to give it a certain frequency support ability. In [8], an improved differential sag plus damping control strategy based on an island microgrid was proposed. The dynamic sag mechanism was established to realize the flexible sharing of transient power among distributed generation, so that the response of the distributed generation to disturbance is flexible and stable. In [9], an optimal droop control of an H-Infinity controller based on an improved artificial bee colony algorithm was proposed, which improves power sharing among parallel inverters, realizes power response optimization and the voltage and frequency stability of a microgrid, and makes the performance of the network more robust to external interference. Compared with traditional droop control and the traditional particle swarm optimization algorithm, the improvement and accuracy of the method were verified.

Based on the above ideas, this paper further studies the feasibility and technical scheme of using the electrode boiler to rapidly adjust the control characteristics and participate in the fast frequency response of the power system. By applying fuzzy control to enhance the control performance of electrode boilers, this paper aims to imbue them with active power–frequency droop characteristics under certain control strategies. This enables them to actively adjust active power based on system frequency deviations, thereby supporting the safe and stable operation of the power system with a high proportion of renewable energy integration.

2. Principle of Operation and Response Characteristics of Electrode Boiler Primary Frequency Control

2.1. The Working Principle of Electrode Boiler Primary Frequency Control

Primary frequency control in a power system refers to the automatic operation of the governing systems of generating units when the system's frequency deviates from its rated frequency. This automatic operation adjusts the active power output of generators to keep the system frequency within an acceptable range. Primary frequency control in a power system is characterized by its rapid response and proportional control, primarily addressing short-term power fluctuations with quick response times. Electrode boilers, as large-capacity loads, can exhibit adjustable droop characteristics through specific control strategies, allowing them to participate in primary frequency control and contribute to frequency stabilization within the system. The working principle of their involvement in primary frequency control is illustrated in Figure 1.

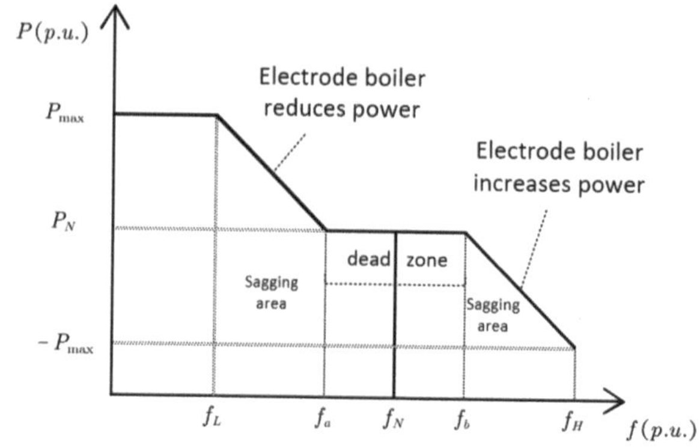

Figure 1. Principle of electrode boiler participation in primary frequency control.

In Figure 1, f_a is the upper limit of the system frequency adjustment dead zone; f_b is the lower limit of the system frequency adjustment dead zone; f_N is the rated frequency; f_L is the minimum frequency that the electrode boiler power can support; f_H is the maximum frequency that the electrode boiler power can support; P_N is the normal operating power of the electrode boiler; and P_{max} is the maximum power of the electrode boiler.

The relationship between the electrode boiler power response and frequency deviation is shown in Equation (1):

$$\Delta P_B = \Delta f \cdot K_D, \tag{1}$$

where Δf represents the frequency deviation; K_D is the frequency regulation effectiveness coefficient of the electrode boiler; and ΔP_B represents the change in power of the electrode boiler.

From the above figure, it can be seen that when $f_a < f < f_b$, the system assumes that the frequency deviation is within the normal range. The electrode boiler operates in a conventional state and does not participate in system frequency regulation. When $f_L < f < f_a$, according to Equation (1), the power setpoint is obtained, and the electrode boiler reduces its corresponding power, delivering power to the grid. When $f_L > f$, the electrode boiler stops working, and the heating is supplied by hot water stored in the heat storage tank, with all the power previously supplied by the electrode boiler now directed to the grid. When $f_b < f < f_H$, based on Equation (1), the power set-point is obtained, and the electrode boiler increases its corresponding power, absorbing power from the grid. When $f > f_H$, the electrode boiler operates at maximum power, absorbing power from the grid. If there is excess heat, it can be stored in the heat storage tank. During this time, it can be used in conjunction with market bidding strategies to lower heating prices, incentivize heat consumption by users, and enhance response flexibility [10].

2.2. The Rapid Response Characteristics of Electrode Boilers

The power of the electrode boiler is primarily influenced by the electrical conductivity of the water and the water level inside the boiler. When the electrical conductivity of the water in the boiler is constant, the power can be controlled through water level adjustments. Based on the deviation between the power and the setpoint, a controller acts on the valve after the circulation pump, regulating the makeup water flow to control the water level inside the inner drum. This achieves the goal of power control. The formula for calculating the power of the electrode boiler is as follows [11]:

$$P = \frac{3I^2 \times (R_0 + R_1)^2 \times L}{0.1835 \times 10^{-3} \times \rho \times \lg \frac{\sqrt{3} \times b \times (1 + \frac{b^2}{4 \times R^4})}{d \times (1 - \frac{b^2}{R^4})}} \tag{2}$$

where P represents the power of the electric boiler; I represents the input current of the electric boiler; L represents the length of the cylindrical electrode in water; ρ represents the resistivity of the solution; b represents the distance from the electrode axis to the center; d represents the diameter of the electrode cross-section; R represents the radius of the cylindrical container; R_0 represents the constant resistance value in the entire circuit; and R_1 represents the resistance of the solution.

By optimizing the control parameters of the electrode boiler, the system can reach the frequency-stable state at about 17 s. According to the "Implementation Rules for Power Grid Operation and Management in the East China Region", in systems primarily consisting of thermal power units, the response time for primary frequency control, i.e., the time it takes for the primary frequency control power increase to reach 90% of the target power increase, should be less than 30 s. Therefore, it can be inferred that the power response time of the electrode boiler meets the timeliness requirements for primary frequency control.

Due to the inherent time lag in thermal systems, when the frequency fluctuations cross the frequency control dead band, making short-term adjustments to the power of the electrode boiler to participate in frequency control does not significantly impact the heating supply of the thermal system. This capability allows the electrode boiler to possess primary frequency control characteristics while still meeting the constraints of the thermal system.

3. Control Strategy of Electrode Boilers Based on Fuzzy Control

3.1. Fuzzy Controller

Fuzzy control is a control method based on fuzzy logic designed to address the complexity, uncertainty, and vagueness in systems. It does not rely on precise mathematical relationships between system inputs and outputs. Instead, fuzzy control defines fuzzy control sets, fuzzy relations, and fuzzy rules to simulate the process of fuzzy reasoning, thereby achieving control over the system. In this paper, the fuzzy control method is employed to achieve the online tuning of control parameters for electrode boilers, enhancing the dynamic performance indicators of the response.

The algorithm flowchart of the fuzzy controller is illustrated in Figure 2.

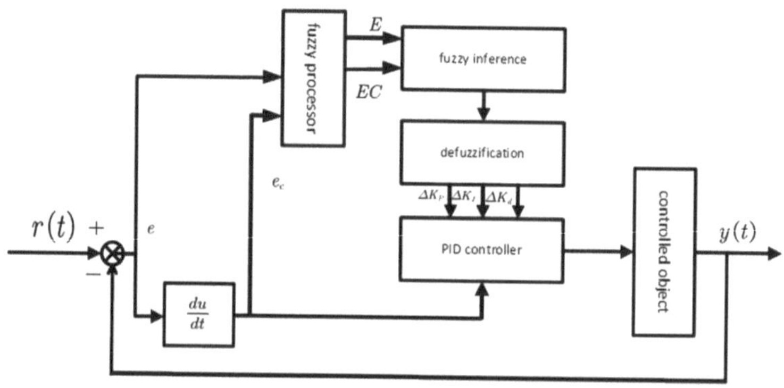

Figure 2. Fuzzy control flowchart.

In Figure 2, it can be seen that the control process of the fuzzy controller mainly includes fuzzification, fuzzy inference, and defuzzification.

This paper takes power deviation e and the rate of change in deviation e_c as inputs. The output variable is obtained through quantification factors, fuzzy processing, and de-fuzzification processing by a fuzzy controller, along with a proportional factor. The calculation formulas for the adjusted K_P, K_I, K_d are as follows [12]:

$$\left\{ \begin{array}{l} K_P = K_P + \Delta K_P \\ K_I = K_I + \Delta K_I \\ K_d = K_d + \Delta K_d \end{array} \right\}, \quad (3)$$

where K_P is the proportionality coefficient; K_I is the integral coefficient; and K_d is the differential coefficient. ΔK_P is the value of the proportional coefficient modification; ΔK_I is the value of the integral coefficient modification; and ΔK_d is the value of the differential coefficient modification. Power deviation and its rate of change are divided into different fuzzy sets, and the membership functions are used to calculate the degree of membership of the input variables to each fuzzy set. The finer the fuzzy set division, the more precise the control. Therefore, variables can be divided into the following levels of fuzzy subsets: $\{NB, NM, NS, ZO, PS, PM, PB\}$. Among them, NB represents strongly negative, NM represents moderately negative, NS represents slightly negative, ZO represents zero, PS represents slightly positive, PM represents moderately positive, and PB represents strongly positive [13]. Their membership functions are Gaussian-type.

Fuzzy control rules are typically summarized based on expert knowledge, with the aim of ensuring a high level of control accuracy. For electrode boiler power control, this paper adopts a total of 49 rules based on expert experience [14,15]. The fuzzy control rules are shown in Table 1.

Table 1. Regulation patterns for ΔK_P, ΔK_I, ΔK_d [16].

E	Ec						
	NB	NM	NS	ZO	PS	PM	PB
NB	PB, NB, PS	PB, NB, NS	PM, NM, NB	PM, NM, NB	PS, NS, NB	ZO, ZO, NM	ZO, ZO, PS
NM	PB, NB, PS	PB, NB, NS	PM, NM, NB	PS, NS, NM	PS, NS, NM	ZO, ZO, NS	NS, ZO, ZO
NS	PM, NB, ZO	PM, NM, NS	PM, NS, NM	PS, NS, NM	ZO, ZO, NS	NS, PS, NS	NS, PS, ZO
ZO	PM, ZM, ZO	PM, NM, NS	PS, NS, NS	ZO, ZO, NS	NS, PS, NS	NM, PM, NS	NM, PM, ZO
PS	PS, NM, ZO	PS, NS, ZO	ZO, ZO, ZO	NS, PS, ZO	NS, PS, ZO	NM, PM, ZO	NM, PB, ZO
PM	PS, ZO, ZB	ZO, ZO, NS	NS, PS, PS	NM, PS, PS	NM, PM, PS	NM, PB, PS	NB, PB, PB
PB	ZO, ZO, PB	ZO, ZO, PM	NM, PS, PM	NM, PM, PM	NM, PM, PS	NB, PB, PS	NB, PB, PB

By using the maximum membership degree method to defuzzify this control rule table, precise control parameter variation values can be obtained. The control parameter modification values are calculated using scaling factors and then combined with the initial parameters to obtain the final control parameters.

3.2. Control Strategy for Electrode Boiler Based on Fuzzy Control

The use of fuzzy control algorithms allows for the rapid adjustment of the power output of an electrode boiler, enabling a quick frequency response. The specific idea behind this is as follows: when a power disturbance occurs, leading to a frequency deviation in the system, the first step is to determine whether the frequency deviation exceeds the system's frequency control dead band. Then, the deviation and the frequency control coefficient K_D are used to calculate the power setpoint value. The electrode boiler takes the power setpoint value as a control signal, and finally, the electrode boiler's rapid action, using the fuzzy control algorithm, balances the system's active power fluctuations in a short time.

The specific process of the electrode boiler control strategy based on fuzzy control is shown in Figure 3.

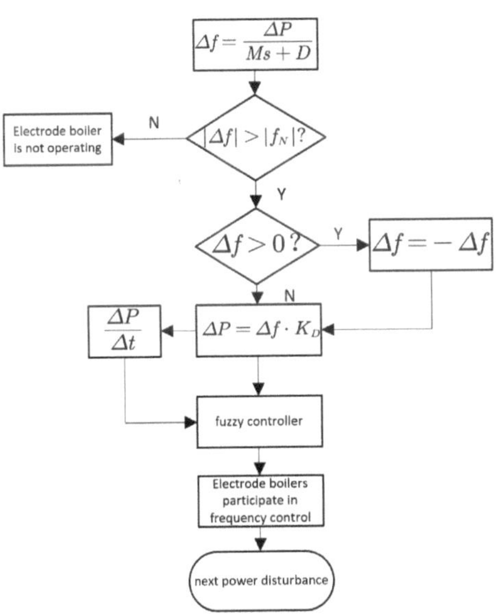

Figure 3. Control flowchart for electrode boiler.

4. System Frequency Response Model

4.1. Electrode Boiler Power Response Model

When an electrode boiler participates in the primary frequency regulation of the system, it rapidly adjusts its output based on the frequency deviation occurring in the system, balancing the active power fluctuations in the power grid to stabilize the frequency within a reasonable range.

Currently, modeling of the electrode boiler mainly involves the study of temperature and power [17]. Since the stable control of temperature takes a long time and cannot meet the time requirement of primary frequency modulation, this study models the power response of electrode boiler. The power simulation model of the electrode boiler can be approximated using a pure lag element and a first-order inertia element, represented by a transfer function as follows:

$$G(s) = \frac{k}{1 + Ts} e^{-\tau s}, \tag{4}$$

where k represents the relationship between the electrode boiler's power and frequency deviation, T represents the inherent characteristics of the electrode boiler, and τ represents the system delay effect. According to the literature [18], the amplification factor is set as $k = 0.8$, the time delay is $T = 15$ s, and $\tau = 2$.

During the frequency regulation process, the grid frequency deviation Δf is used as the control signal, and the active power output ΔP of the electrode boiler is the output signal, as shown in Figure 4.

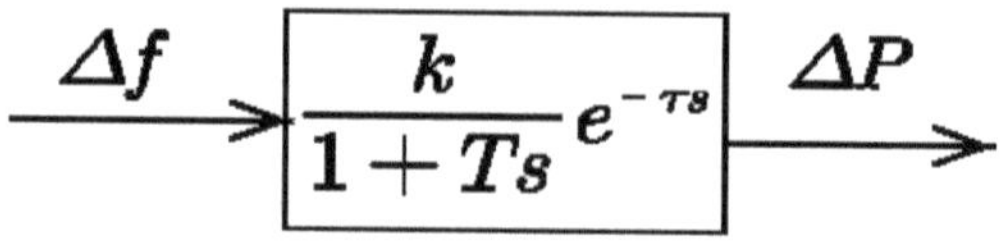

Figure 4. Transfer function diagram of the electrode boiler.

4.2. System Frequency Response Model

The fundamental reason for the frequency deviation in the power system is the mismatch between the active power output of generation units in the system and the power demand on the load side. Power systems experience relatively small power disturbances during stable operation; so, linearization is performed near the system's stable operating point to construct a linear model for dynamic analysis. Following the modeling approach proposed in the literature [19], the dynamic model of the power system, including the electrode boiler, is established as shown in Figure 5. Because the new energy generation with power electronic converters as the interface does not have the ability to participate in primary frequency modulation without an additional control strategy transformation, it is assumed in this paper that the primary frequency modulation of the whole system is mainly provided by the conventional synchronous unit and the modified electrode boiler. In Figure 5, T_g represents the governor's frequency control time constant, T_t represents the boiler's reheating time constant, T_R represents the reheater's time constant, M represents the inertia of the generation unit, D represents the load damping coefficient, Δf represents the frequency deviation, ΔP_B represents the electrode boiler's power variation, ΔP_W represents the load disturbance, K_D represents the frequency regulation effect coefficient of the electrode boiler, and R represents the droop coefficient.

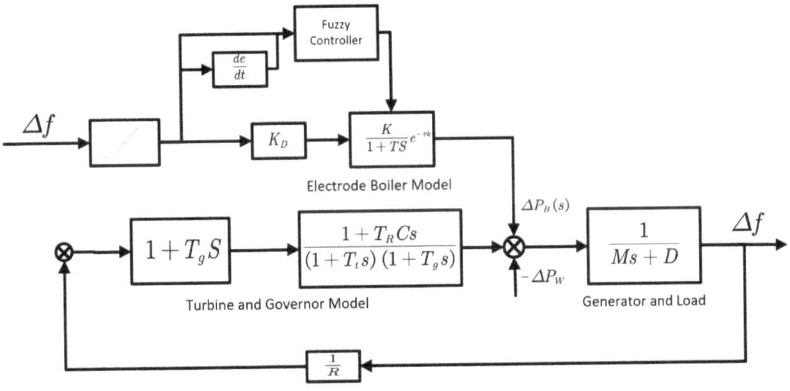

Figure 5. Frequency-response-equivalent model of the power system including the electrode boiler.

5. Simulation Analysis

To validate the impact of a high proportion of new energy sources on the grid frequency stability and the supporting effect of the fast-frequency-response control method based on fuzzy control for electrode boilers in the context of new energy integration, a simulation example system considering different proportions of new energy and electrode boiler frequency regulation is established in Matlab/Simulink-R2022a. The system includes a synchronous generator model, a photovoltaic model, a wind power model, and an electrode boiler model. Nodes 2 and 3 are, respectively, connected to photovoltaic and wind power; node 1 is connected to the synchronous motor and electrode boiler device; and nodes 5, 6, and 8 are connected to the load. Among them, the installed capacity of the synchronous generator is 80 MW, the wind power installed capacity is 10 MW, the photovoltaic installed capacity is 10 MW, it is equipped with a 15 MW electrode boiler, and the maximum heat storage capacity of the heat storage tank is 150 GJ. The operating parameters of the system are shown in Table 2. The system's base frequency is set to 50 Hz. To minimize unnecessary actions of the electrode boiler, the frequency control dead band is set to 50 $\pm$ 0.03 Hz. The model in Figure 6 is subjected to unit scaling, and the parameter selection is as follows: $M = 10$, $T_g = 0.08$ s, $T_t = 0.3$ s, $T_R = 8$ s, $K_D = 5$, and $D = 2$.

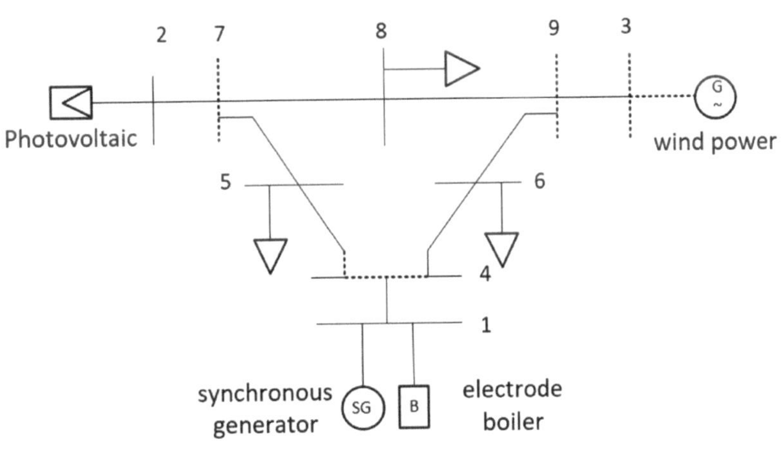

Figure 6. Example simulation system.

Table 2. Running parameters.

Bus	Name	Value
2	Photovoltaic active power	8.7 MW
	Photovoltaic reactive power	2 Mvar
3	Wind active power	9.6 MW
	Wind reactive power	1.5 Mvar
1	Active power of synchronous generator	73.5 MW
	Reactive power of synchronous generator	30 Mvar
	Electrode boiler active power	10 MW
	Electrode boiler reactive power	0 Mvar
5	Load active power	10 MW
	Load reactive power	30 Mvar
6	Load active power	36 MW
	Load reactive power	18 Mvar
8	Load active power	28.7 MW
	Load reactive power	38 Mvar

5.1. Frequency Response Characteristics of Systems with Different Shares of Renewable Energy

According to the method proposed in reference [20] for calculating the penetration rate of renewable energy, different generator inertial constants M can be used to simulate the proportion of renewable energy in the system. It can be calculated that in the model described in this paper, when M = 8 and M = 6 are used, they can simulate the cases where renewable energy accounts for 20% and 40%, respectively. To analyze the frequency response characteristics of systems with different proportions of renewable energy, a step power disturbance of 0.01 p.u. is introduced into the system, and the system's frequency response characteristics are shown in Figure 7.

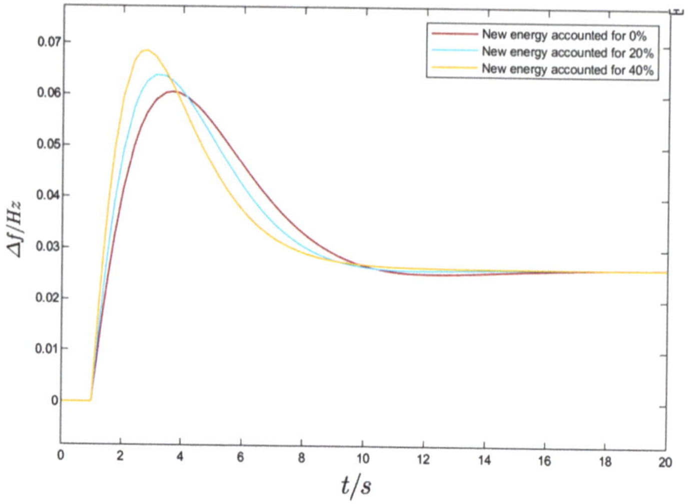

Figure 7. Frequency response characteristics of systems with different renewable energy proportions.

In Figure 7, it can be observed that when a 0.01 p.u. power disturbance occurs in the system, with a 0% proportion of renewable energy, the maximum frequency deviation is

0.06 Hz. With a 20% proportion of renewable energy, the maximum frequency deviation is 0.064 Hz, and with a 40% proportion of renewable energy, the maximum frequency deviation is 0.069 Hz. Moreover, as the proportion of renewable energy increases, the slope of the frequency response curve becomes steeper. It is evident that as the penetration rate of renewable energy increases, the system's frequency rises more rapidly and reaches higher peaks. The frequency response to the same level of power disturbance is more pronounced in high-renewable-energy penetration grids. Therefore, power systems need to have corresponding primary frequency control capabilities to support the continuous integration of renewable energy.

5.2. Frequency Control Characteristics of Electrode Boilers under Step Power Disturbance

In order to analyze the role of the fast frequency response of the electrode boiler and the frequency dynamic response of the system, a power generation disturbance with a duration of 40 s and an amplitude of 0.01 p.u. is added to the frequency simulation model with a new energy penetration rate of 20% in Section 5.1, as shown in Figure 8. The power system frequency and power response curves are shown in Figures 9 and 10 respectively. At the same time, to analyze the optimization effect of the fuzzy control of electrode boiler power, the power response curve of traditional control methods is compared with that of fuzzy control electrode boilers, as shown in Figure 11. The controller parameters for traditional electrode boiler control are selected as follows: $K_P = 4173$, $K_I = 290$, and $K_d = 4390$.

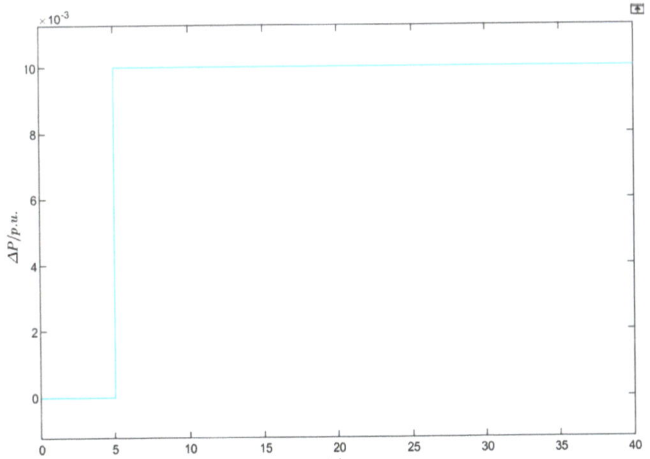

Figure 8. Power disturbance of renewable energy.

When a power disturbance of 0.01 p.u. occurs in the power system, the comparison of the system response characteristics when electrode boilers participate in frequency control or when they do not is shown in Table 3. It can be observed that the participation of electrode boilers in system frequency regulation rapidly provides active support within a short time, suppressing continuous frequency fluctuations. This leads to a reduction in the peak value of system frequency deviation by 0.003 Hz, a 5% decrease; the steady-state value of frequency deviation decreases by 0.005 Hz, a 19% decrease; and the system power deviation decreases by 0.0011 p.u., an 11% decrease. As shown in Figure 11, the overshoot of electrode boiler conventional control is 67%, with a settling time of 20 s. For the fuzzy control of electrode boilers, the overshoot is 64%, with a settling time of 17 s. It is evident that the application of fuzzy control reduces the overshoot of the electrode boiler power response by 3% and reduces the settling time by 3 s, making the response process faster and smoother and causing a lower power impact on the grid.

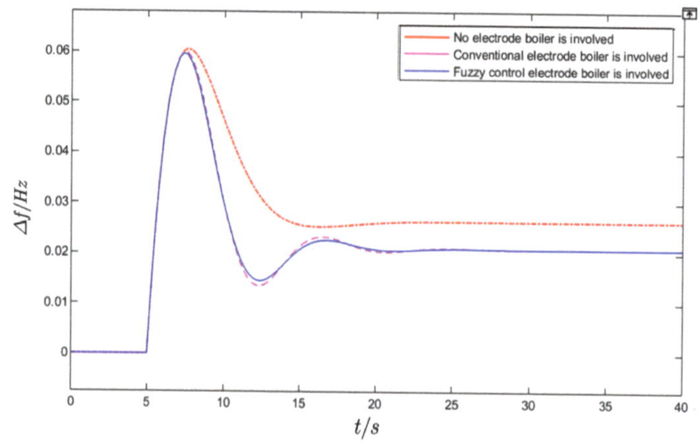

Figure 9. Frequency response of the whole system.

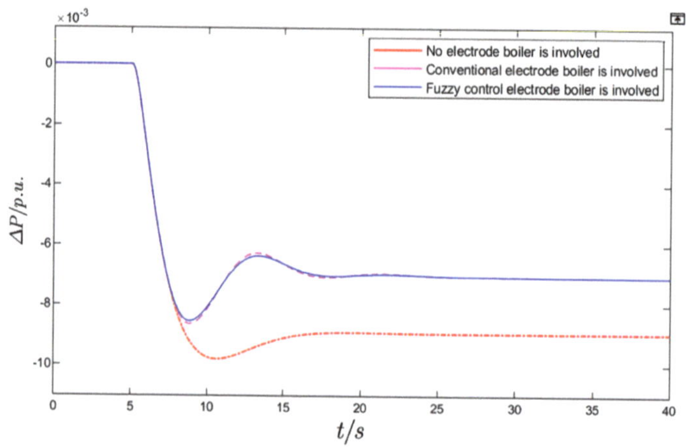

Figure 10. Active power response of the whole system.

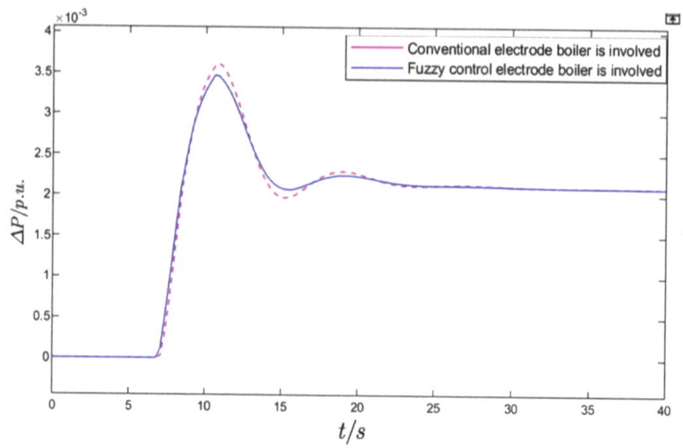

Figure 11. Active power response of electrode boiler.

Table 3. Comparison of electrode boiler participation in frequency control before and after step power disturbance.

	Electrode Boilers Are Not Involved in Frequency Regulation	Electrode Boilers Participate in Frequency Regulation	Percentage Improvement in Performance
maximum frequency deviation	0.061 Hz	0.058 Hz	5%
frequency steady-state value	0.026 Hz	0.021 Hz	19.2%
maximum power deviation	0.0096 p.u.	0.0085 p.u.	11.5%

5.3. Frequency Control Characteristics of Electrode Boilers under Time-Sequential Random Power Disturbance

To validate the rapid frequency response method of electrode boilers in supporting the intermittent output of renewable energy, this section introduces a time-sequential random power disturbance with a maximum amplitude of 0.015 p.u. and a duration of 10 min [21] into the frequency simulation model with a renewable energy penetration rate of 20%, as shown in Figure 12. The grid frequency and power response curves are presented in Figures 13 and 14 respectively. Additionally, the power response curves of electrode boilers with and without the application of fuzzy control algorithms are compared in Figure 15.

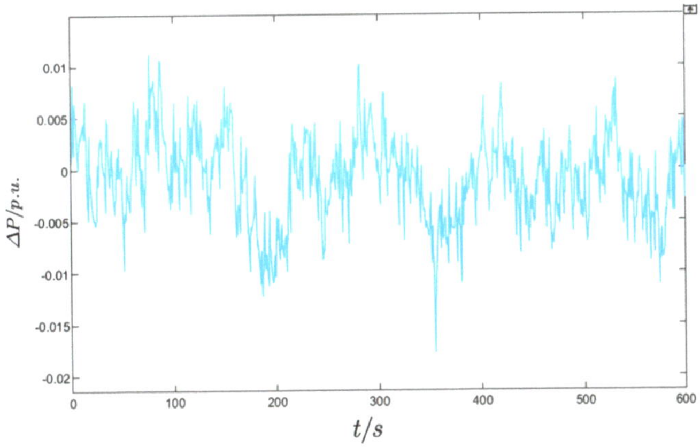

Figure 12. Time-sequential random disturbance.

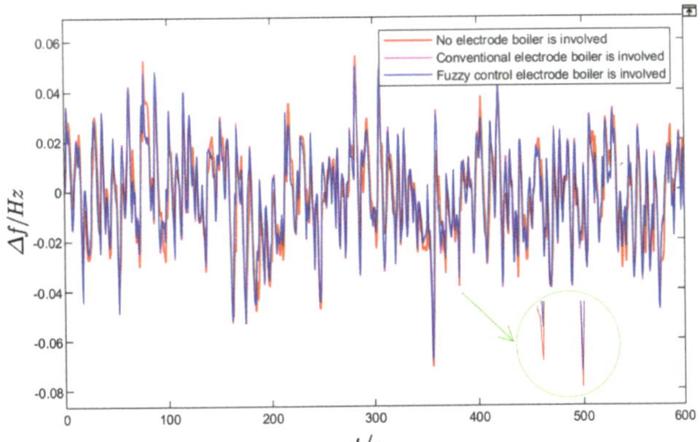

Figure 13. Frequency response of the whole system.

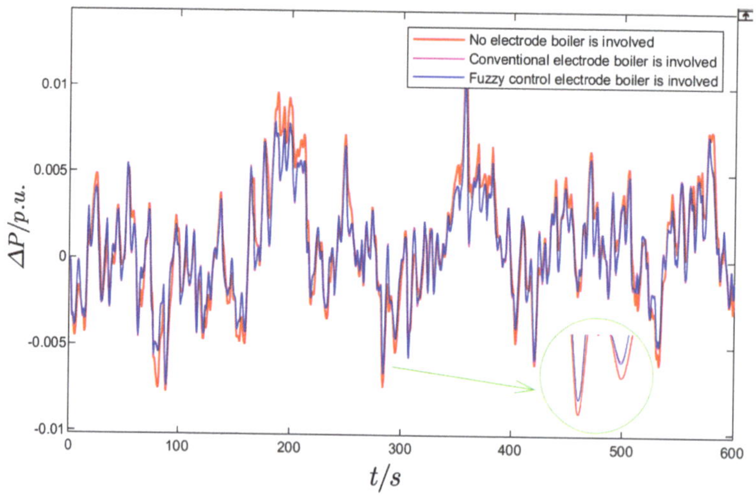

Figure 14. Active power response of the whole system.

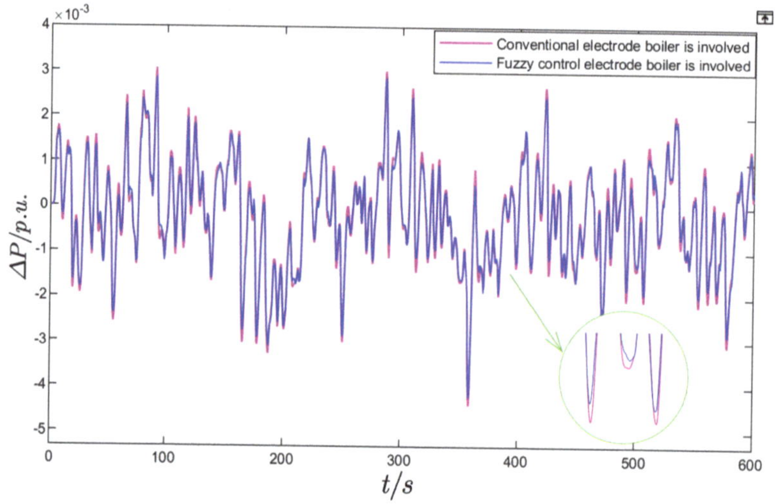

Figure 15. Active power response of electrode boiler.

As shown in Figure 13, the system frequency substantially fluctuates when there is a random power disturbance in the new energy generation. The comparison of the system response characteristics when electrode boilers participate in frequency control or when they do not is presented in Table 4. It can be observed that the application of electrode boilers in frequency control reduces the maximum system frequency deviation by 0.003 Hz, a decrease of 4.2%, and it decreases the active power fluctuation by 0.002 p.u., a reduction of 16.7%. As shown in Figure 15, the power response curve of electrode boilers indicates that the maximum power response of conventional control electrode boilers reaches 0.0031 p.u., while the maximum power response of electrode boilers with fuzzy control is 0.0028 p.u. This suggests that the oscillation amplitude is lower, allowing for more flexible adjustment of the power in response to fluctuations in renewable energy generation.

Table 4. Comparison of electrode boiler participation in frequency control before and after time-sequential random disturbance.

	Electrode Boilers Are Not Involved in Frequency Regulation	Electrode Boilers Participate in Frequency Regulation	Percentage Improvement in Performance
maximum frequency deviation	0.071 Hz	0.068 Hz	4.2%
maximum power deviation	0.012 p.u.	0.001 p.u.	16.7%

6. Conclusions

In this work, considering the impact and challenges brought by the high proportion of new energy access to the power grid on the safe and stable operation of the power system, the fast frequency response control method of the electrode boiler to support the new energy consumption is studied. The improved sag control strategy is applied to the electrode boiler so that it can adjust the output power independently according to the frequency deviation and cope with the fluctuation in the power generated by new energy sources; meanwhile, the fuzzy algorithm is applied to the electrode boiler to optimize the control parameters to shorten the power response time and improve the frequency response performance index. This study shows that the electrode boiler can not only heat the excess power generated by the new energy for heating, increasing the space for new energy consumption, but it can also quickly respond to the instantaneous fluctuation in the power system frequency under the high proportion of intermittent new energy access based on the control method in this paper to provide fast power support and improve the primary frequency regulation support capability of the power system. The follow-up work of this paper is the multipoint distribution of the electrode boiler aggregation control method to meet the high proportion of intermittent new energy access under the secondary frequency regulation needs of power system AGC.

Author Contributions: Conceptualization, T.S. and Z.C.; methodology, T.S.; software, Z.C.; validation, S.G. and D.L.; formal analysis, Z.C.; data curation, D.L.; writing—original draft preparation, Z.C.; writing—review and editing, T.S.; project administration, S.G.; funding acquisition, S.G. All authors have read and agreed to the published version of the manuscript.

Funding: This research was funded by the key R&D and transformation plan project of Qinghai Province "Qinghai Sanjiangyuan Clean Heating Big Data Analysis and Development and Application of Intelligent Management and Control Platform" (No. 2022-GX-C03).

Data Availability Statement: Not applicable.

Acknowledgments: This work was supported by the key R&D and transformation plan project of Qinghai Province "Qinghai San-jiang-yuan Clean Heating Big Data Analysis and Development and Application of Intelligent Management and Control Platform", No. 2022-GX-C03.

Conflicts of Interest: The authors declare no conflict of interest.

Nomenclature

$[f_a, f_b]$	Acceptable frequency deviation range	Hz
f_N	Rated frequency	Hz
f_L	The lowest frequency that the electrode boiler can afford	Hz
f_H	The highest frequency that the electrode boiler can afford	Hz
P_N	Rated power of electrode boiler	MW
P_{max}	Maximum power of electrode boiler	MW
Δf	Frequency deviation	Hz
K_D	Frequency response coefficient of electrode boiler	MW/Hz
ΔP_B	Electrode boiler power change value	MW
P	Electrode boiler power	MW

I	Electrode boiler input current	Ampere
L	Electrode length in water	m
ρ	Solution resistivity	Ω/m
b	Distance from electrode axis to center	m
d	Electrode cross-sectional diameter	m
R	Boiler shell radius	m
R_0	Constant internal resistance of the circuit	Ω
R_1	Solution resistance	Ω
K_P	Proportionality coefficient	--
K_I	Integral coefficient	--
K_d	Differential coefficient	--
ΔK_P	Value of the proportional coefficient modification	--
ΔK_I	Value of the integral coefficient modification	--
ΔK_d	Value of the differential coefficient modification	--
k	Relationship between power and frequency of the electrode boiler	--
T	Inherent characteristics of the electrode boiler	s
τ	Time delay coefficient	s
T_g	Governor's frequency control time constant	s
T_t	Reheating time constant	s
T_R	Reheater's time constant	s
M	Inertia of the generation unit	--
D	Load damping coefficient	--
ΔP_W	New energy power disturbance	MW
R	Droop coefficient	Hz/MW

References

1. Fanglei, L.; Fan, W.; Jiaming, Y.; Guoyi, X.; Tianshu, B. Estimating Maximum Penetration Level of Renewable Energy Based on Frequency Stability Constrains in Power Grid. In Proceedings of the 2020 5th Asia Conference on Power and Electrical Engineering (ACPEE), Chengdu, China, 4–7 June 2020; pp. 607–611.
2. Huang, Y.; Chen, Q.; Zhang, Z.; Liu, X.; Tu, J.; Zhang, L. Optimal Scheduling of Combined Electric and Heating Considering the Control Process of CHP Unit and Electric Boiler. *Processes* **2023**, *11*, 753. [CrossRef]
3. Liu, B.; Li, J.; Zhang, S.; Gao, M.; Ma, H.; Li, G.; Gu, C. Economic Dispatch of Combined Heat and Power Energy Systems Using Electric Boiler to Accommodate Wind Power. *IEEE Access* **2020**, *8*, 41288–41297. [CrossRef]
4. Hu, H.; Wen, L. Multi Source System Unit Commitment Considering Thermal Storage Electric Boiler with High Proportion of Wind Power. In Proceedings of the 2021 IEEE 4th International Conference on Renewable Energy and Power Engineering (REPE), Beijing, China, 9–11 October 2021; pp. 289–293.
5. Cheng, M.; Li, W.; Wang, J.; Tian, Q.; Liu, S.; Xiang, Q. Control of Electric Boiler with Heat Storage for the Mitigation of Wind Curtailment. In Proceedings of the 2020 5th International Conference on Power and Renewable Energy (ICPRE), Shanghai, China, 12–14 September 2020; pp. 602–607.
6. Yang, X.; Ye, T.; Zhang, Y. A Novel Optimization Model for Combined Wind Power Accommodation and Electric Boiler with Thermal Storage. *Int. J. Energy Res.* **2019**, *43*, 6494–6509. [CrossRef]
7. Li, J.; Fu, Y.; Li, C.; Li, J.; Xing, Z.; Ma, T. Improving Wind Power Integration by Regenerative Electric Boiler and Battery Energy Storage Device. *Int. J. Electr. Power Energy Syst.* **2021**, *131*, 107039. [CrossRef]
8. Jouda, M.S.; Kahraman, N. Improved Optimal Control of Transient Power Sharing in Microgrid Using H-Infinity Controller with Artificial Bee Colony Algorithm. *Energies* **2022**, *15*, 1043. [CrossRef]
9. Harasis, S.; Sozer, Y. Improved Transient Power Sharing of Droop Controlled Islanded Microgrids. In Proceedings of the 2020 IEEE Applied Power Electronics Conference and Exposition (APEC), New Orleans, LA, USA, 15–19 March 2020; pp. 2346–2351. [CrossRef]
10. Liu, M. *Electrode Boiler Participates in Day-ahead Optimal Dispatching of Heating System under Frequency Regulation Service*; North China Electric Power University: Beijing, China, 2020. (In Chinese)
11. Yuan, X. *Study on Mechanism Modeling and Dynamic Characteristics of Submerged Electrode Boiler System*; North China Electric Power University: Beijing, China, 2020. (In Chinese)
12. Hu, D.; Liu, H.; Zhu, Y.; Li, Q.; Wang, J.; Zhang, Z.; Sun, J.; Yang, L.; Zhang, Z.; Yuan, J. A Temperature Control Method of Electric Heating Furnace Based on Fuzzy PID. *J. Phys. Conf. Ser.* **2023**, *2503*, 012101. [CrossRef]
13. Mitra, P.; Dey, C.; Mudi, R.K. Fuzzy Rule-Based Set Point Weighting for Fuzzy PID Controller. *SN Appl. Sci.* **2021**, *3*, 651. [CrossRef]
14. Han, J.; Li, Y. *Fuzzy Control Technology*, 5th ed.; Chongqing University Press: Chongqing, China, 2003; pp. 39–40. (In Chinese)
15. Tao, Y. *New PID Control and Its Application*; China Machine Press: Beijing, China, 2002; pp. 43–45. (In Chinese)

16. Yu, F.; Wen, Q.; Lei, H.; Huang, L.; Wen, Z. A Compound Control System for FR4-Based Electromagnetic Scanning Micrograting. *Electronics* **2019**, *8*, 770. [CrossRef]
17. Fan, J. *Electric Boiler Temperature Control Based on The Fuzzy Self-Tuning PID Algorithm*; Ligong University: Shenyang, China, 2013.
18. Jing, H. *The Study on Operation Mode of Electrode Steam Boiler in Heating System of a Thermal Power Plant*; Shenyang Institute of Engineering: Shenyang, China, 2021. (In Chinese)
19. Li, Q. *Research on Frequency Modulation of Thermal Power Unit Assisted by Battery Energy Storage*; Kunming University of Science and Technology: Kunming, China, 2020. (In Chinese)
20. Xiao, Y.; Lin, X.; Wen, Y. Multi-Dimensional Assessment of the Inertia Level of Power Systems with High Penetration of HVDCs and Renewables. *Electr. Power Constr.* **2020**, *41*, 740–741.
21. Michigami, T.; Ishii, T. Construction of Fluctuation Load Model and Dynamic Simulation with LFC Control of DC Power System and Frequency Converter Interconnection. In Proceedings of the IEEE/PES Transmission and Distribution Conference and Exhibition, Yokohama, Japan, 6–10 October 2002; Volume 1, pp. 382–387.

processes

MDPI

Article

Research on Multi-Objective Energy Management of Renewable Energy Power Plant with Electrolytic Hydrogen Production

Tao Shi [1,2,*], Libo Gu [1], Zeyan Xu [1] and Jialin Sheng [1]

[1] College of Automation, Nanjing University of Posts and Telecommunications, Nanjing 210023, China; 1222056621@njupt.edu.cn (L.G.); 1222056624@njupt.edu.cn (Z.X.); 1022051526@njupt.edu.cn (J.S.)
[2] Institute of Advanced Technology for Carbon Neutrality, Nanjing University of Posts and Telecommunications, Nanjing 210023, China
* Correspondence: shitao@njupt.edu.cn

Abstract: This study focuses on a renewable energy power plant equipped with electrolytic hydrogen production system, aiming to optimize energy management to smooth renewable energy generation fluctuations, participate in peak shaving auxiliary services, and increase the absorption space for renewable energy. A multi-objective energy management model and corresponding algorithms were developed, incorporating considerations of cost, pricing, and the operational constraints of a renewable energy generating unit and electrolytic hydrogen production system. By introducing uncertain programming, the uncertainty issues associated with renewable energy output were successfully addressed and an improved particle swarm optimization algorithm was employed for solving. A simulation system established on the Matlab platform verified the effectiveness of the model and algorithms, demonstrating that this approach can effectively meet the demands of the electricity market while enhancing the utilization rate of renewable energies.

Keywords: electrolytic hydrogen; power fluctuation smoothing; peak shaving auxiliary services; fuzzy chance constraints; improved particle swarm algorithm

Citation: Shi, T.; Gu, L.; Xu, Z.; Sheng, J. Research on Multi-Objective Energy Management of Renewable Energy Power Plant with Electrolytic Hydrogen Production. *Processes* **2024**, *12*, 541. https://doi.org/10.3390/pr12030541

Academic Editors: Ferdinando Salata and Virgilio Ciancio

Received: 26 February 2024
Revised: 7 March 2024
Accepted: 7 March 2024
Published: 9 March 2024

1. Introduction

China's "3060 Dual Carbon" target, announced in 2020, has led to a consistent rise in the installed capacity of renewable energy generating units, specifically wind power and photovoltaic systems. As of the end of 2022, China's wind power and photovoltaic capacity reached 7576.1 MW, representing 26.7% of the total installed power capacity. However, the variability of renewable energy output, including intermittency, randomness, and fluctuation, has posed challenges to the stable operation and reliability of the power grid. The lack of dynamic regulation ability of the power system and the issue of growing challenges in renewable energy consumption are becoming more prominent [1,2]. Electrolytic hydrogen production technology, as a flexible resource that can be adjusted, has advantages such as fast response speed and zero carbon emissions [3,4]; moreover, compared with traditional energy storage media, hydrogen energy offers benefits such as high energy density and extended storage duration [5]. The grid-connected operation of the electrolytic hydrogen production system (EHPS) not only helps to improve the fluctuation of wind and solar power output and reduce the rate of wind and solar abandonment (WSA) caused by insufficient power grid consumption capacity but also can participate in peak shaving auxiliary services (PSAS), enhance the dynamic regulation capability of the system, and ensure the secure, stable, and cost-effective operation of the power grid with a high penetration of renewable energy. Therefore, how to effectively use electrolytic hydrogen production technology, enhance the flexibility and regulation capabilities of renewable energy power plants, meet the demand of multiple target scenarios such as smoothing power fluctuation, increasing renewable energy consumption space, and participate in PSAS have emerged as a pressing issue in need of urgent solutions.

There has been a certain research foundation at home and abroad in the coordinated operation of electrolytic hydrogen production systems and renewable energy generating units. The literature [6] investigates a hybrid renewable energy system incorporating hydrogen energy, proposing an efficient hydrogen production system energy management. It considers two different hydrogen production methods and establishes a mathematical model for energy conversion between hydrogen and electricity. The literature [7] explores potential opportunities and risks associated with hydrogen storage systems, assesses the feasibility of using hydrogen storage technology to address the issue of wind and solar energy curtailment in Western China, and proposes two solutions based on fuel cell hydrogen storage systems and hydrogen–natural gas blending. The literature [8] investigates an isolated DC microgrid with a combined electricity–hydrogen storage system. It proposes a cost-minimizing hybrid storage charging and discharging strategy that ensures voltage and power balance within the microgrid. The literature [9] aims to maximize system benefits and reduce environmental costs, constructs a wind–solar complementary hydrogen production system scheduling model, and uses a multi-objective golden eagle algorithm to solve the model. The literature [10] explores refined modeling of hydrogen storage systems and power-to-gas equipment, effectively enhancing the economic performance of integrated energy systems incorporating hydrogen storage. The literature [11] addresses the optimization and scheduling of hybrid systems with hydrogen storage by considering detailed models of flexible loads and hydrogen storage, aiming for an efficient solution. The objective of [12] is to optimize the net income of the microgrid within the service area, with the power balance of the microgrid system as a constraint, construct an optimized dispatching model of the wind and solar power generation service area microgrid including battery and hydrogen energy storage units, and the results show the effectiveness of the optimized dispatching strategy. The literature [13] delves into the capacity configuration of photovoltaic systems combined with hybrid battery–hydrogen storage systems, catering to both electrical and hydrogen loads. It introduces an optimization framework for the energy system servicing the hydrogen load and proposes two energy management strategies to ensure the stable supply of both electricity and hydrogen to the loads. The study in [14] presents an optimized configuration scheme for a wind–solar–hydrogen storage integrated energy system, considering both economic factors and system output fluctuations. The literature [15] studies the control mode of the hydrogen production system, proposes an optimization model of the integrated energy system with hydrogen production, and verifies the role of hydrogen production in system economy and renewable energy consumption in simulation software. The literature [16] based on the wind–hydrogen coupling system develops a capacity optimization model to maximize the revenue of the system and considers the possibility of the system participating in the auxiliary service market to improve system economy. The literature [17] studies integration of the hydrogen production system into the auxiliary peak shaving service of the thermal power unit and proposes a strategy for its participation. The literature [18] investigates integrated electric–gas energy systems incorporating green hydrogen, considering their participation in ancillary services and support for high-proportion renewable energy integration.

The aforementioned research mainly focuses on the operation optimization under a single scenario. However, a single objective is insufficient to meet the demands of the power system; hence, multi-objective optimization scheduling is necessary. The literature [19] proposes a day-ahead optimization scheduling strategy for wind–solar–hydrogen systems, targeting the system's curtailment rate of wind and solar, power deficit rate, and economic efficiency. The feasibility of its day-ahead optimization scheduling strategy was verified through simulation software. The literature [20] crafted an adaptable optimization algorithm focused on reducing expenses and enhancing battery storage capacity. Through the creation and testing of a power flow optimization framework, it facilitated increased adoption of photovoltaic power in standalone microgrids, alleviating the difficulty of managing operational expenses alongside microgrid dependability. The literature [21] proposes an optimized operation focusing on Battery Energy Storage Systems (BESS) and Power-to-Gas (P2G) technologies, aiming to maximize the benefits of Renewable Energy Sources, P2G, and BESS while minimizing facility costs. Optimization is conducted through Mixed-Integer Linear Programming (MILP), and Stochastic Programming (SP) is employed to manage the uncertainty of renewable energy output.

The aforementioned literature discusses multi-objective energy management in microgrid applications and does not focus on renewable energy power plants as the research objective. The multiple objectives in renewable energy power plants mentioned in this paper and the multiple objectives in a microgrid are not the same, but such objectives are necessary for the electricity market and the power system and demonstrate potential in energy management for renewable energy power plants with electrolyte hydrogen units, such as smoothing renewable energy output fluctuations, participating in PSAS. Meanwhile, the random changes in the output of a renewable energy-generating unit make it a random optimization problem under uncertain conditions, requiring special quantification of the randomness of renewable energy output. This paper quantifies the uncertainty of renewable energy output in the form of fuzzy chance constraints and adopts the method of crisp equivalent classes to convert fuzzy chance constraints into deterministic constraints during the solution process.

This paper considers the use of an electrolytic hydrogen production system to smooth the power fluctuations of renewable energy power generation, participate in PSAS, increase the space for renewable energy consumption, reduce the amount of wind and solar discarded, and other application target scenarios, while considering the uncertainty in renewable energy output. It proposes a multi-objective energy management model based on the coordinated optimization of an electrolytic hydrogen production system and renewable energy power generation, and a particle swarm algorithm based on piecewise mapping and improved Levy flight algorithm has been introduced. Finally, the proposed algorithm and model's effectiveness is validated through case simulations. As a whole, the main contributions of this paper are as follows:

1. Multi-objective energy management: this paper presents a multi-objective energy management strategy that utilizes electrolysis hydrogen production technology to mitigate the adverse effects caused by renewable energy generation. This strategy enhances the stability of the power system, improves energy utilization, and offers a viable solution to address the challenges of renewable energy generation.
2. Economic and operational efficiency: this paper's simulation findings deliver an in-depth assessment of the economic and operational advantages associated with a renewable energy power plant including an electrolytic hydrogen production system. It highlights the potential for considerable progress in energy management for renewable energy generation through the optimization of energy usage, involvement in peak-load adjustment services, and participation in carbon trading activities.
3. Uncertain programming: this paper quantifies the uncertainty of renewable energy output in the form of fuzzy chance constraints and adopts the method of crisp equivalent classes to convert fuzzy chance constraints into deterministic constraints during the solution process. The issue of unpredictability in renewable energy generation's output has been addressed, offering a dependable framework for the enhancement of renewable energy systems.
4. Improved particle swarm algorithm: in this paper, the adoption of the particle swarm algorithm based on piecewise mapping and improved Levy flight algorithm is introduced for achieving comprehensive and effective optimization of the proposed model. This technique contributes to the improvement of the precision and dependability of the energy management approach.

2. Basic Principles of Hydrogen Production from Renewable Energy

The renewable energy power-generating system with an electrolytic hydrogen production system mainly uses the EHPS and the surplus power in the grid-connected consumption process of wind power and photovoltaic power generation to store and utilize in the form of hydrogen energy and chemical energy. Its typical structure is shown in Figure 1.

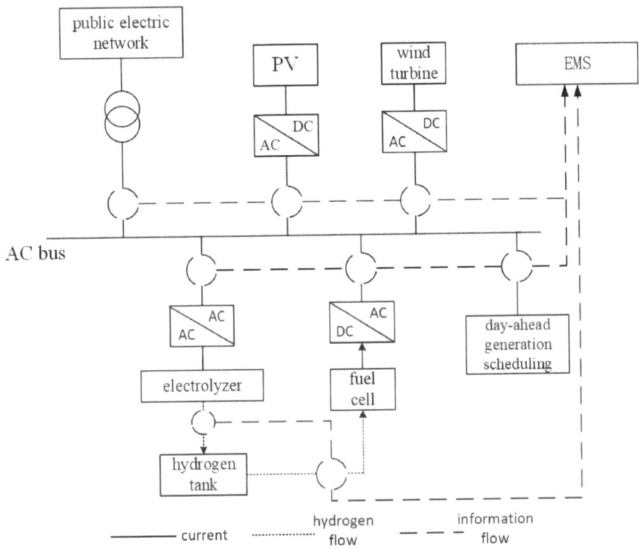

Figure 1. Renewable energy power plant with electrolytic hydrogen production system.

This plant consists of wind power generation, photovoltaic power generation, EHPS, and related power electronic interface equipment. Normally, wind power and photovoltaic power are directly consumed by the grid. However, the intermittent, random, and fluctuating nature of wind power and photovoltaic power generation, influenced by the natural environment, poses challenges, when the flexible regulation ability of the power grid is insufficient, it is necessary to use the EHPS to smooth the output power fluctuations of the renewable energy-generating unit, store the surplus output power of the renewable energy-generating unit, and participate in PSAS under certain price incentives.

The EHPS consists of an electrolyzer, a hydrogen compressor, and a hydrogen storage tank. The EHPS is an important device for electricity conversion and energy storage. It receives direct current from the renewable energy-generating unit and converts electrical energy into chemical energy through the electrolysis process. This conversion process allows the electrolyzer to store electrical energy when the demand of the power system is lower than the supply and to release the stored chemical energy for power supply when the demand is higher than the supply.

3. Technical and Economic Characteristics of Hydrogen Production from Renewable Energy

3.1. Characteristics of Renewable Energy-Generating Unit

3.1.1. Characteristics of Wind Power Generation

The output power P_{wt} of the wind turbine is shown in Equation (1):

$$P_{wt} = \frac{1}{2}C_p(\lambda, \beta)\rho\pi R^2 v^3 \tag{1}$$

In the equation, C_P is the wind energy utilization coefficient; λ is the tip speed ratio; β is the pitch angle; ρ is the air density; R is the radius of the wind turbine rotor; and v is the wind speed.

The output power of the fan can be calculated according to Formula (1). This relationship can be described by the operating parameters of the fan. Commonly used parameters include cut-in wind speed, rated wind speed, cut-out wind speed, and rated power.

$$\begin{cases} 0 & 0 \leq v < v_{in} \\ n(A + Bv + Cv^3)P_{wtN} & v_{in} \leq v \leq v_N \\ nP_{wtN} & v_N \leq v \leq v_{out} \\ 0 & v_{out} < v \end{cases} \tag{2}$$

where:

$$A = \frac{1}{v_{in} - v_N}\left[v_{in}(v_{in} + v_N) - 4(v_{in} \times v_N)\left(\frac{v_{in} + v_N}{2v_N}\right)^2\right] \tag{3}$$

$$B = \frac{1}{(v_{in} - v_N)^2}\left[4(v_{in} + v_N)\left(\frac{v_{in} + v_N}{2v_N}\right)^3 - (3v_{in} + v_N)\right] \tag{4}$$

$$C = \frac{1}{(v_{in} - v_N)^2}\left[2 - 4\left(\frac{v_{in} + v_N}{2v_N}\right)^3\right] \tag{5}$$

When approximated by linearization curve, the piecewise function expression of fan output power characteristics can be written as:

$$\begin{cases} 0 & 0 \leq v < v_{in} \\ nP_{wtN}(v^3 - v_{in}^3)/(v_N^3 - v_{in}^3) & v_{in} \leq v \leq v_N \\ nP_{wtN} & v_N \leq v \leq v_{out} \\ 0 & v_{out} < v \end{cases} \tag{6}$$

In the equation, v_{in} and v_{out} are the cut-in and cut-out wind speeds of the fan, respectively; n is the number of wind turbines; and P_{wtN} is the rated power of the wind turbine.

3.1.2. Characteristics of Photovoltaic Power Generation

The power output of photovoltaic can be calculated using an engineering approximation formula, as shown below:

$$P_{pv} = P_{STC}G_{AC}[1 + k(T_c - T_r)]/G_{STC} \tag{7}$$

In the equation, P_{pv} is the output power of photovoltaic power generation; G_{AC} is the irradiance; P_{STC} is the maximum output power under standard test conditions; G_{STC} is the solar intensity under standard test conditions; k is the power temperature coefficient; T_c is the working temperature of the photovoltaic module; and T_r is the reference temperature.

3.2. *Characteristics of Electrolytic Hydrogen Production and Hydrogen Usage*

3.2.1. Electrolytic

The electrical model of electrolytic water hydrogen production can be represented as follows:

$$\begin{aligned} U_{ele} &= U_r + \tfrac{r_1 + r_2 T_{el}}{S} I_{el} \\ &+ K_{el} \ln\left(\frac{k_{t1} + \frac{k_{t2}}{T_{el}} + \frac{k_{t3}}{T_{el}^2}}{S} I_{el} + 1\right) \end{aligned} \tag{8}$$

In the equation, U_{ele} is the working voltage of the electrolyzer; U_r is the reversible voltage of the electrolyzer that varies with temperature; I_{el} is the working current of the electrolyzer; r_1, r_2 is the ohmic resistance parameter of the electrolyte; T_{el} is the working temperature of the electrolyzer; K_{el}, k_{t1}, k_{t2}, and k_{t3} are the voltage parameters on the electrode of the electrolyzer; and S is the surface area of the electrolyzer motor.

The electrolyzer's hydrogen production rate model is described by the following equation:

$$V_{ele}^{ch}(t) = \eta_{ele} P_{ele}(t) \tag{9}$$

In the equation, $V_{ele}^{ch}(t)$ is the amount of hydrogen output; η_{ele} is the hydrogen production efficiency of the electrolyzer; and P_{ele} is the output power of the electrolyzer.

3.2.2. Hydrogen Storage Tank

According to the different forms of hydrogen, hydrogen storage technology mainly includes solid, gaseous, and liquid hydrogen storage. Industrial hydrogen storage uses gaseous hydrogen storage. The gas pressure $P_{HT}(t)$ in the hydrogen storage tank at time t is related to the amount of hydrogen storage $n_{HT}(t)$, calculated using the ideal gas equation.

$$\begin{cases} P_{HT}(t) = \frac{n_{HT}(t) R T_{H2}}{V_{HT}} \\ n_{HT}(t) = n_{HT}(t-1) + (n_{EL}(t) - n_{FC}(t))\Delta t \end{cases} \tag{10}$$

In the equation, T_{H2} is the temperature of the hydrogen gas; V_{HT} is the volume of the hydrogen storage tank; and $n_{HT}(t-1)$ is the amount of hydrogen gas in the tank at time $t-1$. The hydrogen storage status of the hydrogen storage tank can be calculated as follows:

$$SOHT(t) = \frac{P_{HT}(t)}{P_N} \tag{11}$$

In the equation, P_N is the maximum pressure of the hydrogen storage tank.

3.2.3. Fuel Cell

A fuel cell is similar to the reverse process of hydrogen production in an electrolytic cell. The output voltage can be represented as follows:

$$u_{cell} = u_{nernst} - u_{act} - u_{ohm} - u_{con} \tag{12}$$

In the formula, u_{cell} is the output voltage; u_{nernst} is the thermodynamic electromotive force; u_{act} is the activation polarization overpotential; u_{ohm} is the ohmic polarization overpotential; and u_{con} is the concentration difference overpotential.

4. Optimization Model of Multi-Objective Energy Management

4.1. Optimization Objectives of Energy Management

4.1.1. Smoothing of Power Fluctuations

The integration of renewable energy with significant fluctuations into the grid can cause harm to the grid; thus, certain requirements need to be proposed for the grid-connected renewable energy:

$$|P_{net}(t) - P_{net}(t-1)| \leq Det \tag{13}$$

$$P_{net}(t) = P_{wt}(t) + P_{pv}(t) \tag{14}$$

In the formula, P_{net} represents the actual output of renewable energy; Det represents the fluctuation range that the power grid can withstand, calculated as 10% of the total installed capacity of the plant's renewable energy generating units; P_{wt} represents the output power of wind power generation; and P_{pv} represents the output power of photovoltaic power generation.

4.1.2. Participation in Peak Load Regulation Auxiliary Services

Energy storage facilities absorb power during off-peak periods or during periods of wind, solar, and nuclear power curtailment and release power at other times, thereby providing peak load regulation auxiliary services.

According to the "Compensation Management Method" in the "Operating Rules of the Northeast China Power Auxiliary Service Market" released by the Northeast China Power Auxiliary Service Market, a "step-by-step" compensation strategy is proposed. Peak load regulation pricing details are shown in Table 1. The energy storage facilities built within the metering export of the thermal power plant, in conjunction with the unit, participate in peak shaving and are managed, cost-calculated, and compensated according to the "Compensation Management Method".

Table 1. Peak load regulation pricing details.

Period	Quotation File	Load Rate of Thermal Power Plant	Lower Quotation Limit (CNY/kWh)	Upper Quotation Limit (CNY/kWh)
non-heating season	first gear	40% < load rate ≤ 48%	0	0.4
	second gear	load rate ≤ 40%	0.4	1
heating season	first gear	40% < load rate ≤ 50%	0	0.4
	second gear	load rate ≤ 40%	0.4	1

Due to the current rules and regulations regarding energy storage participation in peak shaving not being fully developed or comprehensive, the energy storage facilities built within the metering export of renewable energy power stations participate in PSAS. The "Compensation Management Method", in which the energy storage facilities are established within the metering export of thermal power plants, is used as a reference to establish a compensation strategy:

$$E_{\text{sup}} = \lambda_{\text{sup}} \sum Q_{\text{sup}} \tag{15}$$

$$\lambda_{\text{sup}} = \begin{cases} 1000 & 0 < f_i \leq 40\% \\ 400 & 40\% < f_i \leq 50\% \\ 0 & 50\% \leq f_i \end{cases} \tag{16}$$

In the formula, E_{sup} represents the compensation for participating in peak load regulation auxiliary services; λ_{sup} represents the subsidized electricity price; Q_{sup} represents the amount of electricity for output peak shaving; and f_i represents the average load rate of the energy storage facilities. The average load rate is calculated as the power generated by the operating unit divided by the capacity of the operating unit, multiplied by 100%.

4.1.3. Increasing the Absorption Space for Renewable Energy

For renewable energy power plants, there are three factors that cause the system to abandon wind and solar. First, the installed capacity of the electrolytic cell is limited. Second, the capacity of the hydrogen storage tank is limited. Third, the installed capacity of electrochemical energy storage is limited.

The power quantity of the plant's abandoned wind and solar is:

$$Q_{\text{loss}}(t) = [P_{net,theory}(t) - P_{net}(t) - P_{ele}(t)|_{V_H < V_H^{\max}} - P_{bat}^{ch}(t)]\Delta t \tag{17}$$

In the formula, Q_{loss} represents the amount of electricity discarded from wind and solar; $P_{net,theory}$ represents the theoretical output of renewable energy; $P_{ele}^{net}(t)|_{V_H < V_H^{\max}}$ represents the operating power of the electrolyzer when absorbing the surplus output of renewable energy and when the storage volume of the hydrogen storage tank is less than the maximum storage volume; P_{bat}^{ch} represents the charging power of electrochemical energy storage; Δt represents the time interval; and Δt equals 15 min.

4.2. Objective Function

Total Benefit Model

$$\text{Max } f = E_{\text{sup}} + E_{\text{co2}} - K_{\text{loss}} - K_{\text{buy}} \tag{18}$$

In the formula, E_{sup} represents the revenue from participating in peak load shifting auxiliary services; E_{co2} represents the revenue from carbon emission reduction; K_{loss} represents the cost of wind and solar power curtailment; and K_{buy} represents the cost of purchasing electricity.

1. Revenue of peak load shifting auxiliary services:

$$E_{\text{sup}} = \lambda_{\text{sup}} \sum P_{\text{sup}}(t)\Delta t \tag{19}$$

2. Revenue of carbon emission reduction:

$$E_{co2} = A_{co2}(B_q - B_p) \tag{20}$$

$$B_p = \lambda_p \sum_{t=1}^{T} P_{\text{buy}}(t)\Delta \text{t} \tag{21}$$

$$B_q = \lambda_q \sum_{t=1}^{T} (P_{wt} + P_{pv} - P_{\text{loss}})\Delta \text{t} \tag{22}$$

In the formula, A_{co2} represents the carbon trading price; B_p represents the plant carbon emissions; B_q represents the allocated carbon emission quota; λ_p represents the equivalent carbon emission coefficient of purchasing electricity from the grid [22]; T represents the optimization period (one day is divided into 96 time periods, that is, $T = 96$); P_{buy} represents the power of purchasing electricity; and λ_q represents the carbon emission quota per unit of electricity generation.

3. Cost of purchasing electricity:

$$K_{\text{buy}} = \sum_{t=1}^{T} A_{\text{buy}} P_{\text{buy}}(t)\Delta t \tag{23}$$

In the formula, A_{buy} represents the time-of-use electricity price and P_{buy} represents the power of purchasing electricity.

4. Cost of wind and solar power curtailment:

$$K_{\text{loss}} = \gamma_{\text{loss}} \sum_{t=1}^{T} Q_{\text{loss}}(t) \tag{24}$$

$$Q_{\text{loss}}(t) = [P_{net,theory}(t) - P_{net}(t) - P_{ele}(t)|_{V_H < V_H^{\max}} - P_{\text{bat}}^{ch}(t)]\Delta t \tag{25}$$

In the formula, γ_{loss} represents the penalty for the unit amount of wind and solar power curtailment.

4.3. Constraint Conditions

4.3.1. Overall Operational Constraints of Renewable Energy Power Stations

1. Power and energy balance constraints:

$$P_{\text{wt}}(t) + P_{pv}(t) = P_{sell}(t) + P_{ele}(t) - P_{buy}(t) - P_{\text{sup}}(t) - P_{fc}(t) + P_{bat}^{ch}(t) - P_{bat}^{disch}(t) + P_{loss}(t) \tag{26}$$

In the formula, P_{sell} represents the power of selling electricity; P_{ele} represents the operating power of the electrolytic cell; P_{sup} represents the power involved in peak shifting; and P_{bat}^{ch} and P_{bat}^{disch} represent the charging power and discharging power, respectively.

2. Hydrogen energy supply and demand balance constraints:

$$V_{hs}(t+1) = V_{hs}(\mathrm{t}) + V_{ele}^{ch}(t) - V_{fc}^{disch}(t) \tag{27}$$

$$\mathrm{V}_{ele}^{ch}(t) = \eta_{ele} P_{ele}(t)\Delta \mathrm{t} \tag{28}$$

$$\mathrm{V}_{fc}^{disch}(t) = \eta_{fc} P_{\mathrm{fc}}(t)\Delta \mathrm{t} \tag{29}$$

In the formula, V_{hs} represents the storage of hydrogen in the hydrogen storage tank; V_{fc}^{disch} represents the amount of hydrogen used by the fuel cell per unit time; η_{fc} represents the hydrogen utilization efficiency of the fuel cell; and P_{fc} represents the operating power of the fuel cell.

3. Power fluctuation constraints:

$$|P_{net}(t) - P_{net}(t-1)| \leq Det \tag{30}$$

4.3.2. Operational Constraints of Each Subsystem

1. Operational constraints of renewable energy-generating unit:

$$P_{wt}^{\min}(t) \leq P_{wt}(t) \leq P_{wt}^{\max}(t) \tag{31}$$

$$P_{pv}^{\min}(t) \leq P_{pv}(t) \leq P_{pv}^{\max}(t) \tag{32}$$

In the formula, $P_{wt}^{\min}$ and $P_{wt}^{\max}$ represent the minimum and maximum output power of wind power, respectively; and $P_{pv}^{\min}$ and $P_{pv}^{\max}$ represent the minimum and maximum output power of photovoltaic power, respectively.

2. Operational Constraints of Hydrogen Energy Systems
   - Operational constraints of electrolytic cell:

$$P_{ele}^{\min} \leq P_{ele}(t) \leq P_{ele}^{\max} \tag{33}$$

In the formula, $P_{ele}^{\min}$ represents the lower limit of the power consumed by the electrolytic cell during normal operation, and there is a risk of explosion when the operating power of the electrolytic cell is below 20~25% of the rated power for a long time [23]; $P_{ele}^{\max}$ represents the upper limit of the power consumed by the electrolytic cell during normal operation.

- Operational constraints of fuel cells:

$$0 \leq P_{fc}(t) \leq P_{fc}^{\max} \tag{34}$$

In the formula, $P_{fc}^{\max}$ represents the upper limit of the power consumed by the fuel cell during normal operation.

- Capacity constraints of hydrogen storage tanks:

$$0 \leq V_{hs}(t) \leq V_{hs}^{\max} \tag{35}$$

In the formula, $V_{hs}^{\max}$ represents the maximum hydrogen storage capacity of the hydrogen storage tank.

3. Operational Constraints of Electrochemical Energy Storage Systems
   - Charging and discharging power constraints:

$$\begin{cases} 0 \leq P_{bat}^{ch}(t) \leq P_{bat}^{rated} \\ -P_{bat}^{rated} \leq P_{bat}^{disch}(t) \leq 0 \\ P_{bat}^{ch}(t) \cdot P_{bat}^{disch}(t) = 0 \end{cases} \tag{36}$$

In the formula, P_{bat}^{rated} represents the rated power of the electrochemical energy storage.

Energy storage state constraints:

- Considering that overcharging and over-discharging of electrochemical energy storage can significantly reduce the cycle life of the battery, it is significant to impose constraints on the state of charge (SOC) of the energy storage system:

$$SOC(t) = \frac{E(t)}{E_{bat}^{rated}} \tag{37}$$

$$SOC(t) - SOC(t-1) = \left(\frac{P_{bat}^{ch}(t)\eta_{bat}^{ch}}{E_{bat}^{rated}} - \frac{P_{bat}^{disch}(t)}{E_{bat}^{rated}\eta_{bat}^{disch}}\right)\Delta t \tag{38}$$

$$SOC\min \leq SOC \leq SOC\max \tag{39}$$

In the formula, $E(t)$ and E_{bat}^{rated} represent the current energy state and rated energy state of the electrochemical energy storage, respectively; η_{bat}^{ch} and η_{bat}^{disch} represent the charging efficiency and discharging efficiency, respectively; $SOC\min$ and $SOC\max$ represent the maximum capacity and minimum capacity of the electrochemical energy storage, respectively.

4.4. Fuzzy Chance Constraints

4.4.1. Fundamental Principle

The modeling idea of fuzzy chance-constrained programming is it allows the decision made to not satisfy the constraint conditions to some extent, but the probability of the fuzzy constraint conditions being established is not less than the preset confidence level.

The single-objective programming with fuzzy chance constraints is as follows:

$$\begin{cases} \min \quad f(x,\zeta) \\ s.t. \quad \Pr\{g(x,\zeta) \leq 0\} \geq \alpha \end{cases} \tag{40}$$

In the formula, x represents the decision variable; ξ represents the fuzzy vector; $\min \quad f(x,\zeta)$ represents the objective function; $g(x,\zeta) \leq 0$ represents the constraint conditions; $\Pr\{\bullet\}$ represents the probability of a certain event occurring; and α represents the confidence level of the system.

4.4.2. Handling of Power and Energy Balance Constraints

Different from traditional optimization methods, due to the randomness of renewable energy output, when a significant amount of renewable energy is integrated into the grid, the error between actual output and predicted output cannot be ignored. In the energy management model, the output of renewable energy has uncertainty. A model is established based on the theory of Chance Constrained Programming, and the fuzzy parameters $\widetilde{P}_{wt}$ and $\widetilde{P}_{pv}$ of renewable energy are introduced to form fuzzy chance constraints. Therefore, Equation (26) can be relaxed to the power and energy balance constraints under a certain confidence level condition, ensuring that the probability of meeting the balance constraints is not less than α.

$$\Pr\left\{ \begin{aligned} \widetilde{P}_{\mathrm{wt}}(t) + \widetilde{P}_{\mathrm{pv}}(t) &= P_{sell}(t) + P_{ele}(t) - P_{\mathrm{buy}}(t) - P_{\mathrm{sup}}(t) \\ &\quad - P_{fc}(t) + P_{bat}^{ch}(t) - P_{bat}^{disch}(t) + P_{\mathrm{loss}}(t) \end{aligned} \right\} \geq \alpha \tag{41}$$

5. Solution Method of Energy Management Model

5.1. Handling Method of Fuzzy Chance Constraints

The methods for handling fuzzy chance-constrained programming are divided into two categories. For simple problems, the fuzzy parameters and decision variables in the constraint conditions can be separated or, when they have a certain linear relationship, they can be transformed into clear equivalent classes for processing and then calculated by traditional methods; for complex problems, random simulation methods can be used for processing. However, the results of random simulation are not accurate and it is not easy to grasp the size of the sample capacity of the simulation, so the first method is chosen [24].

During the operation of the power system in practice, the requirements for safety are high; therefore, the probability of the power and energy balance constraints being established should be above 0.5, that is, $\alpha \geq 0.5$, so:

Assume the form of function $g(x,\zeta)$ is as follows [25]:

$$g(x,\zeta) = \psi_1(x)\zeta_1 + \psi_2(x)\zeta_2 + \cdots + \psi_t(x)\zeta_t + \psi_0(x) \tag{42}$$

In the formula, represents the trapezoidal fuzzy variable $(\gamma_{k1}, \gamma_{k2}, \gamma_{k3}, \gamma_{k4})$, $\psi = 1,2,\ldots, t$. If $\psi_k^+ \vee 0$ and $\psi_k^- \wedge 0$, $k = 1, 2, \ldots, t$. So:

The fuzzy chance constraints in Equation (42) can be clearly equivalent to:

$$\begin{aligned} &(2-2\alpha)\sum_{k=1}^{t}\left(\gamma_{k3}\psi_k^+(x) - \gamma_{k2}\psi_k^-(x)\right) \\ &+(2\alpha-1)\sum_{k=1}^{t}\left(\gamma_{k4}\psi_k^+(x) - \gamma_{k1}\psi_k^-(x)\right) + \psi_0(x) \leq 0 \end{aligned} \tag{43}$$

This paper describes the fuzzy parameters $\widetilde{P}_{\mathrm{wt}}$ and $\widetilde{P}_{\mathrm{pv}}$ of the renewable energy output using trapezoidal fuzzy parameters. Among them, the trapezoidal fuzzy variable is a quadruple $(P_{wt1}, P_{wt2}, P_{wt3}, P_{wt4})$ composed of clear numbers and $P_{wt1} < P_{wt2} \leq P_{wt3} < P_{wt4}$. The quadruple values can be determined based on the predicted renewable energy output:

$$\begin{cases} \widetilde{P}_{wt}(\omega_1,\omega_2,\omega_3,\omega_4) = P_{wt}^{fore}(r_1,r_2,r_3,r_4) \\ \widetilde{P}_{\mathrm{pv}}(\omega_1,\omega_2,\omega_3,\omega_4) = P_{pv}^{fore}(r_1,r_2,r_3,r_4) \end{cases} \tag{44}$$

In the formula, ω_1, ω_2, ω_3, ω_4 represents the membership parameter; r_1, r_2, r_3, r_4 represents the proportionality constant.

The wind and solar prediction output are parameterized in Equation (44):

$$\begin{cases} \widetilde{P}_{wt} \to \frac{\alpha-1}{4}P_{wt1} + \frac{\alpha}{2}P_{wt2} + \frac{3-\alpha}{4}P_{wt3} + \frac{1-\alpha}{2}P_{wt4} \\ \widetilde{P}_{pv} \to \frac{\alpha-1}{4}P_{pv1} + \frac{\alpha}{2}P_{pv2} + \frac{3-\alpha}{4}P_{pv3} + \frac{1-\alpha}{2}P_{pv4} \end{cases} \tag{45}$$

5.2. Improved Particle Swarm Algorithm

5.2.1. Standard Particle Swarm Optimization (PSO) Algorithm

The update formula for the PSO algorithm is given by the following equation:

$$\begin{aligned} v_{\mathrm{i}} &= \omega \times v_{\mathrm{i}} + c_1 \times rand() \times (P_{best_i} - x_i) \\ &+ c_2 \times rand() \times (g_{best_i} - x_i) \end{aligned} \tag{46}$$

$$x_i = x_i + v_i \tag{47}$$

In the formula, $i = 1, 2, \ldots, \mathrm{N}$, where N is the total number of particles in this group; ω is the inertia factor; v_i is the speed of the particle; $rand()$ is a random number between (0, 1); x_i is the current position of the particle; and c_1 and c_2 are learning factors.

5.2.2. Particle Swarm Algorithm Based on Piecewise Mapping and Improved Levy Flight (PLPSO)

(1) Piecewise Mapping

Chaos mapping is a method for generating chaotic sequences. Piecewise mapping, as a typical chaos mapping, has a concise mathematical form and good ergodicity and randomness. The definition of piecewise chaotic mapping is as follows:

$$x(t+1) = \begin{cases} \frac{x(t)}{p} & ,0 \leq x(t) < p \\ \frac{x(t)-p}{0.5-p} & ,p \leq x(t) < 0.5 \\ \frac{1-p-x(t)}{0.5-p} & ,0.5 \leq x(t) < 1-p \\ \frac{1-x(t)}{p} & ,1-p \leq x(t) < 1 \end{cases} \tag{48}$$

In the formula, $p = 1; x(1) = rand$.

(2) Improved Levi Flight

Levy flight is a method of simulating random motion using heavy-tailed distributions. It can describe the dynamic characteristics of some complex systems, such as phenomena in the fields of ecology, geography, physics, etc. The update formula for Levy flight is as follows:

$$\mathrm{X}_{\mathrm{t+1}} = \mathrm{X}_{\mathrm{t}} + a \cdot Levy(\beta) \tag{49}$$

In the formula, X_{t} is the t-th iteration position of X; a is the step length scaling factor; and $Levy(\beta)$ is the Levy random path, satisfying: $Levy \sim u = t^{-\beta}$, $1 \leq \beta \leq 3$.

Levy flight is essentially a random step length, and the distribution of the step length satisfies the Levy distribution. The Levy distribution is complex and is usually implemented using the Mantegna algorithm. For the specific implementation process, please refer to reference [26].

Due to the step length scaling factor of Levy flight usually being set to a fixed value, it leads to slow search speed and low search accuracy during the search process [27]. To address these issues, this paper introduces an enhanced Levy flight approach:

$$\mathrm{X}_{\mathrm{t+1}} = \mathrm{X}_{\mathrm{t}} + a \cdot \sin^2(\frac{\pi}{2} \cdot \frac{1-t}{iter}) \cdot Levy(\beta) \tag{50}$$

In the formula, the fixed step length $a = 0.01$; $\sin^2(\frac{\pi}{2} \cdot \frac{1-t}{iter})$ is the adaptive step length factor, where t is the current iteration number and iter is the maximum number of iterations.

(3) Improved Particle Swarm Algorithm

PSO has some disadvantages, such as easy premature convergence, lack of diversity, and susceptibility to initial values. The improved PSO uses chaos mapping to enhance the diversity of particles and combines with improved Levy flight to enhance the vitality and jumping ability of particles [26], realizing the complementary advantages of PSO, chaos mapping, and Levy flight. Due to the complexity of Levy flight, the algorithm's computation time will experience an increase. However, in the later stage of optimization, the particle swarm algorithm will fall into local optimum, and particles need to perform Levy flight to escape from the local optimum. Therefore, an adaptive random number $(1 + \frac{t}{iter}) \cdot rand$ is introduced. Before each update of particles, the generated adaptive random number is compared with 0.5. If it is greater than 0.5, Levy flight is used to update the particles; otherwise, the particles' velocity and position are updated iteratively through Formulas (46) and (47).

The process description of the improved PSO is as follows:

(a) Population initialization. Randomly initialize the speed of the population and use chaos mapping to initialize the position of the population and find the current optimal particle position and solution.
(b) Calculate fitness. Evaluate the fitness value of each particle and store it in pbest. Compare the fitness values of all particles in the population and store the best individual (gbest) in the population.
(c) Update particle position. Generate an adaptive random number and compare it with 0.5. If it is greater than 0.5, update the particle through Levy flight; otherwise, update the particle through Formulas (46) and (47).
(d) Update individual best position. Compare the current fitness value of the particle with the fitness value before the update, and use the better particle as the particle after the iteration.
(e) Update global optimal solution. Compare pbest with gbest. If pbest is better than gbest, assign the value of pbest to gbest.
(f) Algorithm termination. If the iteration termination condition is met, the algorithm terminates; otherwise, return to step c to continue the search.

The process of the PLPSO algorithm is shown in Figure 2:

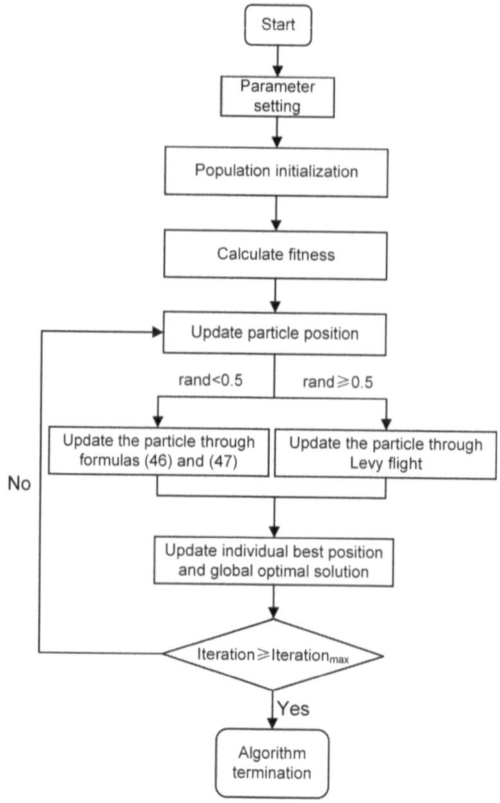

Figure 2. The process of the PLPSO algorithm.

6. Simulation Analysis

6.1. Case Description

Based on the typical daily wind power and photovoltaic output prediction data of a renewable energy power station in a province in Northeast China, with a sampling period of 15 min, the obtained forecast output curve of wind power and photovoltaic on a typical day is shown in Figure 3, and the program output and peak shaving command curve are shown in Figure 4. An EHPS and an electrochemical energy storage system are configured; the installed capacities of wind power and photovoltaics in this region are 25 MW and 200 MW, respectively, the capacity of the electrolytic cell is configured as 25 MW, and the fuel cell capacity is 10 MW; the equipment-related parameters are shown in Table 2 and the economic parameters are shown in Table 3.

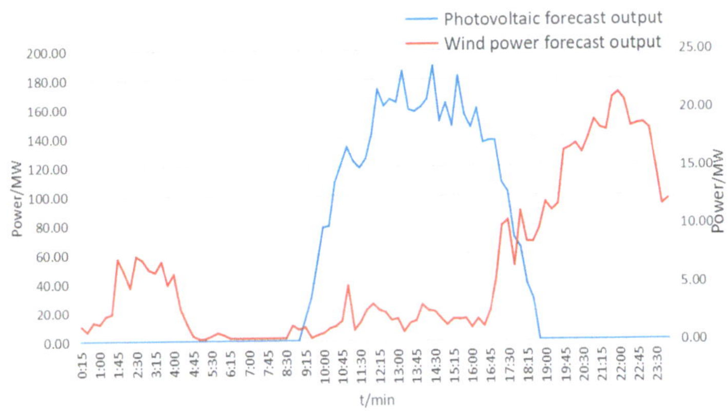

Figure 3. Forecast output curve of wind power and photovoltaic on a typical day.

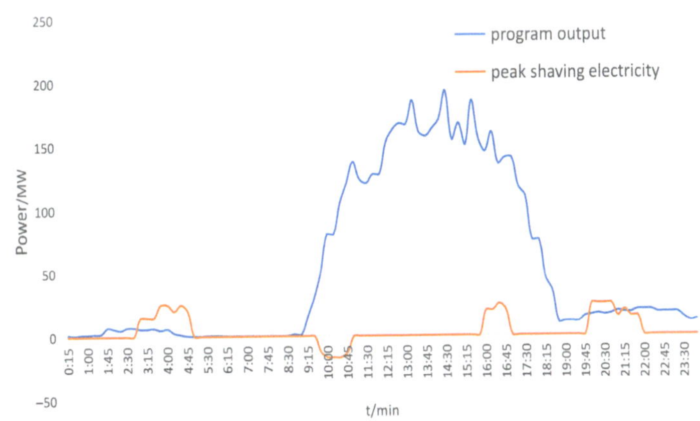

Figure 4. Program output and peak shaving command curve.

Table 2. Operation parameters.

Parameter	Value
$P_{\text{wt}}^{\text{min}}$/MW	0
P_{wt}^{max}/MW	25
$P_{\text{pv}}^{\text{min}}$/MW	0
$P_{\text{pv}}^{\text{max}}$/MW	200
$\eta_{\text{bat}}^{ch}, \eta_{\text{bat}}^{disch}$	0.9
$SOC\text{min}$	0.2
$SOC\text{max}$	0.8
The initial state of SOC	0.5
E_{bat}^{rated}/MW	20
$P_{\text{ele}}^{\text{min}}$/MW	6.25
P_{ele}^{max}/MW	25
$\eta_{\text{ele}}/(\text{Nm}^3/MWh)$	190
$\eta_{\text{fc}}/(\text{Nm}^3/MWh)$	555.56
$V_{hs}^{\text{max}}/\text{Nm}^3$	30,000
$V_{hs}(1)/\text{Nm}^3$	2000

Table 3. Economic parameters.

Parameter	Value
$A_{\text{hyd}}/(\text{CNY}/\text{Nm}^3)$	3
$A_{co2}/(\text{CNY}/\text{kg})$	0.075
$\lambda_{\text{p}}/(\text{kg}/\text{MWh})$	500
$\lambda_{\text{q}}/(\text{kg}/\text{MWh})$	798
$\gamma_{\text{fc}}/(\text{CNY}/\text{MWh}\cdot\text{Day})$	1136
$\gamma_{\text{hs}}/(\text{CNY}/\text{Nm}^3\cdot\text{Day})$	0.5
$\gamma_{\text{ele}}/(\text{CNY}/\text{MW}\cdot\text{Day})$	1748
$\gamma_{\text{bat}}/(\text{CNY}/\text{MW}\cdot\text{Day})$	850
$\gamma_{\text{loss}}/(\text{CNY}/\text{MWh})$	300
A_{buy} (0:00 to 8:00)/(CNY/MWh)	135.0
A_{buy} (9:00 to 12:00 and 18:00 to 23:00)/(CNY/MWh)	521.4
A_{buy} (others)/(CNY/MWh)	328.2

6.2. Comparative Analysis of PSO Algorithm before and after Improvement

The mathematical model proposed in this paper is solved by the traditional PSO and PLPSO algorithms, respectively. The two algorithms have the same parameters: $\omega = 0.8$, $c_1 = c_2 = 1.2$, the population size is 100, the maximum number of iterations is 15,000, and both algorithms run 100 times under the scenario with a confidence level of 0.9. Table 4 gives a comparison of the results of the traditional PSO and PLPSO algorithms.

Table 4. Comparison between traditional PSO and PLPSO.

	PSO	PLPSO
Number of runs	100	100
average run time/s	261.3	248.4
mean value	−654,818.07	−768,875.44
optimal value	−807,479.2	−830,172.1
mean absolute deviation (MAD)	62,614.6	17,792.2

As can be seen from Table 4, the PLPSO algorithm is superior to the PSO algorithm in terms of running time, average value, and optimal value. Through the calculation of the average value, it can be known that the fitness value of the PLPSO algorithm is 114,057.37 smaller than that of the traditional PSO algorithm, and the convergence accuracy is improved by 17.4%; the mean absolute deviation (MAD) measures the average distance between data points and their average value. A smaller average absolute deviation value indicates that the data are more stable. The MAD of the data obtained by the PLPSO algorithm is 71.9% smaller than that of the PSO, indicating that the PLPSO algorithm is more stable. The convergence curves of the fitness values of the two algorithms are shown in Figure 5.

As can be seen from Figure 5, due to the use of piecewise mapping by the PLPSO algorithm, the initial fitness value of the PLPSO algorithm exhibits improvement over the initial fitness value of the traditional PSO algorithm; the PLPSO algorithm terminates convergence around the 5000th generation and the traditional PSO algorithm terminates convergence around the 7500th generation. In comparison to the traditional PSO algorithm, the PLPSO algorithm exhibits an approximately 33.3% faster convergence speed. As evident from the zoomed-in figure, at the end of convergence, the convergence accuracy of the PLPSO algorithm is 14.9% higher than that of the traditional PSO algorithm. Therefore, the PLPSO algorithm is superior to the traditional PSO algorithm.

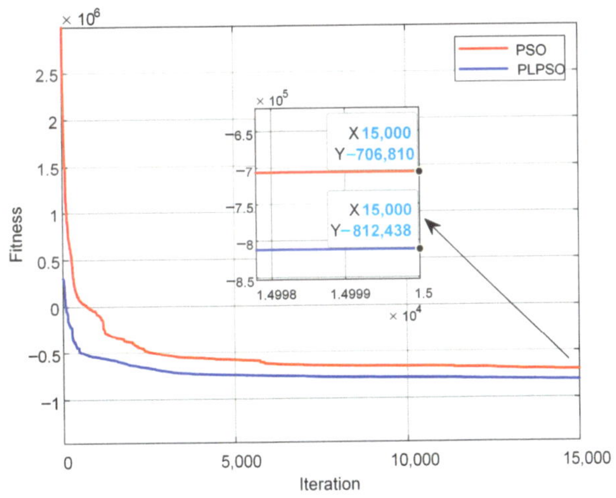

Figure 5. Comparison of convergence curves of two algorithms.

6.3. Analysis of the Impact of Varying Confidence Levels on Decision-Making Outcomes

The confidence level significantly affects the system's economy and safety, and different confidence levels will lead to different scheduling results. As the confidence level increases, the system's rate of WSA and comprehensive benefits are shown in Table 5.

Table 5. Scheduling results under different confidence levels.

Confidence Level	Proportion of the Wind and Solar Energy Abandon	Comprehensive Benefit/CNY
0.55	9.97%	720,723
0.60	7.17%	714,133
0.65	5.36%	731,836.8
0.70	2.06%	736,225.2
0.75	2.32%	730,571.3
0.80	1.98%	738,359.2
0.85	1.25%	748,917.6
0.90	1.42%	762,298.3
0.95	1.16%	774,511.2
1.00	0.67%	784,582.3

As shown in Table 5, as the confidence level continues to increase, the system's wind and light abandonment rate generally shows a downward trend; under the premise that the installed capacity of the EHPS and the electrochemical energy storage system is certain, when the confidence level is low, due to the insufficient installed capacity of the EHPS and the electrochemical energy storage system, the rate of WSA is larger. To mitigate the rate of WSA, a portion of its capacity is allocated to participate in the PSAS, absorbing excess electricity, thereby reducing the economic benefits brought by engaging in the PSAS. Since the plant still has a high rate of WSA, it causes a large WSA penalty. Therefore, when the confidence level is low, the comprehensive benefits are low; when the confidence level is high, due to the sufficient installed capacity of the EHPS and the electrochemical energy storage system, the plant can easily decrease the rate of WSA and has more capacity to engage in PSAS to obtain the highest peak-shaving subsidy. Therefore, as the confidence level increases, the comprehensive benefits show an increasing trend.

The operation of the EHPS and the electrochemical energy storage system is intricately linked to the confidence level. During the scheduling decision-making process, the scheduler must carefully weigh the plant demand, safety, and economy factors to determine the optimal confidence level.

6.4. Analysis of Energy Management Decision-Making Scheme

Taking the confidence level as an example, using the PLPSO algorithm for solving, we can obtain the scheduling situation of the renewable energy output, the EHPS, and the electrochemical energy storage system 24 h before the day. Among them, the peak-shaving subsidy and the peak-shaving command curve are depicted in Figure 6.

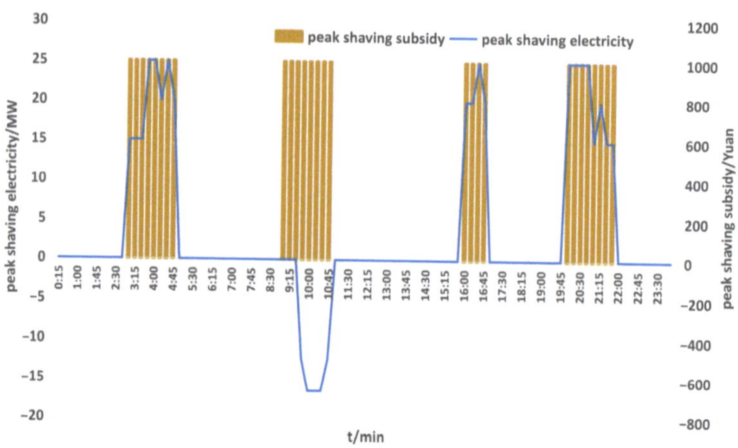

Figure 6. Peak-shaving subsidy and peak-shaving electricity curve.

The renewable energy power plant containing electrolytic hydrogen is derived from the planned output reported to the upper-level power grid, taking into account the time-of-use electricity price, power fluctuation suppression, peak-shaving ladder subsidy, carbon-trading benefits, WSA penalties, and various system operation constraints. It ensures that the actual output fluctuates between 90% and 110% of the planned output. While completing the PSAS, the excess electricity is utilized for hydrogen production (charge) through the EHPS and the electrochemical energy storage system. In the case of insufficient renewable energy output, it discharges. The output curves of the EHPS and the electrochemical energy storage system are displayed in Figures 7 and 8, respectively.

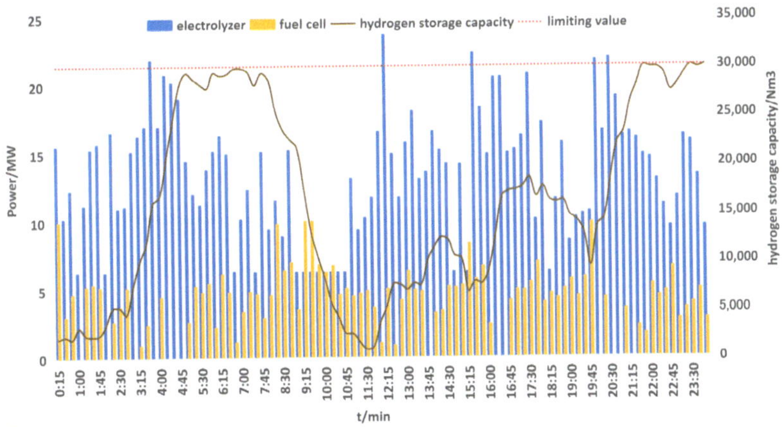

Figure 7. Operation of electrolytic cell, fuel cell, and hydrogen storage tank.

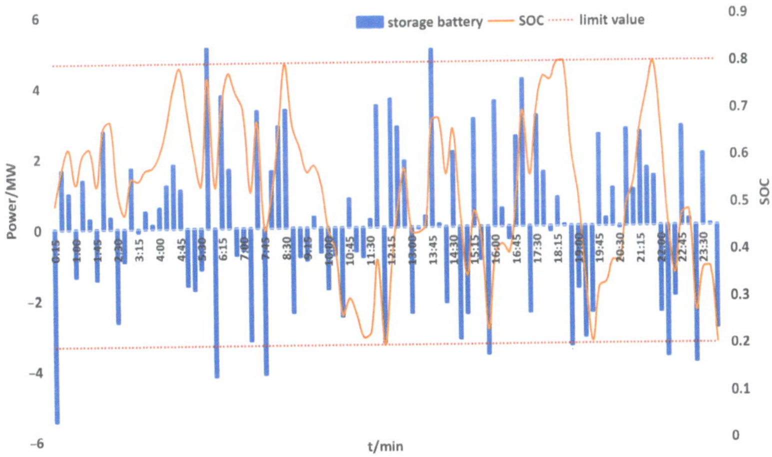

Figure 8. Operation of electrochemical energy storage.

As evident from Figure 7, between 0:00~2:45, 11:00~15:45, and 17:00~19:45, the electrolyzer and fuel cell operate at a certain power to maintain a lower level of hydrogen storage, leaving enough hydrogen storage space for the subsequent periods (3:00~4:45, 16:00~16:45, and 20:00~21:45) to prepare for peak shaving. During the peak-shaving period, the maximum operating power of the electrolyzer is 21.94 MW, the minimum operating power is 14.84 MW, and a total of 68,590 Nm^3 hydrogen is produced; between 5:00 and 8:45, the fuel cell maintains a low-power operation to maintain a higher storage amount of hydrogen in the hydrogen storage tank, preparing for the next period (9:00~10:45) for peak shaving. During the peak-shaving period, the electrolyzer maintains operation at 25% of the rated power (i.e., 6.25 MW) to ensure the electrolyzer operates safely, and the fuel cell is in a high-power working state, with a maximum discharge power of 10 MW, a minimum discharge power of 3.52 MW, a total discharge of 52.02 MW, and a hydrogen consumption of 16,866.8 Nm^3; between 22:00 and 24:00, the electrolyzer operates at high power and the fuel cell operates at low power to ensure that there is a higher amount of hydrogen storage in the hydrogen storage tank at 24:00 to obtain higher hydrogen sales benefits. In one operating cycle, the electrolyzer produces a total of 244,691.03 Nm^3 hydrogen, the fuel cell consumes a total of 218,985.38 Nm^3 hydrogen, and the final hydrogen storage amount is 29,993.85 Nm^3, which is close to the maximum hydrogen storage amount and can obtain higher hydrogen sales benefits.

As shown in Figure 8, during the peak-shaving periods of 3:00~4:45, 16:00~16:45, and 20:00~21:45, the electrochemical energy storage is charged, with a total charge of 23.77 MW in the three peak-shaving periods; during the peak-shaving period of 9:00~10:45, the electrochemical energy storage discharges, with a total discharge of 6.34 MW during the period; the SOC of the electrochemical energy storage is maintained between 0.2 and 0.8, ensuring that the electrochemical energy storage works in a "shallow charge and shallow discharge" state, which helps prolong the service life of the electrochemical energy storage system; the SOC of the electrochemical energy storage is maintained between 0.45 and 0.8 for a total of 66 periods (i.e., 68.75% of a cycle), leaving sufficient power and capacity for scheduling.

Through the operation control of the electrolytic hydrogen production system and the electrochemical energy storage system, the renewable energy power generation system has achieved higher economic benefits under the premise of completing multiple target application scenarios. Among them, the income from PSAS is CNY 524,500, the income from carbon emission reduction is CNY 297,325.2, the income from hydrogen sales is CNY 89,981.55, the cost of electricity purchase is CNY 87,909.82, the penalty for abandoning wind and solar energy is CNY 13,851, and the comprehensive benefit is CNY 810,045.93.

As shown in Figure 9, after calculation, it is found that there is a positive under-compensation of 48.78 MW and a negative under-compensation of 31.18 MW before suppression. The power fluctuates greatly between 9:00 and 7:00, with a total of 14 time periods exceeding the fluctuation limit of renewable energy grid-connected power, and the over-limit probability is 14.58%, which brings a significant negative impact to the power system; after suppression by the EHPS and the electrochemical energy storage system, all the renewable energy outputs participating in the grid connection meet the grid-connected power fluctuation limit.

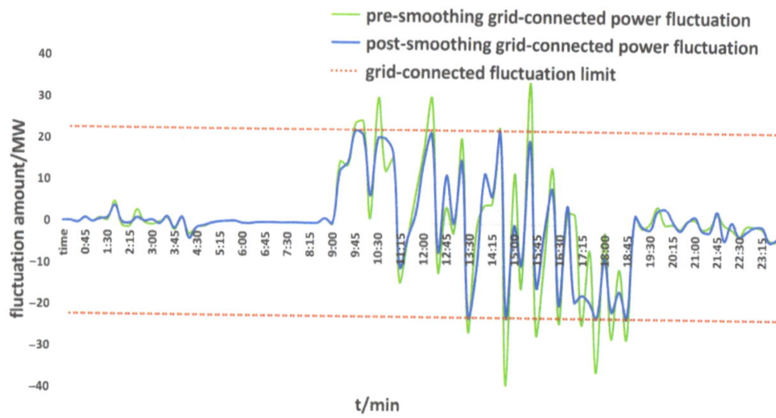

Figure 9. The grid-connected power fluctuation before and after stabilization.

6.5. Effect of Electrolytic Hydrogen Production on Economic Benefits

The effectiveness analysis of the three scenarios—no electrolytic hydrogen with electrochemical energy storage, electrolytic hydrogen without electrochemical energy storage, and electrolytic hydrogen with electrochemical energy storage—are shown in Table 6. The installed capacity of equipment under the three scenarios is as follows:

Table 6. Effectiveness analysis in three scenarios.

	Scenario 1	Scenario 2	Scenario 3
electricity purchasing cost/CNY	28,280.5	71,492.4	87,909.8
abandoned wind power and solar/MW	500.0	308.5	46.2
Hydrogen sales revenue/CNY	0	109,449.0	89,981.6
Peaking revenue/CNY	524,500	512,500	524,500
carbon trading income/CNY	297,789.9	302,926.6	297,325.2
comprehensive benefit/CNY	644,002.6	760,831.5	810,045.9

Scenario 1 (no electrolytic hydrogen with electrochemical energy storage): no electrolyzer, fuel cell, or hydrogen storage tank is set up, and the electrochemical energy storage is 25 MW;

Scenario 2 (electrolytic hydrogen without electrochemical energy storage): in this scenario, the operation and maintenance expenses of the 25 MW electrochemical energy storage in Scenario 3 is proportionally allocated to the EHPS and the fuel cell, among which the electrolyzer is 30 MW, the fuel cell is 10 MW, the hydrogen storage tank is 36,483 Nm^3, and no electrochemical energy storage is set up;

Scenario 3 (electrolytic hydrogen with electrochemical energy storage): the electrolyzer is 25 MW, the fuel cell is 10 MW, the hydrogen storage tank is 30,000 Nm^3, and the electrochemical energy storage is 20 MW. This scenario is the scenario of the proposed mathematical model.

As evident from Table 6, the comprehensive benefit of Scenario 3 proposed in this paper is optimal. Compared with Scenario 1, due to the constraint that electrochemical energy storage cannot perform simultaneous charging and discharging operations, it is impossible to flexibly absorb or release electricity in Scenario 1, and the hydrogen sale benefits brought by electrolytic hydrogen is lacking. Therefore, compared with Scenario 1, Scenario 3 reduces the amount of wind and solar energy discarded by 393.8 MW, reduces the penalty for discarding wind and solar energy by CNY 118,140, increases carbon trading benefits by CNY 10,218.5, and increases comprehensive benefits by CNY 123,247.6. Compared with Scenario 2, due to the lack of electrochemical energy storage and the lower conversion rate of electrolytic hydrogen, Scenario 3 reduces the amount of wind and solar energy discarded by 262.3 MW compared to Scenario 2, reduces the penalty for discarding wind and solar energy by CNY 78,690, and increases comprehensive benefits by CNY 49,214.4. Scenario 3 combines the flexibility of the electrolytic hydrogen production system and the higher conversion rate of the electrochemical energy storage system. Under the premise of completing the multi-objective application scenarios of smoothing power fluctuations, participating in PSAS, and reducing the amount of wind and solar energy discarded, it can still obtain higher comprehensive benefits.

7. Conclusions

This paper focuses on a renewable energy power plant that includes hydrogen production through electrolysis. It considers the use of electrolytic hydrogen to smooth out fluctuations in renewable energy power generation, participate in peak-shaving auxiliary services, and increase the absorption space of renewable energy. It proposes a multi-objective energy management model and algorithm based on the co-ordinated optimization of electrolytic hydrogen and renewable energy power generation. Through simulation, based on the analysis of the operating conditions and safety factors of various facilities in the power station, the advantages of this model for power system safety were carefully studied. With participation in the power market, the system's economics have been greatly improved.

At the same time, the PLPSO algorithm improves the convergence accuracy, convergence speed, and stability of the proposed model compared with the basic algorithm.

The research presented in this paper offers a valuable theoretical reference for energy management in renewable energy-generating plants that include an electrolytic hydrogen production system.

The next step will be to further study the capacity configuration of renewable energy power stations that include an electrolytic hydrogen production system, providing technical support for the installation and construction of renewable energy power stations.

Author Contributions: Conceptualization, T.S.; methodology, T.S. and L.G.; software, L.G.; validation, L.G., J.S. and Z.X.; formal analysis, T.S.; investigation, L.G.; resources, T.S.; data curation, L.G.; writing—original draft preparation, L.G.; writing—review and editing, T.S.; visualization, L.G.; supervision, T.S.; project administration, T.S.; funding acquisition, T.S. All authors have read and agreed to the published version of the manuscript.

Funding: This research was funded by the major project of National Natural Science Foundation of China: "Theory and method of coordinated operation optimization of multi energy supply and demand system containing hydrogen", grant number 62192753.

Data Availability Statement: Data are contained within the article.

Conflicts of Interest: The authors declare no conflicts of interest.

References

1. Li, Z.; Zhang, R.; Sun, H.X. Review on Key Technologies of Hydrogen Generation, Storage and Transportation Based on Multi-Energy Complementary Renewable Energy. *Trans. China Electrotech. Soc.* **2021**, *36*, 446–462.
2. Zhang, P.C.; Xu, J.; Ke, D.P. Optimal Configuration of Energy System in Iron and Steel Park Driven by Hydrogen Energy. *Autom. Electr. Power Syst.* **2022**, *46*, 1–10.
3. Zhang, H.; Yuan, T.J.; Tan, J. Medium and Long-term Forecast of Hydrogen Load in Unified Energy System. *Proc. CSEE* **2021**, *41*, 3364–3372+3662.
4. Cai, G.W.; Kong, L.G.; Xue, Y. Overview of Research on Wind Power Coupled with Hydrogen Production Technology. *Autom. Electr. Power Syst.* **2014**, *38*, 127–135.
5. Jing, P.; Xu, G.Z.; Zhao, B. Large-scale Energy Storage Technology for Global Energy Internet. *Smart Grid* **2015**, *3*, 486–492.
6. Fard, H.H.; Tooryan, F.; Collins, E.R.; Jin, S.; Ramezani, B. Design and optimum energy management of a hybrid renewable energy system based on efficient various hydrogen production. *Int. J. Hydrogen Energy* **2020**, *45*, 30113–30128.
7. Jin, X.; Zhuang, Y.X.; Wang, H.; Zhou, S.Y. Feasibility Analysis Research on Abandoning Wind and Solar Energy with Hydrogen Energy Storage Technology. *Electrotech. Electr.* **2019**, *4*, 63–68.
8. Pu, Y.C.; Li, Q.; Chen, W.R.; Huang, W.Q.; Hu, B.B.; Han, Y.; Wang, X. Energy Management for Islanded DC Microgrid with Hybrid Electric-hydrogen Energy Storage System Based on Minimum Utilization Cost and Energy Storage State Balance. *Power Syst. Technol.* **2019**, *43*, 918–927.
9. Liang, T.; Sun, B.; Mi, D.; Sha, Z. Operation Optimization of Wind-solar Complementary Hydrogen Production System Based on MOGEO. In Proceedings of the 7th International Conference on Power and Renewable Energy (ICPRE), Shanghai, China, 23–26 September 2022. [CrossRef]
10. Cui, Y.; Yan, S.; Zhong, W.Z.; Wang, Z.; Zhang, P.; Zhao, Y.T. Optimal Thermoelectric Dispatching of Regional Integrated Energy System with Power-to-gas. *Power Syst. Technol.* **2020**, *44*, 4254–4264. [CrossRef]
11. Deng, J.; Jiang, F.; Wang, W.Y.; He, G.X.; Zhang, X.H.; Liu, K.C. Low-carbon Optimized Operation of Integrated Energy System Considering Electric-heat Flexible Load and Hydrogen Energy Refined Modeling. *Power Syst. Technol.* **2022**, *46*, 1692–1704. [CrossRef]
12. Shi, R.F.; Ning, J.; Gao, L.Q. Research on Optimal Dispatch Strategy of Wind and Solar Self-consistent Microgrid in Road Transportation System with Hydrogen Energy Storage. *Acta Energiae Solaris Sin.* **2023**, *44*, 513–521. [CrossRef]
13. Mah, A.; Ho, W.S.; Hassim, M.H.; Hashim, H.; Ling, G.H.; Ho, C.S.; Muis, A. Optimization of a standalone photovoltaic-based microgrid with electrical and hydrogen loads. *Energy* **2021**, *235*, 121218. [CrossRef]
14. Wu, F.B. Optimized Configuration of Wind-Solar Hydrogen Storage Integrated Energy System. *Shandong Chem. Ind.* **2020**, *49*, 135–136+138. [CrossRef]
15. Guo, M.J.; Yan, Z.; Zhou, Y. Optimized Operation Design of Integrated Energy System with Wind Power Hydrogen Production. *Electr. Power* **2020**, *53*, 115–123+161.
16. Jiang, Y.W.; Deng, Z.H.; Shi, Y. Size optimization and economic analysis of a coupled wind-hydrogen system with curtailment decisions. *Int. J. Hydrogen Energy* **2019**, *44*, 19658–19666. [CrossRef]
17. Cao, W.; Zhong, X.; Wang, H.H. Optimal Capacity Allocation of Hydrogen Production System Participating Peak Regulation for Auxiliary with Thermal Power Plant. *Distrib. Energy* **2020**, *5*, 15–20. [CrossRef]
18. Cheng, S.; Zhang, J.C.; Han, H.T.; Wei, Z.N.; Sun, G.Q.; Zhou, Y.Z. Optimal Dispatch of Electricity-Gas-Hydrogen Integrated Energy System Considering Auxiliary Service. *Autom. Electr. Power Syst.* **2023**, *47*, 110–120.
19. Jia, C.Z.; Wang, L.M.; Meng, E.L. Optimal Capacity Configuration and Day-Ahead Scheduling of Wind-Solar-Hydrogen Coupled Power Generation System. *Electr. Power* **2020**, *53*, 80–87.

20. Harasis, S.; Sozer, Y.; Elbuluk, M. Reliable Islanded Microgrid Operation Using Dynamic Optimal Power Management. *IEEE Trans. Ind. Appl.* **2021**, *57*, 1755–1766. [CrossRef]
21. Son, Y.-G.; Kim, S.-Y.; Bae, I.-S. Optimal Coordination of Energy Coupling System Considering Uncertainty of Renewable Energy Sources. *Energies* **2024**, *17*, 931. [CrossRef]
22. Zhu, Y.; Lan, Z.B.; Kui, Z. Research on optimal operation of wind-PV-ES complementary system considering carbon emission cost. *Power Syst. Prot. Control* **2019**, *47*, 127–133.
23. Yuan, T.J.; Guo, J.H.; Yang, Z.J. Optimal Allocation of Power Electric-Hydrogen Hybrid Energy Storage of Stabilizing Wind Power Fluctuation. *Proc. CSEE* **2023**, 1–10. [CrossRef]
24. Cui, Y.; Zhou, H.J.; Zhong, W.Z.; Li, H.B.; Zhao, Y.T. Low-carbon scheduling of power system with wind power considering uncertainty of both source and load sides. *Electr. Power Autom. Equip.* **2020**, *40*, 85–93. [CrossRef]
25. Liu, B.D.; Zhao, R.Q.; Wang, G. *Uncertain Programming with Applications*; Tsinghua University Press: Beijing, China, 2003; pp. 178–198.
26. Wang, Q.X.; Guo, X.B. Particle swarm optimization algorithm based on Levy flight. *Appl. Res. Comput.* **2016**, 33, 2588–2591.
27. Zhang, C.; He, X.S.; Ye, Y.R. Particle swarm optimization algorithm based on elite strategy and levy flight. *J. Xi'an Polytech. Univ.* **2018**, *32*, 731–738.

processes

MDPI

Article

Simulation Experiment Design and Control Strategy Analysis in Teaching of Hydrogen-Electric Coupling System

Tao Shi [1,2,*], Jialin Sheng [1], Zhiqiang Chen [1] and Hangyu Zhou [1]

[1] College of Automation, Nanjing University of Posts and Telecommunications, Nanjing 210023, China; 1022051526@njupt.edu.cn (J.S.); 1022051525@njupt.edu.cn (Z.C.); 1022051527@njupt.edu.cn (H.Z.)
[2] Institute of Advanced Technology for Carbon Neutrality, Nanjing University of Posts and Telecommunications, Nanjing 210023, China
* Correspondence: shitao@njupt.edu.cn

Abstract: Hydrogen energy, as a clean and green energy medium, is characterized by large capacity, extended lifespan, convenient storage, and seamless transmission. On the one hand, in the power system, hydrogen can be prepared by the electrolysis of water using the surplus power from intermittent new energy generation, such as photovoltaic and wind power, to increase the space for new energy consumption. On the other hand, it can be used to generate electricity from the chemical reaction between hydrogen and oxygen through the fuel cell and be used as a backup power source when there is a shortage of power supply. In this paper, based on the teaching practice, the conversion mechanism and coupling relationship between various forms of energy, such as photovoltaic energy, hydrogen energy, and electric energy, were deeply analyzed. Further, a hydrogen-electricity coupling digital simulation experimental system, including photovoltaic power generation, fuel cell, and electrolysis hydrogen system, was formed. Simultaneously, considering the synergy between hydrogen production and electricity generation businesses, as well as the demand for the efficient utilization and flexible regulation of multiple energy sources, eight sets of simulation experimental scenarios were designed. A cooperative control strategy for the hydrogen-electric coupling system was proposed and validated through simulation on the MATLAB/SIMULINK-R2023a platform. This study shows that the simulation system has rich experimental scenarios and control strategies, and can comprehensively and accurately demonstrate the multi-energy complementary and cooperative control characteristics of the hydrogen-electric coupling system.

Keywords: teaching experiment; photovoltaic power generation; fuel cell; electrolytic hydrogen production; coordinated control

Citation: Shi, T.; Sheng, J.; Chen, Z.; Zhou, H. Simulation Experiment Design and Control Strategy Analysis in Teaching of Hydrogen-Electric Coupling System. *Processes* **2024**, *12*, 138. https://doi.org/10.3390/pr12010138

Academic Editors: Ferdinando Salata and Virgilio Ciancio

Received: 3 December 2023
Revised: 28 December 2023
Accepted: 3 January 2024
Published: 5 January 2024

1. Introduction

Hydrogen energy, characterized by its cleanliness, high efficiency, large capacity, long lifespan, and ease of storage and transportation, has been extensively applied in diverse fields such as industrial production, energy storage, and transportation. Particularly noteworthy is the capabilities of hydrogen as a chemical energy storage medium, playing a significant role in energy systems. Its elevated specific energy density provides three times greater energy per unit mass compared to gasoline combustion, which offers significant advantages for efficient energy storage and utilization [1]. Meanwhile, as the proportion of wind power and photovoltaic (PV) power in the power system continues to increase, their intermittency, randomness, and volatility have brought challenges and impacts on the security and stability of the power system [2]. By the end of 2022, China's installed wind power capacity reached 760 million kilowatts, accounting for 30% of the total installed capacity for power generation; the PV installed capacity reached 390 million kilowatts, with a year-over-year increase of 28.1%. The increased integration of high proportions of intermittent renewable energy generation has made China's power structure greener and lower in carbon content but has also increased the demand for flexible regulation in the power system. In this context, hydrogen energy, with its flexible characteristics in production, storage, and utilization, has been regarded as a potential flexible regulation resource in the new power system. Comparatively, the traditional gas-thermal power coupling system primarily relies on the conversion and utilization of natural gas, considering the supply demands of both thermal and electrical energy. Natural gas is coupled with the power system in the form of combined heat and power generation, which, to some extent, meets the requirements for the complementary utilization of various forms of energy. However, natural gas is a non-renewable primary energy source, and its combustion process results in carbon dioxide emissions. For hydrogen, the hydrogen-based gas-electric coupling system offers several advantages. Hydrogen can be continuously produced using surplus wind and solar power, and, acting as a crucial industrial product, can be utilized. Additionally, leveraging its characteristics of power generation, heat production, and ease of storage, hydrogen can actively participate in the operation, flexibility, and adjustment of both power and thermal systems. Currently, research on hydrogen energy technologies mainly focuses on fuel cells and electrolysis devices. The literature [3] shows that changing the way hydrogen is produced through renewable electricity can enhance hydrogen–electricity conversion and expand the application range of hydrogen. The literature [4] develops the dynamic modeling and control of proton exchange membrane fuel cells (PEMFCs) and controls the interface components to deliver power to the grid, evaluating the performance of stand-alone power systems. The literature [5] discusses the coupling issue of PEMFCs in scenarios connected to the grid and its impact on efficiency. It analyzes the optimal efficiency conditions and dynamic operational controls using integrated models and iterative algorithms. The literature [6] proposes a power regulation system to ensure high-quality grid injection, applicable to Modified-Y-Source fuel cell inverters. It utilizes LCL filters and PR controllers to optimize the grid injection current quality through parameter adjustment, and simulation experiments demonstrate the effectiveness of the proposed method. Water electrolysis for hydrogen production has advantages such as high current density, high purity, and small footprint [7]. The literature [8] conducted a detailed comparative study on different hydrogen production sources and systems, and its comparisons indicated that photonic options have the highest environmental performance ranking. The literature [9] analyzes two different large-scale electrolytic water technologies for hydrogen production while considering the simulation and determining operational strategies in the market. It compares the levelized hydrogen cost of water electrolysis with SMR factories, evaluating the technical and economic feasibility of water electrolysis in specific regions. The literature [10] presents a hybrid system for hydrogen production, and the modeling simulation is performed in MATLAB to minimize system losses and maximize hydrogen production. The literature [11] introduces a study on a proton exchange membrane electrolysis cell system directly coupled to a 10 kW PV array for hydrogen production. The above literature describes the prospects and technical routes of hydrogen production using renewable energy generation and emphasizes that hydrogen energy will be a key component of the clean energy transition. Further, it also conducts research on the application scenarios, operational characteristics, and operational efficiency improvement methods of hydrogen energy conversion and utilization subsystems, targeting proton exchange membrane fuel cells and electrolytic hydrogen production systems. However, the current research mainly focuses on the energy conversion and utilization technology of hydrogen energy itself, represented by fuel cells and electrolytic hydrogen production systems. There is relatively little research on the coupling modes and synergistic operation mechanisms of hydrogen energy and other energy forms, such as light energy, wind energy, and electric energy.

Therefore, this paper concentrates on the multi-energy complementary and coordinated control characteristics of hydrogen-electric coupling systems, deeply analyzing the conversion mechanisms and coupling relationships among various forms of energy, such as solar energy, hydrogen energy, and electrical energy. A hydrogen-electric coupled digital simulation experimental system was established, which included photovoltaic power generation, fuel cells, and electrolysis hydrogen production systems. Simultaneously, taking into account the synergy between hydrogen production and electricity generation businesses, with goals such as efficient utilization of multiple energy sources, mitigation of power fluctuations, and grid dispatch control, eight sets of simulation experimental scenarios were designed. Corresponding coordinated control strategies for the hydrogen-electric coupling system were proposed, and simulations were conducted using the MATLAB/SIMULINK platform for validation.

2. Typical Structure of Hydrogen-Electric Coupling System

This section presents the design of a typical structure for a hydrogen-electric coupling system that supports the flexible operation of photovoltaic power generation, as shown in Figure 1. In this system, fuel cells and electrolytic hydrogen production devices are used as carriers for hydrogen energy conversion and utilization and are connected in parallel with photovoltaic power generation to the public power grid. When there is an excess of electric energy generated by photovoltaic power generation, it can be converted into hydrogen energy and stored in a hydrogen storage tank through electrolysis. Conversely, when the electricity generated by photovoltaic power generation is insufficient, the hydrogen energy stored in the hydrogen storage tank can be converted back into electricity using a fuel cell. At the same time, because the fuel cell and electrolytic hydrogen production are power electronic interface devices, they have fast dynamic adjustment characteristics. Therefore, they can also be used to suppress the instantaneous fluctuation of photovoltaic power generation output caused by the change of light resources and to execute the instantaneous dispatching control instructions of the superior power grid.

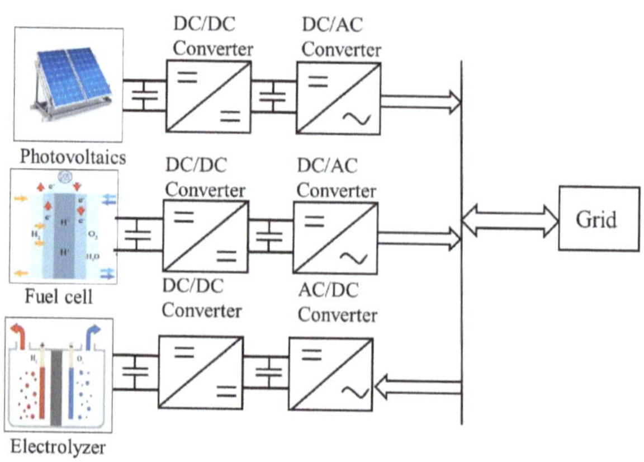

Figure 1. Typical structure of hydrogen-electrode coupling system.

3. Simulation Model of Fuel Cell Power Generation

A fuel cell is a device that converts chemical energy into electricity by electrochemical reactions of hydrogen and oxygen [12]. Taking proton exchange membrane fuel cells (PEMFC) as an example, it supplies hydrogen to the anode and air to the cathode, achieving the separation of hydrogen ions and electrons through the proton exchange membrane. The chemical energy is converted into electrical energy by transferring hydrogen ions from the anode to the cathode via the proton exchange membrane. Simultaneously, electrons travel through an external circuit to the cathode, where they engage in a reaction with oxygen to produce water molecules. The electrochemical reaction formula is as follows:

$$H_2 \rightarrow 2H^+ + 2e^- \tag{1}$$

$$\frac{1}{2}O_2 + 2e^- + 2H^+ \rightarrow H_2O \tag{2}$$

$$2H_2 + O_2 \rightarrow 2H_2O \tag{3}$$

In this paper, a 50 kW PEMFC was taken as an object to establish its dynamic simulation model, as shown in Figure 2. The figure includes two feedback circuits for current feedback and voltage feedback [13,14]. In the figure, response time T_{fc} is used to model the "charge double layer" phenomenon resulting from charge accumulation at the electrode/electrolyte interface [15].

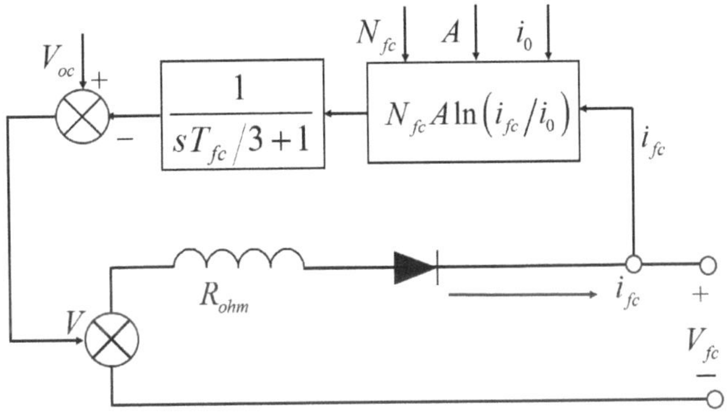

Figure 2. PEMFC equivalent circuit diagrams.

The output voltage of the PEMFC stack is [16]:

$$V_{fc} = V_{oc} - V_{ohm} - V_{act} \tag{4}$$

where: V_{fc} represents the output voltage of the fuel cell; V_{oc} is the open circuit voltage of the fuel cell; V_{ohm} is the ohmic overvoltage loss; and V_{act} represents the absolute polarization voltage loss.

The V_{oc} is [13,16]:

$$V_{oc} = K_c \cdot \left[E_0 + (T_f - 298)\frac{-44.43}{zF} + \frac{RT_{fc}}{zF}\ln\left(\frac{P_{H_2}P_{O_2}^{1/2}}{P_{H_2O}}\right) \right] \tag{5}$$

where: K_c represents the voltage constant under nominal working conditions; E_0 is the electromotive force under standard pressure; T_f represents the operating temperature of fuel cell; z is the number of moving electrons; F is Faraday constant; R is gas constant; P_{H_2} is the partial pressure of hydrogen in the reactor; P_{O_2} represents the partial pressure of oxygen in the reactor; and P_{H_2O} is the partial pressure of water vapor in the reactor.

Considering the ohmic loss caused by electrode and electrolyte resistance, the ohmic overvoltage loss is:

$$V_{ohm} = R_{ohm} \cdot i_{fc} \tag{6}$$

where: R_{ohm} represents the equivalent internal resistance and i_{fc} is the fuel cell current.

Considering the loss caused by slow chemical reaction on the electrode surface, the absolute polarization voltage loss is [15]:

$$V_{act} = N_{fc}A\ln\left(\frac{i_{fc}}{i_0}\right) \cdot \frac{1}{sT_{fc}/3+1} \tag{7}$$

where: N_{fc} represents the number of fuel cells; A is the Tafel slope; i_0 represents the exchange current; and T_{fc} is the response time.

In Equation (7), i_0 and A are [16]:

$$i_0 = \frac{zFk(P_{H_2} + P_{O_2})}{Rh} \bullet \exp(\frac{-\Delta G}{RT}) \tag{8}$$

$$A = \frac{RT}{z\alpha F} \tag{9}$$

where: ΔG is the activation barrier; k represents the Boltzmann constant; α is the charge transfer coefficient; and h is the Planck constant.

Based on the above model, the fuel cell V-I and P-I characteristics are illustrated in Figure 3:

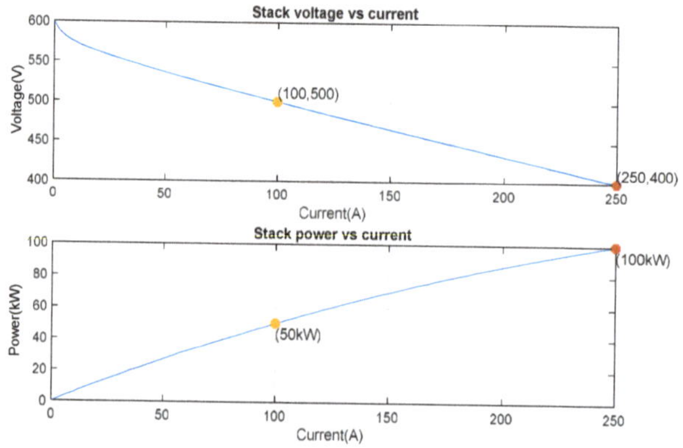

Figure 3. Characteristics of V-I and P-I of fuel cells.

The specific parameters of PEMFC are shown in Table 1:

Table 1. The specific parameters of PEMFC.

PEMFC Parameters	Value
Nominal operating point [Inom (A), Vnom (V)]	[100, 500]
Maximum operating point [Iend (A), Vend (V)]	[250, 400]
Number of cells	300
Nominal Power (kW)	50
Maximal Power (kW)	100
Exchange current (A)	0.54669
Nerst voltage of one cell (V)	1.2041
Fuel cell resistance (ohm)	0.62238
Nominal supply pressure (Fuel (bar))	1.5
Nominal supply pressure (Air (bar))	1
System temperature (Kelvin)	338
Capital cost ($/kW)	400
Lifetime (h)	50,000

4. Simulation Model of Photovoltaic Power Generation

Photovoltaic power generation is a technology that employs the photovoltaic effect to convert solar energy into electrical energy [17]. The current and power output characteristics of photovoltaic systems are influenced by the intensity of illumination and the battery temperature. Photovoltaic cells function similarly to PN junction diodes. Photovoltaic arrays consist of numerous photovoltaic components connected in series and parallel, and their equivalent circuit is depicted in Figure 4.

Based on the equivalent circuit of the photovoltaic array, the relationship between its output voltage V and current I can be expressed as [18,19]:

$$I = I_{pv} - I_d - I_{Rsh} \tag{10}$$

$$I_d = I_0 \left[\exp(\frac{q(V + IR_S)}{\alpha_0}) - 1 \right] \tag{11}$$

where: I_{pv} represents the current generated by the photovoltaic cell due to solar irradiation; I_0 is the saturation current; I represents the load current; V is the output voltage of the photovoltaic cell; R_S represents the series impedance of the photovoltaic cell; and α_0 is the time factor.

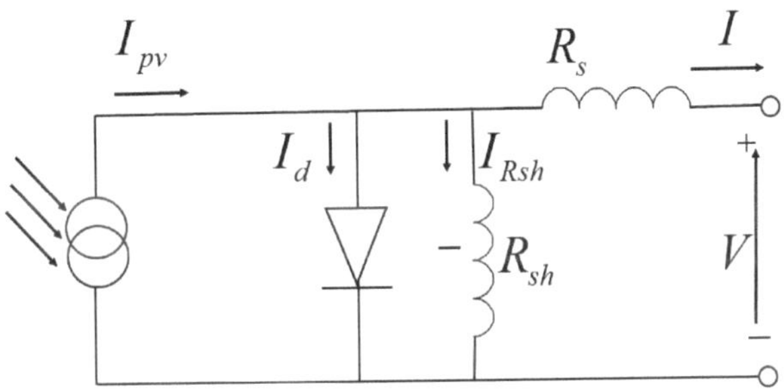

Figure 4. Equivalent circuit diagram of a photovoltaic array.

Since I_{Rsh} is small and can be approximated as 0 [10], Equation (10) can be simplified as follows:

$$I = I_{pv} - I_0 \left[\exp(\frac{q(V + IR_S)}{\alpha}) - 1 \right] \tag{12}$$

The specific parameters of the PV are shown in Table 2:

Table 2. The specific parameters of the PV.

Photovoltaic Parameters	Value
Parallel strings	17
Series-connected modules per string	14
Maximum Power (W)	213.15
Cells per module	60
Open circuit voltage (V)	36.3
Short-circuit current (A)	7.84
Voltage at maximum power point (V)	29
Current at maximum power point Imp (A)	7.35
Nominal temperature (°C)	25

5. Simulation Model of Electrolytic Hydrogen Production

The electrolyzer is a device that utilizes electricity to separate water into hydrogen and oxygen [20,21]. The electrolysis process utilizes an electric current to decompose water molecules into hydrogen and oxygen atoms [22,23]. The electrolyzer is used in the process of electrolyzing water to produce hydrogen. The hydrogen evolution reaction occurs on the cathode side, while the oxygen evolution reaction occurs on the anode side. The cathode and anode undergo the following chemical reactions [24]:

$$H_2O \rightarrow 2H^+ + \frac{1}{2}O_2 + 2e^- \tag{13}$$

$$2H^+ + 2e^- \rightarrow H_2 \tag{14}$$

From a circuit perspective, an electrolyzer is equivalent to a voltage-sensitive nonlinear DC load. Due to the highly nonlinear characteristics of current and voltage, curve fitting is generally used to simulate them, which is applicable to the electrical field. At this point, the voltage-current equation of a single electrolyzer can be calculated [25,26]:

$$U_{el} = U_r + \frac{r_1 + r_2 T_{el}}{A_{el}} I_{el} + s \log\left(\frac{k_{T1} + {k_{T2}}/{T_{el}} + {k_{T3}}/{T_{el}^2}}{A_{el}} I_{el} + 1\right) \tag{15}$$

where: U_{el} represents the working voltage of a single electrolyzer; U_r is the reversible voltage; A_{el} is the electrode area; T_{el} represents the temperature; I_{el} is the current; r_1 and r_2 represent the ohmic resistance parameters; and s, k_{T1}, k_{T2}, and k_{T3} are the coefficient of overvoltage.

The current and voltage characteristics of an electrolyzer are influenced by the operational temperature. In accordance with Faraday's law, the rate of hydrogen production in an electrolytic cell is directly related to the rate of electron transfer at the electrode [27]:

$$N_{H_2} = \eta_e N_e \frac{I_{el}}{zF} \tag{16}$$

$$I_e = \frac{P_e}{N_e U_e} \tag{17}$$

where N_{H_2} represents the rate of hydrogen production, η_e is the electrolyzer efficiency; N_e represents the Number of electrolytic cells; and P_e is the power consumption.

The process of electrolyzing water in an electrolyzer generates hydrogen and oxygen, which are then compressed and stored in a high-pressure hydrogen storage tank. The mathematical model explores the relationship between the pressure of the hydrogen storage tank and its capacity for storing hydrogen [28]:

$$M_{H_2} = \int_{t_1}^{t_2} (N_{H_2} - N_{out}) dt \tag{18}$$

$$P_h = \frac{RT_{el}}{V} M_{H_2} \tag{19}$$

where M_{H_2} represents the storage capacity of hydrogen; P_h is the pressure in the hydrogen storage tank; N_{out} represents the molar hydrogen consumption rate; V is the volume of the hydrogen storage tank; and t_1 and t_2 represent the starting and ending points of hydrogen production.

The modeling data of the electrolysis hydrogen production module are shown in Table 3.

Table 3. The specific parameters of the Electrolysis tank.

Electrolysis Tank Parameters	Value
Reversible Voltage (V)	1.23
Area of Electrode (m^2)	0.1
Number of cells	200
s	0.185
k_{T1}	2.54×10^{-2}
k_{T2}	-0.158
k_{T3}	1.212×10^{3}
r_1	8.232×10^{-5}
r_2	-4.51×10^{-7}
Nominal temperature (°C)	35
Capital cost ($/kW)	100
Lifetime (year)	15

6. Modeling of the Power Electronic Converter

Photovoltaic array, fuel cell, and electrolytic cell grid connection need to use the power electronic converter to realize the transformation of power AC/DC form, to exchange energy and power with the power grid [29]. Because the output voltage of the fuel cell has soft characteristics, its output voltage is affected by the load current, and the photovoltaic power generation has volatility. Therefore, the DC side is required to stabilize the voltage output of the DC side. The Boost circuit topology is shown in Figure 5a. According to the literature [30,31], in terms of complexity, reliability, and cost, the traditional non-isolated DC/DC converter is superior. However, the DC link voltage is typically too high for the electrolytic cell to operate normally. To address this, a Buck circuit is utilized to reduce the voltage to the required operating level. The topology of the Buck circuit is depicted in Figure 5b [30,31].

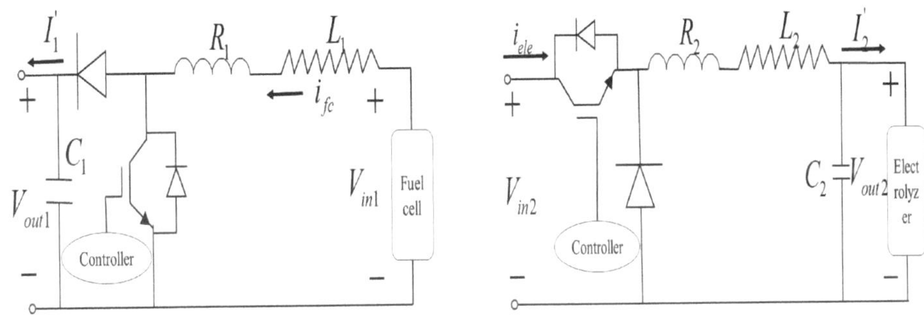

(a) (b)

Figure 5. (**a**) Boost converter topology diagram; (**b**) buck converter topology diagram.

The mathematical model of the Boost circuit can be described using the following differential equation [32]:

$$\begin{cases} \begin{bmatrix} \frac{di_{fc}}{dt} \\ \frac{dV_{out1}}{dt} \end{bmatrix} = \begin{bmatrix} \frac{R_1}{L_1} & \frac{-(1-D_1)}{L_1} \\ \frac{(1-D_1)}{C_1} & \frac{-I_1'}{V_{out1}C_1} \end{bmatrix} \cdot \begin{bmatrix} i_{fc} \\ V_{out1} \end{bmatrix} + \begin{bmatrix} \frac{1}{L_1} \\ 0 \end{bmatrix} V_{in1}, \\ y = \begin{bmatrix} 0 & 1 \end{bmatrix} \cdot \begin{bmatrix} i_{fc} \\ V_{out1} \end{bmatrix} \end{cases} \tag{20}$$

where: i_{fc} represents the inductance current; V_{out1} is the output voltage; V_{in1} is the input voltage; and D_1 represents the control signal.

The mathematical model of the Buck circuit can be described as [32] using the following differential equations:

$$\begin{cases} \begin{bmatrix} \frac{di_{ele}}{dt} \\ \frac{dV_{in2}}{dt} \end{bmatrix} = \begin{bmatrix} \frac{-R_2}{L_2} & \frac{d}{L_2} \\ \frac{d}{C_2} & \frac{-I_2'}{V_{in2}C_2} \end{bmatrix} \cdot \begin{bmatrix} i_{ele} \\ V_{in2} \end{bmatrix} - \begin{bmatrix} \frac{1}{L_2} \\ 0 \end{bmatrix} V_{out2}, \\ y = \begin{bmatrix} 0 & 1 \end{bmatrix} \cdot \begin{bmatrix} i_{ele} \\ V_{in2} \end{bmatrix}, \end{cases} \tag{21}$$

where: i_{ele} is the inductive current; V_{in2} represents the input voltage; V_{out2} is the output voltage; and d represents the control signal.

The inverter and rectifier can be classified according to the voltage source and current source. Due to the similar structure of the inverter and the rectifier in this paper, only the inverter is described, and the inverter topology diagram is shown in Figure 6. The voltage source and inverter in the DC side parallel capacitor are utilized to ensure the DC bus voltage stability. The current source converter is in a series of large inductors on the DC side. Since large inductance reduces the dynamic response of the system, the voltage source inverter is used in this paper.

The grid-type inverter can accurately track the voltage and frequency of the grid to avoid interference with the grid and to ensure that it operates within the normal range. The mathematical model of the inverter in the three-phase coordinate system is shown as follows [33,34]:

$$\begin{cases} U_{aN} = Ri_a + L\frac{di_a}{dt} + e_a \\ U_{bN} = Ri_b + L\frac{di_b}{dt} + e_b \\ U_{cN} = Ri_c + L\frac{di_c}{dt} + e_c \end{cases} \tag{22}$$

where: U_{aN}, U_{bN}, and U_{cN} represent the three-phase output voltage; i_a, i_b, and i_c are the AC side current of the three-phase inverter; R represents the three-phase equivalent resistance; L represents three-phase filter inductance; and e_a, e_b, and e_c are the three-phase power grid voltage.

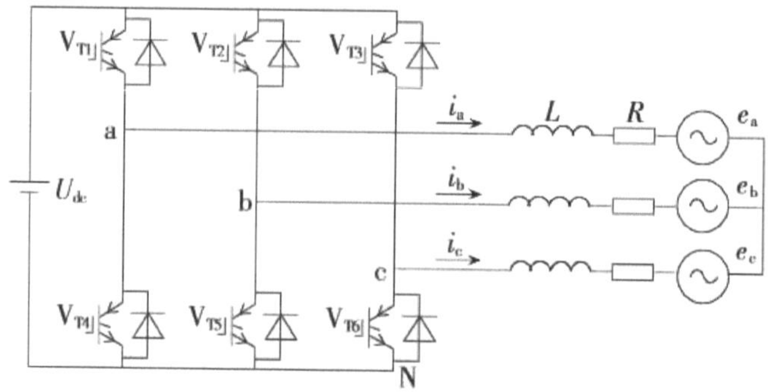

Figure 6. The inverter topology diagram.

The three-phase inverter is inconvenient in designing the inverter control technology. Therefore, it is transformed into the two-phase coordinate system which synchronously rotates at the frequency of the power grid base wave, to facilitate the design of the inverter controller. In the three-phase circuit, the three-phase current is mutually balanced and symmetrical. Based on the principle of power conservation, the current equation of the inverter in the two-phase coordinate system is as follows:

$$\begin{cases} L\frac{di_d}{dt} = u_d^{'} - e_d - Ri_d + wLi_q \\ L\frac{di_q}{dt} = u_q^{'} - e_q - Ri_q - wLi_d \end{cases} \tag{23}$$

where: i_d and i_q represent the current components of the d and q axis; E_d and E_q are the network side voltage components d and q axis; $u_d^{'}$ and $u_q^{'}$ represent the AC output voltage components of the d and q axis; and w is the angular frequency.

The specific parameters of Converter are shown in Table 4:

Table 4. The specific parameters of Converter.

Converter Parameters	Value
Boost converter inductor (H)	0.01
Boost converter capacitor (F)	0.009
Buck converter inductor (H)	0.1
Buck converter capacitor (F)	0.001
Resistor (ohms)	0.1
Filter inductor (H)	0.0015

7. Simulation Scenarios and Control Strategy Design

7.1. Hydrogen-Electric Coupling System Simulation Scenario

To meet the efficient consumption and flexible regulation requirements for grid-connected photovoltaic power generation, this section uses a common AC bus structure based on a hydrogen energy system. Eight simulation scenarios of hydrogen-electric coupling systems are proposed, as shown in Figure 7, where: P_{PV} represents the output power of photovoltaic power generation, P_{load} denotes the power demand of the load, P_s represents the surplus power, P_{grid} represents the power demand from the grid, P_{fc} represents the output power of fuel cell generation, and P_{ele} represents the power consumption of the electrolysis cell.

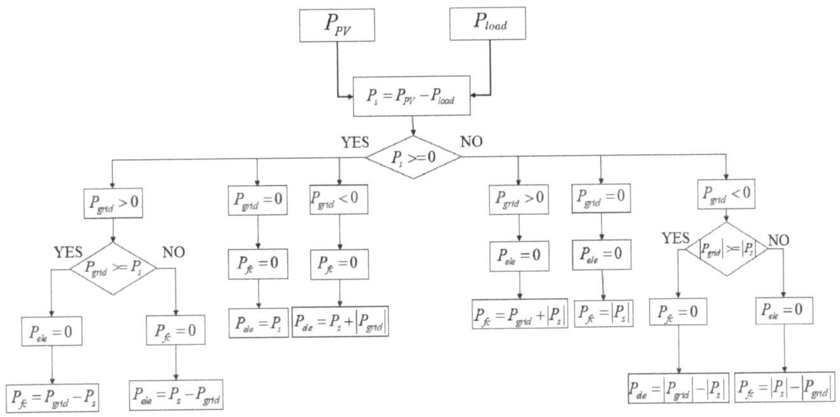

Figure 7. Hydrogen-electric coupling system experimental scenario.

Scenario 1: When photovoltaic power generation can meet the load demand, but not the grid demand, the fuel cell serves as a backup power source to provide the required energy, and the electrolysis cell ceases operation.

Scenario 2: When photovoltaic power generation can simultaneously meet both load and grid demands, the fuel cell stops operating, and the surplus energy is utilized for hydrogen production in the electrolysis cell.

Scenario 3: When photovoltaic power generation satisfies the load demand without grid demand, the fuel cell stops operating, and the surplus energy is used for hydrogen production in the electrolysis cell.

Scenario 4: When photovoltaic power generation meets the load demand and there is an excess of grid electricity, the fuel cell stops operating, and the surplus energy is directed toward hydrogen production in the electrolysis cell.

Scenario 5: When photovoltaic power generation cannot meet both the load and grid demands, the fuel cell serves as a backup power source to provide the required energy, and the electrolysis cell ceases operation.

Scenario 6: When photovoltaic power generation can meet the load demand without grid demand, the fuel cell serves as a backup power source to provide the required energy, and the electrolysis cell stops operating.

Scenario 7: When photovoltaic power generation cannot meet the load demand with an excess of grid electricity in the system and both sources together can meet the load demand, the fuel cell stops operating, and the surplus energy is directed towards hydrogen production in the electrolysis cell.

Scenario 8: When photovoltaic power generation cannot meet both load and grid demands with an excess of grid electricity in the system and the combined operation is still insufficient to meet the load demand, the fuel cell serves as a backup power source to provide the required energy, and the electrolysis cell ceases operation.

7.2. Cooperative Control Strategy of Fuel Cell and Photovoltaic Power Generation

The photovoltaic array adopts the maximum power point tracking (MPPT) technology, and its basic principle is shown in Figure 8 where U_{pv} is the photovoltaic voltage and I_{pv} is the photovoltaic current. After the MPPT control, the reference voltage U_{pvref} of the maximum power operating point is the output, which is compared with the actual photovoltaic voltage to generate the switching signal of the boost converter.

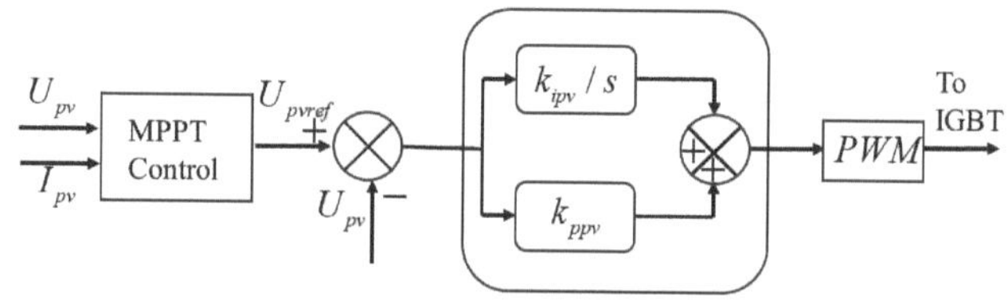

Figure 8. Photovoltaic power generation MPPT control schematic diagram.

In the MPPT control algorithm, the incremental conductance method calculates the maximum power point tracking by comparing the instantaneous conductance of photovoltaic power generation with the variation of conductance. It has the characteristics of anti-interference in the external environment, can respond to changes smoothly and quickly, reduce voltage oscillation, and improve control effect and stability. The flow chart is shown in Figure 9 where: U_k and I_k are the real-time values of photovoltaic voltage and current, dI is the change of current, dU is the change of voltage, dI/dU is the conductance increment, and U_{ref} is the reference voltage.

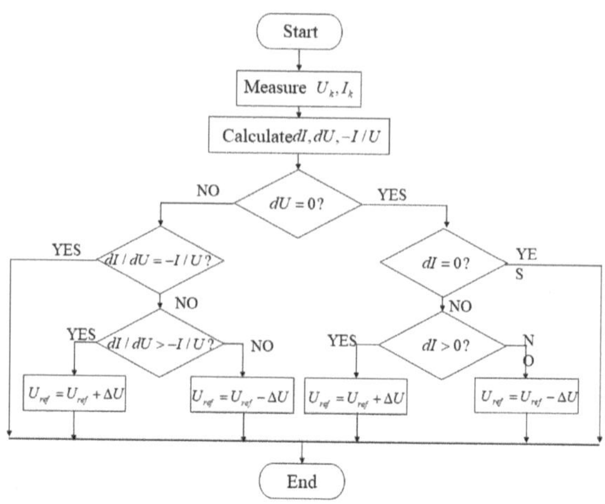

Figure 9. MPPT control flow chart based on incremental conductance method.

The fuel cell system is used as a backup power generation system for the operating conditions when the new energy output cannot meet the load and grid demand.

The V-I and P-I characteristics of the fuel cell, as depicted in Figure 3, show that the port voltage and output power of the fuel cell stack can be adjusted. This is achieved by changing the output current of the fuel cell stack [35].

The fuel cell grid-connected control is mainly divided into power control and inverter control. Figure 10 is the fuel cell grid-connected power control strategy. The basic principle is to compare the reference current with the actual current and use PI control to generate the switching signal of the boost converter, where: P_{fc_ref} is the reference value of the fuel cell output power; V_{fc} is the actual voltage value of the fuel cell; i_{fc_ref} is the reference value of the fuel cell output current; i_{fc} is the actual current of the fuel cell; and k_{ifc} and k_{pfc} are the PI control parameters of the fuel cell power.

In Figure 11, the fuel cell grid-connected inverter control strategy is adopted. The basic principle is to use voltage and current double closed-loop control and complete the current inner loop feedforward decoupling control with the help of traditional PI control. The voltage outer loop control aims at the output voltage of the DC side, stabilizes the DC bus voltage, and provides the current reference value for the inner loop control. In this figure: U_{dcref} is the reference value of DC bus voltage; U_{dc} is the actual DC bus voltage; and i_{qref} is the reference value of q-axis current.

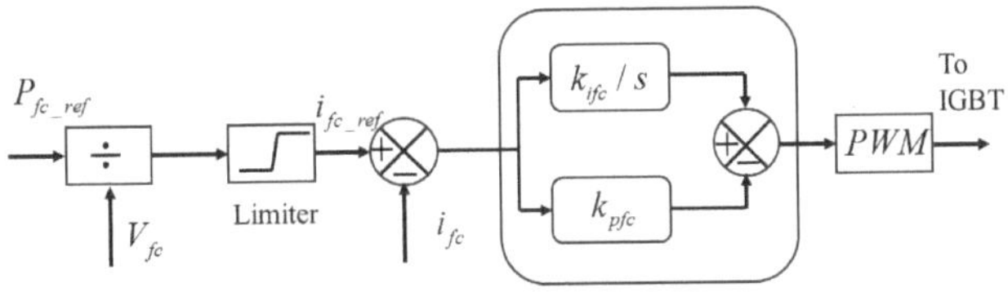

Figure 10. Fuel cell power generation control schematic diagram.

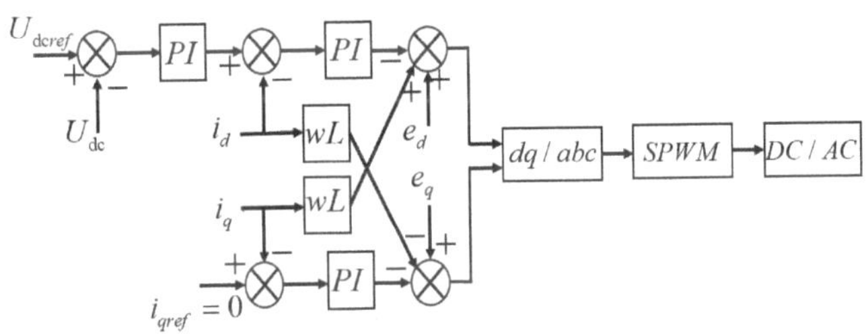

Figure 11. Grid-connected inverter control strategy diagram.

7.3. The Collaborative Control Strategy for Electrolytic Hydrogen Production and Photovoltaic Power Generation

According to the supply and demand situation, the operation of the water electrolysis hydrogen production equipment must be adjusted. When there is abundant electricity and excess supply, the extra electricity will be supplied to the water electrolysis equipment to increase hydrogen production and improve energy utilization. When the electricity supply is tight or even insufficient, the water electrolysis equipment will reduce or suspend hydrogen production. The hydrogen production system in the electrolytic cell converts the excess electricity into hydrogen energy through a power electronic converter, providing fuel for hydrogen fuel cells to achieve regulatory purposes.

The control strategies for hydrogen production through water electrolysis mainly include rectification control and power control. From Figures 10 and 12, it can be seen that both fuel cells and electrolytic cells use the same control principle, but the current direction is different, so both use the same control parameters where: P_{ele_ref} is the reference value of power consumption for electrolytic hydrogen production; V_{ele} is the Actual voltage value of electrolytic hydrogen production; i_{ele_ref} is the reference value of input current for electrolytic hydrogen production; i_{ele} is the actual current of electrolytic hydrogen production; and k_{iele} and k_{pele} are the pi control parameters for electrolytic hydrogen production power.

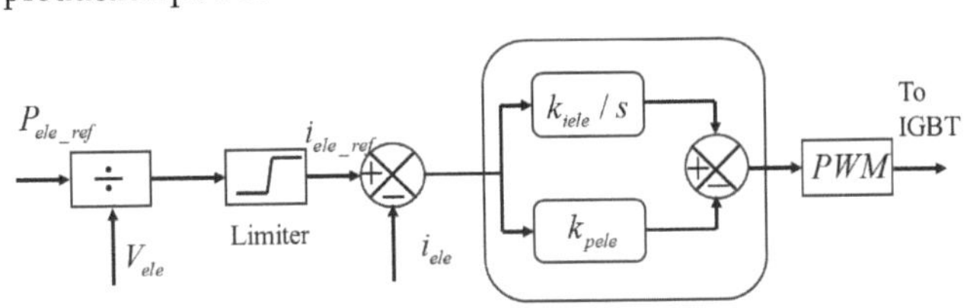

Figure 12. Principle diagram of power control for electrolytic hydrogen production.

8. Simulation Analysis

In order to verify the applicability of the established simulation model and the control strategy in different experimental scenarios, and to analyze the dynamic regulation and response characteristics of the hydrogen-electric coupling system, a hydrogen-electric coupling simulation and experimental system containing the above-mentioned model and the control strategy is established based on the MATLAB/Simulink-R2023a platform in this section, as shown in Figure 13. In this system, the rated capacity of the photovoltaic power generation is 50.73 kW, the rated capacity of the fuel cell is 50 kW, the rated capacity of the electrolysis hydrogen plant is 50 kW, and the hydrogen-electricity coupling system is converged through the AC bus, with a rated voltage of 380 V, and connected to the public power grid with rated frequency of 50 Hz.

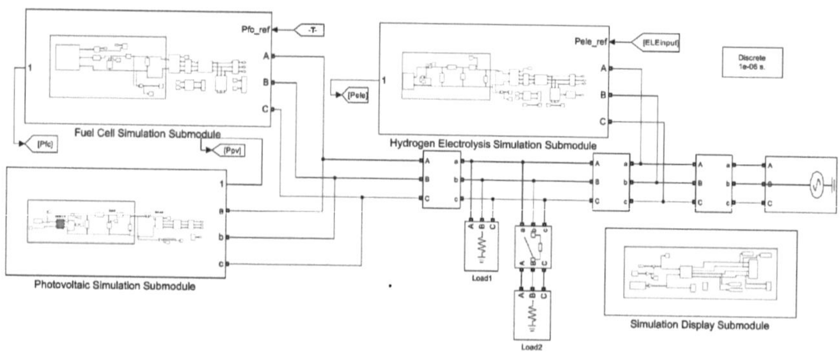

Figure 13. Simulation diagram of the hydrogen-electric coupled system.

8.1. System Response Characteristics under Different Simulation Scenarios

As depicted in Figure 14a–c, the black line represents the reference value, while the red line corresponds to the actual output value. In Figure 14d, the initial load demand was set to 20 kW, and the load increased to 70 kW at t = 2 s. The initial PV output power was 50.73 kW, and the PV output power decreased to 38.395 kW at t = 0.8 s and increased to 50.73 kW at t = 1.2 s.

In the system illustrated in Figure 14d, photovoltaic (PV) and hydrogen systems operate synergistically to achieve optimal performance under different conditions. When surplus electrical energy is generated by a PV, it can be utilized for electrolytic hydrogen production, converting electrical energy into hydrogen, which is stored in the hydrogen storage tank. Conversely, when the electricity generated by a PV is insufficient, a fuel cell can be employed to convert stored hydrogen back into electrical energy. This coupling of hydrogen and electrical systems ensures the complementary utilization of energy forms among solar energy, hydrogen energy, and electrical energy.

At t = 0 s, the grid-connected power was initially set to 70 kW. At this time the system was working in simulation scenario 1. It was difficult to maintain stable operation of the system when only relying on the PV output, and the reference value of the given output power of the fuel cell rose from 0 kW to 39.27 kW.

When t = 0.5 s, the grid power was reduced to 10 kW. At this time, the system was working in simulation scenario 2. Due to the reduced demand for grid connection, the reference value of electrolytic hydrogen consumption power rose from 0 kW to 20.73 kW, and the reference value of the output power of the fuel cell was reduced to 0 kW.

At t = 0.8 s, the grid-connected power remained unchanged and the PV output power dropped to 38.395 kW. At this time, the system was still working in simulation scenario 2, and, due to the reduction of PV output power, the given reference value of electrolytic hydrogen consumption power was reduced from 20.73 kW to 8.395 kW.

At t = 1 s, the grid-connected power decreased to 0 kW. At this time, the system was working in simulation scenario 3, and the given reference value of electrolytic hydrogen consumption power increased from 8.395 kW to 18.395 kW due to the decrease in grid-connected demand.

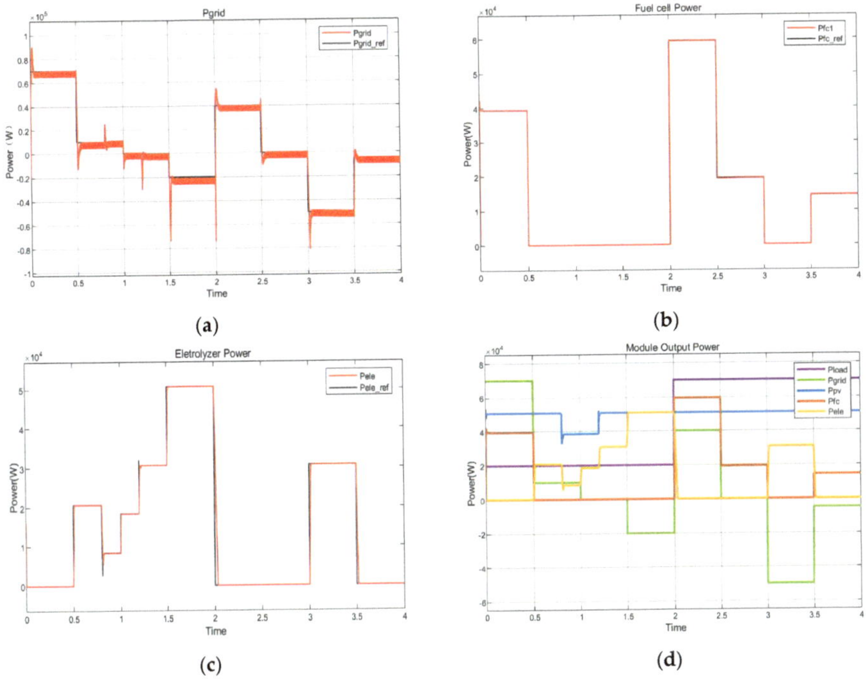

Figure 14. (**a**) Variation curves of grid-connected output power tracking reference value in 8 simulation scenarios; (**b**) Variation curves of fuel cell output power tracking reference value in 8 simulation scenarios; (**c**) Variation curves of electrolysis hydrogen consumption power tracking reference value in 8 simulation scenarios; and (**d**) Variation curves of output power in 8 simulation scenarios for each module.

At t = 1.2 s, the grid power remained unchanged and the PV output power rose to 50.73 kW. At this time, the system was still working in simulation scenario 3, and, due to the rise of the PV output power, the reference value of the given electrolytic hydrogen consumption power rose from 18.395 kW to 30.73 kW.

At t = 1.5 s, the grid-connected power dropped to −20 kW, and then the system worked in simulation scenario 4. The given reference value of electrolytic hydrogen consumption power rose from 30.73 kW to 50.73 kW due to the excess power in the grid.

At t = 2 s, the grid power rose to 40 kW and the load demand rose to 70 kW. At this time, the system was working in simulation scenario 5. Due to the rise in both the grid demand and the load demand, the PV output was not enough to meet the demand, and the reference value of the given fuel cell output power rose from 0 kW to 59.27 kW.

At t = 2.5 s, the grid-connected power decreased to 0 kW. At this time, the system worked in simulation scenario 6. Due to the decrease in the grid-connected demand, the reference value of the given fuel cell output power decreased from 59.27 kW to 19.27 kW.

At t = 3 s, the grid-connected power was reduced to −50 kW. At this time, the system worked in simulation scenario 7, and the reference value of the given electrolytic hydrogen consumption power rose from 0 kW to 30.73 kW due to the surplus power of the grid.

At t = 3.5 s, the grid-connected power rose to −5 kW. At this time, the system worked in simulation scenario 8, and the reference value of the given fuel cell output power rose from 0 kW to 25.27 kW due to the insufficiency of PV output and the surplus of grid power to meet the load demand.

From the above analysis, it can be seen that the fuel cell and electrolyzer could quickly track the demand for flexible regulation of the PV grid-connected system in different scenarios, and accurately respond to the power regulation commands in different scenarios, thus, highlighting the flexible and highly efficient operation of the whole hydrogen-electric coupling system.

8.2. Photovoltaic Cells Energy Conversion

The PV irradiation intensity setting curve and its PV output variation curve are shown in Figure 15.

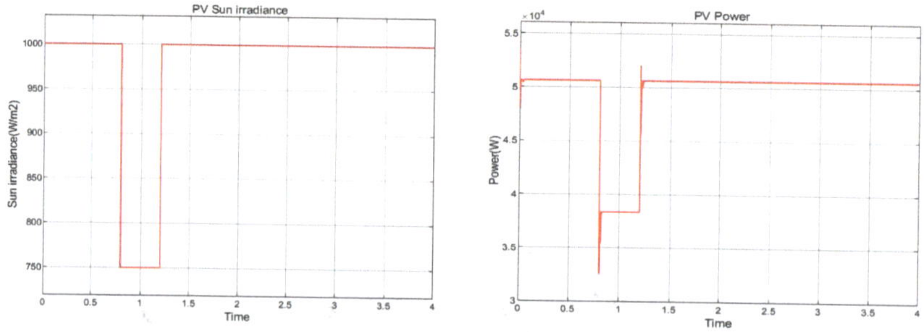

(a) (b)

Figure 15. (**a**) Variation curve of light irradiation intensity; (**b**) Variation curve of photovoltaic output.

As depicted in Figure 15a, the PV irradiation intensity was initially set to 1000 (W/m^2), the PV irradiation intensity decreased to a value of 750 (W/m^2) at t = 0.8 s, and the PV irradiation intensity rose to 1000 (W/m^2) at t = 1.2 s. In Figure 15, the power emitted by the PV rises and falls as the light intensity increases and decreases. At the same temperature, the PV power is consistent with the change in light intensity.

8.3. Fuel Cell Energy Conversion

As shown in Figure 16a, during the dynamic changes of the fuel cell, due to its dynamic characteristics, there will be a certain lag in the process of sudden changes, but it can still enter a steady state within 0.1 s. The dynamic characteristics of the fuel cell voltage were observed at 0 s, 0.5 s, 2 s, 2.5 s, 3 s, and 3.5 s, while the dynamic characteristics of the fuel cell current and power were observed at 0 s and 2 s. Compared to the fuel cell voltage, its current change sensitivity is higher, and the current response is faster. As shown in Figure 16b, the fuel consumption curve of the fuel cell shows a positive correlation with the power curve.

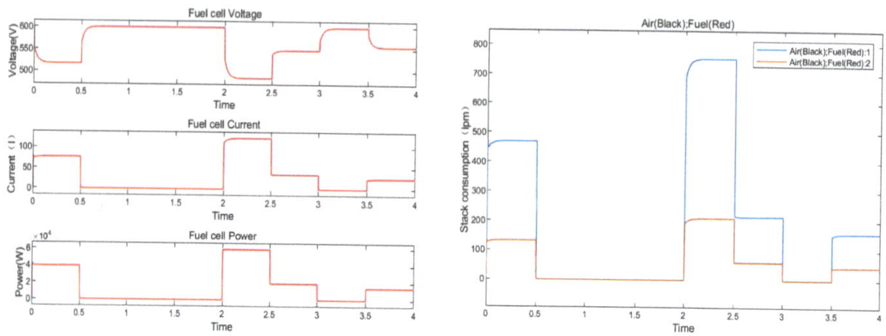

(a) (b)

Figure 16. (**a**) Comparison Chart of V-I-P Changes in Fuel Cells; (**b**) Fuel Consumption Chart of Fuel Cells.

8.4. Electrolysis to Hydrogen Energy Conversion

The rate change curve of electrolytic hydrogen production is shown in Figure 17. When t = 0.5 s, during the startup of the electrolytic hydrogen production unit, the hydrogen production rate increased from 0 (mol/s) to 6.4 (mol/s). When t = 0.8 s, the hydrogen production rate decreased to 2.6 (mol/s). When t= 1 s, t = 1.2 s and t = 1.5 s, the hydrogen production increased to 5.7 (mol/s), 9.5 (mol/s) and 15.6 (mol/s). When t = 1.5 s, the hydrogen production rate reached its maximum value. When t = 2 s and t = 3.5 s, the electrolysis hydrogen production device stopped running and the hydrogen production rate dropped to 0 (mol/s). When t = 3 s, due to the restart of the electrolysis hydrogen production device, the hydrogen production rate rose again from 0 (mol/s) to 9.5 (mol/s).

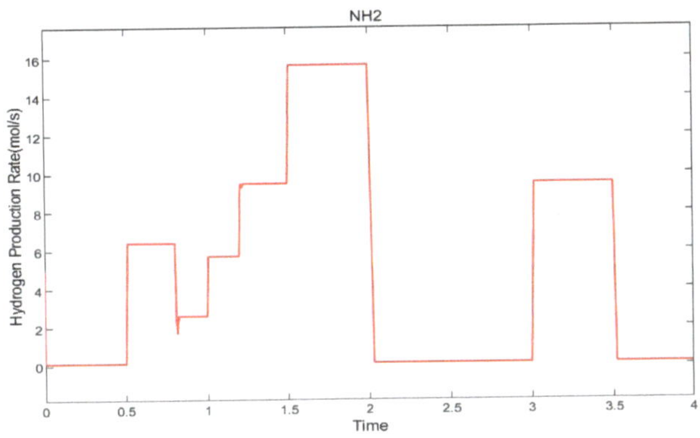

Figure 17. Hydrogen production rate of the electrolyzer.

Figure 18 illustrates the hydrogen production and storage capacity of the electrolyzer, as well as the pressure of the hydrogen storage tank. By comparing these two figures, their changing trends were consistent, and their upward slope was related to the hydrogen production rate. Between 2–3 s and 3.5–4 s, the hydrogen production rate was 0, while the hydrogen storage capacity and hydrogen storage tank pressure remained unchanged. Between 0.5–2 s and 3–3.5 s, the higher the hydrogen production rate, the greater the hydrogen storage capacity and the slope of pressure rise in the hydrogen storage tank. Assuming that the hydrogen produced by electrolytic hydrogen generation in this paper, after meeting the backup demand of fuel cells, remains surplus, it can be sold as an industrial product and applied in other fields.

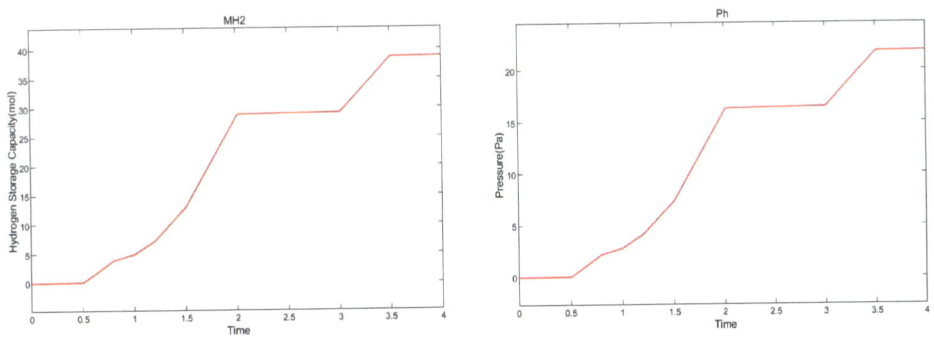

(a) (b)

Figure 18. (**a**) Hydrogen production and storage capacity variation curve of electrolyzer; (**b**) Pressure variation curve of hydrogen storage tank.

8.5. Dynamic Response Characteristics of the Power Electronic Converter

The dynamic response characteristics of the photovoltaic module inverter are shown in Figure 19. Since the outer ring of the inverter was controlled by the DC bus, the set value was 800 V. The reactive power was controlled by q-axis current, and the set value was 0 Var. The figure illustrates that the actual DC bus voltage of the photovoltaic module was effectively maintained at the set value of approximately 800 V. In the dynamic change of t = 0 s, t = 0.8 s, and t = 1 s, there was a short fluctuation. The start-up fluctuation was about 30 V, and the basic fluctuation value was maintained at 10 V. The active power output curve was kept consistent by the photovoltaic power generation curve, and the reactive power was maintained at the set value of 0 Var. The Power losses of the photovoltaic module inverter were maintained at 1100 W, indicating relatively low losses.

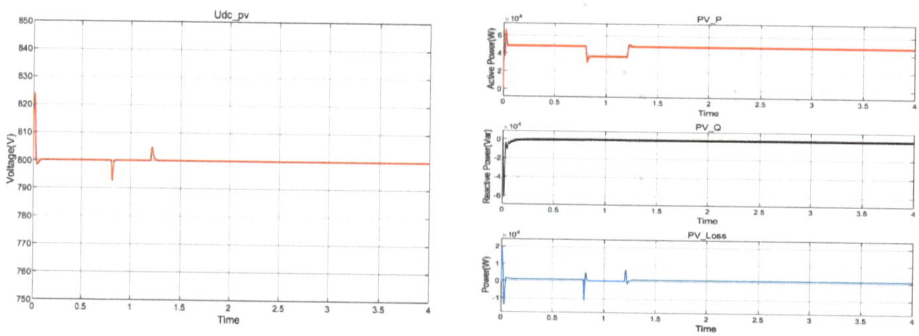

(a) (b)

Figure 19. (**a**) Dynamic voltage response characteristics of DC bus of photovoltaic module; (**b**) Dynamic response characteristics of photovoltaic module power.

The dynamic response characteristics of the fuel cell inverter are illustrated in Figure 20, and its control strategy is consistent with that of the photovoltaic module inverter. The active power output curve of the fuel cell inverter conformed to the reference value of the fuel cell power. Reactive power was maintained at 0 Var. The actual DC bus voltage of the fuel cell module was maintained at the set value of 800 V, and the dynamic fluctuation value was maintained at about ±20 V. The fluctuation is large at t = 0 s, t = 2 s, and t = 2.5 s. The power losses of the fuel cell inverter were maintained at around 1000 W, which was similar to the power losses of the photovoltaic inverter.

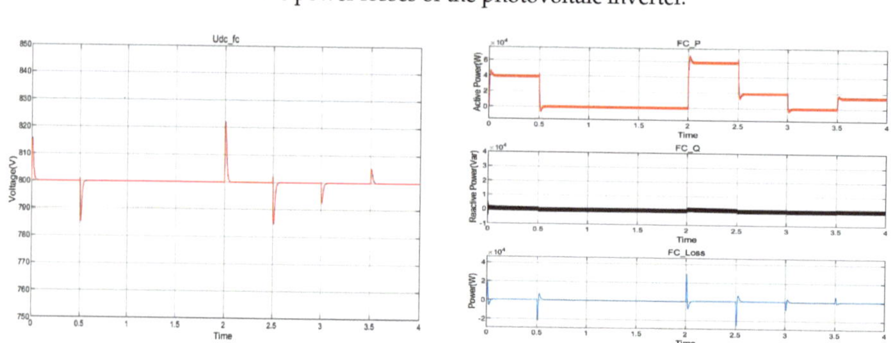

(a) (b)

Figure 20. (**a**) Dynamic voltage response characteristics of DC bus of fuel cell module; (**b**) Dynamic power response of fuel cell module.

The dynamic response characteristics of the electrolytic hydrogen rectifier are shown in Figure 21, and its control strategy was similar to the inverter control strategy. According to the figure, the active power of electrolytic hydrogen production was negative because it absorbed the surplus power from the power grid. The active power curve of the rectification output of electrolytic hydrogen production conformed to the reference value of electrolytic hydrogen production power, and the reactive power was maintained at 0 Var. The actual DC bus bar of the electrolytic hydrogen production module was maintained at the set value of 800 V, and the dynamic fluctuation value was maintained at about ± 20 V. The fluctuation was large at t = 1.5 s, t = 2 s, and t = 3 s. The power losses of the electrolytic hydrogen rectifier were maintained at around 1200 W, indicating relatively low losses.

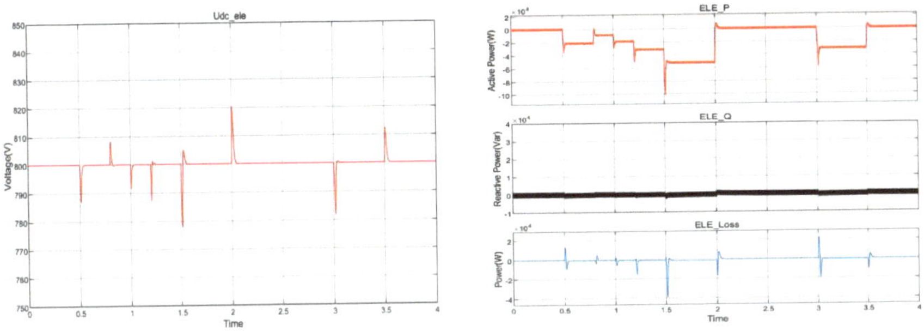

(**a**) (**b**)

Figure 21. (**a**) DC bus voltage dynamic response characteristics of electrolytic hydrogen production module; (**b**) Power dynamic response characteristics of electrolytic hydrogen production module.

The three-phase voltage and current diagram of the photovoltaic module, fuel cell module, and electrolytic hydrogen production module are shown in Figure 22. Because the converters use grid converters, the phase voltage of each module was 311 V, which was consistent with the phase voltage of the power grid.

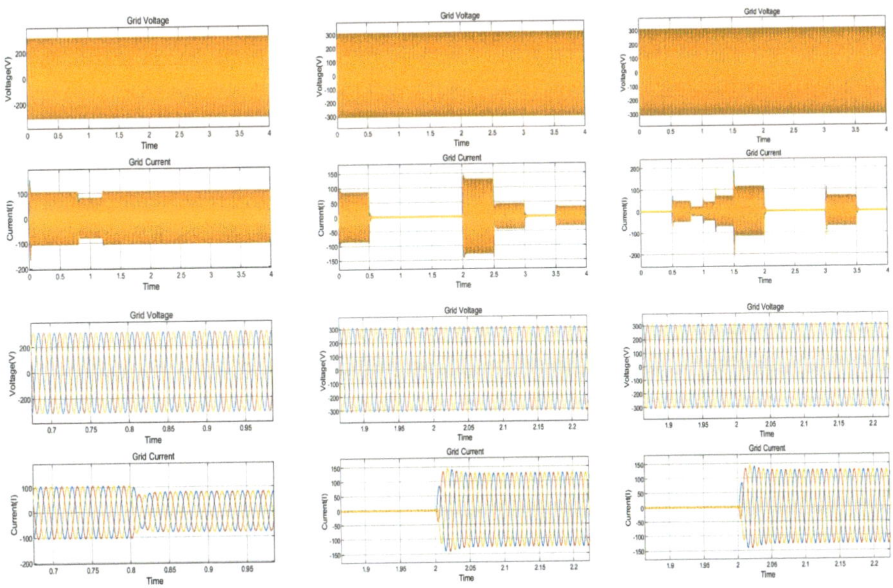

(**a**) (**b**) (**c**)

Figure 22. (**a**) Grid-connected three-phase voltage and current diagram of the photovoltaic module; (**b**) Three-phase voltage and current diagram of fuel cell module; and (**c**) Grid-connected three-phase voltage and current diagram of hydrogen production module of the electrolytic cell.

As shown in Figure 22, in the initial stage of simulation and the mutation process, the grid-connected current output amplitude of the fuel cell, electrolytic cell, and photovoltaic cell all fluctuated because the DC output voltage did not meet the voltage level requirements of the rear inverter in the dynamic. When the system reached a steady state, the inverter output amplitude was a stable sine wave grid current and grid-connected voltage. Because the electrolytic pool absorbed power from the system to produce hydrogen, the grid current was opposite to the voltage phase.

With the fast Fourier transform (FFT) in Simulink, the harmonic distortion (THD) of the grid-connected current can be obtained. As shown in the Figure 23, the grid-connected THD of the photovoltaic module was 1.42%, the THD of the fuel cell module was 1.37%, and the THD of the electrolytic cell hydrogen production module was 1.61%, both of which met less than 5% and met the grid-connected standard.

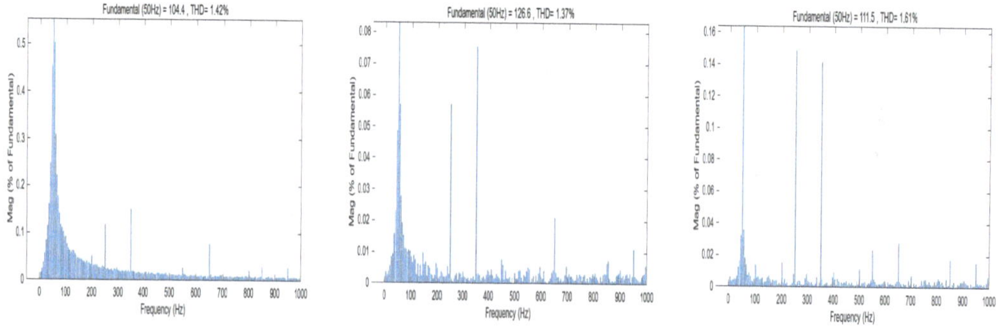

(a) (b) (c)

Figure 23. (**a**) Grid-connected current harmonics of the photovoltaic module; (**b**) Grid-connected current harmonics of the fuel cell module; and (**c**) Grid-connected current harmonics of the hydrogen production module of the electrolytic cell.

9. Conclusions

In this paper, for the multi-energy complementary and cooperative control characteristics of the hydrogen-electricity coupling system, the conversion mechanism and coupling relationship between light energy, hydrogen energy, electric energy, and other forms of energy were analyzed in depth. Further, a hydrogen-electricity coupling digital simulation experimental system, including photovoltaic power generation, fuel cell, and electrolysis hydrogen system, was established, At the same time, taking into account the operational objectives of efficient energy utilization, power fluctuation suppression and grid dispatching control, eight simulation scenarios were designed. Corresponding to cooperative control strategies for the hydrogen-electric coupling system was proposed, and simulation verification was carried out based on the MATLAB/SIMULINK platform. This study showed that the hydrogen energy system, based on fuel cells and electrolytic hydrogen production, had flexible, efficient, and fast energy conversion and regulation characteristics, which met the demand for the flexible regulation of the power system under different scenarios, such as abundant power supply and insufficient power supply. Further, it had good support for the safe and economic operation of the power system under the access of a high proportion of intermittent new energy power generation. The cooperative control model and strategy of the hydrogen-electricity coupling system under different time scales, such as emergency control, primary frequency regulation, AGC control, etc., will be further studied in order to enrich the application functions and scenarios of hydrogen-electricity coupling system and give full play to the energy storage and regulation characteristics of hydrogen energy.

Author Contributions: Conceptualization, T.S., and J.S.; methodology, T.S.; software, J.S.; validation, Z.C., and H.Z.; formal analysis, J.S.; data curation, H.Z.; writing—original draft preparation, J.S.; writing—review and editing, T.S.; project administration, T.S.; funding acquisition, T.S. All authors have read and agreed to the published version of the manuscript.

Funding: This research is funded by the major project of the National Natural Science Foundation of China (Grant No. 62192753).

Data Availability Statement: Data are contained within the article.

Conflicts of Interest: The authors declare no conflict of interest.

References

1. Nicoletti, G.; Arcuri, N.; Nicoletti, G.; Bruno, R. A Technical and Environmental Comparison between Hydrogen and Some Fossil Fuels. *Energy Convers. Manag.* **2015**, *89*, 205–213. [CrossRef]
2. Hossain, M.B.; Islam, M.R.; Muttaqi, K.M.; Sutanto, D.; Agalgaonkar, A.P. Advancement of Fuel Cells and Electrolyzers Technologies and Their Applications to Renewable-Rich Power Grids. *J. Energy Storage* **2023**, *62*, 106842. [CrossRef]
3. Chi, J.; Yu, H. Water Electrolysis Based on Renewable Energy for Hydrogen Production. *Chin. J. Catal.* **2018**, *39*, 390–394. [CrossRef]
4. Amatoul, F.Z.; Er-raki, M. Modeling and Simulation of Electrical Generation Systems Based on PEM Fuel Cell-Boost Converter Using a Closed Loop PI Controller. *Energy Rep.* **2023**, *9*, 296–308. [CrossRef]
5. Sun, L.; Jin, Y.; Pan, L.; Shen, J.; Lee, K.Y. Efficiency Analysis and Control of a Grid-Connected PEM Fuel Cell in Distributed Generation. *Energy Convers. Manag.* **2019**, *195*, 587–596. [CrossRef]
6. Rasekh, N.; Hosseinpour, M. LCL Filter Design and Robust Converter Side Current Feedback Control for Grid-Connected Proton Exchange Membrane Fuel Cell System. *Int. J. Hydrogen Energy* **2020**, *45*, 13055–13067. [CrossRef]
7. Xue, F.; Su, J.; Li, P.; Zhang, Y. Application of Proton Exchange Membrane Electrolysis of Water Hydrogen Production Technology in Power Plant. *IOP Conf. Ser. Earth Environ. Sci.* **2021**, *631*, 012079. [CrossRef]
8. Acar, C.; Dincer, I. Review and Evaluation of Hydrogen Production Options for Better Environment. *J. Clean. Prod.* **2019**, *218*, 835–849. [CrossRef]
9. Nguyen, T.; Abdin, Z.; Holm, T.; Mérida, W. Grid-Connected Hydrogen Production via Large-Scale Water Electrolysis. *Energy Convers. Manag.* **2019**, *200*, 112108. [CrossRef]
10. Ismail, T.M.; Ramzy, K.; Elnaghi, B.E.; Abelwhab, M.N.; El-Salam, M.A. Using MATLAB to Model and Simulate a Photovoltaic System to Produce Hydrogen. *Energy Convers. Manag.* **2019**, *185*, 101–129. [CrossRef]
11. Gaamouche, R.; Abbou, A.; Redouane, A.; Hasnaoui, A.E.; Belhorma, B. Modeling and Simulation of Direct Coupling of an Electrolyzer to PV System for Hydrogen Production. In Proceedings of the 2018 6th International Renewable and Sustainable Energy Conference (IRSEC), Rabat, Morocco, 5–8 December 2018; IEEE: Rabat, Morocco, 2018; pp. 1–5.
12. Kuo, J.-K.; Wang, C.-F. An Integrated Simulation Model for PEM Fuel Cell Power Systems with a Buck DC–DC Converter. *Int. J. Hydrogen Energy* **2011**, *36*, 11846–11855. [CrossRef]
13. İnci, M. Active/Reactive Energy Control Scheme for Grid-Connected Fuel Cell System with Local Inductive Loads. *Energy* **2020**, *197*, 117191. [CrossRef]
14. Bizon, N.; Lopez-Guede, J.M.; Kurt, E.; Thounthong, P.; Mazare, A.G.; Ionescu, L.M.; Iana, G. Hydrogen Economy of the Fuel Cell Hybrid Power System Optimized by Air Flow Control to Mitigate the Effect of the Uncertainty about Available Renewable Power and Load Dynamics. *Energy Convers. Manag.* **2019**, *179*, 152–165. [CrossRef]
15. Njoya, S.M.; Tremblay, O.; Dessaint, L.-A. A Generic Fuel Cell Model for the Simulation of Fuel Cell Vehicles. In Proceedings of the 2009 IEEE Vehicle Power and Propulsion Conference, Dearborn, MI, USA, 7–11 September 2009; IEEE: Dearborn, MI, USA, 2009; pp. 1722–1729.
16. Chen, W.; Han, Y.; Li, Q.; Liu, Z.; Peng, F. Design of Proton Exchange Membrane Fuel Cell Grid-Connected System Based on Resonant Current Controller. *Int. J. Hydrogen Energy* **2014**, *39*, 14402–14410. [CrossRef]
17. Ganiyu, S.O.; Martínez-Huitle, C.A.; Rodrigo, M.A. Renewable Energies Driven Electrochemical Wastewater/Soil Decontamination Technologies: A Critical Review of Fundamental Concepts and Applications. *Appl. Catal. B Environ.* **2020**, *270*, 118857. [CrossRef]
18. Kou, Q.; Klein, S.A.; Beckman, W.A. A Method for Estimating the Long-Term Performance of Direct-Coupled PV Pumping Systems. *Sol. Energy* **1998**, *64*, 33–40. [CrossRef]
19. Bensmail, S.; Rekioua, D.; Azzi, H. Study of Hybrid Photovoltaic/Fuel Cell System for Stand-Alone Applications. *Int. J. Hydrogen Energy* **2015**, *40*, 13820–13826. [CrossRef]
20. Folgado, F.J.; González, I.; Calderón, A.J. Simulation Platform for the Assessment of PEM Electrolyzer Models Oriented to Implement Digital Replicas. *Energy Convers. Manag.* **2022**, *267*, 115917. [CrossRef]
21. Koponen, J.; Ruuskanen, V.; Kosonen, A.; Niemela, M.; Ahola, J. Effect of Converter Topology on the Specific Energy Consumption of Alkaline Water Electrolyzers. *IEEE Trans. Power Electron.* **2019**, *34*, 6171–6182. [CrossRef]

22. Milewski, J.; Guandalini, G.; Campanari, S. Modeling an Alkaline Electrolysis Cell through Reduced-Order and Loss-Estimate Approaches. *J. Power Sources* **2014**, *269*, 203–211. [CrossRef]
23. Ursúa, A.; San Martín, I.; Barrios, E.L.; Sanchis, P. Stand-Alone Operation of an Alkaline Water Electrolyser Fed by Wind and Photovoltaic Systems. *Int. J. Hydrogen Energy* **2013**, *38*, 14952–14967. [CrossRef]
24. Campos Da Paixão, I.; Cardozo, J.C.; Sales Monteiro, M.K.; Gondim, A.D.; Cavalcanti, L.N.; Fabiano De Santana Souza, D.; Martínez-Huitle, C.A.; Vieira Dos Santos, E. A Sustainable Solar-Driven Electrochemical Process for Reforming Lignocellulosic Biomass Effluent into High Value-Added Products: Green Hydrogen, Carboxylic and Vanillic Acids. *RSC Adv.* **2023**, *13*, 35755–35765. [CrossRef]
25. Amores, E.; Rodríguez, J.; Carreras, C. Influence of Operation Parameters in the Modeling of Alkaline Water Electrolyzers for Hydrogen Production. *Int. J. Hydrogen Energy* **2014**, *39*, 13063–13078. [CrossRef]
26. Aboukalam Da Cruz, M.H.A.; Etancelin, M.; Marias, F.; Reneaume, J.-M.; Sochard-Reneaume, S.; Serra, S. Dynamic Modelling of an Alkaline Water Electrolysis System and Optimization of Its Operating Parameters for Hydrogen Production. *Int. J. Hydrogen Energy* **2023**, *48*, 12982–12999. [CrossRef]
27. Tijani, A.S.; Yusup, N.A.B.; Rahim, A.H.A. Mathematical Modelling and Simulation Analysis of Advanced Alkaline Electrolyzer System for Hydrogen Production. *Procedia Technol.* **2014**, *15*, 798–806. [CrossRef]
28. Panahandeh, B.; Bard, J.; Outzourhit, A.; Zejli, D. Simulation of PV–Wind-Hybrid Systems Combined with Hydrogen Storage for Rural Electrification. *Int. J. Hydrogen Energy* **2011**, *36*, 4185–4197. [CrossRef]
29. Saravanan, S.; Ramesh Babu, N. Maximum Power Point Tracking Algorithms for Photovoltaic System—A Review. *Renew. Sustain. Energy Rev.* **2016**, *57*, 192–204. [CrossRef]
30. Valdez-Resendiz, J.E.; Sanchez, V.M.; Rosas-Caro, J.C.; Mayo-Maldonado, J.C.; Sierra, J.M.; Barbosa, R. Continuous Input-Current Buck-Boost DC-DC Converter for PEM Fuel Cell Applications. *Int. J. Hydrogen Energy* **2017**, *42*, 30389–30399. [CrossRef]
31. Guilbert, D.; Collura, S.M.; Scipioni, A. DC/DC Converter Topologies for Electrolyzers: State-of-the-Art and Remaining Key Issues. *Int. J. Hydrogen Energy* **2017**, *42*, 23966–23985. [CrossRef]
32. Armghan, H.; Yang, M.; Ali, N.; Armghan, A.; Alanazi, A. Quick Reaching Law Based Global Terminal Sliding Mode Control for Wind/Hydrogen/Battery DC Microgrid. *Appl. Energy* **2022**, *316*, 119050. [CrossRef]
33. Ansari, S.; Chandel, A.; Tariq, M. A Comprehensive Review on Power Converters Control and Control Strategies of AC/DC Microgrid. *IEEE Access* **2021**, *9*, 17998–18015. [CrossRef]
34. Sharma, R.K.; Mishra, S. Dynamic Power Management and Control of a PV PEM Fuel-Cell-Based Standalone Ac/Dc Microgrid Using Hybrid Energy Storage. *IEEE Trans. Ind. Appl.* **2018**, *54*, 526–538. [CrossRef]
35. Esmaeilian, H.R.; Fadaeinedjad, R. Resolving Power Quality Issues Raised by Aerodynamic Aspects of Wind Turbine in Isolated Microgrids Using Fuel Cell/Electrolyzer System. *IEEE Trans. Sustain. Energy* **2016**, *7*, 1274–1283. [CrossRef]

processes MDPI

Article

Analysis of Enhanced Heat Transfer Characteristics of Coaxial Borehole Heat Exchanger

Lin Sun [1,2], Biwei Fu [2,*], Menghui Wei [2] and Si Zhang [1,2]

1 Cooperative Innovation Center of Unconventional Oil and Gas, Yangtze University (Ministry of Education & Hubei Province), Wuhan 430100, China
2 School of Mechanical Engineering, Yangtze University, Jinzhou 434023, China
* Correspondence: fubiwei@yangtzeu.edu.cn

Abstract: Coaxial borehole heat exchangers provide a practical method for geothermal energy extraction, but heat transfer efficiency is low. In order to address this problem, three coaxial borehole heat exchangers with vortex generators, based on the enhanced heat transfer theory, are proposed in this paper. The author compared and analyzed the heat transfer performance of three coaxial borehole heat exchangers with vortex generators and those of traditional structures, which explains why the new heat exchanger's heat transfer mechanism is enhanced. The results demonstrated that the vortex generator can enhance the fluid flow's turbulent kinetic energy in the coaxial heat exchanger. This generator can also improve the mixing characteristics of the fluid flow and heat transfer. The resultant increase in the inlet flow velocity can decrease the friction coefficient f, increase the Nusselt number and strengthen the coaxial sleeve. As a result, the heat exchange performance of the tubular heat exchanger will also be improved. The thread vortex generator (TVG) heat exchanger outperforms the other three heat exchangers in terms of heat exchange performance, extraction temperature and heat extraction power. The results evidenced that the TVG heat exchanger is better than the smooth tube heat exchanger. The thermal performance coefficient PEC was improved by 1.1 times, and the extraction temperature and heating power were increased by 24.06% and 11.93%, respectively. A solid theoretical foundation is provided by the extracted outcomes for designing and selecting high-efficiency coaxial borehole heat exchangers suitable for geothermal energy extraction.

Keywords: geothermal energy; coaxial heat exchanger; vortex generator; enhanced heat transfer

Citation: Sun, L.; Fu, B.; Wei, M.; Zhang, S. Analysis of Enhanced Heat Transfer Characteristics of Coaxial Borehole Heat Exchanger. *Processes* **2022**, *10*, 2057. https://doi.org/10.3390/pr10102057

Academic Editors: Ferdinando Salata and Virgilio Ciancio

Received: 18 September 2022
Accepted: 9 October 2022
Published: 12 October 2022

1. Introduction

Geothermal energy is sustainable and clean energy with an extremely high utilization value. This energy can effectively address the current global energy crisis and achieve "carbon peaking" and "carbon neutrality". The coaxial borehole heat exchanger is regarded as one of the most efficient methods of geothermal energy extraction [1]. The coaxial borehole heat exchanger is a closed-circulation heat exchange system with numerous advantages, including a simple structure, a large heat exchange area and an easy installation. In addition, there is no geothermal fluid production, which saves energy for reinjection and protects the environment. This heat exchanger has a simple structure, a large heat transfer surface area and a convenient installation procedure, among other advantages. Therefore, it has a wide range of engineering applications. The heat transfer performance of the coaxial borehole heat exchanger directly affects the efficiency of geothermal exploitation. Hence, it is crucial to improve the heat transfer performance of the coaxial borehole heat exchanger. At present, the articles in the literature that have dealt with the heat exchange performance of the coaxial borehole heat exchanger mainly focus on the structural size of the heat exchanger, the geological environment where the heat exchanger is located and the working fluid.

By carrying out numerical calculations and tests, the impact of the heat exchanger's structural size on its heat transfer performance has also been examined in several articles in the literature. Alimonti [2] established a numerical model of wellbore heat exchangers to simulate the effects of different inner diameters of pipes and heat transfer fluids on the performance of heat exchangers. Pan Sheng et al. [3] conducted a sensitivity analysis on the heat exchanger's design parameters, such as the outer diameter, inner diameter, flow rate, outer tube material, grouting material and drilling depth. Consequently, they found that the outer diameter of the pipe, drilling depth and flow rate significantly affected the heat extraction rate. They also proposed an optimization method based on the lowest average energy consumption index. Noorollahi et al. [4] simulated heat transfer between the fluid injected into the well and the surrounding hot rock. Their results demonstrated that thermal gradients, input mass flow, the well casing geometry and size of injection and withdrawal piping are crucial for the thermal recovery output. Song Xianzhi et al. [5] developed the unsteady heat transfer model of a deep coaxial borehole heat exchanger (CBHE) and comprehensively analyzed the essential elements affecting the CBHE's performance.

Simultaneously, some academics have investigated the heat exchanger's geological environment. Renaud et al. [6] used computational fluid dynamics techniques to simulate the 30-year production of a closed-circulation deep borehole heat exchanger (DBHE) and investigated the heat impact and heat recovery associated with drilling into a hypothetical DBHE in the Iceland Deep Drilling Project (IDDP)'s geological environment. Chen Chaofan et al. [7] used OpenGeoSys (OGS) software and the bicontinuous medium method to construct a comprehensive numerical model and examine the effects of pipe material, grouting thermal conductivity, geothermal gradient, soil thermal conductivity and groundwater flow on DBHE. Saedi et al. [8] introduced a novel transient analysis modeling method for geothermal energy production that ensures stable surface fluid temperature by invoking the circulating fluid circulation rate. Simultaneously, geothermal gradients, vertical well depth and thermal inner strength were found to significantly affect the wellhead fluid temperature. Holmberg et al. [1] established the numerical model of the coaxial BHE and conducted a parametric investigation of the coaxial shell with different drilling depths, flow rates and collector characteristics.

The working fluid exerts a significant effect on the heat exchanger's performance. Shi Yan et al. [9] investigated the heat transfer mechanism and optimization method of using CO_2 as a circulating fluid in a mid-depth coaxial heat exchanger geothermal system. They discussed the effects of injection/production pressure differences, mass flow and specific enthalpy changes on the heat production performance. Their results demonstrated that the thermal extraction efficiency of CO_2-based MDCHEs was increased by 31% compared with water-based MDCHEs. Zhang Yiqun et al. [10] investigated the endothermic performance of nine working fluids in a downhole coaxial heat exchanger (DCHE) geothermal system. They evaluated the applicability of the working fluids under the same conditions and studied the effects of parameters on the wellbore, reservoir and fluid. Luo Yongqiang et al. [11] proposed a new analysis model for CBHE. They studied the model's sensitivity to the fluid's flow direction, geothermal gradient and borehole thermal resistance. Bu Xianbiao et al. [12,13] established an equation describing the heat exchange between fluid and rock. They conducted a parametric investigation to determine the optimal values of the main parameters. Their results showed that the geothermal energy generated by abandoned wells strongly depends on fluid flow rates and geothermal gradients.

To improve heat transfer performance in the heat exchanger, it is crucial to increase the heat transfer area, heat transfer temperature difference and heat transfer coefficient. Increasing the heat transfer area is a common and effective approach for improving the heat transfer process. To this end, scholars usually adopt some passive technical devices, such as the use of groove geometry [14], corrugated surfaces [15], twisted belts [16], fins [17], tapered inserts [18] and other types of vortex generators [19]. These devices can significantly increase flow mixing, boundary layer disruption and turbulence to improve the heat transfer performance. Some scholars have also studied the effects of nanofluids on heat transfer [20–23].

Although some studies in the literature have focused on the optimal design of the structure of the coaxial borehole heat exchanger, some research results have achieved improvements in the (application background) heat transfer of the coaxial tube. The strengthening methods mainly include installing ties, coil springs and fin film, among others. Some research work and foundations can be provided for this study. Hamed Arjmandi et al. [24] performed a numerical simulation of the combination of a vortex generator, twisted ribbon vortex generator and Al_2O_3–H_2O nanofluid. They examined the heat transfer rate and pressure drop characteristics of the double tube heat exchanger. The combiner's geometry is optimized based on the response surface method of the central composite design. Mehedi Hasan Tusar et al. [25] used numerical simulations to investigate the heat transfer performance of the coaxial double-tube heat exchanger. Their results demonstrated that the annulus with the spiral belt has a higher heat transfer coefficient (HTC) and friction coefficient than the ordinary annulus. Rishabh Kumar et al. [26] numerically investigated a novel triangular perforated twisted tape with a V-notch inserted in the inner tube of a double-tube heat exchanger. They discovered that its thermal performance exhibited the best hydraulic performance at a lower pitch value of 50 mm. Gnanavel et al. [27] numerically examined the thermal and flow fields of nanofluidic double tube heat exchangers with different coil spring inserts. Furthermore, Do Huu-Quan et al. [28] numerically investigated the turbulent forced convective heat transfer in a new type of inner flat tube double tube heat exchanger. They studied the effects of the inner tube geometry on flow and heat transfer and found that the effect of the inner tube geometry on the thermal properties strongly depends on the value of Re. Iman Bashtani et al. [29] numerically examined the effects of adding alumina to water in a six-gear disk water-to-water double-tube heat exchanger. Their results showed that the turbulator improved the thermal effect. The fluid's collision with the turbulator's surface and the subsequent rupture of the boundary layer increased the heat transfer and local Nusselt number by approximately 70%. Simultaneously, adding nanoparticles to heat exchangers with turbulators increases the average Nusselt number, efficiency and transfer units. Marwa A.M. Ali et al. [30] improved the heat exchange efficiency of the heat exchanger by introducing the rotation of the inner tube of the heat exchanger and adjusting the tube's eccentricity. Their findings demonstrated that the heat transfer rate was significantly increased by 223% because of the variation in the eccentricity by up to 40 mm and the inner tube rotation speed of 500 rpm.

Osama A. Mohsen et al. [31] experimentally analyzed the heat transfer of the double-tube heat exchanger with different fin geometries on the heat exchanger's surface. Fin geometries include interrupted rectangular fins, circular fins and helical ribs. They experimentally measured the HTCs and pressure drops of hot and cold fluids with different Reynolds numbers. Their findings showed that the heat transfer coefficient could be improved by using different fin geometries to expand the surface. Moreover, the improved heat transfer of rectangular fins is the largest, and the improved heat transfer of circular fins is the smallest. Z. Iqbal et al. [32] used the genetic algorithm as the optimizer and the discontinuous Galerkin finite element method (DG-FEM) as the solver of the governing equations to investigate the optimal design problem of the longitudinal perturbation of the outer surface of the double pipe inner pipe. Their findings revealed that the characteristic length, the number of fins, the electrical conductivity of the heating surface material and the number of control points significantly affect the fins' optimal design.

To summarize, vortex generators, such as fins and spiral ties, positively affect the improvement of heat transfer in coaxial tubes. Therefore, three new coaxial borehole heat exchangers with vortex generators (to form a new special structure) were proposed in this work, namely the impeller vortex generator (IVG), the bump vortex generator (BVG) and the thread vortex generator (TVG), for geothermal energy extraction. In addition, numerical simulation was performed to systematically investigate the impact of the vortex generator on the heat transfer enhancement of the heat exchanger by establishing a three-dimensional heat transfer model of a coaxial borehole heat exchanger with a vortex generator. Furthermore, the three heat exchangers under different inlet flow rates were compared and analyzed to strengthen their mechanisms and heat transfer performances. Our work paves the way for the improvement of the extraction efficiency of the coaxial borehole heat exchanger.

2. Geometric Model and Evaluation Index of the Coaxial Borehole Heat Exchanger

2.1. Working Principle of Coaxial Borehole Heat Exchanger

Figure 1 depicts the working principle of the coaxial borehole heat exchanger. The heat exchanger consists of a hot rock mass, an injection channel, an extraction channel and an inner pipe. The well wall forms the annular cavity. The inner pipe was used as the injection channel, the inner pipe was used as the production channel and the bottom of the injection channel was closed. The supercooled fluid continuously absorbs heat from the surrounding hot well wall during its downward injection along the injection channel. After passing through the vortex generator, the supercooled fluid improves the fluid flow and mixing characteristics, destroys the boundary layer near the wall and enhances heat transfer. After this fluid reached the bottom of the well, the high-temperature fluid was produced from the production channel because of the bottom hole pressure, resulting in geothermal energy production.

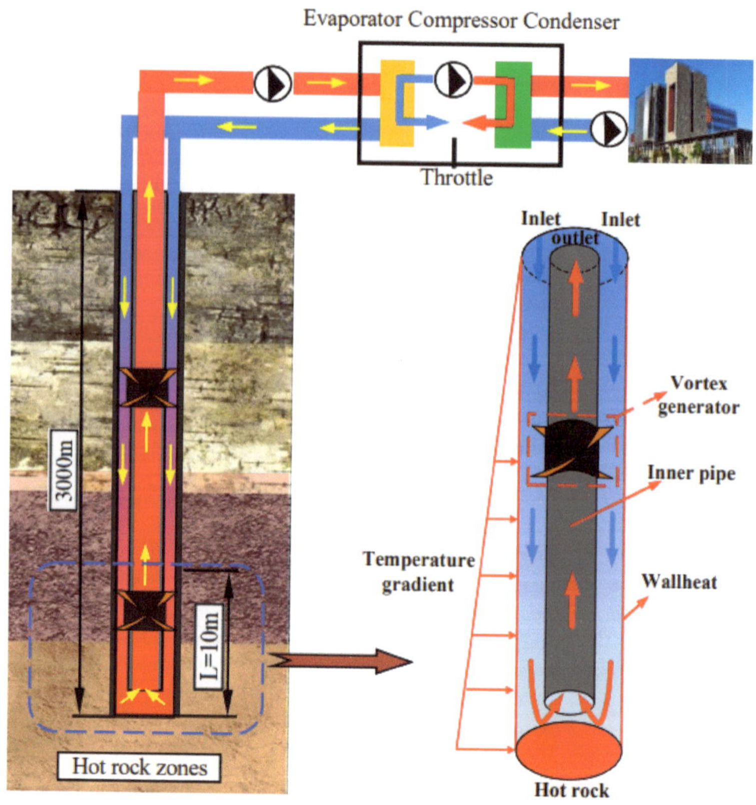

Figure 1. Schematic diagram of working principle of the coaxial borehole heat exchanger.

The coaxial borehole heat exchanger in this study was mainly used for medium and deep geothermal energy extraction, with a well depth from 2000 m to 3000 m. According to the needs of various sites, one or more vortex generators can be added at different well depths to enhance the heat recovery efficiency of coaxial borehole heat exchangers. In order to reveal the enhanced heat transfer mechanism of a single vortex generator, a heat transfer simulation model with an L = 10 m length was constructed based on the coaxial borehole in the hot rock area at the bottom of the well.

2.2. Geometric Model of Coaxial Borehole Heat Exchanger

This paper proposed three types of vortex generators, namely IVG, BVG and TVG structures, to obtain a coaxial borehole heat exchanger structure with improved heat transfer performance. Simultaneously, this paper developed a geometric model of the coaxial heat exchanger with vortex generators (Figure 2) to analyze the effect of different vortex generator structures on the heat transfer performance of the coaxial borehole heat exchanger. The model considers the heat exchanger's bottom hole L = 10 m ($L = L_1 + L_2 + L_3$) to examine the strengthening mechanism of the vortex generator in the heat exchanger in greater detail. Table 1 presents the specific geometric parameters of the model.

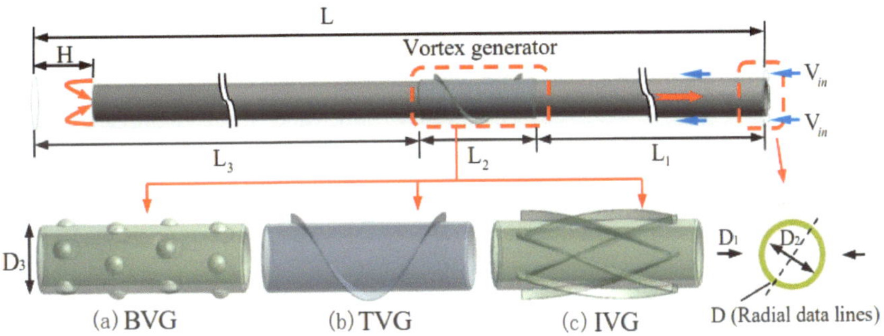

Figure 2. Geometric model of coaxial borehole heat exchanger.

Table 1. Geometrical parameters of coaxial borehole heat exchanger with a vortex generator.

Parameter	Value(mm)
L_1	2850
L_2	300
L_3	7150
D_1	177.8
D_2	100
D_3	130
H	200

2.3. Enhanced Heat Transfer Parameters and Heat Transfer Performance Evaluation Index

2.3.1. Enhanced Heat Transfer Parameters

The local Nusselt number Nu can be expressed as follows [15]:

$$Nu = \frac{hD_h}{k} = \frac{q''}{T_w - T_m}\frac{D_h}{k} \tag{1}$$

q'' represents the heat flux, and T_m can be expressed as follows:

$$T_m = \frac{T_{in} + T_{out}}{2} \tag{2}$$

The hydraulic diameter D_h is expressed as follows:

$$D_h = \frac{4A_c}{P_w} \tag{3}$$

The Darcy coefficient of friction in the fully developed water flow is estimated as follows [25]:

$$f = \frac{2D_h}{L}\frac{\Delta P}{\rho \mu^2} \tag{4}$$

where ΔP is the pressure difference between the inlet and outlet. Importantly, the Darcy coefficient of friction determines the pump's power requirements.

An increased flow resistance accompanies an improved heat transfer. Therefore, most studies [15,18,33] use the thermal performance coefficient, PEC, to assess heat exchangers' overall heat transfer performance. Its specific expression is presented as follows:

$$PEC = \frac{(Nu/Nu_s)}{(f/f_s)^{1/3}} \tag{5}$$

where Nu and Nu_s are the Nusselt numbers of the vortex generator heat exchanger and the smooth tube heat exchanger, respectively.

2.3.2. Heat Transfer Performance Evaluation Index

The extraction temperature T_{out} and the extraction power Q_{out} directly reflect the amount of geothermal energy absorbed by the fluid. These parameters can be used in the evaluation index of the heat exchange performance of the coaxial heat exchanger. The heating power Q_{out} is affected by the fluid's physical parameters, the flow channel's size, the fluid flow rate, and the temperature difference between the inlet and outlet. The heating power is expressed as follows [34]:

$$Q_{out} = c_\rho \rho D_2 V_{out}(T_{out} - T_{in}) \tag{6}$$

3. Numerical Analysis Model and Its Validation

3.1. Model Assumptions

Due to the intricacies of the heat exchange process between the heat exchanger and the rock, the following reasonable assumptions were made to comprehensively analyze the heat exchange performance of the heat exchanger:

(1) The rock and the soil around the geothermal underground coaxial borehole heat exchanger were treated as a homogeneous medium. Moreover, the effect of groundwater seepage is ignored, and the heat transfer in the underground rock and soil is treated as pure heat conduction.
(2) The temperature at the numerical simulation region's radical boundary is considered constant.
(3) The bottom hole's heat source and surface temperature are considered constant.
(4) The temperatures of the rock and the wellbore were considered the same because the wellbore has been attached to the rock for a long time.

3.2. Governing Equations

The (RNG) k-ε turbulence model with good performance in eddies and strong streamline bending was used to solve the flow field. It is based on the above model's assumptions and the Reynolds-averaged Navier–Stokes (RANS) model. The main governing equations for continuity, momentum and energy can be respectively expressed as follows [18]:

$$\frac{\partial}{\partial x_i}(\rho u_i) = 0 \tag{7}$$

$$\frac{\partial}{\partial x_j}(u_j \rho u_i) = -\frac{\partial p}{\partial x} + \frac{\partial}{\partial x_j}\left\{\mu\left[\frac{\partial u_i}{\partial x_j} - \rho \overline{u_i' u_j'}\right]\right\} \tag{8}$$

$$\frac{\partial}{\partial x_j}(\rho u_i T) = \frac{\partial}{\partial x_i}\left\{\left[\frac{\mu}{Pr} + \frac{\mu_t}{Pr_t}\right]\frac{\partial T}{\partial x_i}\right\} \tag{9}$$

where ρ, p, u and μ are the fluid density, pressure, turbulent fluctuation, pressure and dynamic viscosity of water, respectively. The Reynolds stress term $(\overline{\rho u_i' u_j'})$ can be expressed as follows:

$$\overline{\rho u_i' u_j'} = \mu_t\left[\frac{\partial u_i}{\partial x_j} + \frac{\partial u_j}{\partial x_i}\right] - \frac{2}{3}\rho k \delta_{ij} - \frac{2}{3}\mu_t \frac{\partial u_k}{\partial x_k}\delta_{ij} \tag{10}$$

The equations for turbulent kinetic energy (k) and dissipation rate (ε) are respectively expressed as follows [24]:

$$\frac{\partial}{\partial x_i}(\rho u_i k) = \frac{\partial}{\partial x_j}\left\{\left[\frac{\mu_t}{\sigma_k} + \mu\right]\frac{\partial k}{\partial x_j}\right\} + G_k \rho \varepsilon \tag{11}$$

$$\frac{\partial}{\partial x_i}(\rho u_i \varepsilon) = \frac{\partial}{\partial x_j}\left\{\left[\frac{\mu_t}{\sigma_\varepsilon} + \mu\right]\frac{\partial \varepsilon}{\partial x_j}\right\} + \frac{\varepsilon}{k}[C_{1\varepsilon}G_k - \rho C_{2\varepsilon}\varepsilon] \tag{12}$$

In the above equation, $\sigma_k = 1$, $\sigma_\varepsilon = 1.3$, $\sigma_{1\varepsilon} = 1.42$ and $C_{2\varepsilon} = 1.68$. $k = 2$ is a constant for the (RNG) k-ε turbulence model.

3.3. Boundary Conditions

The inlet boundary was set as the velocity inlet, and the outlet boundary was set as the pressure outlet. The heat transfer boundary condition between the circulating fluid in the tube and the tube wall was defined as the coupled heat transfer boundary. The surface temperature was 288.15 K [12], the bottom boundary of the heat exchanger was set at a constant temperature of 408.15 K (Formula 13) and the temperature gradient of the borehole wall was T_g = 4.5 K/100 m [35,36], which was controlled by a user-defined function.

To obtain more accurate numerical calculation results, the rock's temperature and the pressure at the outlet can be expressed as follows:

$$T_W = T_{sur} + T_g z \tag{13}$$

$$P_{out} = \rho g z \tag{14}$$

The fluid medium in the calculation is water, and the casing and inner pipe are steel. The specific physical parameters are shown in Table 2.

Table 2. Physical parameters.

Parameter/Unit	Water	Inner Pipe and Well Wall
Density ρ/kg·m^{-3}	998.2	8030
Specific heat c_p/J·(kg·k)$^{-1}$	4182	502.48
Thermal conductivity k/W·(m·k)$^{-1}$	0.6	16.27
Viscosity μ/kg·(m·s)$^{-1}$	0.001003	0

3.4. Mesh and Model Validation

3.4.1. Meshing and Independence Verification

Figure 3 shows the meshing scheme of the coaxial borehole heat exchanger. Notably, a mixed mesh of structured hexahedrons and unstructured tetrahedrons is used. By considering the impact of the boundary layer near the wall area of heat transfer, the boundary layer and mesh refinements were performed at the interface between the tube wall and the fluid and the surface of the vortex generator to effectively capture the turbulent flow characteristics and address the viscous sublayer effect near the tube wall in question. Moreover,

Table 3 presents the mesh quality parameters for ST and TVG, which are consistent with the literature [25].

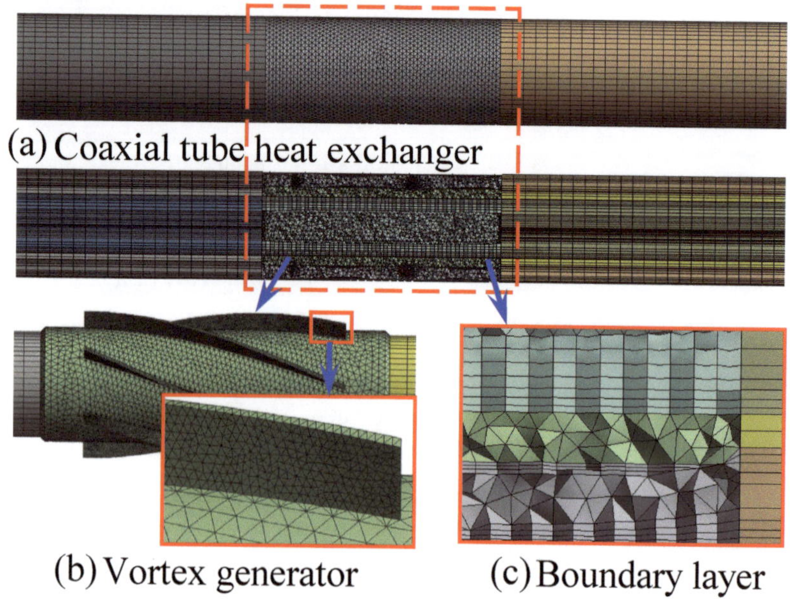

Figure 3. Mesh generation of coaxial borehole heat exchangers.

Table 3. Mesh metrics.

Geometry	Nodes	Average Skewness	Average Orthogonal Quality	Average Aspect Ratio
ST	485,750	0.078	0.985	5.309
TVG	538,195	0.178	0.927	2.108

In order to eliminate the influence of the mesh on the acquired numerical calculation results, the mesh independence verification of the numerical model is performed. As shown in Figure 4, when the inlet speed was $V_{in} = 0.1$ m/s, the Nusselt number Nu, the friction coefficient of the ST heat exchanger and the TVG heat exchanger changed as a function of the number of meshes. Additionally, it can be seen that the Nu and friction factors were almost the same for different mesh nodes. Thus, it can be concluded that the simulations were mesh independent.

3.4.2. Numerical Simulation Model Validation

To verify the extracted numerical simulation results, the Nu and friction factor coefficients of the ST heat exchangers were compared with those obtained by other works in the literature [25,37].

The Dittus–Boelter correlation for Nu is as follows:

$$Nu = 0.023\mathrm{Re}^{0.8}\mathrm{Pr}^{n} \tag{15}$$

where Pr stands for the Prandtl number, $n = 0.4$ for heating (applicable in this study) and $n = 0.3$ for cooling.

$$\mathrm{Pr} = \frac{c_p \mu}{\lambda} \tag{16}$$

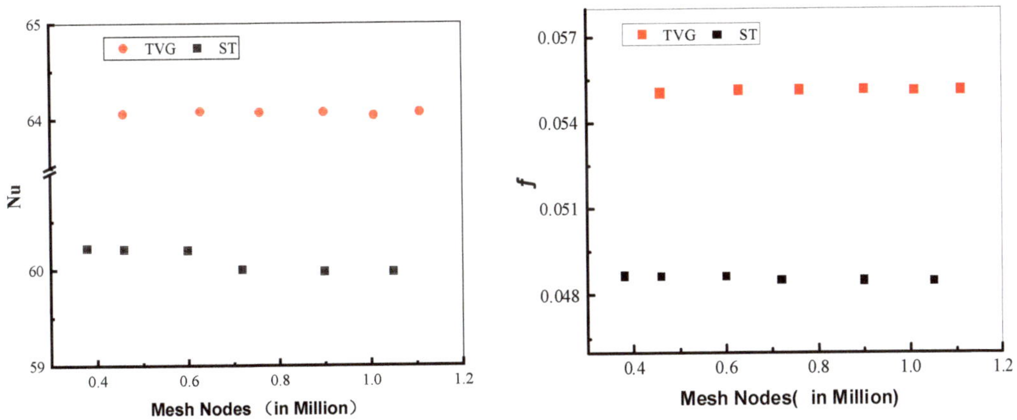

(a) (b)

Figure 4. Grid independency verification; (**a**) grid independency verification of Nu; (**b**) grid independency verification of f.

The Gneilski correlation for Nu is as follows [38]:

$$Nu = \frac{(f/8)(\mathrm{Re}-1000)\mathrm{Pr}}{1+12.7(f/8)^{\frac{1}{2}}(\mathrm{Pr}^{\frac{2}{3}}-1)} \tag{17}$$

where $0.5 \leq \mathrm{Pr} \leq 2000$ and $3000 \leq \mathrm{Re} \leq 10^6$, while the friction factor correlation of Petukhov can be expressed as follows [39]:

$$f = \frac{1}{(-1.64+0.79Ln(\mathrm{Re}))^2},\ 3000 \leq \mathrm{Re} \leq 10^6 \tag{18}$$

Friction factor correlation for Blasius is in the form,

$$= 0.3164\mathrm{Re}^{-0.25}\ 4\times10^3 < \mathrm{Re} < 10^5 \tag{19}$$

Figure 5 displays a comparison of the simulation results of the ST heat exchanger and the empirical formula. As can be ascertained, the error between the simulation results and the empirical correlation formula is less than ±6%, which verifies the rationality of the introduced numerical model.

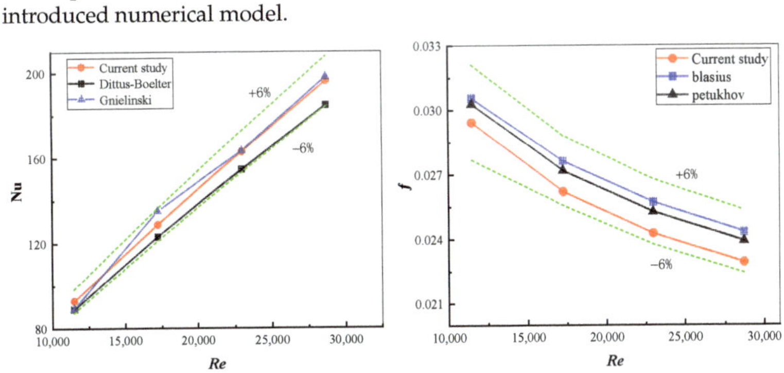

(a) (b)

Figure 5. Validation of the numerical analysis model. (**a**) Comparison of Nu and (**b**) f between the current work and other reports in the literature.

4. Results and Discussions

4.1. Analysis of the Enhanced Heat Transfer Mechanism of Vortex Generator

4.1.1. The Effect of Vortex Generators on the Turbulent Kinetic Energy

Figure 6 shows the turbulent kinetic energy distributions of the four kinds of coaxial heat exchangers in the vortex generator area. As can be observed, the turbulent kinetic energy of the three coaxial heat exchangers with vortex generators is more significant than that of the smooth tube coaxial heat exchanger. Furthermore, the turbulent kinetic energy of the TVG heat exchanger is much larger than that of the other three exchangers. Physically speaking, the vortex generator reduces the cross-sectional area of the fluid flow and increases the average velocity of the flow cross-sectional area as well as Reynolds number, which increases turbulent energy. The guiding and shearing effects of the vortex generator on the fluid led to the following changes: the flow direction and velocity of the fluid changed, the turbulent kinetic energy of the fluid was enhanced and the destructive effect of the thermal boundary layer was enhanced. As a result, the heat transfer between the fluid and the hot rock was enhanced, leading to an improved efficiency in geothermal energy extraction.

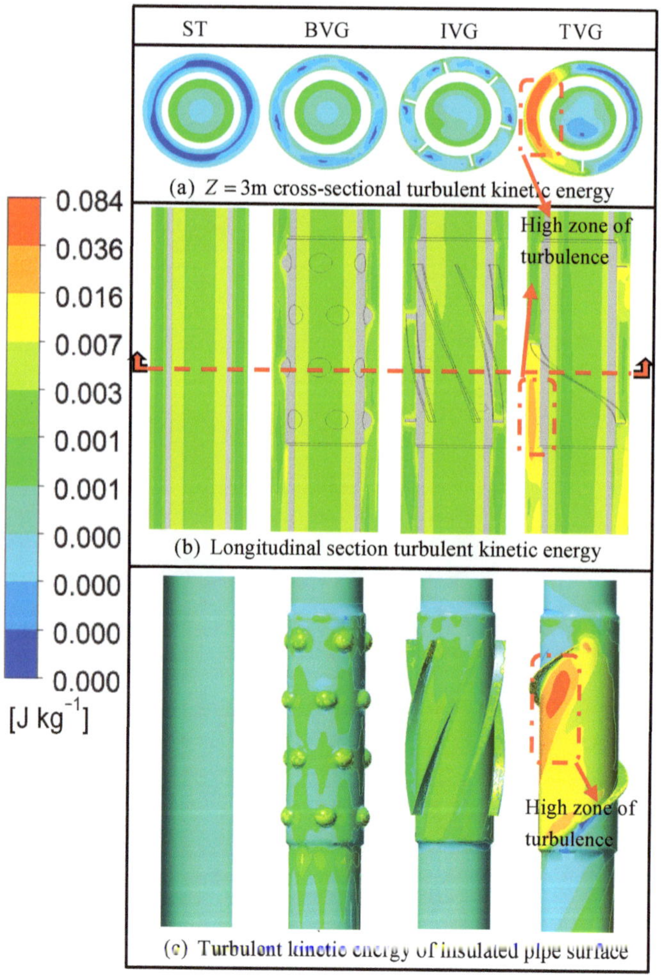

Figure 6. Distributions of the turbulent kinetic energy in the vortex generator area of the four heat exchangers.

Figure 7 shows the distributions of the turbulent energy and velocity vectors at the bottom of the well for the different longitudinal sections of the heat exchanger. This figure thoroughly demonstrates the difference between the turbulent kinetic energy and the velocity vector at the bottom of the well for different heat exchangers. Compared with other heat exchanger structures, the IVG and TVG heat exchangers significantly impact the flow state of the bottom hole fluid. The velocity vector shows a low-velocity zone in the bottom hole cavity of the smooth tube (ST) and BVG heat exchangers. The fluid has more of a slow flow, and the central fluid in the injection channel directly flows into the inner pipe from the end of the inner pipe, resulting in a lower heat exchange effect. The higher turbulent kinetic energy of the IVG and TVG heat exchangers destroys the low-velocity zone in the bottom hole cavity and weakens the boundary layer. This higher energy also allows more slow-flowing high-temperature fluids to mix with the central fluid, thus, improving the heat exchange process. The turbulent kinetic energy is the strongest and has the most substantial effect on the slow-flow region. Simultaneously, a high turbulence area was observed at the inlet of the inner pipe because of the change in the flow channel's diameter, which is conducive to the rapid outflow of the fluid through the inner pipe. It is also conducive to enhancing the heat exchange efficiency of the heat exchanger.

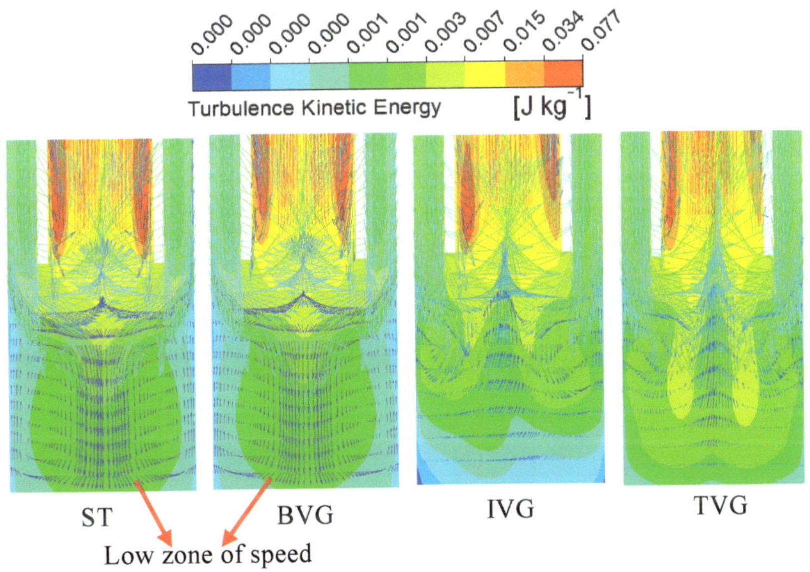

Figure 7. Distributions of turbulent energy and velocity vector at the bottom hole of the different longitudinal sections of the four heat exchangers.

4.1.2. Effect of Vortex Generator on the Velocity Field

Figure 8 depicts the velocity distributions of the vortex generation region along the longitudinal and cross-section Plane1. More specifically, Figure 8a presents the velocity distribution of the Plane1 section. This figure shows that when the fluid passes through the vortex generator of the coaxial heat exchanger, the fluid's velocity increases because of the reduction in the flow cross-sectional area. Figure 8b shows that the fluid velocity first increases and decreases along the injection direction because the fluid enters the vortex generation area, leading to a decrease in the flow cross-sectional area and an increase in the fluid velocity. While flowing out from the vortex generation area, the velocity decreases rapidly because of the increase in the flow cross-sectional area. The massive change in velocity causes an increase in the turbulent kinetic energy of the fluid. As a result, the damage to the thermal boundary layer increases, improving the heat exchange of the coaxial heat exchanger. At the same time, the vortex generator changes the fluid flow from one-dimension to three-dimension. It increases the radial and tangential velocities of the fluid, which destroys the thermal boundary layer and improves the heat exchange performance of the coaxial heat exchanger. Among them, the velocity change in the fluid in the TVG is the most evident, and the fluid's thermal boundary layer destruction effect is the best. To summarize, the TVG heat exchanger exhibited a good heat exchange performance. This structure (TVG), as the vortex generator of the coaxial heat exchanger, is beneficial in improving the mining efficiency of geothermal heat.

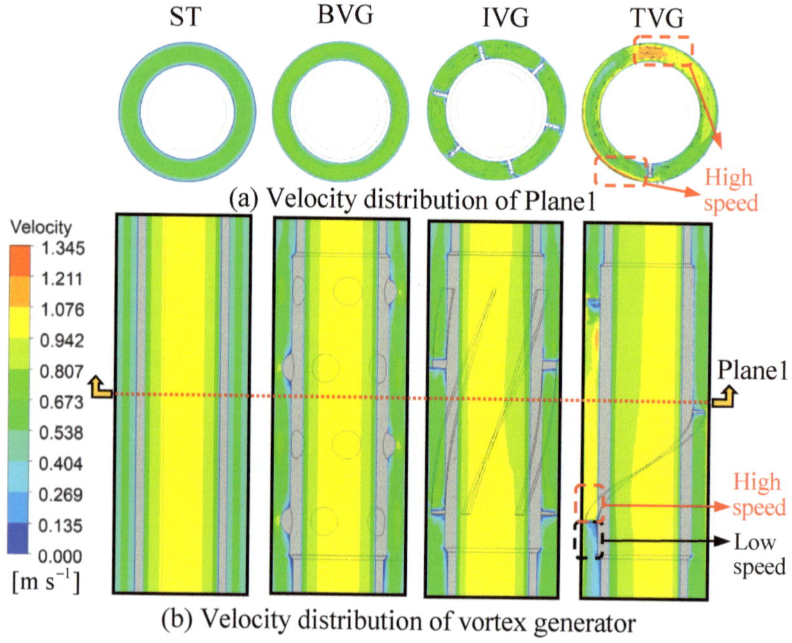

Figure 8. Velocity distributions in the vortex generator area of the four heat exchangers.

Figure 9 presents the velocity variation of fluid in the injection and production channels of different heat exchangers along the well's depth. This figure shows that after the fluid passes through the vortex generator, the fluid velocity increases due to the reduced cross-sectional area of the runner. The increasing velocity strengthens the heat exchange between the fluid and the hot wall. Among them, the TVG heat exchanger exerts the most obvious acceleration effect. Simultaneously, under the disturbance and swirl effect of the vortex generator, the fluid velocity changes from one-dimensional motion to three-dimensional motion. This change increases the time for heat exchange between the fluid and the hot wall and enhances the destructive effect of the boundary layer, thereby improving the heat recovery energy. After the fluid passes through the bottom of the well, the fluid velocity in the inner tube increases because of the reduction in the inner tube's cross-sectional area.

Figure 10 illustrates the velocity variation curve of the cross-section Plane1 along the radial direction. The fluid's velocity in the near-wall boundary layer was relatively low due to fluid viscosity, weakening the heat transfer between the fluid and the hot rock wall. This figure shows that the fluid's flow velocity in the injection channel increases significantly after passing through the vortex generator, and the flow velocity in the area close to the hot rock wall surface increases greatly, which is conducive to enhancing the turbulent flow characteristics of the fluid in the area near the hot rock wall surface, destroying the boundary layer fluid and improving the flow rate of the heat transfer energy between the fluid and the hot rock wall. Compared with the other three heat exchangers, the fluid's flow velocity in the TVG heat exchanger on the Plane1 cross-section fluctuates significantly under the influence of the asymmetric threaded vortex generator's structure. Moreover, the flow velocity in the area near the hot rock wall increases significantly. This increase can effectively reduce the effects of the boundary layer on heat exchange, improve the heat exchange of the fluid diameter in the radial direction and improve the heat recovery effect of the coaxial heat exchanger.

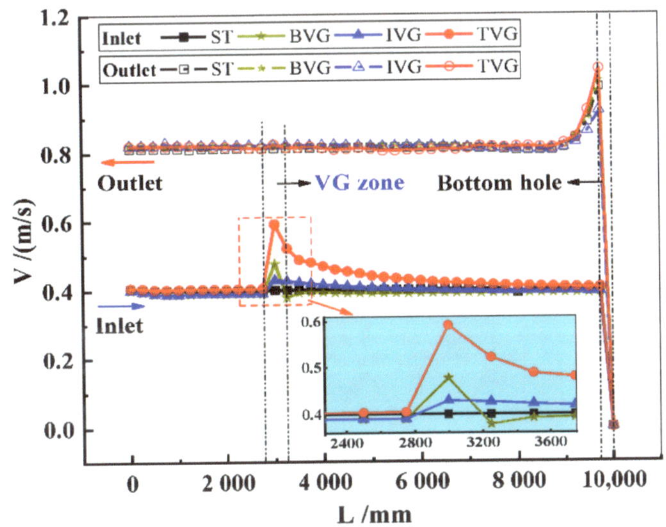

Figure 9. Velocity variation curves along the well's depth in the injection channel and production channel of the four heat exchangers.

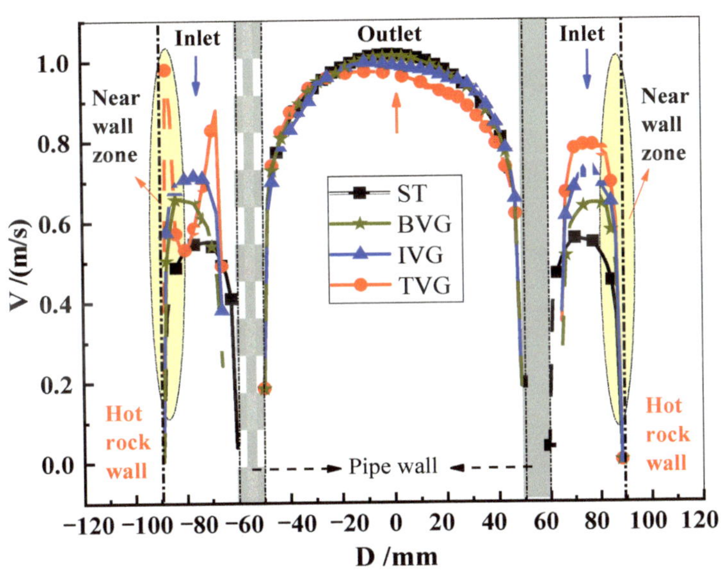

Figure 10. Velocity variation curves along the radial direction of cross-section Plane1 of four heat exchangers.

Figure 11 shows the velocity variation curves of the fluid in the cross-section area injected into the channel at the vortex generator in the *X*, *Y* and *Z* directions, respectively. As can be observed from Figure 11a, the *X*-direction velocity V_x of the ST heat exchanger approaches the value of 0, while V_x of the heat exchangers with vortex generator fluctuates and causes some fluid disturbance in the *X*-direction. Moreover, the velocity fluctuation of the IVG is the most obvious since the flow direction of the fluid is changed by the vortex generator structure. Figure 11b shows that the *Y*-direction velocity V_y of the IVG and TVG heat exchangers was significantly improved compared with the ST heat exchangers. The disturbance and shear effects of the vortex generator structure on the fluid increase the value of V_y, so that the high-speed fluid near the wall can effectively help the fluid to mix. It will destroy the boundary, helping the central cold fluid and the peripheral hot fluid to mix, as well as improving the heat transmission efficiency. The *Y*-direction velocity V_y of the heat exchanger is arranged in order from high to low as follows: TVG > IVG > BVG > ST. As can be ascertained from Figure 11c, the vortex generator also has a promoting impact on the *Z*-direction velocity V_z, whereas the increase in TVG is the greatest. It is conducive to improving the heat transmission effect of the fluid and wall surface.

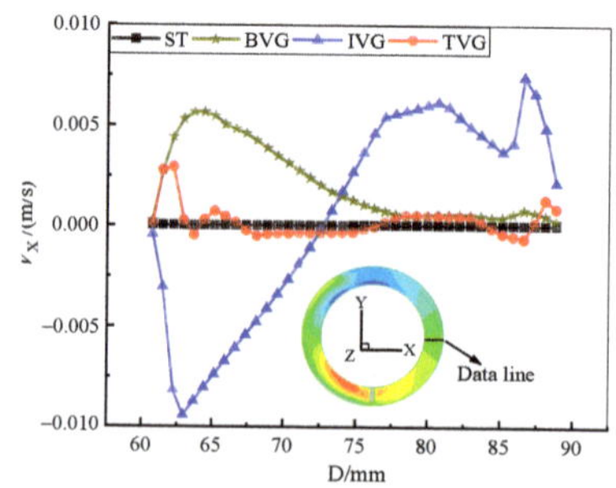

(**a**)

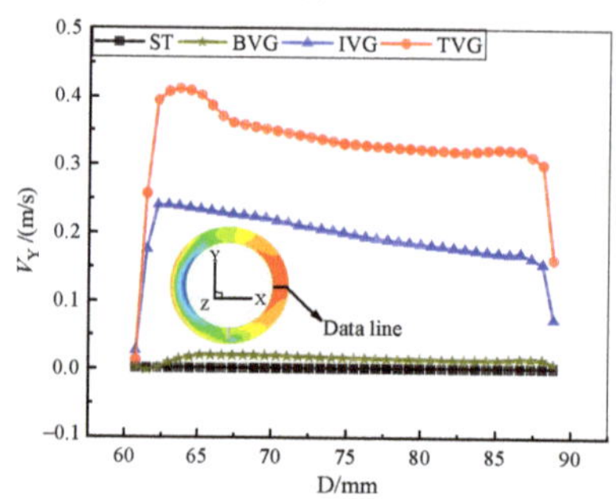

(**b**)

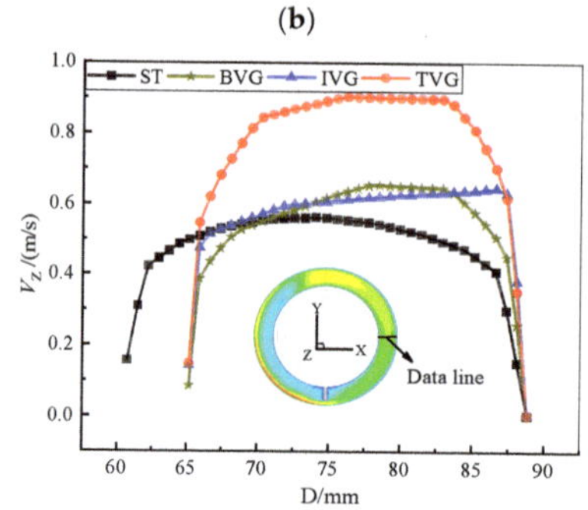

(**c**)

Figure 11. Velocity variation curves of fluid in the injection channel of the cross-section at the vortex generator of the different heat exchangers. (**a**) Variation curve of the velocity in the *X*-direction of the four heat exchangers; (**b**) variation curve of the velocity in the *Y*-direction of the four heat exchangers; (**c**) variation curve of the velocity in the *Z*-direction of the four heat exchangers.

4.1.3. Impact of Vortex Generator on the Temperature Field

Figure 12 shows the cross-sectional temperature distributions of the four coaxial heat exchangers in the vortex generation area. The cross-sectional temperature distribution of the ST heat exchanger reveals that when the low-temperature fluid was injected into the channel, it continued to absorb heat from the hot rock wall, further increasing the fluid temperature in the area close to the hot rock wall. The temperature of the central area of the fluid increases because of the heat transfer inside the fluid in the radial direction. The fluid's temperature near the insulated pipe's wall was still low due to the small thermal conductivity of water. Compared with the ST heat exchanger, the fluid's turbulent flow characteristics are enhanced after it passes through the three vortex generators. Furthermore, the heat exchange between the fluid and the surface of the hot rock was strengthened and the low-temperature area of the fluid was significantly reduced. The low-temperature area basically disappeared in IVG and TVG heat exchangers. These findings demonstrate that the IVG and TVG heat exchangers have a better heat recovery effect.

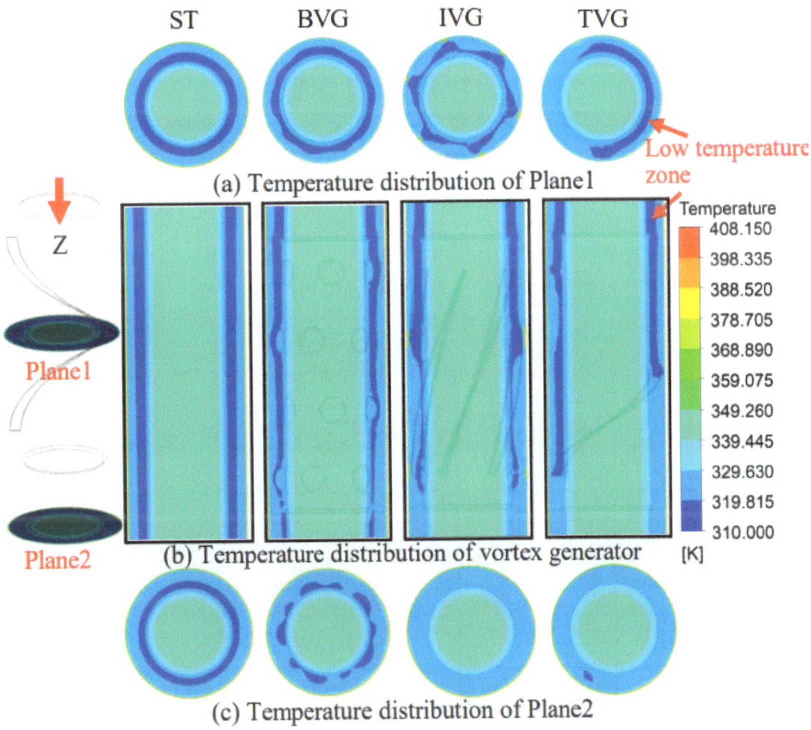

Figure 12. Temperature distributions in the vortex generator area of the four heat exchangers.

Figure 13 presents the average temperature variation curve of the fluid in the injection channel and the production channel of the four heat exchangers along the cross−section of the well's depth. This figure shows that the temperature of the fluid injected into the annulus increases gradually with the well's depth and decreases locally in the region where the vortex occurs. This is because the fluid's velocity is faster in the vortex region and the turbulent characteristics of the fluid are enhanced. In addition, the local high-temperature boundary layer was destroyed and the heat transfer time between the fluid and the hot rock wall was reduced, thus, decreasing the average fluid temperature. Simultaneously, Figure 13 shows that, in the near-bottom region, the high-temperature boundary layer in the injection channel cannot be directly produced from the production channel because of its low velocity. Most of this fluid flows into the bottom hole, and the fluid temperature in the injection channel is higher than that in the production channel.

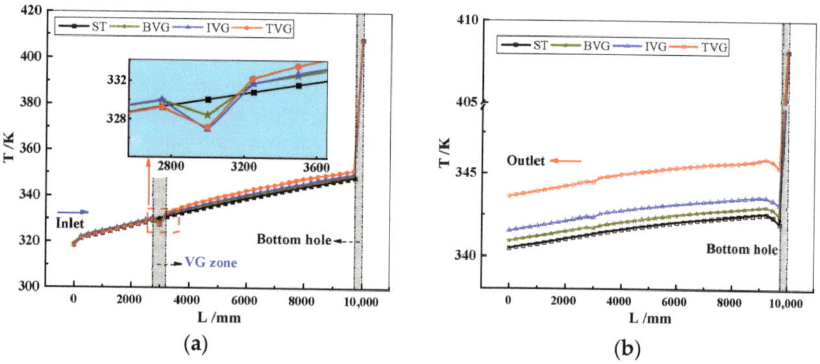

Figure 13. Temperature change curves of fluids in different heat exchanger injection channels and production channels along well depth. (**a**) Inject channel; (**b**) production channel.

Figure 14 shows the temperature distribution curves of the four heat exchangers along the radial direction in the region of the vortex generator. This figure shows that the very high fluid temperature of the boundary layer is attributed to the relatively small fluid velocity of the boundary layer near the hot rock wall surface. Furthermore, more energy is absorbed by the hot rock. The thermal conductivity of water is lower; therefore, the fluid's thermal conductivity in the radial direction is poor and the temperature of the fluid injected into the channel decreases rapidly in the radial direction. According to this figure, the fluid temperature in the production channel is presented as high to low: TVG > IVG > BVG > ST. This arrangement shows that the heat recovery of the TVG heat exchanger effect is the best, and the heat recovery efficiency of the coaxial heat exchanger is improved.

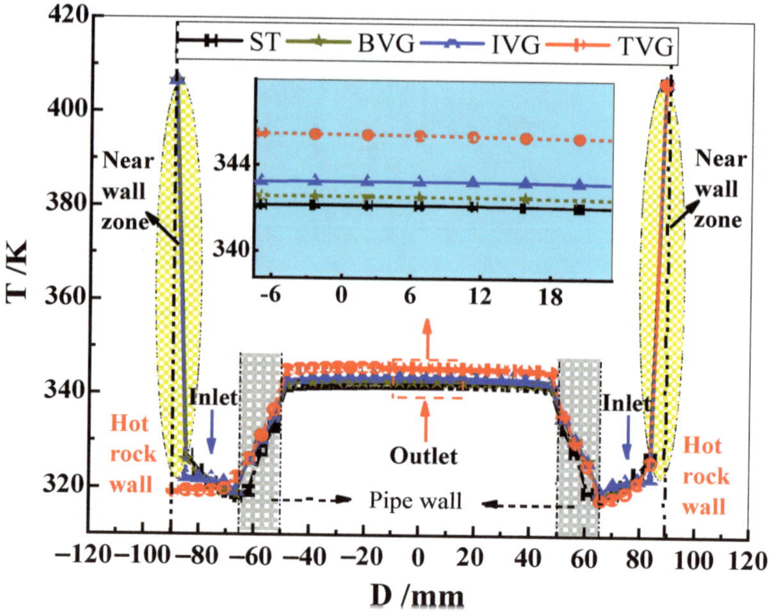

Figure 14. Radial temperature variation curve at the cross-section of the vortex generator region.

4.2. Influence of Vortex Generator on Enhanced Heat Transfer Parameters

Figure 15a shows the variation curve of the friction coefficient of the hot rock wall surface of the four heat exchangers with the inlet flow velocity. The figure shows that the friction coefficient f decreases rapidly when the inlet velocity increases from 0.1 m/s to 0.3 m/s. However, the friction coefficient decreased gradually when the inlet velocity decreased from 0.3 m/s to 1 m/s. Formula (6) shows that reducing the friction coefficient can effectively improve the thermal performance coefficient PEC of the coaxial heat exchanger. Therefore, increasing the inlet flow rate can improve the thermal performance coefficient of the coaxial heat exchanger. When the inlet flow rate is lower than 0.3 m/s, the friction coefficient is higher due to fluid viscosity and the thermal performance coefficient PEC of the coaxial heat exchanger is lower. Simultaneously, Figure 15a shows that under the same flow rate, the friction coefficient f of TVG is significantly higher than that of the other three heat exchangers, which is not conducive to improving the heat exchange performance of the coaxial heat exchanger. Therefore, increasing the inlet flow rate is beneficial in improving the heat exchange performance of the coaxial heat exchanger. Figure 15b shows the variation of the Nusselt number Nu with the inlet flow velocity for four different heat exchangers. The figure shows that the Nusselt number Nu increases linearly with the increase in the flow rate. The increasing speed is presented in descending order as follows: TVG > IVG > BVG > ST. Combined with Formula (6), the thermal performance coefficient PEC of the TVG heat exchanger increases the most with the increase in the inlet velocity. Simultaneously, under the same inlet flow rate, the Nusselt number Nu from large to small is as follows: TVG > IVG > BVG > ST. Compared with the ST heat exchanger, when the inlet velocity is 1 m/s, the Nusselt number Nu in the BVG, IVG and TVG heat exchangers increases by 4.08%, 9.05% and 20.89%, respectively, indicating that the TVG heat exchanger has the best heat transfer performance. These findings show that the TVG heat exchanger exhibits the best heat transfer performance under the same conditions. In addition, increasing the inlet flow rate improves the heat transfer performance of the TVG heat exchanger.

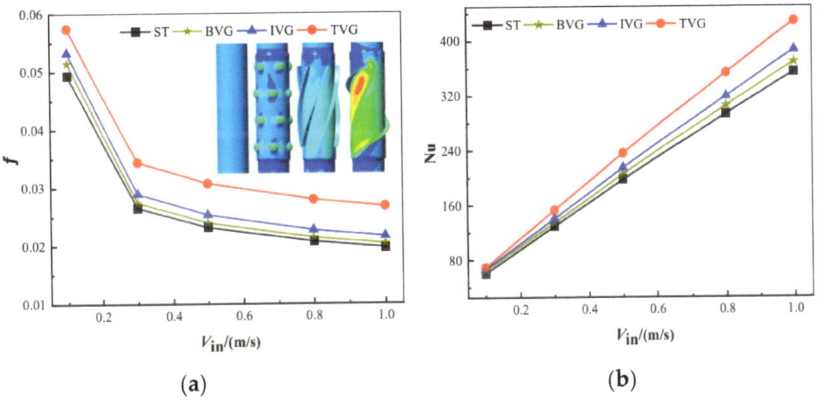

Figure 15. Influence of inlet velocity on strengthening parameters. (**a**) Influence of the inlet velocity on the friction coefficient f; (**b**) influence of the inlet velocity on the Nusselt number Nu.

Figure 16 shows the variation curve of the thermal performance coefficient PEC against the inlet velocity of three kinds of coaxial heat exchangers with vortex generators. The figure shows that the value of PEC is greater than 1.0, which indicates that the proposed new coaxial borehole heat exchanger-enhanced heat transfer design is feasible. The PEC value of TVG increases rapidly with the increase in the inlet velocity. The PEC value increases slowly when the inlet velocity exceeds 0.3 m/s and gradually becomes stable. This result shows that the inlet flow velocity of over 0.3 m/s exerts a minor effect on the heat transfer performance of the TVG heat exchanger. When the inlet velocity is

0.1~0.2 m/s, the PEC values of the IVG and BVG heat exchangers decrease rapidly. Moreover, when the inlet velocity exceeds 0.3 m/s, the PEC values increase slowly and gradually become stable. Simultaneously, under the same conditions, the PEC value of the TVG is much larger than that of the BVG and IVG heat exchangers. When the inlet flow rate was 1 m/s, the PEC of the TVG heat exchangers reached 1.1, whereas that of the BVG and IVG heat exchangers were only 1.03 and 1.06, respectively. This result shows that TVG has a better heat transfer performance.

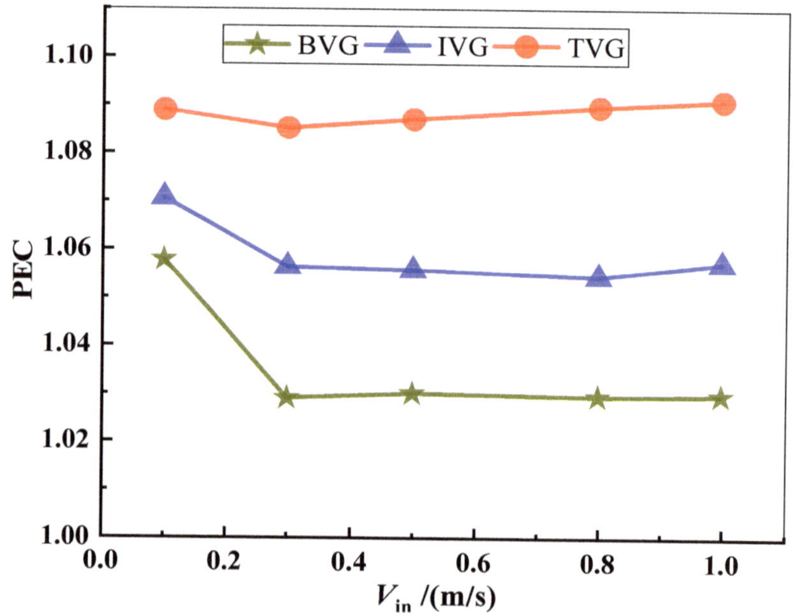

Figure 16. Influence of inlet flow rate on PEC of heat exchangers with vortex generators.

4.3. Heat Transfer Performance Analysis of the Coaxial Borehole Heat Exchanger

4.3.1. Effect of Inlet Flow Rate on Outlet Temperature

Figure 17 shows the variation curves of the production temperature of the four heat exchangers with the inlet flow rate. The figure shows that the production temperature gradually decreases with the increase in the inlet flow rate. When the inlet velocity increased from 0.1 m/s to 1 m/s, the production temperatures of ST, BVG, IVG and TVG heat exchangers decreased by 17.0 K, 17.6 K, 17.3 K and 15.4 K, respectively. Simultaneously, under the same inlet flow rate, the outlet temperature of the heat exchanger is presented as follows: TVG > IVG > BVG > ST. The arrangement shows that the vortex generator can enhance the heat recovery capacity of the coaxial heat exchanger. When the inlet flow rate was 0.5 m/s and the inlet temperature was 310 K, the production temperatures of ST, BVG, IVG and TVG heat exchangers increased by 16.5 K, 17.0 K, 17.8 K and 20.5 K, respectively, compared with the inlet temperature. The production temperature of the TVG heat exchanger is 24.06% higher than that of the ST heat exchanger.

4.3.2. Influence of Inlet Flow Rate on Production Power

Figure 18 shows the variation curves of the heat recovery power of the four heat exchangers with the inlet flow rate. The figure shows that the heat recovery power increases with the increase in the inlet flow rate. When the inlet temperature was 310 K and the inlet velocity increased from 0.1 m/s to 1 m/s, the heat recovery power of ST, BVG, IVG and TVG heat exchangers increased by 6.73 times, 6.68 times, 6.76 times and 7.14 times, respectively. In addition, the heat recovery power of the vortex generator heat exchanger has been improved compared with the ST heat exchanger. When the inlet flow rate was 1 m/s, the heat recovery power of the BVG, IVG and TVG heat exchangers increased by 38.33 KW, 85.28 KW and 241.96 KW, respectively, compared with the ST heat exchanger. Among them, the output power of the TVG heat exchanger increased by 11.93% compared with the ST heat exchanger. In conclusion, the heat recovery effect of the TVG heat exchanger is the highest and the increase in the inlet flow rate is beneficial in improving the heat recovery power of the TVG. However, the fluid flow rate increases with the flow rate, resulting in an increase in the required power of the injection pump, which has higher requirements for the pump's performance. Therefore, the selection of the water injection flow rate in the project can comprehensively consider the amount of heat production, the production temperature and the pump's power consumption.

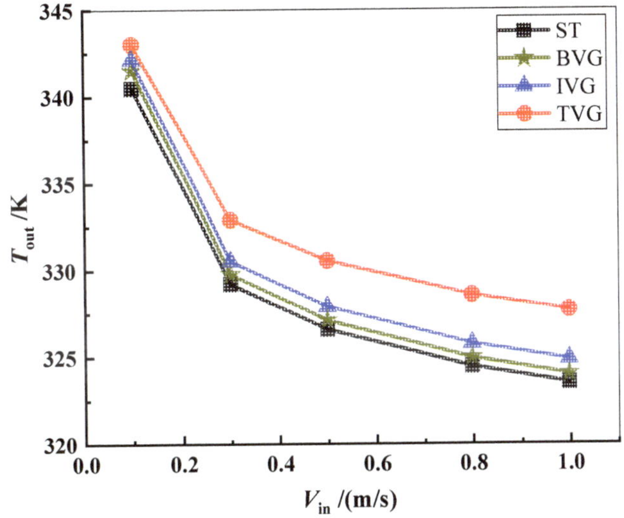

Figure 17. Variation curves of the production temperature of the four heat exchangers with inlet flow velocity.

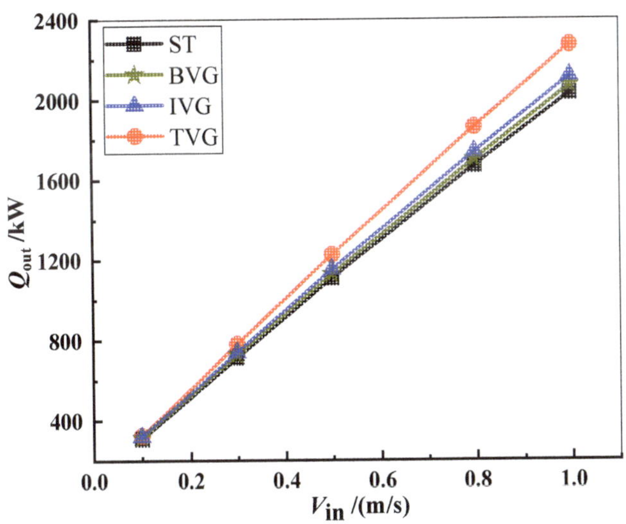

Figure 18. Variation curves of the heat recovery power of the four heat exchangers with inlet flow velocity.

5. Conclusions

Three new structures of the coaxial borehole heat exchanger with vortex generators were proposed in this work to improve the heat transfer efficiency of the coaxial borehole heat exchanger. In addition, the researchers used numerical methods to study the enhanced heat transfer mechanism with vortex generators. Next, the researchers compared and analyzed the heat transfer performance and heat recovery capacity of the four heat exchangers. The conclusions are presented as follows:

(1) The investigation of the enhanced heat transfer mechanism of the vortex generator revealed that the vortex generator can improve the turbulent kinetic energy of the fluid flow, increase the injected fluid velocity and flow velocity in the boundary layer region of the hot rock wall. Similarly, the fluid's velocity in the radial direction fluctuates up and down, thereby destroying the high-temperature boundary layer, strengthening the heat exchange inside the fluid and strengthening heat transfer.

(2) From the investigation of the heat transfer performance of four kinds of coaxial borehole heat exchangers, it was found that increasing the inlet flow rate decreases the friction coefficient and increases the Nusselt number. Compared with the ST heat exchanger, when the inlet velocity was increased by 1 m/s, the Nusselt number in BVG, IVG and TVG heat exchangers increased by 4.08%, 9.05% and 20.89%, respectively. The PEC of the TVG heat exchanger reached 1.1, which is favorable for improving the heat transfer performance of the coaxial heat exchanger. The impact of the pressure drop on the heat exchanger performance needs to be reduced.

(3) The analysis of the heat recovery capacity of four kinds of coaxial borehole heat exchangers showed that the heat recovery power increases linearly with the increase in the inlet flow rate. Increasing the inlet flow rate is beneficial in improving the heat recovery capacity of the coaxial heat exchanger; however, it is necessary to comprehensively consider the working performance of the pump to propose the optimal inlet flow rate.

By comparing the strengthening mechanism, heat exchange performance and heat recovery capacity of the four coaxial borehole heat exchangers, the order of the performance of the four heat exchangers from good to poor is presented as follows: TVG > IVG > BVG > ST. Compared with the ST heat exchanger, the production temperature of the TVG heat exchanger increased by 24.06% and the heating power increased by 11.93%, which is more suitable for geothermal mining.

Studies have shown that vortex generators can greatly improve the heat transfer performance of coaxial borehole heat exchangers, but it will result in pressure loss. Further research on the vortex generator structural parameters, the coaxial borehole structural parameters, process parameters and cycle fluid can be carried out to optimize geothermal energy extraction. It will provide a theoretical basis for the design of a coaxial borehole heat exchanger with vortex generation that has low flow friction and high heat transfer capacity.

Author Contributions: Conceptualization, methodology and validation, B.F.; formal analysis and writing—original draft preparation, L.S.; analysis and writing—part numerical analysis, M.W.; writing—review and editing, S.Z. All authors have read and agreed to the published version of the manuscript.

Funding: This project was supported by the Natural Science Foundation of China (No. 51974036 and No. 51604039), the Yangtze Fund for Youth Teams of Science and Technology Innovation (No. 2016cqt01) and the Open Foundation of Cooperative Innovation Center of Unconventional Oil and Gas, Yangtze University (Ministry of Education & Hubei Province) (No. UOG2022-29).

Data Availability Statement: The data that support the findings of this study are available within the article.

Conflicts of Interest: The authors declare no conflict of interest.

Nomenclature

T_{in}	Inlet temperature, K
T_{out}	Outlet temperature, K
T_g	The geothermal gradient, K/km
V_{in}	Inlet velocity, m/s
V_{out}	Outlet velocity, m/s
c_ρ	Specific heat capacity of water, J/(kg·k)
ρ	Density of water, kg/m^3
K	Thermal conductivity of water, W·(m·k)$^{-1}$
h	Convective heat transfer coefficient of water
T_{sur}	Surface temperature, K
T_w	Local rock temperature, K
T_m	Local fluid temperature, K
H	The distance between the lower end of the inner pipe and the bottom of the well, mm
z	Well depth, m
P	Fluid pressure, Pa
A_c	cross-sectional area, m^2
P_w	Wet perimeter of cross-section, m
g	Gravitational acceleration, m/s^2
μ	Dynamic viscosity of water, kg·(m·s)$^{-1}$
P_{out}	Outlet pressure, Pa
D	Radial dimensions of the heat exchanger, mm
D_1	Heat exchanger's outer tube diameter, mm
D_2	Inner pipe's inner diameter, mm
D_3	Vortex generator's diameter, mm
L_1	Vortex generator's distance from wellhead, mm
L_2	Vortex generator's length, mm
L_3	The distance from the vortex generator to the bottom of the well, mm

References

1. Holmberg, H.; Acuna, J.; Naess, E.; Otto, K.S. Thermal evaluation of coaxial deep borehole heat exchangers. *Renew. Energy* **2016**, *97*, 65–76. [CrossRef]
2. Alimonti, C.; Soldo, E. Study of geothermal power generation from a very deep oil well with a wellbore heat exchanger. *Renew. Energy* **2016**, *86*, 292–301. [CrossRef]
3. Pan, S.; Kong, Y.L.; Chen, C.F.; Pang, Z.H.; Wang, J.Y. Optimization of the utilization of deep borehole heat exchangers. *Geotherm. Energy* **2020**, *8*, 6. [CrossRef]
4. Noorollahi, Y.; Pourarshad, M.; Jalilinasrabady, S.; Yousefi, H. Numerical simulation of power production from abandoned oil wells in Ahwaz oil field in southern Iran. *Geothermics* **2015**, *55*, 16–23. [CrossRef]
5. Song, X.; Wang, G.; Shi, Y.; Li, R.; Xu, Z.; Zheng, R.; Wang, Y.; Li, J. Numerical analysis of heat extraction performance of a deep coaxial borehole heat exchanger geothermal system. *Energy* **2018**, *164*, 1298–1310. [CrossRef]
6. Renaud, T.; Verdin, P.; Falcone, G. Numerical simulation of a Deep Borehole Heat Exchanger in the Krafla geothermal system. *Int. J. Heat Mass Transf.* **2019**, *143*, 118496. [CrossRef]
7. Chen, C.; Shao, H.; Naumov, D.; Kong, Y.; Tu, K.; Kolditz, O. Numerical investigation on the performance, sustainability, and efficiency of the deep borehole heat exchanger system for building heating. *Geotherm. Energy* **2019**, *7*, 18. [CrossRef]
8. Saedi, A.Q.A.; Sharma, P.; Kabir, C.S. A novel cyclical wellbore-fluid circulation strategy for extracting geothermal energy. *Energy* **2021**, *235*, 121385. [CrossRef]
9. Shi, Y.; Bai, Z.; Feng, G.H.; Tian, H.L.; Bei, H. Performance analysis of medium-depth coaxial heat exchanger geothermal system using CO_2 as a circulating fluid for building heating. *Arab. J. Geosci.* **2021**, *14*, 1308. [CrossRef]
10. Zhang, Y.; Yu, C.; Li, G.; Guo, X.; Wang, G.; Shi, Y.; Peng, C.; Tan, Y. Performance analysis of a downhole coaxial heat exchanger geothermal system with various working fluids. *Appl. Therm. Eng.* **2019**, *163*, 114317. [CrossRef]
11. Luo, Y.; Guo, H.; Meggers, F.; Zhang, L. Deep coaxial borehole heat exchanger: Analytical modeling and thermal analysis. *Energy* **2019**, *185*, 1298–1313. [CrossRef]
12. Bu, X.; Ma, W.; Li, H. Geothermal energy production utilizing abandoned oil and gas wells. *Renew. Energy* **2012**, *41*, 80–85. [CrossRef]
13. Bu, X.; Ran, Y.; Zhang, D. Experimental and simulation studies of geothermal single well for building heating. *Renew. Energy* **2019**, *143*, 1902–1909. [CrossRef]

14. Du, J.; Hong, Y. Numerical simulation on fluid flow and heat transfer characteristics in inward sinusoidal rib enhanced tube heat exchangers for waste heat recovery: Comparisons and parametric studies. *Int. J. Therm. Sci.* **2021**, *167*, 107030. [CrossRef]
15. Mohammed, H.A.; Abbas, A.K.; Sheriff, J. Influence of geometrical parameters and forced convective heat transfer in transversely corrugated circular tubes. *Int. Commun. Heat Mass Transf.* **2013**, *44*, 116–126. [CrossRef]
16. Arasteh, H.; Rahbari, A.; Mashayekhi, R.; Keshmiri, A.; Mahani, R.B.; Talebizadehsardari, P. Effect of pitch distance of rotational twisted tape on the heat transfer and fluid flow characteristics. *Int. J. Therm. Sci.* **2021**, *170*, 106966. [CrossRef]
17. Tepe, A.U. Heat transfer enhancement of fin-tube heat exchangers using punched triangular ramp vortex generator on the fin surface. *Int. J. Heat Mass Transf.* **2021**, *174*, 121326. [CrossRef]
18. Nakhchi, M.E.; Esfahani, J.A. Numerical investigation of heat transfer enhancement inside heat exchanger tubes fitted with perforated hollow cylinders. *Int. J. Therm. Sci.* **2020**, *147*, 106153. [CrossRef]
19. Zhang, P.; Rao, Y.; Xie, Y.; Zhang, M. Turbulent flow structure and heat transfer mechanisms over surface vortex structures of micro V-shaped ribs and dimples. *Int. J. Heat Mass Transf.* **2021**, *178*, 121611. [CrossRef]
20. Ghobadi, A.H.; Armin, M.; Hassankolaei, S.G.; Hassankolaei, M.G. A new thermal conductivity model of CNTs/$C_2H_6O_2$–H_2O hybrid base nanoliquid between two stretchable rotating discs with Joule heating. *Int. J. Ambient Energy* **2020**, *43*, 3310–3321. [CrossRef]
21. Zadeh, M.N.; Pourfallah, M.; Sabet, S.S.; Gholinia, M.; Mouloodi, S.; Ahangar, A.T. Performance assessment and optimization of a helical Savonius wind turbine by modifying the Bach's section. *SN Appl. Sci.* **2021**, *3*, 739. [CrossRef]
22. Ghobadi, A.H.; Hassankolaei, M.G. A numerical approach for MHD Al_2O_3–TiO_2/H_2O hybrid nanofluids over a stretching cylinder under the impact of shape factor. *Heat Transf. Asian Res.* **2019**, *48*, 4262–4282. [CrossRef]
23. Ghobadi, A.H.; Hassankolaei, M.G. Numerical treatment of magneto Carreau nanofluid over a stretching sheet considering Joule heating impact and nonlinear thermal ray. *Heat Transf. Asian Res.* **2019**, *48*, 4133–4151. [CrossRef]
24. Arjmandi, H.; Amiri, P.; Pour, M.S. Geometric optimization of a double pipe heat exchanger with combined vortex generator and twisted tape: A CFD and response surface methodology (RSM) study. *Therm. Sci. Eng. Prog.* **2020**, *18*, 100514. [CrossRef]
25. Tusar, M.H.; Bhowmik, P.K.; Salam, B.; Ahamed, J.U.; Kim, J.K. Convective heat transfer and friction factor characteristics of helical strip inserted annuli at turbulent flow. *Int. J. Heat Mass Transf.* **2021**, *176*, 121422. [CrossRef]
26. Kumar, R.; Nandan, G.; Dwivedi, G.; Shukla, A.K.; Shrivastava, R. Modeling of triangular perforated twisted tape with V-Cuts in double pipe heat exchanger. *Mater. Today Proc.* **2020**, *46*, 5389–5395. [CrossRef]
27. Gnanavel, C.; Saravanan, R.; Chandrasekaran, M. Heat transfer enhancement through nano-fluids and twisted tape insert with rectangular cut on its rib in a double pipe heat exchanger. *Mater. Today Proc.* **2019**, *21*, 865–869. [CrossRef]
28. Huu-Quan, D.; Rostami, A.M.; Rad, M.S.; Izadi, M.; Hajjar, A.; Xiong, Q. 3D numerical investigation of turbulent forced convection in a double-pipe heat exchanger with flat inner pipe. *Appl. Therm. Eng.* **2020**, *182*, 116106. [CrossRef]
29. Bashtani, I.; Esfahani, J.A.; Kim, K.C. Effects of water-aluminum oxide nanofluid on double pipe heat exchanger with gear disc turbulators: A numerical investigation. *J. Taiwan Inst. Chem. Eng.* **2021**, *124*, 63–74. [CrossRef]
30. Ali, M.A.M.; El-Maghlany, W.M.; Eldrainy, Y.A. A Attia, Heat transfer enhancement of double pipe heat exchanger using rotating of variable eccentricity inner pipe. *Alex. Eng. J.* **2018**, *57*, 3709–3725. [CrossRef]
31. Mohsen, O.A.; Muhammed, M.A.R.; Hasan, B.O. Heat Transfer Enhancement in a Double Pipe Heat Exchanger Using Different Fin Geometries in Turbulent Flow. *Iran. J. Sci. Technol. Trans. Mech. Eng.* **2020**, *45*, 461–471. [CrossRef]
32. Iqbal, Z.; Syed, K.; Ishaq, M. Fin design for conjugate heat transfer optimization in double pipe. *Int. J. Therm. Sci.* **2015**, *94*, 242–258. [CrossRef]
33. Sani, F.H.; Pourfallah, M.; Gholinia, M. The effect of MoS_2–Ag/H_2O hybrid nanofluid on improving the performance of a solar collector by placing wavy strips in the absorber tube. *Case Stud. Therm. Eng.* **2022**, *30*, 101760. [CrossRef]
34. Fu, B.W.; Sun, L.; Zhang, S. Analysis of Heat Exchange Performance of Coaxial Casing Deep Well Heat Exchanger. *Chin. Hydraul. Pneum.* **2021**, *45*, 95–103. [CrossRef]
35. Caulk, R.A.; Tomac, I. Reuse of abandoned oil and gas wells for geothermal energy production. *Renew. Energy* **2017**, *112*, 388–397. [CrossRef]
36. Xing, L.; Spitler, J.D. Prediction of undisturbed ground temperature using analytical and numerical modeling. Part I: Model development and experimental validation. *Sci. Technol. Built Environ.* **2016**, *23*, 787–808. [CrossRef]
37. Sheikholeslami, M.; Gorji-Bandpy, M.; Ganji, D.D. Effect of discontinuous helical turbulators on heat transfer characteristics of double pipe water to air heat exchanger. *Energy Convers. Manag.* **2016**, *118*, 75–87. [CrossRef]
38. Shaker, B.; Gholinia, M.; Pourfallah, M.; Ganji, D. CFD analysis of Al_2O_3-syltherm oil Nanofluid on parabolic trough solar collector with a new flange-shaped turbulator model. *Theor. Appl. Mech. Lett.* **2022**, *12*, 100323. [CrossRef]
39. Incropera, F.P. *Fundamentals of Heat and Mass Transfer*; Wiley: Hoboken, NJ, USA, 2011.

processes

MDPI

Article

Coordination of Configurations of Technologically Integrated "European Green Deal" Projects

Anatoliy Tryhuba [1,*], Krzysztof Mudryk [2], Inna Tryhuba [1], Taras Hutsol [3,*], Szymon Glowacki [4], Oleksandr Faichuk [5], Nataliia Kovalenko [5], Alona Shevtsova [6], Arkadiusz Ratajski [4], Monika Janaszek-Mankowska [4] and Weronika Tulej [4]

1 Department of Information Systems and Technologies, Lviv National Agrarian University, 80-381 Dublyany, Ukraine
2 Faculty of Production and Power Engineering, University of Agriculture in Krakow, 30-149 Krakow, Poland
3 Department of Mechanics and Agroecosystems Engineering, Polissia National University, 10-008 Zhytomyr, Ukraine
4 Institute of Mechanical Engineering, Warsaw University of Life Sciences-SGGW, 02-787 Warsaw, Poland
5 Faculty of Agrarian management, National University of Life and Environmental Sciences of Ukraine, 03-041 Kyiv, Ukraine
6 Program NAWA, Higher Educational Institution "Podillia State University", 32-300 Kamianets-Podilskyi, Ukraine
* Correspondence: trianamik@gmail.com (A.T.); wte.inter@gmail.com (T.H.)

Citation: Tryhuba, A.; Mudryk, K.; Tryhuba, I.; Hutsol, T.; Glowacki, S.; Faichuk, O.; Kovalenko, N.; Shevtsova, A.; Ratajski, A.; Janaszek-Mankowska, M.; et al. Coordination of Configurations of Technologically Integrated "European Green Deal" Projects. *Processes* **2022**, *10*, 1768. https://doi.org/10.3390/pr10091768

Academic Editor: Maria Gavrilescu

Received: 1 July 2022
Accepted: 30 August 2022
Published: 3 September 2022

Abstract: An analysis of the state of affairs in the theory and practice of implementation of technologically integrated projects in various applied fields was carried out. The peculiarities of the implementation of the technologically integrated projects of the "European Green Deal" for the production of ecologically clean fuel from agricultural waste were analyzed. The expediency of developing a method of coordinating the configurations of technologically integrated "European Green Deal" projects for the production of ecologically clean energy from agricultural waste on the territory of a given region, taking into account their specific project environments, was substantiated. As a result of the conducted research, a method of coordinating configurations of the technologically integrated "European Green Deal" projects for the production of ecologically clean energy from agricultural waste in the territory of a given region, taking into account their specific project environment, was developed. This method involves the implementation of five stages, which ensure consideration of the specific design environment of each region and the type of agricultural raw materials for energy production. This method involves the modeling of individual projects, which makes it possible to increase the accuracy of determining their value indicators, taking into account risk. The balancing of the technologically integrated projects of the "European Green Deal" for the production of clean energy from agricultural waste was carried out on the basis of maximizing value for stakeholders and minimizing risk. On the basis of the proposed method, the computer program "Balancing technologically integrated projects" was developed. The use of this computer program for the given project environment (conditions of LLC "Lutsk Agrarian Company" of the Volyn region, Ukraine) made it possible to forecast the specific value and risk of individual projects involving harvesting raw materials from corn waste. The statistical characteristics of the distribution of the projected specific value of the project of harvesting raw materials from corn waste were established: the estimate of mathematical expectation—EUR 9/ton; dispersion—EUR 25/ton; the estimation of root mean square deviation—EUR 5/ton. The technologically integrated projects of the "European Green Deal" for the production of ecologically clean energy from corn waste with the greatest interconnections in terms of value were identified. The ranking of raw material procurement projects from corn waste was carried out according to their specific values and risks. Among the considered projects, priority was given to project #7 and project #1, which provided the greatest values, 37.6% and 36.6%, respectively, of the total value of the considered projects. The obtained results made it possible to establish priority projects and carry out their balancing by value and risk.

Keywords: integrated projects; configuration; environmentally friendly fuel; agricultural waste; coordination; value; risks

1. Introduction

Providing the population and individual industries in the world with quality and environmentally friendly energy remains an urgent and unresolved problem for mankind [1–4]. Part of this problem is the reduction of greenhouse gas emissions generated by human activities. This is due to the use of fossil fuels, which are the main raw materials for energy production in most countries [5–7]. Combustion of fossil fuels produces carbon dioxide (CO_2). It is known [5] that even a small amount of carbon dioxide can be toxic and cause biochemical changes in the human blood, joint pain, weakness, acid–base imbalance, decreased immunity, kidney disease, and cardiovascular disease.

All the above factors have led to the strengthening of EU requirements for the quality of raw materials for energy production. In particular, they promote the use of environmentally friendly fuels. The EU declared that by 2050, Europe will be the first continent whose economy does not destroy nature. In particular, this was provided for in the European Green Deal policy initiatives initiated by the European Commission.

Regarding environmentally friendly raw materials for energy production, EU countries pay special attention to agricultural waste [8–11]. Some scientific papers [10–12] state that countries that use agricultural waste to produce clean energy are able to ensure their own energy security and achieve the energy policy goals laid out in the European Green Deal.

Integrated projects need to be implemented in individual EU regions to implement European Green Deal initiatives. At the same time, there are a number of scientific and applied problems, the solutions to which will increase the efficiency of European Green Deal projects [1,2,6]. One such task, which unfortunately remains unsolved to this day, is the coordination of the configurations of the technologically integrated "European Green Deal" projects.

Existing scientific publications [13,14] have noted that integrated projects require the coordination of their configurations. Given that the basic projects of the European Green Deal with the use of agricultural waste include projects for the procurement of agricultural waste and the production of clean energy, there is a need to harmonize their configurations.

The analysis performed on the state of the issue in practice showed that currently, the scientific and applied task of coordinating the configurations of the technologically integrated "European Green Deal" projects for the production of ecologically clean energy from agricultural waste in the given region remains unresolved. At the same time, there are no scientific works related to the coordination of the configurations of the technologically integrated "European Green Deal" projects taking into account the characteristics of the project environment where they are implemented. The type of raw materials for energy production is of great importance in the formation of an effective configuration of "European Green Deal" projects. This determines the content of the work performed in the specified projects and the need for certain types of resources (technical, human, material, etc.) [15–18].

All the above factors indicate the need to develop a method of matching the configurations of the technologically integrated "European Green Deal" projects. This should take into account both the changing characteristics of the project environment for the region where they are implemented and the parameters of their configuration objects, which depend on the type of agricultural raw materials used.

2. Analysis of Literature Data and Problem Statement

An analysis of scientific publications showed that some of them relate to the implementation of technologically integrated projects in various fields [14]. However, there are no publications on the coordination of the configurations of technologically integrated "European Green Deal" projects.

Regarding the management of the configuration of projects in various fields of human activity, a number of standards have been developed [19–21], and some scientific papers are devoted to this issue [22–27]. The mentioned research concerns projects of agro-industrial production [24], fire extinguishing [23,26,27], milk procurement [13,17,20], energy [25], and others. Unfortunately, the analyzed scientific works have not used differentiated approaches to the coordination of configurations of technologically integrated projects taking into account the characteristics of the project environment.

The emergence of a practical standard for project configuration management [18] became the basis of an in-depth study of the project configuration management process. Studies on this issue have revealed a number of inconsistencies in this area of knowledge [28,29]. However, these publications, as well as the standard itself [18], have not disclosed the process of matching the configurations of technologically integrated "European Green Deal" projects in a given region, taking into account the characteristics of their design environment and the type of agricultural raw materials used. This determines the specifics and features of European Green Deal projects using agricultural raw materials. To reveal these features, we used both knowledge of project management [19–21] and the results of special studies in projects of various applied industries [30–32], including agricultural production and energy [3,8–13].

An analysis of many scientific papers [6–9] showed that existing methods and models of project configuration management cannot be used to reconcile the configurations of technologically integrated "European Green Deal" projects in a given region with their project environment due to a number of shortcomings. In addition, scientific works on risk management in projects were identified [33–39]. However, they have not taken into account the peculiarities of the changing design environment of the PMC. In addition, they have not provided for the identification of the configuration of projects on the basis of modeling their products, which makes it impossible to obtain maximum systemic value for stakeholders. In order to objectively coordinate the configurations of "European Green Deal" projects, tools should be developed that will take into account the region-specific type of agricultural raw materials for energy production, which will ensure maximum value for stakeholders [40,41].

The aim of the work was to develop a method of coordinating the configurations of technologically integrated "European Green Deal" projects for clean energy from agricultural waste in the region, taking into account their specific design environment; this method is based on individual projects as separate organizational and technical systems and modeling to determine the value indicators in a given scenario of project implementation, which ensures the balancing of their configurations according to the value criterion.

To achieve the goal of the study, the following tasks were completed:

- To propose a method of coordinating configurations in technologically integrated "European Green Deal" projects for the production of clean energy from agricultural waste in the region, taking into account their specific project environments;
- On the basis of the use of the proposed method, to coordinate the configurations of the technologically integrated "European Green Deal" projects for the production of clean energy from corn waste in Ukraine, taking into account the specifics of the project environment of the region.

3. The Method of Coordination of Configurations of Technologically Integrated "European Green Deal" Projects for the Production of Clean Energy from Agricultural Waste in the Region

Coordinating the configurations of technologically integrated "European Green Deal" projects for the production of clean energy from agricultural waste in the region, taking into account their specific project environments, involves the installation of components (projects) that will provide maximum value for stakeholders in a changing project environment [17,18,23]. To propose project configurations, we proposed a method based on the consideration of projects as separate organizational and technical systems. In addition, the proposed method provides for the modeling of individual projects to determine the indicators of their value in a given implementation scenario, which ensures the balancing of their configurations by the criterion of value. The proposed method involved the implementation of five steps, which are presented in Figure 1.

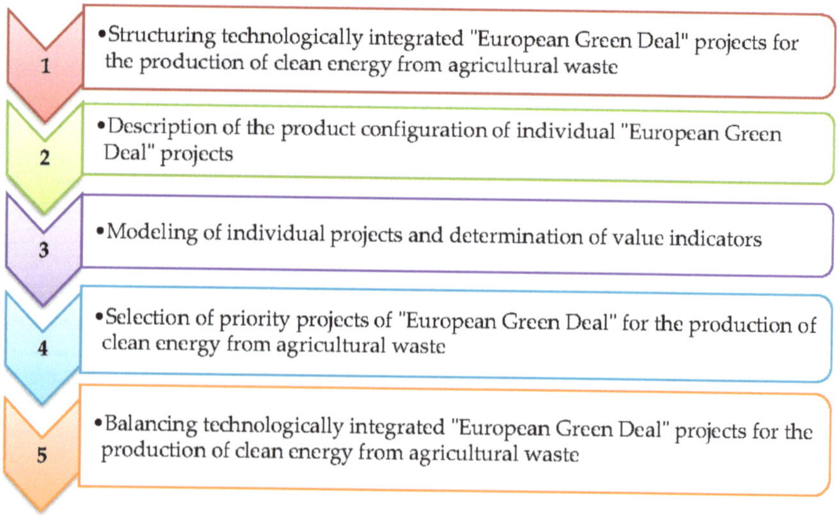

Figure 1. Stages of the method of coordination of configurations of technologically integrated "European Green Deal" projects for the production of clean energy from agricultural waste in the region.

Stage 1. The technologically integrated projects "European Green Deal" for the production of clean energy from agricultural waste include many projects that belong to different levels of their consideration. In particular, as mentioned above, there are two levels of consideration for projects—agricultural waste procurement projects and clean energy projects. Both these types of projects have their own specifics that should be taken into account when agreeing on their configurations. They underlie the process of structuring technologically integrated "European Green Deal" projects for the production of clean energy from agricultural waste [24,32]. This process ensures the definition of the type of project and its belonging in individual regions, as well as changes that occur as a result of these projects.

Stage 2. The next stage involves the description of the product configuration of technologically integrated "European Green Deal" projects. It addresses (1) the type of agricultural waste that will be used for the project; (2) the number and qualification of involved performers; (3) the type and quantity of technical equipment; (4) the characteristics of the project environment (area of fields for growing crops, their soils and fertility, the number of fields, the share of alienation of raw materials for energy production, etc.). At the same time, natural resources, the type of agricultural raw materials, and the share of alienation of raw materials for energy production are of the greatest value in the region of implementation of technologically integrated "European Green Deal" projects.

Stage 3. According to the known configurations of technologically integrated "European Green Deal" project products, project modeling is performed in order to determine the value indicators for each of the formed organizational and technical systems. The most valuable are those projects whose products provide the minimum cost of resources for the production of a unit of product (raw materials or energy). The defining indicator of an individual project is the use of resources ($P^{m}_{i,j}$), i.e., the use of j-th resources in the i-th organizational and technical system of the τ-th level of consideration. To determine this indicator, the simulation model developed in [2] is used, which makes it possible to determine the use of resources ($P^{m}_{i,j}$), i.e., j-th resources in the i-th organizational and technical system:

$$P^{m}_{i,j} = f\left(P^{m}_{i,hr}, P^{m}_{i,tr}, P^{m}_{i,mr}\right) \quad (1)$$

where $P^{m}_{i,hr}, P^{m}_{i,tr}, P^{m}_{i,mr}$—the use of human, technical, and material resources, respectively, in the i-th organizational and technical system of the m-th level of consideration.

Comparing the quantitative values of resource use ($P^{m}_{i,j}$) of the j-th resources in the i-th organizational and technical system obtained on the basis of simulation, the cost levels (R^{m}_{ij}) for the implementation of projects under the given scenario are determined.

Stage 4. To select priority projects "European Green Deal", a model of their value is built (Figure 2).

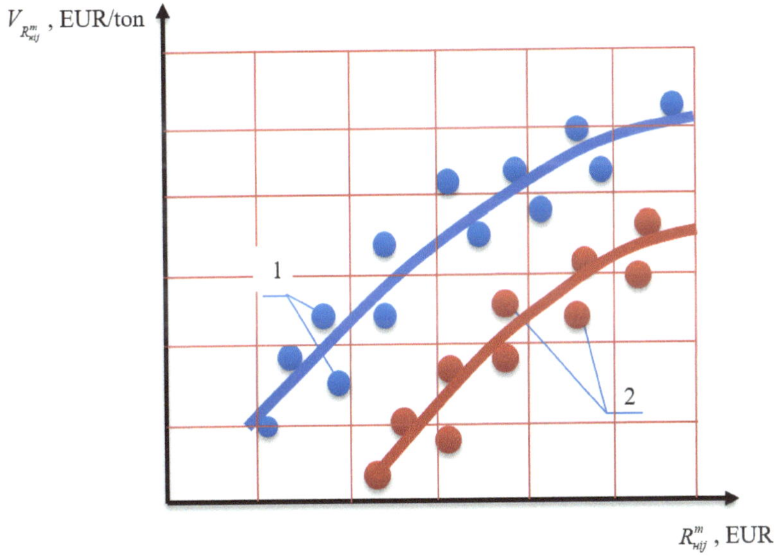

Figure 2. Value model of technologically integrated "European Green Deal" projects that claim to be included in effective technologically integrated projects: R^{m}_{ij}, $V_{R^{m}_{ij}}$ —projects of the *t*-th level of their consideration (1—projects of agricultural waste procurement; 2—projects of environmentally friendly energy production).

Figure 2 presents the value model of projects that claim to be included in effective technologically integrated projects, which is based on known cost levels (R^{m}_{ij}) for project implementation and unit values ($V_{R^{m}_{ij}}$) according to the level of costs and the given level of their consideration.

The authors of the work suggested that the value ($V_{R^{m}_{ij}}$) of the "European Green Deal" projects should be determined by the level of costs (R^{m}_{ij}) of their implementation:

$$V_{R^{m}_{ij}} = \frac{R^{m}_{ij}}{N^{m}_{p}} \quad (2)$$

where $V_{R^{m}_{ij}}$—the specific value by the level of costs (R^{m}_{ij}) for project implementation, EUR/ton; R^{m}_{ij}—the level of costs (R^{m}_{ij}) for the implementation of projects in their given scenario, EUR; N^{m}_{p}—the amount of product obtained from the implementation of individual projects for a given t-th level of their consideration, tons.

Stage 5. At the stage of balancing technologically integrated "European Green Deal" projects for the production of clean energy from agricultural waste, the priority projects selected in the previous stage are ranked according to their specific value in ascending order:

$$V_{R_1^m} \geq V_{R_2^m} \geq \ldots \geq V_{R_n^m} \quad (3)$$

After that, technologically integrated "European Green Deal" projects for the production of clean energy from agricultural waste in the region are selected on the basis of the condition:

$$\{Q_i^m\} \leq Q_{TI}, i = 1, n \quad (4)$$

where $\{Q_i^m\}$—the volume of the received product from the realization of i-th priority projects for the set t-th level of their consideration, tons; Q_{TI}—the total product volume of projects for the production of clean energy from agricultural waste, tons; n—the number of projects included in effective technologically integrated projects, units.

When balancing the technologically integrated "European Green Deal" projects for the production of clean energy from agricultural waste, preference should be given to projects of high value. At the same time, it should be ensured that projects with high levels of implementation costs (R_{ij}^m) and a correspondingly high risk of non-implementation should have as small a share as possible in the structure of technologically integrated projects.

4. Results of Coordination of Configurations of Technologically Integrated "European Green Deal" Projects for the Production of Clean Energy from Corn Waste in Ukraine

First, on the basis of statistical data, we performed an analysis of the state of the raw material base in Ukraine for the production of ecologically clean energy. It was established that today, in the majority of large agricultural enterprises in Ukraine, corn makes up 40–50% of the crop rotation structure. At the same time, the amount of corn sown per grain shows an increasing trend (Figure 3).

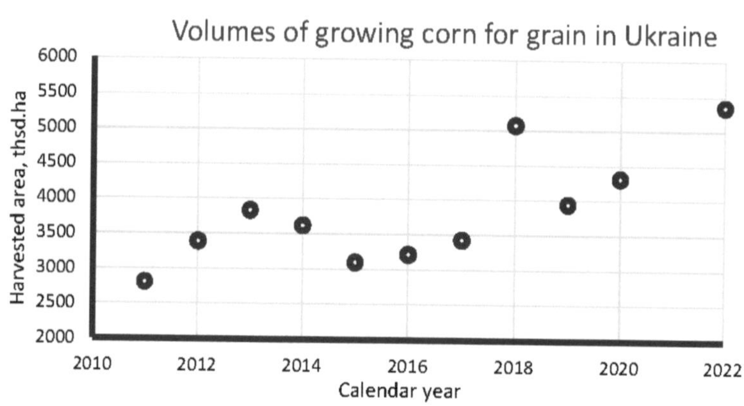

Figure 3. Trends in changes in the amount of corn grown for grain in Ukraine.

Taking into account the above, the production of ecologically clean energy from corn waste was chosen for further research.

On the basis of the disclosure of the content of the stages of the proposed method, which is presented in Figure 1, a corresponding computer program was developed. The specified computer program, "Balancing technologically integrated projects", was developed at the Department of Information Technologies of the Lviv National University of Natural Sciences. It ensured the coordination of configurations of projects for the production of clean energy from agricultural waste in the region. The proposed computer program was tested for adequacy using the generally accepted paired t-test. At the same time, an adequacy check was performed with a comparison of real data for the conditions of "Lutsk Agrarian Company" LLC of the Volyn region (Ukraine) obtained as a result of the use of a computer program for coordinating the configurations of technologically integrated "European Green Deal" projects for the production of ecologically clean energy from agricultural waste. It was determined that the actual data values ($V_{R^{m}_{ij}}$) of the projects of harvesting raw materials from corn waste obtained as a result of the use of the proposed computer program deviated within insignificant limits—1.6 ... 4.9%. This indicates that the proposed computer program adequately ensures the coordination of project configurations for the production of ecologically clean energy from agricultural waste in the given region.

On the basis of the use of the computer program, a study was conducted to coordinate the configurations of technologically integrated "European Green Deal" projects for the production of clean energy from corn waste for a given design environment (conditions Lutsk Agrarian Company LLC, Volyn region, Ukraine). The conducted computer experiments made it possible to predict the specific values ($V_{R^{m}_{ij}}$) of individual projects for the procurement of raw materials from corn waste, which were implemented in different fields. This provided the determination of the specific cost of the disposal of corn waste (Table 1).

Table 1. Results of forecasting the specific values ($V_{R^{m}_{ij}}$) of individual projects for the procurement of raw materials from corn waste.

Indicator	Project Implementation Scenario						
	1	**2**	**3**	**4**	**5**	**6**	**7**
Planned unit cost of maize waste disposal, EUR/ton	14	20	17	18	12	15	11
Market share of maize waste, EUR/ton				24			
Estimated specific value ($V_{R^{m}_{ij}}$) of projects for procurement of raw materials from corn waste	9	14	11	12	8	10	7

The results of the visualization of the distribution of the projected specific value ($V_{R^{m}_{ij}}$) for project No. 1 for the procurement of raw materials from corn waste are presented in Figure 4.

Statistical processing of the obtained data on the distribution of the projected specific value ($V_{R^{m}_{ij}}$) for project No. 1 for the procurement of raw materials from corn waste allowed the determination of the numerical characteristics and the justification of the model (Figure 4), which are described by the normal distribution law with a differential function:

$$f\left(V_{R^{m}_{i1}}\right) = 0.079 \cdot \exp\left(-\frac{\left(V_{R^{m}_{i1}} - 9\right)^2}{50}\right) \quad (5)$$

where $V_{R^{m}_{ij}}$—projected specific value of the project of procurement of raw materials from corn waste, EUR/ton.

The main statistical characteristics of the distribution of the projected specific value ($V_{R^{m}_{i1}}$) for project No. 1 for the procurement of raw materials from corn waste were as follows: estimation of mathematical expectation—EUR 9/ton; dispersion—EUR 25/ton; estimate of standard deviation—EUR 5/ton.

The conducted research provided the construction of covariance and correlation matrices of the forecasted specific values ($V_{R^{m}_{ij}}$) of separate considered projects for the preparation of raw materials from corn waste (Figures 5 and 6).

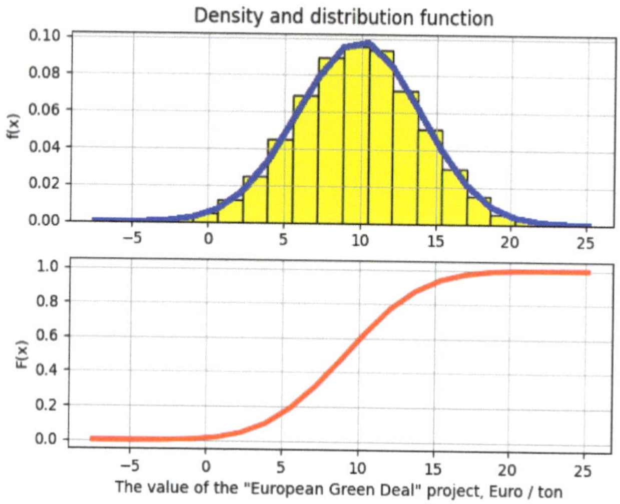

Figure 4. Histogram and theoretical curve of distribution of projected specific value ($V_{R_{ij}^{m}}$) for project No. 1 for the procurement of raw materials from corn waste, EUR/ton.

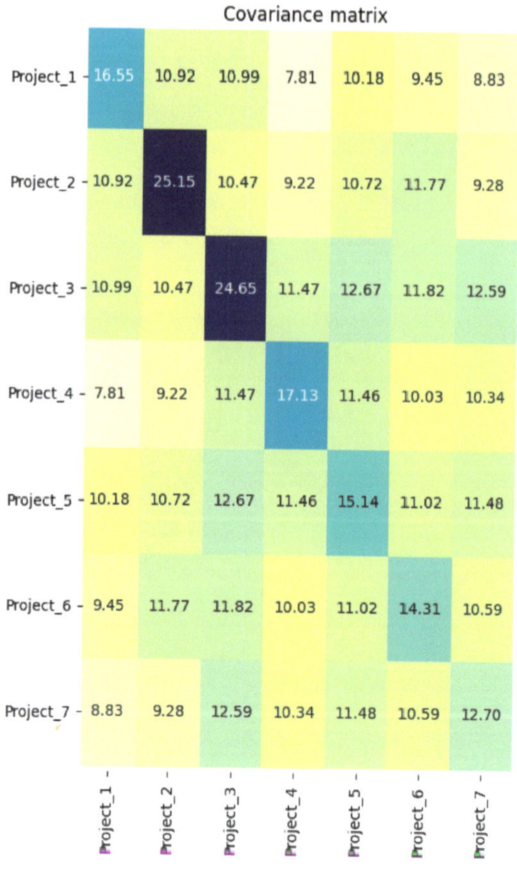

Figure 5. Covariance matrix between the projected specific values ($V_{R_{ij}^{m}}$) of the individual projects under consideration for the procurement of raw materials from corn waste.

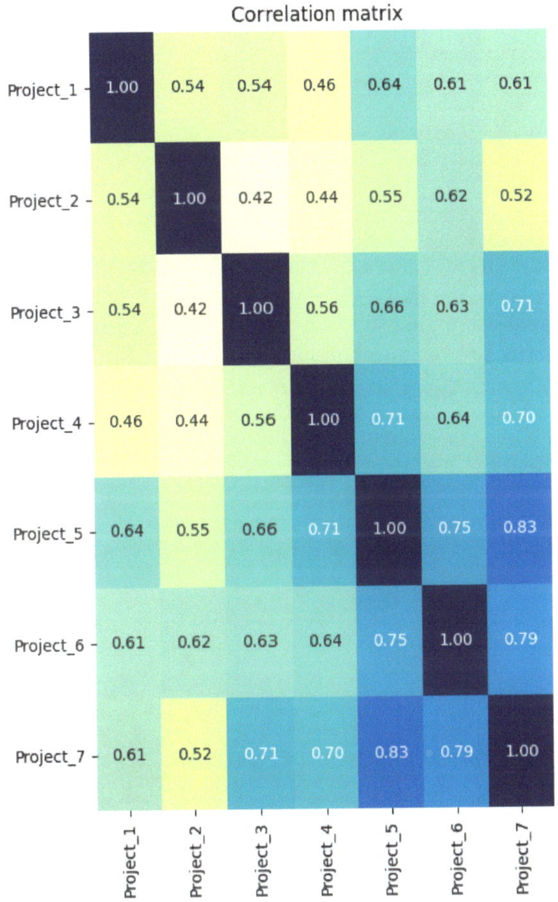

Figure 6. Correlation matrix between the projected specific value ($V_{R^m_{ij}}$) of the individual projects under consideration for the procurement of raw materials from corn waste.

In the resulting covariance matrix of the predicted specific value ($V_{R^m_{ij}}$) of the individual considered projects for harvesting raw materials from corn waste (Figure 5), the diagonal contains the value of the dispersion of the specific value of the considered projects, and the off-diagonal elements characterize the covariances between the specific value of the considered projects. According to the obtained results, we see that the largest deviations of the values of variable specific values from the mathematical expectation were observed for projects No. 2 and No. 3 and were, respectively, $V_{R^m_{ni2}} = 25.15$ EUR/ton and $V_{R^m_{i3}} = 24.65$ EUR/ton. It was for these projects of harvesting raw materials from corn waste that the largest deviations of the obtained values were characteristic, and they are the riskiest for investors. From a risk perspective, they should be the last to be included in technologically integrated "European Green Deal" projects for the production of ecologically clean energy from corn waste.

The obtained correlation matrix of the specific values of projects (Figure 6) shows that a strong relationship according to the specified criterion was observed between projects No. 7 and No. 5 ($r_{75} = 0.83$), and the correlation coefficient between projects No. 7 and No. 6 was $r_{76} = 0.79$. It was established that project #7 had the largest cost relationships with projects on the production of environmentally clean energy from corn waste.

5. Discussion of Research Results

According to the obtained results, we see that the largest deviations of the values of the variable specific value from the mathematical expectation were observed for projects No. 2 and No. 3 and were, respectively, $V_{R^{m}_{ni2}} = 25.15$ EUR/ton and $V_{R^{m}_{i3}} = 24.65$ EUR/ton. It was for these projects that the procurement of raw materials from corn waste was characterized by the largest deviations of the obtained value, and they are the riskiest for investors. From the point of view of risk, they should be included last in technologically integrated "European Green Deal" projects for the production of environmentally friendly energy from corn waste.

In addition, we conducted a rank correlation between the specific value of the considered projects of the procurement of raw materials from corn waste, the results of which are presented in Figure 5. The obtained correlation matrix of the specific value of projects indicates that a strong relationship according to this criterion was observed between projects No. 7 and No. 5, for which the correlation coefficient was $r_{75} = 0.83$, and between projects No. 7 and No. 6, for which the correlation coefficient was $r_{76} = 0.79$. This shows that the No. 7 project had the greatest value relationships in the technologically integrated "European Green Deal" projects for the production of clean energy from corn waste.

Balancing the technologically integrated "European Green Deal" projects for the production of clean energy from agricultural waste made it possible to establish priority projects, that were ranked according to their specific value and risks in ascending order (Table 2).

Table 2. Results of ranking of projects of procurement of raw materials from corn waste by their specific value and risks.

Indicator	Project No.						
	7	1	4	5	6	3	2
Fraction values, %	37.65	36.67	12.96	7.46	1.94	1.7	1.6

As a result of the research, it was established that for a given design environment (conditions of Lutsk Agrarian Company LLC, Volyn region, Ukraine), technologically integrated "European Green Deal" projects for clean energy production from corn waste should be formed in the sequence shown in Table 2. At the same time, the total value of the implementation of technologically integrated projects of harvesting raw materials from corn waste was taken as 100%. Among the projects under consideration, projects No. 7 and No. 1 were prioritized, which provided them the opportunity to obtain the greatest values, which were 37.6% and 36.6% of the projects under consideration, respectively.

The number of corn waste projects to be included in the European Green Deal was selected from condition (4) and depended on the scale and configuration of the clean energy project in the region.

This research allowed the development of a method and a computer program that make it possible to speed up and ensure the accuracy of the process of coordinating the configurations of technologically integrated "European Green Deal" projects for the production of ecologically clean energy from agricultural waste in the territory of the given region. The specified projects were considered as separate organizational and technical systems, which made it possible to take into account their peculiarities and the peculiarities of their project environment. The peculiarity of the developed method is that it involves the modeling of individual projects to determine their value indicators under a given implementation scenario, which ensures the balancing of their configurations according to the value criterion. This ensures accurate results. The performed verification of the adequacy of the computer program for coordinating the configurations of technologically integrated "European Green Deal" projects for the production of ecologically clean energy from agricultural waste confirmed the specified accuracy, as the obtained value $V_{R^{m}_{ij}}$ of the projects of harvesting raw materials from corn waste deviated from the real one within insignificant limits—1.6 ... 4.9%. The possibility of the practical use of the developed toolkit was confirmed by the obtained results of the coordination of the configurations of technologically integrated "European Green Deal" projects for the production of environmentally clean energy from agricultural waste for the conditions of LLC "Lutsk Agrarian Company" of the Volyn region (Ukraine). At the same time, it was established that the value and risks of individual projects for the production of ecologically clean energy from agricultural waste are distinguished within fairly wide limits. This confirms the expediency of using the developed method and computer program to coordinate the configurations of technologically integrated "European Green Deal" projects for the production of ecologically clean energy from agricultural waste, which will make it possible to avoid making erroneous decisions regarding the selection of appropriate projects. Further research should be conducted regarding the coordination of the configurations of technologically integrated "European Green Deal" projects for the production of ecologically clean energy from agricultural waste for other types of raw materials and regions characterizing the project environment. This will provide an opportunity to create relevant knowledge needed by project managers when coordinating the configurations of technologically integrated projects for the production of ecologically clean energy from agricultural waste.

6. Conclusions

The proposed method of coordinating the configurations of technologically integrated "European Green Deal" projects for the production of clean energy from agricultural waste involves the implementation of five stages. These stages take into account the specific project environment for each region, the type of agricultural raw materials for energy production, as well as modeling projects, which allow the determination of their value for each of the existing organizational and technical systems. On the basis of the ranking of projects by their maximum value for stakeholders and the possibility of risk minimization, the balancing of technologically integrated "European Green Deal" projects for the production of clean energy from agricultural waste was performed.

With the use of a computer program developed at the Department of Information Technology of Lviv National University of Nature Management, on the basis of the proposed method, the configurations of the technologically integrated "European Green Deal" projects for clean energy from agricultural waste for a given project environment (conditions Lutsk Agricultural Company, "Volyn region, Ukraine) were determined.

This made it possible to establish numerical characteristics and models of distributions of the predicted specific value ($V_{R^m_{i1}}$) for individual projects of harvesting raw materials from corn waste: the estimation of mathematical expectation—EUR 9/ton; dispersion—EUR 25/ton; the estimation of root mean square deviation—EUR 5/ton. On the basis of the obtained data, covariance and correlation matrices were constructed between the predicted specific value of the individual projects under consideration. The largest deviations of the values of variable specific values from the mathematical expectation were observed for projects #2 and #3 and were, respectively, $V_{R^m_{i2}} = 25.15$ EUR/ton and $V_{R^m_{i3}} = 24.65$ EUR/ton. It was for these projects of harvesting raw materials from corn waste that the largest deviations of the obtained value were characteristic, and they are the riskiest for investors. The ranking of projects by their specific value and risks was carried out, which ensured the identification of priority projects No. 7 and No. 1, which provide the opportunity to obtain the greatest values, which were 37.6% and 36.6% of the projects under consideration, respectively.

Author Contributions: Conceptualization, A.T.; methodology, T.H. and I.T.; database creation, K.M., S.G.; literature review, O.F., N.K.; software, A.S.; validation, W.T.; formal analysis, M.J.-M.; funding acquisition, A.R.; project administration, T.H.; All authors have read and agreed to the published version of the manuscript.

Funding: This research received no external funding.

Institutional Review Board Statement: Not applicable.

Informed Consent Statement: Not applicable.

Data Availability Statement: Data are available from the corresponding author upon reasonable request.

Acknowledgments: We thank the anonymous reviewers for their constructive review, which has greatly improved this manuscript and NAWA.

Conflicts of Interest: The authors declare no conflict of interest.

References

1. Alberti, V.; Caperna, G.; Colagrossi, M.; Geraci, A.; Mazzarella, G.; Panella, F.; Saisana, M. *Tracking EU Citizens' Interest in EC Priorities Using Online Search Data*; The European Green Deal, EUR 30580 EN; Publications Office of the European Union: Luxembourg, 2021; p. JRC123917. [CrossRef]
2. Hutsol, T.; Glowacki, S.; Tryhuba, A.; Kovalenko, N.; Pustova, Z.; Rozkosz, A.; Sukmaniuk, O. *Current Trends of Biohydrogen Production from Biomass–Green Hydrogen*; Libra-Print: Warsaw, Poland, 2021. [CrossRef]
3. Tryhuba, A.; Boyarchuk, V.; Tryhuba, I.; Ftoma, O.; Tymochko, V.; Bondarchuk, S. Model of assessment of the risk of investing in the projects of production of biofuel raw materials. In Proceedings of the 15th International Scientific and Technical Conference on Computer Sciences and Information Technologies (CSIT), Zbarazh, Ukraine, 23–26 September 2020; Volume 2, pp. 151–154.
4. EBA. European Biogas Association. Available online: https://www.europeanbiogas.eu/2020-gas-decarbonisation-pathways-study (accessed on 10 May 2022).
5. Gas Decarbonisation Pathways 2020–2050-Gas for Climate. Available online: https://www.gasforclimate2050.eu (accessed on 11 May 2022).
6. A European Green Deal. Available online: https://ec.europa.eu/info/strategy/priorities/2019-2024/european-green-deal (accessed on 15 May 2022).
7. Von der Leyen, U. My agenda for Europe. Political Guidelines for the Next European Commission 2019–2024. 2019. Available online: https://ec.europa.eu/commission/sites/beta-political/files/political-guidelines-next-commission_en.pdf (accessed on 15 May 2022).
8. Boyarchuk, V.; Tryhuba, I.; Ftoma, O.; Francik, S.; Rudynets, M. Method and Software of Planning of the Substantial Risks in the Projects of Production of raw Material for Biofuel. In Proceedings of the 1st International Workshop IT Project Management (ITPM 2020), Slavsko, Lviv Region, Ukraine, 18–20 February 2020; pp. 116–129. Available online: http://ceur-ws.org/Vol-2565/ (accessed on 25 September 2020).
9. Tryhuba, A.; Boyarchuk, V.; Boiarchuk, O.; Pavlikha, N.; Kovalchuk, N. Study of the impact of the volume of investments in agrarian projects on the risk of their value (ITPM-2021). In Proceedings of the 2nd International Workshop IT Project Management, Lviv, Ukraine, 16–18 February 2021.
10. Tryhuba, A.; Glowacki, S.; Tryhuba, I.; Tabor, S.; Kwasniewski, D.; Sorokin, D.; Yermakov, S. Forecasting Quantitative Risk Indicators of Investors in Projects of Biohydrogen Production from Agricultural Raw Materials. *Processes* **2021**, *9*, 258. [CrossRef]
11. Kondysiuk, I.; Bashynsky, O.; Dembitskyi, V.; Myskovets, I. Formation and risk assessment of stakeholders value of motor transport enterprises development projects. In Proceedings of the International Scientific and Technical Conference on Computer Sciences and Information Technologies, Lviv, Ukraine, 22–25 September 2021; Volume 2, pp. 303–306.
12. Bashynsky, O.; Garasymchuk, I.; Gorbovy, O. Research of the variable natural potential of the wind and energy energy in the northern strip of the ukrainian carpathians. In Proceedings of the 6th International Conference: Renewable Energy Sources (ICoRES 2019). E3S Web of Conferences, Krynica, Poland, 12–14 June 2019; Volume 154, p. 06002.
13. Bashynsky, O. Conceptual model of management of technologically integrated industry development projects. In Proceedings of the 15th International Scientific and Technical Conference on Computer Sciences and Information Technologies (CSIT), Zbarazh, Ukraine, 23–26 September 2020; Volume 2, pp. 155–158.
14. Tryhuba, A.; Komarnitskyi, S.; Tryhuba, I.; Hutsol, T.; Yermakov, S.; Muzychenko, A.; Muzychenko, T.; Horetska, I. Planning and Risk Analysis in Projects of Procurement of Agricultural Raw Materials for the Production of Environmentally Friendly Fuel. *Int. J. Renew. Energy Dev.* **2022**, *11*, 569–580. [CrossRef]
15. Tryhuba, A.; Bashynsky, O.; Hutsol, T.; Rozkosz, A.; Prokopova, O. Justification of Parameters of the Energy Supply System of Agricultural Enterprises with Using Wind Power Installations. *E3S Web Conf.* **2020**, *154*, 04005. [CrossRef]
16. Bioenergy Technologies Office. Available online: https://www.energy.gov/eere/bioenergy/articles/largest-cellulosic-ethanol-plant-world-opens-october-30 (accessed on 20 May 2022).
17. Tryhuba, A.N.; Batyuk, B.V.; Dyndyn, M.L. Coordination of Configurations of Complex Organizational and Technical Systems for Development of Agricultural Sector Branches. *J. Autom. Inf. Sci.* **2020**, *52*, 63–76. [CrossRef]
18. *IEE Std 828-1998*; IEEE Standard for Software Configuration Management Plans. IEEE: New York, NY, USA, 1998.
19. *A Guide to the Project Management Body of Knowledge (PMBOK Guide)*, 4th ed.; Project Management Institute, Four Campus Boulevard: Newtown Square, PA, USA, 2008.
20. *P2M (Project & Program Management for Enterprise Innovation) Guidebook*; Project Management Association of Japan: Tokyo, Japan, 2008.

21. *ISO 21500*; Guidance on Project Management. ISO: Geneva, Switzerland, 2012. Available online: http://www.projectprofy.ru (accessed on 20 May 2022).
22. Ohara, S. *Booklet on P2M/Project & Program Management for Enterprise Innovation*; Project Management Association of Japan: Tokyo, Japan, 2003.
23. Ratushny, R.; Horodetskyy, I.; Molchak, Y.; Grabovets, V. The configurations coordination of the projects products of development of the community fire extinguishing systems with the project environment (ITPM-2021). In Proceedings of the 2nd International Workshop IT Project Management (ITPM 2021), Lviv, Ukraine, 16–18 February 2021.
24. Tryhuba, A.; Hutsol, T.; Kuboń, M.; Tryhuba, I.; Komarnitskyi, S.; Tabor, S.; Kwaśniewski, D.; Mudryk, K.; Faichuk, O.; Hohol, T. Taxonomy and Stakeholder Risk Management in Integrated Projects of the European Green Deal. *Energies* **2022**, *15*, 2015. [CrossRef]
25. Boyarchuk, V.; Koval, N.; Boiarchuk, O.; Pavlikha, N. Risk-Adapted model of the lifecycle of the technologically integrated programs of dairy cattle breeding. In Proceedings of the 16th International Scientific and Technical Conference on Computer Sciences and Information Technologies, Lviv, Ukraine, 22–25 September 2021; Volume 2, pp. 307–310.
26. Tryhuba, A.; Zachko, O.; Grabovets, V.; Berladyn, O.; Pavlova, I.; Rudynets, M. Examining the effect of production conditions at territorial logistic systems of milk harvesting on the parameters of a fleet of specialized road tanks. *East.-Eur. J. Enterp. Technol.* **2018**, *5*, 59–69. [CrossRef]
27. Ratushny, R.; Bashynsky, O.; Ptashnyk, V. Planning of Territorial Location of Fire-Rescue Formations in Administrative Territory Development Projects. In Proceedings of the CEUR Workshop Proceedings, Slavsko, Lviv Region, Ukraine, 18–20 February 2020; pp. 18–20.
28. Tryhuba, A.; Ratushny, R.; Tryhuba, I.; Koval, N.; Androshchuk, I. The Model of Projects Creation of the Fire Extinguishing Systems in Community Territories. *Acta Univ. Agric. Silvic. Mendel. Brun.* **2020**, *68*, 419–431. [CrossRef]
29. Gachie, W. Project risk management: A review of an institutional project life cycle. *Risk Gov. Control. Financ. Mark. Inst.* **2017**, *7*, 163–173. [CrossRef]
30. Gródek-Szostak, Z.; Malik, G.; Kajrunajtys, D.; Szeląg-Sikora, A.; Sikora, J.; Kuboń, M.; Niemiec, M.; Kapusta-Duch, J. Modeling the Dependency between Extreme Prices of Selected Agricultural Products on the Derivatives Market Using the Linkage Function. *Sustainability* **2019**, *11*, 4144. [CrossRef]
31. Ratushnyi, R.; Bashynsky, O.; Ptashnyk, V. Development and Usage of a Computer Model of Evaluating the Scenarios of Projects for the Creation of Fire Fighting Systems of Rural Communities. In Proceedings of the Xl-th International Scientific and Practical Conference on Electronics and Information Technologies (ELIT), Lviv, Ukraine, 16–18 September 2019; pp. 34–39.
32. Faichuk, O.; Voliak, L.; Hutsol, T.; Glowacki, S.; Pantsyr, Y.; Slobodian, S.; Szeląg-Sikora, A.; Gródek-Szostak, Z. European Green Deal: Threats Assessment for Agri-Food Exporting Countries to the EU. *Sustainability* **2022**, *14*, 3712. [CrossRef]
33. Boyarchuk, V.; Ftoma, O.; Padyuka, R.; Rudynets, M. Forecasting the risk of the resource demand for dairy farms basing on machine learning. In Proceedings of the 2nd International Workshop on Modern Machine Learning Technologies and Data Science (MoMLeT+DS 2020), Lviv, Ukraine, 2–3 June 2020.
34. Tryhuba, A.; Bashynsky, O. Coordination of dairy workshops projects on the community territory and their project environment. In Proceedings of the 2019 IEEE 14th International Conference on Computer Sciences and Information Technologies (CSIT), Lviv, Ukraine, 17–20 September 2019; Volume 3, pp. 51–54.
35. Pavlikha, N.; Rudynets, M.; Grabovets, V.; Skalyga, M.; Tsymbaliuk, I.; Khomiuk, N.; Fedorchuk-Moroz, V. Studying the influence of production conditions on the content of operations in logistic systems of milk collection. *East.-Eur. J. Enterp. Technol.* **2019**, *3*, 50–63.
36. Ratushnyi, R.; Khmel, P.; Martyn, E.; Prydatko, O. Substantiating the effectiveness of projects for the construction of dual systems of fire suppression. *East.-Eur. J. Enterp. Technol.* **2019**, *4*, 46–53. [CrossRef]
37. Komleva, N.; Liubchenko, V.; Zinovatna, S.; Kobets, V. Risk management with lean methodology. In Proceedings of the CEUR Workshop Proceedings, Kherson, Ukraine, 15–16 October 2020; Volume 2805, pp. 266–281.
38. DuPont Nevada Site Cellulosic Ethanol Facility Feedstock Collection Program. Available online: http://www.dupont.com/content/dam/dupont/products-and-services/industrial-biotechnology/documents/IB-PDF-04-Feedstock_Collection_Program_2015.pdf (accessed on 22 May 2022).
39. Inequality, Lorenz-Curves and Gini-Index. Available online: https://visualign.org/2011/09/02/inequality-lorenz-curves-and-gini-index/ (accessed on 22 May 2022).
40. Kucher, O.; Hutsol, T.; Glowacki, S.; Andreitseva, I.; Dibrova, A.; Muzychenko, A.; Szeląg-Sikora, A.; Szparaga, A.; Kocira, S. Energy Potential of Biogas Production in Ukraine. *Energies* **2022**, *15*, 1710. [CrossRef]
41. Koval, N.; Kondysiuk, I.; Grabovets, V.; Onyshchuk, V. Forecasting the fund of time for performance of works in hybrid projects using machine training technologies. In Proceedings of the CEUR Workshop Proceedings, Lviv-Shatsk, Ukraine, 5–6 June 2021; Volume 2917, pp. 196–206.

processes

MDPI

Review

Evaluating the Efficacy of Intelligent Methods for Maximum Power Point Tracking in Wind Energy Harvesting Systems

Dallatu Abbas Umar [1,2], Gamal Alkawsi [1,*], Nur Liyana Mohd Jailani [1], Mohammad Ahmed Alomari [3], Yahia Baashar [4], Ammar Ahmed Alkahtani [1], Luiz Fernando Capretz [5] and Sieh Kiong Tiong [1,*]

1 Institute of Sustainable Energy, Universiti Tenaga Nasional, Kajang 43000, Malaysia; dallatu.abbas@uniten.edu.my (D.A.U.); liyana.jailani@uniten.edu.my (N.L.M.J.)
2 Department of Physics, Kaduna State University, Tafawa Balewa Way, PMB 2339, Kaduna 800283, Nigeria
3 Institute of Informatics and Computing in Energy, Department of Informatics, College of Computing and Informatics, Universiti Tenaga Nasional, Kajang 43000, Malaysia; ahmed.alomari@uniten.edu.my
4 Faculty of Computing and Informatics, Universiti Malaysia Sabah (UMS), Labuan 87000, Malaysia
5 Department of Electrical and Computer Engineering, Western University, London, ON N6A 5B9, Canada
* Correspondence: gamal.abdulnaser@uniten.edu.my (G.A.); siehkiong@uniten.edu.my (S.K.T.)

Abstract: As wind energy is widely available, an increasing number of individuals, especially in off-grid rural areas, are adopting it as a dependable and sustainable energy source. The energy of the wind is harvested through a device known as a wind energy harvesting system (WEHS). These systems convert the kinetic energy of wind into electrical energy using wind turbines (WT) and electrical generators. However, the output power of a wind turbine is affected by various factors, such as wind speed, wind direction, and generator design. In order to optimize the performance of a WEHS, it is important to track the maximum power point (MPP) of the system. Various methods of tracking the MPP of the WEHS have been proposed by several research articles, which include traditional techniques such as direct power control (DPC) and indirect power control (IPC). These traditional methods in the standalone form are characterized by some drawbacks which render the method ineffective. The hybrid techniques comprising two different maximum power point tracking (MPPT) algorithms were further proposed to eliminate the shortages. Furtherly, Artificial Intelligence (AI)-based MPPT algorithms were proposed for the WEHS as either standalone or integrated with the traditional MPPT methods. Therefore, this research focused on the review of the AI-based MPPT and their performances as applied to WEHS. Traditional MPPT methods that are studied in the previous articles were discussed briefly. In addition, AI-based MPPT and different hybrid methods were also discussed in detail. Our study highlights the effectiveness of AI-based MPPT techniques in WEHS using an artificial neural network (ANN), fuzzy logic controller (FLC), and particle swarm optimization (PSO). These techniques were applied either as standalone methods or in various hybrid combinations, resulting in a significant increase in the system's power extraction performance. Our findings suggest that utilizing AI-based MPPT techniques can improve the efficiency and overall performance of WEHS, providing a promising solution for enhancing renewable energy systems.

Keywords: MPPT; wind energy harvesting system; artificial intelligence

Citation: Umar, D.A.; Alkawsi, G.; Jailani, N.L.M.; Alomari, M.A.; Baashar, Y.; Alkahtani, A.A.; Capretz, L.F.; Tiong, S.K. Evaluating the Efficacy of Intelligent Methods for Maximum Power Point Tracking in Wind Energy Harvesting Systems. *Processes* **2023**, *11*, 1420. https://doi.org/10.3390/pr11051420

Academic Editors: Ferdinando Salata and Virgilio Ciancio

Received: 10 March 2023
Revised: 12 April 2023
Accepted: 17 April 2023
Published: 8 May 2023

1. Introduction

Energy has played a critical role in driving industrial, commercial, and residential development. However, the increasing demand for energy has led to the need to explore additional resources to boost energy production. While fossil fuels are a common energy source, they also have negative environmental consequences such as air pollution and global warming. In contrast, renewable energy sources such as wind power are clean and do not have a greenhouse effect on the atmosphere, making them ideal for generating electricity without any environmental hazards. Wind power is a viable solution due to its abundance and non-depleting nature, making it an attractive option to address the growing concern for clean and green energy resources [1,2].

Electrical energy conversion from wind energy is achieved by WEHS, which mainly consists of a wind turbine (rotor hub and blades), a generator, and electric power converters [3]. In WEHS, the wind turbine converts the wind kinetic energy into mechanical energy, and the generator further transforms the mechanical energy into electrical energy [4–7].

The electrical power converter connected to the system converts the generated AC power to DC power which the DC load, such as battery charging, can use. For grid-connected WEHS, other devices such as boost converters, inverters, and transformers are required. The boost converter increases the DC output power before passing it to the inverter, which converts the DC power to AC. The step-up transformer boosts the AC power and connects it to the grid. The diagram of a typical grid-connected WEHS is shown in Figure 1.

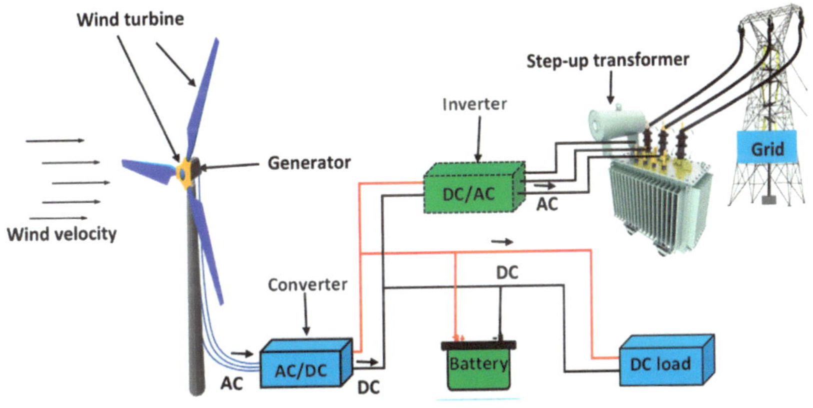

Figure 1. Grid-connected WEHS.

The MPPT controller is an essential component of modern wind energy systems, as it is necessary for optimizing energy conversion and maximizing power generation. Both Photovoltaic Systems (PVS) and WEHS face significant challenges in the implementation of MPPT techniques. These challenges include ensuring the efficiency and accuracy of MPPT, managing environmental factors, maintaining system stability, controlling costs, and overcoming the complexity of implementation.

Despite these challenges, MPPT remains a crucial component of modern energy generation systems. By effectively addressing these challenges, MPPT techniques can improve the overall efficiency and performance of renewable energy systems, making them more viable for widespread adoption and use. Therefore, researchers and engineers continue to work towards developing innovative solutions to overcome these challenges and enhance the implementation of MPPT techniques in both PVS and WEHS systems.

Hence, it is crucial to explore new MPPT techniques and evaluate their performance based on different factors. Recent studies have shown that the hybridization of MPPT techniques with advanced AI methods, such as deep learning, can significantly improve the efficiency and accuracy of MPPT systems. Thus, reviewing and comparing recent MPPT techniques that hybridize with AI methods in both wind and photovoltaic power generation can aid in the development of more efficient and reliable MPPT systems for renewable energy generation.

According to the most recent related reviews on MPPT techniques for the PVS and WEHS systems, as shown in Figure 2, it can be observed that fewer articles are reported on WEHS. Furthermore, a few AI-based MPPT algorithms were reported in only a limited number of review articles, as illustrated in Table 1, and a large number of studies on the MPPT techniques for WEHS focused mainly on the conventional methods, for example, in the studies carried out by Mousa et al. [8] and Pande et al. [9], various types of MPPT algorithm have reviewed, including few of the hybrid and AI-based algorithms. These papers have discussed in detail the application of perturb and observation (P&O) algorithms, followed by the improved version of P&O, such as modified perturb and observation (MPO).

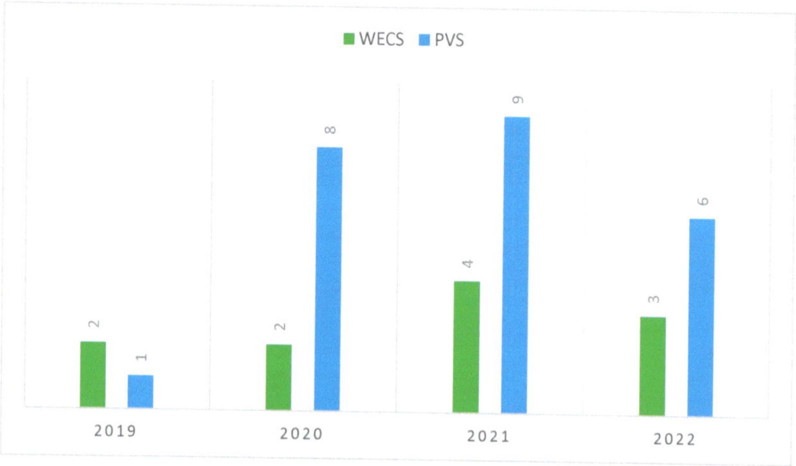

Figure 2. Recently published review articles on MPPT techniques for PVS and WECS.

Table 1. Summary of recent related studies of the MPPT method for WEHSS.

Ref.	Types and Numbers of MPPT Technique Covered				
	Conventional	**AI-Based**	**Hybrid (Conventional + Conventional)**	**Hybrid (AI + Conventional)**	**Hybrid (AI + AI)**
[8]	✓	FLC&NN	P&O+OTC, ORB, PSF	~	×
[9]	✓	FLC&NN	P&O+OTC, ORB, PSF	P&O+FLC, NN, PFS+NN, ORB+PSO	×
[10]	✓	FLC&NN	P&O+PSF	×	×
[11]	P&O, IC	RBN, PSO, ESN ~	RBF+PSF	P&O+FLC (Partially)	×
[12]	✓	ANN, PSO FLC, WOA, GWOA	/	×	×
[13]	✓	FLC&NN	×	PSF+FLC	×
[14]	✓	FLC, NN&GA	×	×	×
[6]	~	FLC, NN	P&O+PSF, OTC+P&O	×	×
[15]	✓	FLC, NN	×	×	×

Symbol: (✓) Covered; (×) Not covered; (~) Partially covered; (/) Not reported.

Table 1 provides an overview of the contemporary research on the efficacy of artificial intelligence based-MPPT techniques applied to wind energy conversion systems.

In summary, the reviews discussed the conventional, hybrid, and AI-based MPPT techniques in WEHS. However, the details provided are not covered enough, especially in standalone and hybrid AI-based MPPT algorithms. Therefore, this paper will focus on the MPPT techniques for WEHS with more emphasis on AI-based MPPT techniques and their performance on WEHS.

The remainder of this paper is organized as follows: Research background is presented in Section 2, modeling of WEHS is presented in Section 3, MPPT algorithms are presented in Section 4, and discussions, future directions and conclusions are presented in Sections 5–7, respectively.

2. Research Background

The operation of WEHS is described by the WT power curve shown in Figure 3. It consists of four main operating regions. In regions one and four, i.e., before cut-in speed and after cut-out speed, the turbine must be stopped and disconnected from the grid so that it is not driven by the generator. In region two, the controllers are used with the MPPT algorithm to track and extract the maximum possible power over the wind speed range. Region three is between the rated power and cut-out speed of the turbine. In this region, the operation of WT must be limited to the rated mechanical power to avoid damage to the electrical generator.

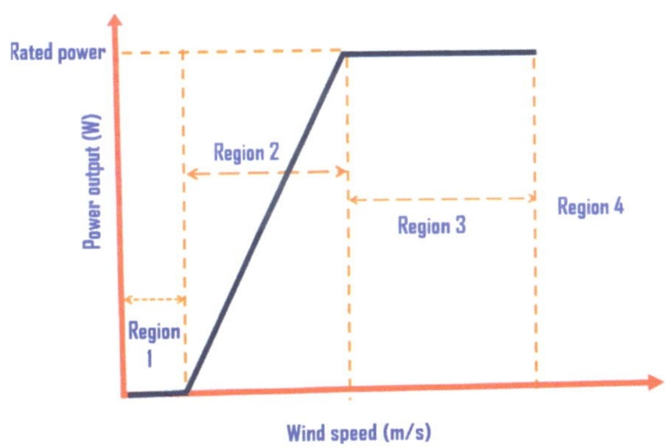

Figure 3. The power curve of the wind turbine.

Due to the intermittent nature of the wind, it is difficult for WEHS to harness the maximum power of the wind over a range of wind speeds. In this context, previous researchers have developed several algorithms to determine the maximum power output of WEHS, as presented by [6,9,16]. These algorithms include the tip speed ratio (TSR), optimal torque control (OTC), and power signal feedback (PSF), which tracks the mechanical power of the WT. Other algorithms, such as the P&O or hill-climb search (HCS) method, incremental conductance (INC), and optimal relation base (ORB), track the maximum converted electrical power from the generator. These traditional methods mentioned above have successfully tracked the MPP of WECS, but they have some drawbacks depending on the method. Therefore, developing an accurate MPPT algorithm to track the MPP is still a challenging task. To solve these problems, some researchers, such as [7,9,16–20], have modified the traditional methods. In [17–19], the TSR method in which the wind speed is measured by a mechanical sensor is replaced with the wind speed estimation method. The issue regarding generator stalling in the PSF method has been resolved in [7] by the concept of a modified PSF algorithm. The variable, adaptive, and hybrid step sizes concepts were proposed by [9,17] as the solution oscillation issue around the MPP by the P&O method. The authors of [16] proposed a method that eliminated the need for a sensor and look-up table as required in the ORB method. In [21], the INC algorithm has been modified for a better system with higher dynamic performance, precision, and fast convergence speed than P&O and the ordinary INC method. Other researchers, such as [22–26], have combined two or more traditional MPPT methods into a hybrid one so that one method could eliminate or reduce the drawback of the other. For instance, the limitation of ORB was addressed in reference [22] by integrating the P&O technique into the algorithm. Reference [23] utilizes a self-rotating P&O-based controller along with ORB to enhance the MPP tracking speed. On the other hand, reference [24] employs OTC in conjunction with P&O to minimize the perturbation step size of the P&O algorithm, thereby facilitating the attainment of MPP.

Recently, researchers have been focusing on implementing AI approaches in MPPT controllers. These approaches have been proposed either in standalone form, such as in references [3,27–33] or in hybrid forms, such as in references [20,34,35]. The use of an artificial neural network (ANN) has been employed in some of these approaches, such as in references [3,27,32,33], resulting in enhanced system performance, resilience, power response, and efficiency. Optimization algorithms such as particle swarm optimization (PSO), ant colony optimization (ACO), Archimedes optimization (AOA), and grasshopper optimization algorithm (GOA) have also been proposed, resulting in improved tracking speed, energy generation, dynamic performance, and global search capability to track the MPP, as demonstrated in references [28–31].

In reference [20], a hybrid AI MPPT algorithm was proposed by integrating radial basis function-neural networks (RBF-NN) and particle swarm optimization algorithms to replace the conventional controller. This hybrid algorithm achieved faster tracking of the MPP, increased system reliability, and a reduction in system losses, size, and cost. In addition, reference [34] proposed a controller that combined fuzzy logic control (FLC) and NN, resulting in improved power harvesting capability in a hybrid renewable energy system (HRES) and shorter simulation time to capture the MPP. Lastly, reference [35] proposed an adaptive neuro-fuzzy inference system (ANFIS) MPPT controller that combines NN and FL approaches, enabling the extraction of maximum power from the wind independently of wind speeds.

3. Modeling of WEHS

The wind power, as seen by the WT blades, is expressed by Equation (1). The WT blade captures the power of the wind and converts it to mechanical power (P_m) of the WT according to Equation (2). The output mechanical torque (T_m) and rotational speed (ω_m) of the WT, which are the inputs to the electrical generator, are given by Equations (3) and (4)

$$P_{wind} = \frac{\rho \pi R^2 V^3}{2} \tag{1}$$

$$P_m = \frac{\rho \pi R^2 C_p(\lambda, \beta) V^3}{2} \tag{2}$$

$$T_m = \frac{P_m}{\omega_m} \tag{3}$$

$$\omega_m = \frac{\lambda V}{R} \tag{4}$$

where $C_p(\lambda, \beta)$, λ, and R are the wind turbine blade efficiency, tip speed ratio, and radius, respectively. Furthermore, V, ρ and β are the wind speed, air density and blade pitch angle.

It is clear from Equation (2), $C_p(\lambda, \beta)$ is a function of λ and β given by Equation (5) [13].

$$C_p(\lambda, \beta) = k_1\left(k_2\frac{1}{\lambda_i} - k_{3\beta} - k_4\beta^{k_5} - k_6\right)Exp\left(-k_7\frac{1}{\lambda_i}\right) \tag{5}$$

$$\text{and } \frac{1}{\lambda_i} = \frac{1}{\lambda + 0.08\beta} - \frac{0.035}{1+\beta^3}, \tag{6}$$

the values for $k_1 - k_7$, λ, and β depend on the wind turbine's type and characteristics.

The mechanical rotating speeds of the wind turbine are transformed into the electrical rotating speed of the generator using Equation (7)

$$\omega_e = \omega_m \times \frac{P}{2} \tag{7}$$

where P is the number of magnetic poles pairs.

And the frequency of rotation is calculated by Equation (8)

$$f = \omega_{rpm} \times \frac{P}{120} \tag{8}$$

where ω_{rpm} is the generator rational speed in revolution per minute.

Considering permanent magnet synchronous generator (PMSG), the dynamic equations for the voltages along the d and q axes are given by Equations (9) and (10)

$$u_d = -r_d i_d + \frac{d\psi_d}{dt} - \omega_e \psi_q \tag{9}$$

$$u_q = -r_q i_q + \frac{d\psi_q}{dt} - \omega_e \psi_d \tag{10}$$

where r_d and r_q represents the stator q and d axes resistance, respectively.

Going with the assumption that there is no rotor flux along the q-axis (i.e., it is only along the d-axis), Equations (11) and (12) are used to determine the currents along the d-axis and q-axis.

$$i_d = \frac{(\psi_{pm} - \psi_d)}{L_d} \tag{11}$$

$$i_q = -\frac{\psi_q}{L_q} \tag{12}$$

where L_d and L_q, ψ_d and ψ_q are the inductances and the flux linkages along the d and q axes, respectively.

The electromagnetic power produced by the PMSG is expressed by Equations (13) or (14)

$$P_e = \frac{3}{2}\left(\omega_e L_q i_q i_d - \omega_e L_d i_d i_q + \omega_e \psi_{PM} i_q\right) \tag{13}$$

$$P_e = \frac{3}{2}\omega_e\left[\psi_{PM} i_q - (L_d - L_q) i_d i_q\right] \tag{14}$$

where ψ_{PM} represent the magnetic flux linkage.

Finally, the electromagnetic torque developed by the PMSG is obtained by Equation (15)

$$T_e = \frac{P_e}{\omega_m} = \frac{3}{2}\,\mathrm{P}\left[\psi_{PM} i_q - (L_d - L_q) i_d i_q\right] \tag{15}$$

The dc current, voltage and electric power, which are the output of the rectifier, are expressed in Equation (16), Equation (17) and Equation (19), respectively.

$$I_{dc} = \frac{\pi I_{ph}}{\sqrt{6}} \tag{16}$$

$$V_{dc} = \frac{3V_{ph}\sqrt{6}}{\pi} \tag{17}$$

$$P_{dc} = V_{dc} I_{dc} \tag{18}$$

where I_{ph} and V_{ph} are the generator stator phase current and voltage of the generator.

The corresponding output current, voltage and power of the boost converter are given by Equations (19), (20) and (21) respectively.

$$I_{out} = \frac{V_{in}XD}{R_{load}} \tag{19}$$

$$V_{out} = \frac{3V_{in}}{1-D} \tag{20}$$

$$P_{out} = V_{out}I_{out} \tag{21}$$

where V_{in} is the converter input voltage and D is the duty cycle which is given by Equation (22)

$$D = 1 - \frac{V_{in}}{V_{out}} \tag{22}$$

4. MPPT Methods for WEHS

The MPPT methods for WEHS can be categorized into traditional MPPT algorithms and intelligence-based MPPT algorithms. The traditional methods are further classified as indirect power control (IPC), which tracks the mechanical power of the WT, and direct power control (DPC), which tracks the maximum electrical power of the generator. The third class of this category of MPPT is the hybrid MPPT method which is the combination of different traditional MPPT algorithms. Smart or Intelligent MPPT algorithms include the MPPT controllers that employ AI algorithms to track the MPP of the WEHS. Therefore, the MPPT methods are broadly classified into four categories such as DPC, IPC), hybrid, and intelligent algorithms [8,9]. Considerable efforts have been dedicated toward the advancement of conventional MPPT controllers, with particular emphasis on enhancing their operational characteristics across various parameters and features. Notably, recent years have witnessed progress in improving the performance of conventional MPPT techniques through the integration or modification thereof alongside traditional and/or AI-based methods. In particular, the utilization of AI techniques in MPPT has garnered significant attention due to its inherent ability to effectively address prevalent issues inherent to these systems.

4.1. Traditional MPPT Methods for WEHS

Examples of IPCs MPPT are TSR, OTC, and PSF. TSR and the other two IPC methods require a mechanical sensor to measure wind speed, while OTC and PSF, in addition, require knowledge of the parameters of WT. In the TSR algorithm, the reference speed of the WT, which corresponds to the MPP, is estimated using Equation (23) and used to control the operation of the WT to the optimal TSR at which the maximum power coefficient is achieved. In OTC, Equation (24), the optimal torque reference relation is used to achieve the MPPT.

$$\omega_{ref}^{*} = \frac{\lambda_{opt}V_r}{R} \tag{23}$$

$$T_{opt}^{*} = 0.5\rho\pi R^5 \frac{C_{p\ max}}{\lambda_{opt}^3}\ \omega_{ref}^2 \tag{24}$$

In the OTC method, the controller maintains a predefined relationship between the electromagnetic torque and rotational speed of the WT in accordance with the maximum power-rotor speed curve such that the rotational speed approaches the optimal value [6]. DPC methods are sensorless and use a precomputed system curve to find the MPP. The MPPT algorithm under DPC includes P&O, ORB, and IC. The P&O algorithm is based on discretizing (perturbing) a control variable, such as generator speed, and observing the resulting effect on generator output. The algorithm compares each successive generator output and adjusts the generator speed in the direction of the MPP [36]. In ORB control, the

MPP is tracked using a look-up table developed based on prior knowledge of the optimal relationship between the WT power and other parameters such as rotational speed, torque, rectifier DC voltage, or current [10]. Classification of the different MPPT algorithms is presented in Figure 4. The comparison between the different traditional methods of MPPT algorithms is summarized together with remarks in Table 2.

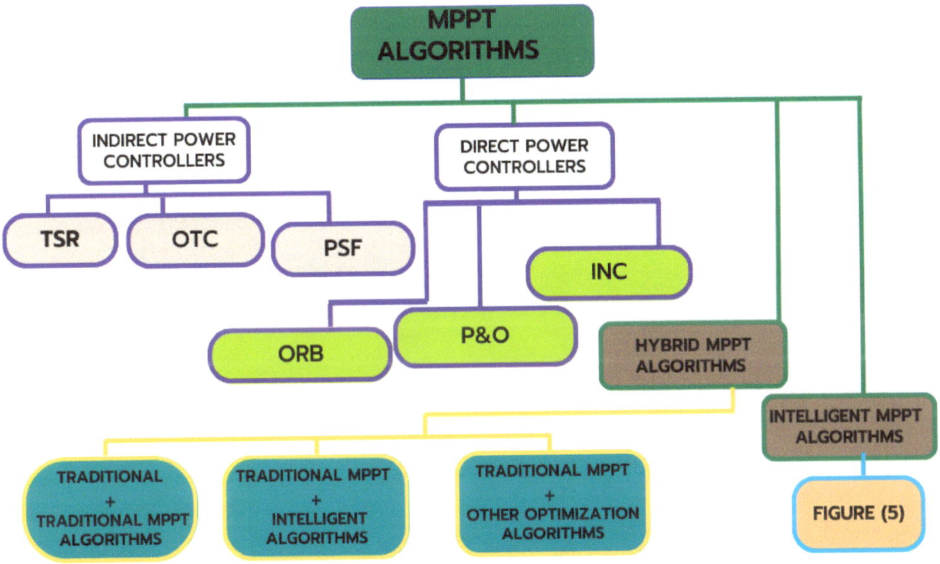

Figure 4. Classification of MPPT algorithms for WECS.

Table 2. Comparison between different type of Traditional method of MPPT algorithms.

MPPT Technique	Advantages	Major Drawbacks	Improvement
TSR	TSR is highly efficient, has high convergence speed, and quickly responds to wind speed changes [37]. Additionally, the TSR method is simple, and no memory is required for the process	The need for a mechanical sensor to measure wind speed results in inaccuracies, leading to increased costs for installation and maintenance.	A novel approach was developed to estimate wind speed, which removed inaccuracies associated with mechanical sensors. The algorithm's speed to track MPP was enhanced, and the technique was made simpler [18–20].
OTC	The OTC method is highly efficient and flexible, with stable torque regulation and easy application. It is also practical as it does not require real-time wind speed measurement, allowing for quick adjustments to changes in wind speed.	The mechanical sensor used for wind speed measurement and knowledge of wind turbine characteristics is necessary. In addition, the turbine's large inertia causes a sluggish response to torque commands, resulting in slow MPP tracking during sudden changes in wind speed. Moreover, measuring electromagnetic torque and turbine speed can increase the system cost and its dependency on generator parameters.	A quantum neural network (QNN) was introduced into the OTC method by [25] and efficiency improvement was recorded more than with conventional OTC and NN. Reference [26] proposed a fuzzy inference-based MPPT method to improve the OTC method. This method enhances the MPPT efficiency under fluctuating wind speeds while ensuring system stability.

Table 2. *Cont.*

MPPT Technique	Advantages	Major Drawbacks	Improvement
PSF	PSF has moderate performance under fluctuating wind speeds and high convergence speeds. The system cost is less compared to the TSR method.	The use of a mechanical sensor to measure wind speed introduces inaccuracies. Additionally, PSF is less efficient and more complex than the TSR and OTC methods. Moreover, it requires knowledge of wind turbine characteristics, and can cause the generator to stall when there are sudden changes in wind speed.	The modified PSF as reported by [7] has solved the issue of generator stalling but raises further issues, such as overshoot of the control variables and greater difficulty in tracking the MPP.
P&O	The proposed method eliminates the need for a mechanical sensor and requires less memory, making it simple to implement. Additionally, it does not require any prior knowledge of the system parameters and characteristics, resulting in lower overall system costs. Although its performance under intermittent wind speed is moderate, it is still a viable option.	Large step size causes oscillation around the maximum power point (MPP) while smaller step size leads to slower response. Both scenarios result in a loss of MPP tracking and reduced efficiency, particularly at varying wind speeds. In addition, the convergence speed is slow.	The drawbacks can be addressed by adopting the followings: Variable, Adaptive, and hybrid step sizes concept [9]. The step size was calculated using trapezoidal rule in [38] which successfully reduce computational complexity of the algorithms and eliminated power oscillation at the MPP.
INC	The benefit of this method is similar to P&O method but with better convergence speed, precision, and MPPT tracking efficiency.	Slow convergence speed. Oscillation at MPP	The INC method proposed by [21] addresses the trade-off between power and convergence speed in P&O methods. Moreover, the modified INC achieves a better system with higher dynamic performance, precision, and fast convergence speed compared to P&O.
ORB	No need for wind speed sensors and look-up tables [16]. Furthermore, high convergence speed than that of P&O and INC. In addition, oscillations around the MPPT are absent in ORB method	Required large memory for pre-obtained optimal relation curve. Required previous knowledge of the system.	Reference [22] improved the ORB method by using the P&O method as an initialization algorithm for online MPP search at local wind speeds. This eliminated the ORB method's drawback by extracting the necessary parameters for its operation.

4.2. Intelligent-Based MPPT Methods

Due to its ability to easily solve problems involving complex mathematical models, AI has proven attractive for applications in WEHS, particularly in the areas of design, modeling, and performance optimization. When applied as a standalone or integrated with the traditional MPPT controllers, AI-based algorithms have shown good results by improving the performance, such as the speed and efficiency of MPPT controllers. The following AI algorithms have been proposed by several researchers to improve the method of tracking the MPP of WEHS. This includes fuzzy logic control (FLC), artificial neural network (ANN), particle swarm optimization (PSO), ant colony algorithm (ACA), Archimedes optimization algorithm (AOA), Cuckoo search (CS), grasshopper optimization algorithm (GOA), multi-objective grasshopper optimization algorithm (MOGOA), electric charge particle optimization (ECPO) and enhanced atom search optimization (EASO) Technique. Figure 5 shows the list of recent AI algorithms that are applied in MPPT controllers for WEHS. Accordingly, refs. [3,27–33] applied the standalone AI-based method to track the MPP of WECS, while references [20,34,35] used the hybrid AI-based method.

The summary of the standalone AI-based (MPPT) algorithms and their contributions to enhancing MPPT tracking in WEHS are provided in Table 3.

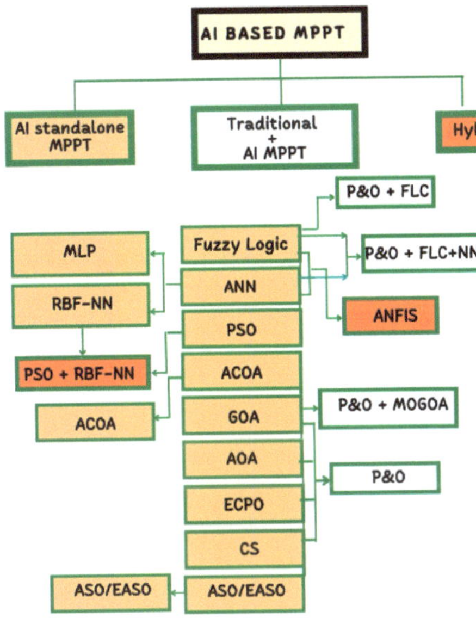

Figure 5. List of recent AI algorithms that are applied for MPPT controllers in WEHS.

Table 3. Role of standalone AI-Based MPPT Algorithms in Enhancing MPP Tracking for WEHS.

Intelligent MPPT Type	Remarks	Algorithm Performance	Reference
FLC	The FLC (fuzzy logic control) controller has demonstrated better power generation and faster response time when compared to other controllers such as P&O and ANN.	A power generation increase of approximately 20 W was achieved when compared to the P&O method.	[34]
ANN (MLP)	An Artificial neural network novel MPPT algorithm was developed Combining an intelligent modular multilayer perceptron (MLP) approach with a simplified model of WEHS.	The model in [3] has achieved an MPPT performance of 99.95% and error of 3% was recorded	[3]
NN (MLP)	Similar approached as in [3], The result obtained showed increased system robustness and fast power response and improved power coefficient.	The system achieved a power coefficient of 0.48 and a time response of 5 s.	[32]
ANN	The use of the ANN as an alternative MPPT algorithm resulted in improved MPP tracking and quicker response times compared to the P&O method.	Increase in power of approximately 30 W was achieved.	[34]
ANN (RBF-NN)	Radial basis function-neural networks (RBF-NN) was proposed to replace the need for measurement instruments, eliminate system errors, and minimize the size and cost of the system. Compared with other AI based methods such as backpropagation of NN and FLC, the proposed RBF-NN method showed better performance.	The response times for FLC, RBF-NN, and BP-NN were 0.47, 0.46, and 0.42, respectively. However, the BP-NN method had the highest ripple factor of 4%	[33]

Table 3. *Cont.*

Intelligent MPPT Type	Remarks	Algorithm Performance	Reference
RNN	A new control strategy for wind power systems was proposed using integral sliding mode control technique based on a recurrent neural network (RNN) where optimal control signals for maximum power extraction is estimated using the RNN. Simulations show that the proposed strategy outperforms existing control strategies in power generation, disturbance rejection, and robustness to parameter variations and uncertainties. The proposed approach can improve the performance and efficiency of wind power systems.	The root mean square error (RMSE) between the optimal power and the tracked power was calculated to be 0.153.	[27]

4.2.1. The Fuzzy Logic-Based MPPT Controllers

FLC is comprised of three main states process, fuzzification, inference, and defuzzification. Fuzzification involves the conversion of physical inputs variables into fuzzy sets (the error and the variation of the error) and the assigning of linguistic variables such as the Negative Big (NB), Negative Medium (NM), Negative Small (NS), Zero (Z), Positive Small (PS), Positive Medium (PM) and Positive Big (PB). The inference stage is the decision-making phase where membership rules are set and also logical relationships between the inputs and outputs variables are constructed, and finally, the fuzzy output is converted to an equivalent numeric value by defuzzification [39].

The concept of FLC based MPPT controller applied to WEHS is explained in Figure 6 and the equivalent MPPT controller circuit is depicted in Figure 7. It consists of two input variables that are fed to the fuzzy toolbox. The first variable is the error ($\epsilon(x)$) which represents the ratio of the current and voltage $\left(\frac{I(x)}{V(x)}\right)$ and their derivatives $\left(\frac{dI(x)}{dV(x)}\right)$, while the second input variable is the change in the error ($\Delta\,\epsilon(x)$), where $\Delta\,\epsilon(x) = \epsilon(x) - \epsilon(x-1)$. The fuzzy toolbox processed the inputs and produced the perturbation parameter, ΔD(x) as its output which in turn is used by the P&O MPPT controller as its input variable. One of the advantages of integrating FLC in MPPT is that the controller can eliminate the oscillation around the MPP, especially in the P&O method. Furthermore, the method does not require the mathematical modeling of the WEHS since the variables of the controller can change in accordance with the dynamic changes of the system. Additionally, wind speed intermittency does not affect the performance of the method.

4.2.2. The Artificial Neural Networks (ANN) Based MPPT Controller

ANN is a numerical and symbolic-based learning technique that uses an arithmetic process rather than logic for pattern recognition, prediction, optimizations, control, system modeling and identification, signal processing, etc. [32,40]. ANN uses feedforward propagation and backpropagation for parameters training, and once the training is performed, the neural network produced almost the same output pattern for similar input data. This ability makes the NN suitable for their applications as intelligence controllers. The concept of ANN applied in the MPPT algorithm for WEHS, as proposed in [3,33], is depicted in Figures 8 and 9. It is an intelligent multilayer perceptron (MLP) structure that is constructed using Kolmogorov's theorem [41], which states that the number of neurons n_i for a hidden layer is obtained by the expression: $2n_i + 1$. Accordingly, the MLP structures contain two inputs, one hidden layer of K- neurons and one neuron output layer. The MLP is integrated into a simple WEHS structure. The MLP is a typical example of a feedforward artificial neural network.

The modular MPL in [3] is used to predict the mechanical rotational speed (ω_m) of the WT, which is used by the model to compute the optimal reference current for the rotor side converter. The MPP of the WEHS is tracked by the control mechanism at the rotor side converter using the optimal current as a reference. In each operation mode of the modular MLP and every dataset, ω_m is estimated as a targeted output variable using the DC current (i_{dc}) and DC voltage (v_{dc}) as input variables.

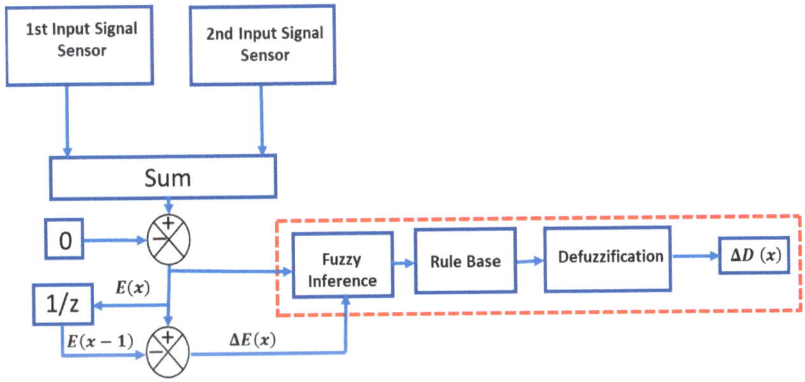

Figure 6. Concept of FLC-based MPPT controller applied to WEHS.

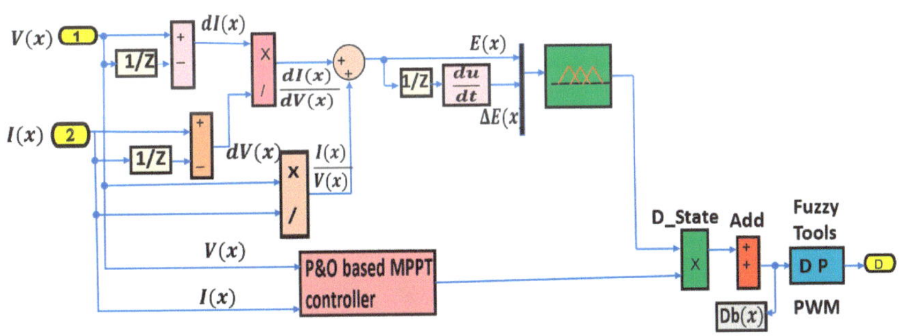

Figure 7. Equivalent circuit of FLC-based MPPT controller [42].

Each training dataset, which comprises the target function and the inputs to the MLP, was normalized according to Equations (25)–(27) using the mean and standard deviation of respective variables.

$$\omega_n^m = \frac{\omega_m - \mu_{\omega_m}}{\sigma_{\omega_m}} \tag{25}$$

$$i_{dc}^n = \frac{i_{dc} - \mu_{i_{dc}}}{\sigma_{i_{dc}}} \tag{26}$$

$$v_{dc}^n = \frac{v_{dc} - \mu_{v_{dc}}}{\sigma_{v_{dc}}} \tag{27}$$

And the final normalized output of the rotational speed of the MLP is calculated using Equations (28) and (29), the hidden layer activation function, and the corresponding linear function of the output neuron.

$$N_i = f_{sgm}(w_{in,1}\omega_n^m + w_{in,2}i_{dc} + w_{in,k}) \tag{28}$$

$$\omega_n^m = w_1N_1 + w_2N_2 + w_3N_3 + w_4N_4 + w_kN_k \ldots + w_{out} \tag{29}$$

where , $w_{\text{in},1} \ldots w_{in,k}$ are weights connecting each successive inputs to the hidden layer neurons, w_1, w_2, $\ldots w_k$ are the weights connecting the output layer from the hidden layer neurons.

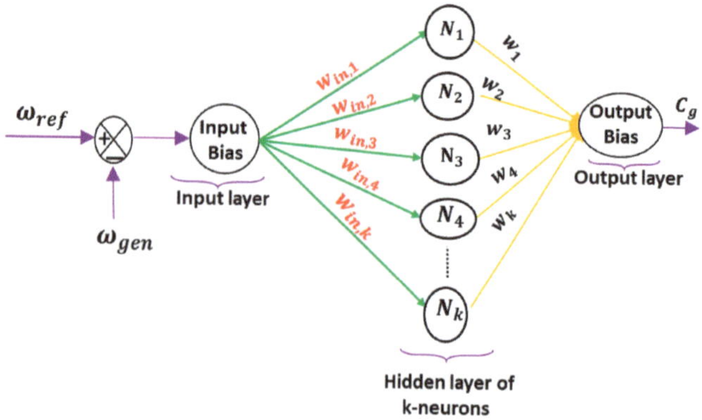

Figure 8. The concept of ANN applied in the MPPT algorithm for WEHS.

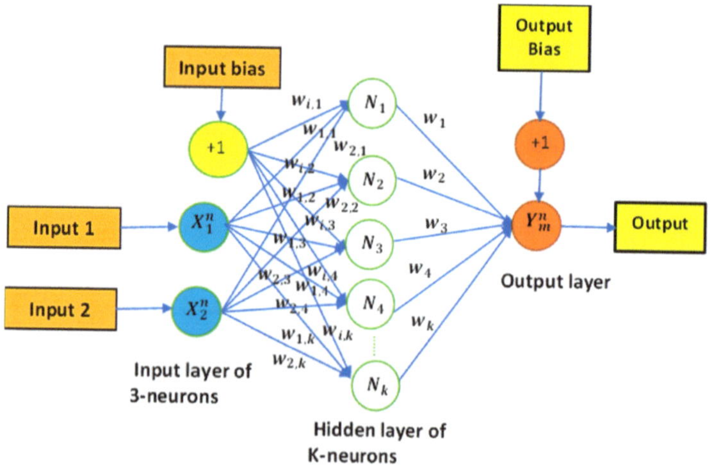

Figure 9. The concept of RBF-NN applied in the MPPT algorithm for WEHS.

4.2.3. MPPT Using PSO Algorithms

The PSO algorithms are a type of intelligent optimization algorithm which belongs to a class of optimization algorithms called metaheuristic algorithms. It is based on swarm intelligence that is inspired by the social behavior of animals (particles) such as fishes or birds (swarm) while searching for food in a physical space. The group of particles moves around in a search space and is guided toward better solutions by a set of rules. The goal of the algorithm is to find the global optimum (global best) of a given objective function by having the particles converge at the optimal solution. The movement of particles toward the optimal solution is influenced by the quality of their current position in the search space as well as the position of other particles in the group and random perturbations. PSO algorithms are often used to solve complex optimization problems that cannot be easily solved using traditional optimization techniques. A simple concept of the PSO algorithm is described in Figure 10.

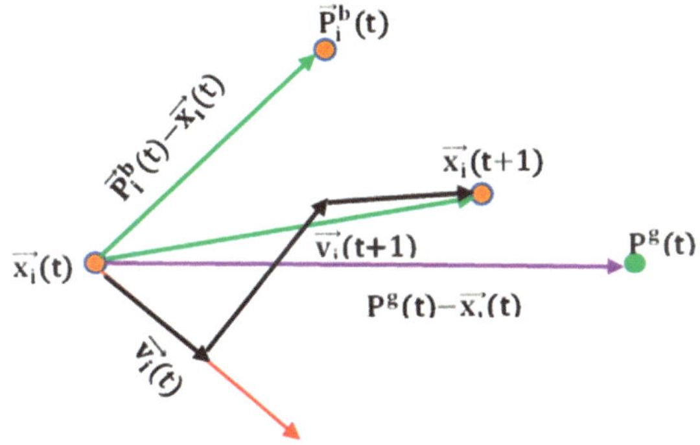

Figure 10. The simple concept of PSO Algorithm.

According to the figure, each particle in the swarm keeps track of its position $\vec{x_i}$(t) in the search space, which signifies the solution to the problem, and the velocity ($\vec{v_i}$(t+1) of each particle specifies its displacement in the searching space. Furthermore, i^{th} particle personal best position is denoted by P_i^b, and the global best position amongst all the particles is denoted by P^g. The general mathematical model of the PSO is described as follows:

x_i(t) is the current position of the particle, $i \leq 1 \leq p$, and p is the swarm population. The new position of each particle is updated using Equation (30)

$$x_i(t+1) \ = x_i(t) + \ v_i(t+1) \tag{30}$$

where $v_i(t + 1)$ is the particle new velocity which is given by Equation (31)

$$v_i \ (t+1) = wv_i \ (t) + c_1r_1\left(P_i^b(t) - x_i\,(t)\right) + c_2r_2(P^g\,(t) - x_i(t)) \tag{31}$$

w, c_1, and c_2 are real values called inertia weight and acceleration coefficients, respectively, and, r_1 and r_2 are uniformly distributed random numbers between 0 to 1.

The above two equations are simple rules to be obeyed by all particles in the swarm for searching for the optimum solution to any given problem.

In a WEHS sense, the fitness function of each particle is calculated by Equation (32)

$$F_{FIT} = \frac{1}{0.1 + abs(\omega_r^* - \omega_r) * abs(P_m^* - p_m)} \tag{32}$$

where ω_r^* and P_m^* are the reference rotor speed and mechanical power of the WT, respectively, ω_r and p_m are the rotational speed and mechanical power of the WT at the wind speed speeds, v.

An effective control MPPT algorithm based on the PSO was proposed by [43] to maximize the efficiency of fixed-pitch wind turbines with double-fed induction generators (DFIGs) by compensating for the errors in the estimation of the circuit parameters of the generator. The MPPT algorithms provide the optimal reference speed that will maximize the mechanical power below the rated speed of the DFIG, while electrical losses of the DFIG are minimized by power management through the optimal rotor current, which is searched by the PSO algorithm. Compared to the results of the conventional methods, the proposed control algorithm has improved the energy generation of the system. Furthermore, in [20], the PSO algorithm was used for RBFNN learning rates and inertia weight adjustment to find their optimum values, as shown in Figure 11. Figure 12 depicts the steps of the PSO algorithm process.

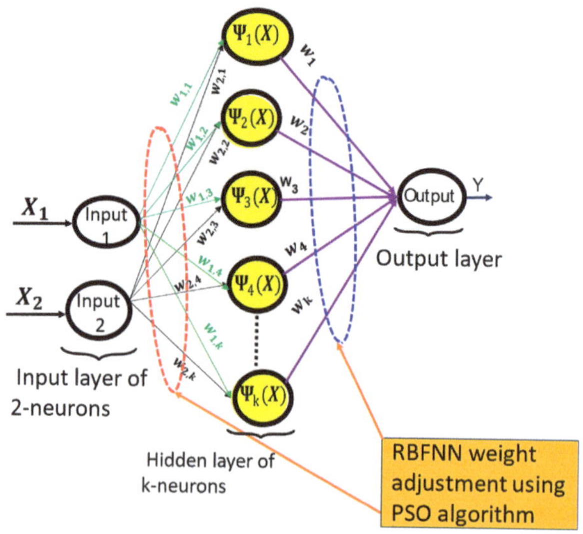

Figure 11. RBFNN-PSO MPPT Algorithm applied to WECS.

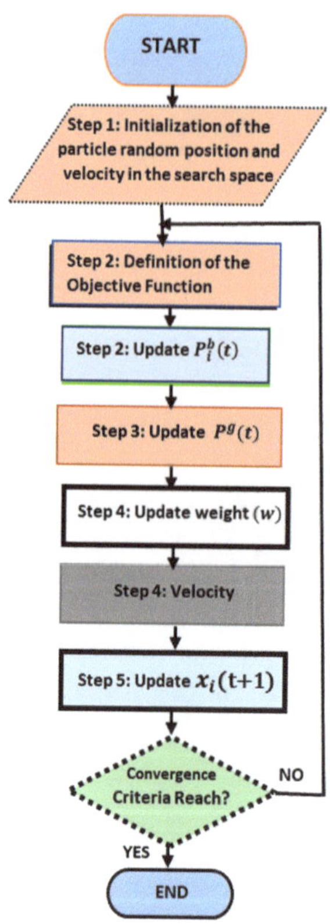

Figure 12. Flowchart for PSO algorithm.

4.2.4. MPPT Method Using Other Optimization Algorithms

Other metaheuristic algorithms such as ant colony optimization algorithm (ACOA), Archimedes optimization algorithm (AOA), Cuckoo search (CS), grasshopper optimization algorithm (GOA), multi-objective grasshopper optimization algorithm (MOGOA), electric charge particle optimization (ECPO), and enhanced atom search optimization (EASO) techniques have been used for MPPT algorithms in WEHS, and the results of their performance have shown very good improvement of the technique in terms of quick searching of the optimum operation point of the WEHS. The different optimization algorithms that are used for MPPT techniques for WEHS are summarized in Table 4.

Table 4. The summary of recent studies of various optimization algorithms used for MPPT in WEHS.

Intelligent MPPT Type	Description	Algorithm Performance Evaluation	Reference
PSO	An effective control MPPT algorithm based on the PSO was proposed to maximize the efficiency of fixed-pitch DFIG WT by compensating the errors in the estimation of the generator circuit parameters. Compared to the results of the conventional methods, the proposed control algorithm has improved the energy generation of the system.	More energy of 1.28% was generated	[43]
ACOA	ACOA was developed and used to tune the PI controller to determine its optimal parameters for speed control. This approach increased the power coefficient and overall performance of the WEHS.	The system achieved a Cp of 0.453, which is slightly higher than the 0.4518 achieved using a PI controller. Additionally, an increase in power output of 150 W was obtained.	[28]
AOA, GOA, CSOA, ECPO,	The shortcomings of the HCS method in terms of MPP tracking speed and efficiency were successfully overcome using the AOA. Compared with other optimization methods such as CSOA, GOA, and ECPO, better performance was obtained with AOA.	The system power generation has increased to 102.20 W, 101.19 W, 78.30 W and 63.52 W respectively with AOA, GOA, ECPO and CSOA algorithm.	[29]
GOA and MOGOA	Fractional order sliding mode controller (FOSMC) based on the traditional P&O method was modified to incorporate the MOGOA. The successful implementation of the MOGOA significantly improved the system's robustness as well as its dynamic performance.	The proposed algorithm achieved an integral of time multiple of absolute error (ITAE) value of 4.18 and 1.48 s for power overshoot and settling time, while the conventional and sliding mode control approaches achieved values of 5.63 and 1.64 s.	[30]
EASO	An optimal solution of high quality and fast system response was achieved using an EASO technique developed for PMSG-based WEHS. According to [30], the technique has a powerful global search capability to track the MPP.	The proposed method achieved an integral absolute error (IAE) control benchmark of 0.1481, which is lower than the values recorded for the PSO and GA algorithms, which were 0.182 and 0.213, respectively.	[31]

4.3. *The Hybrid MPPT Techniques*

This method involves the combination of different kinds of traditional MPPT techniques or involving AI-based algorithms in the MPPT method for tracking the maximum power point of the WEHS. Accordingly, the hybrid MPPT methods are presented as hybrid-traditional MPPT techniques, hybrid-traditional-AI MPPT techniques, or hybrid-AI MPPT techniques. Various hybrid MPPT algorithms that were developed in recent years to improve the performance of the traditional MPPT methods are discussed below.

4.3.1. The Hybrid-Traditional MPPT

Hybrid traditional MPPT algorithms combine various traditional MPPT techniques to track the maximum power point; it is more robust than the other algorithms in their standalone form and can provide a better overall performance of the system. Table 5 provides a summary of hybrid MPPT algorithms based on the traditional MPPT techniques that have been investigated and used in WEHS.

Table 5. Summary of Hybrid-Traditional MPPT Algorithms for WEHS.

Type of MPPT Algorithms	Remarks	Reference
P&O+PSF	Improvement in power efficiency tracking was achieved with the hybrid MPPT	[44]
ORB+P&O	Combines ORB with a self-rotating P&O-based controller that improves the tracking speed of the hybrid MPPT.	[23]
P&O+OTC	The OTC was employed to detect power peak point and reduce the perturbation step size of the P&O algorithm to reach MPP.	[24]

4.3.2. Hybrid Methods (Traditional and Intelligent)

Due to its intelligent ability to solve complex problems, AI-based methods are developed and integrated with the traditional MPPT methods, as presented in Table 6. This will make the technique more robust and less dependent on the machine's characteristics. According to the literature, the combinations have successfully eliminated the drawback of the traditional method and further enhanced the method's reliability and efficiency. Furthermore, the method's tracking speed was improved, and its accuracy was increased.

Table 6. Summary of hybrid methods (traditional MPPT and intelligent) for WEHS.

Hybrid MPPT Name	Description	Algorithm Performance Evaluation	Reference
P&O+FLC	Fuzzy logic controller (FLC) has been integrated into the adaptive P&O MPPT method which increased the computational speed of the MPPT controller. Furtherly, the new hybrid method successfully eliminated the drawbacks of both the standalone conventional adaptive P&O MPPT and the FLC.	The proposed controller yielded a power increase of 37.93% compared to the P&O method and 17.65% compared to the FLC controller. Additionally, 110 W more power was generated with the proposed controller than with the P&O method and 60 W more than the FLC controller [45]. Moreover, 36.38% of energy yield was recorded with the new controller in [46].	[42,45,46]
P&O+ANN	By integrating ANN into the traditional P&O algorithm, an increase in accuracy was achieved.	The new approach effectively monitored the power coefficient at the optimal level of 0.35 and the nominal power generation of 3 MW.	[47]
ORB+ PSO	PSO was used in the ORB algorithm to search for the maximum power coefficient. The resulting hybrid algorithm provided high efficiency	The PSO-ORBMPPT algorithm has a tracking efficiency of up to 99.4%, which is higher than that of conventional OTC and ORB MPPT algorithms. Additionally, the PSO-ORBMPPT algorithm harvests 1.9% more electrical energy than the conventional algorithms.	[48]

4.3.3. Hybrid Methods (Intelligent and Intelligent)

Hybrid intelligent methods, which combine different artificial intelligence techniques, have been used in MPPT control in WEHS to improve the performance of the control system.

More studies were conducted to further enhance the MPPT algorithms' performance by combining several AI algorithms, the results of which are summarized in Table 7.

Table 7. Summary of hybrid methods (intelligent and intelligent) for WEHS.

Types of Intelligent MPPT Algorithm	Description	Performance Metrics and Results	Reference
FLC, NN	The controller in [34] which combined FLC and NN has improved power harvesting capability in a hybrid renewable energy system (HRES) and shorter simulation time to catch the MPP as compared to other standalone methods such as P&O, FLC, and ANN.	Compared to the standalone P&O, FLC, and ANN methods, the hybrid method achieved an increase in power generation of 35 W, 15 W, and 5 W, respectively, in the WEHS.	[34]
ANFIS	An Adaptive Neuro-Fuzzy inference system (ANFIS) (MPPT) controller for grid-connected WEHS was proposed. The method described can extract the MP from the wind by tracking the MPP independently of the wind speeds.	The ANFIS controller resulted in a 37% smaller voltage overshoot compared to the PI controller. Additionally, a power increase of approximately 7.64% was achieved.	[35]
RBF-NN, MPSO	RBF-NN and modified PSO were integrated into MPPT controller in [20], to replace the conventional MPPT controller. The hybrid combination was able to track the MPP of the WEHS in addition to estimating both the effective wind speed and the rotational speed of the WT. Increased in system reliability, and reduced converters loss, size, and system cost were also achieved.	The proposed hybrid method resulted in a 40% reduction in converter size and produced highest power coefficient of 0.498, whereas other methods, such as ENN+PSO, RBNN-GA, and RBFNN, achieved lower power coefficients of 0.475, 0.47, and 0.43, respectively.	[20]

One of the hybrid intelligent methods used in MPPT control is the combination of a neural network and a fuzzy logic controller. In this method, the neural network is used to predict the wind turbine's power output, while the fuzzy logic controller adjusts the rotor speed to ensure that the turbine operates at the maximum power point. The neural network can learn from past wind speed and power output data and use this information to predict the turbine's power output for a given wind speed. The fuzzy logic controller can then adjust the rotor speed based on the predicted power output to ensure that the turbine operates at its maximum power point.

5. Discussions

This section discusses the challenges of hybrid MPPT methods; the main challenges posed by hybrid MPPT are the design and optimization of the hybrid algorithm. In addition, the integration of different techniques can increase system complexity, which can have an impact on the system's reliability and stability. Furtherly, hybrid MPPT algorithm design and optimization necessitate careful consideration of the system's complexity, reliability, and stability, as well as proper validation of the algorithm's robustness and adaptability. The following subsections discuss the different hybrids methods in detail.

5.1. Intelligent-Based MPPT Models

Recent advances in AI have resulted in the development of intelligent-based MPPT algorithms with improved performance. Intelligent-based MPPT models, such as fuzzy logic, neural networks, and genetic algorithms, on the other hand, provide better accuracy and efficiency. These techniques can adapt quickly to changing environmental conditions such as wind speed, temperature, etc., resulting in increased power tracking accuracy and energy conversion efficiency. Furthermore, intelligent-based MPPT models can optimize WEHS control parameters in real-time, resulting in improved system performance under varying wind conditions.

Moreover, intelligent-based MPPT models can overcome traditional MPPT method limitations such as wind speed measurement dependency and system parameter uncertainties. Fuzzy logic-based MPPT methods, for example, can effectively handle uncertainties and nonlinearity, both of which are common challenges in wind energy systems. Similarly, MPPT algorithms based on neural networks can adapt to changing environmental conditions, resulting in improved system performance and increased energy conversion efficiency. Furthermore, the intelligent-based MPPT method using optimization techniques such as PSO and other metaheuristic algorithms such as ACOA, AOA, CS, GOA, MOGOA, ECPO, and EASO techniques have been used for the MPPT in WEHS, and the results of their performance have shown a very good improvement of the technique in terms of quick searching of the WEHS's optimum operation point [28–31,43]. Therefore, in terms of accuracy, efficiency, and adaptability, intelligent-based MPPT methods outperform traditional MPPT techniques, resulting in optimal energy extraction from the wind source. Thus, intelligent-based MPPT techniques are promising approaches for improving WEHS application performance.

5.2. Hybrid Methods (Traditional and Intelligent-Based MPPT)

The traditional MPPT methods, such as P&O and INC techniques, are widely used. Traditional MPPT methods have several performance limitations that can lead to decreased efficiency and energy loss. The sensitivity of traditional MPPT techniques to system parameters and wind speed measurement is one of their primary limitations. Wind turbulence affects the smooth functions of the anemometers, resulting in inaccurate measurement of actual wind speed striking the wind turbine. As a result, traditional MPPT techniques may result in inefficient operation and energy waste.

For improving WEHS performance, MPPT methods that combine traditional and intelligent-based techniques have been proposed. These hybrid approaches seek to overcome the shortcomings of traditional MPPT techniques while retaining their benefits.

For instance, hybrid MPPT methods can optimize WEHS control parameters in real-time by allowing the tracking algorithm to be adjusted based on changing wind speed and direction, resulting in improved system performance and maximum energy extraction from the source under the rapid change in wind conditions.

The hybrid methods proposed by [42,45,46], which combine P&O and FL-based techniques, can adapt to changes in wind speed and direction, resulting in improved tracking accuracy and energy conversion efficiency. Similarly, hybrid MPPT methods that combine P&O and neural network-based techniques yielded a hybrid system that successfully tracked the System MPPT [47]. Therefore, resulting in improved system performance, such as tracking accuracy and increased energy conversion efficiency.

5.3. Hybrid Methods (Intelligent and Intelligent)

Hybrid intelligent MPPT methods combine two or more intelligent techniques to improve the MPPT algorithm's performance. A hybrid approach, for example, that combines FL and ANN-based techniques can overcome the limitations of each technique while retaining their benefits. FL can handle uncertainty in system parameter values, whereas ANN can adapt to changing environmental conditions and provide accurate predictions of wind turbine power output. Similarly, the ANN is used to predict the wind turbine's power output, as in [34], while the FL controller adjusts the rotor speed to ensure that the turbine operates at the maximum power point. The ANN can learn from past wind speed and power output data and use this information to predict the turbine's power output for a given wind speed. The FL controller can then adjust the rotor speed based on the predicted power output to ensure that the WEHS operates at its maximum power point. When compared to other MPPT methods such as P&O, FLC, and ANN in their standalone form, the hybrid method in [34] that combined FLC and ANN achieved improved system power harvesting capability as well as a shorter simulation time to capture the MPP.

Also, the method reported in [35], which utilized the MPPT strategy for grid-connected WEHS based on the ANFIS, was capable of extracting the MP from the wind by tracking the MPP regardless of wind speeds.

Furthermore, in [20], the hybrid combination of RBF-NN and MPSO was able to track the MPP in addition to estimating the effective wind speed and the rotational speed of the WEHS. Additionally, improving system reliability was achieved, and converter size, loss, and cost were all decreased.

Overall, hybrid intelligent methods of MPPT in WEHS can improve the control system's performance and increase the system's energy harvesting efficiency. These methods can learn from past data and use optimization techniques to find the optimal solution for maximizing power output, ensuring that the WEHS operates at their maximum potential.

6. Future Directions

Advancements in AI-based MPPT techniques for wind energy harvesting systems have led to significant improvements in efficiency, accuracy, and overall system performance. However, further research and development are still necessary to explore and implement more advanced AI algorithms, such as reinforcement learning and other metaheuristic optimization techniques, which can improve MPPT techniques even further. Developing real-time MPPT systems that can adapt to rapidly changing environmental conditions such as wind speed and direction can lead to more efficient and reliable wind energy harvesting systems. While simulations are crucial for testing and validating AI-based MPPT techniques, implementing these algorithms in real-world hardware systems can provide valuable insights into their practical feasibility and performance under actual operating conditions.

7. Conclusions

The optimization of wind energy harvesting systems' power output using MPPT has received considerable attention in the research community. Various methods have been proposed for tracking MPPT in WEHS. While traditional techniques have been explored, such as direct and indirect power control, they have certain drawbacks, such as a large convergence speed, the need for system parameter information, the need for wind speed measurement, and low power tracking efficiency. Even though traditional hybrid methods showed an improvement in the WEHS performance, they suffer from other drawbacks, such as the convergence speed of the algorithm, the need for wind speed measurement, and system parameters dependency. AI-based techniques have the ability to swiftly adjust to changes in environmental factors such as wind speed and temperature. Various intelligent-based MPPT techniques, including fuzzy logic-based methods and those based on neural networks, have been developed to improve energy conversion efficiency in wind energy systems. Additionally, optimization techniques such as PSO and metaheuristic algorithms such as ACOA, AOA, CS, GOA, MOGOA, ECPO, and EASO have been utilized, resulting in significant improvements in quickly tracking the optimum operation point of the system. Consequently, results in greater precision in MP tracking and an increase in the efficiency of energy conversion. Hybrid MPPT algorithms comprising the traditional methods and AI-based MPPT algorithms have been proposed. More improvement in system performance has been achieved with the traditional-AI hybrid method by removing the algorithm's dependency on system parameters and the need for wind speed measurement. Furtherly, AI-AI hybrid methods such as FL-ANN, where ANN is used to predict the wind turbine's power output while the FL controller adjusts the rotor speed to its optimal operating point, has proved to be the most efficient method. The AI-based hybrid, in addition to successfully removing the MPPT drawbacks, has also improved the algorithms' robustness and performance. This study focused on the review of AI-based MPPT methods, and their performance was discussed briefly.

The findings revealed that AI-based methods, such as FLC-ANN, have the highest performance in terms of efficiency and accuracy by combining the handling of uncertainty and robustness of FLC with the learning and adaptability of ANN, this hybrid approach benefits from the strengths of both techniques. This synergy enables the FLC-ANN method to better adapt to dynamic and non-linear environments, such as those presented by varying wind speeds and directions, while also demonstrating robustness in the face of uncertainties and system disturbances.

Therefore, it is encouraging to explore and evaluate new MPPT techniques that hybridize with AI methods in wind power generation to improve the efficiency and reliability of renewable energy generation systems.

Author Contributions: Conceptualization, D.A.U., N.L.M.J. and G.A.; methodology and investigation D.A.U., N.L.M.J., G.A. and M.A.A.; resources M.A.A., Y.B. and A.A.A.; writing—original draft preparation, D.A.U., N.L.M.J. and G.A.; writing—review and editing, G.A., Y.B., L.F.C., A.A.A. and S.K.T.; supervision, G.A., A.A.A. and S.K.T.; project administration, A.A.A., L.F.C. and S.K.T.; funding acquisition, A.A.A. and S.K.T. All authors have read and agreed to the published version of the manuscript.

Funding: This research was funded by Universiti Tenaga Nasional (UNITEN).

Data Availability Statement: Data are available from the corresponding author upon request.

Acknowledgments: The authors acknowledge the Ministry of Higher Education, Malaysia for supporting this work under HICoE grant no 2022004HICOE, and deep appreciation for publication support by Low Tuck Kwong International Energy Transition Grant no 202203002ETG at the iRMC of Universiti Tenaga Nasional, Malaysia.

Conflicts of Interest: The authors declare that there is no conflict of interest. The funders had no role in the design of the study; in the collection, analyses, or interpretation of data; in the writing of the manuscript; or in the decision to publish the results.

References

1. Roga, S.; Bardhan, S.; Kumar, Y.; Dubey, S.K. Recent technology and challenges of wind energy generation: A review. *Sustain. Energy Technol. Assess.* **2022**, *52*, 102239. [CrossRef]
2. Famoso, F.; Brusca, S.; D'Urso, D.; Galvagno, A.; Chiacchio, F. A novel hybrid model for the estimation of energy conversion in a wind farm combining wake effects and stochastic dependability. *Appl. Energy* **2020**, *280*, 115967. [CrossRef]
3. Lopez-Flores, D.R.; Duran-Gomez, J.L.; Chacon-Murguia, M.I. A Mechanical Sensorless MPPT Algorithm for a Wind Energy Conversion System based on a Modular Multilayer Perceptron and a Processor-in-the-Loop Approach. *Electr. Power Syst. Res.* **2020**, *186*, 106409. [CrossRef]
4. Tan, J.D.; Chang, C.C.W.; Bhuiyan, M.A.S.; Minhad, K.N.; Ali, K. Advancements of wind energy conversion systems for low-wind urban environments: A review. *Energy Rep.* **2022**, *8*, 3406–3414. [CrossRef]
5. Umar, D.A.; Yaw, C.T.; Koh, S.P.; Tiong, S.K.; Alkahtani, A.A.; Yusaf, T. Design and Optimization of a Small-Scale Horizontal Axis Wind Turbine Blade for Energy Harvesting at Low Wind Profile Areas. *Energies* **2022**, *15*, 3033. [CrossRef]
6. Zhang, X.; Jia, J.; Zheng, L.; Yi, W.; Zhang, Z. Maximum power point tracking algorithms for wind power generation system: Review, comparison and analysis. *Energy Sci. Eng.* **2022**, *11*, 430–444. [CrossRef]
7. Kazmi, S.M.R.; Goto, H.; Guo, H.J.; Ichinokura, O. Review and critical analysis of the research papers published till date on maximum power point tracking in wind energy conversion system. In Proceedings of the 2010 IEEE Energy Conversion Congress and Exposition, Atlanta, GA, USA, 12–16 September 2010; pp. 4075–4082. [CrossRef]
8. Mousa, H.H.H.; Youssef, A.R.; Mohamed, E.E.M. State of the art perturb and observe MPPT algorithms based wind energy conversion systems: A technology review. *Int. J. Electr. Power Energy Syst.* **2021**, *126*, 106598. [CrossRef]
9. Pande, J.; Nasikkar, P.; Kotecha, K.; Varadarajan, V. A review of maximum power point tracking algorithms for wind energy conversion systems. *J. Mar. Sci. Eng.* **2021**, *9*, 1187. [CrossRef]
10. Khan, M.J. Review of Recent Trends in Optimization Techniques for Hybrid Renewable Energy System. *Arch. Comput. Methods Eng.* **2021**, *28*, 1459–1469. [CrossRef]
11. Khan, M.J.; Mathew, L. Comparative Study of Optimization Techniques for Renewable Energy System. *Arch. Comput. Methods Eng.* **2020**, *27*, 351–360. [CrossRef]
12. Kumari, B.; Aggarwal, M. A Comprehensive Review of Traditional and Smart MPPT Techniques in PMSG based Wind Energy Conversion System. In Proceedings of the 2019 International Conference on Power Electronics, Control and Automation (ICPECA), New Delhi, India, 16–17 November 2019; Volume 2019. [CrossRef]
13. Malik, M.Z.; Baloch, M.H.; Gul, M.; Kaloi, G.S.; Chauhdary, S.T.; Memon, A.A. A research on conventional and modern algorithms for maximum power extraction from wind energy conversion system: A review. *Environ. Sci. Pollut. Res.* **2021**, *28*, 5020–5035. [CrossRef]

14. Apata, O.; Oyedokun, D.T.O. An overview of control techniques for wind turbine systems. *Sci. Afr.* **2020**, *10*, e00566. [CrossRef]
15. Xiao, Y.; Li, S.; Xu, M.; Feng, R. Research on the economy of implementing the MPPT for wind-solar hybrid power generation system: A review. In Proceedings of the 2022 41st Chinese Control Conference (CCC), Hefei, China, 25–27 July 2022; Volume 2022, pp. 5339–5344. [CrossRef]
16. Kumar, D.; Chatterjee, K. A review of conventional and advanced MPPT algorithms for wind energy systems. *Renew. Sustain. Energy Rev.* **2016**, *55*, 957–970. [CrossRef]
17. Jia, Y.; Yang, Z.; Cao, B. A new maximum power point tracking control scheme for wind generation. In Proceedings of the International Conference on Power System Technology, Kunming, China, 13–17 October 2002; Volume 1, pp. 144–148. [CrossRef]
18. Wu, S.; Wang, Y.; Cheng, S. Extreme learning machine based wind speed estimation and sensorless control for wind turbine power generation system. *Neurocomputing* **2013**, *102*, 163–175. [CrossRef]
19. Asghar, B.; Liu, X. Adaptive neuro-fuzzy algorithm to estimate effective wind speed and optimal rotor speed for variable-speed wind turbine. *Neurocomputing* **2018**, *272*, 495–504. [CrossRef]
20. Sitharthan, R.; Karthikeyan, M.; Sundar, D.S.; Rajasekaran, S. Adaptive hybrid intelligent MPPT controller to approximate effectual wind speed and optimal rotor speed of variable speed wind turbine. *ISA Trans.* **2020**, *96*, 479–489. [CrossRef] [PubMed]
21. Yang, B.; Zhu, T.; Wang, J.; Shu, H.; Yu, T.; Zhang, X.; Yao, W.; Sun, L. Comprehensive overview of maximum power point tracking algorithms of PV systems under partial shading condition. *J. Clean. Prod.* **2020**, *268*, 121983. [CrossRef]
22. Abdullah, M.A.; Yatim, A.H.M.; Tan, C.W. An online Optimum-Relation-Based Maximum Power Point Tracking Algorithm for Wind Energy Conversion System. In Proceedings of the 2014 Australasian universities power engineering conference (AUPEC), Perth, Australia, 28 September–1 October 2014.
23. Ahmed, S.; Rashid, M.A.; Mohamed, S.B.; Yaakob, S.B. A novel maximum power point tracking algorithm for wind energy conversion system. *Eng. Lett.* **2019**, *27*, 822–830.
24. Kazmi, S.M.R.; Goto, H.; Guo, H.J.; Ichinokura, O. A novel algorithm for fast and efficient speed-sensorless maximum power point tracking in wind energy conversion systems. *IEEE Trans. Ind. Electron.* **2011**, *58*, 29–36. [CrossRef]
25. Ganjefar, S.; Ghassemi, A.A.; Ahmadi, M.M. Improving efficiency of two-type maximum power point tracking methods of tip-speed ratio and optimum torque in wind turbine system using a quantum neural network. *Energy* **2014**, *67*, 444–453. [CrossRef]
26. Liu, J.; Meng, H.; Hu, Y.; Lin, Z.; Wang, W. A novel MPPT method for enhancing energy conversion efficiency taking power smoothing into account. *Energy Convers. Manag.* **2015**, *101*, 738–748. [CrossRef]
27. Yin, X.; Jiang, Z.; Pan, L. Recurrent neural network based adaptive integral sliding mode power maximization control for wind power systems. *Renew. Energy* **2020**, *145*, 1149–1157. [CrossRef]
28. Mokhtari, Y.; Rekioua, D. High performance of Maximum Power Point Tracking Using Ant Colony algorithm in wind turbine. *Renew. Energy* **2018**, *126*, 1055–1063. [CrossRef]
29. Fathy, A.; Alharbi, A.G.; Alshammari, S.; Hasanien, H.M. Archimedes optimization algorithm based maximum power point tracker for wind energy generation system. *Ain Shams Eng. J.* **2022**, *13*, 101548. [CrossRef]
30. Falehi, D. An innovative optimal RPO-FOSMC based on multi-objective grasshopper optimization algorithm for DFIG-based wind turbine to augment MPPT and FRT capabilities. *Chaos Solitons Fractals* **2020**, *130*, 109407. [CrossRef]
31. He, X.; Wei, P.; Gong, X.; Meng, X.; Shan, D.; Zhu, J. Enhanced Atom Search Optimization Based Optimal Control Parameter Tunning of PMSG for MPPT. *Energy Eng. J. Assoc. Energy Eng.* **2022**, *119*, 145–161. [CrossRef]
32. Yaichi, I.; Semmah, A.; Wira, P. Control of doubly fed induction generator with maximum power point tracking for variable speed wind energy conversion systems. *Period. Polytech. Electr. Eng. Comput. Sci.* **2020**, *64*, 87–96. [CrossRef]
33. Kumar, R.; Agrawal, H.P.; Shah, A.; Bansal, H.O. Maximum power point tracking in wind energy conversion system using radial basis function based neural network control strategy. *Sustain. Energy Technol. Assess.* **2019**, *36*, 100533. [CrossRef]
34. Khan, M.J.; Kumar, D.; Narayan, Y.; Malik, H.; Márquez, F.P.G.; Muñoz, C.Q.G. A Novel Artificial Intelligence Maximum Power Point Tracking Technique for Integrated PV-WT-FC Frameworks. *Energies* **2022**, *15*, 3352. [CrossRef]
35. Chhipa, A.A.; Kumar, V.; Joshi, R.R.; Chakrabarti, P.; Jasinski, M.; Burgio, A.; Leonowicz, Z.; Jasinska, E.; Soni, R.; Chakrabarti, T. Adaptive neuro-fuzzy inference system-based maximum power tracking controller for variable speed wecs. *Energies* **2021**, *14*, 6275. [CrossRef]
36. Abdullah, M.A.; Yatim, A.H.M.; Tan, C.W.; Saidur, R. A review of maximum power point tracking algorithms for wind energy systems. *Renew. Sustain. Energy Rev.* **2012**, *16*, 3220–3227. [CrossRef]
37. Abdelqawee, M.; Yousef, A.Y.; Hasaneen, K.M.; Nashed, M.N.F.; Hamed, H.G. Standalone wind energy conversion system control using new maximum power point tracking technique. *Int. J. Emerg. Technol. Adv. Eng. Certif. J.* **2019**, *9*, 95–100.
38. Pande, J.; Nasikkar, P. A Maximum Power Point Tracking Technique for a Wind Power System Based on the Trapezoidal Rule. *Energies* **2023**, *16*, 2799. [CrossRef]
39. Malobe, P.A.; Djondine, P.; Eloundou, P.N.; Ndongo, H.A. A Novel Hybrid MPPT for Wind Energy Conversion Systems Operating under Low Variations in Wind Speed. *Energy Power Eng.* **2020**, *12*, 716–728. [CrossRef]
40. Abdalla, A.; Zakaria, M.N.; Sulaiman, S.; Ahmad, W.F.W. A comparison of feed-forward back-propagation and radial basis artificial neural networks: A Monte Carlo study. In Proceedings of the 2010 International Symposium on Information Technology, Kuala Lumpur, Malaysia, 15–17 June 2010; Volume 2, pp. 994–998. [CrossRef]
41. Kůrková, V. Kolmogorov's theorem and multilayer neural networks. *Neural Netw.* **1992**, *5*, 501–506. [CrossRef]

42. Khan, M.J. An AIAPO MPPT controller based real time adaptive maximum power point tracking technique for wind turbine system. *ISA Trans.* **2022**, *123*, 492–504. [CrossRef] [PubMed]
43. Sompracha, C.; Jayaweera, D.; Tricoli, P. Particle swarm optimisation technique to improve energy efficiency of doubly-fed induction generators for wind turbines. *J. Eng.* **2019**, *2019*, 4890–4895. [CrossRef]
44. Chen, J.; Lin, T.; Wen, C.; Song, Y. Design of a Unified Power Controller for Variable-Speed Fixed-Pitch Wind Energy Conversion System. *IEEE Trans. Ind. Electron.* **2016**, *63*, 4899–4908. [CrossRef]
45. Chaicharoenaudomrung, K.; Areerak, K.; Areerak, K.; Bozhko, S.; Hill, C.I. Maximum Power Point Tracking for Stand-Alone Wind Energy Conversion System Using FLC-P&O Method. *IEEJ Trans. Electr. Electron. Eng.* **2020**, *15*, 1723–1733. [CrossRef]
46. Khan, J.; Mathew, L.; Alotaibi, M.A.; Malik, H.; Nassar, M.E. Fuzzy-Logic-Based Comparative Analysis of Different Maximum Power Point Tracking Controllers for Hybrid Renewal Energy Systems. *Mathematics* **2022**, *10*, 529. [CrossRef]
47. Azzouz, S.; Messalti, S.; Harrag, A. A Novel Hybrid MPPT Controller Using (P&O)-neural Networks for Variable Speed Wind Turbine Based on DFIG. *Model. Meas. Control A* **2019**, *92*, 23–29. [CrossRef]
48. Abdullah, A.; Al-Hadhrami, T.; Tan, C.W.; Yatim, A.H. Towards green energy for smart cities: Particle swarm optimization based MPPT approach. *IEEE Access* **2018**, *6*, 58427–58438. [CrossRef]

processes

MDPI

Article

Game Analysis of the Evolution of Energy Structure Transition Considering Low-Carbon Sentiment of the Decision-Makers in the Context of Carbon Neutrality

Xinping Wang [1], Zhenghao Guo [1], Ziming Zhang [1], Boying Li [2,*], Chang Su [3], Linhui Sun [1] and Shihui Wang [1]

1 School of Management, Xi'an University of Science and Technology, Xi'an 710054, China
2 School of Political Science and International Relations, Tongji University, Shanghai 200092, China
3 School of Safety Science and Engineering, Xi'an University of Science and Technology, Xi'an 710054, China
* Correspondence: liboying@tongji.edu.cn

Abstract: Countries have started to aggressively undertake energy structure transformation strategies in order to reach the objective of carbon neutrality. Both clean and efficient coal energy use and clean energy use will be crucial to the process of changing the energy structure since the two cannot be totally replaced within a short period of time. In this study, we quantify emotions as an irrational factor, combine them with an evolutionary game using RDEU theory, and build an evolutionary game model between government regulators and energy consumers. We then analyze how low-carbon emotions of decision-makers affect their choice of strategy and the transformation of the energy structure. The findings support that by affecting the relative importance of each strategic choice, emotions have a profound impact on the evolutionary steady state of the system. Appropriate stress and anxiety can increase decision-makers' feelings of responsibility, while pleasant emotions frequently support strategic conduct. The main countermeasures are as follows: Allow government regulators and energy consumers to properly release positive information, with government regulators forming subsidies and energy consumers actively cooperating and promoting low-carbon activities. This will properly guide the low-carbon sentiment of game subjects to keep them realistically pessimistic.

Keywords: carbon neutrality; energy structure transition; low-carbon sentiment; RDEU; evolutionary game

Citation: Wang, X.; Guo, Z.; Zhang, Z.; Li, B.; Su, C.; Sun, L.; Wang, S. Game Analysis of the Evolution of Energy Structure Transition Considering Low-Carbon Sentiment of the Decision-Makers in the Context of Carbon Neutrality. *Processes* **2022**, *10*, 1650. https://doi.org/10.3390/pr10081650

Academic Editors: Ferdinando Salata and Virgilio Ciancio

Received: 18 July 2022
Accepted: 17 August 2022
Published: 19 August 2022

1. Introduction

More fossil fuels are being used as science and technology improve, which is a factor in the rising number of environmental problems. Achieving carbon neutrality has been suggested as a solution to the issue of industrial emissions caused by the use of fossil fuels [1]. These emissions have led to issues like the greenhouse effect. By 2050, it is expected that hundreds of countries will be carbon neutral [2] and a number of them have already incorporated carbon neutrality targets into their legal systems. Many nations have begun the shift from fossil fuels to alternative energy sources. Supporting the transition of the changing shape with the advancement of clean fossil fuels and new fuel technologies is critical. As China's economy expands and its CO_2 emissions increase, so does its standard of living, which drives up energy consumption [3]. According to data, China utilized primary energy equal to 3512.8 million tons of oil in 2020, making up 26.1% of the entire amount of energy consumed worldwide [4]. To accomplish carbon neutrality goals and to handle the numerous socio-economic and environmental concerns involved, regional energy transition strategies have been created to achieve low-carbon and sustainable development. The phrase "energy transition" has been referred to in a variety of ways by academics, including "sustainable energy transition" [3], "low carbon energy transition" [4], and "green energy transition" [5]. As a result, reducing greenhouse gas emissions is essential to reducing global warming, its impacts, and the ensuing socioeconomic and environmental issues. Alternative energy sources have been suggested, including examples include the construction of numerous hydroelectric power plants in the Iberian Peninsula and the continued improvement of wind energy generation in Xinjiang [6,7], the gradual advancement of solar energy storage [8], alternative technologies using materials, such as graphene as batteries [9], microbial fuel cell technology [10], and nuclear fission power systems [11,12], all of which have gradually matured, leading to an increase in the share of clean energy in all electricity. As China's economy expands and its CO_2 emissions increase, so does its standard of living, which drives up energy consumption [13]. According to data, China utilized primary energy equal to 3512.8 million tons of oil in 2020, making up 26.1% of the entire amount of energy consumed worldwide [14], it is anticipated that the carbon neutrality goal would be attained by 2060.

China is the country under the most pressure to reduce its emissions and use of energy. Despite fluctuations in economic growth due to the COVID-19 epidemic, China's electricity and energy consumption were still increasing significantly in 2020 [15]. As the main source of energy for power generation, half of fossil energy consumption comes from coal, but the share of clean energy generation has grown significantly [16]. In 2020, 64.7% of China's total power generation came from coal-fired power generation, with hydro-power generation ranking second and accounting for only 16.9% [14]. The country's level of urbanization is still rising, and the industrial structure's impact on the environment, which depends heavily on the combustion of fossil fuels in all areas, is becoming progressively worse [17]. In recent years, smart mines have also been developed with government regulation, emergency response capabilities, and clean coal technologies [18–20], which can operate in low-carbon and in a clean manner while achieving energy conservation [21]. Although green energy sources, such as wind, tidal, and biomass fuels, have relatively little environmental impact, they have the drawback of being unstable and intermittent in their supply [7], which raises the possibility of energy security risk events and jeopardizes the security of the energy supply. This indicates that the use of fossil energy, represented by coal resources, currently dominates the whole energy cycle and is unlikely to alter very soon [22]. Due to technology and other circumstances, the development of new energy applications and development in China is uneven. For example, the development of solar energy, hydrogen energy, and other clean energy sources is still at a very early stage. Due to this, clean energy and coal have been in a constant state of competition, with the latter unable to fully replace the former for an extended length of time. However, clean energy research and utilization have allowed for the safe and effective use of coal. Therefore, coordinating the development of both coal energy and clean energy within the framework of carbon neutrality is a key part of the energy structure change that needs to be dealt with.

The rest of the essay is organized as follows: The literature on the subject of this research is reviewed in Section 2, along with a brief description of the innovation points. The concerns related to research theory are presented in Section 3. Based on the RDEU theory, Section 4 creates an evolutionary game model. The game topics are subjected to stability and asset allocation stability analysis in Section 5. Section 6 runs the model's primary simulation. The study's findings, managerial lessons learned, and inadequacies are presented in Section 7.

2. Literature Review

2.1. Progress of Research on Energy Structure Transition

Analysis of energy strategy games has begun. In order to capture the drivers of the energy transition and in order to simulate and discuss the evolutionary process and evolutionary stabilization strategies to support the development of hydrogen-powered vehicles and solar photovoltaic hydrogen production, Wang et al. [23] analyzed a partnership consisting of an investment company, hydrogen-powered vehicle users, and solar photovoltaic power plants. To replicate and explain the evolutionary process, Wang et al. [24] proposed a partnership of carbon exchange, solar power plants, and coal-fired thermal power plants. Hou et al. [25] proposed a new conceptual model, the institutional economic-technical behavioral framework, to synthesize the similarities and differences in energy transitions in various nations. The German energy transition plan is significant as a framework for fostering the growth of renewable energy, as per Gao et al. [26] 's comprehensive analysis of several strategies for boosting renewable energy during the energy transition. In order to investigate whether the relationship between the government and the public can encourage manufacturers to adopt low-carbon technologies by examining the interaction effects between various stakeholders, Chen et al. [27] developed a three-way game model between the government, manufacturers, and the public under carbon taxes and subsidies. In order to study how financial penalties impact players' tactics and the evolutionary process of optimizing financial penalties, Chang et al. [28] developed a mixed strategy game model and an evolutionary game model for regulators and conventional energy corporations. The application of game theory has been well proven in the study of energy transition, and the theory of evolutionary games can explain how different strategic decisions may have an influence on the entire energy transition system.

2.2. The Use of Evolutionary Games in Energy Structure Transformation

Evolutionary game theory offers a reasonable analytical framework, given that energy structural transformation is a gradual and ongoing process rather than a sudden change that occurs only once. Many academics have conducted pertinent studies on energy transitions using this evolutionary game framework. In a three-way evolutionary game model that included regulators, energy companies, and whistleblowers, Yang et al. [29] discovered a substantial association between the likelihood of a whistleblower, the likelihood of active management by energy companies, and the likelihood of rigorous monitoring. Zhao et al. [30] analyzed the behavioral strategies of generators in connection to renewable energy and the influence of important institutional characteristics on the dynamic evolutionary process of generators. According to Qiao and Yin [31], who used an evolutionary game model based on psychological perceptions, strategic choices made by consumers and companies are essential to the effective implementation of the energy transition. In order to study two different types of power generation enterprises, Liu et al. [32] used an evolutionary game approach and a numerical simulation method of scenario analysis to study two different types of power generation enterprises. In order to deal with the complex relationship between the Chinese government, thermal power producers, and grid companies, Shang et al. [33] used a system dynamic (SD)-based evolutionary game. They came to the conclusion that the Chinese government must strictly enforce the renewable portfolio standard in order to promote green and low-carbon upgrading of energy and electricity (RPS). By examining the many actions each decision-maker in the process takes, these studies have concentrated on how the entire decision-making system changes to support the transition in the energy mix. However, are decision-making processes in government agencies and energy users entirely rational? It is unknown if additional factors have any role in their decision-making.

2.3. Progress of Emotions in Related Research

Nevertheless, research has shown that psychological preferences and feelings have an effect on decision-makers' inclinations and decisions, both in terms of long-term direction and contingency influences [34,35]. As an example, negative emotional states can have a direct impact on risk-taking behavior and decision-making [36]. Since decision-makers are limited in their rationality, in their behavioral choices, and have limited access to information, which can reveal different preferences and subsequently different emotions, emotions play a significant role in decision-making [37]. Decision-makers have varied psychological preferences and risk attitudes as a result of their differing values, interests, and contingent emotions [38]. Both government regulators and energy users experience emotions as a result, and these feelings can further affect their choices. The emotional attitudes of the participants in the energy structure transition towards the transition will have a significant impact on policy practices. However, the traditional evolutionary game does not take into account the psychological preferences and emotions of each player, so further improvement and refinement are required. The rank-dependent expected utility (RDEU) theory and evolutionary games have been merged. To create a co-evolutionary game model of shared manufacturing quality innovation with multi-subject involvement and examine how emotions impact quality improvement motivation, Zhang et al. [39] merged the RDEU theory with evolutionary game theory. To create an RDEU game model for various sectors, Ni et al. [40] merged game theory with the RDEU theory. They came to the conclusion that different emotional states and intensities impact evolutionary outcomes and evolutionary speed. Emotions are less often taken into account as an influencing element in research pertaining to the evolution of energy transitions, however. As a result, this article takes the impact of each game subject's low-carbon sentiment into account and builds an evolutionary game model that takes carbon sentiment into account using the RDEU theory. The mutual synergistic development of coal energy and clean energy is translated into the question of the impact of the share of coal and clean energy in the energy structure transformation on the energy structure transformation based on the issue of the share of coal and clean energy in the integrated energy consumption. Government regulators and energy consumers won't always act rationally in the face of energy structural transformation. In order to promote the efficient use of coal energy, the quick development of clean energy, and new advancements in energy structural transformation, we must first understand the effects that various emotions will have on people's decision-making and behavior, as well as on energy structural transformation.

Major research findings and importance of the study:

(1) In the new condition of energy structure transformation, coal energy and clean energy have been in a position of reciprocal gaming with regard to carbon neutrality. This paper discuss the effects of the shift to a cleaner energy mix after analyzing the decisions chosen by energy consumers and government regulators about the usage of efficient and clean coal.
(2) This paper creatively introduces the factor of low-carbon sentiment, where decision-makers are not in a fully rational state, and analyzes the impact of various low-carbon sentiments of government regulators and energy consumers on decision-making behavior and energy structural transformation. This is in contrast to other traditional studies exploring energy structural transformation.
(3) It contributes to the expansion of evolutionary game theory and applications by building a sentiment model that naturally combines evolutionary game theory and the RDEU theory to explain the tactical decisions chosen by government regulators and energy consumers under various sentiments and risk aversions.

3. Theoretical Assumptions

The Rank-Dependent Expected Utility Theory

Government regulators and energy consumers may have different attitudes toward clean energy options due to their dependence on coal and their usage habits, and thus different emotions in their strategy choices, as a result of the limited information available to them in the face of the energy transition process. Their psychological preferences and risk attitudes may also irrationally shift as a result of interest claims and situational emotions. The psychological interests and emotions of decision-makers may be well explained by the RDEU theory [41], which was put out by Quiggin [42]. It can "satisfactorily" explain the real decision-making behavior in a complicated model and accurately portray the great variability of decision-makers' emotions [43]. A real-valued function V defined by a utility function $U(x)$ and a decision weight function $\pi(x)$ is used to express the level of the decision-maker's preference for various options.

The function expression is:

$$V(x, u, \pi) = \sum_{i=1}^{n} \pi(x_i) U(x_i), i = 1, 2, 3, \cdots, n \quad (1)$$

The method's cumulative probability function is $RP_i = P(X \leq x_i) = \sum_{\tau \geq i}^{n} p_i, i = 1, 2, \cdots n$ for the set of strategies $X = \{x_i; i = 1, 2, \cdots n\}$ and $P = \{X = x_i\} = p_i$. Assuming that the methods x_i are ordered according to the amplitude of the utilitarian calculus $U(x)$ and defining $x_1 > x_2 > \cdots > x_n$, the utilitarian rank of the technique x_i is defined as RP_i. The bigger the strategy's utility, the higher its cumulative probability, and as a result, the more importance the strategy utility will have in the choice.

At this stage, the sentiments function, which is an asymptotic growing function fulfilling $\omega(0) = 0, \omega(1) = 1$, is represented by the decision weight function, $\pi(x) = \omega(p_i + 1 - RP_i) - \omega(1 - RP_i)$, where $\omega(\cdot)$ is the function.

The potential of $X \leq x$ being increased or decreased is possible by the function $\omega(\cdot)$.

Here are the following three situations:

(1) $\omega(\cdot)$ is a concave function when $\omega(p) < p$. $\omega(\cdot)$ reduces the likelihood of $X \leq x$ for any $p \in [0, 1]$, demonstrating the participants' pessimism.
(2) $\omega(\cdot)$ is a convex function when $\omega(p) > p$. $\omega(\cdot)$ widens the likelihood of $X \leq x$ for any $p \in [0, 1]$, demonstrating the participants' optimism.
(3) The possibility is unchanged when $\omega(p) = p$, a sign that people are in a rational mood.

RDEU theory addresses decision weights by non-linearly altering the utility theory in traditional game theory. As a result, the shortcomings of conventional game theory in the attitude dimension can be partially compensated for by the RDEU theory. Additionally, by incorporating this theory, evolutionary game analysis can more objectively and accurately depict each participant's emotional state in the energy structure conversion and the impact of their psychological response on the structure's conversion.

4. Game Model Construction

4.1. Illustration of Relevant Gaming Concepts

Many academics have taken notice of the RDEU theory because it corrects the flaws in conventional game theory that do not adequately account for actual emotions [44] by taking into account how players' emotions may affect their decision-making behavior. An RDEU eagle-dove game model was created by Li et al. [45] to help resolve trade disputes between nations by taking into account the impact of corporate interests, conflict costs, "emotional variables," and "asymmetric elements" on enterprises' strategic decisions. The RDEU game model was developed by Liu et al. [46] after analyzing the equilibrium tactics of subject nations and stakeholders in the marine pollution industry from various interest vantage points. This model is crucial for researching international relations and environmental protection. A conflict-inducing game mechanism-relative expected utility theory (RDEU) evolutionary game model was created by Hong et al. [47] to investigate the evolution of the strategic behavior of expropriated farmers and local governments when emotions are present. In conclusion, the RDEU-based evolutionary game analysis may efficiently investigate how players' emotions affect their behavior while making decisions and may provide a more plausible depiction of the behavioral development of each game topic.

4.2. Model Hypothesis

Hypothesis 1 (H1). *In this study, the evolutionary game model is used for two finitely rational parties: energy consumers and government regulators. In the course of the game, one side alters their strategy in the energy structure transition, while the other side modifies their options in reaction to the other side's decision-making behavior. Each player in the game makes a tactical decision based on the illogical influence of emotion. Government regulators may play the energy structure transformation game in one of two ways: (1) To put green regulations into effect. The regulator adopts the green and low-carbon guiding philosophy and effective supervision with environmental protection as a prerequisite in the process of energy structure reform and development. (2) The use of traditional monitoring. In the process of developing and transforming the energy structure, supervisory measures are implemented using the notion of conventional management as a guiding* *thought. Additionally, there are two energy consumer strategy choices: (1) Put integrated energy utilization into practice and consumption practices that combine renewable energy with coal as a source. (2) The use of coal in a clean and effective manner and the practice of consuming energy only via the clean and effective use of coal as a fuel. In this paper, wind, hydroelectric, solar, and nuclear energy are examples of clean energy since they are all renewable energy sources with minimal environmental impact.*

Hypothesis 2 (H2). *Single coal energy and integrated energy are two categories that are separated based on the amount of energy used. In contrast to clean energy, which is difficult to store, expensive, and unstable but is a new type of energy whose environmental impact is minimal and difficult to cause, China's coal energy reserves are substantial, cheap, easy to transport, and also have the qualities of easy storage and high stability. The fact is that clean energy cannot, in a short amount of time, replace coal energy as the primary source of energy consumption; nonetheless, clean energy utilization and coal energy clean utilization work in tandem and grow together. In this study,* x_1 *stands for the externalized costs caused by the use of coal, such as high levels of air pollution and a high environmental burden, whereas* x_2 *stands for those caused by the use of clean energy, such as supply instability. Positive externalities, such as a reliable supply from the use of coal, are marked by* y_1*, while positive externalities, such as environmental friendliness from the use of renewable energy are denoted by* y_2*.* a *and* b *reflect the respective percentages of coal and clean energy in the total energy consumption.* q *is the probability of occurrence of potential systemic risks in the process of energy structure transformation, and* L *is the hazards/damage of potential systemic risks in the process of energy structure transition.*

Hypothesis 3 (H3). *In order to utilize coal and create clean energy efficiently, green regulations will force government regulators to develop appropriate inputs and outputs, and energy consumers will likewise carry out certain inputs and outputs. The government regulators will implement specific measures as part of the energy restructuring process to encourage the use of clean energy and the clean and efficient use of coal, with the number of subsidies being* l_1 *and* l_2*, and to encourage energy consumers to consider the advantages of environmental friendliness. Only when both government regulators and energy consumers make the right policy decision does the energy usage system reach a state that is secure and stable. Both parties' decision-making practices may influence one another, or a systemic risk event may happen as a result. According to the concept of risk sharing, risk is transferred, and the amount of risk transfer is quantified by a risk transfer factor.*

5. Model Optimization Analysis Considering Heterogeneous Emotions

Under the influence of emotional factors, the subjective probability function becomes $\omega(m_i) = mi^r$. When $r = 1$ the subjective probability value is the same as the objective probability value, and the game subject does not have emotions; when $(r < 1)$ the subjective probability value is greater than the objective probability value and the game subject is optimistic; when $(r > 1)$ the subjective probability value is smaller than the objective probability value, and the game subject is pessimistic.

Based on the assumptions of the underlying model in Section 2 and the relevant principles of RDEU theory, the rank-dependent expected utility models of government regulators and energy consumers under different strategies adopted are constructed, where parameter r_1 denotes the sentiment status of the government regulator, and the emotional intensity of energy consumers is represented by parameter r_2.

5.1. Stability Analysis of Government Regulators' Strategies Considering Low-Carbon Sentiment

Based on the correlation between the dimensions of cost, benefit, reward, and punishment, the utility rating for the government regulators' four strategy options is:

$$\begin{aligned} & a(y_1 - x_1 - S + l_1) + b(y_2 - x_2 - T + l_2) - c_1 > y_1 - x_1 - S + l_1 - c_1 - \beta\eta Lq \\ & > a(y_1 - x_1) + b(y_2 - x_2) - \theta Lq > y_1 - x_1 - Lq \end{aligned} \quad (2)$$

As a consequence, the utility, probability, rank, and decision weights applicable to each of the government regulators' strategies are shown in Table 1.

Table 1. Ranking of the expected benefits of government regulators, taking into account sentiment.

the Effectiveness of Government Regulators' Strategies	Probability	Rank	Decision Weights
$a(y_1 - x_1 - S + l_1) + b(y_2 - x_2 - T + l_2) - c_1$	mn	1	$\omega A(mn)$
$y_1 - x_1 - S + l_1 - c_1 - \beta\eta Lq$	$m(1-n)$	$1 - mn$	$\omega A(m) - \omega A(mn)$
$a(y_1 - x_1) + b(y_2 - x_2) - \theta Lq$	$(1-m)n$	$1 - m$	$\omega A(m + n - mn) - \omega A(m)$
$y_1 - x_1 - Lq$	$(1-m)(1-n)$	$1 - m - n + mn$	$1 - \omega A(m + n - mn)$

The expected benefits of "green regulation" and "regular regulation" by government regulators are U_{1m} and U_{2m}. The average expected return for government regulators is $\overline{U_m}$. Government regulators replicated the dynamic equation as $F(m)$. Specific formula content and derivation data is supported by the Equations (S4)–(S6) are provided in the Supplementary Materials Section.

5.2. Stability Analysis of Energy Consumers' Strategies Considering Low-Carbon Sentiment

Based on how the dimensions of costs, benefits, incentives, and penalties relate to one another, the utility score for the energy consumers' four strategy options is as follows:

$$\begin{aligned} &a(-c_2 + w_1 + S) + b(-c_3 + w_2 + T) > -c_2 + w_1 + S - \eta Lq \\ &> a(-c_2 + w_1) + b(-c_3 + w_2) - \beta\theta Lq > -c_2 + w_1 - Lq \end{aligned} \tag{3}$$

As a consequence, the utility, probability, rank, and decision weights applied to each of the energy consumers' strategies are shown in Table 2.

Table 2. Expected utility of energy consumers' rank dependence considering emotions.

Energy Consumers Strategy Utility	Probability	Rank	Decision Weights
$a(-c_2 + w_1 + S) + b(-c_3 + w_2 + T)$	mn	1	$\omega B(mn)$
$-c_2 + w_1 + R - \eta Lq$	$m(1-n)$	$1 - mn$	$\omega B(m) - \omega B(mn)$
$a(-c_2 + w_1) + b(-c_3 + w_2) - \beta\theta Lq$	$(1-m)n$	$1 - m$	$\omega B(m + n - mn) - \omega B(m)$
$-c_2 + w_1 - Lq$	$(1-m)(1-n)$	$1 - m - n + mn$	$1 - \omega B(m + n - mn)$

The expected benefits of "integrated energy use" and "clean and efficient use of coal" for energy consumers are U_{1n} and U_{2n}. The average expected return for energy consumers is $\overline{U_n}$. The energy consumers replicated the dynamic equation for $F(n)$. Specific formula content and derivation data is supported by the Equations (S7)–(S10) are provided in the Supplementary Materials Section.

5.3. Analysis of Strategy Portfolio Stability

Government regulators may attain local stability by pursuing a green regulatory approach when conditions are $m = 0$, $m = 1$, or $m = m^*$. Consumers of energy in phases $n = 0$, $n = 1$, or $n = n^*$ decide on an integrated energy usage plan to promote localized stability. As a result, the evolutionary game model's five partial equilibria are $E_1(0,0)$, $E_2(0,1)$, $E_3(1,0)$, $E_4(1,1)$, and $E_5(m^*, n^*)$.

Among them:

$$m^* = \frac{[(a-1)(-S+l_1)+b(-T+l_2)+\beta\eta Lq+\theta Lq-Lq]n+2(h_1-d_1)-S+l_1-c_1-\beta\eta Lq-Lq}{[(a-1)(-S+l_1)+b(-T+l_2)+\beta\eta Lq+\theta Lq-Lq]n+a(-S+l_1)-c_1-\beta\eta Lq+Lq}$$

$$n^* = \frac{[(a-1)(-c_2+w_1)+(a-2)R+b(-c_3+w_2)+bD+2\eta Lq-2\beta\theta Lq-Lq]m+R-\eta Lq+Lq}{[(a-1)R+bD+\eta Lq+\beta\theta Lq]m+(1-a)(-c_2+w_1)+R-b(-c_3+w_2)+\beta\theta Lq-\eta Lq}$$

Since the Jacobian matrix's value depends on the values of the model parameters, it changes depending on the players' feelings and emotions, which also affects the equilibrium points that are determined. Therefore, based on the various feelings and emotions of the game subjects, this article analyzes the stability of the strategic portfolios of government regulators and energy consumers, which includes four scenarios: (rational, rational), (rational, emotional), (emotional, rational), and (emotional, emotional).

Scenario 1. *Rational government regulators and rational energy consumers.*

The attitude parameters $r_1 = 1, r_2 = 1$ are now applicable when government regulators and energy customers are both sensible. The strategy portfolio's stability analysis at this stage is shown in Table 3 after the sentiment parameter has been included in each replication dynamic equation.

In Table 3, there are two saddle points in the system when government regulators are rational and energy consumers are rational. When condition $(A_1-2A_2-2A_3-A_4+A_5)(B_1-2B_2-B_3+B_4)-(A_1-A_2)(B_1-B_2)>0$ $A_1-2A_2-2A_3-A_4+A_5+B_1-2B_2-B_3+B_4<0$ is fulfilled,$E_4(1,1)$ is a stable point in the system, when the government regulators' strategy choice is stable in green regulation and the energy consumers' strategy choice is stable in integrated energy use; it also becomes a stable point in the system's evolution when certain conditions are met.

Table 3. Stability analysis of the strategy portfolio under rational government regulators' and rational energy consumers' scenarios.

Equilibrium Point	$\frac{\partial F(m)}{\partial m}$	$\frac{\partial F(m)}{\partial n}$	$\frac{\partial F(n)}{\partial m}$	$\frac{\partial F(n)}{\partial n}$	*Det*(*J*)	*Tr*(*J*)	Consistency
$E_1(0,0)$	0	0	0	0	0	0	Unstable
$E_2(1,0)$	$-A_3-A_4+A_5$	A_1-A_2	0	0	0	×	Unstable
$E_3(0,1)$	0	0	B_1-B_2	$-2B_3+B_4$	0	×	Unstable
$E_4(1,1)$	$A_1-2A_2-2A_3$ $-A_4+A_5$	A_1-A_2	B_1-B_2	$B_1-2B_2-B_3+B_4$	×	×	Saddle Point
$E_5(m^*,n^*)$				Saddle Point			

Scenario 2. *Government regulators' sentiment and energy consumers' sentiment.*

The sentiment parameters are $r_1 \neq 1, r_2 \neq 1$, the sentiment of government regulators and energy customers. The strategy portfolio stability analysis is displayed in Table 4, after adding the sentiment parameters to each replication dynamic equation.

In Table 4, there is a saddle point in the system at the government regulators' sentiment and energy consumers' sentiment. When condition $r_1(A_1-2A_2-2A_3-A_4+A_5)[r_2(B_1-B_2-B_3+B_4)-r_1B_2]-(r_2A_1-r_1A_2)r_1(B_1-B_2)>0$, $r_1[A_1-2A_2-2A_3-A_4+A_5]+[r_2(B_1-B_2-B_3+B_4)-r_1B_2]<0$ is fulfilled,$E_4(1,1)$ is a stable point in the system, when the government regulators' strategy choice is stable at green regulation and the energy consumers' strategy choice is stable at integrated energy use. The specific value of the sentiment parameters r_1, r_2 cannot be determined, and as a result, the stability of the local equilibrium point $E_5(m^*,n^*)$ is dependent on the specific value and sentiment intensity.

At the same time, the intensity of the sentiment of government regulators and energy consumers is unknown.

Table 4. Analysis of the stability of the strategy portfolio under various scenarios, including the attitudes of regulators and consumers of energy.

Equilibrium Point	$\frac{\partial F(m)}{\partial m}$	$\frac{\partial F(m)}{\partial n}$	$\frac{\partial F(n)}{\partial m}$	$\frac{\partial F(n)}{\partial n}$	$Det(J)$	$Tr(J)$	Consistency
$E_1(0,0)$	0	0	0	0	0	0	Unstable
$E_2(1,0)$	$r_1(-A_3 - A_4 + A_5)$	0	0	0	0	×	Unstable
$E_3(0,1)$	0	0	0	$r_2(-B_3 + B_4)$ $-r_2B_3$	0	×	Unstable
$E_4(1,1)$	$r_1(A_1 - 2A_2 - 2A_3$ $-A_4 + A_5)$	$r_2A_1 - r_1A_2$	$r_1(B_1 - B_2)$	$r_2(B_1 - B_2 - B_3$ $+B_4) - r_1B_2$	×	×	Saddle Point
$E_5(m^*, n^*)$	Specific values and emotional intensity are necessary for stability						

Scenario 3. *Government regulators rational, energy consumers emotional.*

The emotional parameters $r_1 = 1, r_2 \neq 1$ occur when government regulators are rational and energy customers are emotional. The strategy portfolio stability analysis is displayed in Table 5 after adding the sentiment parameters to each replication dynamic equation.

In Table 5, there is a saddle point in the system when the government regulators are rational, and the energy consumers are emotional. When condition $(A_1 - 2A_2 - 2A_3 - A_4 + A_5)[r_2(B_1 - B_2 - B_3 + B_4) - B_2] - (r_2A_1 - A_2)(B_1 - B_2) > 0, r_1[A_1 - 2A_2 - 2A_3 - A_4 + A_5] + [r_2(B_1 - B_2 - B_3 + B_4) - r_1B_2] < 0$ is fulfilled, $E_4(1,1)$ is a stable point in the system, when the government regulators' strategy choice is stable at green regulation and the energy consumers' strategy choice is stable at integrated energy use. However, because the emotional intensity of energy consumers is unknown, the particular value of the emotional parameter r_2 cannot be established, and the stability of the local equilibrium point $E_5(m^*, n^*)$ depends on the specific value and emotional intensity.

Table 5. Stability examination of strategy portfolios under scenarios with rational government regulators and optimistic energy consumers.

Equilibrium Point	$\frac{\partial F(m)}{\partial m}$	$\frac{\partial F(m)}{\partial n}$	$\frac{\partial F(n)}{\partial m}$	$\frac{\partial F(n)}{\partial n}$	$Det(J)$	$Tr(J)$	Consistency
$E_1(0,0)$	0	0	0	0	0	0	Unstable
$E_2(1,0)$	$-A_3 - A_4 + A_5$	$-A_2$	0	0	0	×	Unstable
$E_3(0,1)$	0	0	$B_1 - B_2$	$r_2(-B_3 + B_4)$ $-B_3$	0	×	Unstable
$E_4(1,1)$	$A_1 - 2A_2 - 2A_3$ $-A_4 + A_5$	$r_2A_1 - A_2$	$B_1 - B_2$	$r_2(B_1 - B_2 - B_3$ $+B_4) - B_2$	×	×	Saddle Point
$E_5(m^*, n^*)$	Specific values and emotional intensity are necessary for stability						

Scenario 4. *Government regulators' emotions and energy consumers' rationality.*

The rationality of energy consumers, government regulators, and emotional parameters are $r_1 \neq 1, r_2 = 1$. The strategy portfolio stability analysis is displayed in Table 6 after adding the sentiment parameters to each replication dynamic equation.

Table 6. Stability analysis of strategy portfolios under government regulators' sentiment and energy consumers' rationality scenarios.

Equilibrium Point	$\frac{\partial F(m)}{\partial m}$	$\frac{\partial F(m)}{\partial n}$	$\frac{\partial F(n)}{\partial m}$	$\frac{\partial F(n)}{\partial n}$	*Det*(*J*)	*Tr*(*J*)	Consistency
$E_1(0,0)$	0	0	0	0	0	0	Unstable
$E_2(1,0)$	$r_1(-A_3-A_4+A_5)$	A_1	0	0	0	×	Unstable
$E_3(0,1)$	0	0	0	$-B_3+B_4$ $-r_1B_3$	0	×	Unstable
$E_4(1,1)$	$r_1(A_1-2A_2-2A_3$ $-A_4+A_5)$	$A_1-r_1A_2$	$r_1(B_1-B_2)$	$B_1-B_2-B_3$ $+B_4-r_1B_2$	×	×	Saddle Point
$E_5(m^*,n^*)$	Specific values and emotional intensity are necessary for stability						

In Table 6, there is a saddle point in the system when the government regulators are emotional and the energy consumers are rational. When condition $r_1(A_1-2A_2-2A_3-A_4+A_5)[(B_1-B_2-B_3+B_4)-r_1B_2]-(A_1-r_1A_2)r_1(B_1-B_2)>0, r_1[A_1-2A_2-2A_3-A_4+A_5]+[(B_1-B_2-B_3+B_4)-r_1B_2]<0$ is fulfilled,$E_4(1,1)$ is a stable point in the system, when the government regulators' strategy choice is stable at green regulation and the energy consumers' strategy choice is stable at integrated energy use. In addition, the precise value of the sentiment parameter r_1 cannot be identified, and the stability of the local equilibrium point $E_5(m^*,n^*)$ depends on the specific value and sentiment intensity, which makes it impossible to estimate due to the unknown sentiment intensity of government control.

6. Analysis of Simulation

Based on the various emotional states of government regulators and energy customers, the preceding section examines the stability of the strategy portfolios of both groups, which includes four situations: (rational, rational), (rational, emotional), (emotional, rational), and (emotional, emotional). The game subject's emotional state may be further separated into optimistic and pessimistic states while it is in a certain emotional state, and the precise mechanisms by which optimistic and pessimistic emotional states affect the development of the game subject's behavior are likewise distinct. Therefore, on the basis of the previous four scenarios, this section provides further analysis of the evolutionary stability of the system in the nine specific scenarios (rational, rational), (optimistic, pessimistic), (pessimistic, pessimistic), (pessimistic, rational), (rational, optimistic), (rational, pessimistic), and (optimistic, optimistic).

MATLAB was utilized for the simulation study in order to examine the development of the energy structure transition more intuitively. The analysis of the trend of the system's development is not a change in particular values, and evolutionary game simulation is not a quantitative process, but rather a qualitative one. On the basis of the preceding part, it is appropriate to take values within a specified range, and the approach [39] employed is comparable and reasonable, especially mentioning the rules and framework for setting parameters [48,49]. Table 7 details the precise parameter values.

Table 7. Parameter settings.

Parameters	m	n	c_1	c_2	c_3	R	D	l_1	l_2	x_1	x_2	w_1	w_2	y_1	y_2
Initial Value	0.5	0.5	2	2	3	1.5	1.5	2	2	1.5	1.5	5	3	2	2
Parameters	a	b	q	L	r_1	r_2	θ	η	β						
Initial Value	0.7	0.3	0.5	5	1	1	0.5	0.5	0.2						

6.1. (Rational, Rational) State Analysis

Figure 1 depicts the equilibrium plan where energy users and government regulators are both rational actors, when $r_1 = 1, r_2 = 1$. The game has a mixed strategy Nash equilibrium. The likelihood of green regulation by governing bodies is low, ranging between 0.1 and 0.2, while the likelihood of integrated energy usage by energy customers is high, convergent to a steady one.

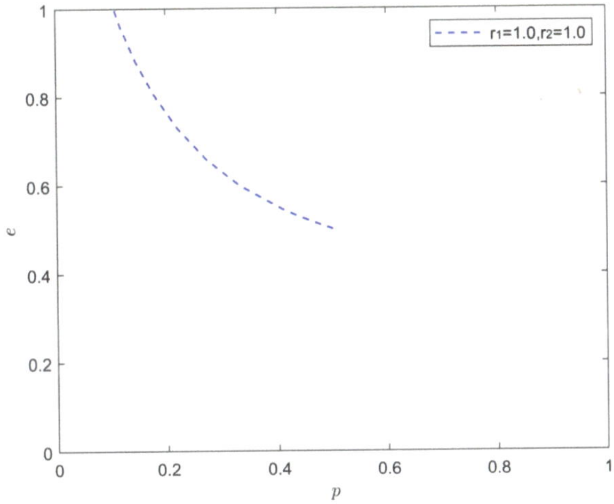

Figure 1. The development of the game's strategy in a (rational, rational) condition.

6.2. (Emotion, Emotion) State Analysis

Figure 2 depicts the optimal approach when both energy users and regulatory agencies are optimistic, i.e., when $r_1 < 1, r_2 < 1$. The likelihood that government regulators would choose green regulation somewhat declines as both government regulators' and energy users' excitement grow. The likelihood that users of energy will choose integrated energy consumption converges and stabilizes at one. The mixed-strategy Nash equilibrium point is moved to the coordinate region's top left corner. According to the aforementioned research, government regulators are more susceptible to optimism than electricity generated, and their regulatory activism lowers when they are unduly enthusiastic about the structural revolution of energy.

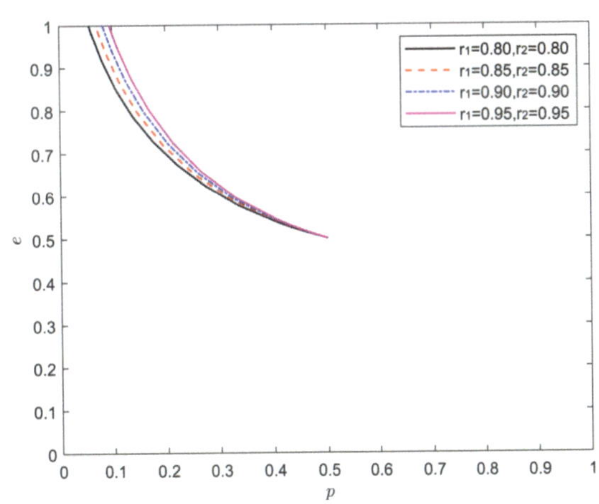

Figure 2. The development of the game's strategy in a (optimistic, optimistic) condition.

Figure 3 depicts the equilibrium plan when both energy users and regulatory agencies are pessimistic, i.e., when $r_1 > 1, r_2 > 1$. The likelihood that government regulators will choose green regulation rises as pessimism among government regulators and energy consumers intensifies. The mixed strategy Nash equilibrium point shifts to the upper right of the coordinate region as the probability of the energy consumers choosing integrated energy use converges to 1, but the probability of the energy consumers choosing integrated energy use declines when the pessimism of government regulators and energy consumers is too great. According to the aforementioned findings, a healthy dose of pessimism can maintain energy consumers' high levels of willingness to integrate energy use and boost government regulators' willingness to enact green regulations, but an excessive amount of pessimism deters both government regulators and energy consumers from making strategic decisions.

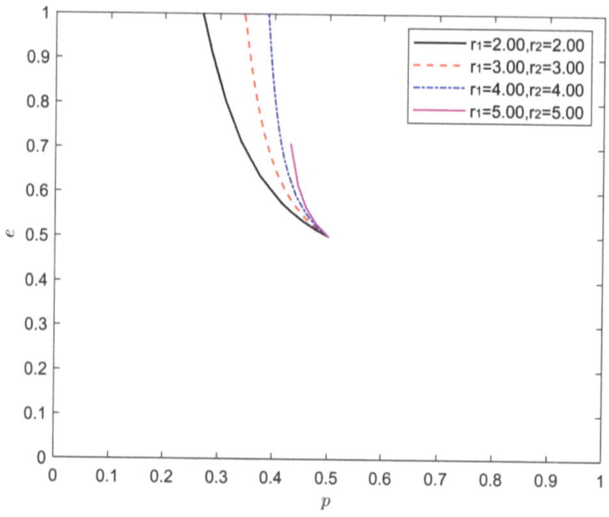

Figure 3. The development of the game's strategy in a (pessimistic, pessimistic) condition.

Figure 4 depicts the optimum approach when energy users are apprehensive and government regulators are optimistic, i.e., when $r_1 < 1, r_2 > 1$. As shown in Figure 4, when watchdogs are optimistic for a period of time, the likelihood that they will choose green regulation continues to decline as the pessimism of electricity generated deepens and the convergence speed of the system's evolution to the stability point (1,0) significantly increases. Further analysis in Figure 4a–c reveal that when the optimism of government regulators increases, the Nash equilibrium point moves to the top left corner of the coordinate area. According to the data above, combinations of (optimistic, pessimistic) diverse sentiments are not the best options for systematic Pareto optimization.

Figure 5 depicts the equilibrium approach when energy users are optimistic and government agencies are pessimistic, i.e., when $r_1 > 1, r_2 < 1$. Figure 5 shows that when there is certainty regarding the optimism of energy consumers, the probability of green regulation by government regulators keeps rising as the pessimism of the government agencies keeps deepening and the mixed strategic plan Nash equilibrium point wants to keep leveling off in the upper right corner of the coordinate region. As demonstrated in Figure 5a–c, the trajectory of strategy development for each game subject is essentially constant, and the growing optimism of the energy consumers has less of an influence on the system's evolution. According to the aforementioned results, the strategy chosen by each game subject is greatly influenced by their pessimism and optimism, with pessimism having a greater impact.

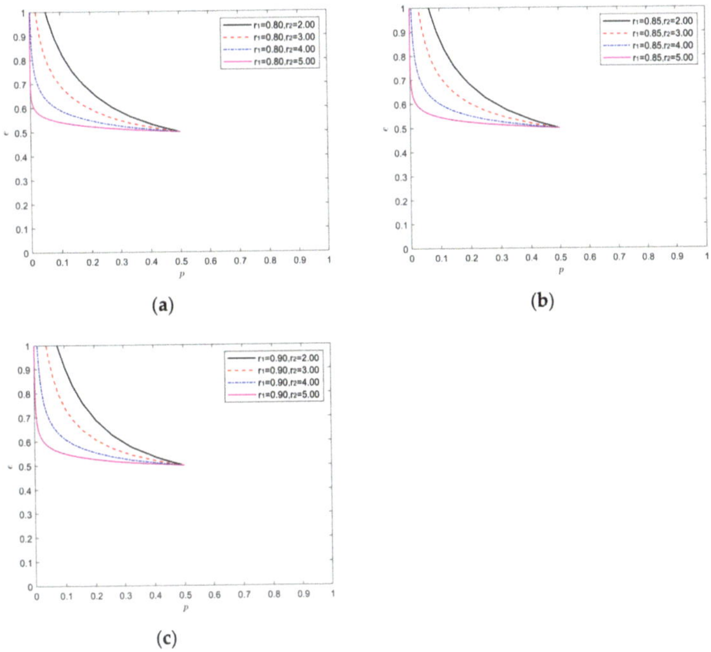

Figure 4. The development of the game's strategy in a (optimistic, pessimistic) condition. (**a**–**c**) represent the results of different parameter values.

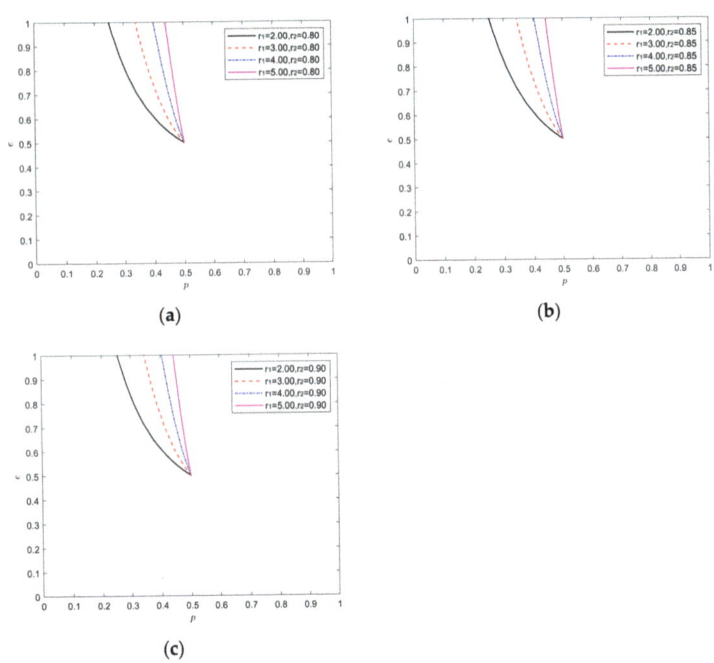

Figure 5. The development of the game's strategy in a (pessimistic, optimistic) condition. (**a**–**c**) represent the results of different parameter values.

6.3. (Emotional, Rational) State Analysis

Figure 6 depicts the equilibrium approach with an optimistic regulatory environment and rational energy users, i.e., when $r_1 < 1, r_2 = 1$. Figure 6 illustrates how the chance of government regulators adopting green regulation continues to decline as their optimism grows, and how the mixed strategy Nash market equilibrium begins to move horizontally toward the top left corner of the coordinate area.

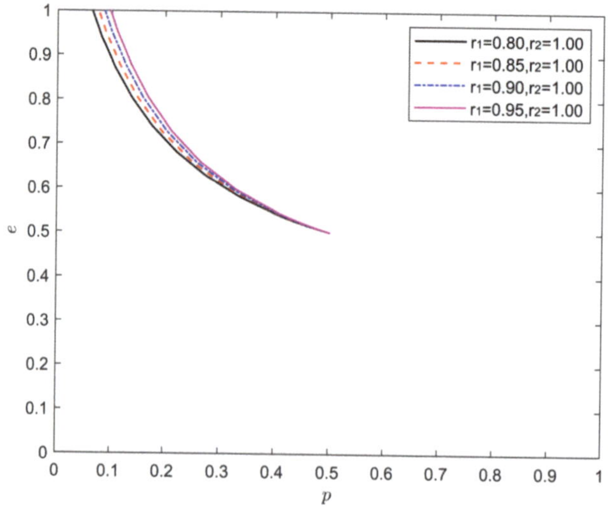

Figure 6. The development of the game's strategy in a (optimistic, rational) condition.

Figure 7 depicts the equilibrium approach where energy customers are rational and government regulators are pessimistic, i.e., when $r_1 > 1, r_2 = 1$. From Figure 7, as government regulators' pessimism deepens, the likelihood that they will choose green regulation rises, and the mixed strategy Nash equilibrium point continues to move horizontally toward the coordinate region's top right corner. According to the facts above, optimism and pessimism have opposing impacts on the technique that government regulators use. Pessimism has a positive impact and may increase the likelihood of green regulation by government regulators. Optimism, on the other hand, has a negative impact and can decrease the likelihood of green regulation by government regulators.

6.4. (Rational, Emotional) State Analysis

Figure 8 depicts the equilibrium plan where energy customers are enthusiastic and government regulators are sensible, when $r_1 = 1, r_2 < 1$. Figure 8 shows that the system's development is less affected by government regulators' growing confidence, and that the general trend of strategy evolution for each game subject is essentially unaltered.

Figure 9 depicts the equilibrium approach where energy consumers are pessimistic and government regulators are sensible, when $r_1 = 1, r_2 > 1$. As energy consumers' pessimism continues to grow, as seen in Figure 9, the pace at which the system's evolution converges to the stability point (0,1) grows noticeably. According to the aforementioned data, a slight increase in the energy of consumers' pessimism may retain the likelihood that they will choose integrated energy usage at a high level.

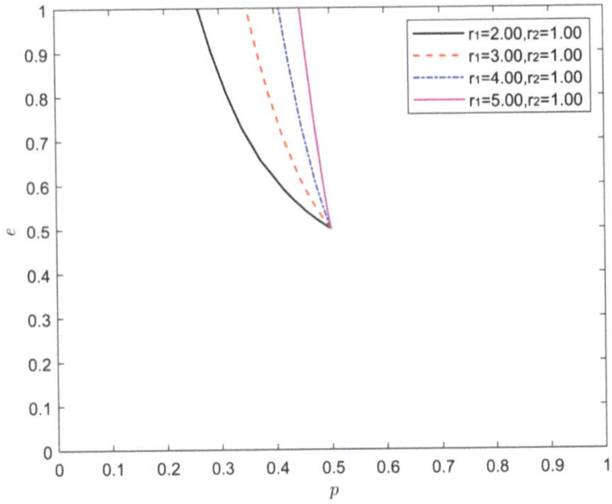

Figure 7. The development of the game's strategy in a (pessimistic, rational) condition.

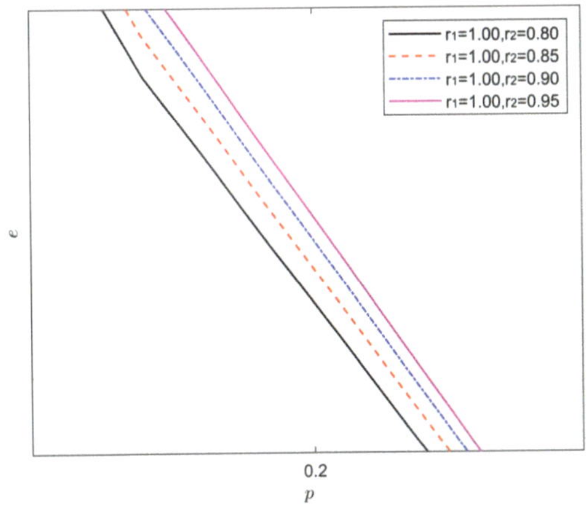

Figure 8. The development of the game's strategy in a (rational, optimistic) condition.

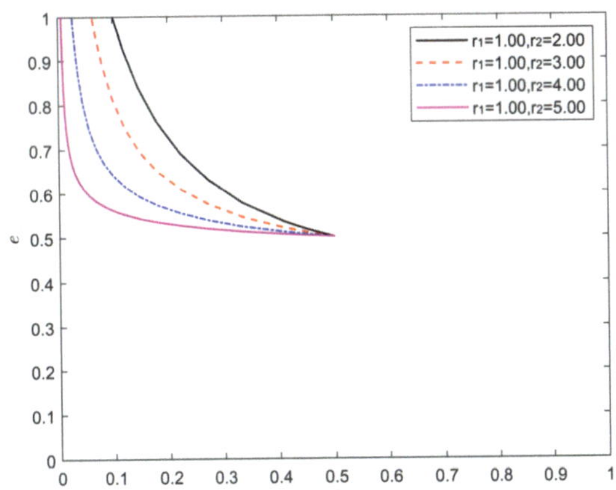

Figure 9. The development of the game's strategy in a (rational, pessimistic) condition.

7. Summary

This essay seeks to examine how various emotions affect the choices and actions of game subjects, as well as how the energy structure transitions. In order to achieve this, given that clean energy cannot quickly replace coal energy, our model quantifies emotions as an irrational element using the RDEU theory and builds an emotion model by fusing the RDEU theory with evolutionary games. The examination of how various emotions affect decision-makers' behavior and the modification of energy structures during strategy decision-making show that incorporating emotions into evolutionary games is feasible. Additionally, in the reality of energy restructuring, government regulators and energy users make irrational decisions. This study serves as a guide for decision-makers' preferred choices when it comes to the clean and effective use of coal energy and the quickening of the development and use of clean energy. China, the greatest developing nation, is a resource-intensive nation with a national situation of "rich in coal, poor in oil, and poor in gas," and it should not aim to "reduce coal," let alone "remove coal." As a result, it is crucial for emerging nations and nations with large coal reserves to understand how China's energy system has changed.

7.1. Conclusions

(1) Emotions have an irrational impact on decision-makers' psychological preferences, which in turn affects their choice of behavioral approach and materially alters the equilibrium state of the model. Government regulators and energy consumers are influenced by emotional and risky attitudes that cause them to make utterly irrational behavioral decisions because they feel dependent on or are unwilling to shift their energy sources too much and too early in favor of coal. The eventual evolution of the system can be impacted by changes in the psychological preferences and feelings of government regulators and energy customers using the same objective decision-making framework.

(2) In terms of the individual consequences of different emotional attitudes, pessimism may raise the decision-maker's psychological preference for the strategy selection process, while optimism can lower this desire. Decision-makers can avoid making judgments with significant mistakes that differ from the system's outcomes by controlling their rational emotions. Pessimism affects decision-makers, increases the likelihood that they will take action, and makes them more likely to choose effective strategies by avoiding acts of omission, but too much pessimism will have a negative impact on the motivation behind decision-making. Appropriate optimism will enable decision-makers to actively adjust to the energy structure transition and intentionally increase the likelihood of attaining the best possible allocation of societal resources. However, overly optimistic decision-makers are prone to arrogance, which can result in motivation loss and inactive behavior.

(3) The interaction of different emotional attitudes results in a more complicated and dynamic synergistic relationship than a simple addition. The effect of pessimism is stronger than that of optimism in this process of synergistic relationships. One of the two sides of the energy structure transformation game displays a rational state, while the other side, regardless of melancholy or optimism, exhibits a state that simply impacts the system's equilibrium state and evolutionary pace rather than the strategy's direction of development. The deepening pessimism of one side will change the evolution rate and equilibrium state of the system as a whole in a relatively positive direction if the optimism of the other side remains unchanged; the deepening optimism of the other side will change the evolution rate and equilibrium state of the system as a whole in a relatively negative direction if the optimism of the other side remains unchanged.

7.2. Revelation

Based on the findings in the previous section, in the context of carbon neutrality, both pessimism and optimism can affect how government regulators and energy consumers make decisions in the face of clean energy and coal energy. However, as compared to optimism, suitable pessimism has a greater impact on government regulators and energy customers. The pessimism of government regulators and energy customers may, thus, be adequately pushed in the context of the reality of the energy structure transition so that both parties can take more proactive steps. The specific recommendations are as follows:

(1) Guiding the appropriate decision-makers to generate appropriate emotional attitudes for decision-makers and regulating the degree of performance of different emotional attitudes to prevent excessive emotions from influencing decision-makers and to strengthen the role of emotions as an irrational factor to support energy structure transformation. Decision-makers are in rational emotions while making judgments about energy structure transformation, and psychological preferences and emotions can influence those decisions in either a positive or a negative way. This is a fact that does exist, and we cannot ignore it. Therefore, we need to take the irrational factor of emotion into full consideration in the overall planning when formulating energy structure transformation strategies and transformation paths. Given that government regulators and energy consumers are adequate at uplifting emotional counseling, preventing the influence of excessive emotions through social monitoring, establishing suitable energy cleanliness standards, etc., they may play a good role.

(2) Increase the pressure, responsibility, and anxiety of both parties appropriately in the energy structure transition in the context of carbon neutrality, so that the pessimism of government regulators and energy consumers gives them the option to take a more proactive approach to opportunities and challenges from all directions. The system for regulating CO_2 emissions can be properly opened up to the public, the relevant data can be made public and clear, the public can be encouraged to get involved, and the public can be asked for their opinions. All stakeholders will naturally cooperate and actively participate in innovation as a result of being made to feel more stressed, anxious, and responsible. This will result in a more reasonable and effective change to the energy system.

(3) Government regulators and energy consumers can release appropriate favorable information to the outside world. As a long-term goal of China's high-quality development, carbon neutrality is not a restriction on energy development, but a strategic goal to achieve carbon neutrality through the transformation and development of the energy structure. Government regulators should, thus, implement relevant regulations, together with subsidies and incentives, during the process of energy structure transition in order to encourage support for the change. Additionally, energy consumers should actively promote low-carbon innovation, actively demonstrate support for pertinent transition policies in regards to implementation and effectiveness, and consistently fortify the linkages and collaboration between various organizations.

7.3. Shortcomings and Prospects

This study creates an evolutionary game model with government regulators and energy customers and creatively incorporates low carbon sentiment into the game of changing the energy structure. This article replicates the various attitudes of government regulators and energy consumers on the reality of energy consumption after determining the Nash equilibrium point and the evolutionary stabilization approach. However, potential game topics, such as whistleblowers, were not considered, and a three-way game could have been played between government regulators, energy customers, and whistleblowers. In the future, a three-way game analysis could be conducted, considering the low carbon sentiment of policymakers. In the upcoming year, we will also consider the possibility that the atypical factor of emotion may have a different salience in the legislative environment than in the social or economic environment. Therefore, it could be interesting to take these various saliences into account in a study session.

Supplementary Materials: The following supporting information can be downloaded at: https://www.mdpi.com/article/10.3390/pr10081650/s1, Formulas S4–S10.

Author Contributions: All authors contributed to the study's conception and design. Material preparation, review, and editing were performed by Z.G., Z.Z., B.L., L.S., C.S. and S.W. The first draft of the manuscript was written by X.W.; all authors commented on previous versions of the manuscript. All authors have read and agreed to the published version of the manuscript.

Funding: This work was supported by the National Social Science Fund of China (grant number, 20BJY109) and "Chenguang Program" from the Shanghai Education Development Foundation and Shanghai Municipal Education Commission (grant number, 19CG24).

Institutional Review Board Statement: Not applicable.

Data Availability Statement: Data can be obtained from the corresponding author upon reasonable request.

Conflicts of Interest: The authors declare no conflict of interest.

References

1. Guan, D.; Klasen, S.; Hubacek, K.; Feng, K.; Liu, Z.; He, K.; Geng, Y.; Zhang, Q. Determinants of stagnating carbon intensity in China. *Nat. Clim. Chang.* **2014**, *4*, 1017–1023. [CrossRef]
2. Wang, D.; Huangfu, Y.; Dong, Z.; Dong, Y. Research Hotspots and Evolution Trends of Carbon Neutrality—Visual Analysis of Bibliometrics Based on CiteSpace. *Sustainability* **2022**, *14*, 1078. [CrossRef]
3. Lange, M.; O'Hagan, A.M.; Devoy, R.R.; Le Tissier, M.; Cummins, V. Governance barriers to sustainable energy transitions—Assessing Ireland's capacity towards marine energy futures. *Energy Policy* **2018**, *113*, 623–632. [CrossRef]
4. Geels, F.W.; Berkhout, F.; van Vuuren, D.P. Bridging analytical approaches for low-carbon transitions. *Nat. Clim. Chang.* **2016**, *6*, 576–583. [CrossRef]
5. Nilsson, M.; Nykvist, B. Governing the electric vehicle transition—Near term interventions to support a green energy economy. *Appl. Energy* **2016**, *179*, 1360–1371. [CrossRef]
6. Kuriqi, A.; Pinheiro, A.N.; Sordo-Ward, A.; Garrote, L. Flow regime aspects in determining environmental flows and maximising energy production at run-of-river hydropower plants. *Appl. Energy* **2019**, *256*, 113980. [CrossRef]
7. Shi, R.-J.; Fan, X.-C.; He, Y. Comprehensive evaluation index system for wind power utilization levels in wind farms in China. *Renew. Sustain. Energy Rev.* **2017**, *69*, 461–471. [CrossRef]
8. Wang, G.; Wang, T.; Jiang, T.; Chen, Z. Influence estimate of liquid lead–bismuth eutectic temperatures on operation and mechanical behaviors of sensible heat storage tank. *Energy Rep.* **2021**, *7*, 4388–4396. [CrossRef]
9. Guo, X.; Zheng, S.; Zhang, G.; Xiao, X.; Li, X.; Xu, Y.; Xue, H.; Pang, H. Nanostructured graphene-based materials for flexible energy storage. *Energy Storage Mater.* **2017**, *9*, 150–169. [CrossRef]
10. Priya, A.D.; Deva, S.; Shalini, P.; Setty, Y.P. Antimony-tin based intermetallics supported on reduced graphene oxide as anode and MnO2@rGO as cathode electrode for the study of microbial fuel cell performance. *Renew. Energy* **2019**, *150*, 156–166. [CrossRef]
11. Gomez, A.; Azzaro-Pantel, C.; Domenech, S.; Pibouleau, L.; Latgé, C.; Haubensack, D.; Dumaz, P. Exergy analysis for Generation IV nuclear plant optimization. *Int. J. Energy Res.* **2010**, *34*, 609–625. [CrossRef]
12. Tsibulskiy, V.F.; Andrianova, E.A.; Davidenko, V.D.; Rodionova, E.V.; Tsibulskiy, S.V. Advantages of Production of New Fissionable Nuclides for the Nuclear Power Industry in Hybrid Fusion-Fission Reactors. *Phys. At. Nucl.* **2017**, *80*, 1220–1226. [CrossRef]
13. Zhao, F.; Liu, X.; Zhang, H.; Liu, Z. Automobile Industry under China's Carbon Peaking and Carbon Neutrality Goals: Challenges, Opportunities, and Coping Strategies. *J. Adv. Transp.* **2022**, *2022*, 1–13. [CrossRef]
14. Huang, S.-Z. The effect of natural resources and economic factors on energy transition: New evidence from China. *Resour. Policy* **2022**, *76*, 102620. [CrossRef]
15. Su, C.; Urban, F. Circular economy for clean energy transitions: A new opportunity under the COVID-19 pandemic. *Appl. Energy* **2021**, *289*, 116666. [CrossRef]
16. BP. *BP Statistical Review of World Energy 2021*; BP: London, UK, 2021.
17. Boucher, O.; Reddy, M. Climate trade-off between black carbon and carbon dioxide emissions. *Energy Policy* **2008**, *36*, 193–200. [CrossRef]
18. Melikoglu, M. Clean coal technologies: A global to local review for Turkey. *Energy Strat. Rev.* **2018**, *22*, 313–319. [CrossRef]
19. Wang, X.; Zhang, C.; Deng, J.; Su, C.; Gao, Z. Analysis of Factors Influencing Miners' Unsafe Behaviors in Intelligent Mines using a Novel Hybrid MCDM Model. *Int. J. Environ. Res. Public Health* **2022**, *19*, 7368. [CrossRef]

20. Xu, H.; Lai, X.; Shan, P.; Yang, Y.; Zhang, S.; Yan, B.; Zhang, Y.; Zhang, N. Energy dissimilation characteristics and shock mechanism of coal-rock mass induced in steeply-inclined mining: Comparison based on physical simulation and numerical calculation. *Acta Geotech.* **2022**, *92*, 22. [CrossRef]
21. Wang, G.; Xu, Y.; Ren, H. Intelligent and ecological coal mining as well as clean utilization technology in China: Review and prospects. *Int. J. Min. Sci. Technol.* **2019**, *29*, 161–169. [CrossRef]
22. Li, Y.; Zhang, B.; Wang, B.; Wang, Z. Evolutionary trend of the coal industry chain in China: Evidence from the analysis of I-O and APL model. *Resour. Conserv. Recycl.* **2019**, *145*, 399–410. [CrossRef]
23. Wang, G.; Chao, Y.; Chen, Z. Promoting developments of hydrogen powered vehicle and solar PV hydrogen production in China: A study based on evolutionary game theory method. *Energy* **2021**, *237*, 121649. [CrossRef]
24. Wang, G.; Chao, Y.; Lin, J.; Chen, Z. Evolutionary game theoretic study on the coordinated development of solar power and coal-fired thermal power under the background of carbon neutral. *Energy Rep.* **2021**, *7*, 7716–7727. [CrossRef]
25. Hou, J.; Zhang, R.; Liu, P.; Zhou, L. A review and comparative analysis on energy transition in major industrialized countries. *Int. J. Energy Res.* **2021**, *45*, 1246–1268. [CrossRef]
26. Gao, A.M.-Z.; Fan, C.-T.; Liao, C.-N. Application of German energy transition in Taiwan: A critical review of unique electricity liberalisation as a core strategy to achieve renewable energy growth. *Energy Policy* **2018**, *120*, 644–654. [CrossRef]
27. Chen, W.; Hu, Z.-H. Analysis of Multi-Stakeholders' Behavioral Strategies Considering Public Participation under Carbon Taxes and Subsidies: An Evolutionary Game Approach. *Sustainability* **2020**, *12*, 1023. [CrossRef]
28. Chang, L.; Song, Y.; Yu, T. Optimization of Financial Penalties for Environmental Pollution by Chinese Traditional Energy Enterprises. *Front. Environ. Sci.* **2020**, *11*, 11. [CrossRef]
29. Yang, Y.; Yang, W.; Chen, H.; Li, Y. China's energy whistleblowing and energy supervision policy: An evolutionary game perspective. *Energy* **2020**, *213*, 118774. [CrossRef]
30. Zhao, X.; Ren, L.; Zhang, Y.; Wan, G. Evolutionary game analysis on the behavior strategies of power producers in renewable portfolio standard. *Energy* **2018**, *162*, 505–516. [CrossRef]
31. Qiao, W.; Yin, X. Understanding the impact on energy transition of consumer behavior and enterprise decisions through evolutionary game analysis. *Sustain. Prod. Consum.* **2021**, *28*, 231–240. [CrossRef]
32. Pingkuoa, L.; Huan, P.; Zhiwei, W. Orderly-synergistic development of power generation industry: A China's case study based on evolutionary game model. *Energy* **2020**, *211*, 118632. [CrossRef]
33. Shang, B.; Huang, T.; Du, X. Impact of government regulation of RPS on China's power market under carbon abatement constraints. *J. Intell. Fuzzy Syst.* **2020**, *39*, 2947–2975. [CrossRef]
34. Stewart, N.; Chater, N.; Stott, H.P.; Reimers, S. Prospect relativity: How choice options influence decision under risk. *J. Exp. Psychol. Gen.* **2003**, *132*, 23–46. [CrossRef] [PubMed]
35. Zhang, Z.; Wang, X.; Su, C.; Sun, L. Evolutionary Game Analysis of Shared Manufacturing Quality Synergy under Dynamic Reward and Punishment Mechanism. *Appl. Sci.* **2022**, *12*, 6792. [CrossRef]
36. Szasz, P.L.; Hofmann, S.G.; Heilman, R.M.; Curtiss, J. Effect of regulating anger and sadness on decision-making. *Cogn. Behav. Ther.* **2016**, *45*, 479–495. [CrossRef]
37. Bandyopadhyay, D.; Pammi, V.C.; Srinivasan, N. Role of affect in decision making. *Prog. Brain Res.* **2013**, *202*, 37–53. [CrossRef]
38. Ghossoub, M.; He, X.D. Comparative risk aversion in RDEU with applications to optimal underwriting of securities issuance. *Insur. Math. Econ.* **2021**, *101*, 6–22. [CrossRef]
39. Zhang, Z.; Wang, X.; Su, C.; Sun, L. Evolutionary Game Analysis of Shared Manufacturing Quality Innovation Synergetic Behavior Considering a Subject's Heterogeneous Emotions. *Processes* **2022**, *10*, 1233. [CrossRef]
40. Ni, S.; Zou, S.; Chen, J. Evolutionary Game Model of Internal Threats to Nuclear Security in Spent Fuel Reprocessing Plants Based on RDEU Theory. *Sustainability* **2022**, *14*, 2163. [CrossRef]
41. Quiggin, J. A theory of anticipated utility. *J. Econ. Behav. Organ.* **1982**, *3*, 323–343. [CrossRef]
42. Ferro, G.M.; Kovalenko, T.; Sornette, D. Quantum decision theory augments rank-dependent expected utility and Cumulative Prospect Theory. *J. Econ. Psychol.* **2021**, *86*, 102417. [CrossRef]
43. Starmer, C. Developments in Non-Expected Utility Theory: The Hunt for a Descriptive Theory of Choice under Risk. *J. Econ. Lit.* **2000**, *38*, 332–382. [CrossRef]
44. Quiggin, J. Comparative statics for rank-dependent expected utility theory. *J. Risk Uncertain.* **1991**, *4*, 339–350. [CrossRef]
45. Li, W.; Huang, S.; Qi, Y.; An, H. RDEU hawk-dove game analysis of the China-Australia iron ore trade conflict. *Resour. Policy* **2022**, *77*, 102643. [CrossRef]
46. Liu, J.; Lyu, Y.; Zhao, H.; Chen, J. Game analysis of nuclear wastewater discharge under different attitudes: Seeking a potential equilibrium solution. *Sci. Total Environ.* **2021**, *801*, 149762. [CrossRef]
47. Hong, K.; Zou, Y.; Zhang, Y.; Duan, K. The Weapon of the Weak: An Analysis of RDEU Game in the Conflict of Farmland Expropriation under the Influence of Emotion. *Sustainability* **2020**, *12*, 3367. [CrossRef]
48. Hartl, B.; Sabitzer, T.; Hofmann, E.; Penz, E. "Sustainability is a nice bonus" the role of sustainability in carsharing from a consumer perspective. *J. Clean. Prod.* **2018**, *202*, 88–100. [CrossRef]
49. Zhang, Y.; Wang, J.; Chen, J.; Liu, W. Does environmental regulation policy help improve business performance of manufacturing enterprises? evidence from China. *Environ. Dev. Sustain.* **2022**, *185*, 1–30. [CrossRef]

processes

MDPI

Article

Thermal, Lighting and IAQ Control System for Energy Saving and Comfort Management

Silvia Maria Zanoli * and Crescenzo Pepe

Dipartimento di Ingegneria dell'Informazione, Università Politecnica delle Marche, Via Brecce Bianche 12, 60131 Ancona, Italy
* Correspondence: s.zanoli@univpm.it

Abstract: The present work proposes a simulation and control framework for home and building automation, focusing on heating, ventilating, and air conditioning processes. Control systems based on different advanced control architectures and different control policies are simulated and compared, highlighting control performances, and energy-saving results in terms of CO_2 emissions reduction. Heat, lighting, and natural ventilation phenomena were modelized through first-principles and empirical equations, obtaining a reliable and flexible simulation framework. Energy-consuming and green energy-supplying renewable sources were integrated into the framework, e.g., heat pumps, artificial lights, fresh air flow, and natural illuminance. Different control schemes are proposed, based on proportional–integral–derivative advanced control architectures and discrete event dynamic systems-based supervisors; different control specifications are included, resulting in a multi-mode control system. The specifications refer to energy savings and comfort management, while minimizing overall costs. Comfort specifications include thermal comfort, lighting comfort, and a good level of indoor air quality. Simulations on different scenarios considering various control schemes and specifications show the reliability and soundness of the simulation and control framework. The simulated control and energy performances show the potential of the proposed approach, which can provide energy-saving results greater or equal to 6 [%] (in each season) and 19 [%] (in one year) with respect to more standard approaches.

Keywords: home and building automation; control system; renewable resources; energy saving; comfort management; indoor air quality; proportional–integral–derivative; discrete event dynamic systems

Citation: Zanoli, S.M.; Pepe, C. Thermal, Lighting and IAQ Control System for Energy Saving and Comfort Management. *Processes* **2023**, *11*, 222. https://doi.org/10.3390/pr11010222

Academic Editors: Ferdinando Salata and Virgilio Ciancio

Received: 4 December 2022
Revised: 31 December 2022
Accepted: 3 January 2023
Published: 10 January 2023

1. Introduction

Energy efficiency and energy performance maximization in buildings represent a challenge, as stated by the European Parliament in the last decade. The main objective is to reduce greenhouse gas emissions further by at least 40% by 2030 as compared with 1990, to increase the proportion of renewable energy consumed, to make energy savings in accordance with European Union level ambitions, and to improve Europe's energy security, competitiveness, and sustainability [1,2]. An efficient and resilient building and construction sector is a fundamental requirement for zero-emission policies and clean energy transition. For example, in 2017, buildings' construction and operations accounted for 36% of the global final energy use and nearly 40% of the energy-related carbon dioxide (CO_2) emissions [3]. In this context, programs have been introduced in the last few years, e.g., Agenda 2030 [4] at the global level and PNRR (Piano Nazionale di Ripresa e Resilienza) at the Italian level [5].

An efficient approach to reaching optimal solutions is to consider the hardware and software aspects of the buildings: hardware may refer to construction technologies and the geometry of the buildings while, software programs can exploit monitoring, modelization, control, and automation strategies [6]. Monitoring, control [7] and automation strategies [8,9] can be cross-fertilized from other areas, e.g., petrochemical and/or energy-intensive industries [10,11].

The processes which require significant energy consumption in buildings are heating, ventilating and air conditioning (HVAC), heating, ventilating, air conditioning and refrigeration (HVACR), and domestic hot water (DHW). The home and building automation (HBA) research area can cover these topics [12]. With regard to HVAC, four main factors have to be taken into account: thermal comfort, lighting comfort, a good level of indoor air quality (IAQ), and energy saving. In this context, comfort can be reached by maximizing the natural illuminance, maintaining a desirable indoor temperature, and ensuring IAQ. Conflicting objectives between comfort and energy saving must be handled [13]. In the last few years, many researchers, engineers, and practitioners have proposed monitoring, simulation, high-level optimization, and control solutions for HBA.

Monitoring and simulation solutions are proposed in [14–19]. In [14], an internet of things-based occupancy monitoring system for energy-efficient smart buildings is proposed. The occupancy monitoring is obtained through a minimally intrusive way, and data fusion techniques are developed to improve the occupancy monitoring accuracy through a multitude of sources. The use of the EnergyPlus tool is discussed in [15], analyzing how simulation can support HBA and how the deployment process of simulation-assisted building control systems can be structured. In [16], the authors accurately describe a building management system (BMS), also known as a building automation and control system (BACS), highlighting the benefits provided by the design, development and implementation of automated step response testing tools. The COVID-19 consequences on buildings' management were analyzed in [17,18], assessing the current problems and difficulties that smart buildings face and the possible future directions of this technology. In [19], a review of machine learning algorithms able to power smart homes is provided.

High-level optimization solutions are proposed in [20–23]. In [20], the bat algorithm is applied for energy optimization in residential buildings. Three environmental parameters, namely temperature, illuminance, and air quality, are the bat algorithm's inputs, and the optimized values of these parameters are the outputs. The error difference between the environmental parameters and the optimized parameters is the inputs of the fuzzy controllers, which return energy as an output, which in turn changes the status of the concerned actuators. In [21], to minimize the daily energy cost of HVAC and lights and maintain occupant comfort, a near-optimal strategy is proposed through a daily cost optimization problem solved by combining Lagrangian relaxation, stochastic dynamic programming, and rollout technique within a surrogate optimization framework. In [22], a real-time demand response strategy based on deep reinforcement learning is proposed as an optimal energy management strategy under the uncertainty of the residents' behavior, outdoor temperature, and renewable generation. In [23], an energy management controller is developed for the demand-side management in smart homes. Fuzzy logic and heuristic optimization techniques for cost, energy consumption, and peak-to-average ratio reduction are used.

The high-level optimization algorithms and monitoring/simulation solutions reported previously are not in charge of the management of the real-time operation of the plants. Real-time operation of the plants can be handled through process controllers. Different HBA control solutions are present in the literature, e.g., based on standard single-input single-output (SISO) proportional–integral–derivative (PID) techniques, fuzzy logic, pole placement, model predictive control (MPC) strategy, and deep learning. Standard PID techniques are proposed in [24–29]. Classical PID control algorithms are exploited in [24] for the control of a HVAC system having two zones with different properties. The PID parameters are tuned to minimize the tracking error, but the steady-state error is not totally eliminated. In [25], a tuning strategy for discharge air temperature control of air handlers in building comfort applications is designed. The proposed method adjusts fewer gain values than other tuning algorithms, and it is tested through a standard SISO loop that manipulates a valve command for the air handler in order to track the defined setpoint for the discharge air temperature. A technique for the fast tuning of the parameters of a standard PID controller of a second-order HVAC system is proposed in [26]. A Big Bang–Big Crunch algorithm is implemented, along with the PID controller, in an FPGA device in order to achieve high tuning speed. In [27], the room temperature and humidity control systems with the conventional PID control using a fixed reset or the modified PID control using adjustable resets, which compensate for the thermal loads upset, are examined through simulations. The optimization of climate conditions in office buildings is proposed in [28] by the use of modeling and simulation tools to define the buildings' energy demand, and the design and implementation of standard PID controls for the different control areas of the HVAC system are also proposed. In [29], the modeling, numerical simulation, and intelligent control of an expert HVAC system having two different zones with variable flow rates were performed by considering the ambient temperature. Standard decoupled SISO PID architectures are used for the control of the temperature of the two zones, and the fuzzy logic is exploited for tuning purposes.

In [30–32], fuzzy logic is employed for control purposes in HBA. The use of smart handheld devices, using MIT App Inventor and fuzzy control, to perform the real-time monitoring and smart control of the designed intelligent windowsill system in a smart home is proposed in [30]. A fuzzy microcontroller exploits information provided by a weather station, which measures indoor illuminance, temperature-humidity, CO_2 concentration and outdoor rain, and wind direction. The control degrees of freedom are fully or partly open to the electric curtain and electric window. Other examples of fuzzy logic implementation are reported in [31,32], pursuing lighting comfort, visual comfort, thermal comfort, and energy-saving objectives. In [33], advanced controllers based on pole placement enhanced with additional variables, namely solar radiation and external temperature, are proposed as climatic control systems. The potential of model predictive control (MPC) for enhancing building and HVAC system energy efficiency is reported in [34], highlighting problem formulation, applications, and opportunities. A recent state-of-the-art review on MPC in the HVAC field is reported in [35], focusing on energy management, energy savings, simulation software, optimization, modelization, and disturbances. In [36], the combination of predictive control and deep thermo-modernization is tackled and studied in a real-world case study: the impact of weather-forecast-based regulation on energy savings for heating in multi-family buildings is assessed. Furthermore, deep learning models, such as convolutional neural networks, are reported.

As a result of the literature review, SISO PID architectures in HBA usually provide decoupled solutions for thermal, lighting, and IAQ control. On the other hand, the main drawbacks of the other analyzed control solutions are the need for in-depth process knowledge for fuzzy logic and the need for accurate models for MPC and pole-placement solutions. The present paper proposes a simulation framework and advanced control approaches for HBA based on PID and discrete event dynamic systems (DEDS), focusing on HVAC processes. Combining different PID/DEDS advanced control architectures, a multi-mode control system is derived. The energy savings and comfort specifications are taken into account and tested through tailored simulations based on different scenarios, assessing control, and energy-saving performances. The paper aims to provide adaptable knobs for engineers, researchers, and practitioners. To the best of the authors' knowledge, in the literature on control systems in HVAC processes, the following aspects have not been explored in depth:

- The design of flexible simulation and control frameworks that allow the modelization and simulation of different environments and the test and comparison of different controllers is not present in the literature. In the design of control systems for energy saving and comfort management in HBA, flexible frameworks can represent a significant tool for designing and prototyping optimal control solutions.
- An assessment of advanced PID control architectures for energy savings and comfort management in HBA is not present in the literature. Exploiting non-standard PID control architectures, coupled control of thermal, lighting, and IAQ subprocesses can

be obtained. In this way, unexpected control margins can be detected and control performance can be improved over standard PID solutions.

- The combination of advanced PID control architectures with DEDS for energy savings and comfort management in HBA is not present in the literature. This combination can result in a significant improvement in energy savings and comfort management performances with respect to more standard control architectures.

Exploiting a multivariable approach with respect to HBA standard PID solutions, further energy savings and comfort margins can be observed. As it will be shown in the paper, the proposed approach can achieve energy-saving results greater than or equal to 6 [%] (in each season) and 19 [%] (in one year) with respect to more standard approaches.

The paper is organized as follows: Section 2 reports the material and methods, focusing on PID controllers, DEDS, and simulation/control frameworks. Section 3 reports the results and discussion, focusing on the control and energy-saving performances. The conclusions are summarized in Section 4.

2. Materials and Methods

2.1. PID Control Architectures

The PID control is exploited in a large number of sectors and at different levels of the automation hierarchy. PID techniques are typically used for SISO feedback control of processes with/without dead times (or delays) [37].

PID controllers are characterized by three main components that process the control error: proportional, integral and derivative. These components are represented by the tuning parameters K_P, T_I, T_D [37]. In order to efficiently control the process, the three terms have to be suitably tuned. In the literature, different tuning methods were proposed, e.g., the Ziegler-Nichols and Cohen-Coon methods. The Ziegler-Nichols frequency response and step response methods are widely used tuning methods for PID controllers [38].

The Ziegler-Nichols frequency response method (close-loop method) assumes to model the controlled process with two parameters: the ultimate gain (K_u) and the ultimate period (T_u). These parameters are determined by forcing the process through a P controller, increasing the proportional gain until the process oscillates critically. The gain yielding marginal stability is the ultimate gain, and the ultimate period is the period of oscillation in correspondence with the ultimate gain. Using the ultimate gain (K_u) and the ultimate period (T_u), the proportional, integral, and derivative gains are typically computed as in Table 1 [37].

Table 1. PID control parameters tuning (Ziegler-Nichols frequency response method) [37].

Controller	K_P	T_I	T_D
P	$0.5 \cdot K_u$		
PI	$0.4 \cdot K_u$	$0.8 \cdot T_u$	
PID	$0.6 \cdot K_u$	$0.5 \cdot T_u$	$0.125 \cdot T_u$

The Ziegler-Nichols step response method (open-loop method) is based on an open-loop step test procedure applied to the process, hence requiring the process to be stable. The unit step response of the process is characterized by two parameters, i.e., α and τ. These can be determined, for example, by drawing a tangent line at the inflexion point of the step response, where the slope of the step response has its maximum value. The intersections of the tangent and the coordinate axes give the parameter α and τ that are used to compute the proportional, integral, and derivative gains (see Table 2) [37].

Table 2. PID control parameters tuning (Ziegler-Nichols step response method) [37].

Controller	K_P	T_I	T_D
P	$1/\alpha$		
PI	$0.9/\alpha$	$3{\cdot}\tau$	
PID	$1.2/\alpha$	$2{\cdot}\tau$	$\tau/2$

As previously described, the PID technique can be exploited for the SISO control of processes with dead times (or delays). Often, industrial processes are modelized by a first-order plus deadtime (FOPDT) model. In the Laplace domain, this model is [39]:

$$y(s) = P(s)u(s) = P_\gamma(s)e^{-Ls}u(s) = \frac{K}{Ts+1}e^{-Ls}u(s) \tag{1}$$

where s is the Laplace variable, $P(s)$ is the input-output transfer function, $P_\gamma(s)$ is the input-output transfer function without delay, y is the output, u is the input, K is the gain of the process, T is the time constant and L is the delay. In this paper, the case where the time constant T and the delay L are comparable is considered, i.e., [37]:

$$0.6 < \frac{L}{T} < 1 \tag{2}$$

Assuming that model (1) is sufficiently reliable, when Equation (2) holds, the Smith Predictor can be used [37]. This PID-based predictive controller overcomes the limitations of a PID controller, which exploits only feedback. A Smith Predictor architecture is reported in Figure 1. In Figure 1, r is the reference to be tracked, w and z are intermediate variables and y' is the feedback variable. The input-output transfer function related to y' is [37]

$$y'(s) = P_\gamma(s)\left(1 - e^{-Ls}\right)u(s) \tag{3}$$

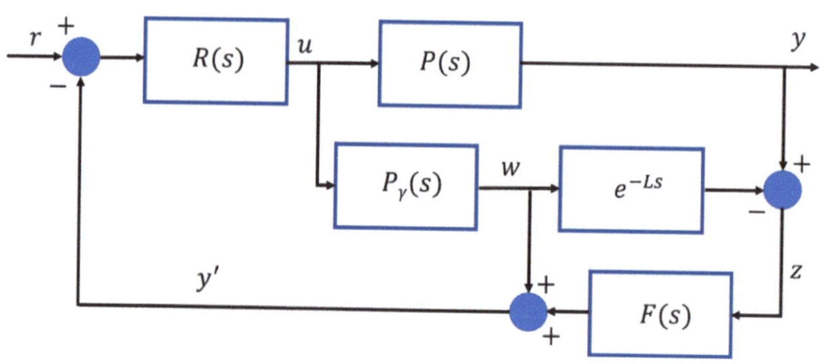

Figure 1. Smith Predictor architecture [37].

In Figure 1, $F(s)$ is a low-pass filter that allows to compensate low frequency errors. $F(s)$ can be defined as [37]:

$$F(s) = \frac{1}{0.5Ls+1} \tag{4}$$

In the scheme of Figure 1, $R(s)$ can be designed as a PID controller.

Different control architectures based on PID controllers were introduced by researchers, engineers, and practitioners. Examples of these architectures are cascade, split-range, valve position, and override control [37].

A cascade control structure involves two or more PID controllers in serial connection and two or more nested control loops. The manipulated variable (MV) of the master controller is connected to the setpoint of the slave controller. The advantage of the cascade control is that disturbance variables (DVs) affecting the inner loop can be compensated much more quickly in the slave (inner) loop than in the slower master (external) loop. The inner controller design is beneficial if its dynamics are faster. Cascade control requires a major effort in the tuning phase with respect to a standard PID controller. Due to the inner loop being an element within the master (external) one, it must be properly tuned before the master controller, with the master controller in manual mode. Then, the master controller should be tuned, with the inner controller in automatic mode [40].

A wide variety of process control problems are characterized by having one controlled variable (CV) and two or more MVs. The challenge from a control perspective is that there is no unique set of input values that can ensure the setpoint tracking of the CV. One of the most common ways of addressing this problem is known as split-range control. A splitter block is used to map the controller output to multiple MVs [41].

In many case studies (e.g., feedstock flow control), the synchronized use of actuators characterized by different dynamics, such as, for example, a small and a large actuation valve, can be required for accurate control over a significant operating range. An ideal solution would be to use the small valve to make fine changes and the large valve to significantly adjust the CV, differently from the split-range control, where valves are sequenced one at a time. Valve position control architecture achieves this type of regulation for the small and large valves [41].

In an override control architecture, two or more CVs share a common actuator (MV). Depending on the current process state, a decision is made on which controller is in charge of the actual manipulation of the actuator; in other words, the various controllers can override each other [37]. Figure 2 reports an override control architecture. In the case reported in Figure 2, the decision is based on a comparison of the computed MV requests of both controllers; for example, the controller that demands the higher MV value takes control of the actuator (high-pass selector). Both controllers run the entire time. This scheme can also be used in applications with more than two CVs and with low-pass selector logic.

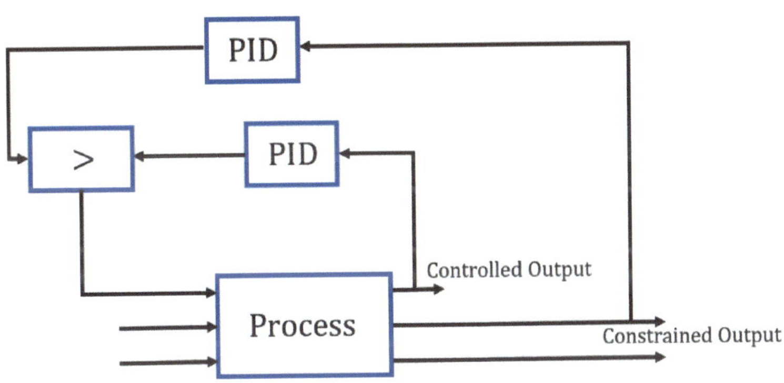

Figure 2. Override control architecture with high-pass selector logic [37].

In order to implement efficient and flexible architectures, the previous control architectures can be enriched with anti-wind-up schemes and bumpless transfer techniques to handle possible transitions between manual and automatic control modes [37].

2.2. HVAC Simulation Framework

In order to design and implement a HVAC simulation framework, first-principles, and empirical models related to an office building were combined. The considered office building is located in Ancona, Italy. The HVAC simulation framework consists of a thermal model, a lighting model and an IAQ model [42,43]. Six walls (two floors and four vertical) and two windows characterize the rectangular layout of the single office room. The considered devices are reported in Figure 3: the actuators are the rolling shutters, the windows, a light dimmer and a heat pump (with a fan coil). The measured variables are the room temperature (through a thermostat) and light (through a luxmeter). Furthermore, a weather station is considered for the measurement of the outdoor temperature, wind speed, and other parameters. Eventually, a presence sensor and IAQ sensors can be present. Table 3 reports the room parameters.

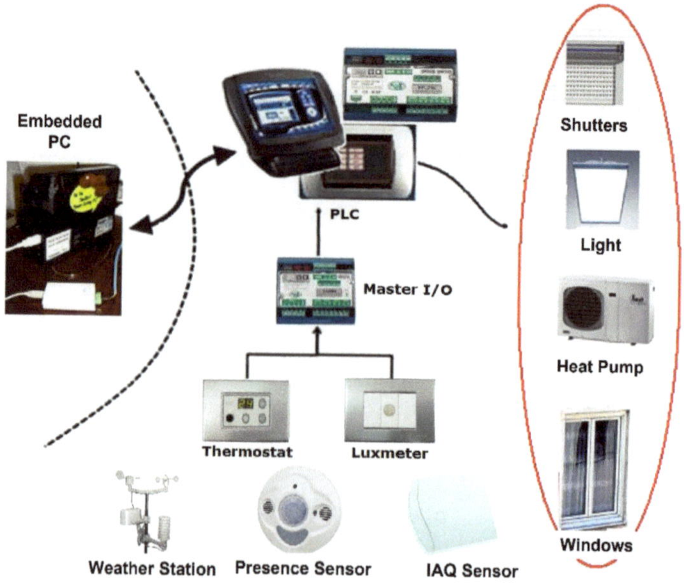

Figure 3. Overall system scheme.

Table 3. Room parameters.

Component/Device	Features	SI Measurement Unit
Room	[5.0 × 4.0 × 2.7]	[m] × [m] × [m]
Wall-SW	[5.0 × 2.7], Vertical	[m] × [m]
Wall-NW	[4.0 × 2.7], Vertical	[m] × [m]
Wall-NE	[5.0 × 2.7], Vertical, (Not Exposed)	[m] × [m]
Wall-SE	[4.0 × 2.7], Vertical, (Not Exposed)	[m] × [m]
Wall-A	[5.0 × 4.0], Horizontal	[m] × [m]
Wall-B	[5.0 × 4.0], Horizontal, (Not Exposed)	[m] × [m]
Window-SW1	[2 × 1.25 × 1], Vertical	[m] × [m] × [m]
Window-SW2	[2 × 1.25 × 1], Vertical	[m] × [m] × [m]
Heat Pump	COP * = 2.8, Max Power = 4	[], [kW]
Artificial Light (Dimmer)	location (2.5, 2.0, 2.4), Flux = 8900	[m], [lumen]

* COP = coefficient of performance.

2.2.1. Thermal Model

The thermal model of the considered office room takes into account the thermal behavior of the room, which can be modelized through an energy balance equation [42] reported in (5). Six main contributions were considered in Equation (5) for the total thermal accumulation, e.g., internal heat sources (Q_{IS}, due to people, lamps, and motors), heat pump source (Q_{HP}), thermal conduction and convection exchanges through walls (Q_W), thermal conduction, convection and radiation exchanges through windows (Q_{GL_S}, $Q_{GL\_S\_I}$), and finally the fluid dynamic exchanges with the outside environment (Q_{VENT}). The thermal contributions are reported in Equations (6)–(9), while the terms involved in the Equations (5)–(9) are reported in Table 4 [42].

$$m_a \cdot c_a \cdot \frac{\partial T_a(t)}{\partial t} = Q_{IS}(t) + Q_{HP}(t) + Q_W(t) + Q_{GL\_S}(t) + Q_{GL\_S\_I}(t) + Q_{VENT}(t) \quad (5)$$

$$Q_W(t) = \sum_{k=1}^{6} A_k \cdot h_{O_k} \cdot \left(T_{W_{k\_L1}}(t) - T_a(t)\right) \tag{6}$$

$$Q_{GL\_S}(t) = \sum_{j=1}^{2} A_{gl_j} \cdot h_{Ogl_j} \cdot \left[(1 - SAF_j(t)) \cdot \left(T_{gl_j}(t) - T_a(t)\right) + SAF_j(t) \cdot \left(T_{gl\_s_j}(t) - T_a(t)\right)\right] \tag{7}$$

$$Q_{GL\_S\_I}(t) = \sum_{j=1}^{2} A_{gl_j} \cdot \left[(1 - SAF_j(t)) \cdot F_{G\_gl_j} + SAF_j(t) \cdot F_{G\_gl\_s_j}\right] \cdot I_{gl_j}(t) \tag{8}$$

$$Q_{VENT}(t) = \sum_{j=1}^{2} \rho_a \cdot c_a \cdot N_j(t) \cdot V_j \cdot (T_e(t) - T_a(t)) \tag{9}$$

Table 4. Thermal model parameters (Equations (5)–(9)).

Symbol	Description	SI Measurement Unit
Q_{IS}	Heat supplied by internal heat sources (people, lamps, and motors)	[W]
Q_{HP}	heat supplied by heat pump source	[W]
Q_W	heat supplied by walls	[W]
Q_{GL_S}, $Q_{GL\_S\_I}$	heat supplied by windows	[W]
Q_{VENT}	heat supplied by the outside environment	[W]
A_k	kth wall area	[m^2]
h_{O_k}	kth wall adduction coefficient	[W/(m^2·K)]
h_{Ogl_j}	jth glass adduction coefficient	[W/(m^2·K)]
$F_{G\_gl_j}$, $F_{G\_gl\_s_j}$	solar gain coefficient (jth glass, jth glass/shutter)	[]
N_j	number of times air is exchanged through the jth window opening	[1/s]
ρ_a	air density	[kg/m^3]
c_a	air specific heat	[J/(kg·K)]
m_a	room air mass	[kg]
V_j	air incoming volume (fixed value) from jth window	[m^3]
$T_{W_{k\_Lj}}$	temperature of jth layer of kth wall	[K]
T_{gl_j}	jth internal temperature of glass	[K]
$T_{gl\_s_j}$	jth internal temperature of glass combined with shutters	[K]
T_a	room temperature	[K]
T_e	outside temperature	[K]
A_{gl_j}	jth glass area	[m^2]
SAF_j	jth shutter actuation factor	[%]
I_{gl_j}	jth glass solar thermal radiation	[W/m^2]

Indoor wall and glass temperatures are included in Equations (6) and (7). These terms are modelized through a dynamic multi-layer model. Figure 4 shows a multi-layer wall composed of five layers. For example, the first wall, i.e., the external wall, composed of five layers, can be modelized by Equations (10)–(14) [42] whose terms are described in Table 5.

$$\frac{\partial T_{W_k}(t)}{\partial t} = A \cdot T_{W_k}(t) + B \cdot u_{W_k}(t) \tag{10}$$

$$T_{W_k}(t) = \begin{bmatrix} T_{W_{k\_L1}}(t) \\ \vdots \\ T_{W_{k\_L5}}(t) \end{bmatrix} \tag{11}$$

$$u_{W_k}(t) = \begin{bmatrix} T_e(t) \\ T_a(t) \\ I_{W_k}(t) \\ I_{gl_1}(t) \\ I_{gl_2}(t) \end{bmatrix} \tag{12}$$

$$A = A_{W_k} \begin{bmatrix} -\frac{K_{e1}+K_{12}}{m_1c_1} & \frac{K_{12}}{m_1c_1} & 0 & 0 & 0 \\ \frac{K_{12}}{m_2c_2} & -\frac{K_{12}+K_{23}}{m_2c_2} & \frac{K_{23}}{m_2c_2} & 0 & 0 \\ 0 & \frac{K_{23}}{m_3c_3} & -\frac{K_{23}+K_{34}}{m_3c_3} & \frac{K_{34}}{m_3c_3} & 0 \\ 0 & 0 & \frac{K_{34}}{m_4c_4} & -\frac{K_{34}+K_{45}}{m_4c_4} & \frac{K_{45}}{m_4c_4} \\ 0 & 0 & 0 & \frac{K_{34}}{m_5c_5} & -\frac{K_{45}+K_{5i}}{m_5c_5} \end{bmatrix} \tag{13}$$

$$B = \begin{bmatrix} A_{W_k}\frac{K_{e1}}{m_1c_1} & 0 & A_{W_k}\frac{\frac{\alpha_e}{(h_e \cdot R_{e1})}}{m_1c_1} & 0 & 0 \\ 0 & 0 & 0 & 0 & 0 \\ 0 & 0 & 0 & 0 & 0 \\ 0 & 0 & 0 & 0 & 0 \\ 0 & A_{W_k}\frac{K_{5i}}{m_5c_5} & 0 & \frac{\Gamma_1 \cdot A_{gl_1} \cdot Tau_1 \cdot F_1}{m_5c_5} & \frac{\Gamma_2 \cdot A_{gl_2} \cdot Tau_2 \cdot F_2}{m_5c_5} \end{bmatrix} \tag{14}$$

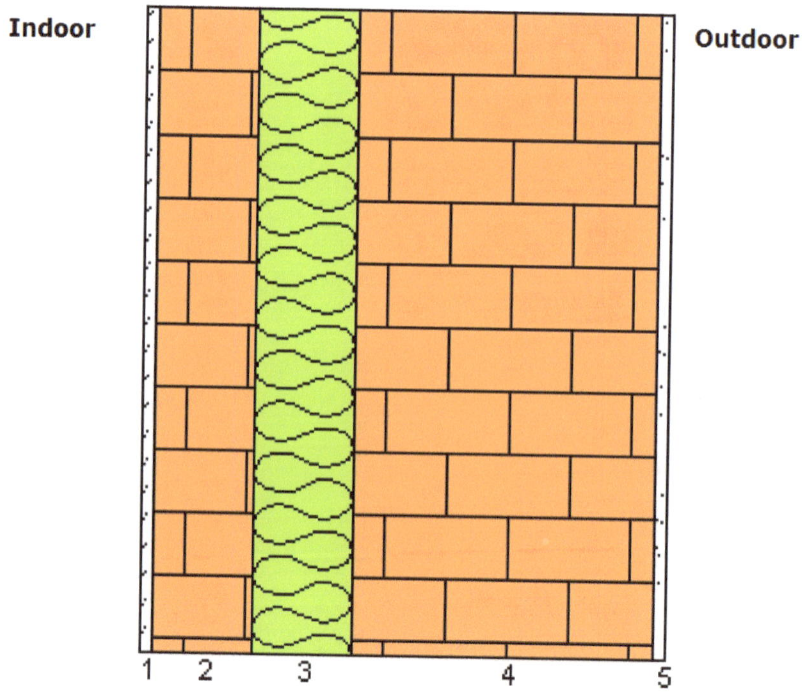

Figure 4. Multi-layer wall [42].

Table 5. Thermal model parameters (Equations (10)–(14)).

Symbol	Description	SI Measurement Unit
$T_{W_{k\_Lj}}$	temperature of jth layer of kth wall	[K]
I_{gl_j}	jth glass solar thermal radiation	[W/m^2]
T_a	room temperature	[K]
T_e	outside temperature	[K]
I_{W_k}	kth wall solar thermal radiation	[W/m^2]
A_{W_k}	kth wall area	[m^2]
K_{ji}	thermal transmittance between layers j and i of the wall	[W/(m^2·K)]
K_{e1}	thermal transmittance between layer one of the wall and outdoor air	[W/(m^2·K)]
K_{5i}	thermal transmittance between layer five of the wall and indoor air	[W/(m^2·K)]
m_i	mass of the layer i of the wall	[kg]
c_i	specific heat of the layer i of the wall	[J/(kg·K)]
α_e	absorption coefficient of the wall	[]
h_e	adduction coefficient of the wall	[W/(m^2·K)]
R_{e1}	thermal resistance of the wall	[(m^2·K)/W]
Γ_j	internal flux parameter	[]
A_{gl_j}	jth glass area	[m^2]
Tau_j	jth glass transparency	[]
F_j	jth shutter shading factor	[]

The window's contribution to the room's thermal behavior is characterized by conduction, convection and radiation phenomena. Equation (7) represents conduction and convection, while Equation (8) reports the radiation contribution. In Equations (7) and (8), two sections characterize the window's contribution (see Figure 5): the first section is related to the lower part of the glass window and takes into account only the contribution of the glass (subscript *gl*), while the second section takes into account the contribution of the glass coupled to the rolling shutter (subscript *gl_s*). These contributions are modelized separately here and not reported for brevity. The density, the thickness, and the specific heat of glass and rolling shutters characterize these contributions [44].

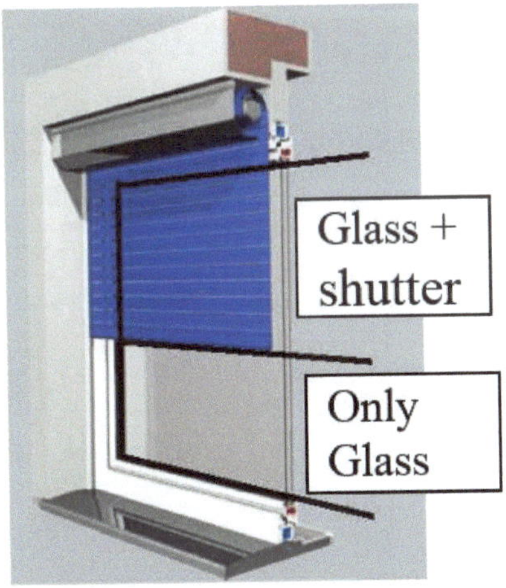

Figure 5. Considered sections of the windows [44].

In order to estimate the thermal radiation related to the windows' glass and to the walls, historical data were selected for the considered location (see Table 3). Different seasons and times of day were taken into account. These inputs, together with the disposition of the considered surfaces, were used in different algorithms, e.g., the Liu and Jordan method or the Collares-Pereira and Rabl method [45].

With regard to the outside temperature used in the thermal model, historical data were selected for the considered location (see Table 3).

The formulated thermal model can be suitably extended to different room configurations and features and to multi-room environments. For example, if a wall with more than five layers were to be considered, a number of equations (similar to Equation (10)) equal to the assumed layers' number would result. In addition, if a multi-room environment is assumed, an equation (similar to Equation (5)) for each room must be included in the model, taking into account possible coupling effects between the different rooms.

2.2.2. Lighting Model

Artificial light and natural illuminance were considered for the designed lighting model. Different assumptions on the room/lights and room/window relative dimensions were made for the lighting model (see Table 3). The superposition principle is applied, so uniformity of the light radiation is assumed. Furthermore, extended light sources and point light sources were considered: natural light sources belong to the first category, while artificial light sources were included in the second one. Equation (15) reports the designed lighting model for a generic point of interest $P(x, y, z)$ [46,47], while Table 6 summarizes the involved parameters.

Table 6. Lighting model parameters (Equation (15)).

Symbol	Description	SI Measurement Unit
E_a	environment illuminance at the point of interest $P(x, y, z)$	[Lux]
$E_{N\_D\_gl,j}$	natural diffuse illuminance on the window	[Lux]
$E_{N\_R\_gl,j}$	natural reflection illuminance on glass	[Lux]
$E_{NAT\_gl,j}$	natural direct illuminance on glass	[Lux]
$C_{j,c}$, $C_{j,r}$	environmental influence of natural diffuse/reflections illuminance at the point of interest $P(x, y, z)$	[]
I_L	artificial light source luminous emission	[cd/klm]
$Lumen$	luminous flux of the artificial light source	[lm]
γ	incidence angle of the light radiation in relation to the point of interest $P(x, y, z)$	[°C]
h	distance between the point of interest and light source	[m]
Tau_j	jth glass transparency	[]
V_j	natural direct illuminance coefficient	[]
A_{gl_j}	jth glass area	[m^2]
ρ	reflection coefficient	[]
$\rho_{weighted}$	average reflection coefficient of the walls	[]
sum_{AREA}	total area of the reflective walls	[m^2]
$E_{NAT\_gl,j}$	jth glass area	[m^2]
η	efficiency of artificial light source	[]
M	maintenance factor	[]

$$
\begin{aligned}
E_a(t) = &\sum_{j=1}^{2} \left| C_{j,c} \cdot E_{N\_D\_gl,j}(t) + C_{j,r} \cdot E_{N\_R\_gl,j}(t) \right| + \frac{I_L(t) \cdot Lumen \cdot 1000 \cdot \cos(\gamma)^3}{h^2} \\
&+ \sum_{j=1}^{2} \frac{Tau_j \cdot V_j \cdot A_{glj} \cdot \rho_{weighted}}{sum_{AREA} \cdot (1 - \rho_{weighted})} E_{NAT\_gl,j}(t) + \frac{Lumen \cdot \eta \cdot M \cdot \rho}{sum_{AREA} \cdot (1 - \rho_{weighted})}
\end{aligned}
\quad (15)
$$

The natural illuminance used in the lighting model was estimated. Historical data were selected for the considered location (see Table 3). Different seasons and times of day were taken into account.

The formulated lighting model can be suitably extended to different room configurations and features and to multi-room environments. For example, if different points of interest were to be considered, an equation (similar to Equation (15)) for each point of interest must be formulated. Furthermore, if more artificial light sources are present, different artificial light source contributions (similar to the second and fourth terms of Equation (15)) must be included. In addition, if more windows are present, different natural light source contributions (similar to the first and third terms of Equation (15)) have to be considered.

2.2.3. IAQ Model

The IAQ model is composed of three main sources of pollution: formaldehyde (HCHO), carbon dioxide (CO_2) and TVOC (total volatile organic compounds). The first one consists of a colorless organic compound with a highly irritating odor, used mainly in the production of wood panels and furniture, paints, and laminate. CO_2 is produced by people in sedentary work in the room. The last source of pollution, i.e., TVOC, includes different chemical compounds. The concentration of toxic waste can be evaluated in terms of the volume of pollution to the volume of standard indoor air [48].

IAQ dynamics are represented in Equations (16)–(19), while Table 7 reports the IAQ model parameters [49,50].

$$\frac{\partial CO_2(t)}{\partial t} = n_{people}(t) \cdot E_p + E_{std_{air}} \cdot G(t) - G(t) \cdot \frac{CO_2(t)}{Vol} \tag{16}$$

$$\frac{\partial HCHO(t)}{\partial t} = \sum_{i=1}^{n_{furn}} A_{furn,i} \cdot E_{furn,i} - G(t) \cdot \frac{HCHO(t)}{Vol} \tag{17}$$

$$\frac{\partial TVOC(t)}{\partial t} = A_{room} \cdot E_{TVOC,\, room} - G(t) \cdot \frac{TVOC(t)}{Vol} \tag{18}$$

$$G(t) = OW_{window} \cdot H_{window} \cdot S_{wind}(t) \tag{19}$$

Table 7. IAQ model parameters (Equations (16)–(19)).

Symbol	Description	SI Measurement Unit
CO_2	room CO_2	$[m^3 CO_2]$
n_{people}	number of people in the room	[]
E_p	CO_2 emissions for each people (sedentary)	$[m^3 CO_2/s]$
$E_{std_{air}}$	standard conditions air CO_2	$[m^3 CO_2/m^3]$
G	natural ventilation flow rate	$[m^3/s]$
Vol	room volume	$[m^3]$
HCHO	room HCHO	[mg HCHO]
n_{furn}	number of the room's furniture	[]
$A_{furn,i}$	room's ith furniture area	$[m^2]$
$E_{furn,i}$	room's ith furniture HCHO emissions per unit area	$[mg\ HCHO/(m^2 \cdot s)]$
$TVOC$	room $TVOC$	[µg]
A_{room}	room area	$[m^2]$
$E_{TVOC,\, room}$	room's ith furniture $TVOC$ emissions per unit area	$[\mu g/(m^2 \cdot s)]$
OW_{window}	opening width of the window	[m]
H_{window}	height of the window	[m]
S_{wind}	wind speed	[m/s]

The wind speed used in the IAQ model was estimated. Historical data were selected for the considered location (see Table 3). Different seasons and times of day were taken into account.

The formulated IAQ model can be suitably extended to different room configurations and features and to multi-room environments. For example, if a multi-room environment is assumed, a set of equations (similar to Equations (16)–(18)) for each room must be included in the model.

2.2.4. Case Study Additional Details

The room planimetry and top view are reported in Figure 6. In the planimetry of Figure 6, only the four vertical walls are reported; the other two horizontal ones (Wall-A and Wall-B in Table 3) are an attic floor and an intermediate floor; Table 3 reports the features of the walls. Walls 1–2 (Wall-SW and Wall-NW in Table 3) are external walls (see Figure 6). Their thickness, which is greater than the thickness of the other walls, is 420 [mm] and their intermediate layer is constituted by extruded expanded polystyrene. Their total thermal transmittance is 0.3 [W/(m^2·K)], i.e., they have a high opposition to thermal radiation. Windows are located on wall one. Walls 3–4 (Wall-SE and Wall-NE in Table 3) are characterized by a thickness equal to 300 [mm] (see Figure 6) and by a lower thermal resistance with respect to the exposed walls. The attic floor wall (Wall-A in Table 3) is similar to walls 1–2, while the intermediate floor wall (Wall-B in Table 3) is characterized by a thermal resistance lower than the attic floor wall [42].

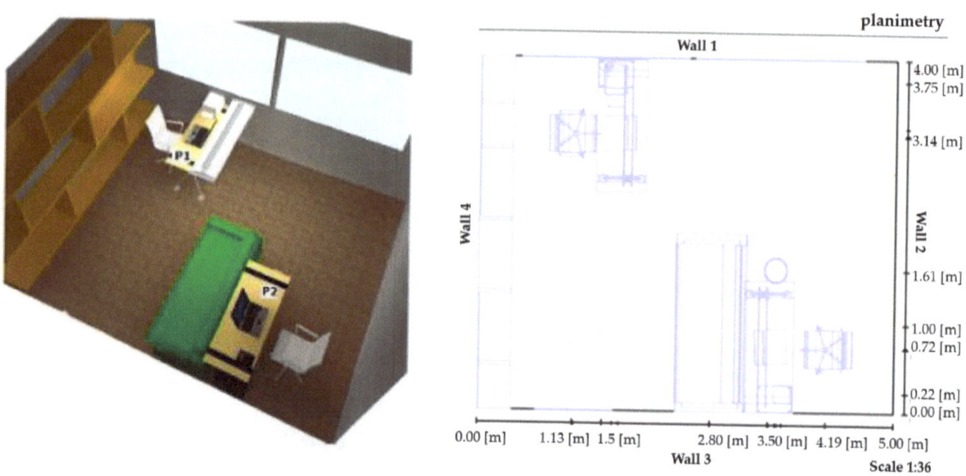

Figure 6. Top view and planimetry of the room (simulated environment).

With regard to the windows, they are characterized by an athermal single glass type, and the transmissivity factor of the glass is equal to 0.46. Through suitable computations, the other parameters related to the thermal behavior of the windows, e.g., solar gain coefficients, were achieved. In addition, tailored computations were performed for achieving the thermal transmittance of the windows, differentiating between the glass-only case and the glass together with shutters case [42].

The room temperature computed by the thermal model is assumed to be the mean temperature of the room (neglecting the temperature layering). The heat pump source contribution will be positive in the case of heating and negative in the case of cooling [51].

With regard to the lighting model, the artificial source is supposed to be located in the middle of the $x - y$ plane of the room (see Table 3), and it is considered a point of light source (see Section 2.2.2 and Figure 7). Furthermore, the artificial source is assumed to be capable of guaranteeing the same illuminance to both work positions (see Figure 6) [46,47]. The point of interest of the lighting model is assumed to be a single point of interest in the room, which was set at the coordinates (2.5, 2.0, 1.0) [m], i.e., in the middle point of the room at a height of 1 [m] above the floor. The choice of using illuminance at the midpoint of the room was mainly because, in most practical applications, only one luxmeter was available for each room [46,47].

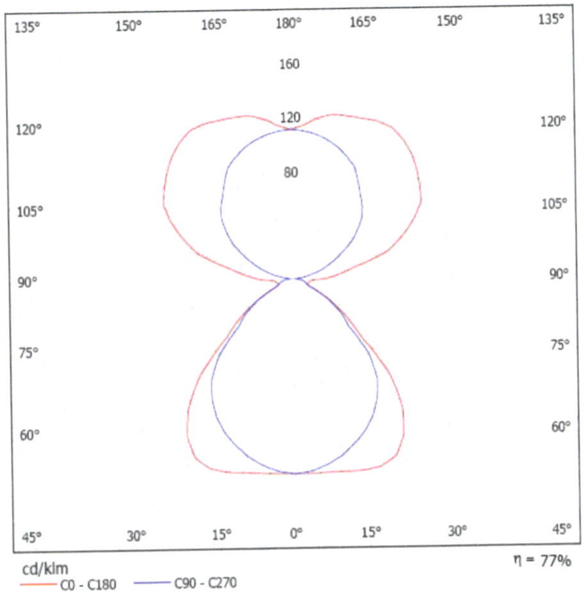

Figure 7. Light emission (simulated environment).

With regard to both thermal and lighting models, the actuation factor of each rolling shutter is a value in the range 0–1 (0 corresponds to an open rolling shutter, 1 corresponds to a closed rolling shutter).

For the IAQ model (see Section 2.2.3), it is assumed that there are two desks composed of laminated panels and a bookcase in the office (see Figure 6). For a work desk height, a horizontal plane with an elevation equal to 0.8 [m] from the floor is assumed. The correspondent average HCHO emission value used in the equation is 3.5 $[mg/(m^2{\cdot}h)]$ and the furniture's TVOC average emission per unit is used in the model is 0.125 $[\mu g/(m^2{\cdot}s)]$. Furthermore, CO_2 average emission produced by two users in sedentary work is assumed to be 0.019 $[m^3CO_2/s]$ [52,53]. Finally, in order to enhance the fit of the framework for the real process, window opening is inhibited if wind speed is greater than a threshold value, i.e., 0.3 [m/s].

2.3. HVAC Control Framework

Based on the HVAC simulation framework described in Section 2.2, a HVAC control framework was designed and implemented. First, the MVs, the CVs, and the DVs were defined. The MVs group includes the heat supplied by the heat pump source, the shutter actuation factor of each of the two rolling shutters, the dimming of the artificial light, and the actuation of the natural ventilation through window opening. The CVs are the mean room temperature, the environment illuminance at the point of interest, and the IAQ parameters. The DVs are the outside temperature, the thermal radiation related to the window's glass and to the walls, the natural illuminance, the heat supplied by the internal heat sources (people, lamps, and motors), and the wind speed.

The control system must guarantee the desired comfort within the room but at the same time must ensure the maximization of energy saving. This last objective can be obtained by maximizing the natural energy sources exploitation. This target can be put into practice, ensuring an efficient multi-variable control of the defined CVs. As previously described, the heat transfer through the glass windows affects the room temperature. The heat is generated by the solar radiation; rolling shutters can be used to regulate this contribution. In addition, the opening of the windows also affects the temperature of the room through the dynamic exchange of fluid. The room illuminance is affected by the rolling shutters' positions, which diffuse and reflect natural contributions. For this reason, the controller must manage possible conflicts between lighting and thermal control objectives [13].

In order to design a flexible control system, Energy Saving and Comfort and Energy Saving control policies were designed. These control policies are defined in the present research work, taking into account different control specifications. For brevity, Comfort and Energy Saving control policy will be indicated as Comfort policy in the following. Energy saving is the main specification in the Energy Saving control policy while fulfilling constraints on user comfort is the main goal in the Comfort control policy. In order to evaluate the CO_2 emissions performances, energy ([kWh]) and CO_2 emissions were related through a linear relationship [54].

The HVAC control framework was based on the use of different PID architectures combined with DEDS (automata) [55] control architectures to obtain a multi-mode control system. Each controller was equipped with anti-wind-up schemes and automatic/manual modes with bumpless transfer techniques (see Section 2.1).

Thanks to the flexibility of the developed simulation and control framework, two main control systems were designed and compared. In the following, the first one will be indicated as the *initial* control system while the second one as the *modified* control system. With regard to the tuning of the controllers included in the *initial* and *modified* control systems (see Sections 2.3.1 and 2.3.2), the methods presented in Section 2.1 were exploited.

To the best of the authors' knowledge, the flexibility of the proposed simulation and control framework represents an innovation in the literature of control systems for energy saving and comfort management in HBA. Thanks to this feature, the frameworks can be adapted based on the environments to be considered as case studies and the controllers to be designed. Different environments can be modelized and simulated in the simulation framework and different control algorithms can be straightly switched in order to compare their performances.

2.3.1. *Initial* Control System

The architecture related to the *initial* control system is shown in Figure 8. A hybrid scheme consisting of PID controllers in a split-range configuration and a DEDS-based (automatic) supervision module performs the thermal control. A split-range configuration is used in order to split the thermal control efforts, considering the three available MVs (rolling shutters, heat pumps, and windows). MVs sequences are assigned based on the energy consumption profile, i.e., a higher priority is assigned to MVs with a lower energy effort (cost). For this reason, heat pumps haave the lowest priority. The DEDS-based (automatic) supervision module represents the splitter of the split range architecture. Based on the current process state, the automata take into consideration all process conditions, such as the malfunction of an actuator or unavailability of an actuator. Six states characterize the designed automata, together with events' set of fifteen elements (see Figure 9 and Table 8). A state represents a defined operating mode. MVs' availability differentiates each operating mode. Table 8 shows the operating modes of the split range configuration ("1" indicates that the considered actuator is available). The states associated with the simultaneous availability of the heat pump and of the windows were assumed to be inadmissible due to energy efficiency purposes. Furthermore, the rolling shutters can be used only if solar radiation is present. States transitions within the automata are triggered by events which are the result of the combination of the following parameters:

- Period of the day (i.e., daytime, nighttime)
- Presence or absence of solar radiation
- Thresholds on the tracking error between the desired reference temperature and the room temperature at different ranges were defined (e.g., tracking error range 0 is associated with a tracking error in the range between -0.2 [°C] and 0.2 [°C])
- Thresholds on the difference between the room temperature and the outside temperature at different ranges were defined (e.g., difference range 0 is associated with a

difference in the range between −2 [°C] and 2 [°C], while a range 1 is associated with a difference greater than 2 [°C])

- Control efforts required for the heat pump
- System switch off

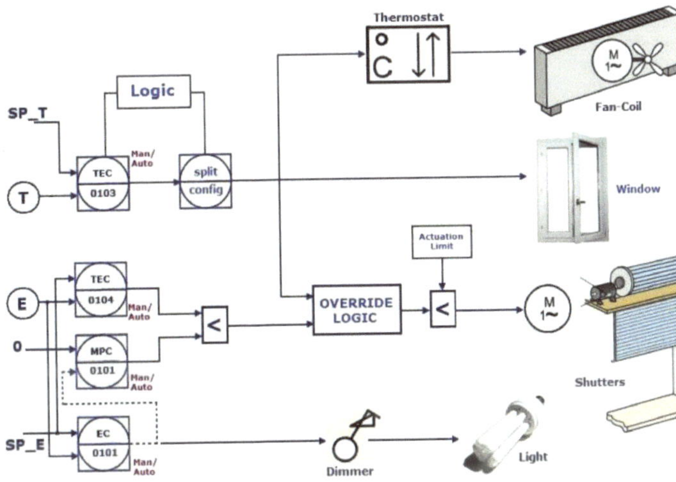

Figure 8. Architecture of the *initial* control system.

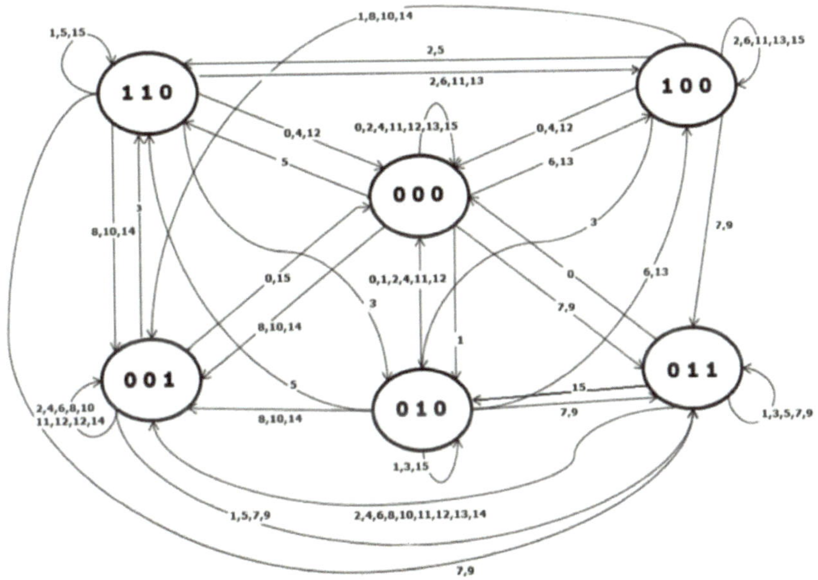

Figure 9. Automata (for the split range of the thermal control) of the *initial* control system.

Table 8. Split range operating modes (thermal control, *initial* control system).

State	Window	Rolling Shutters	Heat Pump
S1	1	1	0
S2	1	0	0
S3	0	1	1
S4	0	1	0
S5	0	0	1
S6	0	0	0

All the events obtained are mutually exclusive. With regard to the control efforts required for the heat pump, if the computed control effort is less than a defined threshold (e.g., 10 [%] of the operating range), the heat pump is switched off. In this way, a major opportunity for intervention is given to the other actuators, e.g., the windows. With regard to the thresholds related to tracking error and the difference between the room temperature and the outside temperature, a suitable hysteresis was added in order to avoid oscillatory behaviors and excessive changes in the asset [38,41]. Table 9 reports the description of some events depicted in Figure 9, while Table 10 reports an example of the state transition matrix related to the first five events.

Table 9. Events description (splitter of the thermal control, *initial* control system).

Event	Description
0	switch-off of the devices
1	daytime, solar radiation, tracking error range 0, difference range 0
2	daytime, no solar radiation, tracking error range 0, difference range 0
3	daytime, solar radiation, tracking error range 0, difference range 1
4	daytime, no solar radiation, tracking error range 0, difference range 1

Table 10. Sub-part of the state transition matrix (splitter of the thermal control, *initial* control system).

Event	Initial State S1	Initial State S2	Initial State S3	Initial State S4	Initial State S5	Initial State S6
0	S6	S6	S6	S6	S6	S6
1	S1	S1	S3	S4	S3	S4
2	S2	S2	S5	S6	S5	S6
3	S4	S4	S3	S4	S3	S4
4	S6	S6	S5	S6	S5	S6

With regard to the lighting control, the dimmer control effort is computed by the controller EC | 0101 (see Figure 8). Since the dimmer control effort can be reduced through the exploitation of natural light, a second controller (MPC | 0101 in Figure 8) is placed in the valve position configuration with a setpoint equal to zero. Finally, a third controller (TEC | 0104) is present; its main function is to not exceed the illuminance threshold acting on the rolling shutters. An override logic handles MPC | 0101 and TEC | 0104 controllers (low-pass mode). MPC | 0101 was tuned as a faster controller with respect to TEC | 0104 in order to limit the dimmer control effort.

In order to handle eventual contrasting requests between thermal and lighting control objectives, an additional override module was added to the control architecture (see Figure 8). This module selects the input signals based on four system configurations, obtained by pairing Energy Saving/Comfort control policies and heating/cooling conditions. The Energy Saving control policy requires the reduction of power consumption as much as possible. If the cooling condition is considered, rolling shutters must be used in order to reduce the use of heat pumps. Conversely, Comfort control policy assigns high priority to the lighting control, and natural illuminance is preferred for the tracking of the desired setpoint. Comfort preferences may be required by the users; for this purpose, specific management constraints were implemented, e.g., for constrained manipulation of the shutter position within defined limits. In this condition, if the rolling shutters saturate the constraints, a request for heat pump activation could be needed.

2.3.2. Modified Control System

The *modified* control system considers the room temperature and the room illuminance together with the IAQ. An enhanced thermal control policy was introduced with respect to the *initial* control system, together with an anti-glare logic and a function that allows for the consideration of solar radiation. The *modified* control system architecture is reported in Figures 10 and 11. Table 11 reports the symbols and the detailed description of the functional blocks depicted in Figures 10 and 11. In addition to the mentioned novelties, IAQ control is introduced; the MV associated with windows' opening and closing is delegated to this task. For this purpose, as can be observed in Figure 10, the window actuation is elevated to a supervisory level and must be considered a DV for the thermal control system.

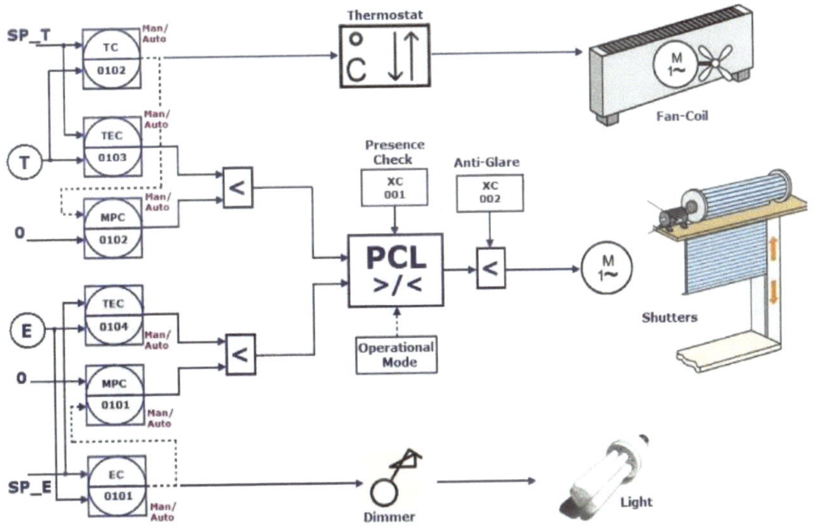

Figure 10. Architecture of the *modified* control system (thermal and lighting control).

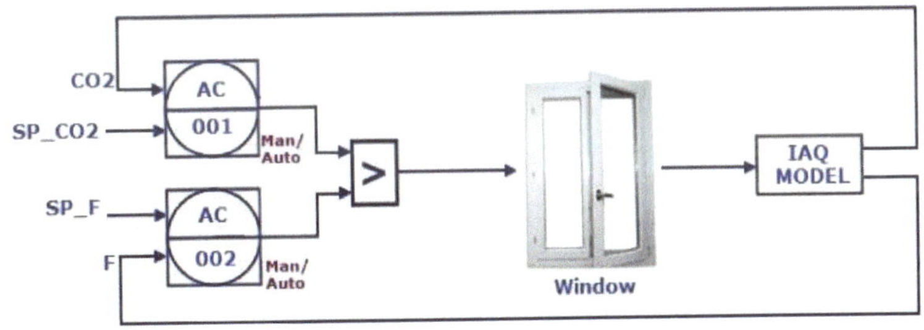

Figure 11. Architecture of the *modified* control system (IAQ control).

Table 11. Functional blocks description (*modified* control system).

Symbol	Description
EC I 0101	PID controller, lighting control
TC I 0102	PID controller, thermal control
TEC I 0103	PID controller, thermal limitation
MPC I 0102	PID controller, motor position control
TEC I 0105	PID controller, lighting limitation
MPC I 0101	PID controller, dimmer position control
MODE	Logic, control mode
XC I 001	Logic, presence radiation
XC I 002	Logic, no excessive brightness
AC I 001	PID controller, CO_2 limitation
AC I 002	PID controller, HCHO limitation

As can be noted in Figure 10, the lighting control of the *modified* control system is the same as the *initial* control system (see Figure 8). Three inter-connected PID controllers characterize the new thermal control system architecture: TC I 0102, TEC I 0103, and MPC I 0102. The activation of the heat pump is managed by TC I 0102, which exploits thermostat information. Heat pump control effort is optimized through MPC I 0102, which is added to the valve position configuration with TC I 0102. A third controller TEC I 0103 is added through an override architecture with MPC I 0102 in order to suggest the rolling shutter actuation for the thermal control.

Possible conflicting requirements on rolling shutters between thermal and lighting control systems are handled through a personal comfort logic (PCL) module (see Figures 10 and 12), depending on the policy (Energy Saving or Comfort) and on the condition (heating or cooling). As can be noted in Figure 12, four states characterize the PCL module: each state is associated with a control action. In state "MIN selection", a coupled thermal and lighting control action with an override architecture configured in low-pass mode is applied; in state "MAX selection" state, a high-pass mode is applied. In the other two states, priority is assigned to the thermal or lighting controller, respectively. When Energy Saving control policy is selected by the user, rolling shutters are used instead of the heat pump when possible. When Comfort control policy is selected by the user, natural illuminance is preferred for the achievement of lighting control objectives. In addition, the same comfort preference considerations given at the end of Section 2.3.1 also apply here.

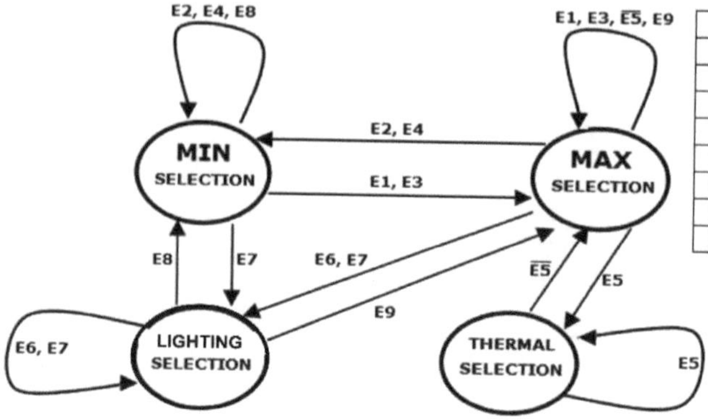

E1	Comfort & Cooling
E2	Comfort & Heating
E3	Saving & Heating
E4	Saving & Cooling
E5	ATTE Alarm Temperature Threshold Exceeded
E6	AITE Alarm Illuminance Threshold Exceeded
E7	Presence Check (True)
E8	$\overline{E7} \wedge (E1 \vee E3)$
E9	$\overline{E7} \wedge (E2 \vee E4)$

Figure 12. Personal comfort logic (PCL) module configuration (*modified* control system).

The PCL module exploits the check "XC I 001–Direct solar radiation Presence Check" (see Figure 10). This check evaluates the presence of direct solar radiation on the glass wall. If direct solar radiation is detected, the selector operates normally. On the other hand, in the case of no solar radiation, the PCL module selects the rolling shutters' values as defined by the lighting control, because in this case, the influence of the solar radiation on room temperature is negligible. The PCL module also takes into account the thermal conduction/convection and the heat radiation contributions. Evaluating all the contributions, the PCL module infers the heat flow direction, i.e., from outside into the environment and vice versa. In this way, this information can be included in the coupled control logic, thus, finalizing the management policies of the rolling shutters.

In addition, the logic "XC | 002–Anti-glare" (see Figure 10) takes into account the sun position to evaluate possible glare effects in the environment and represents an additional check for rolling shutter actuation. For this purpose, azimuth and solar height are considered based on the latitude of the location [56]. This check was introduced in the controller through an override architecture (see Figure 10).

Figure 11 shows the IAQ control system; its functional blocks are described in Table 11. An intelligent logic evaluates the air quality (the presence of CO_2 and HCHO is considered in this section, without loss of generality) and decides the opening or closing of the windows. Suitable hysteresis logics were introduced in order to avoid chattering on the windows' actuation. These logics pre-process the process variables exploited in the computation of tracking errors. IAQ control takes into account two aspects: not increasing energy consumption (such as by excessive window actuation) and not exceeding the imposed pollution limit. The control system works with a virtual sensor of the HCHO and of the CO_2 emissions (or a sensor if available) produced, respectively, by furniture and people. The same control architecture can be extended in order to include other pollutants, e.g., TVOC.

In the control scheme of Figure 11, two regulation loops are present, together with an override selector that allows maintaining the pollution values at fixed levels. Fixed levels that can be considered as maximum constraints are [12,57]:

- CO_2: 1500 [ppm];
- HCHO: 0.1 [ppm];
- TVOC: 300 [$\mu g/m^3$].

In the previous bullet list, 1 [ppm] is equal to approximately 1000 [mg/m^3]. This approach, related to the IAQ control, is known as "performance-based approach" [12,57].

To the best of the authors' knowledge, the design of different advanced PID control architectures, exploiting different configurations, and defining different control specifications, represents an innovation in the literature of control systems for energy savings and comfort management in HBA. This feature allows for bridging the gap between non-optimized solutions and optimized ones, providing a coupled control of thermal, lighting, and IAQ sub-processes. In addition, to the best of the authors' knowledge, the combination of advanced PID control architectures with DEDS for energy savings and comfort management in HBA is not present in the literature; this feature allows for smartly managing the huge number of operating conditions that can occur in HVAC processes. Finally, to the best of the authors' knowledge, the elevation of the IAQ controller to a supervisory level with respect to thermal and lighting controllers represents further innovation that allows for efficiently handling the multi-variable nature of the HVAC processes.

2.4. Software

A MATLAB/Simulink environment was adopted for the HVAC simulation and control frameworks [58]. Furthermore, the MATLAB Identification Toolbox and the MATLAB Control System Toolbox were exploited for process identification and controllers' synthesis [58].

The work has been executed using a laptop computer with the following specifications: Intel(R) Core(TM) i8-3840QM CPU with 3 GHz HDD.

3. Results and Discussion

3.1. Modelization Results

In order to evaluate the proposed HVAC simulation framework (see Section 2.2), tailored simulations were performed. Sampling time of the model was set equal to 60 s.

Figures 13–15 report room temperature and the illuminance of the modelization results assuming no action of the control devices (MVs). The results refer to the effects of the following selected DVs: outside temperature, thermal radiation related to the window glass, and natural illuminance on the walls. In Figure 13, the behavior of the solar radiation for one month is reported. Figure 14 reports the room temperature (blue line) together with the outside temperature (green line) and solar radiation (red line) disturbances. In Figure 15, the room illuminance at the height of the work desk (0.8 [m]) is depicted.

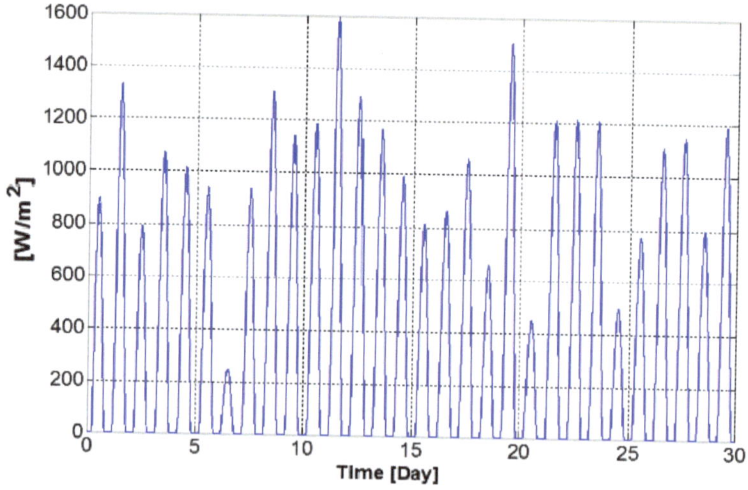

Figure 13. Modelization results: solar radiation.

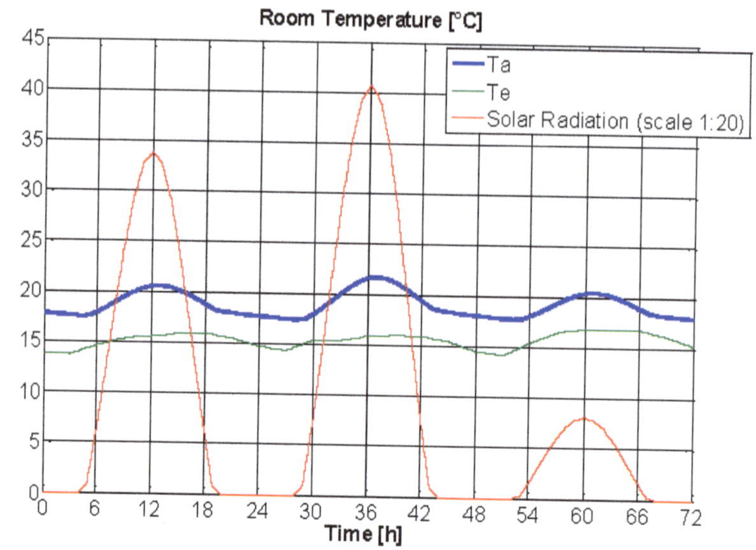

Figure 14. Modelization results with no control actions: room temperature.

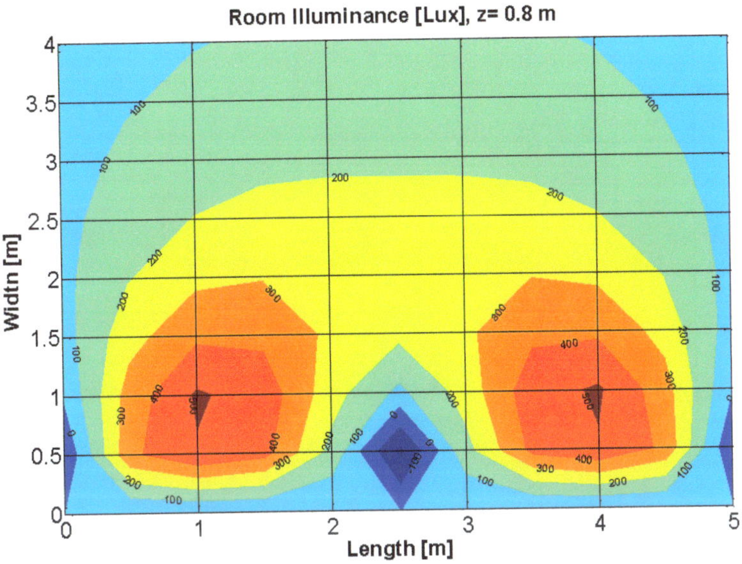

Figure 15. Modelization results with no control actions: room illuminance.

Figures 16–18 report some modelization results on IAQ variables in a typical working day. It is assumed that two people are present in the room during the hours 8–12 and 14–18. The windows are assumed to be closed when the people are in the room. In each figure, the black dashed line indicates the desired maximum constraint of the variable (see Section 2.3), while the variable is depicted by a blue line. Occupancy (suitably scaled) is depicted with a red line in Figure 16. With regard to Figures 16 and 17, as previously stated, 1 [ppm] is equal to approximately 1000 [mg/m^3]. With regard to HCHO and TVOC, zero initial conditions are assumed (see Figures 17 and 18). Observing Figures 16–18, HCHO can be identified as the most critical IAQ variable.

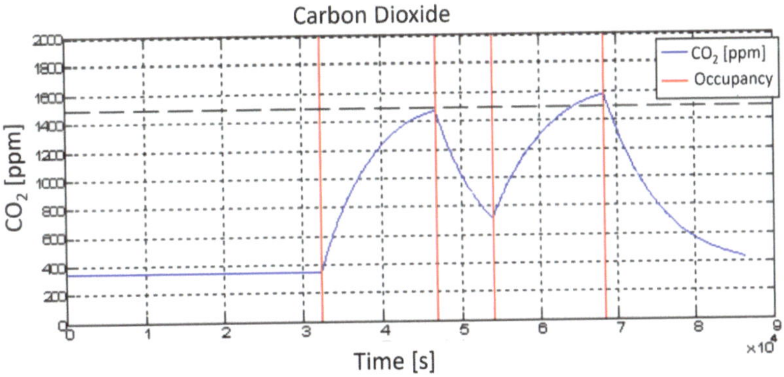

Figure 16. Modelization results on IAQ variables: CO_2.

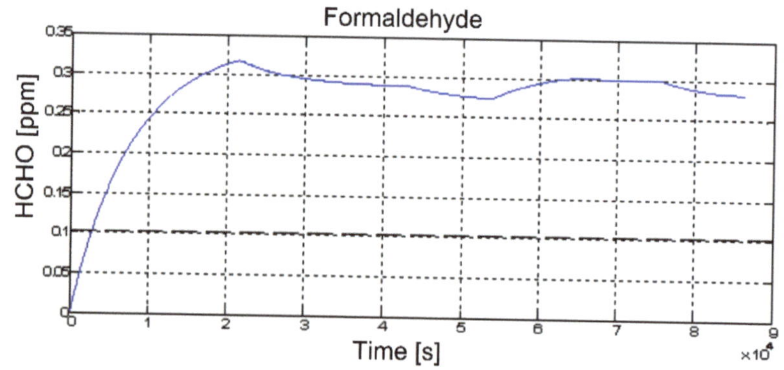

Figure 17. Modelization results on IAQ variables: HCHO.

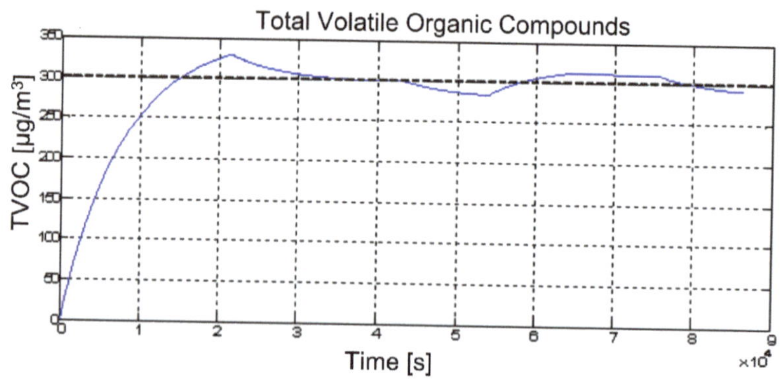

Figure 18. Modelization results on IAQ variables: TVOC.

In order to efficiently tune the controllers described in Section 2.3, step tests were executed in the proposed HVAC simulation framework, assuming a manual management of the involved actuators, thus, setting the PID controllers in manual mode.

With regard to the thermal control through the heat pump (in both the *initial* and *modified* control systems), a unitary step was simulated on the actuator in order to evaluate the response of the room temperature. The other input variables that influenced the temperature were assumed to be constant. The following FOPDT model was assumed to approximate the behavior of the thermal model in the Laplace domain:

$$y(s) = \frac{0.0017}{370s + 1} e^{-345s} u(s) \tag{20}$$

where the time constant and the delay are expressed in seconds.

With regard to the IAQ control through window opening (in the *modified* control system), a step of 1 [m] was simulated on the actuator in order to evaluate the response of the CO_2 and of the HCHO. The step magnitude corresponds to the maximum width of the window opening. A wind speed of 0.15 [m/s] was assumed, which is the average value of wind speed among the values that allow windows to be opened. For the most critical IAQ variable, i.e., HCHO, from the step test, the α and τ parameters were found to be 860 [mg HCHO/m] and 80 [s], respectively (see Section 2.1).

3.2. Control Results

In this section, some tuning results are reported together with tailored simulations of the *initial* and *modified* control systems. Sampling time of the model and of the controllers were set equal to 60 s.

For the heat pump-based thermal controller, the approximate thermal model reported in Equation (20) has been considered in both the *initial* and *modified* control systems. The ratio between the delay and the time constant is:

$$\frac{L}{\tau} = \frac{345\ [\mathrm{s}]}{370\ [\mathrm{s}]} = 0.93 \tag{21}$$

Based on the theoretical assumptions reported in Section 2.1, the Smith predictor architecture reported in Figure 1 was applied to the heat pump-based thermal controller. In order to tune the PID parameters, the Ziegler–Nichols frequency response method (close-loop method; see Table 1) was used (see Figure 19). The ultimate gain (K_u) and the ultimate period (T_u) result in values equal to 0.4 [°C/W] and 1200 [s], respectively (see Figure 19). The PID parameters were computed based on Table 1. Different simulations were performed in order to adjust the parameters starting from the initial tuning. In particular, a faster response was sought while maintaining the overshoot limited to approximately 1.5 [%] of the temperature reference value. Table 12 reports the initial and final tuning parameters. Figures 20 and 21 report the performances related to the room temperature (CV) and the heat pump (MV) with different tuning parameters of the Smith predictor, or in the case of no Smith predictor. The green lines refer to the final tuning parameters in Table 12.

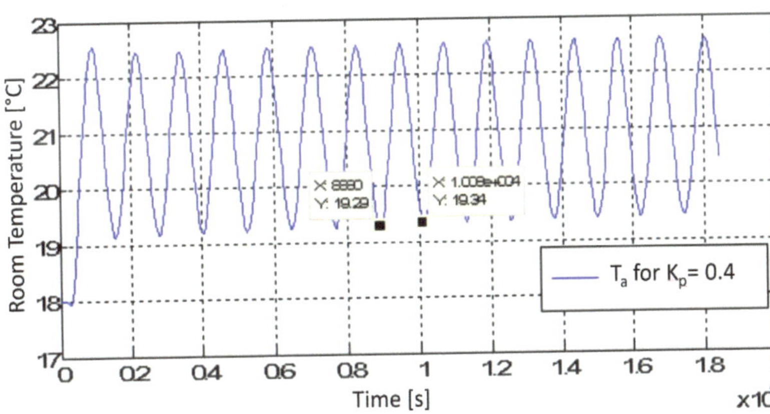

Figure 19. Control results: Smith predictor architecture tuning.

Table 12. Smith predictor architecture tuning.

Parameter	Initial Tuning Value	Final Tuning Value
K_P	0.24 [°C/W]	0.24 [°C/W]
T_I	600 [s]	20 [s]
T_D	150 [s]	150 [s]

In the *modified* control system, with regard to the IAQ controllers, the step tests described in Section 3.1 were exploited for the tuning of the PID parameters, using the Ziegler–Nichols step response method (open-loop method; see Table 2). With respect to the initial tuning parameters, HCHO controller parameters were modified, in order to obtain a faster response to formaldehyde, i.e., the most critical pollutant (see Section 3.1).

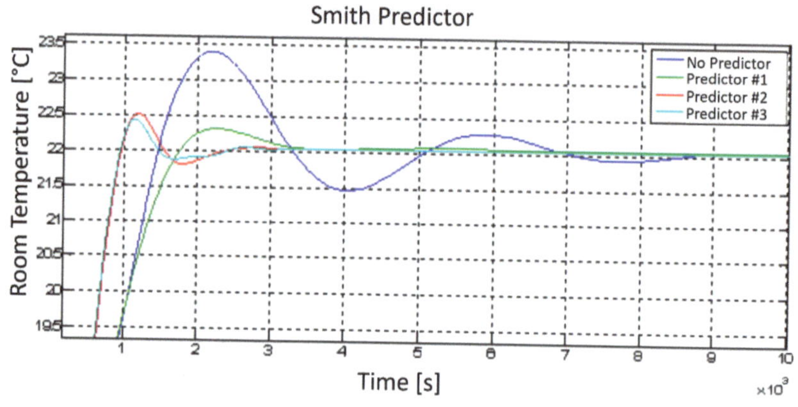

Figure 20. Control results: Smith Predictor architecture performances (CV).

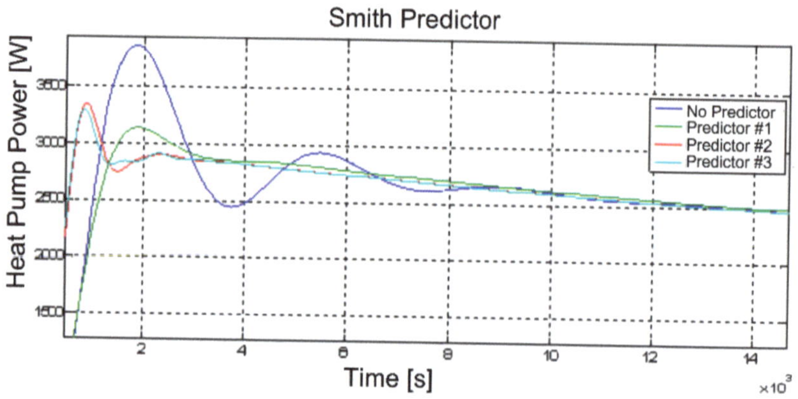

Figure 21. Control results: Smith predictor architecture performances (MV).

Figures 22–25 report the results of the IAQ control over a day achieved by the *modified* control system. In the simulation, the season is assumed to be winter; the difference between room temperature and outside temperature is assumed to be equal to 10 [°C]. It is assumed that two people are present in the room during the hours 8–12 and 14–18. Figures 22–24 report the considered CVs, i.e., CO_2, HCHO, and TVOC. Figure 25 reports the MV, i.e., the window's opening width. In Figures 22–24, a black dashed line indicates the desired maximum constraint of the variable (see Sections 2.3 and 3.1), while the current process variable is depicted by a blue line. Occupancy (suitably scaled) is depicted by a green line in Figure 22. The MV, i.e., the window's opening width, is reported (suitably scaled) through a green line in Figure 23. In Figures 24 and 25, wind speed is also reported (suitably scaled). Looking at the figures, it can be seen that the control system is able to meet IAQ specifications in most cases. The only process condition where constraint violation was observed was when the wind speed was greater than 3 [m/s] (see Figures 24 and 25): windows cannot be opened (see Figure 25), and the HCHO violates its constraint (see Figure 23).

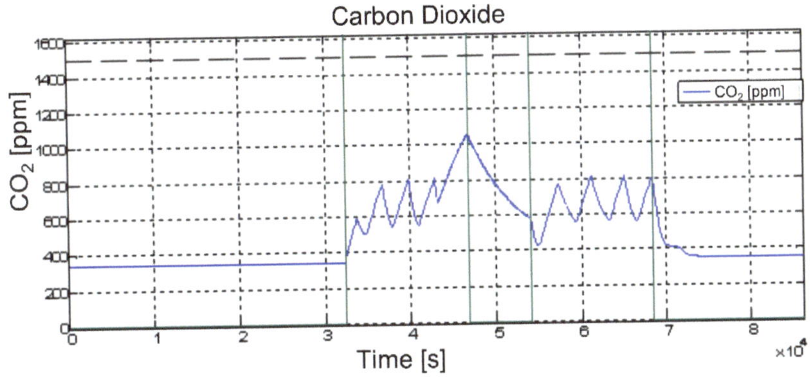

Figure 22. Control results: IAQ control, CO_2 (CV).

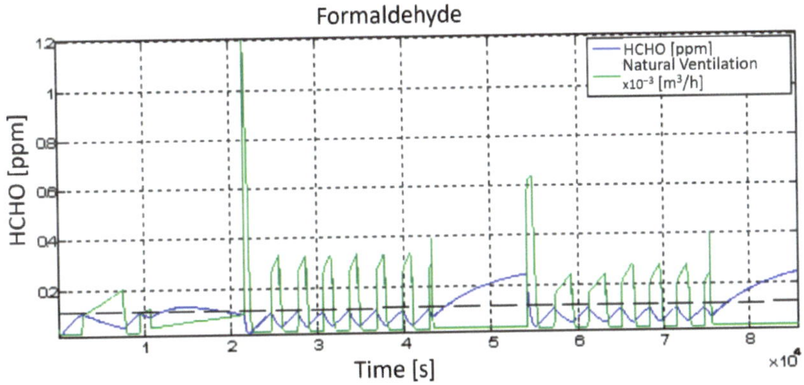

Figure 23. Control results: IAQ control, HCHO (CV).

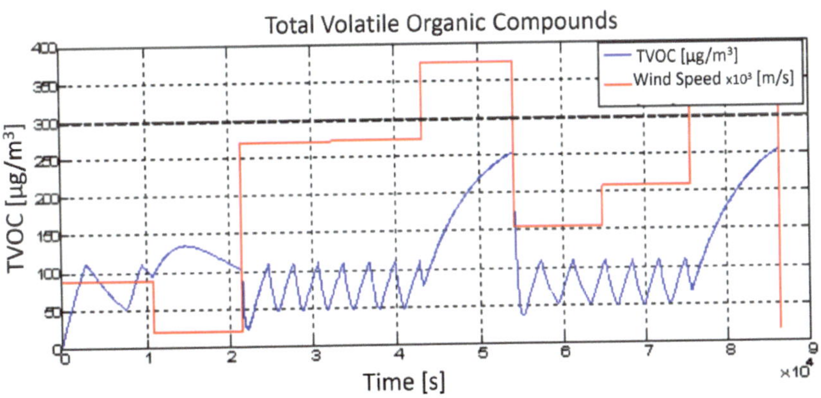

Figure 24. Control results: IAQ control, TVOC (CV).

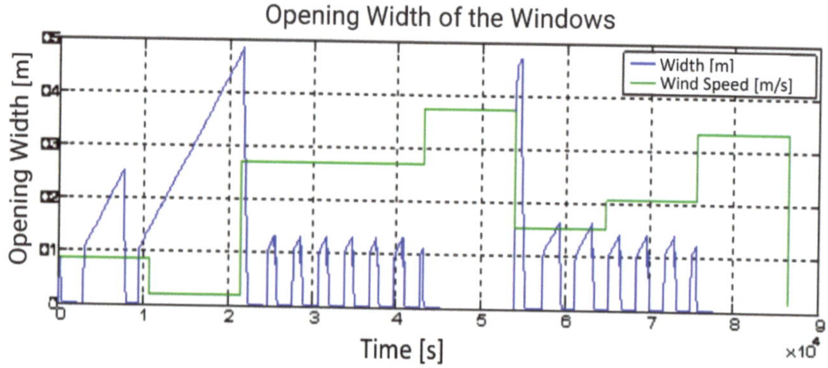

Figure 25. Control results: IAQ control, windows opening (MV).

The performance of the *initial* and *modified* control systems on thermal and lighting control is shown below by simulating three scenarios. An entire day is simulated in each scenario. The first scenario refers to the *initial* control system, while the second and third scenarios refer to the *modified* control system. The simulated scenarios refer to the winter and summer seasons, considering both control policies: Energy Saving and Comfort.

The first scenario is represented in Figures 26–29: it refers to the winter season; the *initial* control system supplies heat to the room in the Energy Saving policy (Figures 26 and 27) and in the Comfort policy (Figures 28 and 29). Figures 26 and 28 report the CVs (room temperature and room illuminance), while the MVs (heat pump, dimmer, and rolling shutters) are depicted in Figures 27 and 29. The room temperature setpoint is set to 22 [°C]. The defined room illuminance setpoint is 200 [Lux]. The controller requires a full opening of the rolling shutters: this action is executed because Energy Saving control policy is active. In this way, thermal solar radiation is exploited and energy saving target is pursued (see Figure 27). In Figure 26, it can be seen that during the middle hours of the day there is a high illuminance in the room, which may be excessive for the user. Comfort control policy performances are reported in Figures 28 and 29. In this case, the highest priority is retained by the room illuminance request: the rolling shutters are closed in the middle part of the day in order to respect the desired illuminance level through artificial light use (see Figures 28 and 29). The Comfort policy tries to search for an optimal tradeoff between energy savings and user comfort specifications: an energy consumption increase is observed in the middle part of the day (see Figure 29). On the day under consideration, thermal radiation is of low intensity: despite this fact, the Energy Saving control policy ensures a reduction in energy consumption compared to the Comfort policy (compare Figures 27 and 29). The transitions required by the controller result in increased control efforts by the heat pump.

The second scenario is represented in Figures 30–35: it refers to the winter season; the *modified* control system supplies heat to the room in the Comfort policy (Figures 30–32) and in the Energy Saving policy (Figures 33–35). Figures 30, 31, 33 and 34 report the CVs (blue lines), together with the defined setpoints (green lines), and two DVs, i.e., the outside temperature (red line) and solar radiation (magenta line). MVs are depicted in Figures 32 and 35, together with the presence of solar radiation. Time-varying setpoints are assigned to room temperature and illuminance. In particular, thanks to the presence check, tailored reference trajectories are imposed, taking into account the occupancy. Similar behaviors can be observed in both the control policies of the room temperature (see Figures 30 and 33); however, MVs behavior is quite different (see Figures 32 and 35): the use of the heat pump is minimized in the Energy Saving policy. This can be clearly observed during periods when solar radiation is present. A significant difference can be observed in room illuminance (see Figures 31 and 34): in the Comfort policy, the tracking of the required CVs setpoint has higher priority, so the rolling shutters are not immediately opened in the middle of the day (when the solar radiation is present, see Figure 32). In this way, a satisfactory tracking of the room illuminance setpoint is obtained (see Figure 31). In the case of the Energy Saving policy, the rolling shutters are immediately opened in the middle of the day so as to allow for exploiting natural heat sources provided by the solar radiation. In this way, considering the higher cost of heat pump activation compared to other MVs, energy savings is achieved at the expense of less comfortable room illuminance.

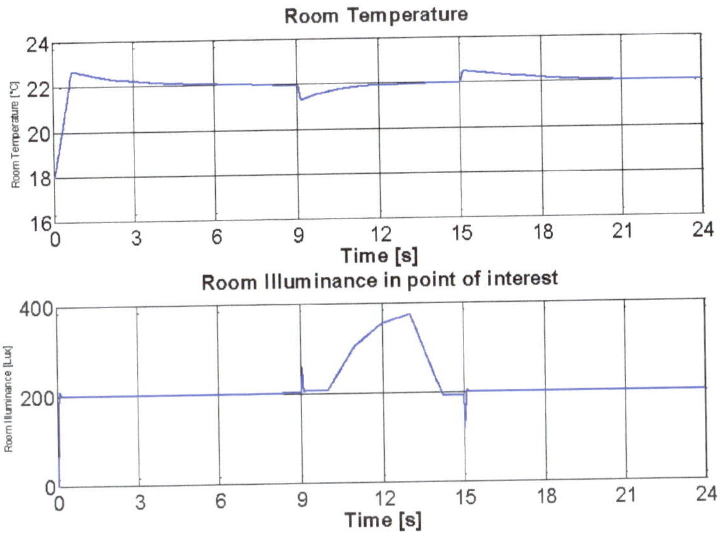

Figure 26. Control results (first scenario, *initial* control system, Energy Saving policy): room temperature and room illuminance (CVs).

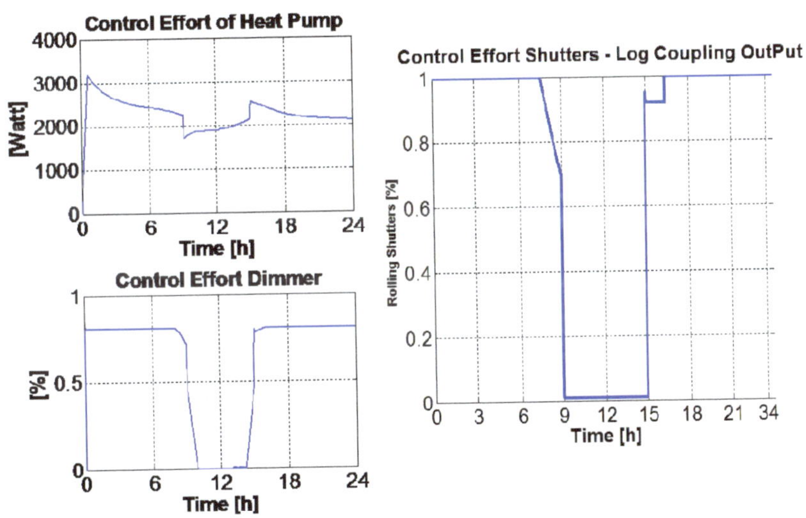

Figure 27. Control results (first scenario, *initial* control system, Energy Saving policy): MVs (heat pump, dimmer, and rolling shutters).

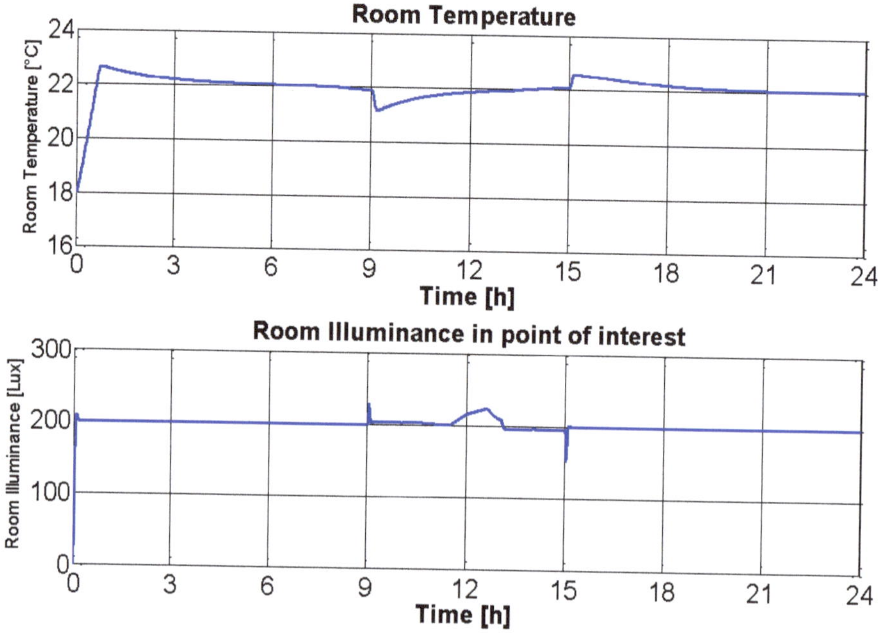

Figure 28. Control results (first scenario, *initial* control system, Comfort policy): room temperature and room illuminance (CVs).

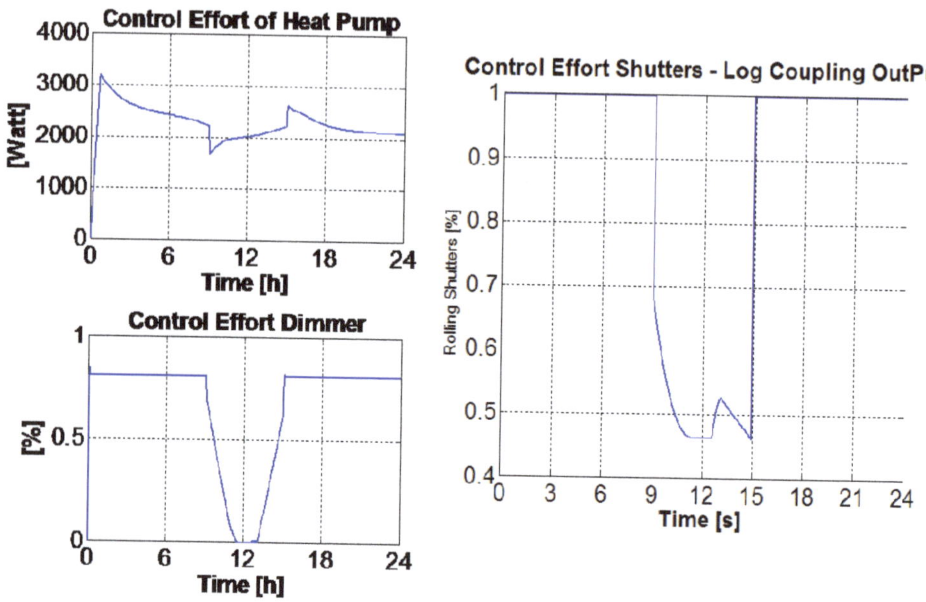

Figure 29. Control results (first scenario, *initial* control system, Comfort policy): MVs (heat pump, dimmer, and rolling shutters).

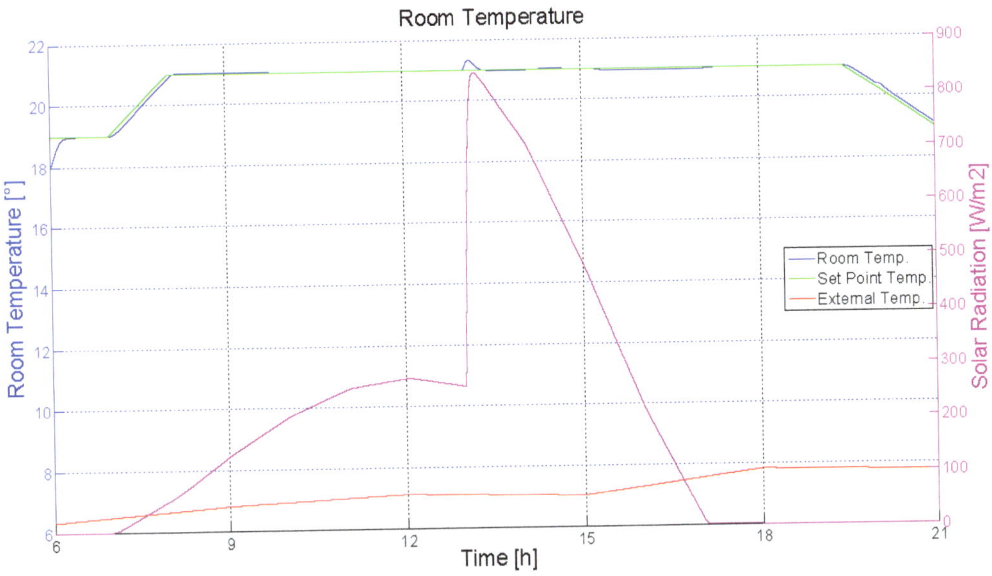

Figure 30. Control results (second scenario, *modified* control system, Comfort policy): room temperature (CV), outside temperature, and solar radiation (DVs).

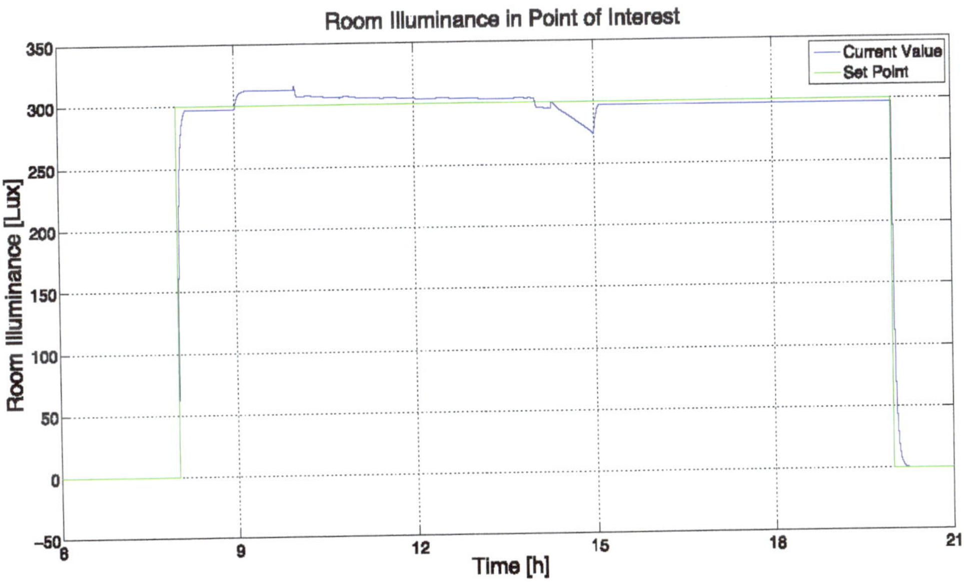

Figure 31. Control results (second scenario, *modified* control system, Comfort policy): room illuminance (CV).

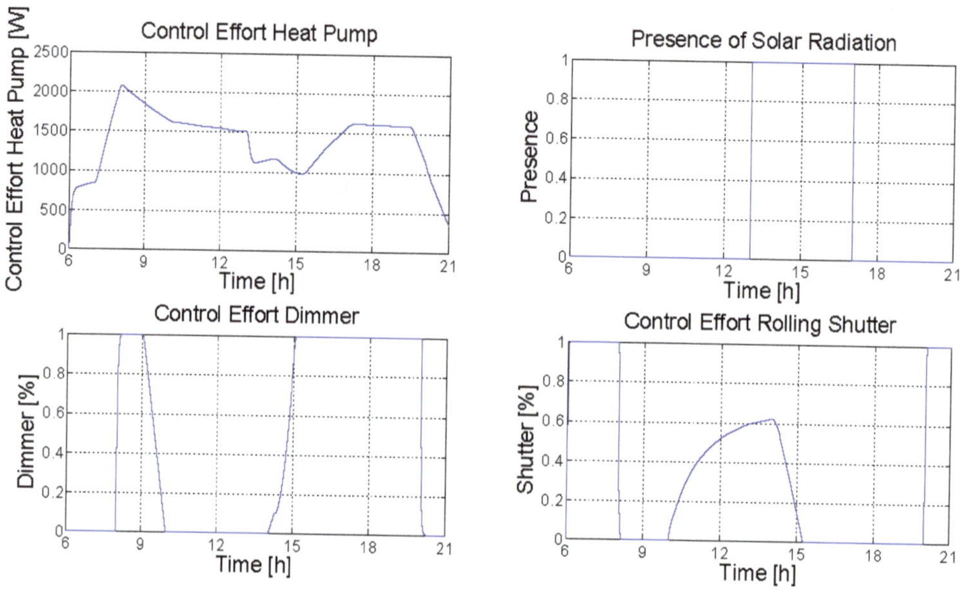

Figure 32. Control results (second scenario, *modified* control system, Comfort policy): MVs (heat pump, dimmer, and rolling shutters) and DV (presence of solar radiation).

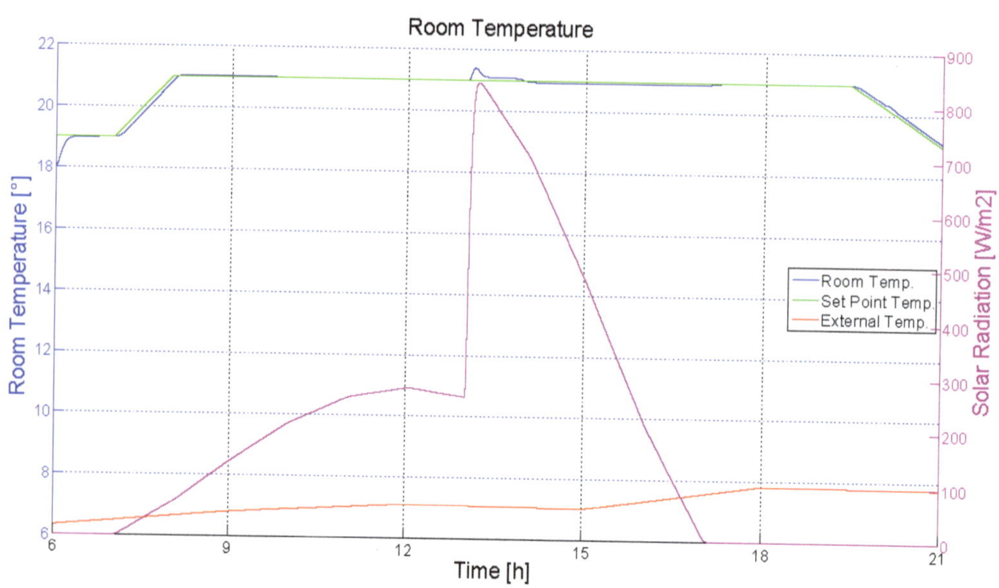

Figure 33. Control results (second scenario, *modified* control system, Energy Saving policy): room temperature (CV), outside temperature, and solar radiation (DVs).

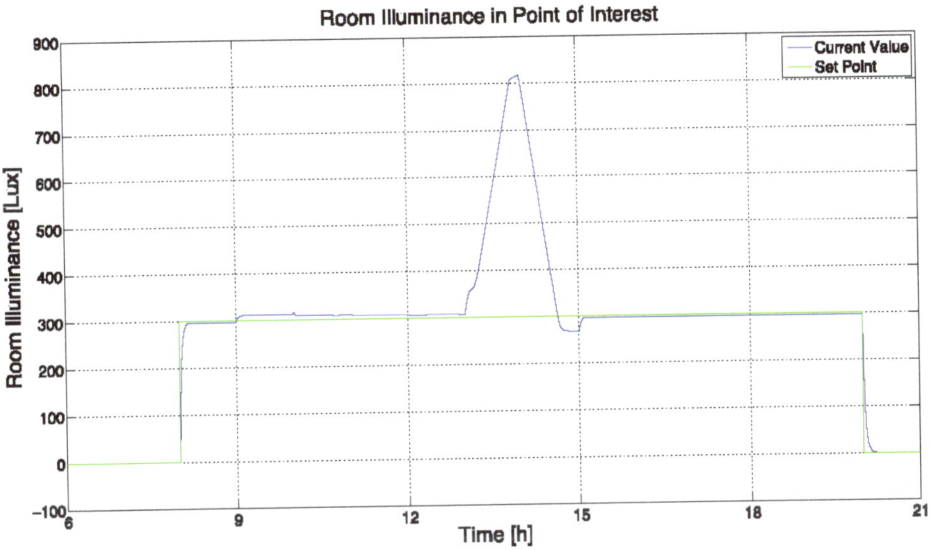

Figure 34. Control results (second scenario, *modified* control system, Energy Saving policy): room illuminance (CV).

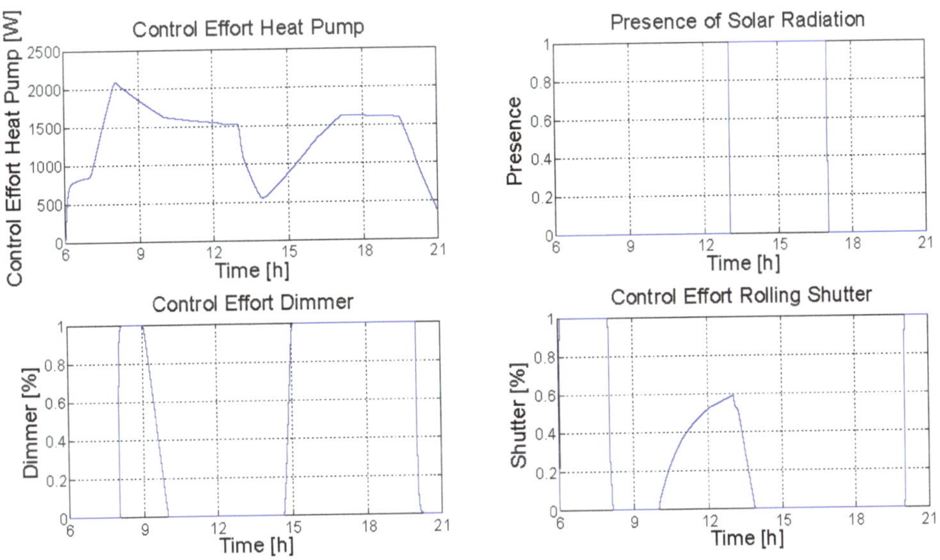

Figure 35. Control results (second scenario, *modified* control system, Energy Saving policy): MVs (heat pump, dimmer, and rolling shutters) and DV (presence of solar radiation).

The third scenario is represented in Figures 36–41: it refers to the summer season; the *modified* control system cools the room in the Comfort policy (Figures 36–38) and in the Energy Saving policy (Figures 39–41). Figures 36, 37, 39, and 40 report the CVs (blue lines), together with the defined setpoints (green lines), and two DVs, i.e., the outside temperature (red line) and solar radiation (magenta line). MVs are depicted in Figures 38 and 41, together with the presence of solar radiation. As in the second scenario, time-varying setpoints are assigned to room temperature and illuminance, considering the presence check. Similar behaviors can be observed in both the control policies on the room temperature (see

Figures 36 and 39); however, the MVs behavior is quite different (see Figures 38 and 41): the use of the heat pump is minimized in the Energy Saving policy. This control action causes a general reduction in the rolling shutters' opening in the Energy Saving policy (see Figures 38 and 41) and, as a consequence, a major increase in the usage of artificial light. These control actions generated some differences in the room illuminance behavior (see Figures 37 and 40).

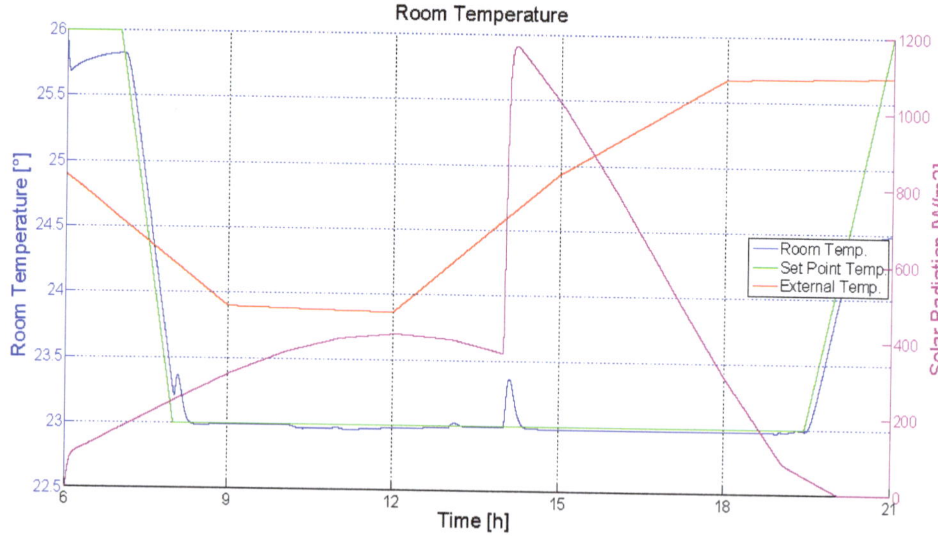

Figure 36. Control results (third scenario, *modified* control system, Comfort policy): room temperature (CV), outside temperature, and solar radiation (DVs).

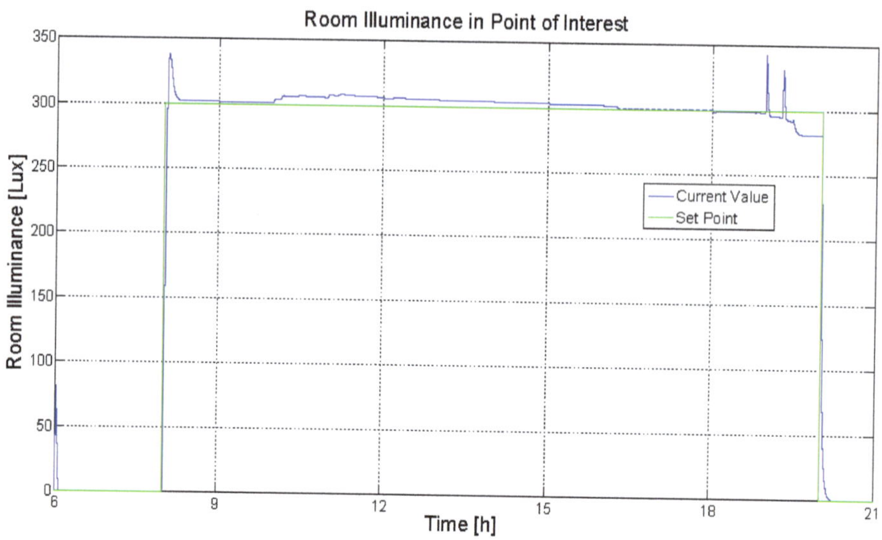

Figure 37. Control results (third scenario, *modified* control system, Comfort policy): room illuminance (CV).

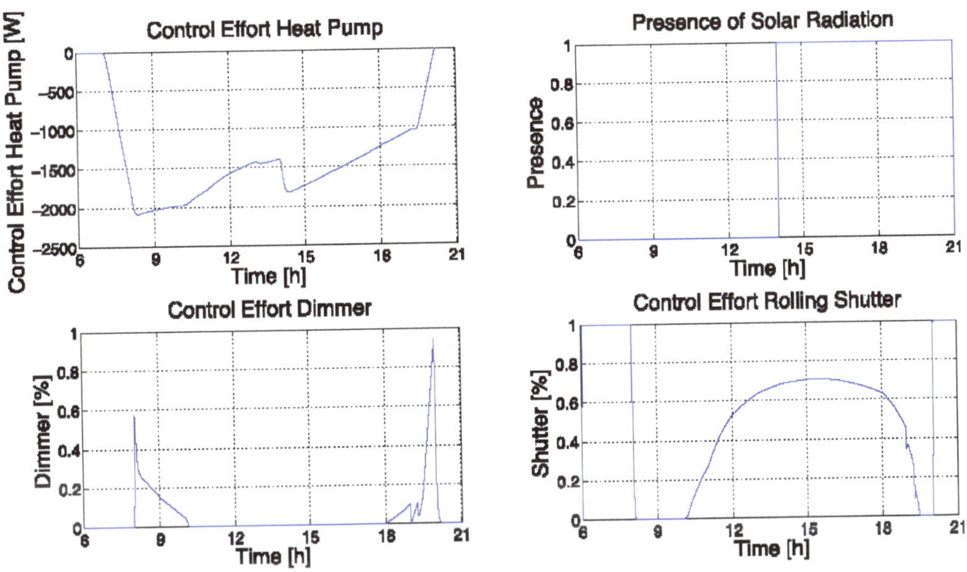

Figure 38. Control results (third scenario, *modified* control system, Comfort policy): MVs (heat pump, dimmer, and rolling shutters) and DV (presence of solar radiation).

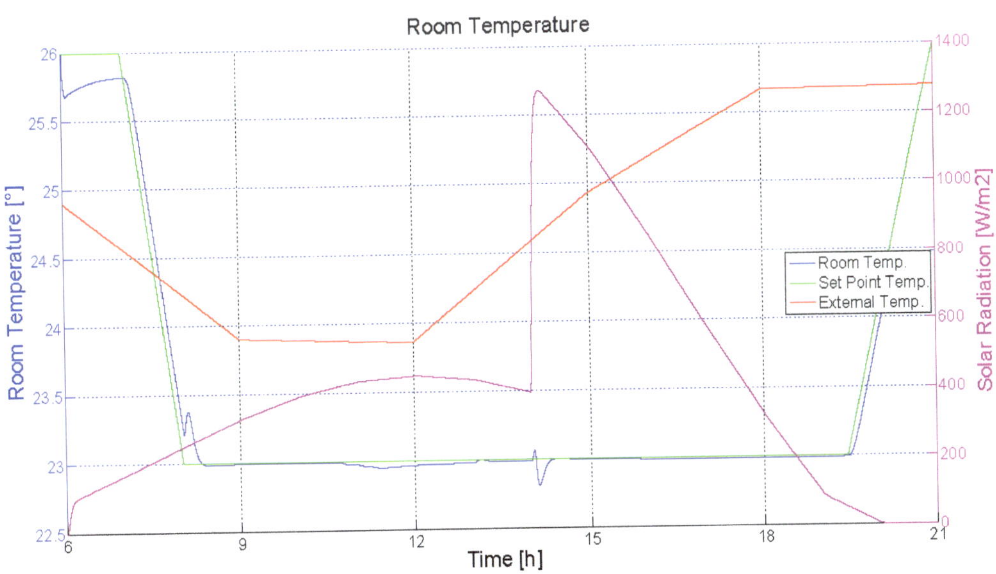

Figure 39. Control results (third scenario, *modified* control system, Energy Saving policy): room temperature (CV), outside temperature, and solar radiation (DVs).

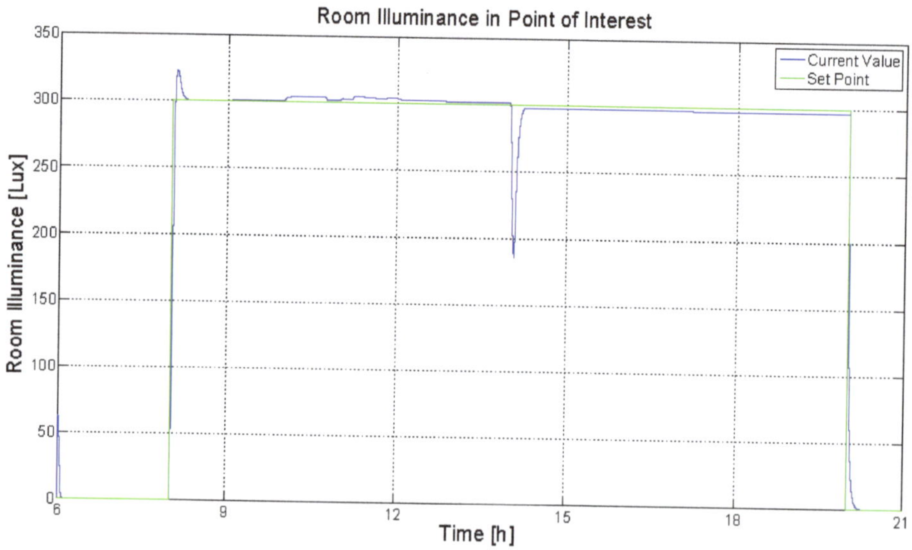

Figure 40. Control results (third scenario, *modified* control system, Energy Saving policy): room illuminance (CV).

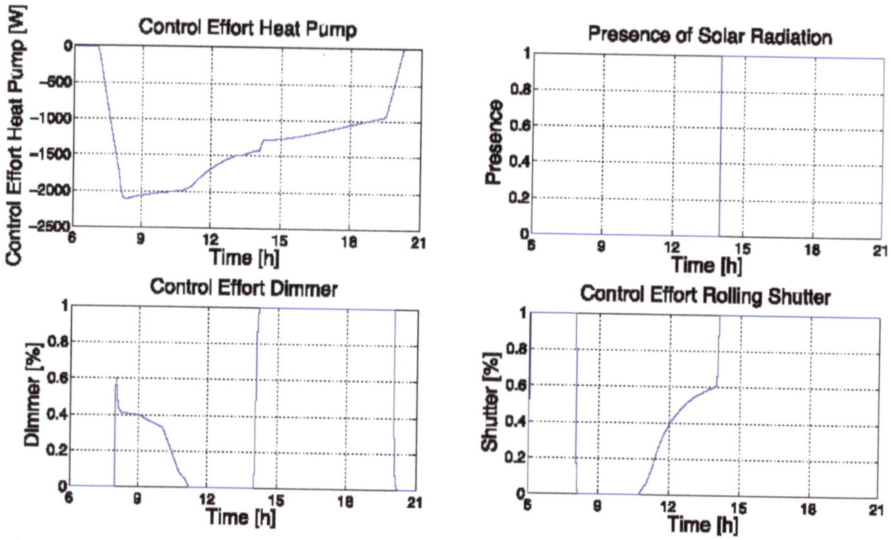

Figure 41. Control results (third scenario, *modified* control system, Energy Saving policy): MVs (heat pump, dimmer, and rolling shutters) and DV (presence of solar radiation).

In order to limit the overshoot of the room illuminance in the presence of solar radiation (main DV), suitable logic can be included to limit the rolling shutter actuation.

Observing the proposed simulations for the *initial* and *modified* control systems, the following conclusions on the control performances can be derived. The *initial* control system is characterized by weak points, mainly in the thermal control due to the presence of bumps on the heat pump control efforts (MV). This is an intrinsic behavior of the MV caused by the adopted thermal control architecture (combination of PID and DEDS). Starting from this consideration, the *modified* control system was designed to implement a different control solution based on the following principle: the thermal control system and lighting have similar requirements and have been designed according to the same criteria of engineering. The *modified* control system was equipped with an enhanced thermal controller based on a more complex architecture. A new automaton has been designed to handle possible conflicting requirements between thermal and lighting controllers, exploiting also additional features, e.g., anti-glare and solar radiation presence. Furthermore, IAQ control was added to the *modified* control system, elevating the actuation of the windows to a supervisory level.

3.3. Energy Saving Results

In order to assess the CO_2 emissions reduction achievable through the proposed control systems, a linear relationship between energy ([kWh]) and CO_2 emissions has been exploited (see Section 2.3). The performances of the *initial* and the *modified* control systems have been compared to the performances of the standard decoupled thermal and lighting PID controllers. The standard PID controllers considered take into account the rolling shutters as measured DVs, while the heat pump and dimmer are the MVs of the thermal and lighting systems, respectively. Since the rolling shutters are measured DVs of the standard PID controllers, two significant conditions are considered in the simulations: the first condition refers to the worst case, i.e., assuming the rolling shutters closed in the winter season and opened in the summer season. The second condition refers to an "average" case, i.e., assuming the half-open rolling shutters. The *initial* and the *modified* control systems are simulated in both the Energy Saving and Comfort policies. Multiple simulations in each season were performed with different weather conditions, obtaining an average energy savings percentage performance for each season of the *initial* and the *modified* control systems with respect to the considered standard PID controllers in both the mentioned rolling shutter cases. Tables 13 and 14 report the results of the simulations performed for the different seasons. In addition, average yearly energy savings and worst yearly energy savings were evaluated for the *initial* and the *modified* control systems with respect to standard decoupled PID controllers (see Figure 42). To calculate the average yearly energy savings, the yearly results obtained based on the experiments shown in Tables 13 and 14 were averaged. On the other hand, to calculate the worst annual energy savings values, the proposed control systems were compared with standard decoupled PID controllers with half-open rolling shutters.

Table 13. Energy saving results: comparison between the *initial* control system and the standard decoupled PID controllers.

	Standard Decoupled PID (Open/Closed Shutters)		**Standard Decoupled PID (Half-Open Shutters)**	
***Initial* control system (Energy Saving)**	Spring 32 [%] Autumn 24 [%]	Summer 47 [%] Winter 21 [%]	Spring 29 [%] Autumn 16 [%]	Summer 35 [%] Winter 13 [%]
***Initial* control system (Comfort)**	Spring 11 [%] Autumn 22 [%]	Summer 41 [%] Winter 19 [%]	Spring 6 [%] Autumn 15 [%]	Summer 25 [%] Winter 11 [%]

As can be noted in Tables 13 and 14 and Figure 42, both the proposed control systems outperform the standard decoupled PID controllers. Furthermore, the *modified* control system outperforms the *initial* one due to the improved control and optimization features. Observing the results reported in Tables 13 and 14, a remarkable aspect can be highlighted: significant energy savings values are obtained in the Comfort policy by the proposed controllers in all the seasons (range 6 [%]–42 [%]). Furthermore, it should be emphasized that the data reported for the winter season refer to a situation where the heat provided by natural sources is minimal due to the unfavorable conditions of average daily solar radiation. In addition, comparing the energy saving results obtained by the proposed controllers in the Energy Saving policy and in the Comfort policy, the Energy Saving policy shows the best performance compared to the Comfort policy. For example, considering the spring season, the high solar radiation forces the rolling shutters to remain almost closed during daylight hours; therefore, the Comfort policy is penalized in terms of energy efficiency.

Table 14. Energy saving results: comparison between the *modified* control system and the standard decoupled PID controllers.

	Standard Decoupled PID (Open/Closed Shutters)		**Standard Decoupled PID (Half-Open Shutters)**	
***Modified* control system (Energy Saving)**	Spring 32.5 [%] Autumn 24.5 [%]	Summer 48 [%] Winter 22 [%]	Spring 30 [%] Autumn 16.5 [%]	Summer 35.5 [%] Winter 14 [%]
***Modified* control system (Comfort)**	Spring 12 [%] Autumn 23 [%]	Summer 42 [%] Winter 20 [%]	Spring 7 [%] Autumn 15.5 [%]	Summer 26 [%] Winter 12 [%]

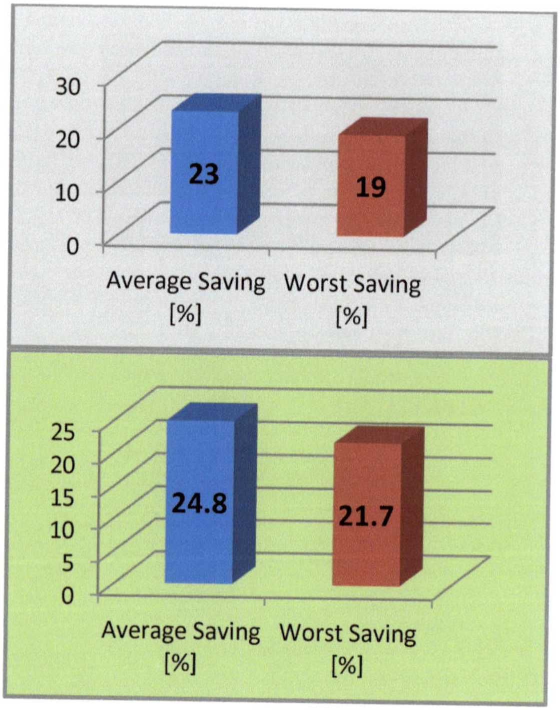

Figure 42. Energy saving results (average yearly energy savings and worst yearly energy savings): *initial* control system (**top**) and *modified* control system (**bottom**).

4. Conclusions

The present work aims at supplying contributions on simulation and control methods for home and building automation, focusing on heating, ventilating, and air conditioning processes. A simulation and control framework was designed and implemented for the thermal, lighting, and indoor air quality (IAQ) subprocesses. The simulation framework was based on first-principles and empirical models, including heat, lighting, and natural ventilation phenomena. Energy-consuming and green energy-supplying renewable sources were integrated into the framework, e.g., heat pumps, artificial lights, fresh air flow, and natural illuminance.

Proportional–integral–derivative (PID) controllers were combined in different advanced architectures. Supervisors based on discrete event dynamic systems (DEDS) methodology were added to the control system, obtaining a multi-mode control framework. Control systems based on different advanced control architectures and different control policies were simulated and compared, highlighting control performances and energy-saving results in terms of CO_2 emissions reduction.

The present work provided qualitative and quantitative results. Qualitative results refer to some methodological innovations provided in the literature on control solutions for thermal, lighting, and IAQ subprocesses. The innovative methods concerned:

- The option to test and simulate different control systems in a flexible framework;
- The assessment of different advanced PID control architectures with the goal of achieving a coupled control of thermal, lighting, and IAQ subprocesses;
- The combination of advanced PID control architectures with DEDS for energy-saving and comfort management.

The proposed control systems achieved significant quantitative results compared with the more standard control approaches. In particular, seasonal and yearly simulations showed that energy-saving results greater than or equal to 6 [%] (in each season) and 19 [%] (in one year) could be achieved compared with the more standard approaches.

Future work will focus on further improving the proposed simulation and control frameworks by adding new features, e.g., multi-room environments, and including other control techniques, e.g., model predictive control. In addition, cost-benefit analysis and feasibility studies for field implementation will be carried out.

Author Contributions: Conceptualization, S.M.Z. and C.P.; methodology, S.M.Z. and C.P.; software, S.M.Z. and C.P.; validation, S.M.Z. and C.P.; formal analysis, S.M.Z. and C.P.; investigation, S.M.Z. and C.P.; data curation, S.M.Z. and C.P.; writing—original draft preparation, S.M.Z. and C.P.; writing—review and editing, S.M.Z. and C.P.; visualization, S.M.Z. and C.P. All authors have read and agreed to the published version of the manuscript.

Funding: This research received no external funding.

Data Availability Statement: Not applicable.

Conflicts of Interest: The authors declare no conflict of interest.

References

1. European Parliament. Directive 2010/31/EU of the European Parliament and of the Council of 19 May 2010 on the energy performance of buildings (recast). *Off. J. Eur. Union* **2010**, *31*, L153/13–L153/35. Available online: https://eur-lex.europa.eu/LexUriServ/LexUriServ.do?uri=OJ:L:2010:153:0013:0035:en:PDF (accessed on 6 September 2022).
2. European Parliament. Directive 2018/844 of the European Parliament and of the Council of 30 May 2018 amending Directive 2010/31/EU on the energy performance of buildings and Directive 2012/27/EU on energy efficiency (Text with EEA relevance). *Off. J. Eur. Union* **2018**, L156/75–L156/91. Available online: https://eur-lex.europa.eu/legal-content/EN/TXT/PDF/?uri=CELEX:32018L0844&from=IT (accessed on 6 September 2022).
3. International Energy Agency, & United Nations Environment Programme. 2018 Global Status Report: Towards a Zero-Emission, Efficient and Resilient Buildings and Construction Sector. 2018. Available online: https://wedocs.unep.org/20.500.11822/27140 (accessed on 6 September 2022).
4. United Nations. Agenda 2030. Available online: https://unric.org/it/agenda-2030/ (accessed on 6 October 2022).
5. Ministero delle Imprese e del Made in Italy. PNRR. Available online: https://www.mise.gov.it/index.php/it/pnrr (accessed on 6 October 2022).
6. Ciardiello, A.; Rosso, F.; Dell'Olmo, J.; Ciancio, V.; Ferrero, M.; Salata, F. Multi-objective approach to the optimization of shape and envelope in building energy design. *Appl. Energy* **2020**, *280*, 115984. [CrossRef]
7. Zanoli, S.M.; Cocchioni, F.; Pepe, C. MPC-based energy efficiency improvement in a pusher type billets reheating furnace. *Adv. Sci. Technol. Eng. Syst. J.* **2018**, *3*, 74–84. [CrossRef]
8. Zanoli, S.M.; Pepe, C.; Rocchi, M. Control and Optimization of a Cement Rotary Kiln: A Model Predictive Control Approach. In Proceedings of the 2016 Indian Control Conference (ICC), Hyderabad, India, 4–6 January 2016. [CrossRef]

9. Zanoli, S.M.; Astolfi, G.; Orlietti, L.; Frisinghelli, M.; Pepe, C. Water Distribution Networks Optimization: A real case study. *IFAC-PapersOnLine* **2020**, *53*, 16644–16650. [CrossRef]
10. Zanoli, S.M.; Pepe, C.; Rocchi, M. Cement Rotary Kiln: Constraints Handling and Optimization via Model Predictive Control Techniques. In Proceedings of the 2015 5th Australian Control Conference (AUCC), Gold Coast, QLD, Australia, 5–6 November 2015; Available online: https://ieeexplore.ieee.org/document/7361950 (accessed on 1 December 2020).
11. Zanoli, S.M.; Pepe, C.; Rocchi, M.; Astolfi, G. Application of Advanced Process Control Techniques for a Cement Rotary Kiln. In Proceedings of the 2015 19th International Conference on System Theory, Control and Computing (ICSTCC), Cheile Gradistei, Romania, 14–16 October 2015. [CrossRef]
12. ASHRAE. Available online: https://www.ashrae.org/ (accessed on 6 October 2022).
13. Mayhoub, M.; Carter, D. A feasibility study for hybrid lighting systems. *Build. Environ.* **2012**, *53*, 83–94. [CrossRef]
14. Akkaya, K.; Guvenc, I.; Aygun, R.; Pala, N.; Kadri, A. IoT-Based Occupancy Monitoring Techniques for Energy-Efficient Smart Buildings. In Proceedings of the 2015 IEEE Wireless Communications and Networking Conference Workshops (WCNCW), New Orleans, LA, USA, 9–12 March 2015. [CrossRef]
15. Leal, S.; Zucker, G.; Hauer, S.; Judex, F. A Software Architecture for Simulation Support in Building Automation. *Buildings* **2014**, *4*, 320–335. [CrossRef]
16. Santos, A.; Liu, N.; Jradi, M. Design, Development and Implementation of a Novel Parallel Automated Step Response Testing Tool for Building Automation Systems. *Buildings* **2022**, *12*, 1479. [CrossRef]
17. Xie, X.; Ramakrishna, S.; Manganelli, M. Smart Building Technologies in Response to COVID-19. *Energies* **2022**, *15*, 5488. [CrossRef]
18. Pedersen, J.M.; Jebaei, F.; Jradi, M. Assessment of Building Automation and Control Systems in Danish Healthcare Facilities in the COVID-19 Era. *Appl. Sci.* **2022**, *12*, 427. [CrossRef]
19. Saleem, A.A.; Hassan, M.M.; Ali, I.A. Smart Homes Powered by Machine Learning: A Review. In Proceedings of the 2022 International Conference on Computer Science and Software Engineering (CSASE), Duhok, Iraq, 15–17 March 2022. [CrossRef]
20. Fayaz, M.; Kim, D. Energy Consumption Optimization and User Comfort Management in Residential Buildings Using a Bat Algorithm and Fuzzy Logic. *Energies* **2018**, *11*, 161. [CrossRef]
21. Sun, B.; Luh, P.B.; Jia, Q.-S.; Jiang, Z.; Wang, F.; Song, C. Building Energy Management: Integrated Control of Active and Passive Heating, Cooling, Lighting, Shading, and Ventilation Systems. *IEEE Trans. Autom. Sci. Eng.* **2012**, *10*, 588–602. [CrossRef]
22. Liu, W.; Wang, Y.; Jiang, F.; Cheng, Y.; Rong, J.; Wang, C.; Peng, J. A Real-time Demand Response Strategy of Home Energy Management by Using Distributed Deep Reinforcement Learning. In Proceedings of the 2021 IEEE 23rd International Conferences on High Performance Computing & Communications; 7th 23rd International Conferences on Data Science & Systems; 19th 23rd International Conferences on Smart City; 7th International Conferences on Dependability in Sensor, Cloud & Big Data Systems & Application (HPCC/DSS/SmartCity/DependSys), Haikou, China, 20–22 December 2021. [CrossRef]
23. Khalid, R.; Javaid, N.; Rahim, M.H.; Aslam, S.; Sher, A. Fuzzy energy management controller and scheduler for smart homes. *Sustain. Comput. Informatics Syst.* **2019**, *21*, 103–118. [CrossRef]
24. Soyguder, S.; Alli, H. Simulation and Modelling of HVAC System Having Two Zones with Different Properties. In Proceedings of the TOK2006 conference, Ankara, Turkey, 31 August–1 September 2006.
25. Zhou, J.Q.; Claridge, D.E. PI tuning and robustness analysis for air handler discharge air temperature control. *Energy Build.* **2012**, *44*, 1–6. [CrossRef]
26. Almabrok, A.; Psarakis, M.; Dounis, A. Fast Tuning of the PID Controller in An HVAC System Using the Big Bang–Big Crunch Algorithm and FPGA Technology. *Algorithms* **2018**, *11*, 146. [CrossRef]
27. Yamazaki, T.; Yamakawa, Y.; Kamimura, K.; Kurosu, S. Air-Conditioning PID Control System with Adjustable Reset to Offset Thermal Loads Upsets. In *Advances in PID Control*; IntechOpen: London, UK, 2011. [CrossRef]
28. Blasco, C.; Monreal, J.; Benítez, I.; Lluna, A. Modelling and PID Control of HVAC System According to Energy Efficiency and Comfort Criteria. In *Sustainability in Energy and Buildings. Smart Innovation, Systems and Technologies*; M'Sirdi, N., Namaane, A., Howlett, R.J., Jain, L.C., Eds.; Springer: Berlin/Heidelberg, Germany, 2012; Volume 12. [CrossRef]
29. Soyguder, S.; Karakose, M.; Alli, H. Design and simulation of self-tuning PID-type fuzzy adaptive control for an expert HVAC system. *Expert Syst. Appl.* **2009**, *36*, 4566–4573. [CrossRef]
30. Wang, J.-M.; Yang, M.-T.; Chen, P.-L. Design and Implementation of an Intelligent Windowsill System Using Smart Handheld Device and Fuzzy Microcontroller. *Sensors* **2017**, *17*, 830. [CrossRef]
31. Liu, J.; Zhang, W.; Chu, X.; Liu, Y. Fuzzy logic controller for energy savings in a smart LED lighting system considering lighting comfort and daylight. *Energy Build.* **2016**, *127*, 95–104. [CrossRef]
32. Ain, Q.-U.; Iqbal, S.; Mukhtar, H. Improving Quality of Experience Using Fuzzy Controller for Smart Homes. *IEEE Access* **2022**, *10*, 11892–11908. [CrossRef]
33. Fontes, F.; Antão, R.; Mota, A.; Pedreiras, P. Improving the Ambient Temperature Control Performance in Smart Homes and Buildings. *Sensors* **2021**, *21*, 423. [CrossRef]
34. Serale, G.; Fiorentini, M.; Capozzoli, A.; Bernardini, D.; Bemporad, A. Model Predictive Control (MPC) for Enhancing Building and HVAC System Energy Efficiency: Problem Formulation, Applications and Opportunities. *Energies* **2018**, *11*, 631. [CrossRef]
35. Yao, Y.; Shekhar, D.K. State of the art review on model predictive control (MPC) in Heating Ventilation and Air-conditioning (HVAC) field. *Build. Environ.* **2021**, *200*, 107952. [CrossRef]

36. Piotrowska-Woroniak, J.; Szul, T.; Cieśliński, K.; Krilek, J. The Impact of Weather-Forecast-Based Regulation on Energy Savings for Heating in Multi-Family Buildings. *Energies* **2022**, *15*, 7279. [CrossRef]
37. Aström, K.J.; Hägglund, T. *PID Controllers: Theory, Design, and Tuning*; ISA: Research Triangle Park, NC, USA, 1995.
38. O'Dwyer, A. *Handbook of PI and PID Controller Tuning Rules*; Imperial College Press: London, UK, 2009. [CrossRef]
39. Ljung, L. *System Identification. Theory for the User*; Prentice-Hall PTR: Upper Saddle River, NJ, USA, 1999.
40. Shinskey, F.G. *Process Control Systems: Application, Design, and Tuning*; McGraw-Hill Professional Publishing: New York, NY, USA, 1996.
41. Morari, M.; Zafiriou, E. *Robust Process Control*; Prentice-Hall PTR: Hoboken, NJ, USA, 1988.
42. Cammarata, G.; Fichera, A.; Forgia, F.; Marletta, L.; Muscato, G. Thermal Load Buildings: General Models and Reduced Models. In Proceedings of the Health Buildings, the 3rd International Conferences, Budapest, Hungary, 22–25 August 1994.
43. Mitsios, I.; Kolokotsa, D.; Stavrakakis, G.; Kalaitzakis, K.; Pouliezos, A. Developing a Control Algorithm for CEN Indoor Environmental Criteria—Addressing Air Quality, Thermal Comfort and Lighting. In Proceedings of the 2009 17th Mediterranean Conference on Control and Automation, Thessaloniki, Greece, 24–26 June 2009. [CrossRef]
44. De Santoli, L. Trasmissione del Calore. In *Fisica Tecnica Ambientale*; CEA: Milan, Italy, 1999; Volume 2.
45. Collares-Pereira, M.; Rabl, A. The average distribution of solar radiation-correlations between diffuse and hemispherical and between daily and hourly insolation values. *Sol. Energy* **1979**, *22*, 155–164. [CrossRef]
46. Bonomo, M. *Illuminazione D'interni*; Maggioli Editore: Santarcangelo di Romagna, Italy, 2009.
47. Paribeni, M.; Parolini, G. *Tecnica Dell'illuminazione*; UTET: Torino, Italy, 2009.
48. Wargocki, P.; Wyon, D.P.; Baik, Y.K.; Clausen, G.; Fanger, P.O. Perceived Air Quality, Sick Building Syndrome (SBS) Symptoms and Productivity in an Office with Two Different Pollution Loads. *Indoor Air* **1999**, *9*, 165–179. [CrossRef]
49. Dasi, H.; Xiaowei, F.; Daisheng, C. On-Line Control Strategy of Fresh Air to Meet the Requirement of IAQ in Office Buildings. In Proceedings of the 2010 5th IEEE Conference on Industrial Electronics and Applications, Taichung, Taiwan, 15–17 June 2010. [CrossRef]
50. Bako-Biro, Z.; Wargocki, P.; Weschler, C.J.; Fanger, P.O. Effects of pollution from personal computers on perceived air quality, SBS symptoms and productivity in offices. *Indoor Air* **2004**, *14*, 178–187. [CrossRef]
51. Kawachi, S.; Hagiwara, H.; Baba, J.; Furukawa, K.; Shimoda, E.; Numata, S. Modeling and Simulation of Heat Pump Air Conditioning Unit Intending Energy Capacity Reduction of Energy Storage System in Microgrid. In Proceedings of the 2011 14th European Conference on Power Electronics and Applications, Birmingham, UK, 30 August–1 September 2011.
52. European Parliament. Directive 2008/50/EC of the European Parliament and of the Council of 21 May 2008 on ambient air quality and cleaner air for Europe. *Off. J. Eur. Union* **2008**, L152/1. Available online: https://eur-lex.europa.eu/legal-content/EN/TXT/HTML/?uri=CELEX:32008L0050&from=IT (accessed on 6 September 2022).
53. European Parliament. Commission Directive (EU) 2015/1480 of 29 August 2015 amending several annexes to Directives 2004/107/EC and 2008/50/EC of the European Parliament and of the Council laying down the rules concerning reference methods, data validation and location of sampling points for the assessment of ambient air quality. *Off. J. Eur. Union* **2015**, L226/4. Available online: https://eur-lex.europa.eu/legal-content/EN/TXT/HTML/?uri=CELEX:32015L1480&from=EN (accessed on 6 September 2022).
54. International Energy Agency. *CO2 Emissions from Fuel Combustion 2019*; IEA: Paris, France, 2019. [CrossRef]
55. Cassandras, C.G.; Lafortune, S. *Introduction to Discrete Event Systems*; Springer Science & Business Media: Berlin/Heidelberg, Germany, 2007.
56. Health and Safety Executive (HSE). *Lighting at Work*; HSE Books: Norwich, UK, 1998.
57. Brager, G.S.; de Dear, R.J. Climate, comfort & natural ventilation: A new adaptive comfort standard for ASHRAE Standard 55. In Proceedings of the Moving Thermal Comfort Standards into the 21st Century, Windsor, UK, 5–8 April 2001. Available online: http://www.escholarship.org/uc/item/2048t8nn. (accessed on 1 December 2020).
58. MathWorks. Available online: https://it.mathworks.com/ (accessed on 24 October 2022).